Geomagnetism, Aeronomy and Space Weather

On the centenary of the International Union of Geodesy and Geophysics, this book reviews the state-of-the-art research in geomagnetism, aeronomy and space weather. Written by eminent researchers from these fields, it summarises the advances in research over the past one hundred years and looks ahead to current and emerging studies on Earth's magnetic field. It provides a comprehensive overview of the generation of Earth's magnetic field, its history and its response to external forces. Starting at the centre of the Earth, the reader is taken on a journey from the interior core and mantle through the upper atmosphere and magnetosphere before reaching the Sun's atmosphere and corona. The applications of this research are also discussed, particularly the societal impact of solar activity on critical infrastructures in our increasingly technologically dependent society. This book is a valuable resource and reference for academic researchers and students in geomagnetism and aeronomy.

Mioara Mandea is Professor and Programme Manager for Solid Earth at the Centre National d'Études Spatiales, France. She is also chair of the Outreach Committee of the European Geoscience Union and General Secretary of the International Association of Geomagnetism and Aeronomy. She has received the International Award of the American Geophysical Union (AGU), the Petrus Peregrinus Medal of the European Geosciences Union (EGU) and the prestigious French Ordre National de Mérite.

Monika Korte is a head of the working group for geomagnetic field evolution at the GFZ German Research Centre for Geosciences, Germany. She is also an associate researcher at the Institute of Geophysics and Planetary Physics, Scripps Institute of Oceanography, University of California. She is Vice President of IAGA and Editor for *Geophysical Research Letters* of the American Geophysical Union.

Andrew Yau is Principal Investigator for the Enhanced Polar Outflow Probe (e-POP) satellite mission and Professor in Physics and Astronomy at the University of Calgary, Canada. He is a member of the IAGA Executive Committee and Editor for *Geophysical Research Letters*. He has worked on several international space science satellite missions.

Eduard Petrovský is Deputy Director at the Institute of Geophysics in the Czech Academy of Sciences, Czech Republic. He is also President of IAGA, Editor of *Geophysical Journal International* and Associate Professor at the University of Helsinki and Charles University, Prague. His current research focuses on basic rock magnetism and its environmental applications.

Special Publications of the International Union of Geodesy and Geophysics Series

The International Union of Geodesy and Geophysics (IUGG) promotes and co-ordinates international scientific studies of Earth (physical, chemical, and mathematical) and its environment in space, aiming to apply this knowledge to societal needs such as mineral resources, mitigation of natural hazards, and environmental preservation. The series is co-published by the IUGG and Cambridge University Press, providing researchers and graduate students with authoritative insights into major scientific developments and state-of-the-art research.

TITLES IN PRINT IN THIS SERIES

A. Ismail-Zadeh, J. Urrutia-Fucugauchi, A. Kijko, K. Takeuchi, and I. Zaliapin, *Extreme Natural Hazards, Disaster Risks and Societal Implications*

J. Li, R. Swinbank, R. Grotjahn, and H. Volkert, *Dynamics and Predictability of Large-Scale, High-Impact Weather and Climate Events*

T. Beer, J. Li, and K. Alverson, *Global Change and Future Earth: The Geoscience Perspective*

M. Mandea, M. Korte, A. Yau, and E. Petrovský, *Geomagnetism, Aeronomy and Space Weather: A Journey from the Earth's Core to the Sun*

Geomagnetism, Aeronomy and Space Weather

A Journey from the Earth's Core to the Sun

EDITED BY

MIOARA MANDEA
Centre National d'Études Spatiales, Paris, France

MONIKA KORTE
GFZ German Research Centre for Geosciences, Potsdam, Germany

ANDREW YAU
University of Calgary, Calgary, Canada

EDUARD PETROVSKÝ
Academy of Sciences of the Czech Republic, Prague, Czech Republic

CAMBRIDGE
UNIVERSITY PRESS

CAMBRIDGE
UNIVERSITY PRESS

University Printing House, Cambridge CB2 8BS, United Kingdom

One Liberty Plaza, 20th Floor, New York, NY 10006, USA

477 Williamstown Road, Port Melbourne, VIC 3207, Australia

314-321, 3rd Floor, Plot 3, Splendor Forum, Jasola District Centre, New Delhi - 110025, India

79 Anson Road, #06-04/06, Singapore 079906

Cambridge University Press is part of the University of Cambridge.

It furthers the University's mission by disseminating knowledge in the pursuit of education, learning and research at the highest international levels of excellence.

www.cambridge.org
Information on this title: www.cambridge.org/9781108418485
DOI: 10.1017/9781108290135

First published 2020

A catalogue record for this publication is available from the British Library

ISBN 978-1-108-41848-5 Hardback

Contents

Contributors

Mangalathayil Ali Abdu
National Institute for Space Research and Aeronautic Technology Institute, São José dos Campos, SP, Brazil

Daniel N. Baker
University of Colorado, Boulder, CO, USA

Balan Nanan
Institute of Geology and Geophysics, Chinese Academy of Sciences, Beijing, China

André Balogh
Imperial College London, London, UK

Harald Böhnel
Universidad Autonoma de Mexico, Queretaro, Mexico

David Boteler
Natural Resources Canada, Ottawa, ON, Canada

Doris Breuer
Institut für Planetenforschung, Deutsches Zentrum für Luft- und Raumfahrt, Berlin, Germany

Arnaud Chulliat
CIRES, University of Colorado, Boulder, CO, USA

C. Robert Clauer
Virginia Polytechnic Institute and State University, Blacksburg, VA, USA

Ed W. Cliver
National Solar Observatory, Boulder, CO, USA

Catherine Constable
University of California, San Diego, La Jolla, CA, USA

Steven Constable
Scripps Institution of Oceanography, La Jolla, CA, USA

Alex Degeling
Physics Department, Shandong University, Shandong, China

Yusuke Ebihara
Kyoto University, Kyoto, Japan

Stephen A. Fuselier
Lockheed Martin Advanced Technology Center, Palo Alto, CA, USA

Stuart A. Gilder
Ludwig Maximilians University, Munich, Germany

Nicolas Gillet
University Grenoble Alpes, University Savoie Mont Blanc, Grenoble, France

Michael Hapgood
STFC Rutherford Appleton Laboratory, Didcot, UK

Bogdan Hnat
Physics Department, University of Warwick, Coventry, UK

Malcolm J. S. Johnston
US Geological Survey, Menlo Park, CA, USA

Andy Kale
Department of Physics, University of Alberta, Edmonton, AB, Canada

David Kerridge
British Geological Survey, Edinburgh, UK

Craig A. Kletzing
University of Iowa, Iowa City, IA, USA

David J. Knudsen
Department of Physics and Astronomy, University of Calgary, Calgary, AB, Canada

Monika Korte
GFZ, German Research Centre for Geosciences, Potsdam, Germany

Elena A. Kronberg
Ludwig Maximilians University, Munich, Germany

Weijia Kuang
NASA, Goddard Space Flight Center, Greenbelt, MD, USA

Jean-Louis Le Mouël
Institut de Physique du Globe de Paris, Paris, France

Vincent Lesur
Institut de Physique du Globe de Paris, Paris, France

Florian Lhuillier
Ludwig Maximilians University, Munich, Germany

Huixin Liu
Department of Earth and Planetary Science, Kyushu University, Fukuoka, Japan

Ruosi Liu
Department of Space Physics, Wuhan University, Wuhan, China

Hermann Lühr
GFZ, German Research Centre for Geosciences, Potsdam, Germany

Mioara Mandea
Centre National d'Études Spatiales, Paris, France

Ian R. Mann
Department of Physics, University of Alberta, Edmonton, AB, Canada

Aurélie Marchaudon
Institut de Recherche en Astrophysique et Planétologie, Université de Toulouse, Toulouse, France

Robert L. McPherron
Department of Earth, Planetary, and Space Sciences, University of California, Los Angeles, Los Angeles, CA, USA

David K. Milling
Department of Physics, University of Alberta, Edmonton, AB, Canada

Kyle R. Murphy
NASA, Goddard Space Flight Center, Greenbelt, MD, USA

Louis G. Ozeke
Department of Physics, University of Alberta, Edmonton, AB, Canada

Ivan P. Pakhotin
Department of Physics, University of Alberta, Edmonton, AB, Canada

Eduard Petrovský
Institute of Geophysics, The Czech Academy of Sciences, Prague, Czech Republic

Viviane Pierrard
Royal Belgian Institute for Space Aeronomy, Brussels, Belgium

Ameya Prabhu
Ludwig Maximilians University, Munich, Germany

I. Jonathan Rae
Mullard Space Science Laboratory, University College London, Dorking, UK

Robert Rankin
Department of Physics, University of Alberta, Edmonton, AB, Canada

Oliver Ritter
GFZ, German Research Centre for Geosciences, Potsdam, Germany

Patricia Ritter
GFZ, German Research Centre for Geosciences, Potsdam, Germany

Terence Sabaka
NASA, Goddard Space Flight Center, Greenbelt, MD, USA

Yoichi Sasai
Tokai University, Tokyo, Japan

Robert J. Strangeway
University of California, Los Angeles, Los Angeles, CA, USA

Andrew Tangborn
NASA, Goddard Space Flight Center, Greenbelt, MD, US

Erwan Thébault
Université de Nantes, Nantes, France

Trond H. Torsvik
University of Oslo, Oslo, Norway

Robert Tyler
NASA, Goddard Space Flight Center, Greenbelt, MD, USA

Jean-Pierre Valet
Institut de Physique du Globe de Paris, Paris, France

Rob van der Voo
University of Michigan, Ann Arbor, MI, USA

Ari Viljanen
Finnish Meteorological Institute, Helsinki, Finland

Michael R. Wack
Ludwig Maximilians University, Munich, Germany

Ingo Wardinski
Université de Nantes, Nantes, France

Colin L. Waters
University of Newcastle, Newcastle, NSW, Australia

Ute Weckmann
GFZ, German Research Centre for Geosciences, Potsdam, Germany

Andrew Yau
Department of Physics and Astronomy, University of Calgary, Calgary, AB, Canada

Jacques Zlotnicki
Observatoire de Physique du Globe de Clermont Ferrand, Clermont Ferrand, France

Preface

The geomagnetic field is generated deep inside the Earth, yet it surrounds our planet in the form of the magnetosphere reaching far into space. As such, it plays an important role in shielding the human habitat against harmful influences from the solar wind and cosmic radiation. Earth's magnetic field signatures and their changes in space and time provide important information on our planet's deep interior dynamics, lithospheric structures and tectonics and the state of Sun–Earth interactions and space weather conditions.

The Earth's core and mantle dynamics are driven by heat flow from the centre to the surface of the Earth. The outer core dynamics, which generate the magnetic field by geodynamo processes, and the coupling between the core and the mantle influence Earth's rotation rate and are linked to mantle dynamics that are responsible for tectonic processes.

The (electro)magnetic signatures of lithospheric rocks and geological formations are interpreted, often in conjunction with other geophysical observables (e.g. gravity and seismics), for the exploration of natural resources, for natural hazard assessment in seismically or volcanic active regions and for understanding geological processes such as plate tectonics or ancient meteor impact craters. The palaeomagnetic signals preserved in volcanic rocks and sediments over geological times not only provide information about past geodynamo action and magnetic field configuration but also find applications in dating and studies of past climatic changes.

Strongly increased solar activity during solar storms can adversely affect critical technological infrastructures, from power grids and rail networks to aviation, telecommunications and satellite navigation, when the prevailing interplanetary conditions lead to the development of geomagnetic storms. The sensitivity of these modern technologies to severe space weather phenomena enhances our vulnerability to their adverse impacts, as society becomes increasingly technology dependent. In order to advance our capability to forecast space weather and mitigate such impacts, an improved and deeper understanding of the geomagnetic field is essential, including its variations on all spatial and temporal scales and its connection to the dynamics of the magnetosphere and solar–terrestrial interactions.

The increased interest in studying the Earth's magnetic field is underlined by the Swarm satellite mission launched by the European Space Agency in 2013. This constellation of three low-Earth-orbiting satellites offers an unprecedented opportunity to achieve an improved separation between the internal and external magnetic field sources, a detailed mapping of the large-scale lithospheric magnetic field and multipoint monitoring of magnetic disturbances and related aeronomic processes for space weather applications.

The International Association of Geomagnetism and Aeronomy (IAGA), one of the eight associations of the International Union of Geodesy and Geophysics (IUGG), brings together scientists from all over the globe working on all aspects of the Earth's magnetic field, the magnetosphere, aeronomy and space weather. Within the Special Publication Series on the one-hundredth anniversary of the IUGG, expert authors from the IAGA community present here reviews of the state of the art and recent progress in the fields of geomagnetism, aeronomy and space weather. This book provides a comprehensive overview of the current knowledge of the Earth's magnetic field; its history, generation and present state; and the way it responds to external – in particular, solar – forces. The book also highlights applications and societal implications of these results. It is structured as follows.

In the first part, the relevance of different aspects of geomagnetic and aeronomy studies is introduced, and the role of IAGA is described. This broad introduction shows the wide range, complexity and heterogeneity of IAGA-related studies, which are not limited to the topics detailed further in this volume. It begins with some historical notes and a general overview of the objectives of geomagnetic and aeronomy studies (Chapter 1). Chapter 2 starts with brief descriptions of how Earth's internal structure can be depicted using electromagnetic induction caused by geomagnetic field variations of external origin and how rocks carrying a magnetic record may help in assessing and untangling Earth's tectonic history, in dating past events and in reconstructing the environment in different geological epochs. This is followed by overviews of applications of studies of geomagnetic field variations, from very slow to rapid and from both the internal geodynamo and the external sources in magnetosphere and ionosphere. Brief reviews of the relationship of magnetosphere and ionosphere with solar variability, which is an important component of space weather, and of hazards associated with space weather are also included in this chapter. Finally, the effects of the geomagnetic field, and in particular, of its changes, on the biosphere are introduced. In the last part of the introduction (Chapter 3), the scientific landscape before IAGA and major achievements obtained either with direct IAGA involvement or representing the most important benchmarks in geomagnetic and aeronomy studies are briefly summarised.

Information on the different sources contributing to the measurable geomagnetic field and on the methods, platforms and instruments used to observe and monitor the field is given in Part II. This part starts with an overview of the geomagnetic field sources from Earth's core to the Sun, in particular focusing on individual current systems interacting in the magnetosphere (Chapter 4). This is followed by a chapter (Chapter 5) on what paleomagnetic data obtained by laboratory experiments from ancient rocks and sediments can tell us about the operation of the outer core geodynamo throughout Earth's history. Chapter 6 describes modern geomagnetic, electromagnetic and magnetotelluric observations at the Earth's surface and gives a comprehensive overview of the available ground-based data providing information for several aspects and applications of geomagnetic field variations. It includes a special discussion on solar cycle responses in geomagnetic records. Chapter 7 discusses data from low-Earth-orbiting satellites and, in particular, their application in data-based modelling of the geomagnetic field. Part II is completed by a chapter (Chapter 8) that provides an exhaustive overview of recent space missions to investigate the solar wind and the terrestrial space environment and magnetosphere.

The different aspects of spatial and temporal geomagnetic field variations and relevant inferences for other scientific fields and society are the topics of Part III. Going from the deep Earth interior out to the magnetosphere, the part starts with an overview of observational constraints on core field variations and their interpretation in terms of outer core fluid dynamics (Chapter 9). A chapter (Chapter 10) on the lithospheric magnetic field signatures in particular focuses on the international efforts to compile the global World Digital Magnetic Anomaly Map and discusses recent advances in interpreting the structures in terms of magnetisation of the Earth's crust. Chapter 11 deals with energy coupling between different regions in the magnetosphere–ionosphere–thermosphere system. Various processes dominating in different geographical regions from equatorial to auroral zones and polar caps and models thereof are discussed, considering both magnetically quiet and solar storm conditions. The next chapter (Chapter 12) provides an overview of recent developments in energy transport from the solar wind into the Earth's magnetosphere and the role of the boundary layer. It describes how charged particles contained in the magnetospheric ring current and Van Allen radiation belts give rise to boundary regions and waves on magnetospheric scale from ultra-low-frequency up to upper-hybrid waves and continuum radiation. The plasmasphere and plasmapause as its associated boundary are also considered in this chapter. The last chapter in this part (Chapter 13) provides a broad overview of temporal geomagnetic variations ranging from reversals of the internal field to rapid variations in the magnetospheric environment.

The highly societally relevant topic of space weather is discussed in detail in Part IV, which consists of three chapters. Chapter 14 discusses the underlying physical processes of weather in the near-Earth space, in terms of the complex system of Sun–Earth interaction processes. Such processes span a wide range of temporal scales, from thousands of years for the hydrological ocean cycles driven by the total solar radiation to scales of minutes and below for particle acceleration in magnetic reconnection. This chapter briefly describes the important component processes, including solar disturbances such as solar flares and coronal mass ejection, and the connection between magnetic storms and magnetic field depressions at the magnetic equator. Chapter 15 discusses the effects of space weather in the ionosphere and thermosphere and at the Earth's surface, respectively, in the context of the three-way coupling between the magnetosphere, ionosphere and thermosphere. This chapter describes the dynamics of the ionosphere during magnetic storms and sub-storms, especially the role of Alfvén waves and field-aligned currents therein, and the resulting magnetic field perturbations. The chapter also outlines the response of the thermosphere to geomagnetic storms and sub-storms in terms of its heating, expansion and composition changes at high and middle latitudes and the surface-level response in terms of geomagnetically induced currents (GIC) and other geomagnetic and geoelectric disturbances. Chapter 16 explains the various potential impact of space weather on our increasingly technology-dependent society, including the adverse effects of GIC, the disruption of satellite communications and navigation and risks of radiation damage both in space and in aviation.

The book concludes in Part V with a brief overview of the magnetic fields of other bodies in the solar system and an outlook on future challenges in geomagnetism and aeronomy research. Firstly, Chapter 17 presents the results of recent studies on the mechanism of dynamo generation and the magnetic field evolution of planetesimals and exoplanets. The general physics of thermal and chemical dynamos are described. Important new scientific results are underlined, including the thermal and magnetic field evolution of Mercury, Mars, the Moon and Ganymede as well as of planetesimals and rocky exoplanets. Chapter 18 describes the physical processes that contribute to the variability of the solar wind flow that expands from its atmospheric and coronal sources to fill (and propagate through) the whole interplanetary medium. A concise view of the different types of phenomena capable of driving space weather events is given, from those intrinsic to the background solar wind flow itself to the impulsive perturbations related to solar activity. The final chapter (Chapter 19) discusses the problems of the geomagnetic field predictability and the techniques applied for forecasting. From the variety of geomagnetic parameters, two representative variables have been chosen for detailed analysis. One is the magnetic signal of gravitational tidal phenomena, which generate relatively weak but periodic magnetic perturbations. Another variable is the secular variation of the main geomagnetic field, which depends on the evolution of geodynamo and changes slowly but chaotically. Thus the first parameter may be related to short-term geomagnetic prediction, while the second one is suitable for long-term prediction.

This volume represents a compilation of different points of view presented by different authors. Thus some of the chapters partly overlap with others, attesting to the complexity and variability of geomagnetism and aeronomy studies.

We thank all the authors for their dedicated efforts to make this book a true community product. We also wish to express our thanks to our associate editors from the IAGA executive committee, who compiled the five parts of the book: Inez Batista, Archana Bhattacharyya, Brian Fraser, Pieter Kotze, Renata Lukianova and Alan Thomson. We are grateful to the following reviewers, who helped ensure the high quality of the individual chapters: Livia Ribeiro Alves, David Boteler, Nicolas Gillet, Andrew Jackson, Larry Kepko, Jan Lastovicka, Vincent Lesur, Stefan Lotz, Robert Lysak, Frederik Menk, Katia Pinheiro, Rui Pinto, Dhananjay Ravat, Craig Rodger, Angelo de Santis, James Slavin, Sabine Stanley and Du Toit Strauss. Last, but not least, we thank VR2Planets and especially the founder of this start-up company, Francois Civet, for preparing the cover figure. The representation of the geomagnetic field is based on a model from satellite and observatory data.

PART I

Introduction

1
Objectives of Geomagnetic and Aeronomy Studies

David Kerridge

Magnetism was discovered in antiquity as the remarkable ability of a type of black rock, lodestone, to attract iron and to both attract and repel other lumps of lodestone. This was followed by the discovery that, when free to rotate, a piece of lodestone turns and aligns itself in a particular direction. The Earth's magnetic field entered human service when this directive property was made useful through the invention of the compass around a thousand years ago, most probably in China. There is evidence that the Chinese also found that the compass direction deviated from true north–south and so discovered what today is termed *magnetic declination*, the angle between true and magnetic north (Needham, 1962). The compass, as a navigational tool, later became invaluable to mariners in their voyages of exploration and trade (Jonkers et al., 2003). (It is interesting to note the geomagnetic field is now used as a directional reference in high-accuracy drilling for production of hydrocarbons and that many of today's ubiquitous smartphones have in-built magnetic sensors and global geomagnetic field models, putting magnetic navigational capability into the hands of people worldwide.)

The invention of the compass was a technological achievement, and the properties of lodestone stimulated curiosity and the beginning of scientific investigation, notably by Petrus Peregrinus, who introduced the concept of magnetic poles in his letter of 1269. William Gilbert (1600) acknowledged Peregrinus' influence when he published the results of years of careful investigation into the properties of a spherical lodestone in his book *De Magnete*. Gilbert vigorously advocated discovery through meticulous experimentation and measurement, and *De Magnete* is often cited as the first scientific textbook. From the evidence of his experiments, he made the revolutionary claim that the Earth itself is magnetic, and it is not unreasonable to suggest that Gilbert's identification of magnetism as a planetary property was the starting point for geomagnetic science.

The pivoted magnetised needle was developed into a scientific instrument with the aim of discovering more about the behaviour of the geomagnetic field. In 1635 Gellibrand noted the slow drift of the direction of the compass in London over several years, so discovering the secular variation of declination. Graham, in London, around 1722, observed diurnal declination changes and, in collaboration with Celsius in Uppsala in 1741, showed that rapid, large changes were observed at the same time in the two locations, and discovered their correlation with the aurora. The connection between solar activity and major geomagnetic disturbances was firmly established in 1859 when Richard Carrington observed a spectacular white-light solar flare, which was followed by a large magnetic storm about 18 hours later (Carrington, 1859). (The term *magnetischer Sturm*, 'magnetic storm', was first used by von Humboldt.)

Knowledge of geographical changes in declination was vital to navigators, who relied on it to make compass corrections as they travelled, particularly on long ocean voyages. Halley was given command of the Royal Navy ship *Paramore* in 1698–1700, surveyed the Atlantic Ocean, and published his famous contour charts of magnetic variation (declination) throughout the Atlantic in 1701.

The nineteenth century saw rapid progress in experimental investigations of electricity and magnetism. Volta reported his invention of the voltaic pile in 1800. Ørsted (1820) established the intimate link between electricity and magnetism, and Ampère (1820) and Biot and Savart (1820) made fundamental discoveries about electrical currents and magnetic fields. Faraday discovered electromagnetic induction in 1831 and introduced the concept of magnetic field lines of force (Faraday, 1852). Maxwell (1873a, 1873b) gave a unified mathematical formulation of the experimental results, and his equations of electromagnetism continue to provide the foundation for theoretical investigations of the geomagnetic (and geoelectric) field.

Gauss played an important role in the development of geomagnetism in a number of ways. With Weber he developed a method of making absolute measurements of the strength of the geomagnetic field (Malin, 1982) and supported von Humboldt's efforts to build a network of worldwide magnetic observatories by establishing the Göttingen Magnetic Union in 1836. Gauss applied the method of spherical harmonic analysis to global observations, showing how the method could be used to separate fields originating above and below the Earth's surface. He concluded that the geomagnetic field originated largely in the Earth's interior (Gauss, 1839). (Spherical harmonic models remain the standard means of representing the global geomagnetic field, as, for instance, in the International Geomagnetic Reference Field regularly revised and published by IAGA.) Gauss also suggested that the time variations of the field could be due to currents in an electrically conducting layer of the atmosphere.

Maxwell wrote, 'Gauss, as a member of the German Magnetic Union, brought his powerful intellect to bear on the theory of magnetism, and on the methods of observing it, and he not only added greatly to our knowledge of the theory of attractions, but reconstructed the whole of magnetic science as regards the instruments used, the methods of observation and the calculation of the results, so that his memoirs on Terrestrial Magnetism may be taken as models of physical research by all those who are engaged in the measurement of any of the forces in nature'.

In quite different experimental investigations starting in the mid nineteenth century, evidence that magnetised rocks may contain a fossil record of the Earth's magnetic field at the time of their formation emerged. This idea was extended to archaeological materials in the early twentieth century. In 1905 Brunhes found baked clays magnetised in the opposite direction to the present-day field in France and came to the conclusion that the Earth's magnetic field had reversed its polarity in the geological past (Brunhes, 1906). This startling discovery was confirmed in rock samples worldwide and became a vital piece of evidence leading to a revolution in understanding the workings of our planet. Under the assumption that the Earth's magnetic field averages to a geocentric dipole over timescales of around 10 ka, the latitude of the site of a rock sample at the time it acquired its magnetisation can be estimated from the direction of its magnetisation. The changes in latitude of a site deduced from measurements at different geological ages provide strong support for Wegener's proposed continental drift (Wegener, 1929). Marine magnetic surveys have revealed the now-familiar barcode of twinned magnetic stripes showing seafloor spreading (Vine and Matthews, 1963), on either side of the mid-Atlantic ridge, for example. The plate tectonic–driven ocean floor conveyor belt creates a reversal 'egg timer' providing a geomagnetic timescale for around 200 Ma.

As the twentieth century approached, systematic mapping and monitoring of the geomagnetic field had begun, cataloguing its variations in space and time; the physics of electricity and magnetism had been established; and Maxwell's equations gave a mathematical toolkit for theoretical developments. A strong foundation had been laid for the rapid developments in research into the geomagnetic field that followed.

One important reason for progress in the twentieth and twenty-first centuries was the establishment of international organisations allowing scientists to present, share and comment on their results and co-operate and co-ordinate activities. Such organisations played important roles in the International Polar Years in 1882–3 and 1932–3, and the International Geophysical Year in 1957–8, all of which provided impetus for research. The year 1957 was also the start of the space age, and the ability to make scientific measurements on board spacecraft has revolutionised research into the geomagnetic field and much of Earth science in general.

Scientific organisations have to reshape in response to changing research interests, and a pertinent example is that following pressure from upper atmospheric scientists to accommodate their interests, the International Association of Terrestrial Magnetism and Electricity expanded its scope and was renamed the International Association of Geomagnetism and Aeronomy in 1954. This followed the suggestions of Chapman to change 'Terrestrial Magnetism' to 'Geomagnetism', a minor change in terminology, and to adopt his personally invented term 'Aeronomy', which he defined as 'the science of the upper region of the atmosphere where dissociation and ionisation are important' (Chapman, 1938, 1946). Modern aeronomy includes studies of the composition, movement and thermal structure and balance of the atmosphere, naturally including the ionosphere created by the ionising effect of solar ultraviolet radiation.

We now understand that the Earth generates its main magnetic field in the fluid part of the core by dynamo action and that the interaction of the field with the solar wind plasma creates a cavity in space, a dynamic magnetic environment, the magnetosphere, with its boundary at a distance of about 10 Earth radii on the sunward side of the Earth and extending well beyond the Moon's orbit on the anti-sunward side. Within the magnetosphere are complex arrangements of electric and magnetic fields controlling a variety of current systems and charged particle populations. Time-varying magnetic fields from within the magnetosphere–ionosphere system induce electric fields in the solid Earth and the oceans, driving electric currents, which, in turn, create secondary magnetic fields. The magnetisation of rocks, which has both induced and remanent components, creates additional fields. Consequently, any instrumental measurement is the composition of fields from all these sources. So, while the core field is the inherent and dominant magnetic field, geomagnetism can be said to encompass the study of geomagnetic fields in the plural.

Geomagnetism has the benefit of long instrumental records, starting with declination observations made more than 400 years ago and observatory records extending over the last 170 years. These data have, among other things, shown that the dipole moment of the geomagnetic field is now decreasing at around 6% per century, provided evidence that the main field is implicated in length-of-day changes through core–mantle coupling (Holme and de

Viron, 2013) and enabled the evolution of flow patterns at the core surface to be estimated (Holme, 2009). The powerful combination of data from ground-based magnetic observatories and a series of magnetic survey satellites over the last few decades has helped to improve mapping the fluid flow at the surface of the core and understanding of the role of waves in generating the secular variation of the field. A particular challenge is to forecast the evolution of the core field based on this understanding.

Probing the Earth's interior is a major challenge for all areas of the Earth sciences, as structure, material properties and processes must be inferred from remote measurements because the interior cannot be instrumented (the deepest drilled borehole penetrates to less than 13 km below the surface). Combining results from different types of observation helps to reduce ambiguities in interpretations in such inverse problems. For the core, seismology, geodesy, geochemistry and thermodynamics and high-pressure and high-temperature laboratory experiments provide information on structure, composition, physical properties and mechanical and thermal boundary conditions relevant to the dynamo problem.

A great deal of progress has been made in recent years in mathematical modelling of the dynamo process. The feasibility of dynamo action in an homogeneous electrical conductor such as the Earth's core was difficult to establish and only proved theoretically through independent work by Backus (1958) and Herzenberg (1958). Early dynamo models prescribed the form of core motions, but, following the pioneering work of Glatzmaier and Roberts (1995), now there are self-consistent models including the equation of motion for the core fluid that attempt to incorporate constraints based on understanding gained from geophysical data and are capable of producing magnetic fields with characteristics similar to the geomagnetic field, including polarity inversions. However, while recent models exhibit Earth-like behaviour in a qualitative sense, working fully in the Earth's parameter regime is beyond current computational ability and remains a major challenge.

Induction studies, based on surface measurements of magnetic and electric fields, allow estimation of the spatial distribution of electrical conductivity within the Earth's interior, from the crust and lithosphere to the deep mantle. This can complement results from seismology, allowing better definition of thermal and geological structures. Joint inversion of data, including links between electrical conductivity and physical parameters controlling seismic wave propagation, is a promising area of research. Conductivity is sensitive to fluid content, and induction studies can play a part in volcanic hazard assessment by using conductivity contrasts to map magma distributions (Desissa et al., 2013).

The ability to make observations in space, with single and, more recently, constellations of satellites, has enabled rapid global magnetic surveys to be made, resulting in advances in mapping the core and crustal fields, making 3-D models of mantle conductivity and even detecting the subtle motional induction signals generated by ocean currents (Sabaka et al., 2018).

Space missions have also allowed 3-D exploration of the magnetosphere with *in situ* measurements of electric and magnetic fields and charged particle populations (Ganushkina et al., 2018). This has greatly advanced understanding of, for example, processes transferring energy and particles from the solar wind into the magnetosphere, the role of the polar cusps, links into the ionosphere, the dynamics of the radiation belts, processes in the magnetotail and mechanisms involved in magnetic storms and sub-storms and production of the aurora. Research is showing the importance of treating the magnetosphere, ionosphere and neutral atmosphere as a coupled system.

The significance of research into magnetospheric processes has gained the attention of governments and the public because of the threat posed by 'space weather' events driven by solar disturbances (Cannon, 2013). During magnetic storms, the rapid changes in the Earth's magnetic field induce electric fields in the solid Earth, driving geomagnetically induced currents (GIC) into the low-resistance paths provided by grounded systems. The GIC risk to modern-day ground-based technology became particularly apparent during the magnetic storm of 13 March 1989, when the whole of the Hydro-Québec power grid collapsed, leaving 6 million people without electricity for nine hours. Since mankind has been able to launch satellites, we have gradually populated near-Earth space, with spacecraft providing positioning, navigation and timing, telecommunications and Earth observation services on which modern society relies. Understanding magnetospheric processes helps to build the description of hazards required for risk assessments for these vital space-based assets. This requires prediction in the context of a diminishing core field and in the sense of probability distributions for future events.

Geomagnetic science is flourishing for a number of reasons. The availability of observations with high resolution in time and space is leading to the discovery of new phenomena and giving the ability to test theories, and modern computing power supports data analysis and the development of complex mathematical models, bringing into range questions that have been too difficult to attack in the past. The geomagnetic field permeates all parts of the solid Earth, extends into space and interacts in processes to some degree wherever there is finite electrical conductivity. Consequently, it is a vital part of the scientific jigsaw puzzle that needs to be assembled to provide a complete picture of the Earth system.

(*Note*: Courtillot and le Mouël (2007) and Stern (2002) are excellent sources of information on the historical development of geomagnetism and palaeomagnetism.)

References

Ampère, A.-M. (1820) Mémoire sur l'action mutuelle entre deux courants électriques, un courant électrique et un aimant ou le globe terrestre, et entre deux aimants, *Annales de chimie et de physique*, **15**, pp. 59–75, 170–218.

Backus, G. (1958) A class of self-sustaining dissipative spherical dynamos, *Annals of Physics*, **4**(4), pp. 372–447. doi: 10.1016/0003-4916(58)90054-X.

Biot, J.-B. and Savart, F. (1820) Note sur le Magnétisme de la pile de Volta, *Annales de chimie et de physique*, **15**, pp. 222–3.

Brunhes, B. (1906) Recherches sur la direction d'aimantation des roches volcaniques, *Journal of Physics: Theories and Applications*, **5**(1), pp. 705–24. doi: 10.1051/jphystap:019060050070500.

Cannon, P. (2013) *Extreme Space Weather: Impacts on Engineered Systems and Infrastructure*. London: Royal Academy of Engineering. Available at: www.raeng.org.uk/spaceweather.

Carrington, R. C. (1859) Description of a singular appearance seen in the Sun on September 1, 1859, *Monthly Notices of the Royal Astronomical Society*, **20**(1), pp. 13–15. doi: 10.1093/mnras/20.1.13.

Chapman, S. (1938) Geomagnetism or terrestrial magnetism?, *Journal of Geophysical Research*, **43**(3), p. 321. doi: 10.1029/TE043i003p00321.

Chapman, S. (1946) Some thoughts on nomenclature, *Nature*, **157**, p. 405. doi: 10.1038/157405b0.

Courtillot, V. and Le Mouël, J.-L. (2007) The study of Earth's magnetism (1269–1950): a foundation by Peregrinus and subsequent development of geomagnetism and paleomagnetism, *Reviews of Geophysics*, **45**(3). doi: 10.1029/2006RG000198.

Desissa, M., Johnson, N. E., Whaler, K. A., Hautot, S., Fisseha, S., Dawes, G. J. K. et al. (2013) A mantle magma reservoir beneath an incipient mid-ocean ridge in Afar, Ethiopia, *Nature Geoscience*, **6**(10), pp. 861–5. doi: 10.1038/ngeo1925.

Faraday, M. (1852) On the physical character of the lines of magnetic force, *The London, Edinburgh, and Dublin Philosophical Magazine and Journal of Science*, **3**(20), pp. 401–28. doi: 10.1080/14786445208647033.

Ganushkina, N. Y., Liemohn, M. W. and Dubyagin, S. (2018) Current systems in the Earth's magnetosphere, *Reviews of Geophysics*, **56**(2), pp. 309–32. doi: 10.1002/2017RG000590.

Gauss, C. F. (1839) Allgemeine Theorie des Erdmagnetismus, Resultate aus den Beobachtungen des magnetischen Vereins im Yahr 1838, pp. 1–57.

Gilbert, W. (1600) *De Magnete*. London: Peter Short. doi: 10.5479/sil.113709.39088016899940.

Glatzmaier, G. A. and Roberts, P. H. (1995) A three-dimensional self-consistent computer simulation of a geomagnetic field reversal, *Nature*, **377**, pp. 203–9. doi: 10.1038/377203a0.

Herzenberg, A. (1958) Geomagnetic dynamos, *Philosophical Transactions of the Royal Society A: Mathematical, Physical and Engineering Sciences*, **250**(986), pp. 543–83. doi: 10.1098/rsta.1958.0007.

Holme, R. (2009) Large-scale flow in the core, in Olson, P. and Schubert, G. (eds) *Treatise on Geophysics: Vol. 8. Core Dynamics*. Amsterdam: Elsevier, pp. 107–30.

Holme, R. and de Viron, O. (2013) Characterization and implications of intradecadal variations in length of day, *Nature*, **499** (7457), pp. 202–4. doi: 10.1038/nature12282.

Jonkers, A. R. T., Jackson, A. and Murray, A. (2003) Four centuries of geomagnetic data from historical records, *Reviews of Geophysics*, **41**(2). doi: 10.1029/2002RG000115.

Malin, S. R. C. (1982) Sesquicentenary of Gauss's first measurement of the absolute value of magnetic intensity, *Philosophical Transactions of the Royal Society A: Mathematical, Physical and Engineering Sciences*, **306**(1492), pp. 5–8. doi: 10.1098/rsta.1982.0060.

Maxwell, J. C. (1873a) *A Treatise on Electricity and Magnetism*, vol. 1, 1st edn. Oxford: Clarendon Press.

Maxwell, J. C. (1873b) *A Treatise on Electricity and Magnetism*, vol. 2, 1st edn. Oxford: Clarendon Press.

Needham, J. (1962) *Science and Civilisation in China: Vol. 4. Physics and Physical Technology, Part 1*, 1st edn. Cambridge: Cambridge University Press.

Ørsted, H. C. (1820) Experiments on the effect of a current of electricity on the magnetic needle, *Annals of Philosophy*, **16**, pp. 273–6.

Sabaka, T. J., Tøffner-Clausen, L, Olsen, N & Finlay, C.C., (2018) A comprehensive model of Earth's magnetic field determined from 4 years of Swarm satellite observations, *Earth, Planets and Space*, **70**(1), p. 130. doi: 10.1186/s40623-018-0896-3.

Stern, D. P. (2002) A millennium of geomagnetism, *Reviews of Geophysics*, **40**(3), p. 1007. doi: 10.1029/2000RG000097.

Vine, F. J. and Matthews, D. H. (1963) Magnetic anomalies over oceanic ridges, *Nature*, **199**(4897), pp. 947–9. doi: 10.1038/199947a0.

Wegener, A. (1929) *Die Entstehung der Kontinente und Ozeane*. 4th edn. Braunschweig: Friedrich Vieweg.

2
Why Study the Geomagnetic Field?

2.1 Understanding the Internal Structure of Our Planet

Steven Constable

External variations in Earth's magnetic field, driven by interaction with the solar wind at periods shorter than the 11 year sunspot cycle and by lightning in the Earth ionosphere cavity at frequencies higher than 8 Hz, induce secondary electric currents in the crust and mantle. Since the magnitude of the induced currents depends on electrical conductivity, geomagnetic induction studies can be used to probe the electrical structure of the crust and mantle, which in turn variously depends on porosity, salinity, melt content, temperature and composition. Applications include applied geophysics (groundwater, geothermal, oil and gas, mining, and environmental studies), tectonics (plate boundary processes, continental evolution), and the deep Earth (water in the transition zone, mantle heterogeneity). There are two basic techniques available: Magnetotelluric (MT) sounding uses coincident measurements of magnetic and electric fields at Earth's surface to produce an electromagnetic response based on the assumption that the incident magnetic field is a vertically incident plane wave, which is a good assumption locally and at periods less than about 100,000 s. Geomagnetic depth sounding (GDS) uses longer period magnetic field measurements at observatories and from satellites along with estimates or assumptions of the spherical harmonic structure of the field to produce a similar response function. See Constable (2015) for more background information.

Estimating the radial conductivity structure of the mantle using magnetic observatory data has not really changed or advanced significantly since the early work by Banks (1969), but the use of magnetic satellites to estimate GDS responses represents a big step forward. The more complete satellite coverage opens up the possibility of estimating three-dimensional models of mantle conductivity, and the nearly continuous measurements since the Oerstedt satellite launched in 1999 allows the work to be extended to long periods. A challenge in using satellite data is separating the spatial and temporal variations in the data, but the current Swarm mission of three simultaneously recording satellites should help to solve this problem (e.g. Puethe et al., 2015). As laboratory studies improve our understanding of the effects of water, temperature, composition, and melt on the electrical conductivity of mantle minerals (e.g. Pommier, 2014), satellite GDS studies have great potential to illuminate the state and dynamics of the upper mantle and perhaps even the lower mantle (e.g. Puethe and Kuvshinov, 2013)

The MT method was developed in the 1960s, and saw limited application in mainly exploration environments in the following decades. Academic use was hampered by poor processing algorithms, cumbersome equipment that was often made in-house, and interpretation of single sites or small arrays using one dimensional parameterized inversion with, at best, blocky 2D forward modeling. This started to change around the turn of the century with the development of robust, multi-station processing algorithms (e.g. Egbert, 1997), 1D and 2D regularized inversion codes (e.g. Constable et al., 1987), and the availability of commercial acquisition equipment. More recent progress has included availability of 3D inversion codes (e.g. Kelbert et al., 2014) and the use of autonomous logging instruments to collect data, which has allowed a big increase in the number of stations that can be collected and the means to interpret them. Higher station density, which improves shallow resolution and avoids aliasing near-surface conductivity structure into deeper parts of the model, and regularized inversion, which allows model complexity to increase without destabilizing the inversions, has resulted in spectacular images of melting and aqueous fluid migration in the crust and upper mantle (e.g. Wannamaker et al., 2014)

The most impressive manifestation of these improvements in MT methodology is the advent of initiatives to cover entire countries and even continents with MT stations on a 50 to 100 km station grid. These include campaigns to cover China (SINOPROBE) (e.g. Dong et al., 2013), Australia (AUSLAMP) (e.g. Robertson et al.,

2016), and the USA (US Array) (e.g. Meqbel et al., 2014). Although there was some justifiable early concern that the site spacing was too large to allow meaningful interpretations, the 3D coverage and large numbers of stations seems to compensate for this. Although all three of these initiatives are on-going and are several years away from completion, there are some very interesting 3D inversion results being published from the data that are currently available. The model resolution is comparable to seismic tomography inversions, yet the sensitivity of the geomagnetic methods to melt fraction, temperature, and water content is superior.

A significant and recent expansion of the geomagnetic induction toolkit involves the extension to offshore data collection. The improvements to the MT method described above, along with a steep increase in industry activity and support, has resulted in a capability to collect offshore MT data with data quality and site spacing comparable to land stations. As with land MT, early development of the current generation of instruments was motivated by commercial targets, but academic applications followed, with some high-profile results being obtained from mid-ocean ridges and subduction zones (e.g. Naif et al., 2013; Key et al., 2013).

References

Banks, R. J. (1969) Geomagnetic variations and the electrical conductivity of the upper mantle. *Geophys. J. R. Astron. Soc.*, **17**, 457–87.

Constable S. (2015) Geomagnetic induction studies. In: Schubert, G. (Ed.), *Treatise on Geophysics*, 2nd edn., Elsevier, Oxford, 219–54.

Constable, S. C., Parker, R. L. and Constable, C. G. (1987) Occam's Inversion: a practical algorithm for generating smooth models from EM sounding data. *Geophysics*, **52**, 289–300.

Dong, S.-W., Li, T.-D., Lu, Q.-T., Gao, R., Yang, J.-S., Chen, X.-H., Wei, W.-B. and Zhou, Q. (2013) Progress in deep lithospheric exploration of the continental China: a review of the SinoProbe. *Tectonophysics*, **606**, 1–13.

Egbert, G. D. (1997) Robust multiple-station magnetotelluric data processing. *Geophys. J. Int.*, **130**, 475–96.

Kelbert, A., Meqbel, N., Egbert, G. D. and Tandon, K. (2014) ModEM: a modular system for inversion of electromagnetic geophysical data. *Comput. Geosci.*, **66**, 40–53.

Key, K., Constable, S., Liu, L. and Pommier, A. (2013) Electrical image of passive mantle upwelling beneath the northern East Pacific Rise. *Nature*, **495**, 499–502.

Meqbel, N. M., Egbert, G. D., Wannamaker, P. E., Kelbert, A. and Schultz, A. (2014) Deep electrical resistivity structure of the northwestern US derived from 3-D inversion of USArray magnetotelluric data. *Earth Planet. Sci. Lett.*, **402**, 290–304.

Naif, S., Key, K., Constable, S. and Evans, R. L. (2013) Meltrich channel observed at the lithosphere-asthenosphere boundary. *Nature*, **495**, 356–9.

Pommier, A. (2014) Interpretation of magnetotelluric results using laboratory measurements. *Surv. Geophys.*, **35**, 41–84.

Puethe, C., Kuvshinov, A., Khan, A. and Olsen, N. (2015) A new model of Earth's radial conductivity structure derived from over 10 yr of satellite and observatory magnetic data. *Geophys. J. Int.*, **203**, 1864–72.

Puethe, C. and Kuvshinov, A. (2013) Determination of the 3-D distribution of electrical conductivity in Earth's mantle from Swarm satellite data: frequency domain approach based on inversion of induced coefficients. *Earth Planets Space*, **65**, 1247–56.

Robertson, K., Heinson, G. and Thiel, S. (2016) Lithospheric reworking at the Proterozoic-Phanerozoic transition of Australia imaged using AusLAMP Magnetotelluric data. *Earth Planet. Sci. Lett.*, **452**, 27–35.

Wannamaker, P. E., Evans, R. L., Bedrosian, P. A., Unsworth, M. J., Maris, V. and McGary, R. S. (2014) Segmentation of plate coupling, fate of subduction fluids, and modes of arc magmatism in Cascadia, inferred from magnetotelluric resistivity. *Geochem. Geophys. Geosyst.*, **15**, 4230–53.

2.2 Tectonic History of the Earth

Rob van der Voo and Trond H. Torsvik

As time goes by, fewer and fewer paleopoles are presented in publications with respect to classical geological, stratigraphic, or structural aspects, including those assigned to the topic 'tectonic history of the Earth'. And yet, paleomagnetism as a field of study and research of tectonic aspects is alive and well and manages to produce more journal articles and databases than ever. Moreover, this is not all: several books have also seen the light in the last decade. One book we are able to highlight is the weighty monograph excellently produced in 2017 by my co-author Trond Helge Torsvik from the University of Oslo and his colleague Robin Cocks of the Natural History Museum in London, entitled *Earth History and Palaeogeography*, and published and marketed by Cambridge University Press.

Data Reliability and Age Precision

One consequence of the substantial shift away from research about paleopoles and continental paleogeography, at least for rocks of Paleozoic age, is that increasing attention is given to innovative ways to establish the reliability of paleopoles, with various analytical tools, such as tilt-tests, conglomerate, angular unconformity tests and more. Equally important are applications of modern radiometric age dating. With typical pole drift of about 0.5°/Ma, an error of some 15 Myr can amount to up to 7.5°. Age determinations of Permian–Triassic and Cambrian-Ediacaran boundaries are examples from the last two decades that have shown that such age inaccuracies can have important consequences (Chen et al., 2015; Bourquin et al., 2011).

Inclination Shallowing

An interesting development has taken place in the determination of primary bedding in coarse clastic sedimentary rocks, and how stratification of red beds may be affected by flattening due to compaction, or by inclination shallowing due to sedimentary fabric changes. Although these notions have been around since the dawn of paleopole determinations, widespread recognition and acceptance of flattening in systematic ways did not take place until well in the present century, as cogently described by Tauxe and Kodama (2009) and Kodama (2012). In a nutshell, a flattening (*f*) can be invoked on the premise that the rocks in question have the right lithology to have suffered a flattening. The problem is, however, that *f* is rarely known and is, instead, assumed and blithely corrected for, using $\tan(Inc_{Observed}) = f\tan(Inc_{Field})$, where *Inc* is the inclination (Torsvik et al., 2012).

Remagnetization

Mentioned above is the issue of reliability, in terms of the age of a set of paleopoles, which is also a long-standing issue that has been around since the publications of Graham (1949), Creer (1968) and McElhinny and Opdyke (1973) and countless others. If rocks acquire a magnetization at low temperatures, in authigenic minerals such as magnetite or hematite, then these minerals are often demonstrably growing at a time clearly younger than the rock's genesis. The remanence of these authigenic minerals is then called a remagnetization. This label is in common paleomagnetic use, regardless of whether there was an original (primary) remanence or not.

Elongation of VGPs or Directions

Primary magnetizations that reflect the geomagnetic field at the time of their acquisition should also reflect distributions of directions that result from the dipole formula ($\tan\lambda = \frac{1}{2}\tan(Inc)$), where λ is the latitude that corresponds to that calculated from a given remanent direction (Dec, Inc). If the virtual geomagnetic poles (VGPs) have a distribution that is circular about the mean ('paleopole'), then the shape of the directions at the sampling site is elongated. Conversely, if the shape of the direction distribution is circular at the sampling site, then the VGP distribution will be elongated. It has been proposed (Deenen et al., 2011) that adherence to this line of reasoning might constitute a reliability criterion; if the VGPs are circularly distributed, the criterion is met, whereas an elongated VGP distribution causes the criterion to be failed. Examples need to be based on substantial numbers of sites. Granted this, some recent results (Tauxe, 2010; Dominguez and Van der Voo, 2014) have borne this out, but it must be noted that some paleomagnetists have doubted the premise of directional elongation (Linder and Gilder, 2012).

Polar Wander

To average out random noise, paleopoles from a continent are commonly used to construct apparent polar wander paths (APWPs) using running mean (moving window) or spherical spline methods. Only primary paleopoles that fulfill certain reliability criteria (Van der Voo, 1990) should be used to calculate APWPs, which visualize the motion of the polar axis relative to a fixed continent. If the relative locations of several continents are known, then paleopoles from these continents can be combined into a Global APWP (GAPWaP; Torsvik et al., 2012). This is illustrated in Figure 2.2.1a, where we rotated the most reliable paleopoles globally to South Africa co-ordinates and then calculated the GAPWaP. This works well back to the assembly of Pangea (~320 Ma) because relative plate circuits at assembly time, and the subsequent break-up history, are reasonably well known.

Movements of APWPs and continents should represent segments of small circles if the Euler pole is kept constant at

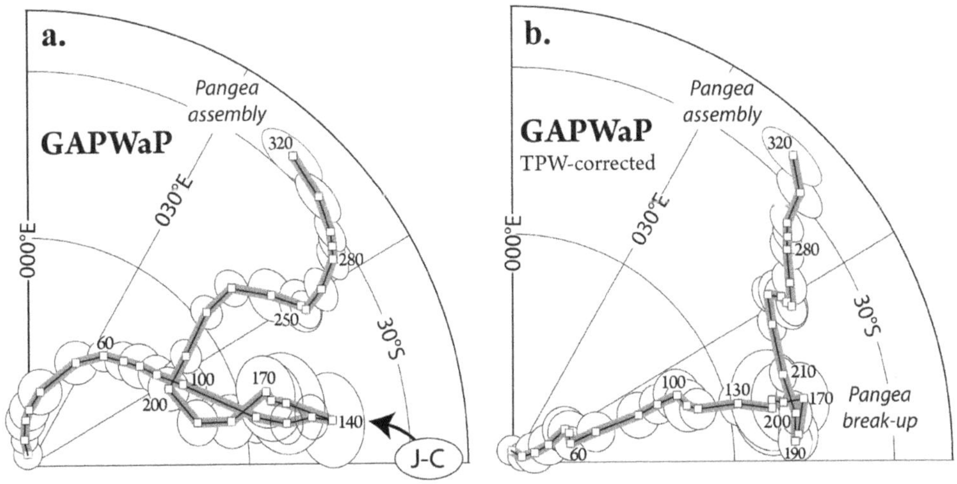

Figure 2.2.1 Global apparent polar wander paths (GAPWaPs) in South African co-ordinates, before and after correction for true polar wander (TPW). The GAPWaP is a running mean path (20 Myr window) and shown with A95 confidence ovals in 10 Myr intervals. The 320–100 Ma TPW corrections are after Torsvik et al. (2012) and Doubrovine et al. (2012) for 100 Ma to recent. J-C, the Jurassic-Cretaceous hairpin (200–100 Ma).

the same location for some time, and sudden changes in the balance of forces driving and resisting plate motions should be reflected in the APWPs as cusps and hairpins (e.g. Gordon et al., 1984). The GAPWaP shows a pronounced Jurassic-Cretaceous hairpin (200–100 Ma; labeled J–C in Figure 2.2.1a), but the motion of the GAPWaP could be due to the drift of a continent ('continental drift'), a rotation of the entire solid Earth (lithosphere and mantle) relative to its spin axis (true polar wander, TPW) or a combination of these causes. TPW is determined by extracting the coherent (mean) rotation of all continents in the paleomagnetic reference frame (Torsvik et al., 2012) but since the Early Cretaceous TPW can also be determined from a direct comparison of paleomagnetic and hotspot frames (Doubrovine et al., 2012). For the past 320 Myrs TPW is characterized by slow (<1°/Myr) oscillations back-and-forth around approximately the same axis (0°N, 11°E). The Early Mesozoic was characterized by systematic counterclockwise rotations of all the continents (250–200 Ma) followed by clockwise rotations until 140 Ma. The Jurassic–Cretaceous hairpin (Figure 2.2.1a) is essentially removed after correction for TPW, and the corrected GAPWaP is dominated by two segments, one from Pangea assembly (320 Ma) to the main break-up between 195 Ma (opening of the Central Atlantic) and 170 Ma (opening of the West Somali Basin between West and East Gondwana), and a second segment from 170 Ma to Recent (Figure 2.2.1b).

Supercontinents

The majority of continents merged to form Pangea at around 320 Ma when Gondwana, Laurussia and intervening terranes collided, and in the process produced the Hercynian-Alleghenian Orogenic Belts in Western Europe and North America. Pangea is the only supercontinent known in some considerable detail but even the initial geometry of Pangea has been vigorously debated (Pangea A vs B) (Irving, 2004; see Domeier et al., 2012). Pangea means 'All land' but did actually not include some of the Asian land-masses (e.g. South China and Annamia), and even though the main break-up phase of Pangea occurred after 200 Ma (opening of the Central Atlantic and splitting Laurussia and Gondwana apart again) some terranes, such as the Cimmerian, rifted off Pangea already in the Early Permian when the Neotethys opened (Figure 2.2.2b).

The reconstruction of older Precambrian supercontinents relies on paleomagnetism in concert with the geological record but the paleomagnetic data are often sparse, and geometric relationships can be ambiguous: at least two Precambrian supercontinents, Rodinia and Nuna (Columbia) have generally been recognized by most workers. From those three supercontinents one may argue for an assembly periodicity of about 750 Myrs based on formation times at 1900 to 1800 Ma (for Nuna), 1100 to 1000 Ma (for Rodinia) and 320 Ma for Pangea (Meert, 2012; Torsvik and Cocks, 2017). Before Nuna, various supercraton connections such as Vaalbara, Superia and Sclavia have been postulated (Evans, 2013).

Paleomagnetic Reconstructions, Longitude and Seismic Tomography

There are many similarities between Wegener (1912) and modern Pangea reconstructions (Figure 2.2.2b) but the most important difference is that we are now able to position Pangea to its ancient latitude and perhaps also its longitude. We are all aware that paleomagnetic reconstructions constrain ancient latitudes and orientations of the continents, and not their longitudes. But this information can be provided by choosing a global reference continent that has remained stable or quasi-stationary with respect to longitude. Africa best meets the criteria and other continents, partnering in the same plate circuit will then occupy their own paleo-longitudinal positions. This is known as the zero-longitude Africa method (Burke and Torsvik, 2004) and GAPWaPs should therefore be constructed in African coordinates (Figure 2.2.1a). From the GAPWaP we calculate Euler rotation poles for southern Africa and rotations for the remaining plates by adding relative plate motions to the 'absolute' rotations of southern Africa in order to make a global reconstruction.

Paleomagnetic reconstructions can also be tested and calibrated in longitude with the positions of subducted material (slabs) which can be observed in tomographic models. Van der Meer et al. (2010) pioneered the concept to interpret the depth and timing of subduction of a total of 28 remnants of slabs in the mantle. Through correlation of those slabs with their geological record, they determined a sinking rate of 12 ± 3 mm/yr, which allowed global correlations of paleo-subduction zones at the surface with seismic tomography images at depth. The longitudes of ancient active margins were constrained by van der Meer et al. (2010) as far back as the Permian (240 Ma) by adjusting a global plate motion model based on paleomagnetic data for improved fits with slab remnants in the tomographic model.

Since the formation of Pangea, plumes that sourced large igneous provinces (LIPs) and kimberlites have been derived from the edges (plume generation zones, PGZ) of two stable thermochemical reservoirs in the lowermost mantle beneath Africa and the Pacific. The remarkable correlation between surface volcanics and lower mantle tomography, and an iterative approach for defining TPW, can be used to position the continents (with their embedded LIPs and kimberlites) in longitude. This is known as the PGZ reconstruction method (Torsvik et al., 2014a) and in Figure 2.2.2b we have used this method to constrain both the Pangean continents and South China (based on the 260 Ma Emeishan LIP) in longitude.

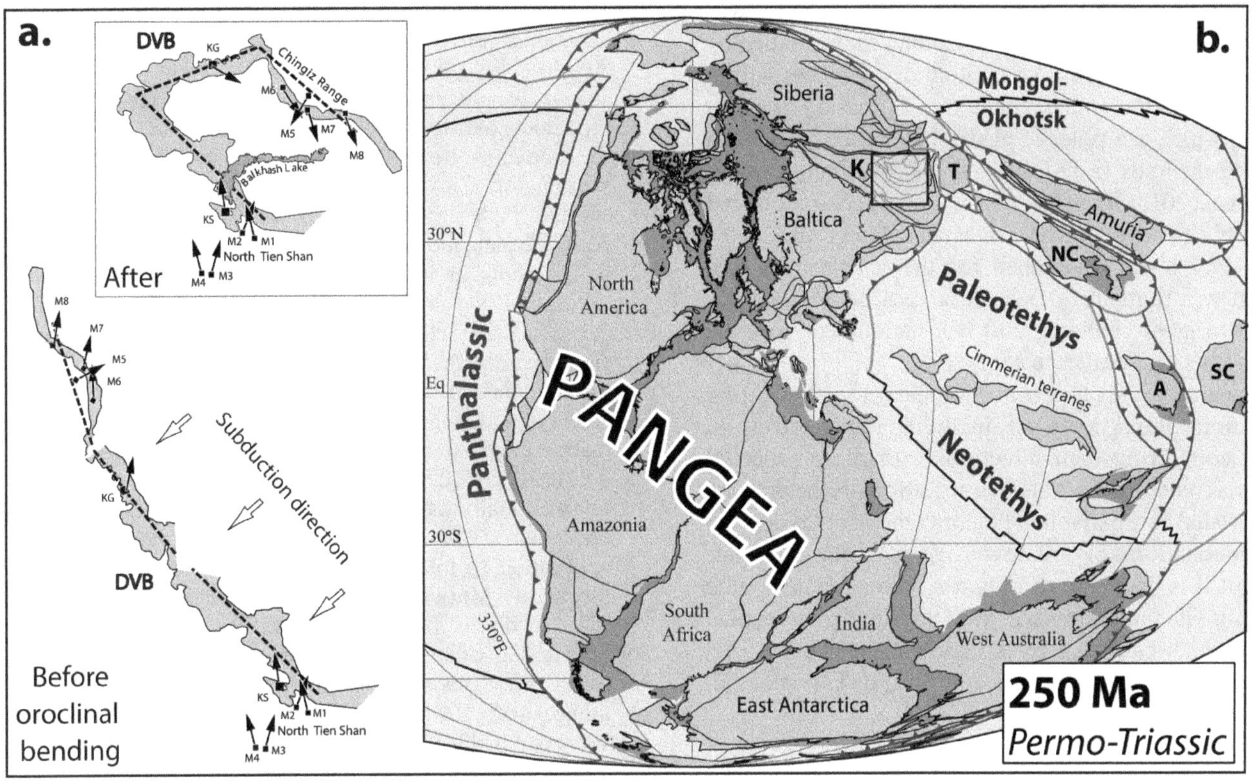

Figure 2.2.2 (a) The relationship between primary Silurian and Devonian declinations (solid arrows) and structural trends in the so-called Devonian Volcanic Belt (DVB) of Kazakhstan in an inferred Middle Devonian configuration before oroclinal bending started and a Permo-Triassic configuration after oroclinal bending (see Abrajevitch et al., 2007). (b) Pangea geography at around 250 Ma, including the opening Neotethys Ocean, the still large but diminishing Paleotethys Ocean, Mongol-Okhotsk and parts of the Panthalassic Ocean. Cimmerian includes terranes from Alborz to Sibumasu, which were parts of the Gondwanan margin in the Early Permian. Solid blue lines are subduction zones with teeth on the upper plate, black lines are spreading centers, and green lines are transform plate margins (Domeier and Torsvik, 2014). Paleomagnetic reconstruction with longitude calibrated with the plume generation zone method (Torsvik et al., 2014a). A, Annamia; K, Kazakhstan (black box approximately the box in panel a, top); NC, North China; SC, South China; T, Tarim. (A black-and-white version of this figure appears in some formats. For the colour version, please refer to the plate section.)

Tectonic Paleomagnetism

Tectonic paleomagnetism comprises studies of deformation (thrusting, faulting, rotations) that have yielded spectacular results in some cases. When mountain belts are curved, while there is evidence for an originally straight (or straighter) configuration, the belt is called an orocline (oros = mountain, cline = tilted, rotated), as first coined by S. Warren Carey (1958). Oroclines of continental scale, hundreds of kilometers, or even a thousand, are incredibly impressive structures that can involve continental margins, displaced continental blocks and belts (Johnston, 2001; Van der Voo, 2004), island arcs and microplates. With the exception of microplates (e.g. Corsica, Sardinia, Tarim), oroclines are features that are defined outside the realm of plate tectonics, which – after all – are premised on crustal rigidity. For a comprehensive review of oroclinal bending and the structures so produced, insufficient space is at this stage available for review. Suffice it to borrow a thousand words from a picture that illustrates the relationship between the structural trends and primary Silurian-Devonian declinations in Kazakhstan, before and after oroclinal bending (Figure 2.2.2a).

Remagnetizations and Apparent Polar Wander Paths

Remagnetizations are very common, and have in past decades deceived many paleomagnetists, these authors included. They can be caused by elevated temperatures (TRM, thermo-remanent) or by rock–fluid interaction and chemical precipitation (CRM). Remagnetizations are generally a nuisance, unless the tests mentioned above allow firm recognition to the contrary. Such is the case for CRMs in carbonate rocks; until the mid 1970s carbonate remagnetizations were thought to be carried by detrital grains of magnetite. Transmission Electron Microscopy (TEM) has revealed some very tiny grains like that, but has hinted rather at authigenesis of magnetite (Weil and Van der Voo, 2002). This has further stimulated a search for the mechanisms of remagnetizations, and has suggested growth of tiny single-domain grains at the time of deformation (folding, faulting), which has been documented to be of a syn-folding nature in many carbonates including those studied by Nemkin et al. (2015, 2016). An exciting development is that carbonate deformation enhances the transformation of smectite to illite, which is a potassic phase that can be dated with argon-argon methods (Fitz-Diaz and van der Pluijm, 2013; Fitz-Diaz et al., 2016). The syn-deformational ages so obtained concur with the ages of the remagnetizations, and allow numerical ages to be established for the remagnetizations. Thus, what was thought of as a nuisance a decade ago has turned into a great challenge and will undoubtedly occupy paleomagnetists for decades to come.

APWPs can serve as a geochronological tool for dating magnetizations of unknown age, including remagnetization events, by comparing their directions with those expected from the reference path. This has traditionally been done through a visual comparison of paleomagnetic poles derived from magnetizations of unknown age with the reference APWP. But it is desirable to obtain a formal age estimate and specify its uncertainty, and a simple statistical approach, which can be used to estimate the 'best-fitting' age and its confidence interval, is described in Torsvik et al. (2014b).

References

Abrajevitch, A. V., Van der Voo, R., Levashova, N. M. and Bazhenov, M. L. (2007) Paleomagnetism of the mid-Devonian Kurgasholak Formation, Southern Kazakhstan: constraints on the Devonian paleogeography and oroclinal bending of the Kazakhstan volcanic arc. *Tectonophysics*, **441**, 67–84.

Bourquin, S., Bercovici, A., López-Gómez, J., Diez, J. B., Broutin, J., Ronchi, A., Durand, M., Arché, A., Linol, B. and Amour, F. (2011) The Permian–Triassic transition and the onset of Mesozoic sedimentation at the northwestern peri-Tethyan domain scale: palaeogeographic maps and geodynamic implications. *Palaeogeogr. Palaeoclimatol. Palaeoecol*, **299**, 265–80.

Burke, K. and Torsvik, T. H. (2004) Derivation of large igneous provinces of the past 200 million years from long-term heterogeneities in the deep mantle. *Earth Planet Sci. Lett.*, **227**, 531–8.

Carey, S. W. (1958) The orocline concept in geotectonics, part 1, *Papers and Proceedings of the Royal Society of Tasmania*, **89**, 255–88.

Chen, D., Zhou, X. Q. and Yong, F. (2015) New U–Pb zircon ages of the Ediacaran–Cambrian boundary strata in South China. *Terra Nova*, **27**, 62–68. doi: 10.1111/ter.12134.

Creer, K. M. (1968) Paleozoic paleomagnetism. *Nature*, **219**, 246–50.

Deenen, M. H. L., Langereis, C. G., van Hinsbergen, D. J. J. and Biggin, A. J. (2011) Geomagnetic secular variation and the statistics of palaeomagnetic directions. *Geophys. J. Int.*, **186**, 509–20.

Domeier, M., Van der Voo, R. and Torsvik, T. H. (2012) Paleomagnetism and Pangea: the road to reconciliation. *Tectonophysics*, **514–17**, 14–43.

Domeier, M. and Torsvik, T. H. (2014) Focus review paper: plate kinematics of the Late Paleozoic. *Geosci. Frontiers*, **5**, 303–50.

Dominguez, A. and Van der Voo, R. (2014) Secular variation of the middle and late Miocene geomagnetic field recorded by the Columbia River Basalt Group in Oregon, Idaho, and Washington, USA. *Geophys. J. Int.*, **197**, 1299–1320.

Doubrovine, P. V., Steinberger, B. and Torsvik, T. H. (2012) Absolute plate motions in a reference frame defined by moving hotspots in the Pacific, Atlantic and Indian oceans. *J. Geophys. Res.*, **117**, B09101. doi: 10.1029/2011JB009072.

Evans, D. A. D. (2013) Reconstructing pre-Pangean supercontinents. *Geol. Soc. Am. Bull.*, **125**, 1735–51.

Fitz-Diaz, E. and van der Pluijm, B. A. (2013) Fold dating: a new Ar/Ar illite dating application to constrain the age of deformation in shallow crustal rocks. *J. Struct. Geol.*, **54**, 174–9. doi: 10.1016/j.jsg.2013.05.011.

Fitz-Diaz, E., Hall, C. M., and van der Pluijm, B.A. (2016). XRD-based 40Ae/39Arage correction for fine-grained illite with application to folded carbonates in the Monterrey Salient (northern Mexico). *Geochim. Cosmochim. Acta*, **181**, 201–216, doi: 10.1016/j.gca.2016.02.004.

Gordon, R. G., Cox, A. and O'Hare, S. (1984) Paleomagnetic Euler poles and the apparent polar wander and absolute motion of North America since the Carboniferous. *Tectonics*, **3**, 499–537. doi: 10.1029/TC003i005p00499.

Graham, J. W. (1949) The stability and significance of magnetism in sedimentary rocks. *J. Geophys. Res.*, **54**, 131–67.

Irving, E. (2004) The case for Pangea B, and the Intra-Pangean Megashear: timescales of the paleomagnetic field. *Geophys. Monogr.*, **145**, 13–27.

Johnston, S. T. (2001) The Great Alaskan Terrane Wreck: reconciliation of paleomagnetic and geological data in the northern Cordillera. *Earth Planet. Sci. Lett.*, **193**, 259–72.

Kodama, K. P. (2012) *Paleomagnetism of Sedimentary Rocks: Process and Interpretation*. Wiley-Blackwell, London.

Linder, J. M. and Gilder, S. A. (2012) Latitude dependency of the geomagnetic secular variation S parameter: A mathematical artifact. *Geophys. Res. Lett.*, **39**, L02308. doi: 10.1029/2011GL050330.

McElhinny, M. W. and Opdyke, N. D. (1973) Remagnetization hypothesis discounted: a paleomagnetic study of the Trenton Limestone, New York State. *Bull. Geol. Soc. Amer.*, **84**, 3697–3708. doi: 10.1130/0016/7606(1973)84<3697:rhdaps>2.0.CO;2.

Meert, J. G. (2012) What's in a name? The Columbia (Palaeopangea/Nuna) Supercontinent, *Gondwana Res.*, **21**, 987–93.

Nemkin, S. R., Fitz-Diaz, E., van der Pluijm, B. A. and Van der Voo, R. (2015). Dating syn-folding remagnetization: approach and field application (central Sierra Madre Oriental, Mexico). *Geosphere*, **11**, 1–12. doi: 10.1130/GES01187.1.

Nemkin, S. R., Lageson, D., van der Pluijm, B. A. and Van der Voo, R. (2016) Remagnetization and folding in the frontal Montana Rocky Mountains. *Lithosphere*, **8**, 716–28. doi: 10.1130/L579.1.

Tauxe, L. (2010) *Essentials of Paleomagnetism*. University of California Press, Berkeley.

Tauxe, L. and Kodama, K. P. (2009). Paleosecular variation models for ancient times: clues from Keweenawan lava flows. *Phys. Earth Planet. Inter.*, **177**, 31–45.
Torsvik, T. H., Van der Voo, R., Preeden, U., Mac Niocaill, C., Steinberger, B., Doubrovine, P. V., van Hinsbergen, D. J. J., Domeier, M., Gaina, C., Tovher, E., Meert, J. G., McCausland, P. J. and Cocks, L. R. M. (2012) Phanerozoic polar wander, paleogeography and dynamics. *Earth Sci. Rev.*, **114**, 325–68.
Torsvik, T. H., Van der Voo, R., Doubrovine, P. V., Burke, K., Steinberger, B., Ashwal, L. D., Trønnes, R., Webb, S. J. and Bull, A. L. (2014a) Deep mantle structure as a reference frame for movements in and on the Earth. *PNAS*, **111**, 24, 8735–40.
Torsvik, T. H., Doubrovine, P. V. and Domeier, M. (2014b) Continental drift (paleomagnetism). In: *Encyclopedia of Scientific Dating Methods*. Springer Science, Berlin. doi: 10.1007/978-94-007-6326-5_107-1.
Torsvik, T. H. and Cocks, L. R. M. (2017) *Earth History and Palaeogeography*. Cambridge University Press, Cambridge.
van der Meer, D., Spakman, W., van Hinsbergen, D. J. J., Amaru, M. L. and Torsvik, T. H. (2010) Absolute plate motions since the Permian interred from lower mantle slab remnants. *Nature Geosci.*, **3**, 36–40. doi: 10.1038/NGEO708.
Van der Voo, R. (1990) Phanerozoic paleomagnetic poles from Europe and North America and comparisons with continental reconstructions. *Rev. Geophys.*, **28**, 167–206.
Van der Voo, R. (2004) Paleomagnetism, oroclines, and growth of the continental crust. *GSA Today*, **14**. doi: 10.1130/1052-5173-(2004)014<4:poagot>2.0.C0;2.
Wegener, A. (1912) Die Entstehung der Kontinente. *Petermann's Mittelungen aus Justus Perthes' Geographischer Anstalt*, **58**, 185–95, 253–6, 305–9.
Weil, A. B. and Van der Voo, R. (2002) Insights into the mechanism for orogen-related carbonate remagnetization from growth of authigenic Feoxide: a scanning electron microscopy and rock magnetic study of Devonian carbonates from northern Spain. *J. Geophys. Res.*, **107**(B4), 2063. doi: 10.1029/2001JB000200.

2.3 Paleomagnetic Dating

Harald Böhnel

For paleomagnetic dating, the record of the paleomagnetic field in a rock or rock sequence is compared to reference curves which describe the variations of this field during the past, and which are already known reliably. This knowledge may be subdivided into four categories and the consequent reference data types: changes of paleopoles over time defining apparent polar wander paths (APWP) for tectonic plates, polarity changes of the Earth's field and the magnetic polarity timescale (MPTS), geomagnetic excursions and events and the geomagnetic instability timescale (GITS), and the short-term changes of the magnetic field over centuries to thousands of years (secular variation, SV). These reference data resolve time periods covering tens of millions of years, hundred thousands of years and tens to thousands of years, respectively, and thus are used in very different contexts of paleomagnetic dating.

Paleomagnetic Dating Using Apparent Polar Wander Paths (APWP)

Starting in the 1950s, APWP have been developed for many of the main tectonic plates known today on Earth. The details of these curves is variable and still changing, particularly for early geological periods (e.g. Torsvik et al., 2012). Nevertheless, reference paleopoles are available for many plates with a resolution of 10 Ma and with 95% uncertainty angles A95 of about 5° on average (e.g. Kent and Irving, 2010). This allows comparing a paleopole obtained by a paleomagnetic study with such an APWP with a typical uncertainty between about 10–30 Ma, depending on the uncertainties of observed and reference paleopoles, but also on the speed of change of the APWP. For example, for the Cretaceous–Paleogene an extended standstill of the APWP is observed for North America plate, over a period of about 80 Ma (e.g. Kent and Irving, 2010).

Paleomagnetic dating using APWP was more common decades ago, when absolute ages by means of geochronology methods were more difficult to obtain or in the context of mineralizations (e.g. Cochran and Elmore, 1987). More recently, it has been used to assign ages to remagnetization processes which may be more difficult to date otherwise (e.g. Henry et al., 2001).

Paleomagnetic Dating Using the Geomagnetic Polarity Timescale (GPTS)

Polarity changes of the Earth's magnetic field have been encountered early in paleomagnetism, and soon their potential for stratigraphic studies e.g. in lava flow sequences in Iceland or volcanic fields in Central France was recognized (Roche, 1953; Hospers, 1954). Since the 1960s, the GPTS has been continuously refined in the context of marine magnetic anomalies produced by seafloor spreading (Vine and Matthews, 1963), and the development of K-Ar (e.g. McDougall et al., 1966) and Ar-Ar (McDougall and Harrison, 1999) geochronology methods. Finally, it has much benefitted from the combination with orbital tuning (e.g. Hilgen et al., 1995; Langereis et al., 2010) and this combination is also called astronomical polarity timescale (APTS). Orbital tuning allows to define the boundaries between consecutive polarity chrons often with a precision of 0.01 Ma or less, normally impossible for absolute geochronology methods. The GPTS is therefore defined with high precision and accuracy over large periods of the Mesozoic and Cenozoic (e.g. Ogg, 2012), and its use may therefore provide age data with high resolution. Magnetostratigraphy has been early and widely applied for correlation and dating to marine (e.g. Opdyke, 1972) and continental sediments (e.g. Fang et al., 2007) and lava flow sequences (e.g. Watkins and Walker, 1977), and also to cave deposits in the context of hominine studies (e.g. Herries and Shaw, 2011). Finally, it was successfully used in volcano-sedimentary sequences, where the paleomagnetic record was thought to be rather unreliable (Lenhardt et al., 2010), but it indeed provided a precise and high-resolution magnetostratigrapy.

Paleomagnetic Dating Using the Geomagnetic Instability Timescale (GITS)

During a polarity chron, geomagnetic excursions and events define moments when the field deviated strongly from that of the geocentric axial dipole. Commonly this was accompanied by much reduced field intensities. The duration of excursions and events is typically shorter than 10 ka, and therefore they are difficult to encounter. During the last decades, large efforts have been undertaken to define excursions and events, by studying marine sediments of high deposition rates and lava flow sequences. The results of such efforts have been concluded in a geomagnetic instability timescale GITS (Singer et al., 1999) and according to Singer (2014), about 19 (14) sub-chrons, events and excursions may have occurred during the Matuyama (Brunhes) chron. These present high-resolution markers in stratigraphic sections, and are therefore of interest in studies of marine sediments (e.g. Nowaczyk et al., 1994). Conversely, such sediment studies have also much contributed to recognize events and excursions.

Paleomagnetic Dating Using the Secular Variation of the Earth's Magnetic Field

The secular variation (SV) produces irregular changes of the direction (inclination and declination) and intensity of the Earth's magnetic field, as has been observed since the earliest geomagnetic observatory measurements, with rates between decades and thousands of years. These changes are known globally in excellent detail only for historic periods (last 400 years; Jackson et al., 2000), and going back in time have to be determined indirectly from paleomagnetic archives like archaeologic and volcanic materials, and lake and cave sediments. Such data are available in detail from some regions (e.g. Bulgaria; Kovacheva et al., 2014), and for others only sporadically or incompletely (e.g. Hagstrum and Champion, 2002). Particularly, only few full vector SV curves of high resolution are available beyond the archaeological period. In regions where high-resolution SV curves are available, paleomagnetic dating of a material can be achieved by comparison of its paleomagnetic record (e.g. Speranza et al., 2008). Elsewhere, local SV curves may be

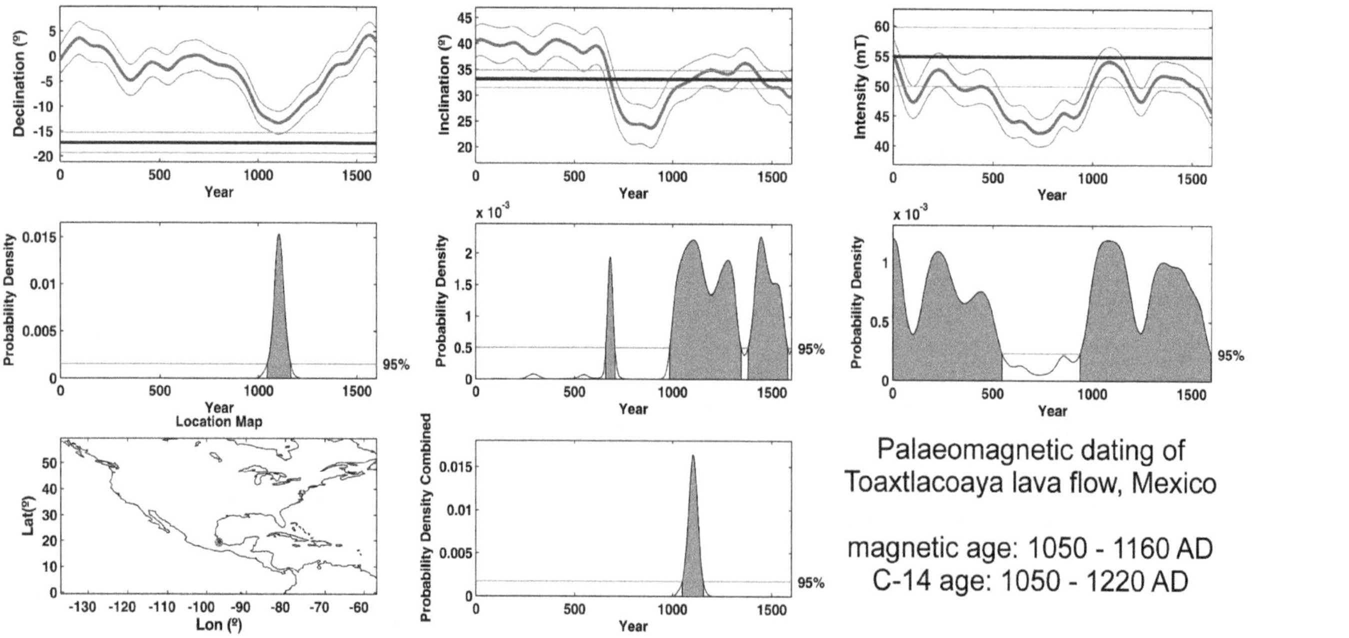

Figure 2.3.1 Paleomagnetic dating of a Holocene lava flow using the archaeo_dating Matlab tool (Pavón-Carrasco et al., 2011). Thick red curves are SV model curves with their 95% uncertainty shown as thin red lines; blue lines are measured paleomagnetic declination, inclination and paleointensity with their 95% uncertainties. Probability density curves indicate the possible ages, with 95% confidence limits. Note how the different field elements provide different possible age ranges but only one combined probability density. Also note the good coincidence between paleomagnetic and C-14 ages. (A black-and-white version of this figure appears in some formats. For the colour version, please refer to the plate section.)

calculated from global field models which have been developed for the last 10–14 ka, using methods like spherical harmonic analysis (CALS10k.1b, Korte et al., 2011; SHA.DIF.14k, Pavón-Carrasco et al., 2014; pfm9k, Nilsson et al., 2014; ARCH10k.1, Constable et al., 2016). Such SV curves will be more deterministic if sufficient regional or close-by paleomagnetic data were involved in the construction of the model. In many regions, the availability of data decreases going back in time, and thus the curves derived from such models loose resolution (e.g. Pavón-Carrasco et al., 2014). Such limitations have to be taken into account when carrying out a paleomagnetic dating. Programs like archaeo_dating (Pavón-Carrasco et al., 2011) and the availability of global paleomagnetic databases like GEOMAGIA50 have much facilitated carrying out paleomagnetic datings (Figure 2.3.1). This way an age range is provided with 1σ or 2σ uncertainty limits, which are comparable to the uncertainties provided by the

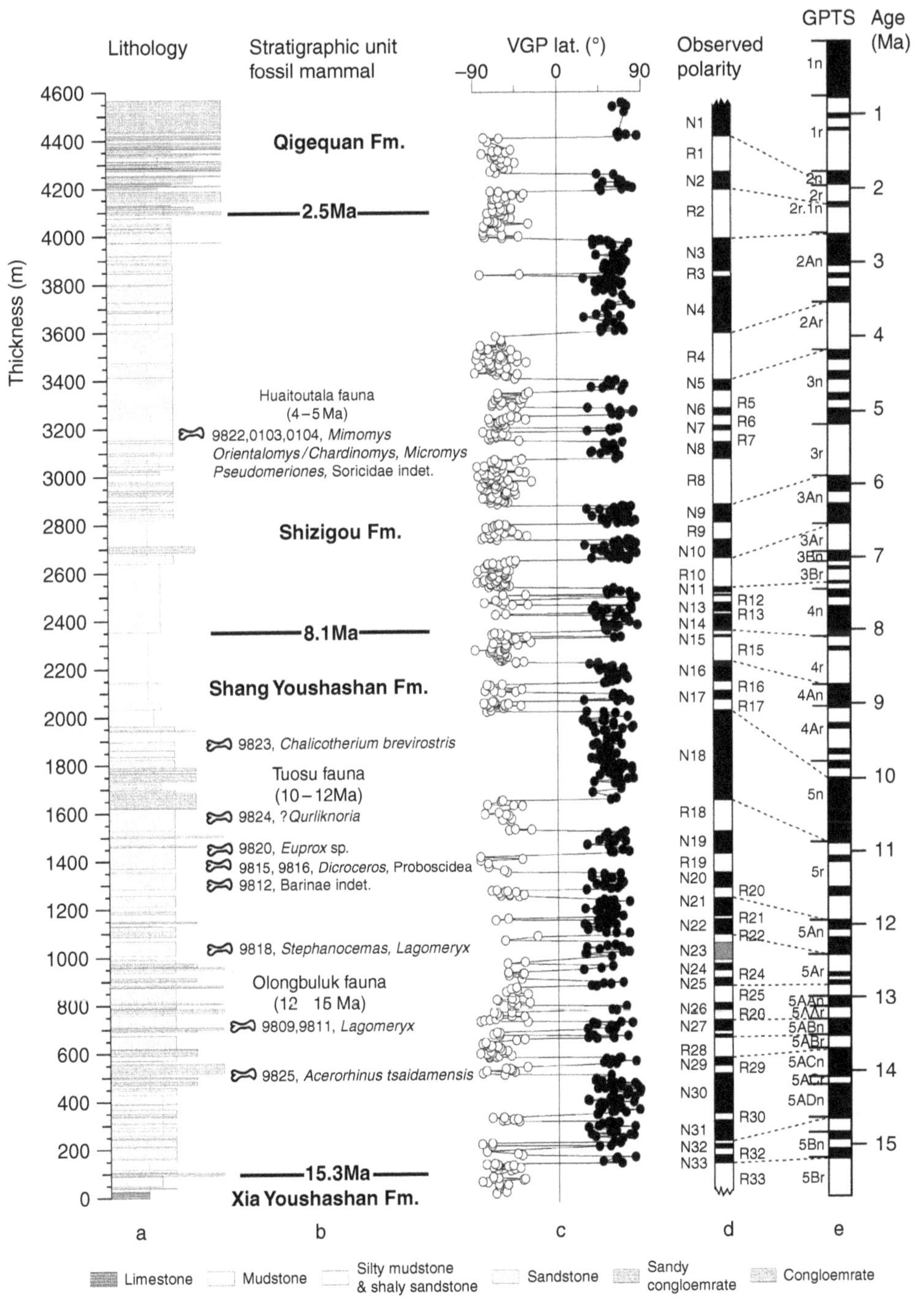

Figure 2.3.2 Magnetostratigraphy of fluvio-lacustrine sediments from the NE Tibetan Plateau: (a) lithology, (b) mammal fauna, (c) VGP latitude, (d) observed polarity and (e) GPTS. From Fang et al. (2007), © Elsevier. This high-resolution magnetostratigraphy allowed the interpretation of sedimentation rate changes and intervals of rapid uplift, not possible by other geochronology methods.

radiocarbon method. Secular variation has produced repetitive field directions and intensities over time, and the dating may thus produce several possible age ranges. Only when additional information is available, like stratigraphic restrictions, some of these age options may be excluded. It must be noted here that a paleomagnetic dating may result in one or several Holocene age ranges, even if the rock indeed is significantly older than the available SV reference curves.

Figure 2.3.2 shows an example of a paleomagnetic dating of a lava flow in Mexico, which has been C-14 dated at 1050–1216 cal. AD. The paleomagnetic age of 1040–1162 AD coincides well within the uncertainty limits. This example also shows the repeating behavior of the magnetic field elements and how the experimental uncertainties impact on the correlations between measured declination, inclination and paleointensity with the SV reference curves; the resulting age here is exclusively determined by the declination.

Paleomagnetic dating based on SV curves has been successfully applied in archaeology (e.g. Catanzariti et al., 2008), where often a broad age range is known by the properties of the piece of interest, and the dating then may provide a much more restricted age. Often archaeological materials are un-oriented and thus only provide paleointensities, which then limits the application of paleomagnetic dating. It has also been very successful in the context of volcanic activity during the Holocene, where in general the full vector may be recovered. This is of major interest for volcanic risk assessment, but in cases also for analyzing the impact of eruptions on the local ancient civilizations (e.g. Mahgoub et al., 2018).

References

Catanzariti, G., McIntosh, G., Soares, A. M. M., Díaz-Martínez, E., Kresten, P. and Osete, M. L. (2008) Archaeomagnetic dating of a vitrified wall at the Late Bronze Age settlement of Misericordia (Serpa, Portugal). *J. Archaeol. Sci.*, **35**, 1399–1407.

Cochran, K. A. and Elmore, R. D. (1987) Paleomagnetic dating of Liesegang bands. *J. Sedim. Res.*, **57**, 701–8.

Constable, C., Korte, M. and Panovska, S. (2016) Persistent high paleosecular variation activity in southern hemisphere for at least 10 000 years. *Earth Planet. Sci. Lett.*, **453**, 78–86.

Fang, X., Zhang, W., Meng, Q., Gao, J., Wang, X., King, J., Song, C., Dai, S. and Miao, Y. (2007) High-resolution magnetostratigraphy of the Neogene Huaitoutala section in the eastern Qaidam Basin on the NE Tibetan Plateau, Qinghai Province, China and its implication on tectonic uplift of the NE Tibetan Plateau. *Earth Planet. Sci. Lett.*, **258**, 293–306.

Hagstrum, J. T. and Champion, D. E. (2002) A Holocene paleosecular variation record from 14C-dated volcanic rocks in western North America. *J. Geophys. Res. Solid Earth*, 107 (B1), 2025. doi: 10.1029/2001JB000524.

Henry, B., Rouvier, H., Le Goff, M., Leach, D., Macquar, J. C., Thibieroz, J. and Lewchuk, M. T. (2001) Palaeomagnetic dating of widespread remagnetization on the southeastern border of the French Massif Central and implications for fluid flow and Mississippi Valley-type mineralization. *Geophys. J. Int.*, **145**, 368–80.

Herries, A. I. and Shaw, J. (2011) Palaeomagnetic analysis of the Sterkfontein palaeocave deposits: Implications for the age of the hominin fossils and stone tool industries. *J. Human Evol.*, **60**, 523–39.

Hilgen, F. J., Krijgsman, W., Langereis, C. G., Lourens, L. J., Santarelli, A. and Zachariasse, W. J. (1995) Extending the astronomical (polarity) time scale into the Miocene. *Earth Planet. Sci. Lett.*, **136**, 495–510.

Hospers, J. (1954) Magnetic correlation in volcanic districts. *Geol. Mag.*, **91**, 352–60.

Jackson, A., Jonkers, A. R. and Walker, M. R. (2000) Four centuries of geomagnetic secular variation from historical records. *Philos. Trans. R. Soc. London A*, **358**, 957–90.

Kent, D. V. and Irving, E. (2010) Influence of inclination error in sedimentary rocks on the Triassic and Jurassic apparent pole wander path for North America and implications for Cordilleran tectonics. *J. Geophys. Res. Solid Earth*, **115**, B10103. doi: 10.1029/2009JB007205.

Korte, M., Constable, C., Donadini, F. and Holme, R. (2011) Reconstructing the Holocene geomagnetic field. *Earth Planet. Sci. Lett.*, **312**, 497–505.

Kovacheva, M., Kostadinova-Avramova, M., Jordanova, N., Lanos, P. and Boyadzhiev, Y. (2014) Extended and revised archaeomagnetic database and secular variation curves from Bulgaria for the last eight millennia. *Phys. Earth Planet. Inter.*, **236**, 79–94.

Langereis, C. G., Krijgsman, W., Muttoni, G. and Menning, M. (2010) Magnetostratigraphy – concepts, definitions, and applications. *Newsl. Stratigr.*, 43(3), 207–33.

Lenhardt, N., Böhnel, H., Wemmer, K., Torres-Alvarado, I. S., Hornung, J. and Hinderer, M. (2010) Petrology, magnetostratigraphy and geochronology of the Miocene volcaniclastic Tepoztlán Formation: implications for the initiation of the Transmexican Volcanic Belt (Central Mexico). *Bull. Volcanol.*, **72**, 817–32.

Mahgoub, A. N., Reyes-Guzmán, N., Böhnel, H., Siebe, C., Pereira, G. and Dorison, A. (2018) Paleomagnetic constraints on the ages of the Holocene Malpaís de Zacapu lava flow eruptions, Michoacán (México): implications for archeology and volcanic hazards. *Holocene*, **28**, 229–45.

McDougall, I., Allsopp, H. L. and Chamalaun, F. H. (1966) Isotopic dating of the Newer Volcanics of Victoria, Australia, and geomagnetic polarity epochs. *J. Geophys. Res. Soild Earth*, **71**, 6107–18.

McDougall, I. and Harrison, T. M. (1999) *Geochronology and Thermochronology by the 40Ar/39Ar Method.* Oxford University Press, New York.

Nilsson, A., Holme, R., Korte, M., Suttie, N. and Hill, M. (2014) Reconstructing Holocene geomagnetic field variation: new methods, models and implications. *Geophys. J. Int.*, **198**, 229–48.

Nowaczyk, N. R., Frederichs, T. W., Eisenhauer, A. and Gard, G. (1994) Magnetostratigraphic data from late Quaternary sediments from the Yermak Plateau, Arctic Ocean: evidence for four geomagnetic polarity events within the last 170 Ka of the Brunhes Chron. *Geophys. J. Int.*, **117**, 453–71.

Ogg, J. G. (2012) Geomagnetic polarity time scale. In: Gradstein, F. M., Ogg, J. C., Schmitz, M. D. and Ogg, G. M. (Eds), *The Geologic Time Scale 2012*. Elsevier, Amsterdam, 85–113.

Opdyke, N. D. (1972) Paleomagnetism of deep-sea cores. *Rev. Geophys.*, **10**, 213–49.

Pavón-Carrasco, F. J., Rodríguez-González, J., Osete, M. L. and Torta, J. M. (2011) A Matlab tool for archaeomagnetic dating. *J. Archaeol. Sci.*, **38**, 408–19.

Pavón-Carrasco, F. J., Osete, M. L., Torta, J. M. and De Santis, A. (2014) A geomagnetic field model for the Holocene based on archaeomagnetic and lava flow data. *Earth Planet. Sci. Lett.*, **388**, 98–109.

Roche, A. (1953) Sur l'origine des inversions d'aimantation constatés dans les roches d'Auvergne. *C. R. Acad. Sci. Paris*, **236**, 107–9.

Singer, B. S. (2014) A Quaternary geomagnetic instability time scale. *Quat. Geochronol.*, **21**, 29–52.

Singer, B. S., Hoffman, K. A., Chauvin, A., Coe, R. S. and Pringle, M. S. (1999) Dating transitionally magnetized lavas of the late Matuyama Chron: toward a new 40Ar/39Ar timescale of reversals and events. *J. Geophys. Res. Solid Earth*, **104**, 679–93.

Speranza, F., Pompilio, M., D'Ajello Caracciolo, F. and Sagnotti, L. (2008) Holocene eruptive history of the Stromboli volcano: constraints from paleomagnetic dating. *J. Geophys. Res. Solid Earth*, **113**, B09101. doi: 10.1029/2007JB005139.

Torsvik, T. H., Van der Voo, R., Preeden, U., Mac Niocaill, C., Steinberger, B., Doubrovine, P. V., van Hinsbergen, D. J., Domeier, M., Gaina, C., Tohver, E. and Meert, J. G. (2012) Phanerozoic polar wander, palaeogeography and dynamics. *Earth Sci. Rev.*, **114**, 325–68.

Vine, F. J. and Matthews, D. H. (1963) Magnetic anomalies over oceanic ridges. *Nature*, **199**, 947–9.

Watkins, N. D. and Walker, G. P. L. (1977) Magnetostratigraphy of eastern Iceland. *Am. J. Sci.*, **277**, 513–84.

2.4 Surveying and Exploration

Jacques Zlotnicki, Malcolm J. S. Johnston, Yoichi Sasai and Jean-Louis Le Mouël

The Total Magnetic Field

The Earth is a complex and fascinating planet. Its main dipolar magnetic field results from convective movement of the electrically conducting fluid iron–sulfur mix that forms the liquid outer core at depths between 3480 and 5150 km. This field is tilted by about 10.5° from the Earth's rotational axis and varies smoothly in space and time (the secular variation) due to changing current interactions in the core. Transient and highly variable contributions in the main magnetic field result from the continuous bombardment by solar flares, solar winds and cosmic rays that disturb the magnetosphere and ionosphere down to a few hundreds of kilometers above the Earth's surface by interactions with resident electric charges. Depending on the electrical conductivity of the crust and mantle, these external disturbance magnetic fields penetrate the Earth to varying depths. The greater the disturbance field period, the deeper the penetration. At the Earth's surface, the total magnetic field is then the sum of the main dipole field, the induced and remanent fields from crustal and mantle materials and transient disturbance fields of ionospheric/magnetospheric origin.

Surface observations show that internal magnetic fields and external transient fields at mid-latitudes are about 45,000 nT and as much as 200 nT, respectively. The time variability of the transient magnetic fields generates electric fields in the crust where these tightly linked electromagnetic (EM) fields follow Maxwell laws (Reitz et al., 2008).

The Earth is a living planet where tectonics, hydrogeology, thermal transfers and different mineralogies may drastically distort the simple pattern described above but measurements of these magnetic fields continuously provide new insights and knowledge about the planet. In the following, we illustrate some of the findings obtained from the study of the magnetic field and more generally from the electromagnetic field.

Magnetism and the Earth's Crust and Mantle

The Earth may be explored with either studies of the main field or the effects of external transient fields penetrating the Earth.

Contributions from Dipole Field Behavior and Paleomagnetism

Ductile convection in the mantle combined with the Earth's rotation produces the well-known crustal plate tectonics that transports magnetic minerals to the ground surface. Progressive magnetization occurs as ambient temperatures decreases below the Curie temperature and freezes in the magnitude and direction the magnetic field at that point in geologic time. Deep magma emerging from volcanoes and cooling at the ground surface creates the same magnetic fingerprint. Thus, the newly magnetized rocks become a marker of the main magnetic field at that epoch.

This mechanism has given rise to a number of findings. The study of magnetic rocks of different epochs defines polar wander, seafloor spreading and the continued movement of the tectonic plates. The reversal in the magnetization of rocks during different geologic periods has shown that the main dipole field has reversed many times in the past

(Cox et al., 1964). Movement of different tectonic plates with similar magnetizations provided a way to determine the duration of the magnetic field reversals. More specifically, the history of volcanic episodes may be approached by stratigraphic studies of the magnetized volcanic products (Gialanella et al., 1993).

Contributions from Magnetotellurics

In contrast to the slow secular variation of the main magnetic field of the Earth, the highly variable external magnetic fields induced by solar activity penetrate the Earth, inducing a secondary transient magnetic field that depends on the electrical conductivity of geological structures (Chave and Jones, 2012). At the ground surface, the synchronous records of transient magnetic and electric fields are linked by an impedance tensor that characterizes the geological structures beneath. For a homogenous structure, the depth penetration or skin depth p (in meters) obeys the simple relationship

$$p \approx 503\sqrt{\rho T},$$

where $1/\rho$ (in S/m) is the electrical conductivity and T is the period (in seconds) of the variations of the transient magnetic field. This is the basis of the magnetotelluric methodology that allows scanning and imaging the Earth from the ground surface to several hundred kilometers depth. Thus, the panel of studies is prodigious. Ore deposits, geothermal fields, gas and oil resources may be found and analyzed for industrial purposes. Roots of faults and magma plumbing systems of volcanoes may be approached for mitigating natural hazards.

Surveying the Earth

Magnetic observatories. Since the beginning of the nineteenth century, the number of magnetic observatories on Earth has grown steadily with the main objective to explore and survey the dynamics of the Earth's interior. At present more 150 observatories distributed over the world contribute to monitor the total magnetic field with a higher and higher accuracy (<0.1 nT) (www.intermagnet.org/). It has been shown that the secular variation due to the convection in the core exhibits a variation from decades to thousands millions of years, which induces a nonregular variation in time and direction at any location on the Earth (Love, 2008). Mapping the secular variation at least every 5 years was necessary when compass-based equipment was used for surveying and navigation but now is not necessary with the advent of GPS (Global Positioning System). The shift of the magnetic declination and temporary accelerations of the secular variation are related to variations in the global length-of-day. At the present time, satellites provide most global surveying information. While global studies were first made using single satellites (CHAMP, DEMETER missions; e.g. Tyler et al., 2003), global surveying is now mostly completed using constellations of interconnected satellites. An example of a global surveying result is the mapping of the Earth magnetic field based on Swarm satellites constellation (Figure 2.4.1) (https://directory.eoportal.org/web/eoportal/satellite-missions/s/swarm).

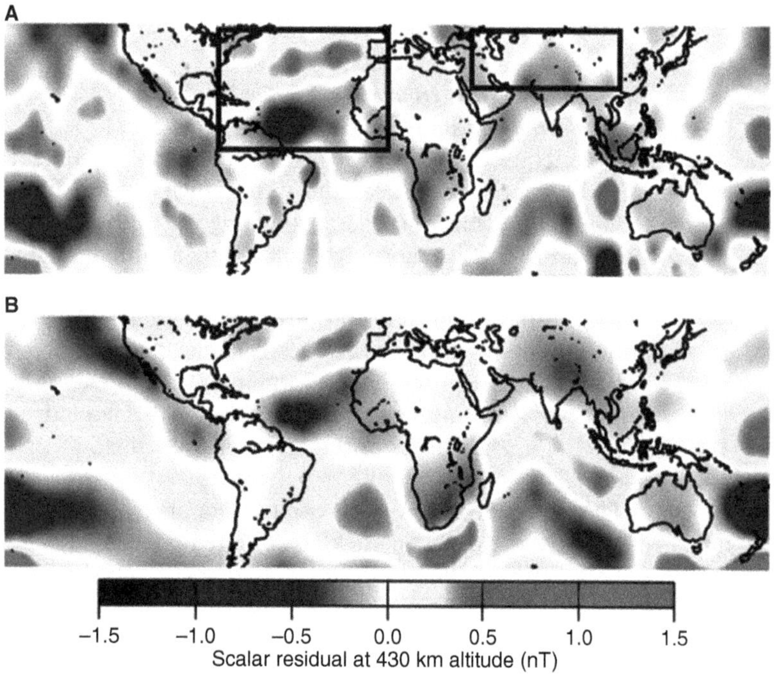

Figure 2.4.1 Earth's magnetic field in June 2014 as observed by the Swarm constellation. Image from ESA/DTU Space, 'Swarm reveals Earth's changing magnetism', ESA, 19 June 2014, www.esa.int/Our_Activities/Observing_the_Earth/Swarm/Swarm_reveals_Earth_s_ changing_magnetism. (A black-and-white version of this figure appears in some formats. For the colour version, please refer to the plate section.)

Monitoring Crustal Activity

The behavior of the magnetic field is intrinsically associated with fault rupture and volcanic activity as a result of active tectonics. Continuous monitoring the magnetic and electric fields allows detection of abnormal temporal and spatial variations linked to these violent disruption phenomena. These phenomena have a great impact on human life since several billions of people live along active faults and on volcanoes. These events have killed more than 1 million people in the last 15 years.

Earthquakes

Seismo-electromagnetic effects refer to electromagnetic (EM) signals generated by fault failure processes in the Earth's crust. These may occur slowly (when associated with plate tectonic loading, slow earthquakes, post-seismic slip, etc.) or rapidly during and following earthquakes. Several different physical processes related to crustal failure can contribute to the generation of seismo-electromagnetic effects and unambiguous observations of these effects provide new independent information about the physics of fault failure. Causal relations between co-seismic magnetic field changes and earthquake stress release have been clearly documented. However, despite many decades of high quality monitoring, clear demonstration of the existence of one-to-one precursory EM signals has not been achieved (Park et al., 1993; Johnston, 1997, 2002, 2007; Uyeda et al., 2002, 2009; Varotsos et al., 2011).

The primary mechanisms for generation of electric and magnetic fields with crustal deformation and earthquake related fault failure include piezomagnetism, stress/conductivity, electrokinetic effects, charge generation processes, charge dispersion, magnetohydrodynamic effects and thermal remagnetization and demagnetization effects (Figures 2.4.2 and 2.4.3). Physical limitations, constraints and frequency limitations placed on these processes are discussed in Johnston (2002).

The precision of local magnetic and electric field measurements on active faults varies as a function of frequency, spatial scale, instrument type and site location. Most measurement systems on the Earth's surface are limited more by noise generated by ionosphere, magnetosphere and cultural

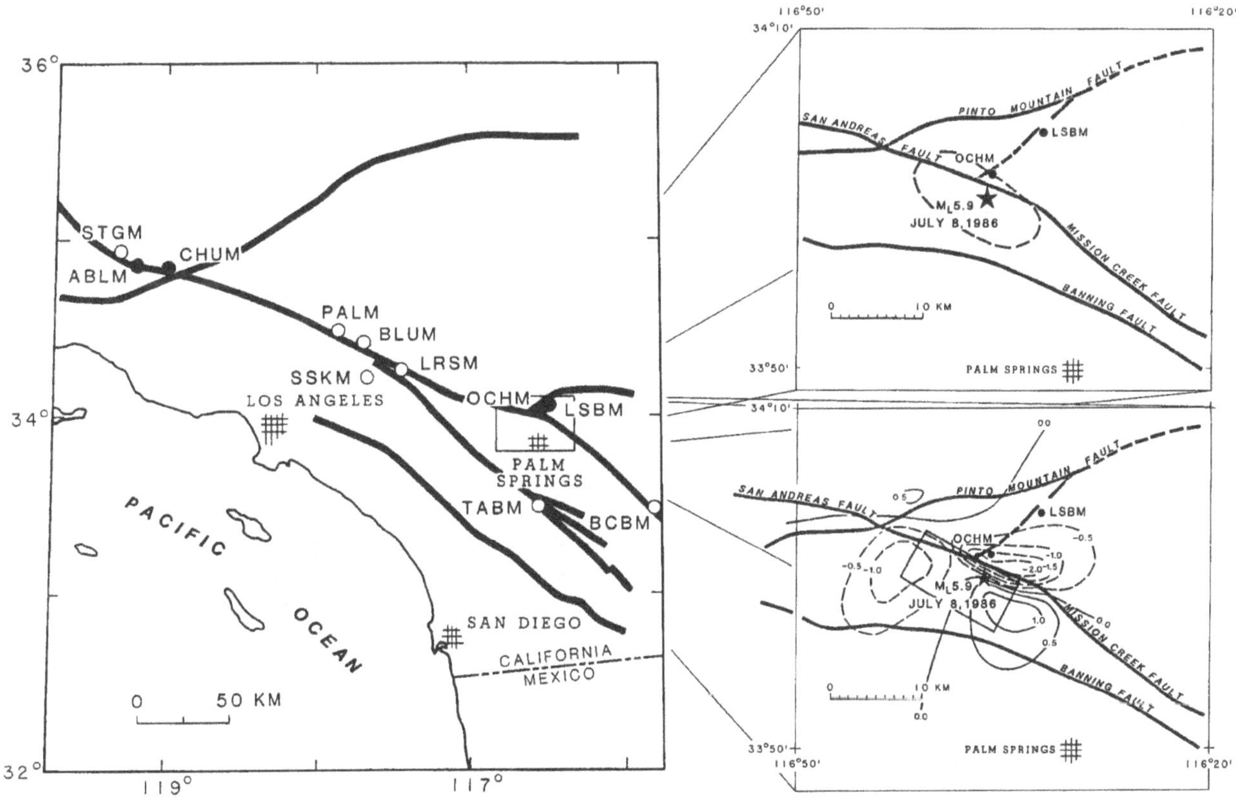

Figure 2.4.2 (left) Recording magnetometer network along the San Andreas fault system in southern California. Filled dots show locations of instruments operating at the time of the 8 July 1986 North Palm Springs Earthquake. (upper right) Expanded section shows the location of the two nearest recording magnetometers, OCHM and LSBM, relative to the epicenter (star) and the aftershock zone (dashed line) of the earthquake. (lower right) Expanded section shows contours of the magnetic field change expected from the earthquake. The surface projection of the fault model is shown as a solid rectangle. Adapted from Figure 1 of Johnston and Mueller (1987).

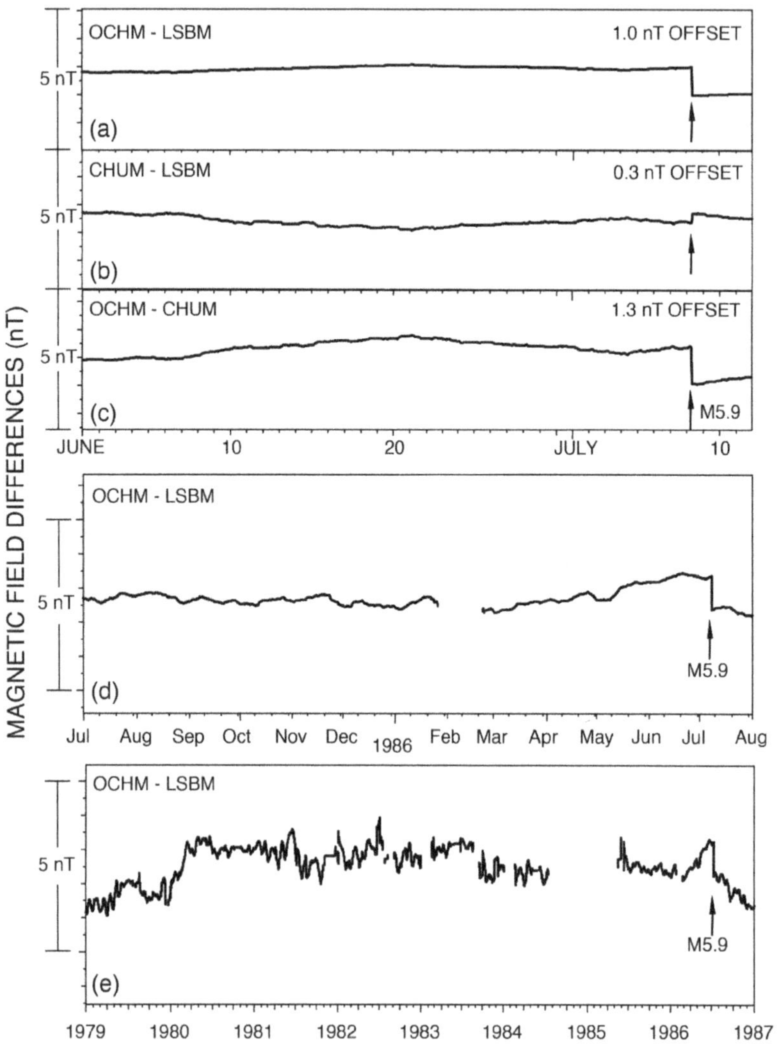

Figure 2.4.3 (a) Magnetic field differences between the two local sites, OCHM and LSBM, for more than a month before and a few days after the North Palm Springs earthquake (occurrence time indicated by an arrow). (b) Magnetic field differences between the sites CHUM and LSBM about 260 km to the northwest for the same time period. (2c) Magnetic field differences between the sites OCHM and CHUM also for the same time period. (2d) Magnetic field differences between the sites OCHM and LSBM for the 12-month period before the earthquake. (2e) Magnetic field differences between the sites OCHM and LSBM for the 7.5 year period before the earthquake. Adapted from Figure 2 of Johnston and Mueller (1987).

disturbances than by instrumental noise. Thus, systems for quantifying these disturbance sources are of crucial importance if changes in electromagnetic fields related to crustal activity are to be uniquely identified.

EM noise increases approximately linearly with site separation. Cultural noise further complicates measurement capability because of its inherent unpredictability and useful electromagnetic field measurements in urban areas are difficult to perform. At lower frequencies (less than some tens of hertz) for both electric and magnetic field measurements, the most common technique involves the use of reference sites with synchronized data sampling in arrays using site spacing comparable to the expected source sizes of a few tens of kilometers. Adaptive filtering and use of multiple variable-length sensors in the same and nearby locations can further reduce noise by about a factor of three.

These same techniques can be applied to electromagnetic field measurements at higher frequencies (100 Hz to MHz)

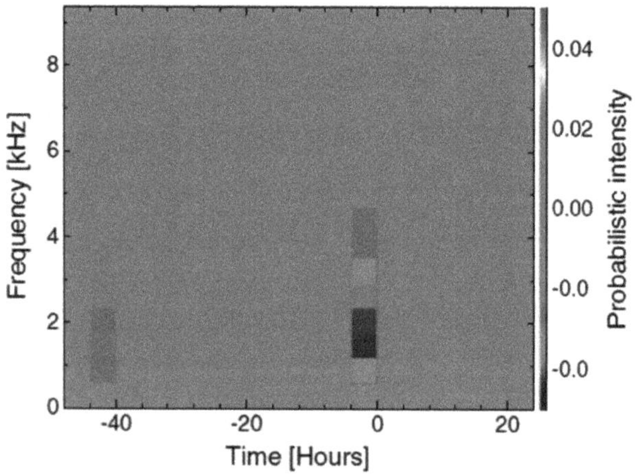

Figure 2.4.4 Frequency-time dependence of the probabilistic intensity obtained from the night-time electric field data measured within 440 km of the epicenters of earthquakes with magnitudes >5.0 during DEMETER mission. After Pisa et al. (2013). (A black-and-white version of this figure appears in some formats. For the colour version, please refer to the plate section.)

but much less is known about the scale and temporal variation of these signals since basic physics precludes simple propagation of high frequency EM signals from seismogenic depths (5–100 km) on active faults in the Earth's crust where the electrical conductivity is more than 0.1 S/m.

Dedicated satellites are now contributing to the Earth's monitoring. DEMETER micro-satellite (2004–10 mission, http://demeter.cnrs-orleans.fr/) recorded the EM field on polar and circular orbits at the altitude of about 700 km. Spatial shift between consecutive orbits only allowed resurveying of specific target every few days. However, it was found that earthquakes of magnitude above 5 may be preceded (few hours) by a decrease of the wave electric field intensity at frequencies of about 1.7 kHz (Parrot et al., 2006; Pisa et al., 2013; Figure 2.4.4). Results were sufficiently promising that the Chinese Earthquake Administration in China has launched one satellite in February 2018 devoted to the monitoring of earthquakes and plans to launch several more (www.leos.ac.cn/data-getData.action?gjx=2&backimage=1).

Volcanoes

Monitoring volcanoes is an easier task since the size of the source area is limited and typically located below the edifice. A focused design of multi-parameter electromagnetic networks that record both magnetic and electric fields can be located around this source. Many mechanisms may produce volcano-EM signals: pore pressure and stress field changes, crystallization process in magma reservoirs, release of thermal fluids and gases from depth, reactivation of hydrothermal systems, eruption of charged ash and gas into the atmosphere, etc. All of these are related to specific physical well-known mechanisms such as piezomagnetic (changes in the magnetization of rocks with the stress fields), electrokinetic (transfers of fluids flow carrying electrical charges), resistivity (electrical changes in bedrocks), induced transient (disturbance of the transient magnetic field) and thermomagnetic (changes in the magnetization of rocks with the temperature) effects (i.e. Stacey and Johnston, 1972; Ishido and Mitzutnai, 1981; Yukutake, 1990; Johnston, 2002). The different space and time distributions of signals from different sources allows separation of these diverse mechanisms. For example, thermal effects take much longer time (months to years) (Kanda et al., 2010) to develop compared to piezomagnetic (months to seconds) (Sasai et al., 2002) or electrokinetic (weeks to seconds) (Zlotnicki et al., 2018) effects related to stress changes or fluids flow changes. Thus, accurate monitoring and data processing may clearly define volcano-EM signals, prior to volcanic activity (Figures 2.4.5 and 2.4.6).

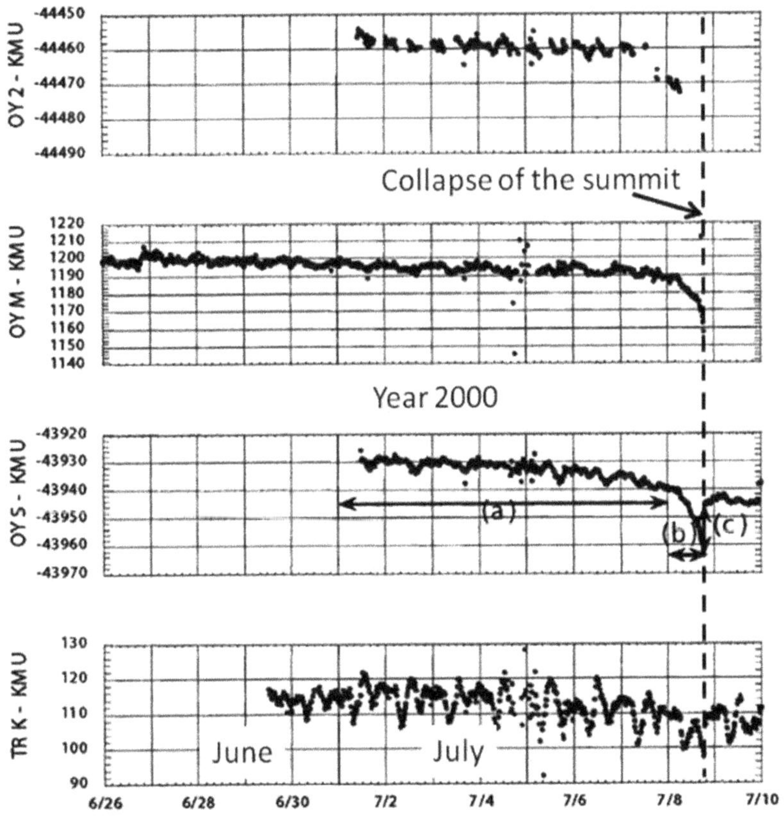

Figure 2.4.5 Total magnetic field changes preceding the 8 July 2000 summit collapse observed at stations along a NS line (top to bottom) crossing Miyake-jima volcano. These short-term signals appear 5 days before the caldera formation. After Sasai et al. (2002).

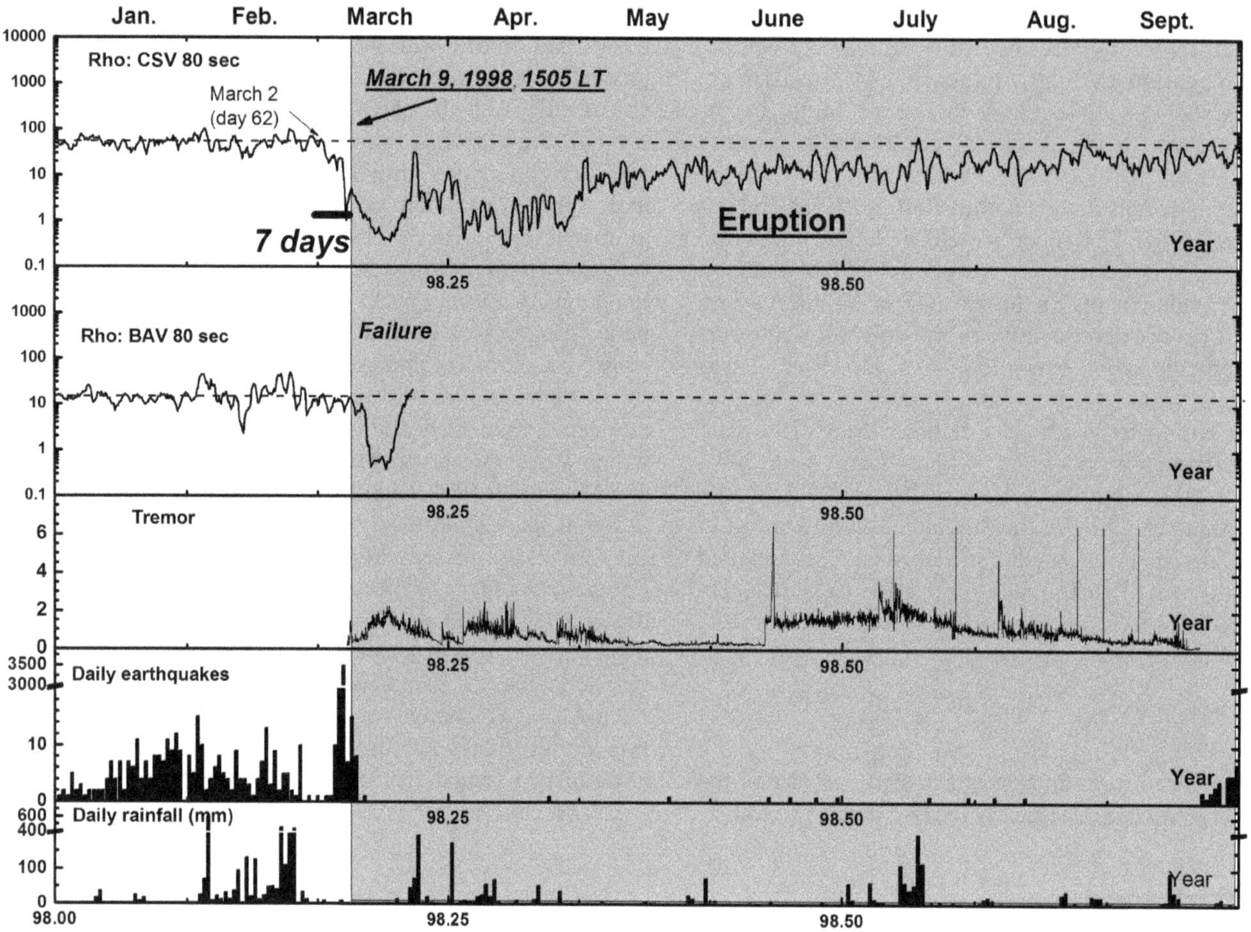

Figure 2.4.6 (from top to bottom) Variations of the resistivity at two stations CSV and BAV before and during the 9 March 1998 eruption at la Fournaise volcano for the 80 s period computed by single MT method. Tremor activity, daily number of earthquakes and daily rainfall. After Wawrzyniak et al. (2017).

Monitoring volcanoes from space could also be very promising. The DEMETER mission has provided records of transient electromagnetic signals during the final two months before eruptions. Seventy-three volcanoes erupted during the 6.5 years of recordings. For 57 of them, 269 anomalies in the ionosphere were apparently related to 89 eruptions. The maximum number of anomalies were recorded between 30 and 15 days before a burst of surface activity, corresponding to the period of accelerating phenomena (Zlotnicki et al., 2013). More than 1,300 volcanoes have erupted in the last 10,000 years, and fewer than one-fourth are monitored by a geophysical instrumentation. Therefore, monitoring volcanoes by a flotilla of satellites could greatly contribute to monitor volcanic eruptions.

References

Chave, A. D. and Jones, A. G. (2012) *The Magnetotelluric Method: Theory and Practice*. Cambridge University Press, New York.

Cox, A., Doell, R., Brent, R. and Dalrymple, G. (1964) Reversals of the Earth's magnetic field. *Science*, **144**, 1537–43.

Gialanella, P. R., Incoronato, A., Russo, F. and Nigro, G. (1993) Magnetic stratigraphy of Vesuvius products. I. 1631 lavas. *J. Volcanol. Geotherm. Res.*, **58**, 211–15, doi: 10.1016/0377-0273(93)90109-5.

Ishido, T. and Mizutani, H. (1981) Experimental and theoretical basis of electrokinetic phenomena in rock-water systems and its applications to geophysics. *J. Geophys. Res.-Sloid Earth*, **86**, 1763–75.

Johnston, M. J. S. and Mueller, R. J. (1987) Seismomagnetic observations during the 8 July 1986, magnitude 5.9 North Palm Springs, California, earthquake. *Science*, **237**, 1201–3.

Johnston, M. J. S. (1997) Review of electric and magnetic fields accompanying seismic and volcanic activity. *Surv. Geophys.*, **18**, 441–76.

Johnston, M. J. S. (2002) Electromagnetic Fields Generated by Earthquakes. In: *International Handbook of Earthquake and Engineering Seismology*, vol. 81A, Academic Press, New York, 621–35.

Johnston, M. J. S. (2007) Seismo-electromagnetic effects. In: *Encyclopedia of Geomagnetism and Paleomagnetism*, Springer, The Netherlands, 908–10.

Kanda, W., Utsugi, M., Tanaka, Y., Hashimoto, T., Fujii, I., Hasenaka, T. and Shigeno, N. (2010) A heating process of Kuchi-erabu-jima volcano, Japan, as inferred from geomagnetic field variations and electrical structure. *J. Volcanol. Geotherm. Res.*, **189**, 158–71. doi: 10.1016/j.jvolgeores.2009.11.002.

Love, J. J. (2008) Magnetic monitoring of Earth and space. *Physics Today*, Feb., 31–6.

Park, S. K., Johnston, M. J. S., Madden, T., Morgan, R., Dale, F. and Morrison, H. F. (1993) Electromagnetic precursors to earthquakes in the ULF band: a review of observations and mechanisms. *Rev. Geophys.*, **31**, 117–32. doi: 10.1029/93RG00820.

Parrot, M., Berthelier, J. J., Lebreton, J. P., Sauvaud, J. A., Santolík, O. and Blecki, J. (2006) Examples of unusual ionospheric observations made by the DEMETER satellite over seismic regions. *Phys. Chem. Earth*, **31**, 486–95.

Piša, D., Němec, F., Santolík, O. R., Parrot, M. and Rycroft, M. (2013) Additional attenuation of natural VLF electromagnetic waves observed by the DEMETER spacecraft resulting from preseismic activity. *J. Geophys. Res. Space Phys.*, **118**, 5286–95. doi: 10.1002/jgra.50469.

Reitz, J. R., Milford, F. J. and Christy, R. W. (2008) *Foundations of Electromagnetic Theory*. 4th edn. Addison-Wesley, New York.

Sasai, Y., Uyeshima, M., Zlotnicki, J., Utada, H., Kagiyama, T., Hashimoto, T. and Takahashi, Y. (2002) Magnetic and electric field observations during the 2000 activity of Miyake-jima volcano, Central Japan. *Earth Planet. Sci. Lett.*, **203**, 769–77. doi: 10.1016/S0012-821X(02)00857-9.

Stacey, F. D. and Johnston, M. J. S. (1972) Theory of the piezomagnetic effect in titanomagnetite-bearing rocks. *Pure Appl. Geophys.*, **97**, 146–55.

Tyler, R. H., Maus, S. and Lühr, H. (2003) Satellite observations of magnetic fields due to ocean tidal flow. *Science*, **299**, 239–41. doi: 10.1126/science.1078074.

Uyeda, S., Hayakawa, M., Nagao, T., Molchanov, O., Hattori, K., Orihara, Y., Gotoh, K., Akinaga, Y. and Tanaka, H. (2002) Electric and magnetic phenomena observed before the volcano-seismic activity in 2000 in the Izu Island Region, Japan. *Proc. Natl. Acad. Sci. USA*, **99**, 7352–5. doi: 10.1073_pnas.072208499.

Uyeda, S., Nagao, T. and Kamogawa, M. (2009) Short-term earthquake prediction: current status of seismo-electromagnetics. *Tectonophysics*, **470**, 205–13.

Varotsos, P., Sarlis, N. V. and Skordas, E. S. (2011) *Natural Time Analysis: The New View of Time: Precursory Seismic Electric Signals, Earthquakes and Other Complex Time Series*. Springer, New York. doi: 10.1007/978-3-642-16449-1.

Wawrzyniak, P., Zlotnicki, J., Sailhac, P. and Marquis, G. (2017) Resistivity variations related to the large March 9, 1998 eruption at La Fournaise volcano inferred by continuous MT monitoring. *J. Volcanol. Geotherm. Res.*, **347**, 185–206, doi: 10.1016/j.jvolgeores.2017.09.011.

Yukutake, T. (1990) An overview of the eruptions of Oshima Volcano, Izu, 1986–1987 from the 1149 geomagnetic and geoelectric standpoints. *J. Geomagn. Geoelectr.*, **42**, 141–50.

Zlotnicki, J., Li, F. and Parrot, M. (2013) Ionospheric disturbances detected by DEMETER satellite over active volcanoes: August 2004 to December 2010. *Geophys. J. Int.*, **183**, 1332–47. doi: 10.1155/2013/530865.

Zlotnicki, J., Sasai, Y., Johnston, M., Fauquet, F., Villacorte, E. and Cordon, J. M., Jr (2018) The 2010 seismovolcanic crisis at Taal Volcano (Philippines). *Earth Planets Space*, **70**, Article 159. doi: 10.1186/s40623-018-0925-2.

2.5 Geospace Environment: State of the Magnetosphere and Ionosphere

Mangalathayil Ali Abdu

The geospace environment may be defined as the region surrounding the Earth in which solar–terrestrial interaction shapes the coupled system of the magnetosphere and ionosphere plasma domains and the neutral atmosphere that underlies them. For practical purposes this region extends from the Earth's mesosphere or the lower ionosphere up to the Sun's vicinity. The geospace environment, which may also be called the geoplasma environment, is a key factor influencing a variety of space-based activities on which the modern high-tech based society depends, on a day-to-day basis. An example is the denser plasma domain of the environment, the ionosphere, in which the plasma structuring at wide-ranging spatial and temporal scales can cause disruptive impact on space application systems (such as the navigation and communication systems operated by global navigation satellite systems (GNSS)). Rapid changes in this environment occurring due to solar and space weather disturbances can further cause serious disruptions to high technology infrastructure of modern life. For these reasons an in-depth understanding of the complex processes controlling the state of the geospace environment, and especially of the magnetosphere–ionosphere system, is a fundamental requirement for developing capability to mitigate the harmful impact of such disturbances. Such understanding is achieved through continuous monitoring of the geospace environmental properties from space- and ground-based platforms, and by using physics-based model simulations on the complex processes that govern the dynamic state of the environment.

The forces controlling the dynamic state of the geospace environment vary from being externally imposed in the form of solar mass and energy emission carried by the solar wind interacting with the upper reaches of the environment to being internally imposed ones from the entire atmospheric system, such as upward propagating wave forces driven by the absorption of solar electromagnetic radiation in the atmosphere. The most distinctive aspect of the geoplasma environment is its control by geomagnetic field. The geomagnetic field uniquely shapes the dynamic states of the plasma distribution and current flow in the entire system, except at its lower boundary, where collisional interaction of the plasma with background neutrals also plays role in shaping such characteristics. Different plasma domains

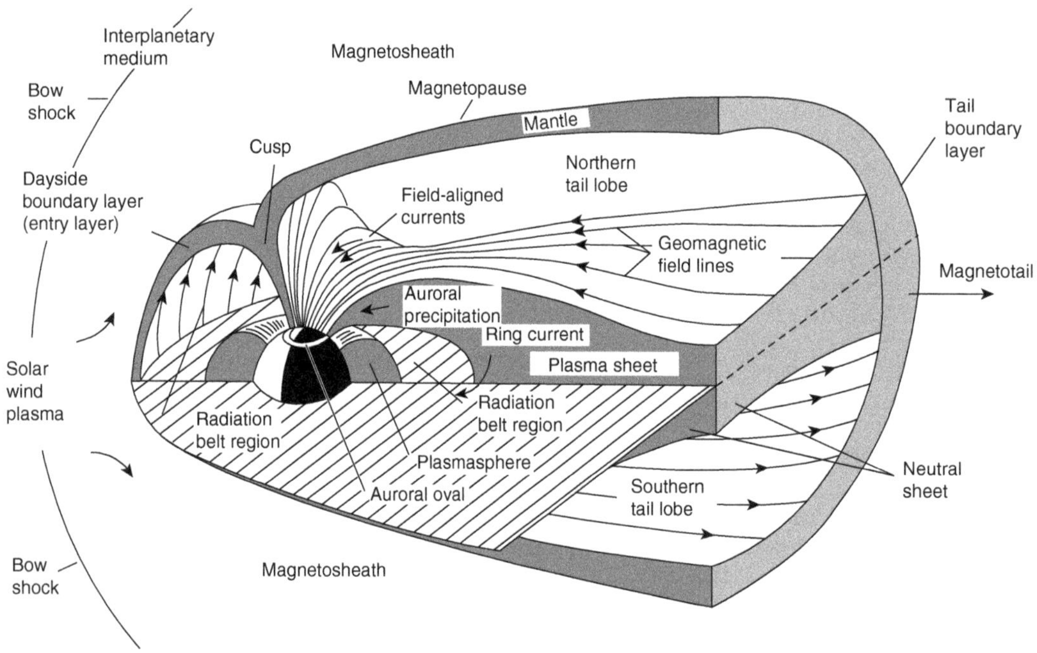

Figure 2.5.1 Schematic of the plasma domains and current systems of the geospace system. After Kelley (1989), courtesy of J. Roederer.

and current systems are present in the system as identified by direct measurements from space-borne detectors as well as by remote diagnostics from ground-based observations. Broadly classified in terms of the ionosphere and magnetosphere, the geoplasma environment includes also the plasmasphere that connects the two regions as well as different other plasma regions produced by the solar wind interaction with the magnetosphere. The constant impact of the streaming plasma of the solar wind and its magnetic field (the interplanetary magnetic field) with the magnetosphere, results in the compression and elongation of the magnetosphere on the dayside and nightside, respectively, as well as in the formation of the highly dynamic plasma domains: the bow shock and magnetopause on the dayside, and the magneto tail and plasma sheet on the night side, of the Earth (Kivelson and Russell, 1995; Southwood et al., 2015).

Figure 2.5.1 (after Kelley, 1989) illustrates the different plasma domains and current systems that shape the geoplasma environment. The auroral oval and precipitation regions, radiation belt regions, field-aligned currents, plasma sheet, ring current and plasmasphere are specifically indicated in the figure. The outer and inner radiation belts encircle the geoplasma environment (Van Allen and Frank, 1959). The complex interactive processes operative in structuring the different plasma domains also lead to formations of the current systems of the geoplasma environment: magnetopause current, ring current, partial ring current, the region 1 and 2 field-aligned currents, tail current, neutral sheet current and auroral substorm current systems, all of which (though not specifically indicated in the figure) play important roles in the electrodynamics of this environment. The magnetic fields due to these currents, as also that of the horizontal currents in the ionosphere, modify the geomagnetic field measured by ground-based and space-borne magnetometers (Luhr et al., 2017; Yamazaki and Maute, 2017). The separation of the different sources of magnetic field components from the geomagnetic field measurements is an important requirement for development of high precision models of the core geomagnetic field. Such separation is a highly complex and challenging task, especially under 'quiet' conditions of the geospace environment. In this context it needs to be noted that significant progress is being achieved in interpreting the different sources of the magnetic field from the dedicated magnetic satellite missions, such as the Swarm (Luhr et al., 2017).

Large-scale disturbances in the geoplasma environment occur following rapid enhancement in solar mass and energy emission, in the form of coronal mass ejections and fast coronal streams producing co-rotating interaction regions, which impact with the magnetosphere. The resulting disturbances in the state of the magnetosphere and ionosphere are manifestations of space weather disturbances, which also include hazardous radiations present in the entire Sun–Earth space and heleosphere. The solar wind shock front carrying high-speed and high-density magnetized plasma cloud (ICME) striking the Earth's bow shock promptly initiates a series of interactive processes that results in drastic modifications of the dynamic state of the magnetosphere–ionosphere system. When the interplanetary magnetic field (Bz) is directed southward, magnetic field line reconnection occurs on the dayside permitting solar wind energy access to high-latitude magnetosphere and ionosphere (Dungey, 1961; Hansen et al., 1971; Tsurutani et al., 1988; Gopalaswamy, 2016). The motional electric field due to the interplanetary magnetic field is mapped directly to high-latitude ionosphere causing build-up of polar cap potential drop, with the associated dawn-dusk electric field driving large-scale plasma convection. The field lines pulled out to the extended night side magnetosphere contribute to enhanced magnetic flux of the tail region, cross tail currents and intensified plasma sheet. Further field line reconnection taking place in this region leads to energized particle precipitation into the auroral oval causing break out of substorms. Concurrent injections of solar wind particle and high-latitude ionospheric ion into the magnetosphere result in ring current intensification defining the development of a global scale magnetic storm. While the development phase of an intense storm is marked by successive substorms, weaker category substorms can occur unaccompanied by magnetic storms (Gonzalez et al., 1994). In all events the solar mass and momentum deposited at high latitude during active periods produce profound changes in the global atmosphere–ionosphere–magnetosphere system. In particular, severe, or super, storms pose potential threat of disruption to ground- and space-based technological systems. The storm induced ionospheric plasma enhancements, and their structuring and dynamics, are of practical interest to the community, due to their disruptive impacts on space application and operational systems of our day-to-day lives, which therefore demand detailed investigations. The degree and extend of such impacts vary around globe, the effects being more intense at equatorial and aurora latitudes.

Furthermore, during the onset and development phases of substorms/storms, energetic particle precipitation in the auroral region causes enhanced ionospheric conductivity so that the dawn-dusk convection electric field drives enhanced auroral electrojet current system and the associated region 1 and region 2 field-aligned currents (FAC), with the accompanying horizontal DP current loops extending to low latitudes (Zmuda et al., 1966; Nishida, 1968; Kamide, 1974; Ijima and Potemra, 1978). Under these situations the convection electric field unshielded by inner magnetosphere penetrates to subauroral and equatorial latitudes as undershielding prompt penetration electric field (PPEF) with eastward (westward) polarity on the dayside (nightside). The region 2 FAC in the inner magnetosphere of closed field lines provides shielding of the convection electric field so that its build-up (in tens of minutes) results in rapid decrease in the PPEF intensity. A subsequent northward turning of the Bz that marks the storm recovery, therefore results in overshielding electric field penetration to low latitude with reverse (dusk–dawn) polarity (Rastogi and Patel, 1975; Kelley et al., 1979). These electric fields occurring in sequence can produce drastic modifications in the plasma density and structuring, and in the current systems, of the equatorial–low latitude ionosphere (Fejer, 2011). A PPEF of eastward polarity associated with the storm development can drive, on the dayside, enhanced equatorial electrojet current (Kikuchi et al., 2008) and large uplift of the ionosphere resulting in large increase of the total electron content (TEC) of the ionosphere, accompanied also by simultaneous pole-ward expansion of the equatorial ionization anomaly (where the plasma density has a global maximum) (Abdu, 1997; Mannucci et al., 2005). These changes can impose enhanced propagation delay on satellite signals leading to errors in positioning applications operated by space-based navigation systems. The same electric field (PPEF) occurring in sunset sector can cause enhanced generation of equatorial plasma bubble irregularities that can disrupt satellite communications due to scintillation effects. When the storm recovery occurs the associated overshielding electric field may reverse some of the detrimental effects, but its eastward polarity in the night sector can cause enhanced development of plasma bubble irregularities.

The recovery phase of an intense storm is marked by delayed global-scale disturbances in thermospheric circulation and composition changes resulting from the high-latitude heating and ion-drag acceleration created during the storm development phase by the energy deposition in the auroral region. The heating can produce upwelling of the auroral ionosphere–thermosphere (IT) system, and depending upon the dynamics and spatial extension of the heated region large-scale atmospheric gravity waves are generated. Their equatorward propagation as fast speed traveling ionospheric disturbances (TID) leaves the middle and low-latitude ionosphere in highly perturbed state. Furthermore, the storm associated disturbance wind directed equatorward and westward interacts with the nearly vertical geomagnetic field lines over mid-latitude resulting in the generation of disturbance dynamo electric field that has dusk–dawn polarity (Blanc and Richmond, 1980; Richmond et al., 2003). As a result the ionospheric phenomena are suppressed during the day while they are enhanced during late night hours. In this way we note that a part of the solar wind and magnetospheric energy deposited at high latitude is dissipated through different channels at equatorial and lower latitude regions of the Earth.

Even under quiescent state of the geoplasma environment representing the absence of activity on the Sun, when the magnetosphere–ionosphere coupling process is in steady balance, with the related magnetic signatures being therefore less conspicuous, the Earth's near space dense plasma environment, the ionosphere, can often be in highly dynamic or perturbed state. Such perturbations arise from vertical coupling processes involving interaction among the atmospheric and ionospheric layers through upward propagating atmospheric wave disturbances, such as tides, planetary waves and gravity waves, originating from the meteorological domain of the lower atmosphere (Forbes, 2000).

The geomagnetic field measurements by ground-based instruments and dedicated magnetic satellite missions (such as CHAMP and Swarm), complemented by model representations of the core field have demonstrated the capability for identifying the sources of magnetic field contributions arising from the different current systems of the magnetospheric and ionospheric domains as well as from those connecting the two (Luhr et al., 2017). Achieving precise results in this regard is a complex task, and is a fundamental requisite in the study of the variables states of the magnetosphere–ionosphere system, and especially in understanding the ionospheric weather in relation to and as a subset of the space weather disturbances. Such studies are already contributing to our in-depth scientific understanding of the interactive processes of the interplanetary–magnetosphere–ionosphere system needed to develop prediction models designed to mitigate possible hazards and disruptions in the technological and space-based application systems of modern life. Related aspects of the key points touched in this brief introduction will be discussed further in the follow-up chapters.

References

Abdu, M. A. (1997) Major phenomena of the equatorial ionosphere-thermosphere system under disturbed conditions. *J. Atmos. Sol. Ter. Phys.*, **59**, 1505–19.

Blanc, M. and Richmond, A. D. (1980) The ionospheric disturbance dynamo. *J. Geophys. Res. Space Phys.*, **85**, 1669–86. doi: 10.1029/JA085iA04p01669.

Dungey, J. W. (1961) Interplanetary magnetic fields and the auroral zones. *Phys. Rev. Letts.*, **6**, 47–8.

Fejer, B. G. (2011) Low latitude ionospheric electrodynamics. *Space Sci. Rev.*, **158**, 145–66. doi: 10.1007/s11214-010-9690-7.

Forbes, J. F. (2000) Wave coupling between the lower and upper atmosphere: case study of an ultra-fast Kelvin Wave. *J. Atmos. Sol. Terr. Phys.*, **62**, 1603–21.

Gonzalez, W. D., Joselyn, J. A. Kamide, Y., Kroehl, H. W., Rostoker, G., Tsurutani, B. T. and Vasyliunas, V. M. (1994) What is a geomagnetic storm. *J. Geophys. Res. Space Phys.*, **99**, 5771–92.

Gopalaswamy, N. (2016) History and development of coronal mass ejections as a key player in solar–terrestrial relationship. *Geosci. Lett.*, **3**, UNSP 8. doi: 10.1186/s40562-016-0039-2.

Hansen, R. T., Garcia, C. J., Grognard, R. J. M. and Sheridan, K. V. (1971) A coronal disturbance observed simultaneously with a white-light corona-meter and the 80 MHz Culgoora radio heliograph. *Proc. Astron. Soc. Austr.*, **2**, 57–60.

Ijima, T. and Potemra T. A. (1978) Large-scale characteristics of field-aligned currents associated with substorms. *J. Geophys. Res. Space Phys.*, **83**, 599–615.

Kamide, Y. (1974) Association of DP and DR fields with the interplanetary magnetic field variation. *J. Geophys. Res. Space Phys.*, **79**, 49–55. doi: 10.1029/JA079i001p00049.

Kelley, M. C. (1989) *The Earth's Ionosphere, Plasma Physics and Electrodynamics*. Academic Press, New York.

Kelley, M. C., Fejer, B. G. and Gonzales, C. A. (1979) An explanation for anomalous equatorial ionospheric electric fields associated with a northward turning of the interplanetary magnetic field. *Geophys. Res. Letts.*, **6**, 301–4.

Kikuchi, T., Hashimoto, K. K. and Nozaki, K. (2008) Penetration of magnetospheric electric fields to the equator during a geomagnetic storm. *J. Geophys. Res. Space Phys.*, **113**, A06214. doi: 10.1029/2007JA012628.

Kivelson, M. G. and Russell C. (Eds) (1995) *Introduction to Space Physics*. Cambridge University Press, Cambridge.

Luhr, H., Xiaong C., Olsen, N. and Le, G. (2017) Near-Earth magnetic field effects of large-scale magnetospheric currents. *Space Sci. Rev.*, **206**, 521–45. doi: 10.1007/s11214-016-0267-y.

Mannucci, A. J., Tsurutani, B. T., Iijima, B. A., Komjathy, A., Saito, A., Gonzalez, W. D., Guarnieri, F. L., Kozyra, J. U. and Skoug, R. (2005) Dayside global ionospheric response to the major interplanetary events of October 29–30, 2003 'Halloween storms'. *Geophys. Res. Lett.*, **32**, L12S02.

Nishida, A. (1968) Geomagnetic DP-2 fluctuations and associated magnetospheric phenomena. *J. Geophys. Res. Space Phys.*, **73**, 1795–1803.

Rastogi, R. G. and Patel, V. L. (1975) Effect of interplanetary magnetic field on the ionosphere over the magnetic equator. *Proc. Indian Acad. Sci. A*, **82**, 121–41.

Richmond, A. D., Peymirat, C. and Roble, R. G. (2003) Long-lasting disturbances in the equatorial ionospheric electric field simulated with a coupled magnetosphere–ionosphere–thermosphere model. *J. Geophys. Res. Space Phys.*, **108**, 1118. doi: 10.1029/2002JA009758.

Southwood, D., Stanley, W. H., Cowley, F. R. S. and Simon, M. (Eds) (2015) *Magnetospheric Plasma Physics: The Impact of Jim Dungey's Research*. Springer, Berlin. doi: 10.1007/978-3-319-18359-6.

Tsurutani, B. T., Gonzalez, W. D., Tang, F., Akasofu, S.-I. and Smith, E. J. (1988) Origin of interplanetary southward magnetic fields responsible for major magnetic storm near solar maximum (1978–1979). *J. Geophys. Res. Space Phys.*, **93**, 8519–31.

Van Allen, J. A. and Frank, L. A. (1959) Radiation around the Earth to a radial distance of 107,400 km. *Nature*, **183**, 430–34.

Yamazaki, Y. and Maute, A. (2017) Sq and EEJ – a review on the daily variation of the geomagnetic field caused by ionospheric dynamo currents. *Space Sci. Rev.*, **206**, 299–405.

Zmuda, A. J., Martin, J. H. and Heuring, F. T. (1966) Transverse magnetic disturbances at 1100 kilometers in the auroral region. *J. Geophys. Res.*, **71**, 5033–45.

2.6 Space Weather Hazard

David Boteler

Studies of the geomagnetic field have always had a practical application through the use of compasses for navigation (Turner, 2011). The early focus was on measurements of the main field of the Earth with Halley publishing the first map of declination in 1701. However, occasionally there were irregular variations which coordinated measurements in 1741 by Graham in London, UK, and Celsius and Hiorter in Uppsala, Sweden, showed were simultaneous over large areas. The Swedish scientists also noted the coincidence of the magnetic variations with the northern lights, heralding a field of study that continues to this day. In the next century the establishment of magnetic observatories enabled continuous recording of the magnetic fluctuations just in time for the introduction of the telegraph, the first electrical system with long conductors deployed across the Earth's surface. Thus when problems occurred because of spurious voltages in the telegraph system it was soon realized that these coincided with magnetic field fluctuations (Prescott, 1866). Measurements of the magnetic field variations at the newly established Toronto magnetic observatory showed an 11 year pattern of occurrence coincident with the sunspot cycle suggestive of a connection (Sabine, 1852). More dramatic indication of a solar source for the magnetic disturbances and their effects on technology came with the solar flare observed by the astronomer Richard Carrington on 2 September 1859 and the subsequent magnetic disturbance that caused worldwide problems with the telegraph (Boteler, 2006).

Since these beginnings in the mid nineteenth century the connections between solar activity and geomagnetic disturbances have been the subject of extensive investigations and the collection of processes is now referred to as 'space weather'. There has also been a growing recognition of the hazard that space weather represents for a wide range of man-made technology ranging from ground systems to radio communications and navigation systems affected by the ionosphere and satellites and astronauts affected by the space environment. On the ground magnetic disturbances continue to be a problem not only for navigation but also in specialized applications like directional drilling and aeromagnetic surveys. However, the most widespread concern about the geomagnetic effects of space weather has been with regard to their effects on systems such as phone cables, pipelines and power systems that involve the use of long conductors.

Problems with the telegraph continued throughout the nineteenth century with problems reported during each major magnetic disturbance. Changes in the technology used with communication cables revealed new problems for both land and undersea cables. Even the modern use of fiber-optics for undersea cables has not removed the problem as these cables still need a conductor to carry power to the repeaters used to boost the signals and geomagnetically induced electric fields interfere with the power feed for the cable. Oil and gas pipelines represent another type of long conductor affected by geomagnetic disturbances. Here the induced voltage fluctuations interfere with the cathodic protection systems used to inhibit the electrochemical processes that cause corrosion of the pipeline: magnetic disturbances interrupt pipeline surveys for monitoring the pipeline protection systems and the pipeline voltage fluctuations take the pipeline outside the safe range for cathodic protection increasing the possibilities of corrosion. The use of higher resistance coatings on modern pipelines has increased the size of the geomagnetically induced (telluric) voltage fluctuations and they are now a factor that has to be taken into account in pipeline design.

Geomagnetic effects on power systems were first reported during the Easter Sunday magnetic storm of 1940. During subsequent magnetic storms there were reports of power system effects in North America and Scandinavia but it was not until March 1989 that the full range of effects became apparent. In this event, the largest magnetic storm of the twentieth century the geomagnetically induced currents (GIC) produced in power systems caused both damage to equipment (transformers) and interfered with system operation (tripping lines out of service), and also led to a blackout of the Hydro-Québec power system (Guillon et al., 2016). Nowadays there is concern that a magnetic storm of the size of the Carrington event, or larger, could cause even more widespread and damaging effects on today's power systems. With modern society's reliance on all thing electrical, anything that has the potential to disrupt the supply of electrical power is a cause for concern and geomagnetic effects on power systems are now regarded as the biggest threat from space weather. Considerable work has been done over the last 150 years to understand the space weather phenomena that lead to the effects on modern technology, but there is more to be done if we are to be able to properly evaluate the risks to these systems and to be able to accurately forecast space weather conditions to maintain operation of the systems on which society depends.

References

Boteler, D. H. (2006) The super storms of August/September 1859 and their effects on the telegraph system. *Adv. Space Res.*, **38**, 159–72.

Guillon, S., Toner, P., Gibson, L. and Boteler, D. (2016) A colorful blackout. *IEEE Power Energy Mag.*, **14**, 59–71.

Prescott, G. B. (1866) *History, Theory, and Practice of the Electric Telegraph*. Ticnor and Fields, Boston.

Sabine, E. (1852) On periodical laws discoverable in the mean effects of the larger magnetic disturbances – No. II. *Philos. Trans. R. Soc. London*, **142**, 103–24.

Turner, G. (2011) *North Pole, South Pole: The Epic Quest to Solve the Great Mystery of Earth's Magnetism*. The Experiment, New York.

2.7 Geomagnetic Field and Biosphere

Jean-Pierre Valet

Magnetism has fascinated since the early discovery of the magnet. Despite its relatively weak strength, the Earth's magnetic field does not escape this rule. It is generated by somehow mysterious motions within the Earth's liquid core and interacts with Sun activity yielding amazing and splendid events like aurorae. It is also remarkable that the geomagnetic field history has been punctuated by large and puzzling convulsions that reach their apogee with the polarity reversals. These extraordinary phenomena incite to speculate on the existence of causal links between the Earth's magnetic field and the biosphere.

The geomagnetic field strength and its geometry impose the shape of the magnetosphere surrounding the planet that is usually regarded as a shield which protects us against penetration of highly energetic cosmic and solar particles. The possibility of links between the geomagnetic field and the biosphere has always generated a large interest which has been recently reactivated by the acquisition of long and detailed time series. The first studies emerged after two major discoveries. The first one was the observation of radiation belts (Van Allen and Franck, 1959), while the second one was the existence of geomagnetic reversals that were rapidly seen as events associated with weak field periods and therefore poor protection against harmful radiations that could affect the biosphere.

There are two issues behind the question of a link between the biosphere and the geomagnetic field. The first one is to find evidence for suitable correlations between significant evolution or extinctions of species and large geomagnetic field changes, particularly during episodes of very low field intensity. The first studies were motivated by the fact that the first detailed paleomagnetic records from marine sedimentary cores that recorded the succession of the last reversals revealed systematic intensity drops during the transitions. Several studies focused on apparent correlations between evolutions of species, particularly radiolarians and field reversals but were disputed and disregarded on statistical grounds. The second issue is to find a suitable mechanism that could account for the impact of low field strength. This is not straightforward because the atmosphere is much more efficient than the geomagnetic field to protect us against harmful radiations.

There is no reason to consider reversals only without taking into account other large intensity drops that occurred during periods of stable field polarity and mostly led to excursions and other similar events referred to as aborted reversals. If we assume that the same processes are at work for successful and unsuccessful transitions, there is no reason for treating the two cases separately. These frequent events are well documented from detailed and multiple records of relative paleointensity and recently from large peaks of beryllium ten production. The existing composite records of geomagnetic field intensity do not go much further than the past 2 Ma. They tell us that intensity dips were very frequent during the past 2 Ma with occurrences going from 20 kyr to 100 kyr. This relatively high frequency may not be fully compatible with appearance or disappearance of species.

Another link between biosphere and geomagnetic field could result from the interaction of cosmic rays with atmospheric compounds (Glassmeier and Vogt, 2010). We know that field reversals and short events are always associated with very low field strength and a complex geometry which is likely multipolar. In such conditions, the entire atmosphere becomes accessible to relatively low energetic solar protons. Ionization in the middle atmosphere initiates a chain of chemical reactions that yields production of nitric oxide (Crutzen et al., 1975; Jackman et al., 2001) which is an ozone depleting chemical. Simulations have shown that, in the absence of field, the total ozone low can reach 40% at medium to high latitudes, particularly in the Northern Hemisphere. The resulting UV-B flux increases by up to 20%, which is three to five times as large as the present annual Southern Hemisphere ozone hole. Potential health effects of UV-B radiations include skin cancer, tissue damage to eyes and effects on cellular immunity.

Two studies suggest that this scenario may not be unrealistic. The first one (Valet and Valladas, 2010) pointed out the synchronism between the Laschamp event (the most recent aborted reversal) that occurred only 40 kyr ago and the demise of Neanderthal during the same period. Conditions for generating significant ozone losses between the geographic poles and the mid-latitudes were met during the Laschamp and persisted during centuries of reduced geomagnetic intensity. The large doses of UV-B may have caused deleterious effects particularly for the young population while the modern men had black skin and were therefore much less affected. There was also a large reserve of population in the African territories that was not concerned by ozone loss because nitric oxide lifetime is considerably shortened by solar radiation.

The second study (Meert et al., 2016) links the terminal Ediacaran extinction with geomagnetic reversal frequency. From around 550 Ma and during much of the Cambrian, the Earth's magnetic field was rapidly reversing. The authors assume that the averaged field was low enough to generate increased dosing of UVB radiation into the shallow marine environment and provided selective pressure favoring organisms that could detect and avoid UVB damage. If there was a significant increase of UV-B at the Earth's surface during periods of low geomagnetic intensity, it might be possible in future to test it through biological indices. Glassmeier et al. (1999) suggest that the density of colored organic matter could be one approach.

Another link between geomagnetic reversals and evolution has also been recently suggested (Wei et al., 2014). The concept is that oxygen escape from the Earth's atmosphere can increase by 3–4 orders during periods of low geomagnetic field that happen during reversals. The authors noticed an apparent link between reversal frequency increase, oxygen level decrease in parallel with the pattern of marine species mass extinctions.

Attempts were also made to detect significant long-term variations in the evolution that would be associated with pseudo-periodicities of the geomagnetic field. A relationship between climatic changes generated by Earth's orbital changes and geomagnetic field has been frequently suggested. This assumption implies that the fluid core motions that drive the geodynamo would be sensitive to the Earth's orbital changes. After being disregarded on experimental and theoretical grounds (Kent and Carlut, 2001), the hypothesis of a precession-driven dynamo has recently been advocated (Tilgner, 2007; Wu and Roberts, 2008). In the meantime links between climatic boundaries and geomagnetic lows were put forward (Worm, 1997; Fuller, 2006; Thouveny et al., 2008), but they are not systematic.

On a much longer timescale, the most striking evidence for indirect but related changes between the biosphere and the geomagnetic field involves thermal and chemical couplings between the successive envelopes from the Earth's core to the atmosphere. Three of the four most recent large mass extinction events in the Phanerozoic – the Cretaceous–Tertiary, the Triassic–Jurassic, the Permo-Triassic and the Guadalupian–Tatarian were preceded by major superchrons (the oldest one being not magnetically documented yet) (Courtillot, 1994). It is proposed that exceptional volcanism was related to thermals released after the unusually long periods with no magnetic reversals. Hot material raised from the core–mantle boundary to the surface where it caused widespread basaltic volcanic eruptions. The long time taken for the ascent of material through the mantle explains the 10–20 Myr delay between the end of each superchron and the subsequent mass depletion event. These major eruptions released large amounts of CO_2 and sulfates which have a pronounced effect on the climate and biota. This is also compatible with recent modeling (Mussard et al., 2014) which defends that continuous release of volcanic CO_2 was not sufficient to generate long-term warming and that it was enhanced by large-scale SO_2 injections.

Summarizing, the past two decades have seen the emergence of plausible models linking the history of the geomagnetic field and the biosphere which involve successions of complex and mostly nonlinear processes. They either rely on secondary effects resulting from formations of atmospheric compounds that affected the ozone layer, on climatic changes generated by intense volcanic eruptions, or changes in the geodynamo generated by Earth's orbital variations. These assumptions are sustained by observations, but in most cases they remain to be tested further using multiple approaches. There is no doubt that the upcoming years will see substantial progress on these aspects.

References

Courtillot, V. (1994) Mass extinctions in the last 300 million years: one impact and seven flood basalts. *Isr. J. Earth Sci.*, **43**, 255–66.

Crutzen, P. J., Isaksen, I. S. and Reid, G. C. (1975) Solar proton events: Stratospheric sources of nitric oxide. *Science*, **189**, 457–9.

Fuller, M. (2006) Geomagnetic field intensity, excursions, reversals and the 41,000-yr obliquity signal. *Earth Planet. Sci. Lett.*, **245**, 605–15.

Glassmeier, K. H., Othmer, C., Cramm, R., Stellmacher, M. and Engebretson, M. (1999) Magnetospheric field line resonances: a comparative planetology approach. *Surv. Geophys.*, **20**, 61–109.

Glassmeier, K. H. and Vogt, J. (2010) Magnetic polarity transitions and biospheric effects. *Space Sci. Rev.*, **155**, 387–410.

Jackman, R. J., Floyd, T. M., Ghodssi, R., Schmidt, M. A. and Jensen, K. F. (2001) Microfluidic systems with on-line UV detection fabricated in photodefinable epoxy. *J. Micromech. Microeng.*, **11**, 263–9.

Kent, D. V. and Carlut, J. (2001) A negative test of orbital control of geomagnetic reversals and excursions. *Geophys. Res. Lett.*, **28**, 3561–4.

Meert, J. G., Levashova, N. M., Bazhenov, M. L. and Landing, E. (2016) Rapid changes of magnetic field polarity in the late Ediacaran: linking the Cambrian evolutionary radiation and increased UV-B radiation. *Gondwana Res.*, **34**, 149–57.

Mussard, M., Le Hir, G., Fluteau, F., Lefebvre, V. and Goddéris, Y. (2014). Modeling the carbon-sulfate interplays in climate changes related to the emplacement of continental flood basalts. *Geol. Soc. Am. Spec. Pap.*, **505**, 339–52.

Thouveny, N., Bourlès, D. L., Saracco, G., Carcaillet, J. T. and Bassinot, F. (2008) Paleoclimatic context of geomagnetic dipole lows and excursions in the Brunhes, clue for an orbital influence on the geodynamo? *Earth Planet. Sci. Lett.*, **275**, 269–84.

Tilgner, A. (2007) Kinematic dynamos with precession driven flow in a sphere. *Geophys. Astrophys. Fluid Dyn.*, **101**, 1–9.

Valet, J. P. and Valladas, L. (2010) The Laschamp-Mono lake geomagnetic events and the extinction of Neanderthal: a causal link or a coincidence? *Quat. Sci. Rev.*, **29**, 3887–93.

Van Allen, J. A. and Frank, L. A. (1959) Radiation around the Earth to a radial distance of 107,400 km. *Nature*, **183**, 430–34.

Wei, Y., Pu, Z., Zong, Q., Wan, W., Ren, Z., Fraenz, M. and Hong, M. (2014) Oxygen escape from the Earth during geomagnetic reversals: implications to mass extinction. *Earth Planet. Sci. Lett.*, **394**, 94–8.

Worm, H. U. (1997) A link between geomagnetic reversals and events and glaciations. *Earth Planet. Sci. Lett.*, **147**, 55–67.

Wu, C. C. and Roberts, P. H. (2008) A precessionally-driven dynamo in a plane layer. *Geophys. Astrophys. Fluid. Dyn.*, **102**, 1–19.

3

Major Scientific Contributions of the International Association of Geomagnetism and Aeronomy (IAGA) during the Past 100 Years

Edward W. Cliver and Eduard Petrovský

3.1 Geomagnetism and Aeronomy circa 1919

The landscape of geomagnetic and aeronomic science one hundred years ago when IAGA was founded bore little resemblance to that of today. The scientific highlights in geomagnetism for the century preceding 1919 reveal both the level of understanding of geomagnetism and aeronomy and the pace of progress in these fields at the time of the founding of IAGA's predecessor, the Section for Terrestrial Magnetism and Electricity (STME)[1] which became the International Association of Terrestrial Magnetism and Electricity (IATME) in 1930 and IAGA in 1954. These highlights include: (1) Gauss's absolute measurement of the geomagnetic field in 1832, his establishment, with Weber, of the Göttingen Magnetic Union during the early 1830s, and his application of spherical harmonic analysis to the data obtained therefrom in 1839 that refuted Halley's resilient quadrupolar picture of the Earth's magnetic field (Garland, 1979); (2) the development of the British Colonial Observatories, at the suggestion of von Humboldt and under the direction of Sabine, in the 1840s (Cawood, 1979); (3) the first observation of a solar flare (Carrington, 1860; Hodgson, 1860) and first association of a solar flare with geomagnetic disturbances (Stewart, 1861); (4) the inference by Stewart (1883), Kennelly (1902) and Heaviside (1902) that the regular diurnal geomagnetic variability originated in an electrically conducting region of the atmosphere (although the existence of the ionosphere was not confirmed until 1925; Appleton and Barnett, 1925), with the first rudimentary depiction of the magnetosphere (the Chapman-Ferraro cavity; Chapman and Ferraro, 1931a, 1931b, 1933) following in the next decade; (5) mapping of the northern auroral zone by Loomis (1860) and Fritz (1881); (6) the introduction of the first geomagnetic index, the Ci planetary character figure, in 1906 (Lincoln, 1967); (7) the discovery of a causative role of the Sun for Earth's magnetic variability (Schwabe, 1844; Sabine, 1852; Wolf, 1852a, 1852b; Gautier, 1852; Maunder, 1904) even though the mechanism(s) remained elusive (Stern, 1989; Cliver 1994a, 1994b, 1995); and (8) Birkeland's (1901, 1908) experiments with cathode rays and a terella (magnetized sphere) which provided support for the 'corpuscular' hypothesis (e.g. Fitzgerald, 1892, 1900; Cliver 1994b) of solar–terrestrial interaction. In 1918 Chapman presented his first theory of geomagnetic storms, positing streams of charged particles (thought to be mainly of one sign) as the responsible agent (Chapman, 1918). This model was challenged by Lindemann (1919) because of the deleterious effect of the repulsion of like charges. Lindemann suggested instead that the solar stream would have to be electrically neutral, i.e. what we today call a plasma. Such was the state of the art for geomagnetic and upper atmospheric science ca. 1919. In terms of observations, Merlin and Somville (1910) compiled information on eighty stations with reported geomagnetic programs/instrumentation at that time (Figure 3.1.1a) versus the current list of ~130 stations worldwide (Figure 3.1.1b) in the International Real-time Magnetic Observatory Network (INTERMAGNET) consortium (Love and Chulliat, 2013). The principal difference between the two figures is the geographic distribution, evolving from a strong European concentration circa 1910 to more balanced partition today. The other prominent development is the transition from analog to digital recording, with all INTERMAGNET stations providing access to 1 min data within 72 hours of acquisition.

In this chapter, we review the salient institutional contributions from 1919 to 2019 of IAGA and its predecessors within the International Union of Geodesy and Geophysics (IUGG) to our understanding and specification of Earth's magnetism and its upper atmosphere. In this review, we draw attention to the outstanding geomagnetic challenges/questions for scientists during the coming hundred years.

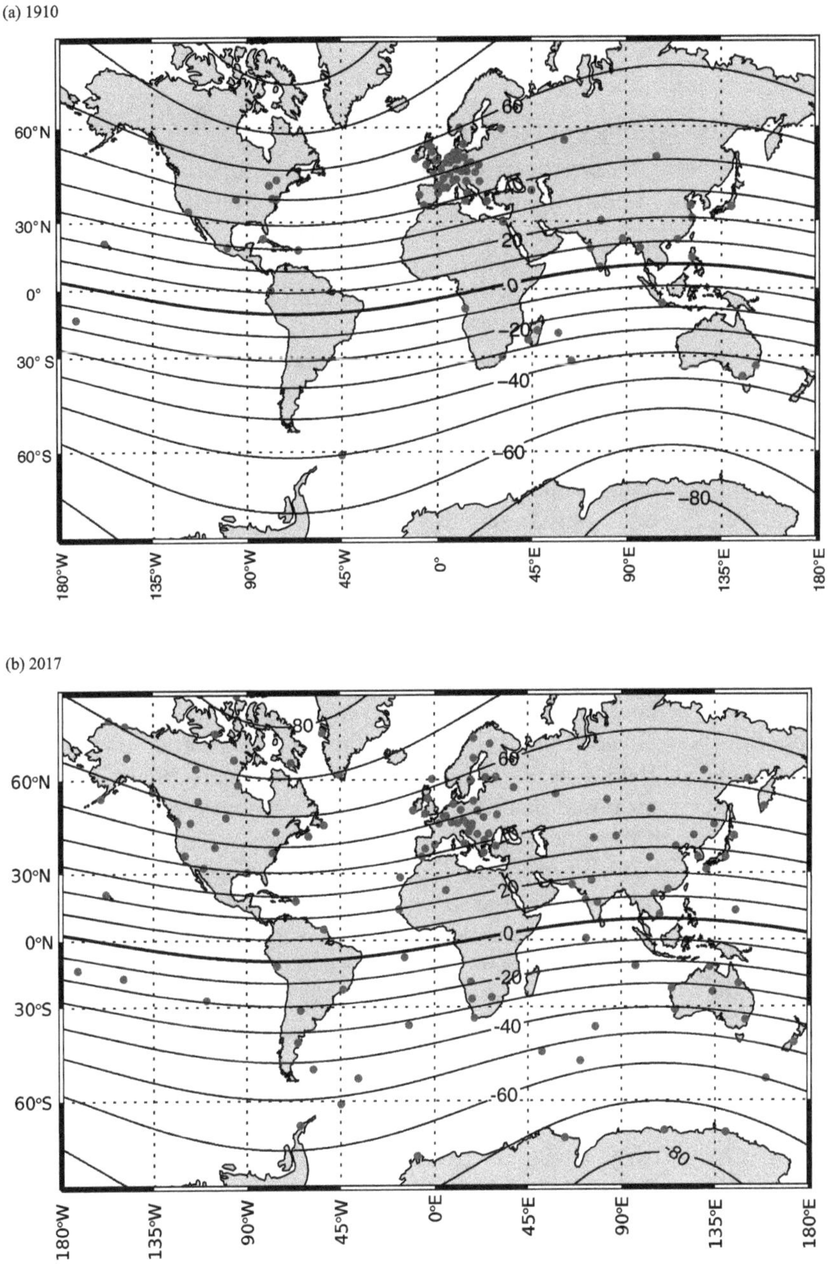

Figure 3.1.1 (a) Map showing locations of 80 magnetic observatories listed by Merlin and Somville (1910). (b) Locations of current INTERMAGNET Observatories (www.intermagnet.org/imos/imomap-eng.php). Five high-latitude stations are beyond the range of the figure.

3.2 IAGA's Institutional Contributions, 1919–2019

Global Infrastructure: Biennial Assemblies and International Years

Geomagnetism is intrinsically a worldwide phenomenon, requiring global observations. The need for international cooperation in geomagnetism was formally recognized in 1873 with the establishment of the precursor of STME/IATME/IAGA, the Commission for Terrestrial Magnetism and Atmospheric Electricity of the International Meteorological Organization. The broad objectives of IAGA are promotion of research in geomagnetism and upper atmospheric physics, facilitation of international coordination of such research, and provision of an international forum for presentation and discussion of research results via biennial meetings (www.iaga-aiga.org/about/statutes/). During the last century IAGA has provided moral (via resolutions) and organizational support for several high-profile scientific campaigns including the Second International Polar Year (IPY) of 1932–3 (Laursen, 1984), the International Geophysical Year (IGY) of 1957–8 (Fukushima, 1995), the International Quiet Sun Year (IQSY) of 1964–5 (e.g. Anonymous, 1963), and more recently, two international efforts in 2007–8 to commemorate the fiftieth anniversary of the IGY: International Heliophysical Year (IHY) and Electronic Geophysical Year (eGY) (Ismail-Zadeh, 2016).

Geomagnetic Indices

Of the eight scientific associations in the IUGG, IAGA is the de facto approval authority for geomagnetic indices (http://isgi.unistra.fr/about_indices.php; Love and Remick, 2007). Quoting from Mayaud (1980), 'Within the frame of the International Association of Geomagnetism and Aeronomy …, geomagneticians are entrusted with a particular and difficult task: to derive geomagnetic indices for the whole scientific community investigating geophysical phenomena in the Earth's environment'. Thus IAGA issues resolutions endorsing calculation of new (and existing) geomagnetic indices, e.g. Kp (IAGA Resolution 6/1954; Menvielle et al., 2011); Dst (IAGA Resolutions 12/1963, 2/1969, 15/1969 and 6/1995); AE (IAGA Resolutions 2/1969, 15/1969, 6/1995 and 4/2005); aa and am (IAGA Resolutions 2/1969, 3/1975); and PC (IAGA Resolution 3/2013). Responsibility for validation, calculation, dissemination, and archival of the various geomagnetic indices is overseen by the International Service of Geomagnetic Indices (ISGI; http://isgi.unistra.fr/) with the work carried out by several Collaborating Institutes (aa, am and CK days, Ecole et Observatoire des Sciences de la Terre, France; Kp and QD days, Helmholtz-Zentrum Potsdam Deutsches GeoForschungsZentrum, Germany; Dst and AE, World Data Center for Geomagnetism, Japan; PC, by both Arctic and Antarctic Research Institute, Russian Federation and Denmark Technical University National Space Institute, Denmark; and SC and SFE, Observatori de l'Ebre, Spain).

From 1940 to 1969, the geomagnetic Ci, Kp, and Ap indices were published in IAGA Bulletin 12 (1948–71; for Kp and Ap from 1932 to 1939, see Bulletin 18). For the years 1970–95, these indices and Dst were published in IAGA Bulletin 32 from 1972 to 2001, with aa added and Ci dropped beginning for the year 1976, and AE added in 1981. The aa values from 1968 to 1975 are contained in IAGA Bulletin 39 (1977) and Dst values from 1957 to 1986 are given in Bulletin 40 (1991). Geomagnetic data were archived at World Data Centers (WDCs) in Boulder (USA), Copenhagen (Denmark), Edinburgh (UK), Kyoto (Japan) and Mumbai (India). World Data Centers were established by the International Council of Scientific Unions (ICSU) in preparation for the IGY. In 2009, the World Data Centers were reformed under an ICSU-established World Data System (WDS). The data repositories are no longer federated into the WDC organization but through WDS certification of Member Organizations – holders and providers of data, data products or services – from wide-ranging fields by using internationally recognized standards. In this context, in late 2017, the ICSU World Data System (WDS) and the Data Seal of Approval (DSA) announced the launch of a new certification organization: CoreTrustSeal.

INTERMAGNET

Consistently from the outset in 1919, two goals, whether implicit or explicit (e.g. Fukushima, 1995), of IAGA and its predecessors have been to standardize observational programs and facilitate interchange of data. Both of these objectives were strongly advanced in the late 1980s with the creation of INTERMAGNET (Love and Chulliat, 2013), a voluntary consortium of standardized digital geomagnetic observatories. As noted in Section 3.1, INTERMAGNET presently encompasses ~130 geomagnetic observatories. Quoting from St. Louis (2011), 'The possibility of worldwide data communication between magnetic observatories was first raised seriously at the Workshop on Magnetic Observatory Instruments, held in Ottawa, Canada, in August 1986. Further discussions, particularly between the British Geological Survey (BGS) and US Geological Survey (USGS) took place in May 1987 at the Nordic Comparison Meeting at Chambon la Forêt, France. A pilot scheme between BGS and USGS was described at the sessions of Division of IAGA during the XIXth General Assembly of IUGG in Vancouver, Canada, in August 1987,[2] with the proposal that the geomagnetic community should adopt automatic observatories with satellite communications as its mode of operation for the future. INTERMAGNET embodies the proposal to extend worldwide the network of observatories communicating in this way'. In regard to IAGA's role in the creation of INTERMAGNET, Kerridge (2001) wrote, 'The initiative to establish INTERMAGNET was supported by [IAGA] and close contact is maintained with the IAGA bodies with interest in magnetic observatory practice, data, data derivatives, and applications'. The need for digital recording at magnetic observatories was recognized as early as 1967 in IAGA Resolution 2/1967 from the IUGG General Assembly in St. Gallen, Switzerland.

Aeronomy: What's in a Name?

An important, and perhaps under-appreciated, contribution of IAGA to the science of the upper atmosphere was the incorporation of the term 'aeronomy' in the name of the association. In 1931, the arrival of the then as yet unnamed science of aeronomy was signaled by the publication of two landmark studies (Chapman, 1931a, 1931b) on the formation of an ionized layer by solar irradiation of Earth's atmosphere (see papers by J. A. Ratcliffe, D. R. Bates, M. Nicolet and W. B. Hanson in Akasofu et al., 1968). Work in this emerging field was divided among the (then) International Association of Meteorology and IATME. The term aeronomy was coined in 1946 by Sydney Chapman as a replacement for 'meteorology'. When this suggestion did not take root; Chapman (1953) restricted the term to the science of the upper atmosphere (Chamberlain, 1989; Nagy, 2008). At the 1954 meeting of the IUGG in Rome, Chapman defined aeronomy as 'the science of the upper atmospheric regions where dissociation and ionization are important'. At the same meeting, with Chapman's advocacy, the IATME national representatives voted to change the name of the association to the International Association of Geomagnetism (another Chapman coinage; Chapman, 1938) and Aeronomy (Coulomb, 1982; Fukushima, 1994). Kaplan (1977) wrote, 'With this change in the name of one of the seven associations that made up the IUGG, aeronomy finally found an international home that it has occupied with increasing activity and distinction'. In Kaplan's opinion, 'aeronomy was principally responsible for the interest in and eventual great success of the International Geophysical Year and its remarkable offspring, science in space'.[3]

Space Weather

Interest in space weather was spawned in the mid twentieth century because of its effect on radio wave propagation (for defense communication and radar systems) and the potential hazards it posed for humans in space and space vehicles. More recently, the increased susceptibility of technological society, particularly the electrical grid, to extreme space weather events (e.g. US National Academy of Sciences; Lloyds of London; JASON; UK Royal Academy of Engineering)[4] has driven research in this area. The need for IAGA to increase focus on Sun–Earth relationships was first addressed in 1963 when the 11 IAGA Committees were transformed into nine IAGA Commissions, one of which was Commission V on Solar–terrestrial and Cosmic–terrestrial Relations (Fukushima, 1994). In 1974, this Commission became IAGA Division IV, Solar Wind and Interplanetary Field. The current emphasis on Space Weather is reflected by the fact that of the four IAGA Interdivisional Commissions, one is devoted to Space Weather, with the other three addressing Developing Countries, History, and Education and Outreach.

The Interdivisional Commission on Space Weather provides the most direct scientific link among the eight scientific associations of the IUGG to the operational International Space Environment Service (ISES). ISES is composed of 15 Regional Warning Centers and four Associated Warning Centers (www.spaceweather.org/ISES/intro/intro.html).

It is commonly estimated that the state of space weather prediction lags that of terrestrial weather forecasting by about 50 years. The perceived lag is primarily due to a lack of observations, both remote of the Sun and in situ in interplanetary space. In part because of the dearth of observations relative to terrestrial weather, space weather models are also in their infancy. Given the increasing reliance of humans on space technology and the potential threats posed by extreme space weather to such technology and the electrical grid, improvement in this area is a high IAGA priority for the coming century (Kerridge, 2007; IAGA Resolutions 1 and 2/2017).

International Geomagnetic Reference Field and World Digital Magnetic Anomaly Map

The International Geomagnetic Reference Field (IGRF) is a series of mathematical models of the Earth's main magnetic field and its secular variation. The models are used to calculate the large-scale, internal part of the Earth's magnetic field at times between 1900.0 AD and the present, at locations on or above Earth's surface. IGRF is maintained since 1968 by a working group of volunteer scientists from several international institutions, which was initiated by discussions started in the early 1960s (IAGA Resolutions 25/1971, 18/1973, 14/1975, 13/1981 and 6/1983; cf. Anonymous, 1969a, 1969b; Peddie, 1983). It is derived from observations collected by satellites, at magnetic observatories and during magnetic surveys. It is used for example in studies of space weather, investigations of local magnetic anomalies, and also by commercial organizations and private individuals who often use the geomagnetic field as a source of orientation information. Temporal variations of the internal part of the geomagnetic field, which are on timescales of years to decades, require revisions of the IGRF to remain up to date and as accurate as possible. The first generation IGRF (IGRF-1) for the period 1955.0–1975.0 was published by Zmuda (1971). At present, IGRF-12 (released in December 2014 at www.ngdc.noaa.gov/IAGA/vmod/igrf.html) extends and updates previous versions (summarized in Finlay et al., 2010) and provides a new Definitive Geomagnetic Reference Field model for epoch 2010.0. Moreover, it proposes a provisional reference field model for epoch 2015.0 and a predictive part for epochs ranging from 2015.0 to 2020.0 (Thébault et al., 2015a). A special issue of Earth, Planets and Space documents the construction of IGRF-12 (Thébault et al., 2015b).

Marine and airborne magnetic anomaly data have been collected for more than half a century, providing global coverage of the Earth. Due to the changing main field from the Earth's core, and due to differences in quality and coverage, combining these data to a consistent global magnetic grid is challenging. The World Digital Magnetic Anomaly Map (WDMAM) project is an international effort, coordinated by IAGA (IAGA Resolutions 16/1977, 16/1979, 5/1989, 4/1993, 4/1995, 3/1997, 4/2011), to integrate all available near-surface and satellite magnetic anomaly data. In 2003, a task force was created with the aim to compile the WDMAM, combining aeromagnetic and marine data worldwide on a global, 5 km cell size, grid. The map is the product of years of work, research and coordination. The first version of the WDMAM was released in 2007 (Korhonen et al., 2007; Hemant et al., 2007). Following this work, Maus et al. (2009) and Maus (2010) continued collecting data and proposed their own map and associated magnetic field model covering spherical harmonic (SH) degrees from 16 to 720. The main limitation of this original grid was the way the oceanic data gaps were filled. Thus, a second version of the map was produced (Lesur et al., 2016). It has been built by two groups that were originally working independently in the 'Institut de Physique du Globe' in Paris (IPGP) and in the 'GeoForschungsZentrum' in Potsdam (GFZ). This version has been approved by IAGA during the twenty-sixth IUGG assembly in 2015 and publicly released. It can be downloaded at www.wdmam.org.

Geodynamo

The geomagnetic field has attracted humans' attention for centuries. In 1088, Shen Kuo, a Chinese scientist and statesman of the Song dynasty, first described a controlled experiment with a compass needle and postulated the concept of declination (Needham, 1962; Smith and Needham, 1967). While it is not our aim to provide a detailed summary of the whole history of geomagnetism here (see Courtillot and le Mouël, 2007, for a very nice historical excursion), we have to mention Gilbert's (1600) work *De Magnete*, often regarded as one of the first 'modern' treatises in physics. Gilbert's main idea was that the compass needle pointed north because the Earth itself attracted it as a giant magnet would. The most challenging scientific question for geomagnetism is the origin of Earth's magnetic field. In fact, Albert Einstein once ranked the source of Earth's magnetism among the most important unsolved problems in classical physics (Busse, 1978). A key development in this area, in the same year as the founding of IAGA's STME predecessor, was Larmor's (1919) proposal that the magnetic field of the Earth (and Sun) could be maintained by a self-excited dynamo. Although Larmor's work was subsequently questioned by Cowling (1934), works by Elsasser (1939) and Bullard (1949a, 1949b) revitalized the theory of geodynamo. In the following 50 years, mostly due to computational facilities, this theory developed significantly, resulting in the first demonstration of a successful ('Earthlike') numerical dynamo by Glatzmeier and Roberts (1995a, 1995b). Various problems involving convection, driven either thermally or by other forces, have been addressed with the aim of modeling certain aspects of the observed behavior of geomagnetic field (e.g. secular variations and reversals). This development of dynamo theory was accompanied and supported by several laboratory experiments, dating back to the work of Steenbeck et al. (1967, 1968), who demonstrated the α effect in a liquid sodium flow. It was obvious that such an experiment requires a large fluid volume and high flow rates. Only recently, two dynamos ran successfully with liquid sodium flows (Gailitis et al., 2000; Müller and Stieglitz, 2000, 2002; Stieglitz and Muller, 2001). In particular the latter one, developed in Karlsruhe (Müller and Stieglitz, 2000, 2002; Stieglitz and Muller, 2001), can be well understood as a mean-field dynamo of α^2. A valuable review of the genesis and progress in geomagnetic dynamos was published by Roberts and King (2013). Another comprehensive review of the main concept and some recent developments for the mean-field dynamo is presented by Radler (2014). Despite substantial progress thus far, the geodynamo and dynamos of other planets remain a significant challenge for the future.

Paleomagnetism and Databases

Paleomagnetism is aimed at deciphering the orientation and intensity of magnetization recorded in rock samples, either during their formation or subsequent genesis. This knowledge, often requiring demanding and time consuming laboratory work, is applied to several problems, from paleogeographic reconstruction through polar wander path to magnetostratigraphy. Again, we can't review the whole rich history of paleomagnetism, and refer the reader to the excellent review of Courtillot and Mouël (2007). However, with respect to IAGA (and IUGG), we need to mention Mercanton (1926), who postulated that reversals (of magnetic inclination) found in the Northern Hemisphere would be found also in the other hemisphere, thus providing evidence that magnetic poles have undergone enormous displacements. Moreover, he also asked the International Union of Geodesy and Geophysics to extend its observational (paleomagnetic) base and pointed to the need for centralized sample archiving and unification of methods to study magnetic properties of rocks. Fortunately, this great idea has been materialized over the last decades, during which the number of works in paleomagnetism has exponentially increased. We note a few examples here. Standardized methods based on step-wise heating of samples are used in paleointensity studies (Thellier and Thellier, 1941, 1959; for review, see e.g. Dunlop, 2011), and Zijderveld (1958) reported on a method for depicting individual components of remanent magnetization (direction and intensity) stored in a rock sample during laboratory demagnetization. Jelinek (1981) developed a statistical procedure for analyzing the anisotropy of magnetic susceptibility that is widely used in tectonic studies. These and other methods, further improved, are still used as 'gold standards' in paleomagnetism. The same applies to rock magnetism, the basic and great provider of experimental approaches in paleomagnetism, archeomagnetism and environmental magnetism. From Curie point determination through analysis of the viscous component of remanence to first-order reversal curves (FORC diagrams), all these methods further improve the quality of data needed for reliable interpretation of magnetic information stored in natural samples. The increasing number of laboratories, methods, instruments and above all data was realized at the 1987 IUGG Assembly, when IAGA issued the first resolution urging international coordination for paleomagnetic database(s)

(IAGA Resolution 2/1987). Since then, several databases, regional and global, have been developed. The first one of great importance was the IAGA Global Paleomagnetic Database (McElhinny and Lock, 1990a, 1990b, 1993, 1996). Later on, this database was followed up by a MagIC project (Tauxe, 2010; Koppers et al., 2010). Other databases include the paleointensity database PINT (Perrin and Schnepp, 2004; Biggin et al., 2010), the database of Precambrian paleomagnetic data PALEOMAGIA (Veikkolainen et al., 2014) and most recently the HISTMAG database (Arneitz et al., 2017), combining historical, archeomagnetic and volcanic data. Despite the significant progress made since early (medieval) studies, paleomagnetism still faces many challenges and remains an attractive field of research.

Acknowledgements

Eduard Petrovský is grateful for the support from the Ministry of Education, Youth and Sports of the Czech Republic through an Inter-Vector project LVT19011.

References

Akasofu, S.-I., Fogle, B. and Haurwitz, B. (Eds), 1968. Sidney Chapman, Eighty. University of Colorado Press, Boulder, pp. 27–30, 31–4, 35–8, 39–41.

Anonymous [IAGA-IQSY Committees], 1963. IAGA Newsletter 1, pp. 17–20.

Anonymous, 1969a. International Geomagnetic Reference Field 1965.0: IAGA Commission 2 Working Group 4, Analysis of the geomagnetic field. *J. Geophys. Res.*, 74, 4407–8. doi: 10.1029/JB074i017p04407.

Anonymous, 1969b. International Geomagnetic Reference Field 1965.0. *J. Geomagn. Geoelectr.*, 21, 569–71.

Appleton, E. V. and Barnett, M. A. F., 1925. On some direct evidence for downward atmospheric reflection of electric rays. *Proc. Royal Soc. London. Series A*, 109, 621–41.

Arneitz, P., Leonhardt, R., Schnepp, E., et al., 2017. The HISTMAG database: combining historical, archaeomagnetic and volcanic data. *Geophys. J. Int.*, 210, 1347–59. doi: 10.1093/gji/ggx245.

As, J. A. and Zijderveld, J. D. A., 1958. Magnetic cleaning of rocks in palaeomagnetic research. *Geophys. J. R. Astron. Soc.*, 1, 308–19.

Biggin, A. J., McCormack, A. and Roberts, A., 2010. Paleointensity database updated and upgraded. *Eos Trans. AGU*, 91(2), 15. doi: 10.1029/2010EO020003.

Birkeland, K., 1901. Expédition Norvegienne de 1899–1900. Résultats magnétiques. Vidensk Skrifter, 1. Mat. naturv. Kl.

Birkeland, K., 1908. The Norwegian Aurora Polaris Expedition, 1902–1903, vol. 1, section 1, H. Aschehoug, Oslo.

Bullard, E. C., 1949a. The magnetic field within the Earth. *Proc. R. Soc. London, Ser. A*, 197, 433–53.

Bullard, E. C., 1949b. Electromagnetic induction in a rotating sphere. *Proc. R. Soc. London, Ser. A*, 199, 413–43.

Busse, F. H., 1978. Magnetohydrodynamics of the Earth's dynamo. *Ann. Rev. Fluid. Mech.*, 10, 435–62.

Carrington, R. C., 1860. Description of a singular appearance seen in the Sun on September 1, 1859. *Mon. Not. R. Astron. Soc.*, 20, 13–15.

Cawood, J., 1979. The magnetic crusade: science and politics in early Victorian Britain. *Isis*, 70, 493–518.

Chamberlain, J. W., 1989. Aeronomy: The word. *Eos Trans. AGU*, 70, 1544.

Chapman, S., 1918. An outline of a theory of magnetic storms. *Proc. R. Soc. London, Ser. A*, 97, 61–83.

Chapman, S., 1931a. The absorption and dissociative or ionizing effect of monochromatic radiation in an atmosphere on a rotating earth. *Proc. Phys. Soc.*, 43, 26–45.

Chapman, S., 1931b. The absorption and dissociative or ionizing effect of monochromatic radiation in an atmosphere on a rotating earth part II. Grazing incidence. *Proc. Phys. Soc.* 43, 483–501.

Chapman, S., 1938. Geomagnetism or terrestrial magnetism? *Terr. Magn. Atmos. Electr.*, 43, 321.

Chapman, S., 1946. Some thoughts on nomenclature. *Nature*, 157, 405.

Chapman, S., 1953. Nomenclature in meteorology. *Weather*, 7–8, 62.

Chapman, S. and Ferraro, V. C. A., 1931a. A new theory of magnetic storms. *Terr. Magn. Atmos. Electr.*, 36, 77–97.

Chapman, S. and Ferraro, V. C. A., 1931b. A new theory of magnetic storms, Part I – The initial phase (continued). *Terr. Magn. Atmos. Electr.*, 36, 171–86.

Chapman, S. and Ferraro, V. C. A., 1933. A new theory of magnetic storms, Part II – The main phase. *Terr. Magn. Electr.*, 38, 79–96.

Cliver, E. W., 1994a. Solar activity and geomagnetic storms: the first 40 years. *Eos Trans. AGU*, 75, 569, 574–5.

Cliver, E. W., 1994b. Solar activity and geomagnetic storms: the corpuscular hypothesis. *Eos Trans. AGU*, 75, 609, 612–13.

Cliver, E. W., 1995. Solar activity and geomagnetic storms, From M regions and flares to coronal holes and CMEs. *Eos Trans. AGU*, 76, 75, 83.

Coulomb, J., 1982. From IATME to IAGA (1951–1954). *IAGA News*, 21, 114–16.

Courtillot, V. and le Mouël, J. L., 2007. The study of Earth's magnetism (1269–1950): a foundation by Peregrinus and subsequent development of geomagnetism and paleomagnetism. *Rev. Geophys.*, 45, RG3008. doi: 10.1029/2006RG000198.

Cowling, T. G., 1934. The magnetic field of sunspots. *Mon. Not. R. Astron. Soc.*, 94, 32–48.

Dunlop, D. J., 2011. Physical basis of the Thellier-Thellier and related paleointensity methods. *Phys. Earth Planet. Inter.*, 187, 118–38. doi: 10.1016/j.pepi.2011.03.006.

Elsasser, W., 1939. On the origin of the Earth's magnetic field. *Phys. Rev.*, 60, 876–83.

Finlay, C. C., Maus, S., Beggan, C. D., et al., 2010. International Geomagnetic Reference Field: the eleventh generation. *Geophys. J. Int.*, 183, 1216–30. doi: 10.1111/j.1365-246X.2010.04804.x.

Fitzgerald, G. F., 1892. Sunspots and magnetic storms. *The Electrician*, 30, 48.

FitzGerald, G. F., 1900. Sunspots, magnetic storms, comet tails, atmospheric electricity, and aurorae. *The Electrician*, 46, 287–8.

Fritz, H., 1881. *Das Polarlicht*. Brockhaus, Leipzig.

Fukushima, N., 1994. Some topics and historical episodes in geomagnetism and aeronomy. *J. Geophys. Res.*, 99, 19113–42.

Fukushima, N., 1995. History of the International Association of Geomagnetism and Aeronomy (IAGA). *IUGG Chronicle*, 226, 73–87.

Gailitis, A., Lielausis., O., Dementev, S., et al., 2000. Detection of a flow induced magnetic field eigenmode in the Riga dynamo facility. *Phys. Rev. Lett.*, 84, 4365–8. doi: 10.1103/PhysRevLett.84.4365.

Garland, G. D., 1979. The contributions of Carl Friedrich Gauss to geomagnetism. *Historia Mathematica*, 6, 5–29.

Gautier, A., 1852. Relation entre les taches du Soleil et les phénomènes magnétiques. *Arch. Sci.*, 21, 194–5.

Gilbert, W., 1600. *De Magnete*. Excudebat Petrus Short, London. (English translation by P. Fleury Mottelay, Dover, Mineola, New York, 1958.)

Glatzmeir, G. H. and Roberts, P. H., 1995a. A 3-dimensional self-consistent computer-simulation of a geomagnetic-field reversal. *Nature*, 377, 203–9. doi: 10.1038/377203a0.

Glatzmeir, G. H. and Roberts, P. H., 1995a. A 3-dimensional convective dynamo solution with rotating and finitely conducting inner-core and mantle. *Phys. Earth Planet. Inter.*, 91, 63–75. doi: 10.1016/0031-9201(95)03049-3.

Heaviside, O., 1902. Telegraphy. I. Theory. In: *Encyclopedia Britannica*, 10th ed., pp. 213–18.

Hemant, K., Thébault, E., Mandea, M., Ravat, D. and Maus, S., 2007. Magnetic anomaly map of the world: merging satellite, airborne, marine and ground-based magnetic data sets. *Earth Planet. Sci. Lett.*, 260, 1–2, 56–71.

Hodgson, R., 1860. On a curious appearance seen in the Sun. *Mon. Not. Roy. Astron. Soc.*, 20, 15.

Ismail-Zadeh, A., 2016. Geoscience international: the role of scientific unions. *Hist. Geo. Space Sci.*, 7, 103–23.

Jelinek, V., 1981. Characterization of the magnetic fabric of rocks. *Tectonophysics*, 79, T63–7. doi: 10.1016/0040-1951(81)90110-4.

Kaplan, J., 1977. The aeronomy story: a memoir. In: *Washington Essays on the History of Rocketry and Astronautics*, vol. 2, Hall, R. C. (Ed). NASA, Hampton, VA. pp. 423–7.

Kennelly, A. E., 1902. On the elevation of the electrical conducting strata of the Earth's atmosphere. *Elec. World. Eng.*, 39, 473.

Kerridge, D. J., 2001. INTERMAGNET: World-wide near real-time geomagnetic observatory data, paper presented at Space Weather Workshop: Looking towards a European Space Weather Programme, Eur. Space Res. and Technol. Cent., Noordwijk, Netherlands, 17–19 Dec. Available at www.esa-spaceweather.net/spweather/workshops/SPW_W3/index.html.

Kerridge, D., 2007. IAGA, International Association of Geomagnetisim and Aeronomy. In: *Encyclopedia of Geomagnetism and Paleomagnetism*, Gubbins, D. and Herrero-Bervera, E. (Eds). Springer, Dordrecht, pp. 407–8.

Koppers, A. A. P., Minnett, R., Tauxe, L. and Constable, C., 2010. MagIC database: comprehensive archiving and visualization of rock- and paleomagnetic data using Web 2.0 technology. *Geochim. Cosmochim. Acta*, 74, A531.

Korhonen, J. V., Fairhead, J. D., Hamoudi, M., Hemant, K., Lesur, V., Mandea, M., Maus, S., Purucker, M., Ravat, D., Sazonova, T. and Thébault, E., 2007. *Magnetic Anomaly Map of the World*. Geological Survey of Finland, Espoo.

Larmor, J., 1919. How could a rotating body like the Sun become a magnet? *Rep. Br. Assoc. Adv.*, pp. 159–60.

Laursen, V., 1984. IATME and the second international polar year. *IAGA Newsletter*, 22, 104–8.

Lesur, V., Hamoudi, M., Choi, Y., Dyment, J. and Thébault, E., 2016. Building the second version of the World Digital Magnetic Anomaly Map (WDMAM). *Earth Planets Space*, 68, 27. doi: 10.1186/s40623-016-0404-6.

Lincoln, J. V., 1967. Geomagnetic indices. In: *Physics of Geomagnetic Phenomena*, vol. 1, Matsushita, S. and Campbell, W. H. (Eds). Academic Press, New York, pp. 67–100.

Lindemann, F. A., 1919. Note on the theory of magnetic storms. *Philos. Mag.*, 38, 669.

Loomis, E., 1860. The great auroral exhibition of August 28 to September 4, 1859, and the geographical distribution of auroras and thunder storms. *Am. J. Sci.*, 2, 30, 79–100.

Love, J. J. and Chulliat, A., 2013. An international network of magnetic observatories, *Eos Trans. AGU*, 94, 42, 373–4.

Love, J. J. and Remick, K. J., 2007. Magnetic indices. In: *Encyclopedia of Geomagnetism and Paleomagnetism*, Gubbins, D. and Herrero-Bervera, E. (Eds). Springer, Dordrecht, pp. 509–12.

Maunder, E. W., 1904. Magnetic disturbances, 1882 to 1903, as recorded at the Royal Observatory, Greenwich, and their association with sun-spots. *Mon. Not. Roy. Astron. Soc.*, 65, 2–34.

Maus, S., 2010. An ellipsoidal harmonic representation of earth's lithospheric magnetic field to degree and order 720. *Geochem Geophys Geosyst.*, 11, Q06015. doi: 10.1029/2010GC003026.

Maus, S., Barckhausen, U., Berkenbosch, H., et al., 2009. EMAG2: a 2–arcmin resolution earth magnetic anomaly grid compiled from satellite, airborne, and marine magnetic measurements. *Geochem. Geophys. Geosyst.*, 10, Q08005. doi: 10.1029/2009GC002471.

Mayaud, P. N., 1980. *Derivation, Meaning, and Use of Geomagnetic Indices*. American Geophysical Union, Washington, DC.

McElhinny, M. W. and Lock, J., 1990a. IAGA global paleomagnetic database. *Geophys. J. Int.*, 101, 763–6. doi: 10.1111/j.1365-246X.1990.tb05582.x.

McElhinny, M. W. and Lock, J., 1990b. Global paleomagnetic database project. *Phys. Earth Planet. Inter.*, 63, 1–6. doi: 10.1016/0031-9201(90)90053-Z.

McElhinny, M. W. and Lock, J., 1993. Global paleomagnetic database supplement number one – update to 1992. *Surv. Geophys.*, 14, 303–29. doi: 10.1007/BF00690947.

McElhinny M. W. and Lock, J., 1996. IAGA paleomagnetic databases with access. *Surv. Geophys.*, 17, 575–91. doi: 10.1007/BF01888979.

Menvielle, M., Iyemori, T., Marchaudon, A. and Nosé, M., 2011. Geomagnetic indices. In: *Geomagnetic Observations and Models*, Mandea M. and Korte, M. (Eds). Springer, Dordrecht, p. 204.

Mercanton, P. L. 1926. Inversion de l'inclinaison magnétique terrestre aux ages géologiques. *J. Geophys. Res.*, 31, 187–90.

Merlin, E., Somville, O., 1910. *Liste des observatoires magnétiques et des observatoires séismologiques*. Observatoire Royal de Belgique, Bruxelles.

Muller, U. and Stieglitz, R., 2000. Can the Earth's magnetic field be simulated in the laboratory? *Naturwissenschaften*, 87, 381–390, DOI: 10.1007/s001140050746.

Muller, U. and Stieglitz, R., 2002. The Karlsruhe dynamo experiment. *Nonlin. Process. Geophys.*, 9, 165–70.

Nagy, A. F., 2008. Preface (to Comparative Aeronomy). *Space Sci. Rev.*, 139, 1–2.

Needham, J., 1962. Science and Civilisation in China, Vol. 4, Physics and Physical Technology, Part 1, *Physics*. Cambridge University Press, New York.

Peddie, N. W., 1983. International Geomagnetic Reference Field – its evolution and the difference in total field intensity between new and old models for 1965–1980. *Geophysics*, 48, 1691–6. doi: 10.1190/1.1441450.

Perrin, M. and Schnepp, E., 2004. IAGA paleointensity database: distribution and quality of the data set. *Phys. Earth Planet. Inter.*, 147, 255–67. doi: 10.1016/j.pepi.2004.06.005.

Radler, K. H., 2014. Mean-field dynamos: the old concept and some recent developments Karl Schwarzschild Award Lecture 2013. *Astro. Nachr.*, 335, 459–69. doi: 10.1002/asna.201412055.

Roberts, P. H. and King, E. M., 2013. On the genesis of the Earth's magnetism. *Rep. Prog. Phys.*, 76, 096801. doi: 10.1088/0034-4885/76/9/096801.

Sabine, E., 1852. On periodical laws discoverable in the mean effects of the larger magnetic disturbances. *Philos. Trans.*, 142, 103–24.

Schwabe, S. H., 1844. Solar observations during 1843. *Astron. Nach.*, 21, 495.

Smith, P. J. and Needham, J., 1967. Magnetic declination in Mediaeval China. *Nature*, 214, 1213–14. doi: 10.1038/2141213b0.

St. Louis, B., 2011. *INTERMAGNET Technical Reference Manual*, version 4.5. British Geological Survey, Edinburgh.

Steenbeck, M., Kirko, I. M., Gailitis, A., Klawina, A. P., Krause, F., Laumanis, I. J. and Lielausis O. A., 1967. Der experimentelle Nachweis einer elektromotorischen Kraftlängs einesäußeren Magnetfeldes, induziert durch eine Strömung flüssigen Metalls (α-Effekt). *Mber. Dtsch. Akad. Wiss. Berlin*, 9, 714–19.

Steenbeck, M., Kirko, I. M., Gailitis, A., Klyavinya, A. P., Krause, F., Laumanis, I. Ya. and Lielausis, O. A., 1968. Experimental discovery of the electromotive force along the external magnetic field induced by a flow of liquid metal (α-effect). *Soviet Phys. Dokl.*, 13, 443–5.

Stern, D. P., 1989. A brief history of magnetospheric physics before the spaceflight era. *Rev. Geophys.*, 27, 103–14.

Stewart, B., 1861. On the great magnetic disturbance which extended from August 28 to September 7, 1859, as recorded by photography at Kew Observatory. *Philos. Trans.*, 151, 423–30.

Stewart, B., 1883. Terrestrial magnetism. In: *Encyclopedia Britannica*, 9th ed., 16, pp. 159–84.

Stieglitz, R. and Muller, U., 2001. Experimental demonstration of a homogeneous two-scale dynamo. *Phys. Fluids*, 13, 561–4. doi: 10.1063/1.1331315.

Tauxe, L., 2010. *The MagIC Database: Essentials of Paleomagnetism*. University of California Press, Oakland, pp. 391–5.

Thébault, E., Finlay, C. C., Beggan, C. D., et al., 2015a. International Geomagnetic Reference Field: the 12th generation. *Earth Planets Space*, 67, 79. doi: 10.1186/s40623-015-0228-9.

Thébault, E., Finlay, C. C. and Toh, H., 2015b. Preface to special issue 'International Geomagnetic Reference Field – the twelfth generation'. *Earth Planets Space*, 67, 158. doi: 10.1186/s40623-015-0313-0.

Thellier, E. and Thellier, O., 1941. On the thermic variations of thermoremanent magnetisation of burned earth. *Comptes Rendus Hebdomadaires des Seances de L'Academie des Sciences*, 213, 59–61.

Thellier, E. and Thellier, O., 1959. Sur l'intensité du champ magnétique terrestre dans le passé historique et géologique. *Ann. Géophys.*, 15, 285–376.

Tsurutani, B. T., 1991. SPR name change to 'Space Physics and Aeronomy'. *Eos Trans. AGU*, 72, 172.

Veikkolainen, T., Pesonen, L. J. and Evans, D. A. D., 2014. PALEOMAGIA: A PHP/MYSQL database of the Precambrian paleomagnetic data. *Stud. Geophys. Geod.*, 58, 425–41. doi: 10.1007/s11200-013-0382-0.

Wolf, R., 1852a. Sonnenflecken – Beobachtungen in der ersten Hälfte desJahres 1852; Entdeckung des Zusammen-hanges zwischen den Declinationsvariationen der Magnetnadel und den Sonnenflecken. *Mitt. Naturf. Ges.*, 224–64, 179–84.

Wolf, R., 1852b. Liaison entre les taches du Soleil et les variations en declinaison de l'aiguille aimantée. *Compt. Rend.*, 35, 364.

Zmuda, A. J., 1971. The International Geomagnetic Reference Field: introduction. *Bulletin of the International Association of Geomagnetism and Aeronomy*, 28, 148–52.

Notes

1 Fukushima (1994) recounts the role of the American Geophysical Union (AGU) in the founding of STME within the International Geodetic and Geophysical Union, the predecessor of the International Union of Geodesy and Geophysics (IUGG). Other valuable resources on IAGA history include Fukushima (1995), Kerridge (2007), IAGA Newsletters (www.iaga-aiga.org/index.php?id=newsletter), IAGA Resolutions (www.iaga-aiga.org/resolutions/) and IAGA Bulletins (www.iaga-aiga.org/publications/bulletins/).

2 This pilot program was successfully executed with an exchange of data between Hartland Observatory in the UK and Boulder Observatory in the US via a Geostationary Operational Environmental Satellite (Kerridge, 2001).

3 The Terrestrial Magnetism and Electricity section of the AGU (the US analog of IAGA within the IUGG) changed its name to the Geomagnetism and Aeronomy section in 1958 and then to Solar-Terrestrial Relationships in 1968 and Solar-Planetary Relationships in 1970 before adopting its present name, Space Physics and Aeronomy, in 1992 (Tsurutani, 1991).

4 US National Academy of Sciences: www.nap.edu/catalog/12507/severe-space-weather-events-understanding-societal-and-economic-impacts-a; Lloyds of London: www.lloyds.com/news-and-insight/risk-insight/library/natural-environment/space-weather; JASON: https://fas.org/irp/agency/dod/jason/space weather.pdf; Royal Academy of Engineers: www.raeng.org.uk /publications/reports/space-weather-full-report.

PART II

Geomagnetic Field

Sources and Observations

4

Geomagnetic Field Sources

From Earth's Core to the Sun

Daniel N. Baker

The Earth's main magnetic field originates in the molten outer core of our planet. It arises through the action of convection currents that drive the geomagnetic dynamo (see Chapter 9). Earth's interior is modelled to have a solid inner core with a geocentric radius of ~1200 km and a liquid outer core extending from the inner core boundary out to about 3400 km geocentric radial distance. Heat flow from the inner core (~6000 K) to this core-mantle boundary maintains the outer core in its liquid state. The iron-nickel composition of the outer core allows generation of strong electrical currents due to the convective flow caused by Earth's rotation. This flow is further shaped by viscous interaction with the inner core as well (see Figure 4.0.1).

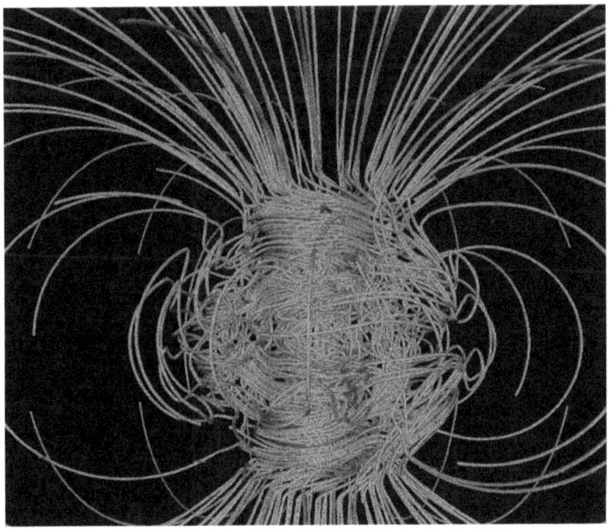

Figure 4.0.1 Computer simulation of the Earth's magnetic field in a period of normal polarity between magnetic reversals. The lines represent magnetic field lines: blue when the field points toward the centre of the Earth and yellow when it points away. The rotation axis of the Earth is centred and vertical in the diagram. From Glatzmaier and Roberts (1995). (A black-and-white version of this figure appears in some formats. For the colour version, please refer to the plate section.)

The geomagnetic field extends from the deep interior outward through the mantle and crust beyond the Earth's surface (Gauss, 1839, in Glassmeier and Tsurutani, 2014). The geomagnetic field often is approximated as an offset, tilted dipole magnetic field with a magnetic moment of 7.94×10^{22} A-m^2. However, as reported in subsequent chapters, the field is immensely more complex and time-varying than this simple picture might suggest. In the dipole approximation, the magnetic field is tilted by about 11° relative to the Earth's rotational axis. At Earth's surface (1 R_E = 6372 km), the magnetic field intensity varies from about 0.25 Gauss (G) to about 0.65 G. In SI units, the field varies from 2.5×10^4 nano tesla (nT) in the South Atlantic region near South America to about 6.5 $\times 10^4$ nT in the north polar regions over Siberia in the north and over Antarctica in the South (Figure 4.0.2).

The geomagnetic field extending outward beyond Earth's solid surface encounters strong, highly variable flow of hot ionised gas from the Sun. This flow of magnetised solar plasma (called the solar wind) compresses and shapes the dayside of Earth's magnetic field as shown in Figure 4.0.3. On the night (anti-sunward) side of the Earth, the magnetic field gets drawn out into a long, comet-like tail. Present evidence is that this magnetotail region extends to hundreds or thousands of Earth radii. The paper by Cowley (1991) gives both theoretical and observational estimates of the tail length, providing arguments that the tail extends to ~5000 R_E. Research over the past six decades of the Space Age has revealed much about the various current systems that shape the Earth's 'magnetosphere' as illustrated schematically in Figure 4.0.3.

The remainder of this chapter is devoted to providing a broad overview of the individual current systems (Campbell, 2003) that, acting together, generate the complex and fascinating geomagnetic field. More details of all those constituent elements are provided in subsequent Chapters 5–13. In the remainder of Part II of this book, Chapter 6 describes magnetic field measurements made from ground observatories around the world. Chapter 7 discusses amazing new observations of the Earth field made from LEO (low-Earth orbit) spacecraft such as the Swarm mission. Chapter 8 describes recent measurement at higher altitudes made with

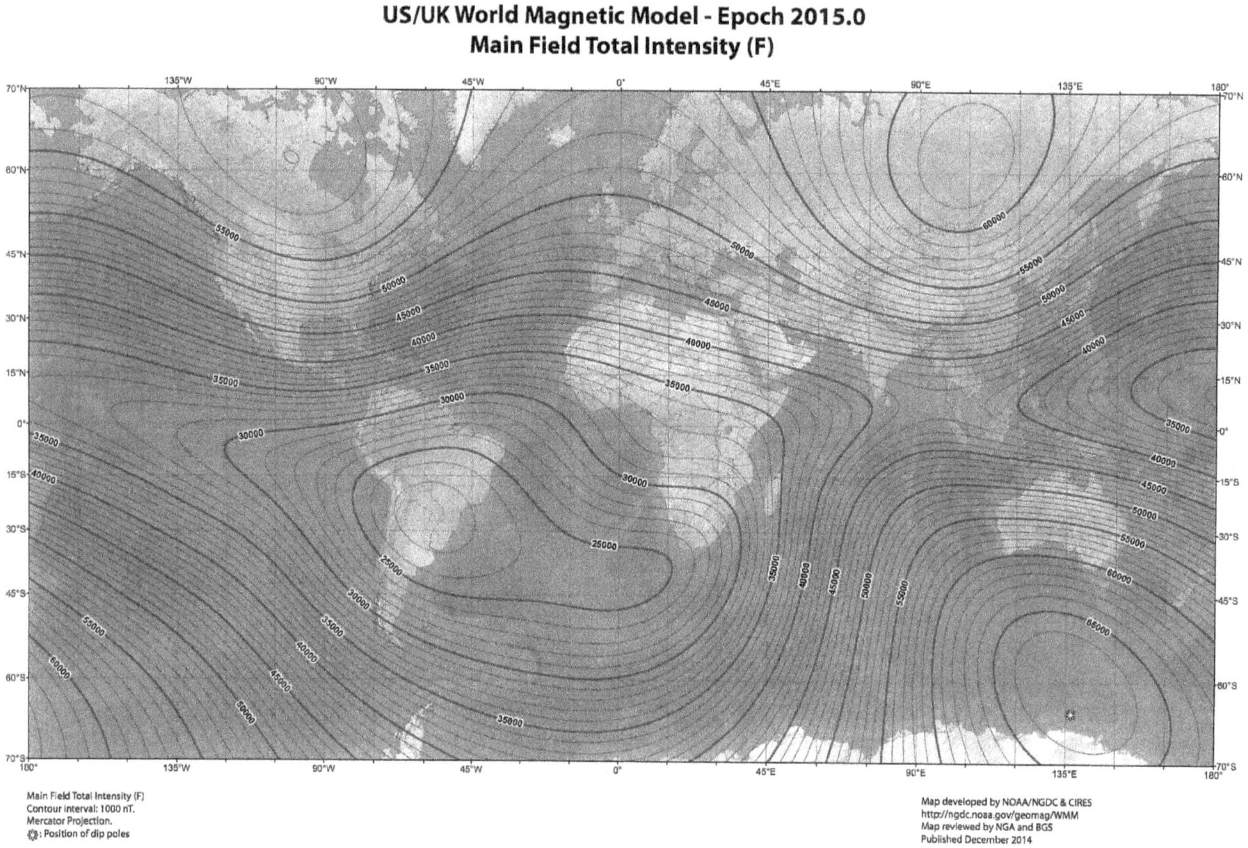

Figure 4.0.2 World magnetic model – Epoch 2015.0 showing contours of the main magnetic field total intensity. From NOAA National Geophysical Data Center (NGDC) and Cooperative Institute for Research in Environmental Science (CIRES).

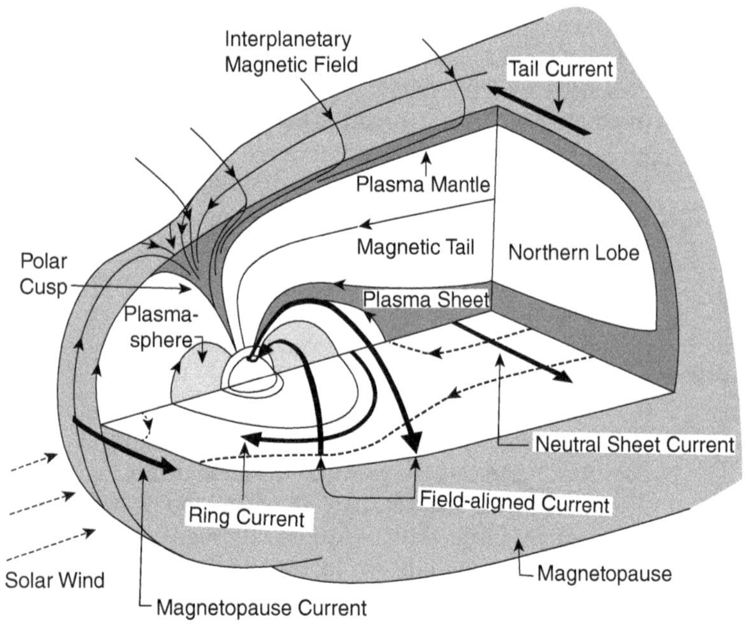

Figure 4.0.3 A 3-D cutaway diagram showing the principal current systems that together shape and modify the Earth's magnetic field and magnetospheric system.

the four-spacecraft Cluster mission as well as several other modern satellite observatories.

4.1 Ring Current

As shown in Figure 4.0.3, a key current system in the equatorial plane is the extraterrestrial ring current. The ring current plays a central role by adding to the Earth's dipole field in order to shape the overall magnetosphere (Cole, 1966). It circles the Earth in the equatorial plane and is generated by the azimuthal drift of energetic (~10 to ~200 keV) charged particles trapped on field lines between $r \sim 2R_E$ and $r \sim 7\ R_E$. During geomagnetic storms, ring current energetic particle populations are significantly increased, with peak fluxes occurring in the inner ring current region (at $r \lesssim 4R_E$). The quiet-time ring current comprises predominantly H^+ ions, while the storm-time ring current also contains a significant amount of ionospheric O^+. The oxygen contribution to ring current energy density can exceed that of H^+ for brief periods near the maximum of particularly intense geomagnetic storms.

The formation of the storm-time ring current is generally attributed to two distinct processes: (1) the injection of plasma into the inner magnetosphere during the active phase of magnetospheric 'substorms' (see Chapter 12) and (2) increased convective transport of energetic charged particles from the nightside plasma sheet deep into the inner magnetosphere ($r < 4\ R_E$) due to intensification of the Earth's dawn–dusk convection electric field during extended periods of strong southward IMF. The present understanding of ring current formation favours the enhanced convection model over the substorm plasma injection model; however, it is recognised that magnetospheric substorms (Baker et al., 1996), while not the dominant direct driver, nonetheless play a key role in the growth of the storm-time ring current (e.g. by energising ions in the near-Earth plasma sheet prior to their transport into the inner magnetosphere).

The storm-time growth of the ring current typically lasts from 3 to 12 hours and constitutes the 'main phase' of a magnetic storm (see Chapter 12). Following this main phase, the ring current begins to decay, returning to its pre-storm state in a period of two to three days (Jacobs, 1987, 1991). During the storm recovery phase, particle transport into the ring current slows. This allows loss processes to reduce ring current particle fluxes back to their quiet-time level. The principal loss process during both the storm main and recovery phases is ring current particle charge exchange with neutral (low-energy) hydrogen atoms in the geocorona. Another loss process, affecting principally low-energy ring current ions, involves Coulomb collisions with the thermal plasma of the plasmasphere. Yet another process contributing to ring current decay is the precipitation loss of ring current particles into the atmosphere as a result of wave-particle interactions. The volume by Hultqvist et al. (1999) provides a comprehensive overview of source and loss processes.

The role of all these loss processes in the evolution of the ring current is still the subject of ongoing magnetospheric research (Chapter 12). The growth and recovery of the ring current is measured by changes in the *Dst* (disturbance storm-time) index, the geomagnetic index that serves as the usual indicator of ring current activity.

4.2 Field-Aligned (or Birkeland) Currents

So-called Birkeland currents are large-scale electron (and ion) flows along magnetic field lines that connect the higher altitude parts of Earth's magnetosphere with the high-latitude ionosphere (Iijima and Potemra, 1978). These currents are driven directly within the magnetosphere by interactions with the magnetised solar wind flow (see Figure 4.0.3) and also by the large-scale plasma flow through the magnetosphere (convection that is indirectly driven by the interplanetary interaction). The strength of the Birkeland currents changes with activity in the magnetosphere (e.g. during magnetospheric substorms). Small-scale variations in the upward current sheets (downward flowing electrons) accelerate magnetospheric electrons (see Figure 4.2.1) which, when they reach the upper atmosphere, create the northern and southern lights (Aurora Borealis and Australis). In the high-latitude ionosphere (i.e. the auroral zones), the Birkeland currents close through the region of the auroral electrojet, which flows perpendicular to the local magnetic field in the ionosphere. These perpendicular currents are known as Pedersen currents (Figure 4.2.1). The Birkeland currents occur in two pairs of field-aligned current sheets. One pair extends from noon through dusk to the midnight sector. The other pair extends from noon through dawn to the midnight sector. The sheet on the high-latitude side of the auroral zone is referred to as the region 1 current sheet and the sheet on the low-latitude side is called the region 2 current sheet (see Figure 4.2.1).

Auroral Birkeland currents carry some 100,000 amperes (A) during quiet times and can carry more than 1 million A during geomagnetically disturbed times. Birkeland's original work estimated currents 'at heights of several hundred kilometres, and strengths of up to a million amperes' (Birkeland, 1908). The Pedersen currents that connect the field-aligned currents heat the upper atmosphere due to the finite conductivity of the ionosphere. This heat (known as Joule dissipation) is transferred from the ionospheric plasma to the neutral gas of the upper atmosphere. This neutral atmosphere consequently rises buoyantly and can greatly increase drag on low-altitude satellites. This is a key space weather concern (see Chapter 14).

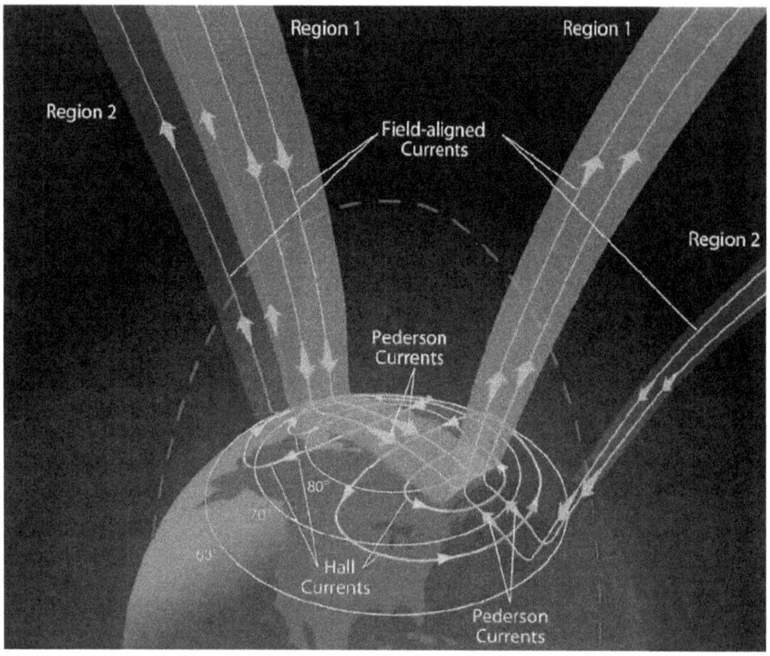

Figure 4.2.1 A 3-D diagram illustrating the field-aligned (Birkeland) and perpendicular (Pedersen) current systems in Earth's high-latitude magnetosphere–ionosphere system. The Region 1 and Region 2 current sheets of the Birkeland system are illustrated.

4.3 Magnetotail Currents

The tail current sheet is a major shaper of the Earth's external magnetic field, especially at large radial distances (Tsyganenko and Stern, 1996). This current system forms one of the principal boundaries in the magnetosphere, defining its global geometry. Predicting the position of the tail current sheet is important for the planning of the science of magnetospheric missions and accurately interpreting their results. It is also crucial for development of empirical models of the magnetospheric fields and plasma distributions. The shape of the tail current sheet as a function of the magnetic dipole tilt and interplanetary conditions has been addressed empirically in numerous studies using both magnetometer data and plasma data. These studies, based on different datasets and techniques, established the key features of the current sheet geometry and quantified its location using various analytical approximations (Tsyganenko, 1989). The fundamental effect of the seasonal and diurnal oscillation of the dipole tilt angle is a periodic warping of the neutral sheet surface, so that it bends northward (or southward) near the midnight meridian, but moves in the opposite direction or stays at rest near the tail's flanks (see Figure 4.3.1). The IMF component is found to exert a torque on the Earth's magnetotail, resulting in a left- or right-handed twisting of the cross-tail current sheet for positive or negative IMF, respectively. A good recent review concerning the size, shape, and predicted location of the distant magnetotail is given by Sibeck and Lin (2014).

Most of the volume of the Earth's magnetotail is taken up by two bundles of oppositely directed, nearly parallel magnetic field lines (see Figure 4.3.2). The field lines north of the equatorial plane point sunward and connect to a roughly circular region in Earth's ionosphere including the northern magnetic pole (the north 'polar cap'). The southern bundle of field lines points away from the Sun and is linked to the southern polar cap region. The two tail volumes relatively

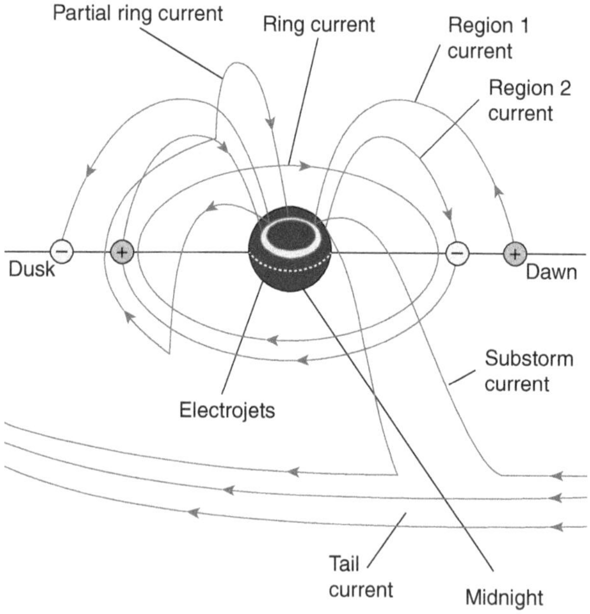

Figure 4.3.1 The principal near-Earth current systems in the terrestrial magnetosphere and their relationship to the nightside cross-tail current system that forms Earth's magnetotail. Courtesy of Robert McPherron.

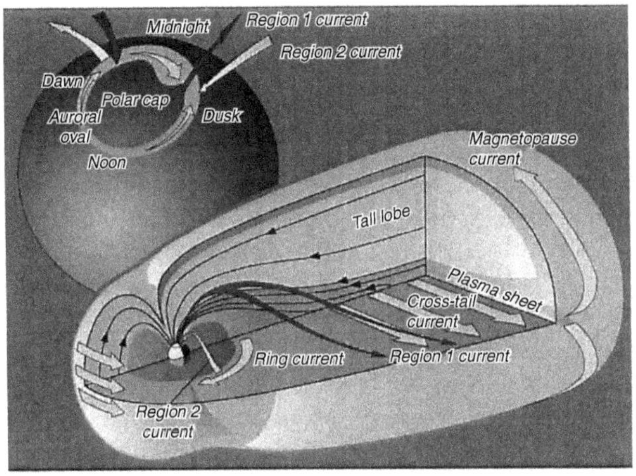

Figure 4.3.2 A diagram showing current systems from Earth's ionosphere and polar cap extending to the deep magnetotail. The plasma sheet and tail current systems are emphasised to the right of the figure. From Baker (2000).

devoid of plasma are known as the 'tail lobes'. They extend far from Earth (having been observed many hundreds of R_E downstream). At those large downstream distances the lobes can have considerable solar wind plasma within them, but near Earth they are almost empty. This extremely low density suggests that field lines of the lobe ultimately connect to the distant solar wind, somewhere far downstream from Earth. These are called 'open' magnetic field lines. Ions and electrons can easily flow away from Earth along lobe field lines until they are swept into the solar wind. In general, few solar wind ions can flow back upstream toward Earth and thus the lobes have strong magnetic field but low plasma density.

Separating the two magnetotail lobes is the 'plasma sheet'. This is a layer of weaker magnetic field and higher density plasma. It is centred on the equatorial plane and is typically 2–6 R_E thick. Unlike field lines of the tail lobes, those of the plasma sheet pass across the equator (though they are quite stretched out anti-sunward). The weak magnetic field in the plasma sheet means that the plasma is less restrained here than nearer to Earth so the plasma can dominate more over the magnetic field. The current flowing across the plasma sheet from dawn to dusk closes over the top and bottom portions of the magnetotail (Figure 4.3.2) to form a solenoidal configuration that supports the stretched out magnetic field lines of the magnetotail.

4.4 Magnetopause Currents

The magnetopause is the relatively sharp boundary between the magnetosphere and the onrushing plasma of the shocked solar wind. In general, the magnetopause is the demarcation between Earth's magnetic field and the solar wind plasma (see Figure 4.0.3). The location of the magnetopause is determined by the balance between the pressure of the terrestrial magnetic field and the dynamic pressure of the external (solar wind) flow. As the solar wind pressure increases or decreases, the magnetopause moves inward and outward in direct response. Waves (ripples and flapping motion) along the magnetopause move in the direction of the solar wind flow in response to small-scale variations in the solar wind pressure and to Kelvin–Helmholtz instability (see Chapters 12 and 13).

The solar wind is supersonic and passes through a standing bow shock wave where the direction of flow is changed so that most of the solar wind plasma is deflected to either side of the magnetosphere (Figure 4.0.3). The zone of shocked solar wind plasma is called the magnetosheath. At Earth (and all the other planets with intrinsic magnetic fields), some solar wind plasma penetrates and enters the magnetosphere, becoming trapped within the magnetosphere. At Earth, the solar wind plasma that enters the magnetosphere largely forms the nightside plasma sheet. The amount of solar wind plasma energy that enters the magnetosphere is regulated by the north-south orientation of the interplanetary magnetic field (IMF), embedded in the solar wind (e.g. Baker et al., 1996).

Chapman and Ferraro (1931) proposed that a plasma was emitted by the Sun in a burst as part of a solar flare event. This solar blast was presumed to disturb the Earth's magnetic field so as to cause a geomagnetic storm. The collision frequency of particles in the plasma in the interplanetary medium is very low and the electrical conductivity is so high that this solar 'cloud' could be approximated as an infinite conductor. Chapman and Bartels (1940) illustrated this concept by postulating a plate with infinite conductivity placed on the dayside of the terrestrial dipole. The field lines on the dayside are distorted due to the magnetopause currents flowing in the boundary. At low latitudes, the magnetic field lines are pushed inward. At high latitudes, the magnetic field lines are pushed backwards and over the polar regions. The boundary between the region dominated by intrinsic magnetic field (i.e. the magnetosphere) and the plasma in the interplanetary medium was taken to be the magnetopause current layer. Since the solar wind is continuously flowing anti-sunward, the magnetopause above, below and to the sides of the Earth is swept backwards into the geomagnetic tail as shown in Figure 4.0.3. Thus, magnetopause currents have effects on shaping the geomagnetic field over both the dayside and the nightside of the Earth. Solar wind particles can fairly readily enter the magnetosphere through the cusp region. Because the solar wind exists at all times and not just times of solar flares, the magnetopause really is a permanent feature of the space near Earth, not just a storm-time phenomena (contrary to the original Chapman–Ferraro picture).

4.5 Magnetospheric Dynamics

The magnetic field lines of the geomagnetic field are not stationary. They are continuously joining or merging with magnetic field lines of the interplanetary magnetic field. The joined field lines are swept back over the poles into the Earth's magnetic tail. In the tail, the field lines from the Earth's magnetic field are re-joined and start moving back toward the Earth (see Chapter 12). The physics of this process (Figure 4.5.1) was first explained by Dungey (1961). This picture of IMF interconnection with the geomagnetic field is the fundamental cause of sporadic geomagnetic activity such as magnetospheric substorms and geomagnetic storms (Baker et al., 1996).

Inside the magnetosphere is the plasmasphere, a donut-shaped region containing low-energy plasma. This region begins at a height of 60 km, extends up to 3 or 4 R_E. It includes the ionosphere. This plasmasphere rotates with the Earth and supports a wide variety of plasma waves. There are also two concentric torus-shaped regions, called the Van Allen radiation belts, with high-energy ions and electrons. The electrons dominate the outer belt (energies from 0.1 to over 10 million electron volts (MeV)) while ions (from MeV to GeV energies) dominate the inner Van Allen zone. The inner belt is 1–2 R_E out while the outer belt is at ~3–7 R_E. The plasmasphere and Van Allen belts have partial overlap, with the extent of overlap varying greatly with solar activity (Chapter 12).

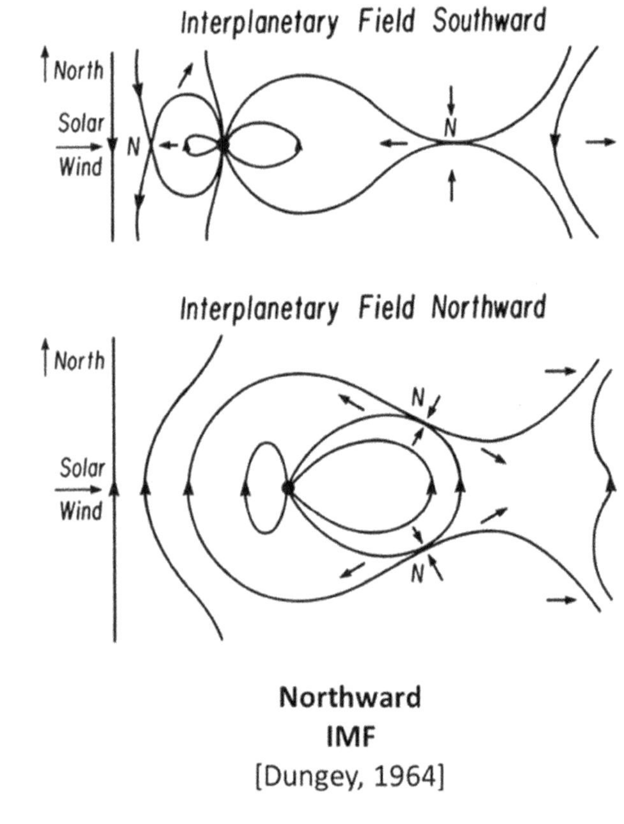

Figure 4.5.1 Schematic of coupling between the interplanetary magnetic field and the Earth's magnetic field by reconnection. This coupling leads to energy input to the magnetosphere and to enhance substorm activity. From Dungey (1961).

The varying conditions in the magnetosphere, known as space weather, are largely driven by solar activity (Chapters 14 and 15). If the solar wind is weak, the magnetosphere expands; if it is strong, it compresses the magnetosphere and more of the solar wind gets in. Periods of particularly intense activity (i.e. geomagnetic storms) can occur when a coronal mass ejection erupts near the Sun and sends a shock wave through the interplanetary space. Such a wave can take just 1–3 days to reach the Earth and are the cause of the most intense episodes of geomagnetic storm activity.

4.6 Summary

This chapter has provided a brief and very high-level overview of the geomagnetic field and its interactions with the magnetic field emanating from our Sun. Through careful study, we have come to a quite deep understanding of the history of Earth's magnetic field (palaeomagnetism) and we have also developed great insight into the many manifestations of the complex, present-day geomagnetic field. Subsequent chapters in this book will describe in remarkable ways the changes occurring in Earth's magnetic environment on timescales ranging from seconds to millennia. When considered in the context of life on Earth and space weather threats to our modern technological systems, the geomagnetic field is – and always has been – crucial to our very existence. The remainder of this book will explore the many fascinating aspects of our magnetically active planet.

References

Baker, D. N. (2000) Effects of the Sun on the Earth's environment, *J. Atmos. Sol. Terr. Phys.*, **62**, 1669–81.

Baker, D. N., et al. (1996) The neutral line model of substorms: Past results and present view, *J. Geophys. Rev.*, **101**, 12,995–13,010.

Birkeland, K. (1908) *The Norwegian Aurora Polaris Expedition*, 1902–1903, H. Aschehoug, Christania, Norvège.

Campbell, W. N. (2003) *Introduction to Geomagnetic Fields*, 2nd edn., Cambridge University Press, Cambridge.

Chapman, S. and Bartels, J. (1940) *Geomagnetism*, Oxford University Press, New York.

Chapman, S. and Ferraro, V. C. A. (1931) New theory of magnetic storms, *Terr. Magn. Atm. Elec.*, 36, 77.

Cole, K. D. (1966) Magnetic storms and associated phenomena, *Space Sci. Rev.*, **5**, 699–770.

Cowley, S. W. H. (1991) The structure and length of tail-associated phenomena in the solar wind downstream from the Earth, *Planet. Space Sci.*, **39**, 1039–43.

Dungey, J. W. (1961) Interplanetary magnetic field and the auroral zones, *Phys. Rev. Lett.*, **6**, 47.

Gauss, C. F. (1839) Allgemeine Theorie des Erdmagnetismus, in *Resultate aus den Beobachtungen des magnetischen Vereins im Jahre 1838*, Ed. Gauss, C. F. and Weber, W., 1–57, Weidmannsche Buchhandlung, Leipzig.

Glassmeier, K.-H. T. and Tsurutani, B. (2014). Carl Friedrich Gauss – *General Theory of Terrestrial Magnetism* – a revised translation of the German text. *Hist. Geor. Space Sci.*, 5, 11-62. doi: 10.5194/hgss-5-11-2014.

Glatzmaier, G. A. and Roberts, P. H. (1995) A three-dimensional self-consistent computer simulation of a geomagnetic field reversal, *Nature*, **377**, 203–9.

Hultqvist, B., Oieroset, M., Paschmann, G. and Treumann, R. (Eds.) (1999) *Magnetospheric Plasma Sources and Losses*, Space Science Series 6, Kluwer Academic, Dordrecht.

Iijima, T. and Potemra, T. A. (1978) Large scale characteristics of field-aligned currents associated with substorms, *J. Geophys. Res.*, **83**, 599–615.

Jacobs, J. A. (Ed.) (1987) *Geomagnetism*, vols. 1 and 2, Academic Press, New York.

Jacobs, J. A. (Ed.) (1991) *Geomagnetism*, vol. 4, Academic Press, New York.

Sibeck, D. G. and Lin, R.-Q. (2014) Size and shape of the distant magnetotail, *J. Geophys. Res.*, 119, 1028–43.

Tsyganenko, N. A. (1989) Magnetospheric magnetic field with a warped tail current sheet. *Planet. Space Sci.*, **73**, 5.

Tsyganenko, N. A. and Stern, D. P. (1996) Modeling the global magnetic field of the large scale Birkeland current system, *J. Geophys. Res.*, **101**, 187–98.

5

Can Paleomagnetism Distinguish Dynamo Regimes?

*Stuart A. Gilder and Florian Lhuillier**

This section describes a paleomagnetic approach to test whether the geodynamo has existed in distinct operating modes through Earth's history. A central question surrounds how the geodynamo functions during superchrons, where periods lasting tens of millions of years passed with no reversals, versus times closer to the present, when reversals happened four to five times per million years. Five numerical dynamos with characteristic reversal frequencies ranging from 2 to 13 Myr^{-1} were studied to derive the parameters that discriminate them using data sets routinely acquired by paleomagnetists. Once defined, the feasibility of the approach was tested through field and laboratory work.

5.1 Background

Geomagnetic field reversals constitute a persistent feature of Earth's dynamo. Reversal frequency has fluctuated through Earth's history from roughly five times per million years over the past 5 million years to periods lasting over 30 million years with no reversals. The latter are termed superchrons, of which three have been documented in the Phanerozoic: the Cretaceous normal superchron (CNS), which lasted from 124 Ma to 83 Ma; the Permo-Carboniferous (Kiaman) reversed superchron, which lasted from 317 Ma to 265 Ma; and the Ordovician (Moyero) reversed superchron, which lasted from 485 to 463 Ma (Pavlov and Gallet, 2005). Superchrons have also been proposed in the Proterozoic (Elston et al., 2002; Pavlov and Gallet, 2010; Gallet et al., 2012; Driscoll and Evans, 2016), indicating that they could be a persistent feature of the magnetic field over the whole of Earth's history.

Superchrons and changes in reversal frequency are thought to be influenced by the thermal structure at the core–mantle boundary. Modification in the heat flux at the core–mantle boundary occurring over plausible mantle convection rates can alter conditions at the boundary layer where the dynamo is produced thereby causing perturbations in its convective pattern (Larson, 1991; Glatzmaier et al., 1999). Heat can be removed from the core–mantle boundary by mantle plumes (Loper and McCartney, 1986; Courtillot and Besse, 1987; Courtillot and Olson, 2007; Olson and Amit, 2015) or the boundary layer can be cooled by subducting cold lithosphere (Eide and Torsvik, 1996; Gallet and Hulot, 1997). Altering convective processes within the outer core, such as growth of the inner core, could also alter magnetohydrodynamic patterns (McFadden and Merrill, 1986; Roberts and Glatzmaier, 2001; Heimpel et al., 2005; Landeau et al., 2017).

Several theories have been put forth to explain the origin of superchrons (Jacobs, 2001). End-member hypotheses are high and low energy states whose origins differ by convection vigor in the outer core. Larson and Olson (1991) proposed that superchrons result from enhanced core convection that stabilizes the geodynamo. This model predicts less geomagnetic secular variation, and potentially higher field strength, during superchrons. In contrast, Loper and McCartney (1986) argued that chaotic flow from stronger core convection results in geomagnetic reversals whereby superchrons represent times of weaker core convection, characterized by greater variations in direction and potentially lower field intensity. Alternatively, superchrons are perhaps a continuum of a single process, arising from a non-linear transition of the geodynamo between reversing and stable states (Hulot and Gallet, 2003).

The question becomes how to distinguish geodynamo regimes in the geologic record. The typical way is to study the rate of change of the geomagnetic field, called paleosecular variation (Cox, 1975; Bloxham and Gubbins, 1985; Jackson and Bloxham, 1991). Because paleointensity data are more difficult to acquire than directional data, paleosecular variation is most often quantified via the angular standard deviation of virtual geomagnetic poles (VGPs). VGPs are calculated from the site mean directions (declination and inclination) at the geographic location of the site using the law of cosines. The mean angular distance Δ of an

* We thank the Alexander von Humboldt Foundation for funding, an anonymous reviewer for helpful suggestions and Pieter Kotze and Inez Batista for editorial handling.

individual VGP away from the mean of the VGPs establishes the S-value, defined as $S = \sqrt{\frac{1}{N-1}\sum_{i=1}^{N}\Delta_i^2}$. High S reflects high (fast) geomagnetic activity while low S, low (slow) activity.

S-values vary with latitude, being two times higher for sites at polar regions than at the equator. Several workers explained the latitudinal dependency using a combination of dipole and non-dipole components that vary in proportion, intensity and direction (Creer et al., 1959; Irving and Ward, 1964; Cox, 1962, 1970; McElhinny and Merrill, 1975; McFadden and McElhinny, 1984; McFadden et al., 1988; Constable and Parker, 1988; Harrison, 1995; Tauxe and Kent, 2004). Although the S parameter has been used to draw geomagnetic inferences in the paleorecord (e.g. Cronin et al., 2001; Riisager et al., 2001; Tarduno et al., 2002; Biggin et al., 2008); in practice, describing secular variation with S-values can be problematic. If S indeed varies with latitude and time, then using S to indicate geomagnetic activity necessitates defining the slope in S as a function of latitude for each independent time interval. This requires not only a large time investment in the laboratory, but also geologic conditions must be appropriate in time and space (e.g. Linder and Gilder, 2011).

5.2 Distinguishing Numerical Dynamos with Simulated Paleomagnetic Data

What is the best way to distinguish dynamo regimes in the paleomagnetic record? To answer this, we studied numerical dynamos in order to define parameters that paleomagnetists could use to better constrain secular variation models – especially those parameters that are invariant with respect to latitude. In other words, what can we learn from dynamo simulations that have distinctly different properties (reversal frequency, dipole strength, etc.) that paleomagnetism can adequately distinguish them?

To this aim, we used five dynamo models that are characterized by fixed Ekman ($E = 6.5 \times 10^{-3}$), hydrodynamic Prandtl ($Pr = 1$) and magnetic Prandtl ($Pm = 20$) numbers (Lhuillier et al., 2013). The first model had an insulating inner core and resembled the dynamos studied by Olson (2007) and Olson et al. (2009); the other four had conducting inner cores and operated in the same parameter range as the dynamos investigated by Driscoll and Olson (2011) and Olson et al. (2011, 2013). The models varied by their modified Rayleigh numbers (Ra*), which led to polarity chron durations ranging from ~517 to 83 kyr (~2–13 reversals/Myr) (Table 5.2.1; Lhuillier and Gilder, 2013).

Each model was integrated over the equivalent of 40–50 Myr. From these runs, three-component magnetic data were generated in the local reference frame at the Earth's surface,

Table 5.2.1 *Properties of the five numerical models investigated by Lhuillier and Gilder (2013)*

Model no.	Inner core type	Ra* × 10^{-3}	Rm	f_{13_dip}	f_{oct_dip}	μ_{chron} kyr
1	Insulating	5.5	110	0.58	0.14	517
2	Conducting	5.1	108	0.57	0.09	176
3	Conducting	5.5	117	0.54	0.09	086
4	Conducting	5.9	124	0.54	0.10	083
5	Conducting	6.3	133	0.50	0.11	094

Note. Ra* = the modified Rayleigh number was used as an input parameter with inner cores that were either insulating or conducting as specified. Rm = the magnetic Reynolds number, where Rm = UD/η, with U being the time-averaged root mean square velocity over the outer core shell. D = the shell thickness assuming an aspect ratio of 0.35. η = the magnetic diffusivity. f13_dip = the time-averaged root mean square amplitude of the dipole relative to the total field up to spherical harmonic degree 13 at the core–mantle boundary. f_{oct_dip} = the time-averaged root mean square amplitude of the octupolar component (g_3^0) to that of the axial dipole (g_1^0) at the Earth's surface. μ_{chron} = the mean chron length in kyr rescaled by the magnetic diffusion time.

from which the total intensity F, declination D and inclination I were derived along longitudinal bands at 35 geographic latitudes distributed every 5° from 85°S to 85°N. With some caveats, we found that F was Gaussian distributed and that virtual geomagnetic poles (VGPs), and directions to a lesser extent, were approximately Fisherian distributed (Lhuillier and Gilder, 2013). For example, in each of the five models, the average intensity (μ_F) was approximately twice as large at the poles than at the equator, which is the expected result for a predominantly dipolar field (Figure 5.2.1a). After studying the full gamut of the output data (components and field elements), we found that the relative variability in intensity (ε_F), which is equal to the standard deviation of F (σ_F) divided by μ_F ($\varepsilon_F = \sigma_F / \mu_F$), was the least latitude-dependent parameter (Figure 5.2.1). The humps visible at ~±45° latitude are probably the consequence of a ~10% octupolar component produced by the investigated dynamos (Table 5.2.1). We anticipate they should vanish in a more dipolar-dominated dynamo, such as the Earth's dynamo (e.g. Courtillot and Besse, 2004).

From these indices, we found that the best way to discriminate the models was to combine both the intensity and directional dispersion information, with the latter being defined by either the Fisher (1953) precision parameter of the directions (k_D) or the S-value of the poles (S_P). Figure 5.2.2 plots ε_F versus k_D and S_P for each 5° latitudinal increment produced from each dynamo model. The average value from all latitudes is also shown. In this way, each dynamo model can be discriminated either from the overall averages or from a single observation at a given latitude. They predict that dynamos with lower reversal frequencies

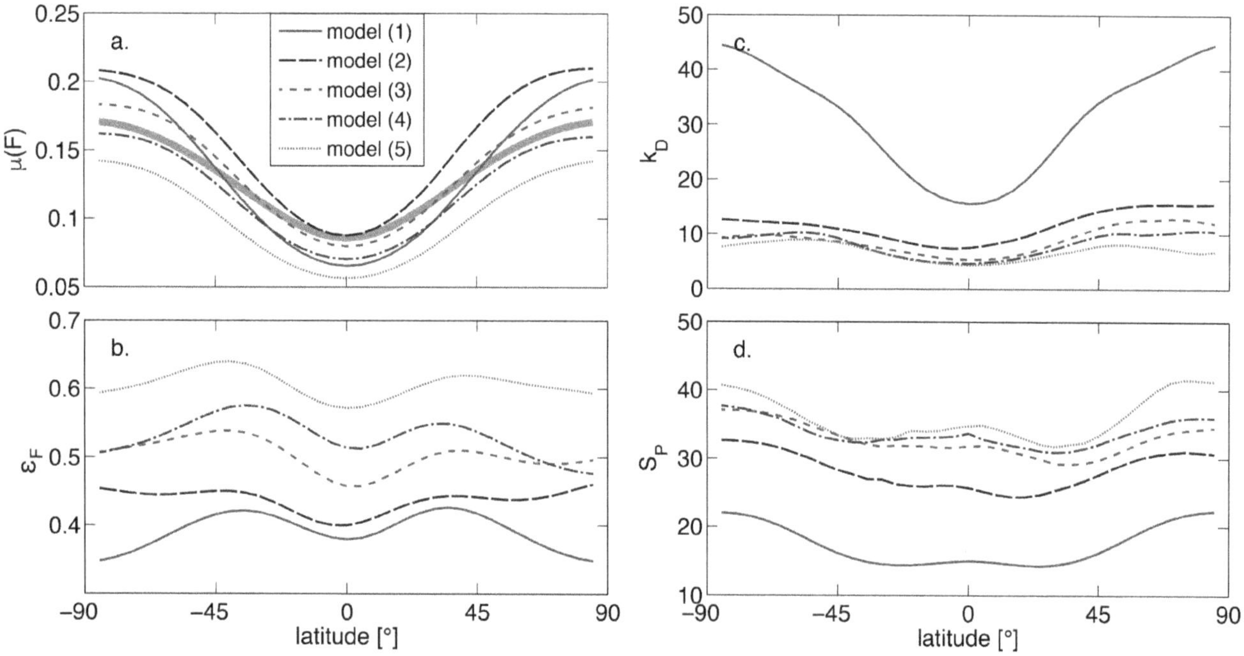

Figure 5.2.1 Latitudinal variation of (a) the average intensity μ_F, (b) the relative variability in intensity ε_F, (c) the precision parameter of directions k_D and (d) the angular dispersion of poles S_P for the five investigated models. The thick gray curve in panel (a) is the dipole formula fit to the mean of the five models (after Lhuillier and Gilder, 2013). (A black-and-white version of this figure appears in some formats. For the colour version, please refer to the plate section.)

will be more stable with slower, less variable secular variation rates than dynamos with higher reversal frequencies.

From a practical viewpoint, evaluating ε_F from absolute paleointensity determinations is unrealistic due to the method's low success rate, which makes relative paleointensity determinations advantageous to obtain ε_F. In optimal cases, the pseudo-Thellier (relative) method (Tauxe et al., 1995) applied to volcanic rocks (e.g. de Groot et al., 2013, 2016) often leads to success rates greater than 50%, thereby allowing ε_F to be more routinely calculated (e.g. Lhuillier et al., 2016, 2017). This technique is similar to Thellier-style experiments (Thellier and Thellier, 1959; Coe, 1967; Tauxe and Staudigel, 2004) but relies on the acquisition of an anhysteretic remanent magnetization and its subsequent demagnetization with an alternating field. As the method circumvents thermal demagnetization, it avoids alteration of the magnetic remanence carriers, which often leads to failed absolute paleointensity determinations.

5.3 Testing against Real Data

We went to western China (Tuoyun: 75.3°E, 40.2°N) ~30 km south of the border of Kyrgyzstan in the Chinese Tianshan Mountains, which contain two sequences of basaltic lava flows that are separated in time by ~50 Myr: one set lies within the Cretaceous normal superchron (CNS) at 112–115 Ma and a second at 59–70 Ma spanning the Cretaceous–Paleogene boundary (Figure 5.3.1) (Lhuillier et al., 2016). This locality provides a unique geological setting where secular variation can be observed at the same site during a superchron and during a time when the field was actively reversing. Any difference with respect to latitude could be ignored, as the paleolatitude of Tuoyun did not change significantly between 115 and 59 Ma (Gilder et al., 2008).

At Tuoyun we collected 405 samples in 76 sites of lava flows in a 541 m thick section that spanned the Cretaceous–Paleogene boundary (59–70 Ma) and 285 samples at two Cretaceous (112–115 Ma) sections: 22 sites of lava flows in 110 m and 35 sites of lava flows in 198 m. The ages from the two Cretaceous sections were distinct in time, so we combined their data for subsequent analyses. Figure 5.3.1 includes stereonet plots of the site mean directions of the two age populations. The inter-site precision parameter was 1.5 times higher for the 56 sites from the Cretaceous normal superchron (k_D = 19.4) than after the superchron, when the reversal frequency was ~2/Myr (k_D = 12.8, N = 75), implying faster secular variation after the superchron than during it, consistent with the prediction from the numerical models. The inclination distributions are distinctly different between the two age groups, which may also be important for distinguishing field behavior although we do not discuss it here (Lhuillier et al., 2016).

Relative paleointensity data were successfully obtained from the 112–115 Ma rocks but not from the K-T section. To compare the data from the superchron with a period when reversals were frequent, we carried out a paleomagnetic study on more than 80 eruptive units ranging in age from 3200 to 60 ka from the Boring (Oregon, USA) volcanic field (Lhuillier et al., 2017; see also Hagstrum

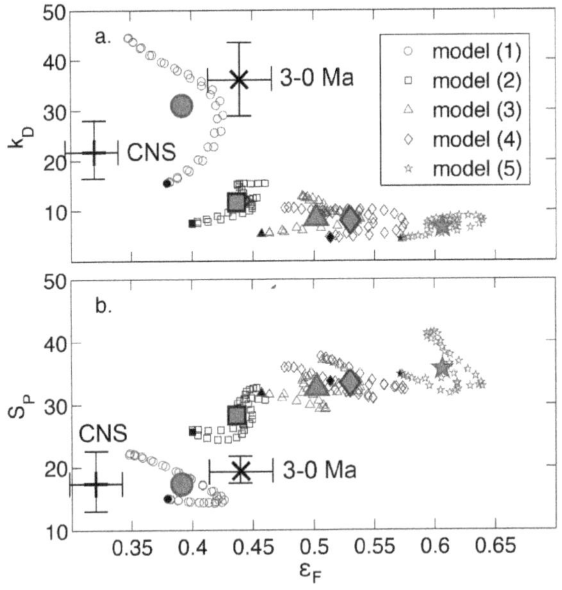

Figure 5.2.2 The precision parameter for directions (k_D) and the angular dispersion of poles (S_P) as a function of the relative variability in intensity (ε_F) for each of the 35 latitudes for each of the five dynamo models. Equatorial values are in black; large symbols represent the overall averages (after Lhuillier and Gilder, 2013). Model 1 has the lowest reversal rate (2/Myr) and 5 the highest (13/Myr). These models predict that dynamos with lower reversal frequencies will be more stable with slower, less variable secular variation rates than dynamos with higher reversal frequencies. The plot includes experimental results from Tuoyun (Xinjiang, China) for the Cretaceous normal superchron (CNS) and from Boring (Oregon, USA) for the last 3 Ma (3–0 Ma). Uncertainties on ε_F (1σ) were derived from bootstrapping the data; 95% confidence limits for k_D and S_P were calculated according to Cox (1969). (A black-and-white version of this figure appears in some formats. For the colour version, please refer to the plate section.)

et al., 2017). The Boring and Tuoyun data are plotted on Figure 5.2.2 to compare with the output from the dynamo models.

5.4 Conclusions

Extrapolating k_D, ε_F and S_P from the dynamo simulations in Figure 5.2.2 to lower reversal frequencies predicts superchons should possess relatively low ε_F and S_P and high k_D. This is consistent with the results from the CNS section from Tuoyun (Figure 5.2.2). Although intensity data were not obtained from the K-T section that had reversals, its k_D was 34% lower than the k_D during the CNS in agreement with the modeling results. The 3–0 Ma Boring volcanic rocks yielded higher ε_F than the CNS, also as predicted from the models with higher reversal frequencies. On the other hand, k_D was higher than expected, S_P too low. The k_D and S_P values from Boring are somewhat contradictory. An important difference between the Tuoyun and Boring studies is that the Tuoyun samples were collected from a few continuous sections, whereas the Boring samples were collected at a large number of independent flows that were assumed to be emplaced on horizontal surfaces. Thus, any deviation from horizontal that went unobserved in the field was not accounted for in Boring and would be mitigated at Tuoyun. Normally, random differences in dip direction would lower k_D; however, a regional tilt could bias the flow directions in a way that changed the geometrical distribution of directions without changing the degree of clustering. Because VGPs are distorted when translating from directions using the law of cosines, a systematic change in directions by either elongating them in declination or

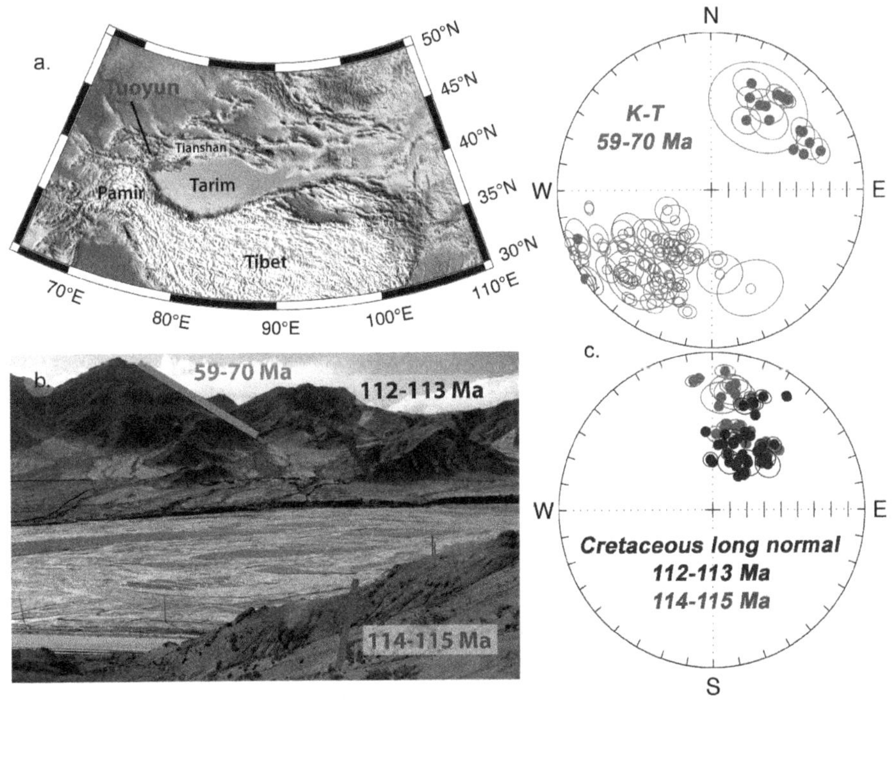

Figure 5.3.1 (a) Topographic locality map and (b) photographic overview of the two Cretaceous long normal (112–115) sections and the Cretaceous–Tertiary (K-T) boundary section investigated by Lhuillier et al. (2016). Stereonet plots on the right are the 75 and 56 site mean directions from the K-T and superchron sections, respectively (two sites were omitted from the original data set). (A black-and-white version of this figure appears in some formats. For the colour version, please refer to the plate section.)

inclination space, could lead to under- or over-estimated S_P values (Linder and Gilder, 2012).

Although the geodynamo is certainly different than the chemically driven dynamos we studied, these dynamos surprisingly yielded Earth-like paleomagnetic data distributions. An important conclusion is that both directional and intensity information should be combined to define secular variation. Considering that the two points in Figure 5.2.2 represent two ends of the spectrum (superchron vs 4–5 reversals/Myr), it appears the geodynamo lies within a fairly restricted area in k_D–ε_F–S_P space, with the hint that superchrons and reversing regimes are part and parcel of the same process. It remains to be seen whether future data will define a broad cluster of points or separate into distinct regimes in regards to Figure 5.2.2. Further studies are under way to generate more data – particularly for the Ethiopian traps at around 29 Ma and 15 Ma.

References

Biggin, A., van Hinsbergen, D., Langereis, C., Straathof, G., & Deenen, M. (2008). Geomagnetic secular variation in the Cretaceous Normal Superchron and in the Jurassic. Physics of the Earth and Planetary Interiors, 169(1–4), 3–19.

Bloxham, J. & Gubbins, D. (1985). The secular variation of Earth's magnetic field. Nature 317, 777–81.

Coe, R. S. (1967). The determination of paleo-intensities of the Earth's magnetic field with emphasis on mechanisms which could cause nonideal behavior in Thellier's method. Journal of Geomagnetism and Geoelectricity, 19(3), 157–79.

Constable, C. G. & Parker, R. L. (1988). Statistics of the geomagnetic secular variation for the past 5 m.y. Journal of Geophysical Research, 93(B10), 11569–81.

Courtillot, V. & Besse, J. (1987). Magnetic field reversals, polar wander, and core–mantle coupling. Science, 237(4819), 1140–47.

Courtillot, V. & Besse, J. (2004). A long-term octupolar component in the geomagnetic field? (0–200 million years BP), in *Timescales of the Paleomagnetic Field*, edited by J. E. T. Channell, D. V. Kent, W. Lowrie & J. G. Meert, AGU, Washington, DC.

Courtillot, V. & Olson, P. L. (2007). Mantle plumes link magnetic superchrons to phanerozoic mass depletion events. Earth and Planetary Science Letters, 260(3–4), 495–504.

Cox, A. (1962). Analysis of present geomagnetic field for comparison with paleomagnetic results. Journal of Geomagnetism and Geoelectricity, 13, 113–19.

Cox, A. (1969). Confidence limits for the precision parameter *k*. Geophysical Journal of Royal Astronomical Society, 18, 545–9.

Cox, A. (1970). Latitude dependence of the angular dispersion of the geomagnetic field. Geophysical Journal of the Royal Astronomical Society, 20(3), 253–69.

Cox, A. (1975). The frequency of geomagnetic reversals and the symmetry of the nondipole field. Reviews of Geophysics and Space Physics, 13(3), 35–51.

Creer, K., Irving, E. & Nairn, A. (1959). Palaeomagnetism of the Great Whin Sill. Geophysical Journal of the Royal Astronomical Society, 2, 306–23.

Cronin, M., Tauxe, L., Constable, C., Selkin, P. & Pick, T. (2001). Noise in the quiet zone. Earth and Planetary Science Letters, 190(1–2), 13–30.

de Groot, L. V., Biggin, A. J., Dekkers, M. J., Langereis, C. G. & Herrero-Bervera, E. (2013). Rapid regional perturbations to the recent global geomagnetic decay revealed by a new Hawaiian record. Nature Communications, 4, 1–7.

de Groot, L. V., Pimentel, A. & di Chiara, A. (2016). The multimethod palaeointensity approach applied to volcanics from Terceira: Full-vector geomagnetic data for the past 50 kyr. Geophysical Journal International, 206(1), 590–604.

Driscoll, P. E. & Evans, D. A. D. (2016). Frequency of Proterozoic geomagnetic superchrons. Earth and Planetary Science Letters, 437, 9–14.

Driscoll, P. E. & Olson, P. L. (2011). Superchron cycles driven by variable core heat flow. Geophysical Research Letters, 38(9), L09304.

Eide, E. & Torsvik, T. (1996). Paleozoic supercontinental assembly, mantle flushing, and genesis of the Kiaman Superchron. Earth and Planetary Science Letters, 144(3–4), 389–402.

Elston, D. P., Enkin, R. J., Baker, J. & Kisilevsky, D. K. (2002). Tightening the belt: Paleomagnetic-stratigraphic constraints on deposition, correlation, and deformation of the Middle Proterozoic (ca. 1.4 Ga) Belt-Purcell Supergroup, United States and Canada. Geological Society of America Bulletin, 114(5), 619–38.

Fisher, R. A. (1953). Dispersion on a sphere. Proceedings of the Royal Society of London, Series A, 217, 295–305.

Gallet, Y. & Hulot, G. (1997). Stationary and nonstationary behaviour within the geomagnetic polarity time scale. Geophysical Research Letters, 24(15), 1875–8.

Gallet, Y., Pavlov, V. E., Halverson, G. & Hulot, G. (2012). Toward constraining the long-term reversing behavior of the geodynamo: A new 'Maya' superchron ~1 billion years ago from the magnetostratigraphy of the Kartochka Formation (southwestern Siberia). Earth and Planetary Science Letters, 339–40, 117–26.

Gilder, S. A., Gomez, J., Chen, Y. & Cogné, J. P. (2008). A new paleogeographic configuration of the Eurasian landmass resolves a paleomagnetic paradox of the Tarim Basin (China). Tectonics, 27, TC1012, doi: 10.1029/2007TC002155.

Glatzmaier, G., Coe, R., Hongre, L. & Roberts, P. (1999). The role of the Earth's mantle in controlling the frequency of geomagnetic reversals. Nature, 401(6756), 885–90.

Hagstrum, J. T., Fleck, R. J., Evarts, R. C. & Calvert, A. T. (2017). Paleomagnetism and $^{40}Ar/^{39}Ar$ geochronology of the Plio-Pleistocene Boring volcanic field: Implications for the geomagnetic polarity time scale and paleosecular variation, Phys. Earth Planet. Inter., 262, 101–15.

Harrison, C. (1995). Secular variation of the earth's magnetic field. Journal of Geomagnetism and Geoelectricity, 47, 131–47.

Heimpel, M. H., Aurnou, J. M., Al-Shamali, F. M. & Gomez-Perez, N. (2005). A numerical study of dynamo action as a function of spherical shell geometry. Earth and Planetary Science Letters, 236(1–2), 542–57.

Hulot, G. & Gallet, Y. (2003). Do superchrons occur without any palaeomagnetic warning? Earth and Planetary Science Letters, 210(1–2), 191–201.

Irving, E. & Ward, M. (1964). A statistical model of the geomagnetic field. Pure and Applied Geophysics, 57(1), 47–52.

Jackson, A. & Bloxham, J. (1991). Mapping the fluid flow and shear near the core surface using the radial and horizontal components of the magnetic field. Geophysical Journal International, 105(1), 199–212.

Jacobs, J. (2001). The cause of superchrons. Astronomy & Geophysics, 42(6), 6–30.

Landeau, M., Aubert, J. & Olson, P. L. (2017). The signature of inner-core nucleation on the geodynamo. Earth and Planetary Science Letters, 465, 193–204.

Larson, R. (1991). Latest pulse of Earth: Evidence for a mid-Cretaceous superplume. Geology, 19(6), 547.

Larson, R. & Olson, P. (1991). Mantle plumes control magnetic reversal frequency. Earth and Planetary Science Letters, 107 (3–4), 437–47.

Lhuillier, F. & Gilder, S. A. (2013). Quantifying paleosecular variation: Insights from numerical dynamo simulations. Earth and Planetary Science Letters, 382, 87–97.

Lhuillier, F., Gilder, S. A., Wack, M., He, K., Petersen, N., Singer, B. S., Jicha, B. R., Schaen, A. J. & Colon, D. (2016). More stable yet bimodal geodynamo during the Cretaceous superchron? Geophysical Research Letters, 43(12), 6170–7.

Lhuillier, F., Hulot, G. & Gallet, Y. (2013). Statistical properties of reversals and chrons in numerical dynamos and implications for the geodynamo. Physics of the Earth and Planetary Interiors, 220, 19–36.

Lhuillier, F., Shcherbakov, V., Gilder, S. A. & Hagstrum, J. T. (2017). Variability of the 0–3 Ma palaeomagnetic field observed from the Boring Volcanic Field of the Pacific Northwest. Geophysical Journal International, 211, 69–79.

Linder, J. & Gilder, S. A. (2011). Geomagnetic secular variation recorded by sediments deposited during the Cretaceous normal superchron at low latitude. Physics of the Earth and Planetary Interiors, 187, 245–60.

Linder, J. M. & Gilder, S. A. (2012). Latitude dependency of the geomagnetic secular variation S parameter: A mathematical artifact. Geophysical Research Letters, 39, L02308, doi: 10.1029/2011GL050330.

Loper, D. & McCartney, K. (1986). Mantle plumes and the periodicity of magnetic field reversals. Geophysical Research Letters, 13(13), 1525–8.

McElhinny, M. & Merrill, R. (1975). Geomagnetic secular variation over the past 5 My. Reviews of Geophysics, 13(5), 687–708.

McFadden, P. & McElhinny, M. (1984). A physical model for palaeosecular variation. Geophysical Journal of the Royal Astronomical Society, 78(3), 809–30.

McFadden, R. & Merrill, R. (1986). Geodynamo energy source constraints from palaeomagnetic data. Physics of the Earth and Planetary Interiors, 43(1), 22–33.

McFadden, P., Merrill, R. & McElhinny, M. (1988). Dipole/quadrupole family modeling of paleosecular variation. Journal of Geophysical Research, 93(B10), 11583–8.

Olson, P. L. (2007). Gravitational dynamos and the low-frequency geomagnetic secular variation. Proceedings of the National Academy of Sciences of the United States of America, 104 (51), 20160–66.

Olson, P. L. & Amit, H. (2015). Mantle superplumes induce geomagnetic superchrons. Frontiers in Earth Science, 3, 1–11.

Olson, P. L., Driscoll, P. E. & Amit, H. (2009). Dipole collapse and reversal precursors in a numerical dynamo. Physics of the Earth and Planetary Interiors, 173, 121–140.

Olson, P. L., Deguen, R., Hinnov, L. A. & Zhong, S. (2013). Controls on geomagnetic reversals and core evolution by mantle convection in the Phanerozoic. Physics of Earth and Planetary Interiors, 214, 87–103.

Olson, P. L., Glatzmaier, G. A. & Coe, R. S. (2011). Complex polarity reversals in a geodynamo model. Earth and Planetary Science Letters, 304(1–2), 168–79.

Pavlov, V. & Gallet, Y. (2005). A third superchron during the Early Paleozoic. Episodes, 28(2), 78–84.

Pavlov, V. E. & Gallet, Y. (2010). Variations in geomagnetic reversal frequency during the Earth's middle age. Geochemistry, Geophysics, Geosystems, 11(1), Q01Z10, doi: 10.1029/2009GC002583.

Riisager, J., Perrin, M., Riisager, P. & Vandamme, D. (2001). Paleomagnetic results and paleointensity of Late Cretaceous Madagascan basalt. Journal of African Earth Sciences, 32(3), 503–18.

Roberts, P. H. & Glatzmaier, G. A. (2001). The geodynamo, past, present and future. Geophysical and Astrophysical Fluid Dynamics, 94(1–2), 47–84, doi: 10.1080/03091920108204131

Tarduno, J., Cottrell, R. & Smirnov, A. (2002). The Cretaceous superchron geodynamo: Observations near the tangent cylinder. Proceedings of the National Academy of Sciences of the United States of America, 99(22), 14020–25.

Tauxe, L. & Kent, D. V. (2004). A simplified statistical model for the geomagnetic field and the detection of shallow bias, in *Timescales of the Paleomagnetic Field*, AGU, Washington, DC, pp. 101–15.

Tauxe, L., Pick, T. & Kok, Y. S. (1995). Relative paleointensity in sediments: A pseudo-Thellier approach. Geophysical Research Letters, 22(21), 2885–8.

Tauxe, L. & Staudigel, H. (2004). Strength of the geomagnetic field in the Cretaceous normal superchron: New data from submarine basaltic glass of the Troodos Ophiolite. Geochemistry, Geophysics, Geosystems, 5(2), doi: 10.1029/2003GC000635.

Thellier, E. & Thellier, O. (1959). Sur l'intensité du champ magnétique terrestre dans le passé historique et géologique. Annales de Géophysique, 15, 285–376.

6

Geomagnetic and Electromagnetic Observations at Ground Level

6.1 Geomagnetic Measurements

*Arnaud Chulliat**

6.1.1 Introduction

The geomagnetic field has a wide variety of sources (Chapter 4), from liquid metal flows in the Earth's outer core to magnetized rocks, oceanic tides and currents and electric currents in the ionosphere and magnetosphere. At or near the Earth's surface, the core field (also referred to as the main field) represents about 95% of the total field on average, with amplitudes in the 30,000–70,000 nanotesla (nT) range. Contributions from other sources generally do not exceed a few thousand nT, except at high latitudes during very intense magnetic storms, and in some locations where subsurface rocks carry strong magnetization.

Some sources are static or quasi-static on historical timescales (e.g. magnetized rocks), while others vary within seconds or less (e.g. electric currents in the magnetosphere). As a result, the directly observable part of the geomagnetic temporal spectrum is broad and encompasses all timescales from seconds to centuries (Figure 6.1.1). Longer timescales can be indirectly inferred from archeomagnetic and paleomagnetic measurements (Chapter 5). Shorter timescales (e.g. Schumann resonances, magnetospheric waves beyond the ultra-low-frequency band) are generally considered outside geomagnetism proper and are measured by different instruments and methods than those reviewed in this chapter.

Despite much progress throughout the twentieth century, no single existing instrument is able to measure the geomagnetic field over the entire spectrum represented in Figure 6.1.1 to the accuracy needed for research applications. Data used in Figure 6.1.1 were obtained by combining measurements from several instruments. These measurements were collected at a magnetic observatory, a facility designed to allow collecting homogeneous and accurate geomagnetic measurements at a single location for decades and even centuries. The concept of geomagnetic observatories was invented in the early nineteenth century and has led to numerous discoveries from immediately after the first observatory was installed until today. Data collected at today's observatories are generally considered accurate enough when investigating geomagnetic signals in the frequency range shown in Figure 6.1.1 at a single location.

Spatial variations of the Earth's magnetic field also encompass a broad spectrum. While fields originating in the outer core and distant magnetosphere are predominantly dipolar at the Earth's surface, other fields generated by less distant sources such as magnetized rocks in the crust and electric currents in the ionosphere can vary over much smaller spatial scales. Magnetic observatory data can recover the largest spatial scales fairly accurately, but have inherent limitations due to the geographical distribution of emerged land masses. Smaller scales are observed by various other methods, each method being generally designed for a particular source or set of sources. For example, crustal field variations in the 1–300 km range are typically measured by airborne magnetic surveys over dense, regular grids, and high-latitude magnetic fields are monitored by dense arrays of magnetometers. However, these other measurement methods are neither permanent, nor do they cover the entire globe, leaving many unobserved gaps in the spatio-temporal spectrum. This limitation is partially addressed by low-Earth orbit (LEO) satellite measurements (Chapter 7), which provide a nearly global coverage, but limited universal time coverage and at a distance of several hundreds of kilometers from crustal source.

This chapter will first briefly review the main instruments used in today's ground-based geomagnetic observations, focusing on their performances and working principles (from a user's point of view). Next, the major measurement methods and systems currently in use will be introduced, with a focus on the latest developments in the field.

6.1.2 Instruments

Various types of instruments have been used since the sixteenth century to measure the Earth's magnetic field and its temporal variations (e.g. Jankowski and Sucksdorff, 1996, and references therein). Many of these instruments,

* I thank the editors for inviting me to write this section and an anonymous reviewer for useful comments and suggestions.

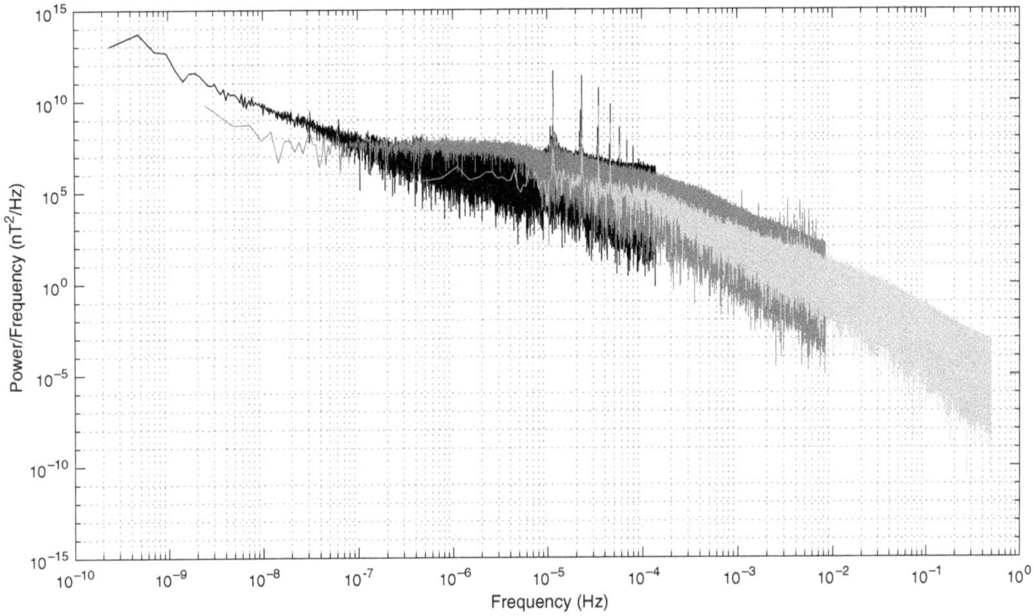

Figure 6.1.1 Periodogram of magnetic field data (Y component) recorded at the Chambon-la-Forêt magnetic observatory between 1883 and 2014 (hourly data, in black), 1996 and 2008 (minute data, in dark gray) and in January 2014 (1 s data, in light gray). The power generally decreases with frequency, as longer periods such as the secular variation have larger amplitudes than smaller periods. Peaks at the 24 hour period (about 1.16×10^{-5} Hz) and harmonics are caused by the ionospheric Sq field. Peaks at higher frequencies are caused by geomagnetic pulsations. Minute data cover only one solar cycle (cycle 23), while hourly data cover 12 solar cycles, hence the level differences below 10^{-5} Hz. (A black-and-white version of this figure appears in some formats. For the colour version, please refer to the plate section.)

although marvels of craftsmanship, were cumbersome to use and could only be operated by well-trained observers. Modern instruments started to be developed in the mid twentieth century, building upon progress in quantum mechanics and electronics. Most instruments currently used in ground-based and satellite geomagnetism fall into two broad categories: (1) fluxgate magnetometers, which measure the field along one or more directions using the ferromagnetic properties of some materials, and (2) scalar magnetometers, which rely on quantum mechanical properties of liquid and gas to measure the modulus of the field. When coupled with a theodolite and forming a so-called DI-flux, a fluxgate magnetometer can indirectly provide very accurate measurements of the magnetic inclination and declination. Other geomagnetism instruments, not reviewed in this section, include induction coil magnetometers, which record field variations at higher temporal frequencies, and cryogenic, Superconducting Quantum Interference Device (SQUID) magnetometers, mostly used in paleomagnetism (Chapter 5).

6.1.2.1 Fluxgate Magnetometers

Fluxgate magnetometers are widely used in geomagnetism, both for ground-based and space-based measurements, but also in many other applications, e.g. navigation and submarine detection. The first fluxgate was developed by Aschenbrenner and Goubau (1936). Since then, fluxgate magnetometry has become a mature field (see, e.g., the review by Primdahl, 1979). A fluxgate magnetometer measures the ambient magnetic field in one or more directions, thus acting as a 'vector' magnetometer. Each direction is sampled by a sensor composed of a highly permeable ferromagnetic core, an excitation coil and a pickup (or sense) coil. When an alternating current is passed through the excitation coil, the core becomes alternately magnetized in opposite directions, which in turn induces an electromotive force (emf) in the pickup coil. Because the magnetization curve of a ferromagnetic material is nonlinear, the output when adding a non-zero ambient field (along the sensor axis) doesn't exactly have the same frequency characteristics as the input field. This property is used to infer the magnitude of the ambient field. The core can be viewed as acting like a 'gate' for the ambient magnetic field flux, gating it in and out of the pickup coil, hence the name 'fluxgate'.

A simple mathematical model (inspired from Jankowski and Sucksdorff, 1996) can be used to illustrate the working principle of a fluxgate. Assume the magnetization curve of the core is the piecewise polynomial function represented in Figure 6.1.2a, where

$$B/\mu_0 = aH^3 + cH \tag{6.1.1}$$

for $|H| \leq \sqrt{-c/(3a)}$, $c = 1$, $a \approx -0.15$ and $B/\mu_0 = 1$ at saturation. (Units are not specified on purpose, as this is not a realistic model.) Here B is the magnetic field, H the excitation field and μ_0 the vacuum permeability. When

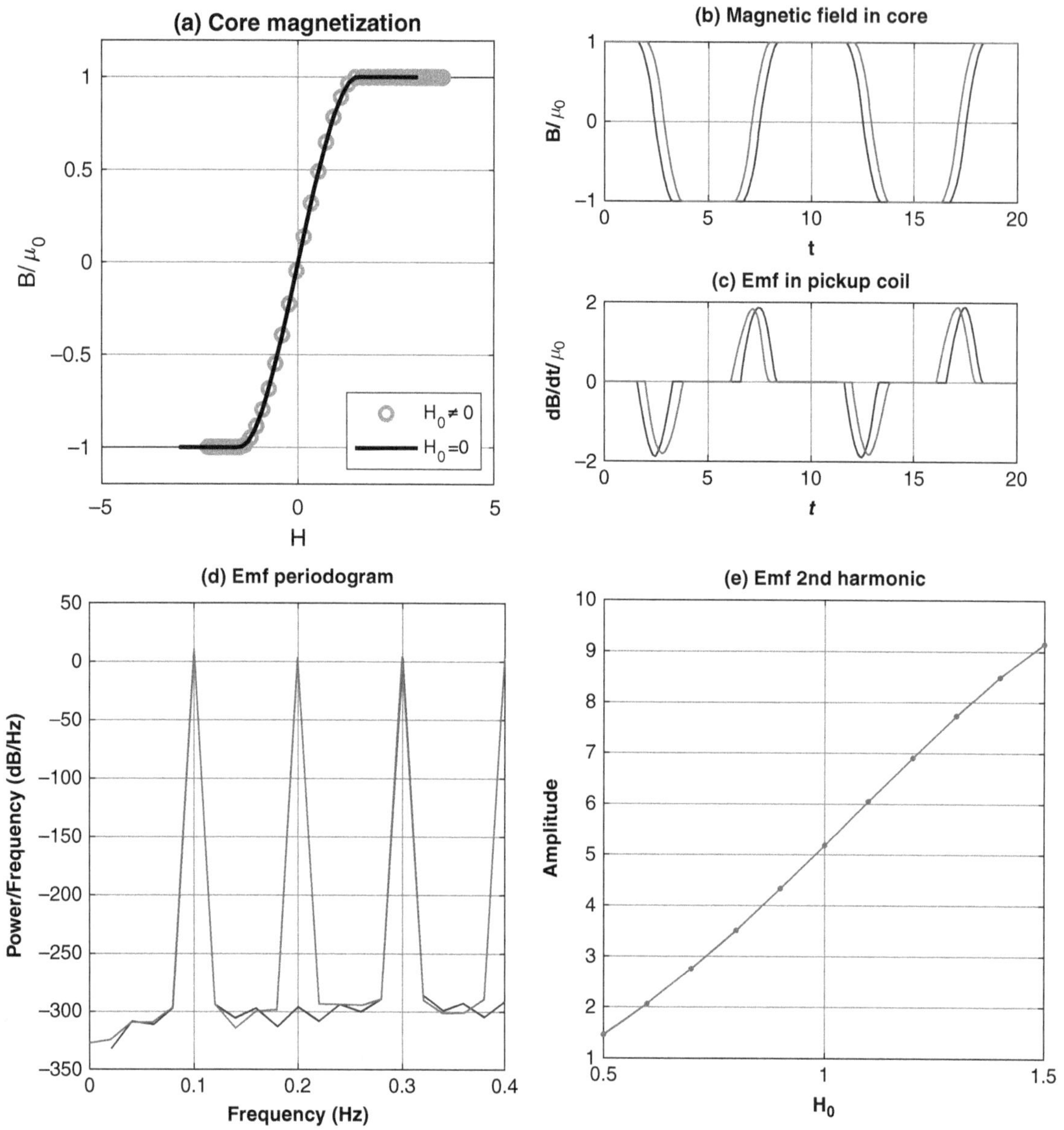

Figure 6.1.2 (a) Illustration of the working principle of a fluxgate magnetometer using a simple model of core magnetization, as given by Equation (6.1.1). Time variations of (b) the magnetic field in the core and (c) the electromotive force (emf) in the pickup coil. Adding a non-zero ambient field H_0 leads to asymmetrical flux variations (b, light gray curve) and even harmonics in emf variations (d). (e) The amplitude of the second harmonic and the magnitude of the ambient field are almost linearly related.

driving the core with $H = H_1 \cos 2\pi t/T$ of period $T = 10$, the curve representing flux variations inside the core (Figure 6.1.2b, dark gray curve) is symmetrical with respect to its zeros; as a result, emf variations in the pickup coil (Figure 6.1.2c) contain odd harmonics only (Figure 6.1.2d). When adding a non-zero ambient field H_0, flux variations become asymmetrical (Figure 6.1.2b, light gray curve), and emf variations contain even harmonics (Figures 6.1.2c and 6.1.2d, light gray curves). (Note that this effect is due to the cubic term in Equation (6.1.1) and doesn't necessarily require the core to saturate; in the absence of saturation, the emf curves have the same isolated zeros but different minima and maxima.) The amplitude of the second harmonic is almost linearly related to the magnitude of the ambient field (Figure 6.1.2e), and therefore can be used to infer it. In most modern instruments, a feedback coil is used to cancel the ambient field and the fluxgate sensor is used as a null-detector, allowing a larger dynamic range.

Fluxgate magnetometers come in a variety of designs. Rod core fluxgates are made of either one core, as discussed above, or two parallel cores permeated by excitation fields in opposition of phase. The twin core design (either Vacquier or Förster type) cancels the odd harmonics in the pickup coil emf and is prevalent in most modern rod core fluxgates. The ring core design (which was the original design used by Aschenbrenner and Goubau, 1936) is an extension of the twin core design where the two rods are connected to each other to form a ring. Both twin and ring core fluxgates are used in ground-based measurements, while ring core fluxgates are preferred for space magnetometry due to their lower noise and smaller power consumption (see e.g. Nielsen et al., 1995, for a description of the magnetometer used on board the LEO Ørsted satellite).

Long-term stability is an important requirement for fluxgate use in magnetic observatories. The INTERMAGNET global network currently requires that vector magnetometers in observatories do not drift by more than 5 nT/yr (St Louis, 2012). In fluxgates, drifts affect the instrument sensitivity, offset (i.e. output in the presence of a zero field) and sensor axis direction. They may be caused by various small changes in the electronics, the coil system or even the sensor material itself. Some of these changes are related to temperature, as all sensor materials (core, coils, etc.) are subject to thermal expansion, and temperature has a well-known effect on electrical conductivity. The temperature dependence can be mitigated to some extent by adding a temperature feedback loop. Today's best observatory fluxgates have temperature coefficients of the order of 0.2 nT/deg (e.g. Hrvoic and Newitt, 2011). Another external source of instability is the slow movement of the instrument itself, for example caused by a tilting pillar. Tilt (but not rotation) can be mitigated by using a suspension system (e.g. Rasmussen and Kring Lauridsen, 1990).

Noise is another important characteristic of a fluxgate magnetometer, as it places a limit on the smallest (hence fastest, see Figure 6.1.1) geomagnetic signal that can be recorded. Requirements in that regards used to be less stringent for observatories than magnetometer arrays and space-based observations. However, this is no longer the case as many observatories have started recording 1 s data (e.g. Chulliat et al., 2009). Although INTERMAGNET has not published an official noise requirement yet, it is generally believed that fluxgates should have a noise no larger than 10 pT/sqrt(Hz) at 1 Hz, and ideally as close as possible to 1 pT/sqrt(Hz), to being able to accurately measure geomagnetic activity in that part of the spectrum. This is achieved by some observatory instruments, but not all. The noise performance of the amorphous alloy generally used as core material is key to the overall instrument noise (Narod et al., 1985) and has been the subject of recent research (e.g. Marusenkov, 2017).

6.1.2.2 Scalar Magnetometers

Scalar measurements of the geomagnetic field are usually performed with either a proton precession magnetometer, an Overhauser magnetometer or an optically pumped magnetometer. These three types of magnetometers have in common the use of quantum mechanical properties of liquids or gas, and the measurement of a frequency linearly related to the modulus of the ambient field. Scalar magnetometers can be made very accurate (better than 0.5 nT) and, unlike fluxgate magnetometers, are not subject to long-term drifts. For this reason, they are used as absolute references in magnetic observatories and onboard LEO satellites (Chapter 7). They are also widely used in aeromagnetic and ground surveys.

The proton precession magnetometer was invented by Varian (1951) and Packard (1958), a few years after the discovery of the nuclear magnetic resonance (NMR) in the 1940s. Let us briefly summarize its working principle (see e.g. Freude, 2006 for a presentation of NMR basics). Every proton has a spin angular momentum $\vec{L}$ and a spin magnetic moment $\vec{\mu}$. Both quantities are related through

$$\vec{\mu} = \gamma_p \vec{L} \tag{6.1.2}$$

where γ_p is the gyromagnetic ratio of the proton. The standard value recommended by the International Association of Geomagnetism and Aeronomy (IAGA, www.bgs.ac.uk/iaga/vobs/) is $\gamma_p = 2.675153362 \cdot 10^8\ \mathrm{T^{-1}s^{-1}}$ (Mohr et al., 2008). When protons are placed in a magnetic field $\vec{B}_0$, their energy level is split into two Zeeman levels, one for each proton spin quantum number −½ and ½. Both levels correspond to different magnetic moments and occupation probabilities (in thermal equilibrium). As a result, level splitting leads to a (paramagnetic) macroscopic magnetization

$$\vec{M}_0 = \chi \vec{B}_0 / \mu_0 \tag{6.1.3}$$

in the same direction as $\vec{B}_0$, where χ is the magnetic susceptibility.

In a proton precession magnetometer, an artificial magnetic field $\vec{B}_0$ is first applied perpendicularly to the Earth's magnetic field $\vec{B}_E$. The modulus of $\vec{B}_0$ is typically 100–1000 times larger than that of $\vec{B}_E$ so that the effect of $\vec{B}_E$ is negligible. When $\vec{B}_0$ is switched off, $\vec{B}_E$ exerts a magnetic torque on the protons, which then start precessing at the Larmor angular frequency

$$\omega_L = \gamma_p \vec{B}_E \tag{6.1.4}$$

around the direction of $\vec{B}_E$. At the macroscopic level, the magnetization also experiences a precession at the frequency ω_L. However, due to interactions at the subatomic level, it simultaneously decays (and thus spirals down) from its initial value $\vec{M}_0$ toward the (much smaller) magnetization $\vec{M}_E$ induced by $\vec{B}_E$. The decay is exponential and involves two timescales known as the longitudinal relaxation time T_1 and the transverse relaxation time T_2 in the nuclear magnetic resonance literature. A coil (usually the same one that generated $\vec{B}_0$) picks up the rotating magnetic field generated by the spiraling magnetization, and a frequency counter measures ω_L, hence the magnitude of $\vec{B}_E$.

In practice, the proton precession magnetometer sensor is a bottle filled with a proton rich liquid such as kerosene, alcohol or water, and placed within a coil. The signal to be detected typically oscillates at a few kHz (2.13 kHz for a 50,000 nT field) for a few seconds (i.e. the transverse relaxation time T_2 of the liquid). The liquid needs to be polarized by $\vec{B}_0$ before each new measurement, which typically also takes a few seconds (i.e. the longitudinal relaxation time T_1, which is usually larger than T_2). The longer the polarization, the stronger $\vec{M}_0$ and the better the sensitivity. Typically, a sensitivity of 0.1 nT is achieved with readings every 3 s (Hrvoic and Newitt, 2011).

The low sampling frequency of proton precession magnetometers is a serious limitation. In so-called Overhauser magnetometers, the magnetization is enhanced by adding free electrons (e.g. in free radicals) to the liquid and applying a suitable radiofrequency radiation that saturates selected energy levels. This effect was first discovered in metals by Overhauser (1953). An external magnetic field is still required, but it can be shorter and less intense, resulting in higher sampling frequency, lower power consumption and less stray fields in the vicinity of the instrument. A modern Overhauser can achieve a sensitivity of 10–20 pT with readings every second (Hrvoic and Newitt, 2011). Because of their superior performance and relative affordability, Overhauser magnetometers have become the preferred absolute instruments in observatories. An Overhauser was also used as reference onboard the Ørsted and CHAMP satellites; the total error of that particular instrument was estimated to be as low as 0.25 nT (Duret et al., 1996).

In optically pumped magnetometers, the sensor liquid is replaced by a gas, either helium or an alkali vapor such as caesium, rubidium or potassium. Optical pumping is used to enhance the effects of an electronic magnetic resonance by raising electrons to higher energy levels. The resonance occurs between the Zeeman sublevels created by the ambient magnetic field, and is induced by a radiofrequency radiation. As the pumping light is transmitted through the sensor cell containing the gas, its intensity changes are used to detect the resonance (see e.g. the classic review by Stuart, 1972). The source of light can be a gas discharge lamp of the same element as the gas, or a laser. The latter allows a higher degree of polarization and therefore an improved sensitivity (e.g. Kernevez et al., 1992). This property has recently been used to design an helium magnetometer for the Swarm satellite mission, known as the Absolute Scalar Magnetometer (ASM). This instrument achieves 1 pT/sqrt(Hz) resolution over the DC to 100 Hz bandwidth, and its accuracy is better than 50 pT (Léger et al., 2009). Optically pumped magnetometers in general have higher measurement rates and lower noise than Overhauser magnetometers, making them suitable for measurements from fast moving platforms, e.g., airplanes and satellites. Some recently developed instruments reach subfemtotesla sensitivity (Sheng et al., 2013). Research in optically pumped magnetometers is very active and is largely motivated by applications requiring highly sensitive magnetic measurements such as biomedical applications (e.g. Knappe et al., 2014).

6.1.2.3 DI-Flux

Long-term drifts of fluxgate magnetometers make them unable to accurately monitor the geomagnetic secular variation for scientific purposes. Onboard satellites, this problem is addressed by coupling a fluxgate and a scalar magnetometer, and utilizing the rotation of the Earth's magnetic field in the instrument reference frame to periodically (typically once a day) recalibrate the fluxgate. However, this approach is not applicable to ground-based measurements, as the Earth's magnetic field at a given location doesn't change by more than a few percent at most. Today's magnetic observatories rely on a specifically designed instrument, the declination–inclination fluxgate magnetometer or DI-flux, to periodically correct whatever long-term drift exists in fluxgate measurements. An early version of the DI-flux was initially developed for high-latitude measurements, including the determination of the North Magnetic Pole (Serson and Hannaford, 1956), for which use of traditional instruments such as quartz horizontal magnetometers (QHMs) was unpractical. When coupled with a scalar magnetometer, a DI-flux can make vector measurements to an accuracy better than 1 nT (Bitterly et al., 1984). DI-flux are also used in ground surveys and repeat station measurements, i.e. whenever an accurate ground-based vector measurement is needed.

A DI-flux is composed of a non-magnetic theodolite and a single-axis fluxgate magnetometer mounted on top of the theodolite telescope, with its axis parallel to that of the telescope (Figure 6.1.3). The measurement is done by a trained observer. It consists in determining the declination (D), i.e. the angle between geographic North and the horizontal component of the geomagnetic field, and the inclination (I), i.e. the angle between the horizontal plane and the geomagnetic field vector. D is measured with respect to a known azimuth mark and is obtained by finding the angle for which the field is zero in the horizontal plane. I is obtained by finding the zero in the previously determined magnetic meridian plane. There are two methods, the null method and the residual method; each method relies on flipping the telescope in four positions so that telescope-fluxgate misalignments and fluxgate offsets are cancelled out (e.g. Bitterly and Lalanne, 2003, for step by step descriptions of both procedures). Although there are many sources of error (Hrvoic and Newitt, 2011), trained observers are able to make consistent measurements over a long period of time and achieve an accuracy better than 5 arcsec (which corresponds to the aforementioned vector accuracy of 1 nT at mid-latitudes).

6.1.3 Measurement Methods and Systems

The magnetometers described in the previous section are used in various configurations to measure the Earth's magnetic field and its variations on the ground. In magnetic observatories, well-trained observers combine the three

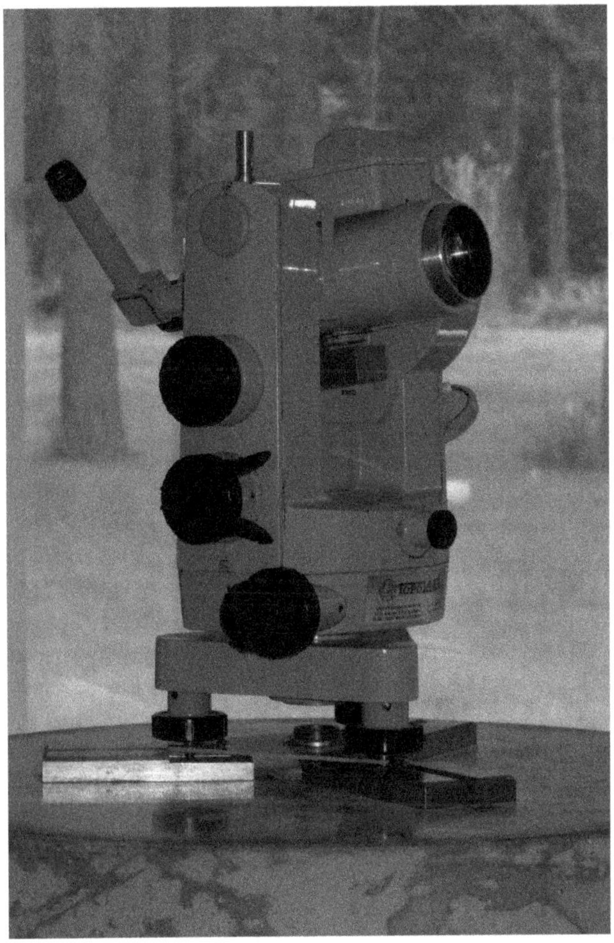

Figure 6.1.3 Declination–inclination fluxgate magnetometer or DI-flux in use at the Chambon-la-Forêt observatory (France). The fluxgate magnetometer is inside the gray box on top of the theodolite telescope.

types of magnetometers to provide the most accurate observations possible over a wide range of temporal scales. Magnetometer arrays usually rely on automatically operating fluxgate magnetometers to observe external field variations with a higher spatial resolution than observatories. In marine and airborne magnetic surveys, scalar magnetometers are towed or flown over vast oceanic and land areas to recover the small spatial scales of the lithospheric field that cannot be detected from satellites. For each of these observation methods, operations and data distribution are coordinated to some extent at the national and/or global level and are constantly evolving thanks to new and improved technology.

6.1.3.1 Magnetic Observatories

Modern magnetic observatories aim at measuring geomagnetic variations with timescales ranging from centuries down to seconds. Historically, magnetic observatories focused on the longest timescales; the typical sampling frequency was one measurement per hour and many observatories only published annual mean values on a regular basis. However, most modern observatories now publish 1 min data, and a growing number of them publish 1 s data. This evolution was largely driven by INTERMAGNET, the International Real-time Magnetic Observatory Network (e.g. Love and Chulliat, 2013). Since its founding in the late 1980s, INTERMAGNET has required its member observatories to meet strict equipment and data quality standards (St Louis, 2012), including recording 1 min vector data with a 0.1 nT resolution and full absolute control. In recent years, INTERMAGNET has been developing new specifications for acquisition of 1 s data and has promoted the acquisition of such data among its members.

IAGA maintains a full list of magnetic observatories (including non-INTERMAGNET ones), with support from the British Geological Survey (www.bgs.ac.uk/iaga/vobs/). Figure 6.1.4 shows a map of the 190 currently operating observatories registered with IAGA. Among them, 133 are INTERMAGNET observatories (see full list at www.intermagnet.org) and 13 have the IAGA 'provisional' status, meaning that they have not yet distributed a full year of definitive data. In recent years, about 70 observatories produced 1 s data (Chulliat et al., 2017).

Although different types of instruments can be found in magnetic observatories, current operation of a magnetic observatory follows a well-established set of practices (e.g. Jankozski and Sucksdorff, 1996; Turner et al., 2009). Most observatories rely on a triaxial, generally fluxgate, magnetometer to continuously record geomagnetic variations. This instrument is placed on a stable, half-buried pillar in a magnetically clean and temperature-controlled environment, and is typically operating for several years without interruption. Accurate timing is achieved using a GPS receiver. Absolute control of the vector magnetometer long-term drifts is provided by a DI-flux and a scalar, generally proton precession or Overhauser, magnetometer, both located a few tens of meters away from the vector magnetometer. Generally, these two instruments are placed in a separate, magnetically clean shelter, and the DI-flux sits on top of a stable pillar with an azimuth mark within sight. Magnetic gradient is preferably low in the vicinity of the instruments, but this requirement is not always met as some observatories are located in areas such as volcanic islands where highly magnetized rocks are prevalent.

Absolute measurements are typically made once a week, following the procedure outlined above. These measurements are then used to correct vector magnetometer measurements by calculating the so-called baseline for each component. A baseline is a smooth mathematical function with the same sampling rate as the vector instrument, obtained by fitting the differences between the vector instrument output and the absolute measurements on a given component. The more absolute measurements are taken, the more accurate the baseline. (The term 'baseline' dates back to the era when continuous vector measurements were made by torsion variometers and a baseline needed to be applied to each daily photographic recording.) Because time variations of the geomagnetic field are small compared to the main magnetic field vector, changes in the nine calibration parameters of a fluxgate magnetometer can be corrected by only three baseline values (one per component) at any single time.

Per INTERMAGNET policy, 1 min and 1 s data have to be made available from the INTERMAGNET website within less than 72 hours from acquisition (near–real time). The time delay is actually smaller at a growing number of observatories; it ranges from 5 min to 1 hour at about 40 INTERMAGNET observatories. Real-time observatory data are used in various operational applications, including space weather monitoring and forecasting (Love and Finn, 2011) and directional drilling (Buchanan et al., 2013). When distributed in real or near–real time, observatory data are of the so-called 'variation' or 'provisional' types, according to the classification of the IAGA2002 data exchange format (www.ngdc.noaa.gov/IAGA/vdat/IAGA2002/iaga2002format.html).

Every year, observatories produce so-called definitive data by correcting all 1 min and (if applicable) 1 s data with final baselines calculated over the whole previous year. Definitive data are observatories' highest quality data product. INTERMAGNET definitive data undergo a further quality check by third party experts before being released on INTERMAGNET website. Definitive 1 min data are then used to calculate hourly mean values and annual means, which are distributed by the World Data Center for Geomagnetism in Edinburgh (www.wdc.bgs.ac.uk/). Other geomagnetic data centers part of the World Data System (WDS, www.icsu-wds.org/) maintain extensive archives of recent and older definitive observatory data, including scanned magnetograms. Since the beginning of the Swarm satellite mission, quality-controlled and jump-corrected hourly means are also made available by the European Space Agency as a Swarm auxiliary data product (Macmillan and Olsen, 2013). Monthly means are routinely calculated for secular variation studies (e.g. Chulliat and Telali, 2007), although no continuously updated repository currently exists.

Since the late 2000s, a growing number of observatories have started distributing so-called 'quasi-definitive' data, defined by INTERMAGNET as '(H, D, Z) or (X, Y, Z) 1 min data:

- Corrected using temporary baselines
- Made available less than 3 months after their acquisition
- Such that the difference between the quasi-definitive and definitive (X, Y, Z) monthly means is less than 5 nT for every month of the year'

Currently between 60 and 70 observatories produce quasi-definitive data on a routine basis, using various processing methods (Peltier and Chulliat, 2010; Matzka, 2012; Clarke et al., 2013), and distribute them via INTERMAGNET. Quasi-definitive data provide information on the most recent geomagnetic secular variation, without having to wait until the release of definitive data which typically occurs with a one year delay. As such they have been widely used in recent years to calculate and validate geomagnetic models inferred from Swarm data (e.g. Finlay et al., 2016), as well as investigate sudden changes of trend or 'geomagnetic jerks' occurring in secular variation records (e.g. Chulliat and Maus, 2014; Torta et al., 2015).

Until recently, many countries used to make additional absolute measurements at predefined locations and regular time intervals in order to calculate higher resolution models of the geomagnetic secular variation on their territory, mostly for navigation purposes. Such repeat stations used to be a very important part of a country's observatory operations (Newitt et al., 1996). However, the utility of repeat stations has declined since 2000, due to the nearly continuous availability of high-quality magnetic satellite measurements (see Chapter 7). Today, repeat stations measurements are still performed in some countries, but are generally considered as a safety net in case satellite measurements become unavailable, or are used for model validation purposes.

A long-standing limitation of magnetic observatories is their heterogeneous geographical distribution (Figure 6.1.4). While some regions (e.g. Europe, North America, Australia, parts of Asia) are covered by dense networks of observatories, others (e.g. Africa, Pacific Ocean) have large coverage gaps. A sustained effort was made in recent decades to upgrade old observatories and install new observatories in remote locations such as oceanic islands (e.g. Matzka et al., 2010, and references therein). However, despite this effort, the global geographical coverage of observatories has remained nearly constant since the International Geophysical Year in 1958, at about 60% of a 92 cells equal area grid (Chulliat et al., 2017). The reason is that, while new observatories were opened, old ones were closed at a nearly equal rate. For example, the Bangui observatory in Central African Republic, which opened in 1952 and was one of a few observatories located in Africa, had to close in 2011 due to safety concerns. One way to improve the global geographical coverage would be to install automated observatories (Alldredge and Saldukas, 1964) in remote locations. The recently developed AUTODIF instrument (Van Loo and Rasson, 2007; Poncelet et al., 2017), which is an automated DI-flux, could provide a technical solution to this problem. Another way would be to install magnetic observatories at the bottom of the oceans. Since the early 2000s, a few seafloor observatories have been successfully installed in the Pacific Ocean (Toh et al., 2006), allowing the detection of tsunami (Toh et al., 2011) and tidal (Schnepf et al., 2014) magnetic signals.

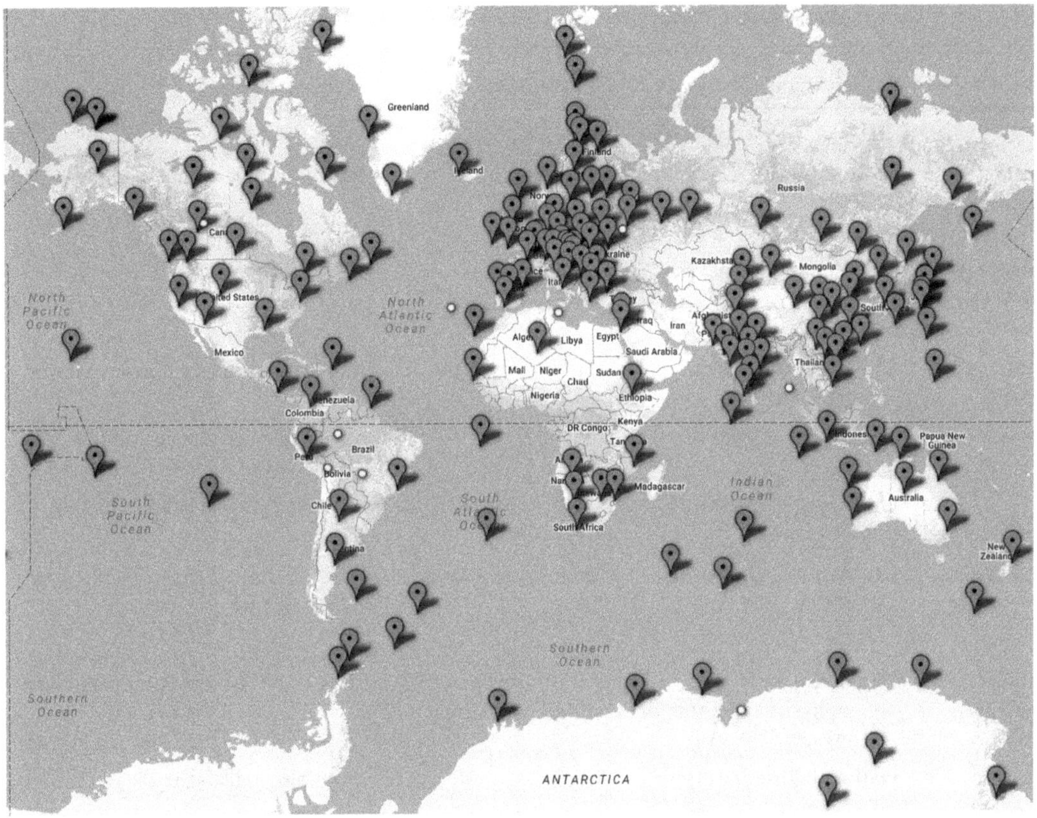

Figure 6.1.4 Operating magnetic observatories registered with IAGA, as of January 2018. Observatories with 'open' status (i.e. having delivered at least one year of definitive data to a world data center) are marked as gray pins. Observatories with 'provisional' status are marked as white circles.

6.1.3.2 Magnetometer Arrays

Absolute control is generally not required when investigating geomagnetic variations with timescales of days or smaller, i.e. variations caused by rapidly varying electric currents in the ionosphere, magnetosphere and oceans and their induced counterparts. Therefore, all what is needed to record such variations at a given location is a vector magnetometer (or 'variometer') with a known orientation and placed in a magnetically clean environment. This setup can be fully automated and is relatively low cost compared to a full observatory. It is referred to as a magnetometer station. The vector magnetometer is often a fluxgate magnetometer, sometimes supplemented with an induction coil magnetometer to record fast variations up to 100 Hz.

Most currently operating magnetometer stations were installed as part of regional or global arrays of magnetometers aimed at investigating specific science questions. Unlike observatories, such arrays are not meant to be permanent, although some of them have been operating for decades. Also, magnetometer stations do not register with IAGA or any other organization, which makes listing past and current magnetometer arrays and their stations challenging. Table 6.1.1 provides a partial list of selected arrays, including institution information and links to relevant websites. The most instrumented areas are North America and Europe. Both regions are covered by several large arrays, mostly at high latitudes, complementing dense distributions of observatories (Figure 6.1.5). Smaller arrays cover Antarctica and parts of Africa and South America. A truly global array, MAGDAS (Yumoto, 2001, 2006) includes stations along the 210° magnetic meridian from New Zealand to Northeastern Siberia, and along the dip-equator from South America to Africa, Asia and some Pacific Islands. It was made possible by decades-long partnerships between Kyushu University and about 30 institutions worldwide.

Magnetometer arrays come in different sizes and shapes, depending on the targeted scientific application. A common design is that of a chain of magnetometers along a magnetic meridian. It allows investigating various magnetosphere-ionosphere processes having a latitudinally dependent effect on the ground magnetic field such as, for example, variations in plasmaspheric mass density (e.g. Chi et al., 2013, McMAC; Lichtenberger et al., 2013, EMMA), geomagnetic storms (e.g. Xu et al., 2017, GMA) and the equatorial electrojet and electric field (e.g. Yizengaw et al., 2014, SAMBA and AMBER; Alken et al., 2013, WAMNET). Closely spaced two-dimensional arrays, alone or in combination with chains, are used to investigate, e.g. electric currents in the auroral ionosphere (Weygand et al., 2011, most North American arrays), substorms (Tanskanen,

Table 6.1.1 *Selected Magnetometer Arrays Providing Data to SuperMAG*

Acronym	Full name	PI institution	Geographical location	Website
North America				
AUTUMN	Athabasca University THEMIS UCLA Magnetometer Network	Athabasca University	Canada	http://autumn.athabascau.ca/
CARISMA	Canadian Array for Realtime Investigations of Magnetic Activity	Canadian Space Agency, University of Alberta	Canada	www.carisma.ca/
GIMA	Geophysical Institute Magnetometer Array	University of Alaska Fairbanks	Alaska	www.gi.alaska.edu/magnetometer/
MACCS	Magnetometer Array for Cusp and Cleft Studies	Augsburg College	Canada	http://space.augsburg.edu/maccs/index.html
McMAC	Mid-continent Magnetoseismic Chain	University of California Los Angeles	North America	http://spc.igpp.ucla.edu/mcmac/
THEMIS GBO	THEMIS Ground Based Observatories	University of California Berkeley	Canada and Alaska	http://themis.ssl.berkeley.edu/gbo/display.py
Europe/Russia				
EMMA	European Magnetometer Array for PLASMON	Geological and Geophysical Institute of Hungary	Eastern Europe, Southern Africa	http://geofizika.canet.hu/plasmon/
ENIGMA	HellENIc GeoMagnetic Array	National Observatory of Athens	Greece	http://enigma.space.noa.gr/
GMA	Greenland magnetometer array	Danish Technical University	Greenland	www.space.dtu.dk/English/Research/Scientific_data_and_models/Magnetic_Ground_Stations.aspx
IMAGE	International Monitor for Auroral Geomagnetic Effects	Finnish Meteorological Institute	Northern Europe/Scandinavia	http://space.fmi.fi/image/
RapidMag	Russian Auroral and Polar Ionospheric Disturbance Magnetometers	Johns Hopkins University Applied Physics Laboratory	Northern Russia	http://sd-www.jhuapl.edu/rapidmag/
SAMNET	Sub-Auroral Magnetometer Network	Lancaster University	United Kingdom	https://aurorawatch.lancs.ac.uk/
TGO	Tromsø Geophysical Observatory	UiT The Arctic University of Norway	Norway, Svalbard, Greenland	http://flux.phys.uit.no/geomag.html
Antarctica				
BAS LPM	Low Power Magnetometer Network	British Antarctic Survey	Antarctica	www.bas.ac.uk/project/low-power-magnetometer-lpm-network/
PENGUIn	Antarctic Low-Powered Magnetometers	Virginia Tech	Antarctica	http://mist.nianet.org/index.html
Global/Equatorial				
AMBER	African Meridian B-Field Education and Research	Boston College, University of Michigan	Global (along dip-equator)	http://magnetometers.bc.edu/
MAGDAS/CPMN	MAGnetic Data Acquisition System/Circum-pan Pacific Magnetometer Network	Kyushu University	Global	http://magdas2.serc.kyushu-u.ac.jp/tbrtgut 12pt

Table 6.1.1 (*cont.*)

Acronym	Full name	PI institution	Geographical location	Website
SAMBA	South American Meridional B-Field Array	NASA, University of Michigan	South America, Antarctica	http://magnetometers.bc.edu/
WAMNET	West African Magnetometer NETwork	Institut de Physique du Globe de Paris	Western Africa	www.bcmt.fr/wamnetnetwork.html
210 MM	210 Magnetic Meridian Chain	Kyushu University, Nagoya University	Western Pacific	http://stdb2.stelab.nagoya-u.ac.jp/mm210/

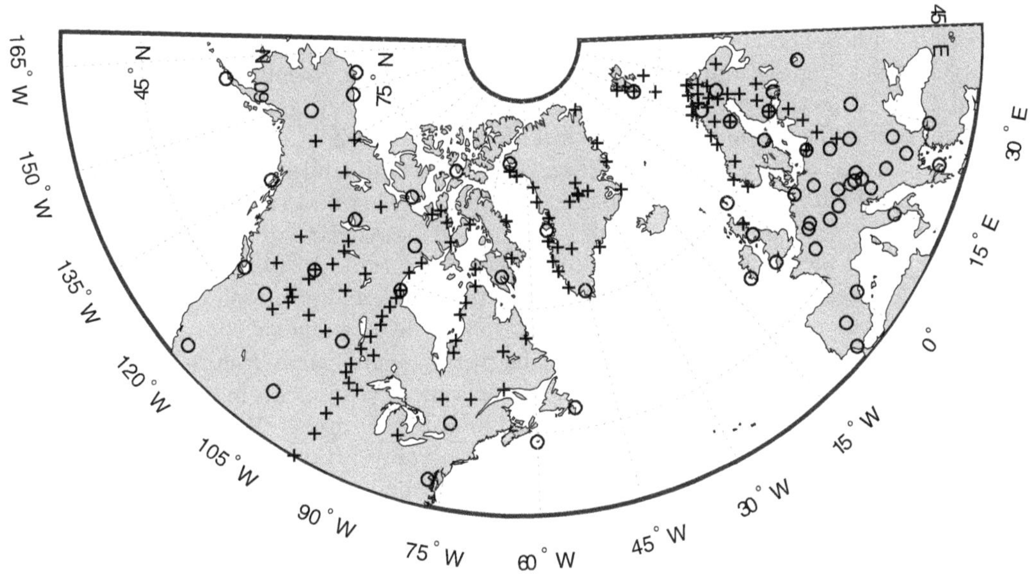

Figure 6.1.5 Selected magnetometer arrays (AUTUMN, CARISMA, GMA, IMAGE, MACCS, McMAC, SAMNET – crosses) and INTERMAGNET magnetic observatories (circles) in Europe and North America as of January 2018.

2009, IMAGE) and geomagnetic pulsations (Engebretson et al., 1995, MACCS). Arrays in the both hemispheres can be used to investigate geomagnetic activity at conjugate points (e.g. Kim et al., 2013, MACCS, GMA, PENGUIn, BAS). These are only a few examples among the many uses of magnetometer arrays, some of which are discussed in detail in later chapters of this book.

Until recently, each magnetometer array project was distributing its own data using its own format, coordinate system, and distribution method (e.g. a website). The situation improved a few years ago when SuperMAG (Gjerloev, 2009, 2012) started distributing data from different arrays through a single website (http://supermag.jhuapl.edu). SuperMAG now includes data from about 300 magnetometer stations and observatories, including all INTERMAGNET observatories. Before redistributing data, SuperMAG cleans them, rotates them in the same coordinate system (local magnetic North, East and vertical), resamples them using an identical time resolution (1 min) and removes a baseline including the static field, slow temporal variations (attributed mostly but not solely to the core field variation), and diurnal variations (attributed mostly to the Sq ionospheric current system). The whole procedure is aimed at facilitating the use of data from different magnetometer arrays for research into ionospheric and magnetospheric current systems. This is a strikingly different approach than the one used by INTERMAGNET, which doesn't modify any data from its member observatories but instead focuses on helping observatories meet strict data quality standards and redistributing their data in near–real time.

A recent workshop in the United States (Engebretson and Zesta, 2017) reaffirmed the scientific value of magnetometer arrays for space weather and identified several problems that the US community is facing. These include a lack of consistency in operations and data availability, magnetometer aging and a limited number of higher-level data products. Various recommendations were made by workshop participants to address these issues, such as changes in how arrays are organized and funded at the national level, the development of new data products and visualizations based upon array data, and the specific allocation of funding to the development of improved instruments.

6.1.3.3 Marine and Airborne Magnetic Surveys

Magnetized rocks in the Earth's lithosphere generate magnetic fields with a broad spectrum of spatial scales, from thousands of kilometers to a few meters. At scales greater than 3000 km, the lithospheric field is overlapping with the more intense core field and is largely unknown. At intermediate scales (from 3000 to 250–300 km), it dominates the core field and can be globally mapped from LEO satellite data (Chapter 7). Recently, gradient data from the Swarm constellation were used to infer the global lithospheric field down to about 250 km length scales (Olsen et al., 2017), and slightly smaller scales are expected to be detected when Swarm satellites orbit at lower altitudes later in the mission. At scales smaller than 250 km, the lithospheric field becomes too small to be detected at satellite altitude with existing instrument and field modeling technology. Near surface measurements are needed to recover this part of the spectrum. There are broadly two types of such measurements: marine magnetic surveys, where a magnetometer is towed by or attached to a ship or an undersea vessel; and airborne magnetic (aeromagnetic) surveys, where a magnetometer is attached to a flying object such as an airplane, a helicopter or a balloon. Scales as small as a few hundreds of meters can be recovered by low altitude, airborne surveys. Even smaller scales can be recovered by other means, e.g. bicycling or pedestrian surveys, but the geographical extent of such measurements is very limited.

The first airborne magnetic surveys were carried out in the 1920s and 1930s, first using balloons, then aircrafts. During World War II, airborne surveys were used to detect undersea objects such as submarines and mines, and their development was closely linked to that of the fluxgate magnetometer. Today, aeromagnetic surveys are used for a wide range of purposes, including the search of minerals, geological mapping and oil and gas exploration (e.g. Nabighian et al., 2005). They usually rely on scalar magnetometers, which provide an accuracy better than 1 nT. The overall survey accuracy also depends on how well the position of the magnetometer is determined. The advent of the Global Positioning System has led to dramatically improved positioning accuracy, with errors less than a few meters horizontally and vertically when using single-frequency receivers. Airborne surveys typically consist of regular-spaced flight lines, with spacing between 50 and 500 meters, and perpendicular tie lines with 10 times larger spacing (e.g. Hamoudi et al., 2011). Most surveys are flown at constant ground clearance, between 150 and 1000 m. The closer the spacing and the lower the survey, the higher the achievable spatial resolution. A base station (e.g. a nearby observatory or a station specifically installed for that purpose) is often used to provide a reference record of magnetic time variations in the survey area, thus allowing for accurate external field correction. Measurements made on tie lines are also used for data correction, a technique known as magnetic leveling. Aeromagnetic data are usually gridded for visualization purposes, but also as a first step of further processing (e.g. Reeves, 2005). Recent developments in the field include the use of Unmanned Aerial Vehicles (e.g. Miles et al., 2008) to acquire data at low altitude on land or over water without putting a flight crew at risk, and the use of ultra-sensitive instruments such as Superconducting Quantum Interference Device (SQUID) magnetometers to do full tensor gradient measurements at low altitude (e.g. Schiffler et al., 2014).

Because compasses were prevalent for centuries in marine navigation, magnetic measurements made on board ships form the bulk of historical geomagnetic observations (Jonkers et al., 2003; Matzka et al., 2010). Marine records were successfully used with land-based, historical measurements (e.g. in observatories and repeat stations) to model the global geomagnetic secular variation back to the sixteenth century (Jackson et al., 2000). However, neither type of data has the resolution and positioning accuracy needed to detect lithospheric field anomalies.

Modern marine magnetic surveys started in the 1950s, prompted by the same developments in magnetometry that were exploited in airborne surveys. Early magnetic profiles were among the key observations supporting the seafloor spreading hypothesis and plate tectonics, generating tremendous interest for such surveys. As a result, the bulk of marine magnetic surveys were carried out during the 1960s and 1970s (e.g. Hamoudi et al., 2011). A typical setting consists in a scalar magnetometer towed by a ship. A 200 meters minimum separation between the magnetometer and the ship is typically required to avoid contamination by the ship magnetic field, which can be quite large and varies with the direction of the cruise (Bullard and Mason, 1961). Another important source of error is the positioning error, which can reach 100 m in surveys predating the advent of the GPS. Some recent surveys include vector magnetic measurements, combined with inertial attitude measurements, in order to improve the interpretation of magnetic anomalies (e.g. Engels et al., 2008). A deep tow magnetometer or a deep-sea vehicle can be used to make measurements closer to the sea floor, in order to resolve smaller scale magnetic anomalies (see the review by Hamoudi et al., 2011, and references therein).

Various governmental and private institutions hold data collected by marine and airborne surveys. The NOAA National Centers for Environmental Information maintain an archive known as the Geophysical Data System (GEODAS), which includes data provided by over 100 institutions worldwide over 50 years (Figure 6.1.6). Recent additions to GEODAS were documented by Meyer et al. (2017) and included in NOAA's Earth Magnetic Anomaly

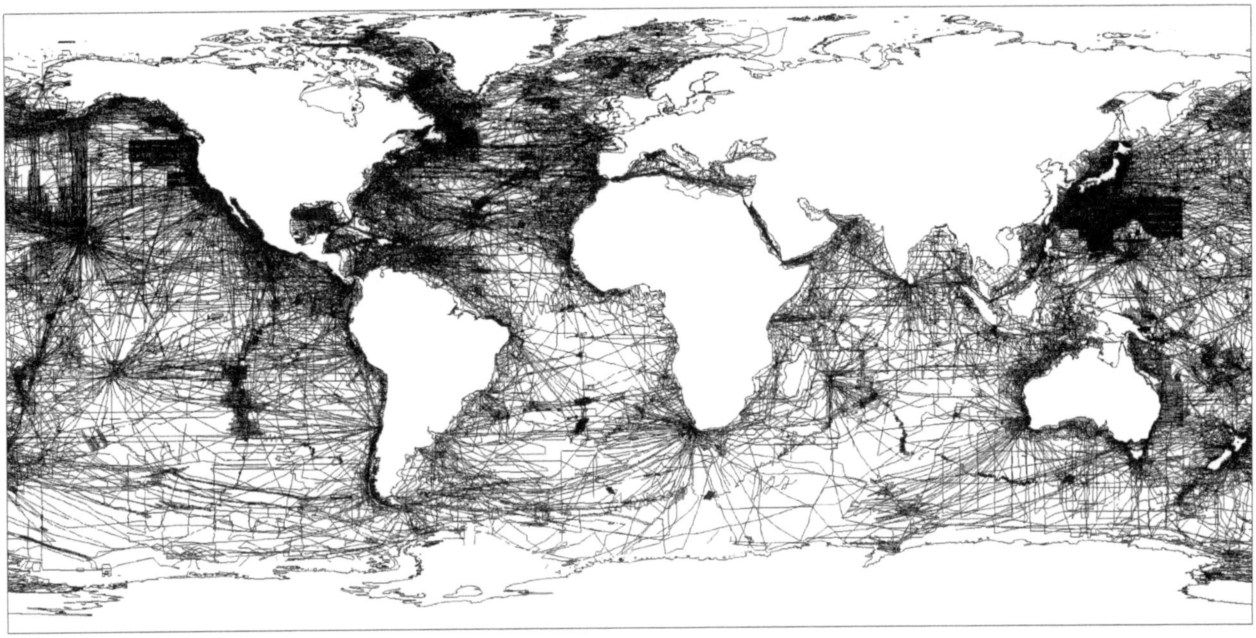

Figure 6.1.6 Marine (track-line) magnetic data archived in NOAA's Geophysical Data System (GEODAS) as of January 2018. Data can be accessed at https://maps.ngdc.noaa.gov/viewers/geophysics/.

Grid at 2 arc-min resolution – version 3. GEODAS data was also used to produce the World Digital Magnetic Anomaly Map (see Chapter 7). It is worth noting that survey data can be privately archived in GEODAS and only distributed at a reduced resolution, if that is the wish of the data owner. This capability addresses a long-standing issue in survey data availability for scientific purposes, as large amounts of survey data (e.g. Fairhead and Green, 2015) are not in the public domain due to their high commercial value.

6.1.4 Conclusion and Outlook

The data collected by the three broad measurement methods presented in the previous section have led to numerous scientific discoveries and provided tremendous economic value. Magnetic observatories record the broadest range of timescales, from seconds to centuries, and are used in virtually all branches of geomagnetism, either alone or jointly with other ground-based and/or space-based observation systems. Magnetometer arrays complement the global observatory network in areas most relevant to external magnetic field and space weather studies, such as the polar, auroral and equatorial zones. Marine and airborne surveys allow studying the lithospheric magnetic field at spatial scales smaller than what can be detected by LEO satellites, i.e. about 250 km, and provide key observations for geophysical exploration and a wide range of geological and tectonic studies.

Each of these methods faces some specific challenges that hinder its development. However, with advances in instrument, teleoperation and data management technologies, new options exist to overcome these challenges. Automated land and seafloor observatories could soon allow expanding the magnetic observatory network into oceans and remote land areas, thus bringing it closer to a truly global coverage. Robust teleoperation of flying drones and deep-sea vehicles provides new ways to survey the lithospheric field at small spatial scales over regions of interest. Modern data management tools make it possible to archive and distribute all types of geomagnetic data in a consistent way, such that scientific data usage is maximized, while addressing privacy issues often invoked to limit data availability.

Scarcity of funding is another issue, especially for magnetic observatories and magnetometer arrays, as securing recurrent funding for long-term observations is always difficult. Private companies have a stake in such observations, and some of them provide financial contributions to selected programs. Improved coordination and organization at the national and global levels can certainly help with funding issues, as was recently noted by the US magnetometer array community. It can also catalyze the adoption of new technologies, as was shown by INTERMAGNET over the past three decades. INTERMAGNET led to more robust operations, higher data quality and more consistent data distribution, while not receiving any significant direct funding. It provides a model that could be emulated in other types of ground-based observations and in future developments of the global observatory network. IAGA, as the main international scientific organization in the field of geomagnetism, could facilitate the emergence of new and innovative organizational structures aimed at further improving ground-based geomagnetic observations.

References

Alken, P., Chulliat, A. & Maus, S. (2013). Longitudinal and seasonal structure of the ionospheric equatorial electric field. *J. Geophys. Res. Space Physics*, **118**, 1298–1305, doi: 10.1029/2012JA018314.

Alldredge, L. R. & Saldukas, I. (1964). An automatic standard magnetic observatory. *J. Geophys. Res.*, **69**(10), 1963–70.

Aschenbrenner, H. & Goubau, G. (1936). Ein Anordnung zur Registrierung rascher magnetischer Störungen. *Hochfrequenztechnik Elektroakustik*, **47**, 177–81.

Bitterly, J., Cantin, J. M., Schlich, R. & Folques, J. (1984). Portable magnetometer theodolite with fluxgate sensor for Earth's magnetic field component measurements. *Geophys. Surv.*, **6**, 233–9.

Bitterly, J. & Lalanne, X. (2003). *Observatoire Magnétique Planétaire – Manuel d'opérations / Operation Manual* (English translation). Institut de Physique du Globe de Paris, Paris.

Buchanan, A., Finn, C. A., Love, J. J., Worthington, E. W., Lawson, F., Maus, S., Okewunmi, S. & Poedjono, B. (2013). Geomagnetic referencing – the real-time compass for directional drillers. *Oilfield Rev.*, **25**(3), 32–47.

Bullard, E. C. & Mason, R. G. (1961). The magnetic field astern of a ship. *Deep Sea Res.*, **8**(1), 20–27, doi: 10.1016/0146-6313(61)90012-0.

Chi, P. J., et al. (2013). Sounding of the plasmasphere by Mid-continent MAgnetoseismic Chain (McMAC) magnetometers. *J. Geophys. Res. Space Physics*, **118**, 3077–86, doi: 10.1002/jgra.50274.

Chulliat, A., Matzka, J., Masson, A. & Milan, S. E. (2017). Key ground-based and space-based assets to disentangle magnetic field sources in the Earth's environment. *Space Sci. Rev.*, **206**, 123–56, doi: 10.1007/s11214-016-0291-y.

Chulliat, A. & Maus, S. (2014). Geomagnetic secular acceleration, jerks, and a localized standing wave at the core surface from 2000 to 2010. *J. Geophys. Res. Solid Earth*, **119**, 1531–43, doi: 10.1002/2013JB010604.

Chulliat, A., Savary, J., Telali, K. & Lalanne, X. (2009). Acquisition of 1-second data in IPGP magnetic observatories. *US Geol. Surv. Open-File Rep.*, 2009–1226, 54–9.

Chulliat, A. & Telali, K. (2007). World monthly means database project. *Publs. Inst. Geophys. Pol. Acad. Sc.*, **C-99**(398), 268–74.

Clarke, E., Baillie, O., Reay, S. J. & Turbitt, C. W. (2013). A method for the near real-time production of quasi-definitive magnetic observatory data. *Earth Planets Space*, **65**, 1363–74, doi: 10.5047/eps.2013.10.001.

Duret, D., Léger, J. M., Francès, M., Bonzom, J. & Alcouffe, F. (1996). Performances of the OVH magnetometer for the Danish Oersted satellite. *IEEE Trans. Magnetics*, **32**(5), 4935–7.

Engebretson, M. J., Hughes, W. J., Alford, J. L., Zesta, E., Cahill, L. J., Jr., Arnoldy, R. L. & Reeves, G. D. (1995). Magnetometer array for cusp and cleft studies observations of the spatial extent of broadband ULF magnetic pulsations at cusp/cleft latitudes. *J. Geophys. Res.*, **100**(A10), 19371–86, doi: 10.1029/95JA00768.

Engebretson, M. J. & Zesta, E. (2017). The future of ground magnetometer arrays in support of space weather monitoring and research. *Space Weather*, **15**, 1433–41, doi: 10.1002/2017SW001718.

Engels, M., Barckhausen, U. & Gee, J. S. (2008). A new towed marine vector magnetometer: methods and results from a Central Pacific cruise. *Geophys. J. Int.*, **172**(1), 115–29, doi: 10.1111/j.1365-246X.2007.03601.x.

Fairhead, J. D. & Green, C. M. (2015). Generating a high-resolution global magnetic model for oil and mineral exploration. *The Leading Edge*, **34**(9), 1096–1102, doi: 10.1190/tle34091096.1.

Finlay, C. C., Olsen, N., Kotsiaros, S., Gillet, N. & Tøffner-Clausen, L. (2016). Recent geomagnetic secular variation from Swarm and ground observatories as estimated in the CHAOS-6 geomagnetic field model. *Earth Planets Space*, 68, 112, doi: 10.1186/s40623-016-0486-1.

Freude, D. (2006). *Spectroscopy for Physicists*. http://home.uni-leipzig.de/energy/freuse.html.

Gjerloev, J. W. (2009). A global ground-based magnetometer initiative. *Eos Trans. AGU*, **90**(27), 230–31, doi: 10.1029/2009EO270002.

Gjerloev, J. W. (2012). The SuperMAG data processing technique. *J. Geophys. Res.*, **117**, A09213, doi: 10.1029/2012JA017683.

Hamoudi, M., Quesnel, Y., Dyment, J. & Lesur, V. (2011). Aeromagnetic and marine measurements. In M. Mandea & M. Korte, eds., *Geomagnetic Observations and Models*. Springer, pp. 57–103.

Hrvoic, I. & Newitt, L. R. (2011). Instruments and methodologies for measurement of the Earth's magnetic field. In M. Mandea & M. Korte, eds., *Geomagnetic Observations and Models*. Springer, pp. 105–26.

Jackson, A., Jonkers, A. R. T. & Walker, M. R. (2000). Four centuries of geomagnetic secular variation from historical records. *Philos. Trans. R. Soc. Lond. A*, **358**, 957–90.

Jankowski, J. & Sucksdorff, C. (1996). *Guide for Magnetic Measurements and Observatory Practice*. International Association of Geomagnetism and Aeronomy, Warsaw.

Jonkers, A. R. T., Jackson, A. & Murray, A. (2003). Four centuries of geomagnetic data from historical records. *Rev. Geophys.*, **41**, 1006, doi: 10.1029/2002RG000115.

Kernevez, N., Duret, D., Moussavi, M. & Léger, J. M. (1992). Weak field NMR and ESR spectrometers and magnetometers. *IEEE Trans. Magnetics*, **28**(5), 3054–9.

Kim, H., Cai, X., Clauer, C. R., Kunduri, B. S. R., Matzka, J., Stolle, S. & Weimer, D. R. (2013). Geomagnetic response to solar wind dynamic pressure impulse events at high-latitude conjugate points. *J. Geophys. Res. Space Physics*, **118**, 6055–71, doi: 10.1002/jgra.50555.

Knappe, S., Sander, T. & Trahms, L. (2014). Optically-pumped magnetometers for MEG. In S. Supek & C. J. Aine, eds., *Magnetoencephalography*. Springer, Berlin, pp. 993–9.

Léger, J. M., Bertrand, F., Jager, T., Le Prado, M., Fratter, I. & Lalaurie, J. C. (2009). Swarm absolute scalar and vector magnetometer based on helium 4 optical pumping. *Procedia Chem.*, **1**, 634–7, doi: 10.1016/j.proche.2009.07.158.

Lichtenberger, J., Clilverd, M., Heilig, B., Vellante, M., Manninen, J., Rodger, C., Collier, A., Jørgensen, A., Reda, J., Holzworth, R. & Friedel, R. (2013). The plasmasphere during a space weather event: first results from the PLASMON project, *J. Space Weather Space Clim.*, **3**, A23, doi: 10.1051/swsc/201304.

Love, J. J. & Chulliat, A. (2013). An international network of magnetic observatories. *Eos Trans. AGU*, **9**(4), 373–4, doi: 10.1002/2013EO420001.

Love, J. J. & Finn, C. A. (2011). The USGS geomagnetism program and its role in space weather monitoring. *Space Weather*, **9**, S07001, doi: 10.1029/2011SW000684.

Macmillan, S. & Olsen, N. (2013). Observatory data and the Swarm mission. *Earth Planets Space*, **65**, 1355–62, doi: 10.5047/eps.2013.07.011.

Marusenkov, A. (2017). Possibilities of further improvement of 1 s fluxgate variometers. *Geosci. Instrum. Method. Data Syst.*, **6**, 301–9, doi: 10.5194/gi-6-301-2017.

Matzka, J. (2012). Preparation of quasi-definitive (QD) data for the observatories Narsarsuaq, Qeqertarsuaq and Tristan da Cunha. In P. Hejda, A. Chulliat & M. Catalan, eds., *Proceedings of the XVth IAGA Workshop on Geomagnetic Observatory Instruments, Data Acquisition, and Processing*, Real Instituto Y Observatorio de la Armada en San Fernando, San Fernando, Boletin Roa No. 03/13, pp. 50–53.

Matzka, J., Chulliat, A., Mandea, M., Finlay, C. C. & Qamili, E. (2010). Geomagnetic observations for main field studies: From ground to space. *Space Sci. Rev.*, **155**, 29–64, doi: 10.1007/s11214-010-9693-4.

Meyer, B., Chulliat, A. & Saltus, R. (2017). Derivation and error analysis of the Earth magnetic anomaly grid at 2 arc min resolution Version 3 (EMAG2v3). *Geochem. Geophys. Geosys.*, **18**, 4522–37, doi: 10.1002/2017GC007280.

Miles, P. J., Partner, R. T., Keeler, K. R. & McConnell, T. J. (2008). Unmanned airborne vehicle for geophysical surveying. U.S. patent US2008/0125920 A1, published 29 May.

Mohr, P. J., Taylor, B. N. & Newell, D. B. (2008). CODATA recommended values of the fundamental physical constants: 2006. *Rev. Mod. Phys.*, **80**, 633–730, doi: 10.1103/RevModPhys.80.633.

Nabighian, M. N., Grauch, V. J. S., Hansen, R. O., LaFehr, T. R., Li, Y., Peirce, J. W., Phillips, J. D. & Ruder, M. E. (2005). The historical development of the magnetic method in exploration. *Geophysics*, **70**(6), 33–61, doi: 10.1190/1.2133784.

Narod, B. B., Bennest, J. R., Strom-Olsen, J. O., Nezil, F. & Dunlap R. A. (1985). An evaluation of the noise performance of Fe, Co, Si, and B amorphous alloys in ring-core fluxgate magnetometers. *Can. J. Phys.*, **63**(11), 1468–72, doi: 10.1139/p85-246.

Newitt, L. R., Barton, C. E. & Bitterly, J. (1996). *Guide for Magnetic Repeat Stations*. International Association of Geomagnetism and Aeronomy, Boulder, CO.

Nielsen, O. V., Petersen, J. R., Primdahl, F., Brauer, P., Hernando, B., Fernandez, A., Merayo, J. M. G. & Ripka P. (1995). Development, construction and analysis of the 'Ørsted' fluxgate magnetometer. *Meas. Sci. Technol.*, **6**, 1099–1115.

Olsen, N., Ravat, D., Finlay, C. C. & Kother, L. K. (2017). LCS-1: A high-resolution global model of the lithospheric magnetic field derived from CHAMP and *Swarm* satellite observations. *Geophys. J. Int.*, **211**(3), 1461–77, doi: 10.1093/gji/ggx381.

Overhauser, A. W. (1953). Paramagnetic relaxation in metals. *Phys. Rev.*, **89**, 689–700, doi: 10.1103/PhysRev.92.411.

Packard, M. E. (1958). Gyromagnetic resonance magnetometer. U.S. patent 2,856,579, issued 14 October.

Peltier, A. & Chulliat, A. (2010). On the feasibility of promptly producing quasi-definitive magnetic observatory data. *Earth Planets Space*, **62**, e5–e8, doi: 10.5047/eps.2010.02.002.

Poncelet, A., Gonsette, A. & Rasson, J. (2017). Several years of experience with automatic DI-flux systems: theory, validation and results. *Geosci. Instrum. Method. Data Syst.*, **6**, 353–60, doi: 10.5194/gi-6-353-2017.

Primdahl, F. (1979). The fluxgate magnetometer. *J. Phys. E Sci. Instrum.*, **12**, 241–53.

Rasmussen, O. & Kring Lauridsen, E. (1990). Improving baseline drift in fluxgate magnetometers caused by foundation movements, using band suspended fluxgate sensors. *Phys. Earth Planet. Inter.*, **59**, 78–81.

Reeves, C. (2005). *Aeromagnetic Surveys – Principles, Practice and Interpretation*. Geosoft (http://www.geosoft.com).

Schiffler, M., Queitsch, M., Stolz, R., Chwala, A., Krech, W., Meyer, H.-G. & Kukowski, N. (2014). Calibration of SQUID vector magnetometers in full tensor gradiometry systems. *Geophys. J. Int.*, **198**, 954–64, doi: 10.1093/gji/ggu173.

Schnepf, N., Manoj, C., Kuvshinov, A., Toh, H. & Maus, S. (2014). Tidal signals in ocean-bottom magnetic measurements of the Northwestern Pacific: observation versus prediction. *Geophys. J. Int.*, **198**, 1096–110, doi: 10.1093/gji/ggu190.

Serson, P. H. & Hannaford, W. L. W. (1956). A portable electrical magnetometer. *Can. J. Technol.*, **34**, 232–43.

Sheng, D., Li, S., Dural, N. & Romalis, M. V. (2013). Subfemtotesla scalar atomic magnetometry using multipass cells. *Phys. Rev. Lett.*, **110**, 160802, doi: 10.1103/PhysRevLett.110.160802.

St Louis, B., ed. (2012). *INTERMAGNET Technical Reference Manual*, version 4.6.

Stuart, W. F. (1972). Earth's field magnetometry. *Rep. Prog. Phys.*, **35**, 803–81.

Tanskanen, E. I. (2009). A comprehensive high-throughput analysis of substorms observed by IMAGE magnetometer network: Years 1993–2003 examined. *J. Geophys. Res.*, **114**, A05204, doi: 10.1029/2008JA013682.

Toh, H., Hamano, Y. & Ichiki, M. (2006). Long-term seafloor geomagnetic station in the northwest Pacific: A possible candidate for a seafloor geomagnetic observatory. *Earth Planets Space*, **58**, 697–705.

Toh, H., Satake, K., Hamano, Y., Fujii, Y. & Goto, T. (2011). Tsunami signals from the 2006 and 2007 Kuril earthquakes detected at a seafloor geomagnetic observatory. *J. Geophys. Res.*, **116**, B02104, doi: 10.1029/2010JB007873.

Torta, J. M., Pavón-Carrasco, F. J., Marsal, S. & Finlay, C. C. (2015). Evidence for a new geomagnetic jerk in 2014. *Geophys. Res. Lett.*, **42**, 7933–40, doi: 10.1002/2015GL065501.

Turner, G. M., Rasson, J. L. & Reeves, C. V. (2009). Observation and measurement techniques. In G. Schubert & M. Kono, eds., *Treatise on Geophysics – Geomagnetism*. Elsevier, Amsterdam, pp. 93–146.

Van Loo, S. A. & Rasson, J. L. (2007). Presentation of the prototype of an automated DIFlux. *Publs. Inst. Geophys. Pol. Acad. Sc.*, **C-99**(398), 77–86.

Varian, R. H. (1951). Method and means for correlating nuclear properties of atoms and magnetic fields. US patent 2,561,490, issued 24 July.

Weygand, J. M., Amm, O., Viljanen, A., Angelopoulos, V., Murr, D., Engebretson, M. J., Gleisner, H. & Mann, I. (2011). Application and validation of the spherical elementary currents systems technique for deriving ionospheric equivalent currents with the North American and Greenland ground magnetometer arrays. *J. Geophys. Res.*, **116**, A03305, doi: 10.1029/2010JA016177.

Xu, Z., Hartinger, M. D., Clauer, C. R., Peek, T. & Behlke, R. (2017). A comparison of the ground magnetic responses during the 2013 and 2015 St. Patrick's Day geomagnetic storms. *J. Geophys. Res. Space Physics*, **122**, 4023–36, doi: 10.1002/2016JA023338.

Yizengaw, E., Moldwin, M. B., Zesta, E., Biouele, C. M., Damtie, B., Mebrahtu, A., Rabiu, B., Valladares, C. F. & Stoneback, R. (2014). The longitudinal variability of equatorial electrojet and vertical drift velocity in the African and American sectors. *Ann. Geophys.*, **32**, 231–8, doi: 10.5194/angeo-32-231-2014.

Yumoto, K. (2006). MAGDAS project and its application for space weather. ILWS Workshop 2006, Goa, 19–24 February.

Yumoto, K., & CPMN Group (2001). Characteristics of Pi 2 magnetic pulsations observed at the CPMN stations: A review of the STEP results. *Earth Planets Space*, **53**, 981–92.

6.2 Electromagnetic and Magnetotelluric Measurements

Ute Weckmann and Oliver Ritter

Electromagnetic (EM) methods form a discipline of geophysics, applied to study the electrical conductivity structure within Earth in a wide depth range. As the name suggests, EM measurements always depend on recordings of magnetic and electric field variations. Observed electromagnetic fields can be of natural origin i.e. generated by worldwide thunderstorm activity or ionospheric current systems (Constable, 2016), generated by a manmade (controlled) sources (Streich, 2016), or a combination of both. These electromagnetic fields penetrate into the Earth and depending on the electric properties within, secondary fields are induced that can be measured at the Earth's surface by magnetometers and telluric electrodes.

The depth of investigation depends on the so-called skin effect, which describes that for a given conductivity (or its inverse electrical resistivity) low-frequency EM fields penetrate deeper into the Earth than their higher frequency counterparts. EM field experiments can be designed to cover sounding depths from a few meters to a couple of hundred kilometers. Resistivity values of Earth's materials vary over several orders of magnitude, typically in the range between 10^{-1} to 10^{6} Ωm.

The electrical conductivities of rocks are generally less controlled by the matrix, which may define a rock, but by constituents which can conduct electrical currents and as a secondary effect by temperature. The bulk conductivity of a rock increases if conductive constituents form interconnected networks, for instance containing brines, fluids, melts, graphite or other (metallic) mineralization.

With EM methods we strive to obtain high-resolution information on lateral and vertical changes in electrical resistivity at depth that can be related to geological units, tectonic processes or deposition of geo-resources. The following examples give a brief glimpse of the broad range of EM applications, including its variety magnetotellurics (MT). More details and many more references for further reading can be found in the review papers given at the international EM workshops of IAGA Division VI 'Electromagnetic Induction into the Earth and Planetary Bodies'.

6.2.1 Land Measurements and Observations

Magnetotelluric measurements are based on naturally occurring EM field variations and can cover a frequency range from 10^{-5} to 10^{4} Hz. MT recordings contain the lowest frequency signals and provide the largest sounding depths of all EM methods. MT is therefore routinely used for regional-scale investigations on crustal and upper mantle depth levels, such as plate boundaries, mountain ranges and large fault zones. Profiles are often interpreted using two-dimensional (2-D) inversion (Ledo, 2006), while arrays typically require three-dimensional (3-D) inversion approaches (Miensopust, 2017).

A prominent example of a large scale tectonic target for MT is the San Andreas Fault (SAF) in northern America. The crustal and upper mantle resistivity structure across the Parkfield–Cholame segment of the SAF was imaged using a regional-scale MT array (Becken and Ritter, 2012). Broad band MT data were collected in the frequency range of 1 kHz – 0.1 mHz and interpreted with 2-D and 3-D inversion (Figure 6.2.1). The MT data were collected along several profiles crossing creeping (e.g. profile 2) and locked (e.g. profile 5) segments of the SAF. The most prominent feature

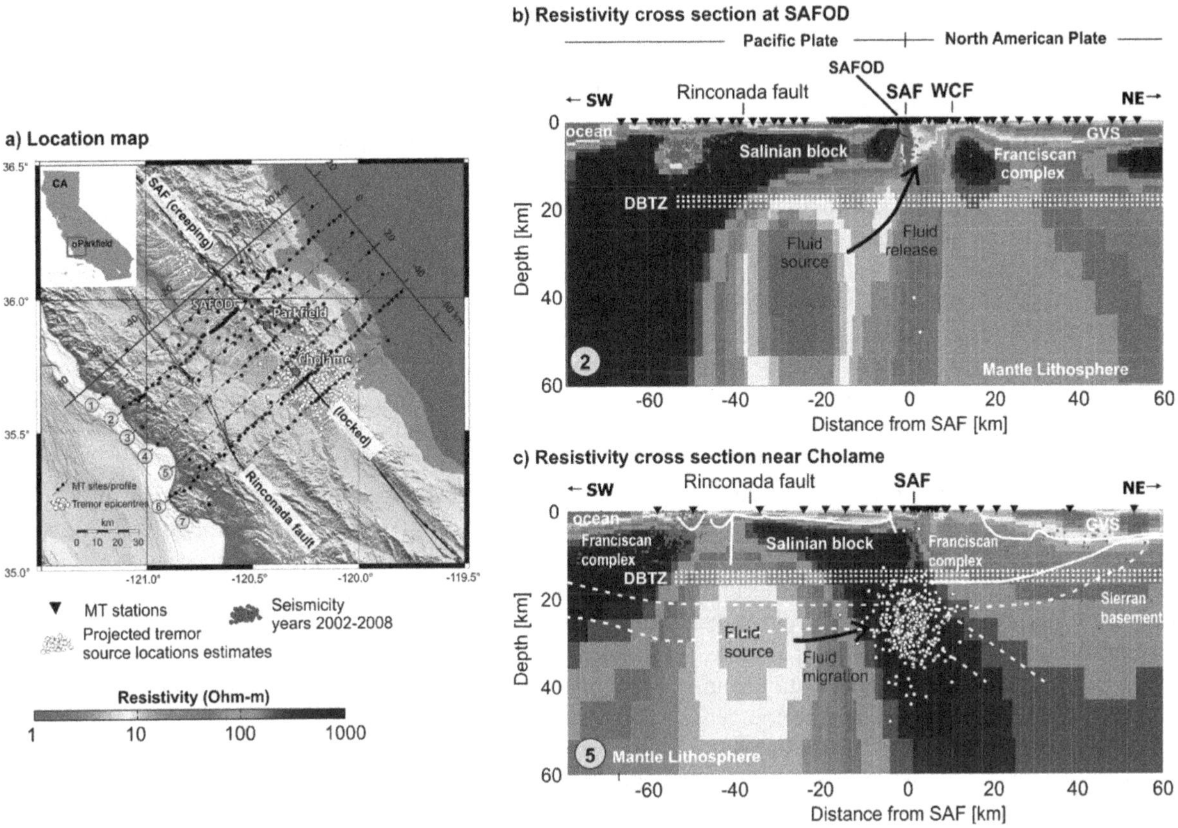

Figure 6.2.1 Map view. Black dots mark MT stations along seven ~130 km long profiles. Black lines indicate the San Andreas and Rinconada faults. The northwestern part of the SAF is described as creeping, while the southeastern section is locked. Regional-scale resistivity models across the SAF (a) at SAFOD near Parkfield and (b) at the source region of tremor near Cholame. WCF = Waltham canyon fault. BDTZ = brittle-ductile transition zone. Superimposed are boundaries of geological units (white lines). Modified after Becken and Ritter (2012). (A black-and-white version of this figure appears in some formats. For the colour version, please refer to the plate section.)

within the sections is a deep highly conductivity zone (HCZ, 1–5 Ωm), centered 30–40 km southwest of the SAF at middle crustal levels. Becken and Ritter (2012) suggested mineralized fluids within an interconnected network of high porosities as a possible cause. However, the upper crustal structure surrounding the SAF exhibits distinct along-strike variation related to the occurrence of upper crustal seismicity and deep non-volcanic tremor (NVT). Where the fault is creeping and characterized by microseismicity, a conductive pathway connects with the HCZ; but, where the fault is locked with the episodes of NVT, the upper parts around the SAF are resistive and void of a conductive channel into the HCZ. The HCZ may therefore correspond to the source region of mantle-derived fluids, which find pathways through the crust into the creeping fault segment, while these fluids may be trapped at greater depths in the locked segment of the SAF, where NVT is observed.

Besides faults, entire tectonic units or even continents are nowadays studied with MT to examine their evolution history. Prominent ongoing continental-scale research includes for example the USArray, AusLamp or Sinoprobe initiatives. Regional-scale experiments have a long tradition and were typically acquired along profiles and transects. Recent reviews include the African (Weckmann, 2012) and Asian continents (Zhang, 2017) as well as Central Europe (Neska, 2016). Historic studies already linked crustal conductors with tectonic boundaries and detachment zones and anomalies in the upper mantle with partial melts. But with high-resolution and array measurements the internal structures of these tectonic blocks could be imaged with MT (Selway, 2014) and modern studies fostered a revision of 'old' anomalies that were initially attributed to deep lithospheric sources which are now associated with crustal anomalies or deep conductive basins. Many new insights could be obtained, for instance, in the Himalayas, where a large number of MT experiments revealed lithospheric zones of high electrical conductivity which triggered the idea of lower crustal flow (Unsworth, 2010).

Long period MT studies have been used to obtain conductivity images of the Earth's lithosphere and mantle down to depths of 1000 km, typically based on 1-D interpretations (Kuvshinov, 2012). Since deep EM studies exhibit a substantial level of lateral heterogeneity in the upper and middle to lower mantle, recent research efforts attempt to derive global models using 3-D inversion. These models are usually not based on MT data but utilize satellites and data from geomagnetic observatories (see section 6.1). Up to now, different approaches often result in disparate models though. Causes are the unevenly distributed MT data sets (on a global scale), variability in data quality and applied data interpretation tools, numerical and computational costs and the inherent non-uniqueness of the problem.

Electrical conductivity is a crucial parameter for the characterization of geothermal systems which are composed of faults and/or fractures filled with geothermal fluids and altered rocks (Muñoz, 2014; Patro, 2017). In hydrothermal systems, electrical resistivity is dominated by alteration mineralogy, which can be used as a proxy for temperature. In non-volcanic reservoirs, saline geothermal fluids and their pathways along faults and fractures can be imaged with EM methods. Provided that the electrical conductivity of these fluids is known, it is possible to infer porosity and/or permeability of the system, which are among the most important reservoir parameters.

While MT is nowadays routinely used for geothermal exploration, methods based on controlled sources, such as frequency or time domain EM, are also applicable. The main benefit of controlled source EM methods is generally better signal to noise ratios, particularly when working in inhabited or industrialized regions, with the disadvantage of more limited sounding depths and the additional logistical overhead for operating sources. For shallow targets (a few hundred meters), transient EM methods (TEM) are often applied, which interpret the decay of the measured magnetic field (transient) to determine the conductivity distribution in the subsurface. TEM has proven to be very efficient in mapping e.g. clay and saltwater intrusions or in detecting polluted areas and delineating lower boundaries of deep waste deposits (Tezkan, 1999). Long offset TEM (LOTEM) is designed for exploration depths down to several kilometers and uses long horizontal electric dipoles to generate the transient source fields.

Land CSEM has been applied commercially in mining applications, where resistivity contrasts between targets and host rocks are commonly large (Smith, 2014) and well logging, where the lack of a pronounced conductivity contrast poses a challenge for EM methods (Strack, 2014). MT measurements are generally very useful to delineate conductive structures (e.g. ore bodies), but cannot delineate well (thin) resistive features (e.g. hydrocarbons), for which controlled source methods are much better suited. This is largely due to differences in the physics of the respective source fields and their diffusion into subsurface. For MT and other plane wave source methods, such as CSAMT or VLF, the induced current systems are predominantly horizontal, while those associated with controlled sources (e.g. transient EM (TEM) or modern CSEM) also contain vertical components. The geometry of controlled sources has to be taken into account, however, which adds additional complexity to subsequent modeling and inversion (Streich, 2016). Depending on the environment and the target, different sources such as horizontal or vertical electric (HED, VED) or magnetic (HMD, VMD) dipoles, and vertical magnetic dipole (VMD) sources can be applied. Of similar importance is the layout of magnetic and electric receivers since they also have different sensitivity to conductive or restive target structures (Streich, 2016).

Very shallow EM investigations are mainly applied in archeology, for soil characterization and prospecting for gemstones. Very popular are *Slingram*-type systems, dual-coil electromagnetic profiling equipment with transmitter and receiver at a fixed distance. These methods are highly efficient in terms of ease of operation and measurement progress, as they can be used in an almost continuous way for profiling. *Slingram*-type data are mostly used for a qualitative interpretation, however, as the secondary field that carries information on subsurface conductivity is measured in the presence of the larger primary field. Without adequate compensation, these data can only be used in a relative sense, which is often sufficient for environmental applications.

6.2.2 Marine Measurements

Marine MT and EM measurements started to attract a larger interest in the early 2000s when controlled source EM (CSEM) methods were applied for off-shore hydrocarbon exploration (Key, 2012). For CSEM, an electric or magnetic dipole transmitter system is deep-towed from a ship to generate a time-varying current and the related EM fields diffuse through the ocean, seafloor and air. The EM field is modified by the conductivity of the media it passes through. The resulting attenuation and phase shift can then be recorded with EM receivers (magnetometers and/or electric dipoles) either at seafloor or via a towed line of electrodes (streamer). It should be noted that marine CSEM was first applied to determine the conductivity of the ocean lithosphere, although nowadays hydrocarbon reservoir detection and monitoring are more common applications. Similar to land (controlled source) EM measurements it is of utmost importance to develop a target-adapted experimental strategy since receiver layout, transmitter-receiver offsets and type of EM signals need to ensure highest resolution to the target structures.

As its counterpart on land, marine MT is based on naturally occurring low-frequency plane waves that diffuse into seabed and which can be measured at the bottom of the ocean with magnetometers and electric dipoles. In the marine environment, the conductive sea water damps away the shortest periods of the natural EM fields so that a typical frequency range extends from 10 Hz to 0.1 mHz. Marine MT is useful for sedimentary basin exploration and deeper tectonic studies and joint inversions of CSEM and MT data can result in more robust conductivity models (Key, 2012). Amphibious MT projects, a combination of off- and onshore stations are an interesting field of emerging research to investigate continental shelfs and the transition from oceanic to continental lithosphere. The land–ocean transition is obviously a first order conductivity contrast with strong topographical gradients, affecting both land and marine stations and which is a challenge for inversion codes.

6.2.3 Airborne Measurements

Airborne measurements are often applied in mining geophysics, mineral exploration and ground water prospection, where large areas are examined for shallow anomalies. Industry and academia have developed a huge variety of configurations, sources, sensors and concepts for airborne EM measurements to optimize efficiency. In this context, only a few approaches can be mentioned, but good reviews were given by Smith (2014) and Everett (2012). Most airborne systems use controlled source methods, particularly for detection of small scale features that have a very large conductivity contrast with the host rock. Both frequency and time domain systems exist. VTEM, SkyTEM and HELITEM are examples of controlled source airborne systems that have a transmitter and receiver below an aircraft or helicopter and work in time domain. An example of a natural-field system is ZTEM (Z tipper EM). The ZTEM system measures the natural-field vertical magnetic field component on a sensor carried below a helicopter. The vertical magnetic field measurements are then related as in MT, to the horizontal magnetic field components measured at a base station at surface.

References

Becken, M. & Ritter, O. (2012). Magnetotelluric studies at the San Andreas Fault Zone: Implications for the role of fluids. *Surv. Geophys.*, 33, 65–105, doi: 10.1007/s10712-011-9144-0.

Constable, C. (2016). Earth's electromagnetic environment. *Surv. Geophys.*, 37, 27–45, doi: 10.1007/s10712-015-9351-1.

Everett, M. (2012). Theoretical developments in electromagnetic induction geophysics with selected applications in the near surface. *Surv. Geophys.*, 33, 29–63, doi: 10.1007/s10712-011-9138-y.

Key, K. (2012). Marine electromagnetic studies of seafloor resources and tectonics. *Surv. Geophys.*, 33, 135–67, doi: 10.1007/s10712-011-9139-x.

Kuvshinov, A. V. (2012). Deep electromagnetic studies from land, sea, and space: Progress status in the past 10 years. *Surv. Geophys.*, 33, 169–209, doi: 10.1007/s10712-011-9118-2.

Ledo, J. (2006). 2-D versus 3-D magnetotelluric data interpretation. *Surv. Geophys.*, 27, 511–43, doi: 10.1007/s10712-005-1757-8.

Miensopust, M. (2017). Application of 3-D electromagnetic inversion in practice: Challenges, pitfalls and solution approaches. *Surv. Geophys.*, 38, 869–933, doi: 10.1007/s10712-017-9435-1.

Muñoz, G. (2014). Exploring for geothermal resources with electromagnetic methods. *Surv. Geophys.*, 35, 101–22, doi: 10.1007/s10712-013-9236-0.

Neska, A. (2016). Conductivity anomalies in Central Europe. *Surv. Geophys.*, 37, 5–26, doi: 10.1007/s10712-015-9349-8.

Patro, P. (2017). Magnetotelluric studies for hydrocarbon and geothermal resources: Examples from the Asian region. *Surv. Geophys.*, 38, 1005–41, doi: 10.1007/s10712-017-9439-x.

Selway, K. (2014). On the causes of electrical conductivity anomalies in tectonically stable lithosphere. *Surv. Geophys.*, 35, 219–57, doi: 10.1007/s10712-013-9235-1.

Smith, R. (2014). Electromagnetic induction methods in mining geophysics from 2008 to 2012. *Surv. Geophys.*, 35, 123–56, doi: 10.1007/s10712-013-9227-1.

Strack, K. (2014). Future directions of electromagnetic methods for hydrocarbon applications. *Surv. Geophys.*, 35, 157–77, doi: 10.1007/s10712-013-9237-z.

Streich, R. (2016). Controlled-source electromagnetic approaches for hydrocarbon exploration and monitoring on land. *Surv. Geophys.*, 37, 47–80, doi: 10.1007/s10712-015-9336-0.

Tezkan, B. (1999). A review of environmental applications of quasi-stationary electromagnetic techniques. *Surv. Geophys.*, 20, 279–308, doi: 10.1023/A:1006669218545.

Unsworth, M. (2010). Magnetotelluric studies of active continent-continent collisions. *Surv. Geophys.*, 31, 137–61, doi: 10.1007/s10712-009-9086-y.

Weckmann, U. (2012). Making and breaking of a continent: following the scent of geodynamic imprints on the African continent using electromagnetics. *Surv. Geophys.*, 33, 107–34, doi: 10.1007/s10712-011-9147-x.

Zang, L. (2017). A review of recent developments in the study of regional lithospheric electrical structure of the Asian continent. *Surv. Geophys.*, 38, 1043–96, doi: 10.1007/s10712-017-9439-x.

6.3 Solar Cycle Variations in Differential Instrumental Responses from Ground-Based Geomagnetic Records

Stuart A. Gilder, Michael R. Wack, Elena A. Kronberg and Ameya Prabhu

6.3.1 Introduction

A small fraction of the magnetic field measured at the Earth's surface originates predominantly from electromagnetic processes extending tens to hundreds of Earth radii into space. Protons and electrons in the magnetosphere, together with ionized chemical compounds from the atmosphere, are guided by the internally generated magnetic field to create a complex system of electrical currents above the rotating Earth (see Baumjohann and Treumann, 2012, for a review). Interaction of the solar wind with the magnetosphere produces a measurable effect on the Earth's surface, such as magnetic storms or substorms with amplitudes that reach several hundreds of nanoteslas (nT), or as much lower amplitude, higher frequency fluctuations (e.g. Tamao, 1964; Southwood, 1974; Menk, 1988; Allan and Poulter, 1992; McPherron, 2005).

The effects of the solar wind on the Earth's magnetic field have been studied intensely from ground-based observatories. Several geomagnetic indices help quantify the magnetic field's response to solar activity (Mayaud, 1980), and most workers find that they vary in concert with changes in the solar wind including its electric field, velocity, and magnitude of the southward component of the interplanetary magnetic field (IMF) (e.g. Wilcox et al., 1967; Hirschberg and Colburn, 1969; Garrett et al., 1974; Crooker et al., 1977; Feynman, 1982; Le Mouël et al., 2005; Weimer et al., 2010). Much attention has been directed to understanding the magnetic spectrum concerning pulsations or field line resonances from ground-based measurements of the magnetic field as well as from instruments on board satellites (e.g. Russell and Fleming, 1976; Takahashi and McPherron, 1982; Engebretson et al., 1987; Kivelson et al., 1997; Chi and Russell, 2001; Kepko et al., 2002; Vellante et al., 2004; Takahashi et al., 2006; Heilig et al., 2007; Keiling et al., 2016). These studies typically utilized short time intervals (minutes to a few days) that specifically focused on particular frequencies like those predicted for cavity modes (e.g. 1.3, 1.9, 2.6 and 3.3 mHz; Walker, 2002), whereas time intervals that did not contain specific, sought after frequencies were usually neglected. Below, we introduce a new technique based on data acquired from ground-based magnetic observatories that can be used to study space weather phenomena over long time periods (days to decades) and over a broad frequency band, simultaneously from several locations on the globe. We describe how the technique works and demonstrate that similar effects can be observed from most observatories located at all geomagnetic latitudes, thereby opening a vast dataset to the study of magnetic field oscillations ultimately derived from solar activity and its interaction with the Earth environment.

6.3.2 International Real-Time Magnetic Observatory Network (INTERMAGNET) Protocol

Each of the ~125 geomagnetic observatories participating in the worldwide INTERMAGNET (*Inter*national *R*eal-time *Mag*netic Observatory *Net*work) consortium house both vectorial and scalar magnetometers (Campbell, 1997; St Louis, 2008; Auster, 2008). Prior to 2009, the INTERMAGNET protocol for vector (fluxgate-type) magnetometers uses a Gaussian-weighted filter centred on the minute that incorporates 19 readings of three-component data acquired every 5 s (nine values before the minute in question, nine after, and one at the minute round). Each individual measurement takes about a tenth of a second to be made. The Gaussian-filtered data are known as 1 min data. INTERMAGNET observatories started switching to 1 s data in 2009, yet as of 2017 about half of the participating observatories still produce 1 min values.

Most observatories use Overhauser or non-Overhauser effect proton precession magnetometers to obtain scalar field data. The exact specifications at each observatory can be obtained from the yearly definitive data published by INTERMAGNET concerning the different instruments and acquisition methods. Measurements from non-Overhauser effect magnetometers take about 5 s – 3 s for polarization followed by an exponentially decaying signal over roughly 2 s. The measurement time stamp is taken at the end of the 5 s interval; Overhauser effect proton precession magnetometers measure at rates up to 5 Hz. Most observatories do not apply Gaussian-weighted filters to the scalar magnetometer data. Thus, although the measurement cadence of vectorial and scalar magnetometers is fixed as identical according to the same clock (GPS, DCF77, etc.), the actual measurements are not carried out at precisely the same time and/or do not average over the same amount of time.

Each participating observatory furnishes INTERMAGNET the minute values of the total field data from the scalar magnetometer together with the filtered three-component vector data. A simple test of magnetic observatory data quality compares the difference in total field (F) values between the scalar magnetometer (F_s) and the vectorial magnetometer (F_v) (i.e. the square root of the sum of the squares of the three components), which is known as ΔF ($\Delta F = F_s - F_v$). Because both magnetometers lie several meters apart, ΔF is non-null, but it should ideally be constant in time. While observing data from several INTERMAGNET observatories run by the Bureau Central de Magnétisme Terrestre (France), we noticed a faint yet repeatable signal that ΔF was usually 'noisier' (greater variation about a constant ΔF) around local noon, yet such variability was not noticeable in the component data. This observation led to the present study suggesting that the variance about ΔF stems from magnetic field line oscillations.

6.3.3 Methods and Results

6.3.3.1 Development and Numerical Modeling of the New Method

To test whether ΔF undergoes a diurnal cycle, we calculated the average and single sample standard deviation (1σ) of the minute data for each hour (typically $n = 60$) of each day for an entire year, and then from these values ($n = 365$ or 366), calculated the average ΔF and average 1σ for each hour of the day for that year. Figure 6.3.1 gives two examples for the Chambon-la-Forêt (France) observatory for the years 2000 and 2004 where two effects are noted. First, ΔF exhibited a

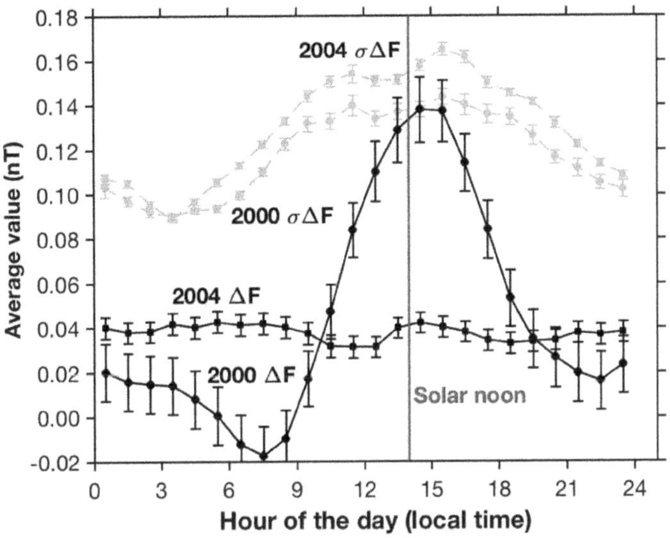

Figure 6.3.1 Daily variation of the hourly averaged value (average of 60 min values averaged over 366 days) of the difference in total field intensity (ΔF) and the single standard deviation (1σ) about ΔF ($\sigma_{\Delta F}$), together with their standard errors, measured with vectorial and scalar magnetometers at the Chambon-la-Forêt (CLF, France) observatory in 2000 and 2004.

diurnal variation amplitude of about 0.2 nT in 2000 with a maximum at around solar noon, whereas insignificant diurnal variation in ΔF was observed in 2004. Second, the scatter (1σ) about ΔF also exhibited a diurnal variation, independent of the magnitude or rate of change of ΔF. By examining ΔF curves month by month, we found that the change in the character of ΔF occurred in October 2002 when the electronics for the vectorial magnetometer were moved into a thermally controlled environment.

Despite the change in character of ΔF, the average 1σ variation about ΔF, hereafter called $\sigma_{\Delta F}$, was invariant to ΔF (Figure 6.3.1); moreover, we found no significant correlation between $\sigma_{\Delta F}$ and F, dF/dt or $d\Delta F/dt$. Rather, $\sigma_{\Delta F}$ showed some dependence to magnetic field behavior and time of day. For example, a proxy for local field variability is given by the geomagnetic K index, which ranges from 0 (quiet) to 9 (extremely disturbed) and is calculated every three-hours. A comparison of three-hour averages of $\sigma_{\Delta F}$ as a function of the K index during a day (Figure 6.3.2) showed that for a given level of disturbance (e.g. constant K), the three-hour average $\sigma_{\Delta F}$ was systematically greater around local noon than at night, further confirming that field oscillation amplitudes reached a maximum at local solar noon, consistent with previous observations (Menk, 1988; Lee and Lysak, 1991; Allan and Poulter, 1992; Sakurai et al., 1999; Menk et al., 2000; Verö et al., 2000; Chi and Russell, 2001; Heilig et al., 2007; Takahashi and Ukhorskiy, 2008).

To understand the origin of the $\sigma_{\Delta F}$ variations, we first used fast Fourier transform routines to analyze the frequency content of the magnetic field time series from the

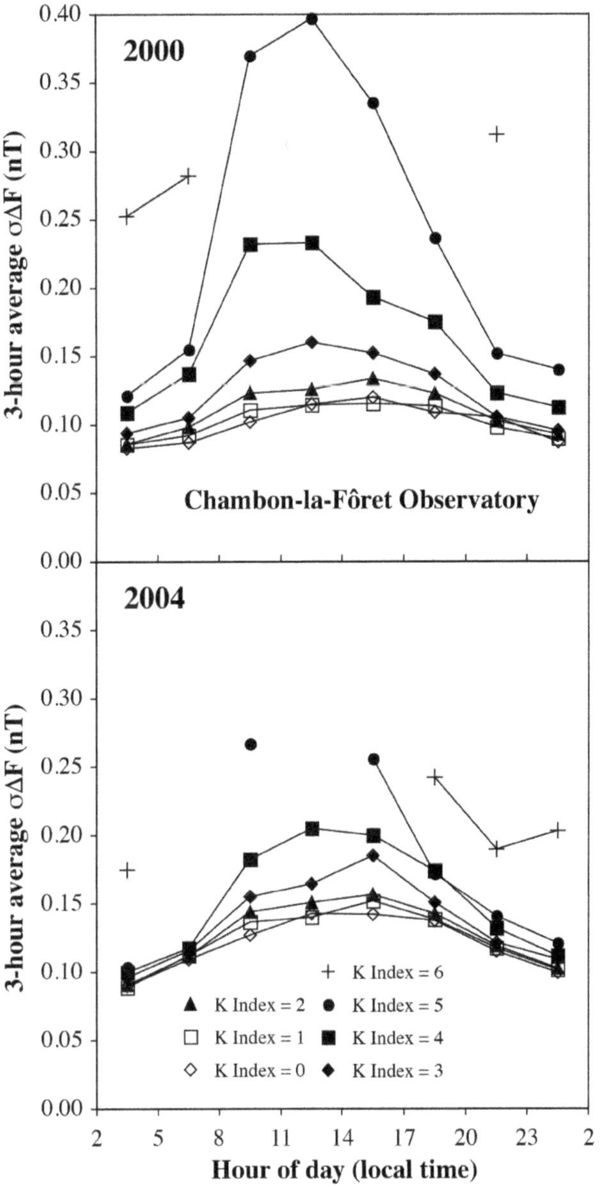

Figure 6.3.2 Daily variation of the 3 hour average $\sigma_{\Delta F}$ as a function of K Index for the Chambon-la-Forêt Observatory in 2000 and 2004. In 2000 (2004), there were 151 (254), 792 (830), 1006 (952), 652 (597), 229 (210) and 17 (24) values for K indices from 0 to 6, respectively. K indices of 7 and 8 [6 (7)] were not considered.

Chambon-la-Forêt (CLF, France) observatory in April 2004 (a relatively quiet month) for the minute F data from the proton magnetometer as well as for ΔF and $\sigma_{\Delta F}$ (Figure 6.3.3). A prominent peak at a 1.16×10^{-5} Hz frequency (one-day periodicity) was apparent in the F scalar field readings and in $\sigma_{\Delta F}$, but not in ΔF, in good agreement with Figure 6.3.1. The fact that there was no diurnal frequency visible in the ΔF spectrum demonstrated that there was either no, or the exact same bias, between magnetometers caused by parasitic, diurnal signals like

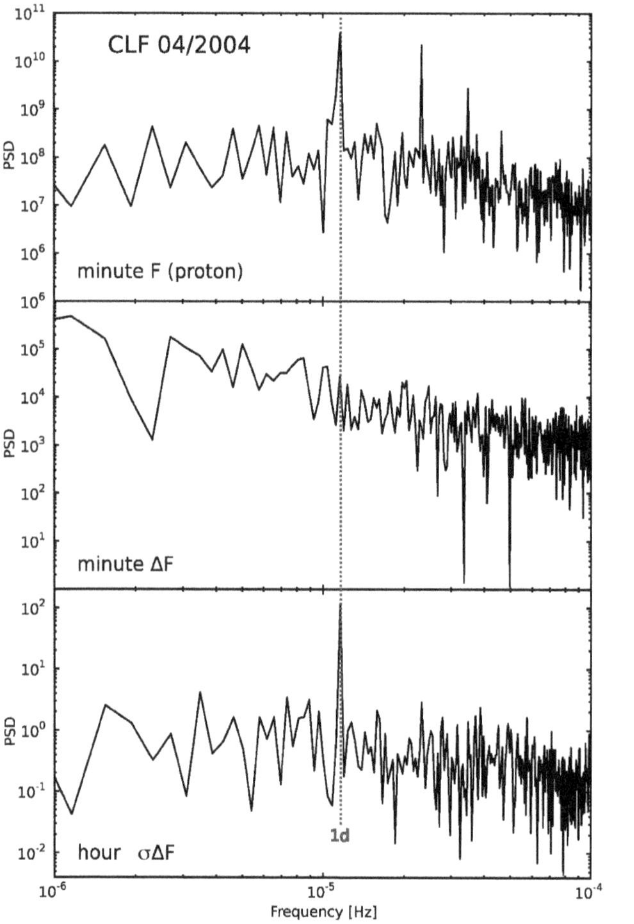

Figure 6.3.3 Fast Fourier transform (FFT) spectra of data from the Chambon-la-Forêt Observatory in April 2004. No tapering, windowing, stacking or linear regression was applied to the data; using a Hamm taper resulted in fairly identical plots. The analyzed time series are: total magnetic field (F) recorded with a proton magnetometer each minute; difference between total field values (ΔF) from the scalar and vectorial magnetometers for each minute; hourly average standard deviation of 60 min ΔF values (= $\sigma_{\Delta F}$). PSD is power spectrum density; 1d represents a 1 day period. Subdividing the data into weeks, applying FFT and stacking the data also showed a peak at a 1 day frequency.

temperature or voltage variations. It is highly unlikely that totally different and independent instruments (fluxgate and proton magnetometers) will show the same response to a parasitic signal, which argues strongly against influences on the measured values stemming from digital filtering noise, rounding of data, etc. The diurnal frequency seen in $\sigma_{\Delta F}$ can only come from filtering inherent in the method. In other words, $\sigma_{\Delta F}$ arises from the difference in timing between the datasets of the two magnetometers and is not intrinsic to the magnetometers, otherwise one would see an effect in ΔF. Moreover, it is shown below that the one-day periodicity in F and $\sigma_{\Delta F}$ stem from two independent sources.

We developed a numerical model to understand the frequency content and source of $\sigma_{\Delta F}$ by using a monochromatic sinusoidal waveform representing the ambient magnetic field consisting of 3.6 million data points (1000 data points per second over a one-hour duration) (Figure 6.3.4). This resolution ensured that even the highest frequencies (50 Hz) were still well represented by 20 points per period to avoid aliasing affects in the source waveform. Monotonic sinusoids were generated for 10^5 discrete frequencies. At each input frequency, we calculated the total field value for each minute using the INTERMAGNET protocol for both the simulated proton (Overhauser and non-Overhauser with no Gaussian filter) and fluxgate (19 point Gaussian filter) magnetometers, as well as their mean hourly values, ΔF and $\sigma_{\Delta F}$. Figures 6.3.5a and 6.3.5b show a summation of these sinusoids with periods spanning from 10^3 to 10^{-2} s resulting from the typical INTERMAGNET protocol.

The difference in total field values (ΔF) for both magnetometers at each frequency over the one-hour interval was essentially zero despite spikes arising from aliasing effects (i.e. the input waveform was always sampled at the same phase). Much more interesting was the systematic deviation in $\sigma_{\Delta F}$ in the frequency range from approximately 10 Hz to 10 mHz (periods of 0.1 s to 100 s). The signal was about 65% of the amplitude of the input signal over a large swath of this frequency range. A similar result was obtained with a non-Overhauser effect proton magnetometer except the frequency range diminished by a factor of 20 (red line in Figure 6.3.5b). Hence, the models revealed no signal in ΔF, but did show observable signals in $\sigma_{\Delta F}$, meaning that $\sigma_{\Delta F}$ depends solely on temporal differences in data acquisition and processing and not on magnetometer working principles, noise from the magnetometers, etc.

6.3.3.2 *Verification and Application of the Method*

To explain the daily variation of ~0.06 nT in the $\sigma_{\Delta F}$ Chambon-la-Forêt data (Figure 6.3.1) with our model using an Overhauser effect proton magnetometer, a sinusoidal signal with an amplitude of 0.06 nT / 0.65 = 0.09 nT during daytime within the 10 mHz to 10 Hz frequency range is sufficient. Such a small signal would be hardly detectable with direct readings from the magnetometers.

A test of the $\sigma_{\Delta F}$ method from the year 2000 with a selection of five observatories demonstrated when the method did or did not work (Figure 6.3.6). The ΔF data from Nacengyk (NCK, Hungary) showed that on 25 August, ΔF was offset by ~1 nT from all other days (Figure 6.3.6a). Including this data in the analysis resulted in no diurnal $\sigma_{\Delta F}$ signal; whereas recomputing $\sigma_{\Delta F}$ after omitting the day of offset ΔF values revealed a diurnal $\sigma_{\Delta F}$ signal with a maximum near solar noon (Figure 6.3.6b). The Niemeck (NGK, Germany) observatory had no diurnal

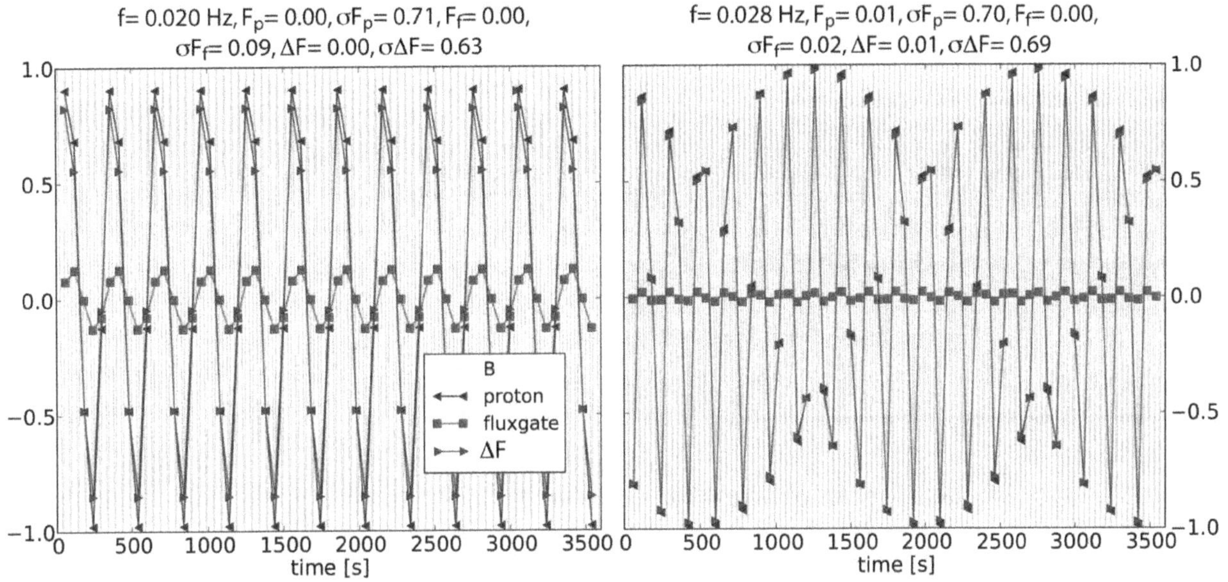

Figure 6.3.4 Examples of numerical simulations for input frequencies of 0.020 Hz (left) and 0.028 Hz (right) over a 1 hour period. All numerical values are relative to the amplitude (±1) of the artificial input magnetic field (B, light gray), which is a sinusoidal waveform consisting of 1000 data points per second over a 1 hour duration. One minute instrument readings are calculated based on the Intermagnet protocol for non-filtered, non-Overhauser effect proton (blue triangles) and Gaussian-filtered fluxgate (red squares) magnetometers. Observed measurement oscillations are due to aliasing effects and are smaller for the fluxgate readings due to the Gaussian filter. Green triangles show the scalar differences of the measurements for each minute (ΔF). Listed above each plot are the hourly mean values of F_p, F_f and ΔF as well as their standard deviations (σF_p, σF_f, $\sigma \Delta F$) (subscripts p = proton, f = fluxgate). This protocol was performed for 10^5 discrete frequencies to produce Figure 6.3.5. (A black-and-white version of this figure appears in some formats. For the colour version, please refer to the plate section.)

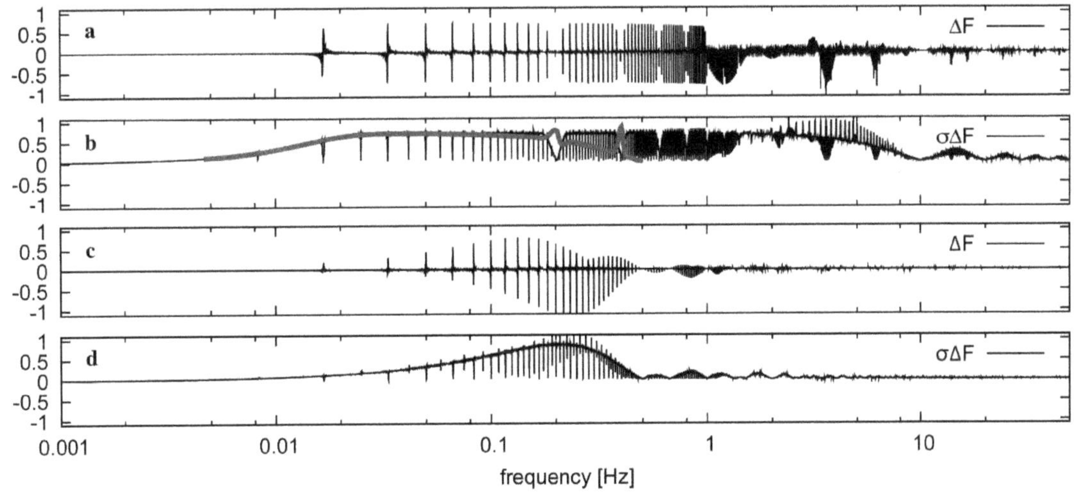

Figure 6.3.5 Numerical model results for an Overhauser effect proton and fluxgate magnetometer summed over 10^5 different input signal frequencies using the INTERMAGNET protocol (see Figure 6.3.4 for examples). The input amplitude is one unit. The (a) ΔF and (b) $\sigma_{\Delta F}$ of the hourly average and standard deviation of minute values are calculated for the Overhauser effect proton and fluxgate magnetometers. For comparison, the red line in $\sigma_{\Delta F}$ (b) was calculated using a non-Overhauser effect proton magnetometer that integrates over 2 s. (c and d) Numerical model results of ΔF and $\sigma_{\Delta F}$ for two non-Overhauser effect proton magnetometers whose measurement times were staggered by 1.5 s (0.5 s of overlap in the measurement).

signal because a Gaussian filter was applied to both the proton and vectorial magnetometer data, which eliminated the $\sigma_{\Delta F}$ signal (Figure 6.3.6c). Boulder (BOU, USA), Honolulu (HON, USA) and Resolute Bay (RES, Canada) all showed diurnal $\sigma_{\Delta F}$ curves (Figures 6.3.6d–6.3.6f); the truncated peak around local noon at the Boulder observatory was unlike any other observatory we tested (Figure 6.3.6d).

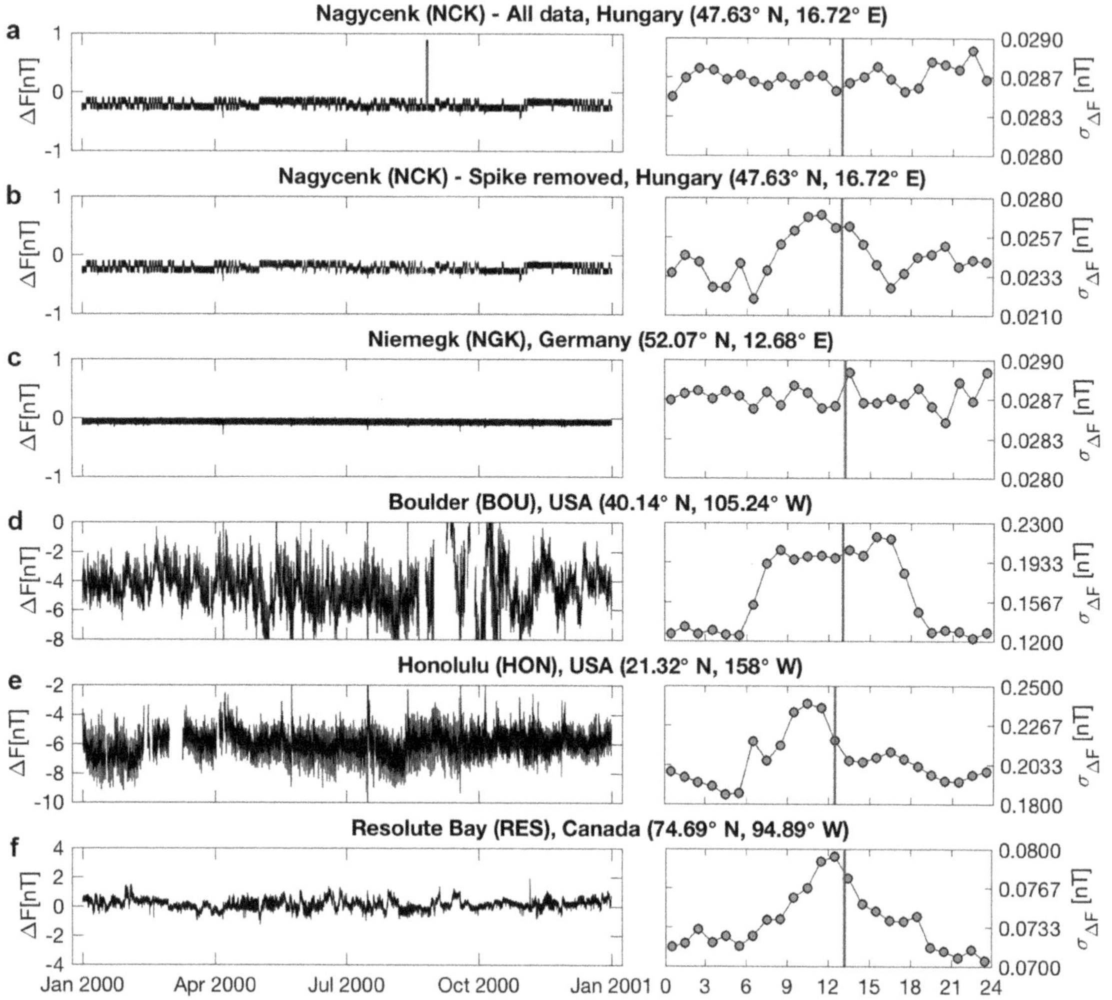

Figure 6.3.6 Comparison of yearly ΔF and stacked daily $\sigma_{\Delta F}$ for 2000 from five observatories (solar noon indicated as the orange line). The top figure (a) from Nagycenk shows that a spike of ~0.85 nT in 1 day's worth of ΔF values obliterates a diurnal $\sigma_{\Delta F}$ variation. Removing the data from this day results in an $\sigma_{\Delta F}$ whose maximum is around local noon (b). (c) Niemegk $\sigma_{\Delta F}$ shows no diurnal variation because this observatory applies Gaussian filters to the data from scalar and vectorial magnetometers. Despite the much greater yearly variability in ΔF, stacked daily $\sigma_{\Delta F}$ for (d) Boulder, (e) Honolulu and (f) Resolute Bay all exhibit maxima around local noon.

We then turned to the 40 Hz data from the Parkfield, California earthquake experiment, which provides publicly accessible, three-component, electromagnetic measurements from two stations (e.g. Kappler et al., 2010). The average power calculated for each hour in the 0.01 to 10 Hz frequency range for the data in April 2004 at both stations show distinct maxima centered at local noon, broadly consistent with $\sigma_{\Delta F}$ (Figure 6.3.7). Hence, when used carefully, our technique allows one to access the 0.01 to 10 Hz frequency range in the INTERMAGNET database, consistent with the modeling shown in Figure 6.3.5.

We next focused on an observatory near the geomagnetic equator Huancayo (HUA, Peru) with a field inclination of approximately −1° and another near the south geomagnetic pole Dumont d'Urville (DRV, Antarctica) with a field inclination of approximately −89°. For HUA, the analysis was applied to a 4 year period from 2002 to 2005; diurnal variations exist for both ΔF and $\sigma_{\Delta F}$ (Figure 6.3.8), again $\sigma_{\Delta F}$ attains a maximum value around local noon. Average ΔF diurnal amplitudes range from 0.6 to 0.8 nT and $\sigma_{\Delta F}$ reaches nearly 2 nT.

Our study from Dumont d'Urville commenced in 1994, after the scalar magnetometer was changed from a Geometrics G618 to a Geomag SM90R (Overhauser), which, together with a change in the grounding and connector cables and the application of a 10 s filter (Jean-Jacques Schott, personal communication, 2011), affected the character of the data. If solar phenomena influenced $\sigma_{\Delta F}$, then one would expect there to be a seasonal variation in $\sigma_{\Delta F}$ in the polar region with a maximum in summer and minimum in winter. This is precisely what was observed (Figure 6.3.9); $\sigma_{\Delta F}$ amplitudes were roughly five times

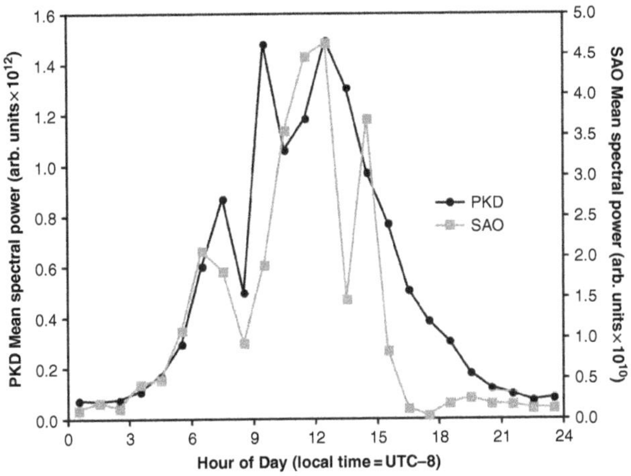

Figure 6.3.7 Hourly mean spectral power in the 0.01 to 10 Hz band from the Parkfield (PKD, Lat.: 35.95°N, Lon.: 120.54°W) and San Andreas Geophysical Observatory (SAO, 36.76°N, Lon.: 121.45° W) electromagnetic stations for the month of April 2004. The US Geological Survey and UC Berkeley Seismological Lab operate the three-component, ElectroMagnetics Inc. (now Schlumberger) BT4 sensors.

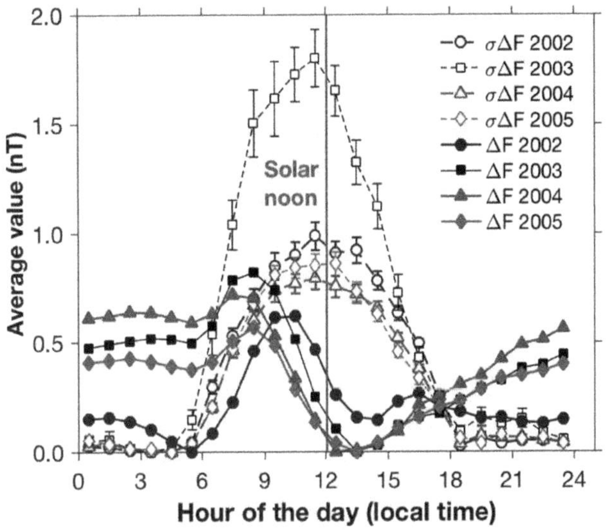

Figure 6.3.8 Daily variation of the hourly averaged value of ΔF and $\sigma_{\Delta F}$ for the Huancayo (HUA, Peru) observatory from 2002 to 2005. The data are scaled so that the minimum value = 0; standard errors for $\sigma_{\Delta F}$ are shown. (A black-and-white version of this figure appears in some formats. For the colour version, please refer to the plate section.)

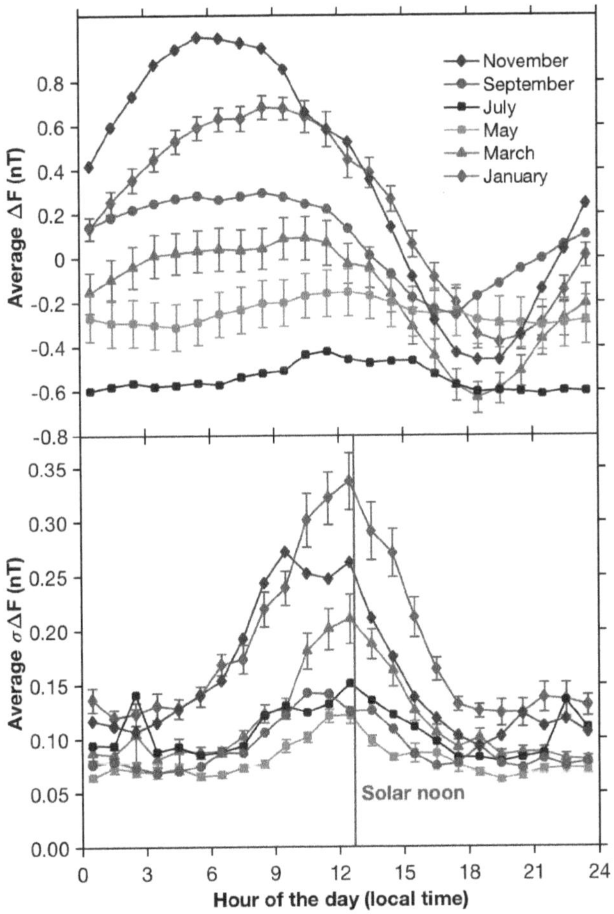

Figure 6.3.9 Daily variation of the hourly averaged value of (a) ΔF and (b) $\sigma_{\Delta F}$ for selected months at the Dumont d'Urville (DRV, Antarctica) observatory in 2004. Standard errors for select months are shown. (A black-and-white version of this figure appears in some formats. For the colour version, please refer to the plate section.)

greater in summer than in winter. However, we note that high-latitude observatories within the auroral oval: Abisko (ABK, Sweden) and Poste de la Baleine (PBQ, Canada) in the Northern Hemisphere and Macquarie Island (MCQ, Australia) in the Southern Hemisphere, yielded maxima during spring and fall for certain years. Besides the data from the 11 observatories presented herein, we also studied $\sigma_{\Delta F}$ from AAE (Ethiopia), BNG (Central African Republic) KOU (French Guiana), FUR (Germany), PPT (Tahiti) and TUC (USA), which all displayed diurnal variations.

We then explored whether a seasonal dependence could be discerned in the Chambon-la-Forêt data by comparing $\sigma_{\Delta F}$ from January and July from 1988 to 2007 (Figure 6.3.10a). Taking the average of all these years for each of the months yielded virtually identical curves (Figure 6.3.10b), despite that the amount of daylight (sunrise to sunset) at CLF lasts 8.8 hours on 15 January and 15.7 hours on 15 July. On the other hand, the diurnal amplitude of F itself for 2000 and 2004 did show a distinct seasonal variation with a maximum in summer and minimum in winter (Figure 6.3.10c). This suggested that the mechanism responsible for generating the current producing the diurnal variation in F was distinct from that which produces the field oscillations seen in $\sigma_{\Delta F}$. Specifically, at this mid-latitude observatory, $\sigma_{\Delta F}$ displayed no seasonal variation, including at the equinoxes, hence independent of the Sq (solar quiet) current system (e.g. Pham Thi Thu et al., 2011).

Figure 6.3.10a suggests that the diurnal amplitude of $\sigma_{\Delta F}$ appeared to change from year to year. To examine whether the long-term trend in $\sigma_{\Delta F}$ contained information related to the solar cycle, for each year from Chambon-la-Forêt, we

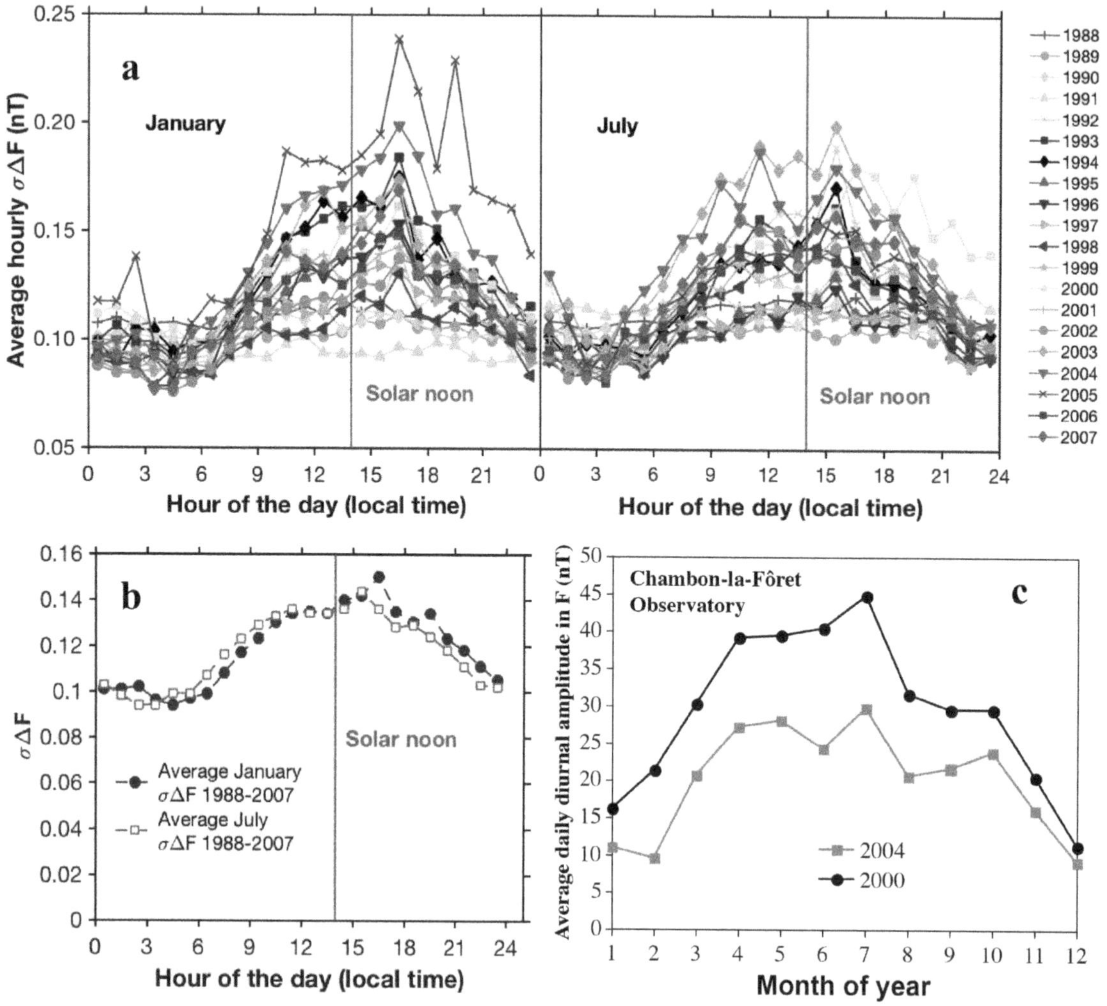

Figure 6.3.10 (a) Daily variation of the hourly averaged value of $\sigma_{\Delta F}$ for January and July for the years 1988 to 2007 for the Chambon-la-Forêt (CLF, France) observatory (January 1999 data not included). (b) Average daily variation of the hourly averaged value of $\sigma_{\Delta F}$ for January and July for the years 1988 to 2007 for the Chambon-la-Forêt (CLF, France) observatory. (c) Average daily diurnal amplitude in total field (*F*) per month in 2000 and 2004 from the Chambon-la-Forêt (CLF, France) observatory. (A black-and-white version of this figure appears in some formats. For the colour version, please refer to the plate section.)

calculated the $\sigma_{\Delta F}$ amplitude in January and July by defining the maximum as the average value between 9:30 and 15:30 (local time) and the minimum as the average value between 21:30 and 3:30 and took the difference between the two. We then plotted these amplitudes, which ranged from 0.003 to 0.081 nT, against the average yearly number of sunspots (Wolf number [*W*]; compiled by SILSO, Royal Observatory of Belgium, Brussels) and observed a seemingly robust correlation albeit phase shifted by about 3 years (Figure 6.3.11a). We did the same for Dumont d'Urville for each month of January since 1994, when the amplitude in $\sigma_{\Delta F}$ was often the highest during the year; the maximum value was taken as the average between 9:45 and 13:45 (local time) and the minimum as the average between 19:45 and 1:45 (Figure 6.3.11b). Again, there existed a phase shift of about three years. To better understand this, we calculated the time derivative of *W*, took its negative ($-dW/dt$), and normalized all data sets to their maximum values (Figure 6.3.11c). A cross-correlation function analysis (Figure 6.3.11d) of the data in Figure 6.3.11c showed that the maxima clearly coincided with the waning phase of the sunspot cycle, which is largely consistent with pulsation activity measured at the mid-latitude Nagycenk observatory (Zieger, 1991). This further suggested that $\sigma_{\Delta F}$ from ground-based magnetic observatory data contained information originating from the solar wind.

6.3.4 Discussion and Conclusions

Solar wind forcing on the magnetopause from wave-particle instabilities or from fluctuations in solar wind pressure generates compressional mode waves that enter the magnetosphere on the dayside and drive measurable magnetic field fluctuations at the Earth's surface (see Menk and Waters, 2013, for a review). Several studies have shown correlations

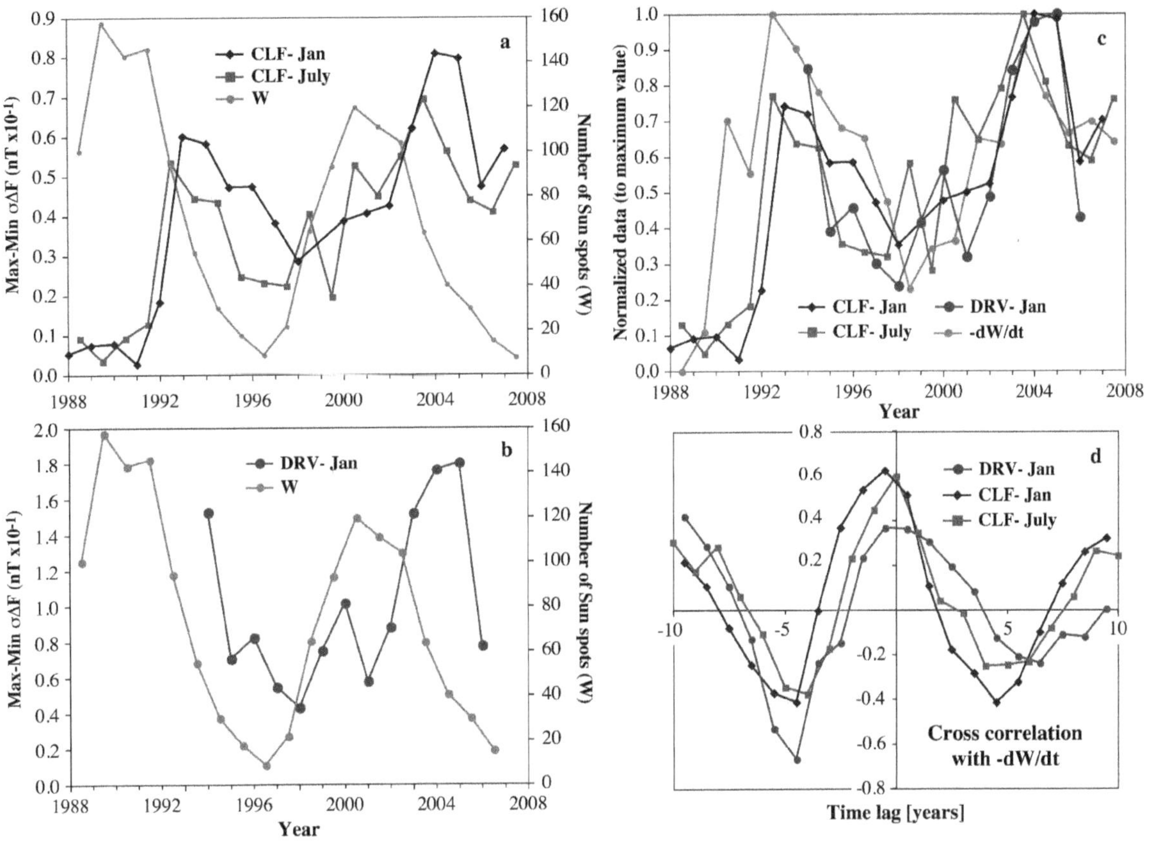

Figure 6.3.11 Difference between the average maximum and minimum variation of $\sigma_{\Delta F}$ compared against sunspot (Wolf, W) number. (a) January and July from 1988 to 2007 for the Chambon-la-Forêt (CLF) observatory. (b) January from 1994 to 2006 for the Dumont d'Urville (DRV) observatory. (c) Same observatory data as in (a) and (b), normalized by the maximum value, and plotted with the negative rate of change of the yearly averaged W ($-dW/dt$), also normalized to the maximum value. (d) Cross-correlation functions of the average maximum-minimum variation of $\sigma_{\Delta F}$ with the negative rate of change of the sunspot (Wolf, W) number. Cross-correlation functions were calculated with Python's xcorr algorithm using time series shifted to zero mean and normalized to a maximum value of one. That the cross correlations maximize for all three time series around zero time lag demonstrates that the observed signals are in phase with the negative rate of change of the sunspot number. (A black-and-white version of this figure appears in some formats. For the colour version, please refer to the plate section.)

between field line resonance and broadband compressional spectra (e.g. Engebretson et al., 1987; Potemra et al., 1989; Lee and Lysak, 1991; Goldstein et al., 2005; Clausen et al., 2009). We therefore tested $\sigma_{\Delta F}$ by correlating it against solar wind parameters (Figure 6.3.12). A good correlation existed using the IMF vector σ_B/B where B is the average magnetic field magnitude for one year and σ_B is its standard deviation; σ_B/B signifies the level of magnetic field turbulence (*Zimbardo* et al., 2010). Comparing $\sigma_{\Delta F}$ from CLF for January with σ_B/B yields a Pearson correlation coefficient of $r = 0.49$ and a 2 tailed p-value of 0.027. On the other hand, both cone angle (the angle between the IMF direction and the Sun–Earth line) and solar wind velocity (V_{SW}) are associated with the penetration of Alfvén waves into the magnetosphere (e.g. Allan et al., 1986; Allan and Poulter, 1992; Clausen et al., 2009; Dimmock et al., 2014). Indeed we found a robust correlation between $\sigma_{\Delta F}$ and V_{SW} from DRV for the January amplitudes (Pearson statistics $r = 0.75$; $p = 0.003$).

Determining which solar wind parameter correlated with ground-based observations from each observatory simultaneously was challenging. For example, Dimmock et al. (2014) found that for southward IMF conditions during the twenty-third solar cycle, magnetic turbulence was enhanced in the magnetosheath. They noted that high solar wind velocity associated with high-speed solar wind streams correlated with increased bow shock and magnetosheath turbulence. Although we found a good correlation with the IMF B_z component during the twenty-third solar cycle, this was not so for the twenty-second, as the twenty-second and twenty-third solar cycles differ with respect to the behavior of the IMF and the solar wind speed. Because a complete solar cycle is 22 years, it seems that both the solar wind speed and the direction of the IMF influence the propagation of the solar wind magnetic turbulence into the magnetosphere down to the ground. As solar wind speed and the IMF vary with time and are not necessarily correlated, we considered a turbulence penetration parameter

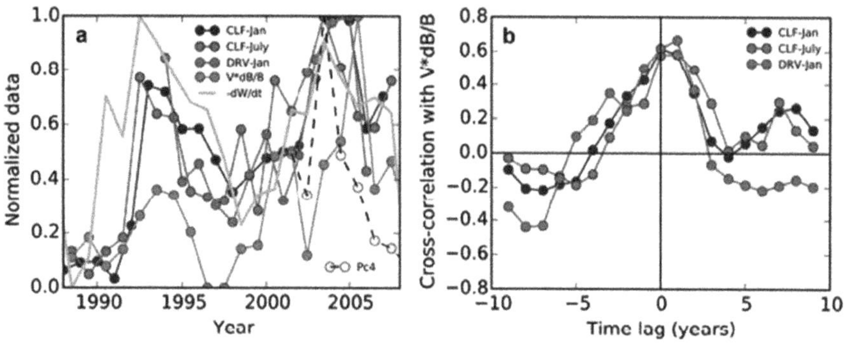

Figure 6.3.12 (a) Normalized proxy of magnetic field fluctuations at the ground ($\sigma_{\Delta F}$, Figure 6.3.11) compared to the turbulence penetration parameter V_{ay}•σ_B/B (light blue, 1 min data from the OMNI database: http://omniweb.gsfc.nasa.gov) and wave power of the Pc4 magnetic field fluctuations in the inner magnetosphere (black dashed line, from the MAARBLE project, www.maarble.eu/, with data obtainable at the Cluster Science Archive, www.cosmos.esa.int/web/csa). Scaling uses the normalization $X' = \frac{X-X_{\min}}{X_{\max}-X_{\min}}$. (b) Cross-correlation functions between $\sigma_{\Delta F}$ and the turbulence penetration parameter V•σ_B/B. Cross-correlation is done to a time series shifted by subtracting the mean values of the variables and normalizing to a maximum value of one. (A black-and-white version of this figure appears in some formats. For the colour version, please refer to the plate section.)

V_{ay}•σ_B/B where V_{ay} is the average yearly velocity. This parameter significantly correlated with magnetic fluctuations at the ground (CLF-Jan: $r = 0.61, p = 0.0039$; CLF-July: $r = 0.57, p = 0.0078$; DRV-Jan: $r = 0.61$, $p = 0.024$; Figure 6.3.12b), which implies that magnetic field variations at the ground were influenced by the solar wind, with the strongest variations occurring during the declining phase of the solar cycle that are associated with high-speed streams and turbulent Alfvénic waves (Potapov, 2013). Figure 6.3.12a also plots the wave power of the Pc4 magnetic field fluctuations in the inner magnetosphere (inside 10 Earth radii) where fluctuations peaked during the declining phase of the solar cycle. This demonstrated that the strength of magnetic field fluctuations in the solar wind correlated with magnetic field fluctuations in the inner magnetosphere, which were detected at the ground.

Allan and Poulter (1992) showed how dayside boundary effects with mHz frequency hydromagnetic waves can excite field line resonances. Variations across the plasmapause also help generate cavity resonances (Allan et al., 1986) that produce field oscillations measurable on the ground. The typical way to detect field line resonances (pulsations) on the ground has been through the analysis of magnetic data limited to very specific time windows when the pulsations were noticeable (e.g. Kepko and Spence, 2003; Lee and Lysak, 1991; Menk et al., 2000, 2003). Fast Fourier transforms were typically applied to the data to derive the frequency content (e.g. Mathie et al., 1999; Plaschke et al., 2009). However, one must justify the use of Fourier transforms and other techniques to construct power spectra, and insure that stationary conditions were satisfied (e.g. Goldstein et al., 2005). Equally or even more likely is that turbulence in the solar wind and generation of compressional or Alfvén waves are produced in an extremely dynamic magnetosphere that could conceivably produce a continuum of frequencies (Troitskaya and Gul'elmi, 1967; Poulter and Nielsen, 1982; Allan and Poulter, 1992; Walker, 2002). As $\sigma_{\Delta F}$ defined maxima at solar noon across most latitudes and times, it seems likely a continuum in frequencies had been recorded as opposed to specific frequencies produced by cavity modes, for example. It seems plausible that field line oscillations can be excited by the Alfvén waves embedded into the solar wind. They are associated with high-speed streams (Desai et al., 2008), which are most frequently observed during the waning phase of the solar cycle (Xystouris et al., 2014). Figure 6.3.12 suggests that the solar wind turbulence leads to field line oscillations in the inner magnetosphere that are observable with $\sigma_{\Delta F}$.

When considering only the diurnal $\sigma_{\Delta F}$ record, one could conclude that the effect comes from a response of the ionosphere under direct Sun exposure from solar irradiance (X-ray, ultra-violet, total luminosity, etc.). This appears unlikely, however, if the diurnal and long-term variations in $\sigma_{\Delta F}$ have a common cause since solar irradiance varies in phase with the solar cycle (Willson and Hudson, 1991; Qian et al., 2012). On the other hand, the descending phase of the solar cycle corresponds to the time when coronal holes, which are areas of open field lines dominated by large-amplitude Alfvén waves, migrate from the pole to the equator (Tsurutani et al., 1994; Gonzalez et al., 1999). Thus turbulence in the solar wind, and specifically large-amplitude, non-linear Alfvén waves within bodies of co-rotating streams associated with coronal holes (Tsurutani et al., 1995, 2006), likely contribute to producing the oscillations, which eventually reach the ground (Menk and Waters, 2013, and references therein).

Turbulence is pervasive in space plasmas (e.g. Chen et al., 2010; Horbury et al., 2011). Anisotropy of the magnetic field turbulence (the fluctuating part) can be described by the distribution of magnetic energy in wave vector space, by the magnetic variances in the different directions, etc. The anisotropy changes when passing from the solar wind to the magnetosheath (Perri et al., 2009). It seems plausible that the turbulence, which is always present and greatest during the waning phase of the solar cycle, enters the magnetosphere on the dayside (Figure 6.3.12). Just how the solar wind transmits turbulence at the shock front depends on the cone angle – the relative orientation between the IMF direction with respect to the magnetopause (Allan and Poulter, 1992; Plaschke et al., 2009; Baumjohann and Treumann, 2012). Although we do not know the exact transfer mechanism, Figure 6.3.12 shows that the solar wind turbulence leads to field line oscillations in the inner magnetosphere that are observable with $\sigma_{\Delta F}$.

One needs to keep in mind that each observatory has its own protocol that might deviate from INTERMAGNET specifications (e.g. vector magnetometers that measure at 1 Hz and use a Gaussian filter for 1 s data to generate 1 min data). This would alter the frequency spectra, which would mostly affect the relative amplitude of $\sigma_{\Delta F}$ but not the long-term (diurnal, annual, etc.) waveform. On the contrary, it seems highly unlikely that instruments function differently at different times of the day (Figure 6.3.1) and under different geomagnetic field conditions (Figure 6.3.2). Moreover, the effects shown herein can be observed at most INTERMAGNET observatories irrespective of geographic location, and the oscillations observed on the ground correlate with solar wind parameters (Figure 6.3.12). This makes the $\sigma_{\Delta F}$ method useful because INTERMAGNET data span a wide geographic and temporal range and are instantly accessible. In addition, $\sigma_{\Delta F}$ is independent of spectral methods thereby avoiding inconsistencies with stationarity assumptions.

Without the need for new instrumentation, ground-based magnetic observatories can measure different frequencies using the $\sigma_{\Delta F}$ method without changing the operational frequency of the magnetometers themselves. For example, if two non-Overhauser proton magnetometers operate 1.5 s out of phase, so that there is a 0.5 s overlap in the measurements between the two, the sampled frequency spectra is restricted to periods of roughly 2 and 20 s (Figures 6.3.5c and 6.3.5d). Similar studies using 1 s data, available since 2009, will also measure a different frequency content. Finally, we recommend that individual observatories supply raw, unfiltered data to INTERMAGNET, which can provide routines that filter the data for those requiring it.

Acknowledgement

We thank Vincent Courtillot for making this study possible, Yuri Khotyaintsev for useful discussions concerning Pc4 data and Rudolf Treumann for helpful advice. François Truong participated in the early stages of this work at the CLF observatory. We appreciate the helpful comments from an anonymous reviewer and the editorial handling by Pieter Kotze. All ground-based observatory data presented in this paper can be downloaded at: http://www.intermagnet.org/index-eng.php, the 40 Hz data can be found at: http://seismo.berkeley.edu/bdsn/em.overview.html and the interplanetary magnetic field vector and the solar wind speed are taken from the OMNI data base: http://omniweb.gsfc.nasa.gov/form/omni_min.html. Data from the MAARBLE project (http://www.maarble.eu) can be found at http://www.cosmos.esa.int/web/csa.

References

Allan, W., Poulter, E. M. & White, S. P. (1986). Hydromagnetic wave coupling in the magnetosphere: Plasmapause effects on impulse-excited resonances. *Planet. Space Sci.*, 34, 1189–1200.

Allan, W. & Poulter, E. M. (1992). ULF waves-their relationship to the structure of the Earth's magnetosphere. *Rep. Prog. Phys.*, 55, 533–98.

Auster, H. U. (2008). How to measure Earth's magnetic field. *Phys. Today*, 61, 76–77.

Baumjohann, W. & Treumann, R. A. (2012). *Basic Space Plasma Physics*, rev. edn., Imperial College Press, London.

Campbell, W. H. (1997). *Introduction to Geomagnetic Fields*, Cambridge Univ. Press, Cambridge.

Chen, C. H. K., Horbury, T. S., Schekochihin, A. A., Wicks, R. T., Alexandrova, O. & Mitchell, J. (2010). Anisotropy of solar wind turbulence between ion and electron scales. *Phys. Rev. Lett.*, 104, 255002, doi: 10.1103/PhysRevLett.104.255002.

Chi, P.J. & Russell, C. T. (2001). On two methods using magnetometer-array data for studying magnetic pulsations. *Terrestrial Atmos. Ocean Sci.*, 12, 649–62.

Clausen, L. B. N., Yeoman, T. K., Fear, R. C., Behlke, R., Lucek, E. A. & Engebretson, M. J. (2009). First simultaneous measurements of waves generated at the bow shock in the solar wind, the magnetosphere and on the ground. *Ann. Geophys.*, 27, 357–71.

Crooker, N. U., Feynman, J. & Gosling, J. T. (1977). On the high correlation between long-term averages of solar wind speed and geomagnetic activity. *J. Geophys. Res.*, 82, 1933–7.

Desai, M. I., Mason, G. M., Müller-Mellin, R., Korth, A., Mall, U., Dwyer, J. R. & von Rosenvinge, T. T. (2008). The spatial distribution of upstream ion events from the Earth's bow shock measured by ACE, Wind, and STEREO. *J. Geophys. Res.*, 113, A08103, doi: 10.1029/2007JA012909.

Dimmock, A. P., Nykyri, K. & Pulkkinen, T. I. (2014). A statistical study of magnetic field fluctuations in the dayside magnetosheath and their dependence on upstream solar wind conditions. *J. Geophys. Res.*, A119, 6231–48, doi: 10.1002/2014JA020009.

Engebretson, M. J., Zanetti, L. J., Potemra, T. A., Baumjohann, W., Lühr, H. & Acuña, M. H. (1987). Simultaneous observation of Pc 3–4 pulsations in the solar wind and in the Earth's magnetosphere. *J. Geophys. Res.*, 92, 10,053–62.

Feynman, J. (1982). Geomagnetic and solar wind cycles, 1900–1975. *J. Geophys. Res.*, 87, 6153–62.

Garrett, H. B., Dessler, A. J. & Hill, T. W. (1974). Influence of solar wind variability on geomagnetic activity. *J. Geophys. Res.*, 79, 4603–10.

Goldstein, M. L., Eastwood, J. P., Treumann, R. A., Lucek, E. A., Pickett, J. & Décréau, P. (2005). The near-Earth solar wind. *Space Sci. Rev.*, 118, 7–39.

Gonzalez, W. D., Tsurutani, B. T. & Clúa de Gonzalez, A. L. (1999). Interplanetary origin of geomagnetic storms. *Space Sci. Rev.*, 88, 529–62.

Heilig, B., Lühr, H. & Rother, M. (2007). Comprehensive study of ULF upstream waves observed in the topside ionosphere by CHAMP and on the ground. *Ann. Geophys.*, 25, 737–54.

Hirshberg, J. & Colburn, D. S. (1969). Interplanetary field and geomagnetic variations – a unified view. *Planet. Space Sci.*, 17, 1183–1205.

Horbury, T. S., Wicks, R. T. & Chen, C. H. K. (2011). Anisotropy in space plasma turbulence: solar wind observations. *Space Sci. Rev.*, 172, 325–42, doi: 10.1007/s11214-011-9821-9.

Kappler, K. N., Morrison, H. F. & Egbert, G. D. (2010). Long-term monitoring of ULF electromagnetic fields at Parkfield, California. *J. Geophys. Res.*, 115, B04406, doi: 10.1029/2009JB006421.

Keiling, A. D., Lee, H. & Nakariakov, V. (eds.) (2016). *Low-Frequency Waves in Space Plasmas*. American Geophysical Union, Washington, DC.

Kepko, L., Spence, H. E. & Singer, H. J. (2002). ULF waves in the solar wind as direct drivers of magnetospheric pulsations. *Geophys. Res. Lett.*, 29, 1197, doi: 10.1029/2001GL014405.

Kepko, L. & Spence, H. E. (2003). Observations of discrete, global magnetospheric oscillations directly driven by solar wind density variations. *J. Geophys. Res.*, 108, 1257, doi: 10.1029/2002JA009676.

Kivelson, M. G., Cao, M., McPherron, R. L. & Walker, R. J. (1997). A possible signature of magnetic cavity mode oscillations in ISEE spacecraft observations. *J. Geomagn. Geoelectr.*, 49, 1079–98.

Lee, D. H. & Lysak, R. L. (1991). Monochromatic ULF wave excitation in the dipole magnetosphere. *J. Geophys. Res.*, A96, 5811–17.

Le Mouël, J. L., Kossobokov, V. & Courtillot, V. (2005). On long-term variations of simple geomagnetic indices and slow changes in magnetospheric currents: The emergence of anthropogenic global warming after 1990? *Earth Planet. Sci. Lett.*, 232, 273–86, doi: 10.1016/j.epsl.2004.07.046.

Mathie, R. A., Menk, F. W., Mann, I. R. & Orr, D. (1999). Discrete field line resonances and the Alfvén continuum in the outer magnetosphere. *Geophys. Res. Lett.*, 26, 659–62.

Mayaud, P. N. (1980). Derivation, Meaning, and Use of Geomagnetic Indices. Geophysical Monograph 22. American Geophysical Union, Washington, DC.

McPherron, R. L. (2005). Magnetic pulsations: Their sources and relation to solar wind and geomagnetic activity. *Surv. Geophys.*, 26, 545–92.

Menk, F. W. (1988). Spectral structure of mid-latitude Pc3-4 geomagnetic pulsations. *J. Geomagn. Geoelectr.*, 40, 33–61.

Menk, F. W., Waters, C. L. & Fraser, B. J. (2000). Field line resonances and waveguide modes at low latitudes 1. Observations. *J. Geophys. Res.*, A105, 7747–61.

Menk, F. W., Yeoman, T. K., Wright, D. M., Lester, M. & Honary, F. (2003). High-latitude observations of impulse-driven ULF pulsations in the ionosphere and on the ground. *Ann. Geophys.*, 21, 559–76.

Menk, F. W. & Waters, C. L. (2013). *Magnetoseismology*. Wiley-VCH, Weinheim, Germany.

Qian, L. Y., Burns, A. G., Solomon, S. C. & Chamberlin, P. C. (2012). Solar flare impacts on ionospheric electrodyamics. *Geophys. Res. Lett.*, 39, L06101, doi: 10.1029/2012GL051102.

Perri, S., Yordanova, E., Carbone, V., Veltri, P., Sorriso-Valvo, L., Bruno, R. & André, M. (2009). Magnetic turbulence in space plasmas: Scale-dependent effects of anisotropy, *J. Geophys. Res.*, 114, A02102, doi: 10.1029/2008JA013491.

Pham Thi Thu, H., Amory-Mazaudier, C. & Le Huy, M. (2011). Sq field characteristics at Phu Thuy, Vietnam, during solar cycle 23: Comparisons with Sq field in other longitude sectors. *Ann. Geophys.*, 29, 1–17, doi: 10.5194/angeo-29-1-2011.

Plaschke, F., Glassmeier, K. H., Sibeck, D. G., Auster, H. U., Constantinescu, O. D., Angelopoulos, V. & Magnes, W. (2009). Magnetopause surface oscillation frequencies at different solar wind conditions. *Ann. Geophys.*, 27, 4521–32.

Poulter, E. M. & Nielsen, E. (1982). The hydromagnetic oscillation of individual shells of the geomagnetic field. *J. Geophys. Res.*, 87, 10432–8.

Potapov, A. S. (2013). ULF wave activity in high-speed streams of the solar wind: Impact on the magnetosphere. *J. Geophys. Res.*, 118, 6465–77, doi: 10.1002/2013JA019119.

Potemra, T. A., Lühr, H., Zanetti, L. J., Takahashi, K., Erlandson, R. E., Marklund, G. T., Block, L. P., Blomberg, L. G. & Lepping, R. P. (1989). Multisatellite and ground-based observations of transient ULF waves. *J. Geophys. Res.*, 94, 2543–54.

Russell, C. T. & Fleming, B. K. (1976). Magnetic pulsations as a probe of the interplanetary magnetic field: A test of the Borok B Index. *J. Geophys. Res.*, 81, 5882–6.

Sakurai, T., Tonegawa, Y., Kitagawa, T., Yumoto, K., Yamamoto, T., Kokubun, S., Mukai, T. & Tsuruda, K. (1999). Dayside magnetopause Pc 3 and Pc 5 ULF waves observed by the GEOTAIL Satellite. *Earth Planets Space*, 51, 965–78.

Southwood, D. J. (1974). Some features of field line resonances in the magnetosphere. *Planet. Space Sci.*, 22, 483–491.

St Louis, B. (2008). *INTERMAGNET Technical Reference Manual*, version 4.4. Internal publication. INTERMAGNET, Edinburgh.

Takahashi, K. & McPherron, R. L. (1982). Harmonic structure of Pc 3–4 pulsations. *J. Geophys. Res.*, A87, 1504–16.

Takahashi, K., Chi, P. J., Denton, R. E. & Lysak, R. L. (eds.) (2006). *Magnetospheric ULF Waves: Synthesis and New Directions*, American Geophysical Union, Washington, DC.

Takahashi, K. & Ukhorskiy, A. Y. (2008). Timing analysis of the relationship between solar wind parameters and geosynchronous Pc5 amplitude. *J. Geophys. Res.*, A113, A12204, doi: 10.1029/2008JA013327.

Tamao, T. (1964). The structure of three-dimensional hydromagnetic waves in a uniform cold plasma. *J. Geomagn. Geoelectr.*, 16, 89–114.

Troitskaya, V. A. & Gul'elmi, A. V. (1967). Geomagnetic micropulsations and diagnostics of the magnetosphere. *Space Sci. Rev.*, 7, 689–768.

Tsurutani, B. T., Ho, C. M., Smith, E. J., Neugebauer, M., Goldstein, B. E., Mok, J. S., Arballo, J. K., Balogh, A., Southwood, D. J. & Feldman, W. C. (1994). The relationship between interplanetary discontinuities and Alfvén waves: Ulysses observations. *Geophys. Res. Lett.*, 21, 2267–70.

Tsurutani, B. T., Gonzalez, W. D., Gonzalez, A. L. C., Tang, F., Arballo, J. K. & Okada, M. (1995). Interplanetary origin of geomagnetic activity in the declining phase of the solar cycle. *J. Geophys. Res.*, A100, 33.

Tsurutani, B.T., Gonzalez, W. D., Gonzalez, A. L. C., Guarnieri, F. L., Gopalswamy, N., Grande, M., Kamide, Y., Kasahara, Y., Lu, G., Mann, I., McPherron, R., Soraas, F. & Vasyliunas, V. (2006). Corotating solar wind streams and recurrent geomagnetic activity: A review. *J. Geophys. Res.*, A111, A07S01, doi: 10.1029/2005JA011273.

Vellante, M., Lühr, H., Zhang, T. L., Wesztergom, V., Villante, U., De Lauretis, M., Piancatelli, A., Rother, M., Schwingenschuh, K., Koren, W. & Magnes, W. (2004). Ground/satellite signatures of field line resonance: A test of theoretical predictions. *J. Geophys. Res.*, A109, A06210, doi: 10.1029/2004JA010392.

Verö, J., Zieger, B., Szendröi, J., Vellante, M., Stresgtik, J., Lühr, H., Best, A., Körmendi, A., Lichtenberger, J., Ménesi, T., Bencze, P., Märcz, F. & Wesztergom, V. (2000). Connections between whistlers and pulsation activity. *Ann. Geophys.*, *18*, 866–73.

Walker, D. M. (2002). Excitation of field line resonances by MHD waves originating in the solar wind. *J. Geophys. Res.*, A107, 1481, doi: 10.1029/2001JA009188.

Weimer, D. R., Clauer, C. R., Engebretson, M. J., Hansen, T. L., Gleisner, H., Mann, I. & Yumoto, K. (2010). Statistical maps of geomagnetic perturbations as a function of the interplanetary magnetic field. *J. Geophys. Res.*, 115, A10320, doi: 10.1029/2010JA015540.

Wilcox, J. M., Schatten, K. H. & Ness, N. F. (1967). Influence of interplanetary magnetic field and plasma on geomagnetic activity during quiet-Sun conditions. *J. Geophys. Res.*, 72, 19–26.

Willson, R. C. & Hudson, H. S. (1991). The Sun's luminosity over a complete solar cycle. *Nature*, 351, 42–4, doi: 10.1038/351042a0.

Xystouris, G., Sigala, E. & Mavromichalaki, H. (2014). A Complete Catalogue of High-Speed Solar Wind Streams during Solar Cycle 23. *Solar Phys.*, 289, 995–1012, doi: 10.1007/s11207-013-0355-z.

Zieger, B. (1991). Long-term variations in pulsation activity and their relationship to solar wind velocity, geomagnetic activity and F2 region electron density. *J. Geophys. Res.*, 96, 21,115–23.

Zimbardo, G., Greco, A., Sorriso-Valvo, L., Perri, S., Vörös, Z., Aburjania, G., Chargazia, K. & Alexandrova, O. (2010). Magnetic turbulence in the geospace environment. *Space Sci. Rev.*, *156*, 89–134, doi: 10.1007/s11214-010-9692-5.

7 Modelling Internal and External Geomagnetic Fields Using Satellite Data

*Ingo Wardinski and Erwan Thébault**

7.1 Introduction

Since the availability of precise geomagnetic measurements from satellites such as Ørsted and CHAMP, which flew between 1999 and 2010, a new era began that changed our understanding of the different sources which contribute to Earth's magnetic field. This progress mutually developed with advancements in geomagnetic field modelling and new theoretical derivations in describing the various source processes. Prominent achievements and findings of this era are the identification of magnetic signatures caused by tidal oceanic currents (Tyler et al., 2003; Maus and Kuvshinov, 2004; Sabaka et al., 2016), the detection of an accelerating jet motion in the liquid outer core (Livermore et al., 2017), the assembly of a highly resolved crustal magnetic field model by integration of airborne and satellite data (Maus et al., 2008; Lesur et al., 2016) and implementations of data assimilation techniques to infer the dynamical regime within Earth's core through numerical dynamo simulations (e.g. Fournier et al., 2010, 2011).

As an example, Figure 7.1.1 shows the lithospheric field described by the MF7 model (Maus et al., 2008). This model is often used as a benchmark for lithospheric field models because of its high spatial resolution.

A further example is shown in Figure 7.1.2. The amplitude of motional induced magnetic fields related to M2-tides reaches at satellite altitude only a few nT. Although this is a very weak signal, it can be retrieved from geomagnetic satellite data because of its periodic nature. Fainter magnetic signals of the O1 and N2 tides can also be resolved in satellite data

Recently, Finlay et al. (2017) provided a detailed description of methods that are currently used to construct mathematical models of the internal magnetic field and to handle magnetospheric and ionospheric signatures in spherical harmonics. Magnetic fields generated by magnetospheric and ionospheric current systems are the main obstacles to resolve short spatial and temporal variations of the internal magnetic field, as they show a variability ranging from large to small spatial scales and from a few months to centuries. Yet, a better description of the short temporal and spatial variability of the internal field would provide insights on the processes generating the internal field and the dynamical regime in Earth's liquid outer core. These findings may also contribute to the discussion about the structure of Earth's outer core. For example, seismic observations tend to indicate the presence of a stratified layer below the core surface (Helffrich and Kaneshima, 2010; Kaneshima and Matsuzawa, 2015), while geomagnetic evidence suggests that, if there is a stratification, it must be weak (Whaler, 1980; Lesur et al., 2015).

Lithospheric magnetic field modelling is another area of geomagnetic research which largely benefits from satellite-based observations. Magnetic signatures of the lithospheric field are small compared to other sources of the main field, and therefore dedicated studies of the lithospheric magnetic field require a thorough analysis of the data to extract this signal. Thébault et al. (2017) discuss commonly used methods to identify magnetic signatures of the lithosphere, such as filtering and line levelling techniques. To this end, Thébault et al. (2017) showed that such empirical corrections can be effective in correcting data for effects of the disturbance field. But these approaches remain limited because of their simplistic assumptions about the source geometry and statistical properties of the disturbance field that is dynamic and poorly described.

The description of both magnetic field sources, internal and lithospheric, is pledged to our limited knowledge of the disturbance field generated by ionospheric and magnetospheric current systems. Studies by Maus et al. (2006) and Lesur et al. (2013) provide valuable approaches towards the incorporation of information on spatial structures of disturbance fields and may also reduce uncertainties related to these magnetic fields. ESA's Swarm satellite mission, which is a constellation of three satellites in low Earth orbit launched in November 2013, offers with its high-precision and high-resolution measurements the essential requirements

* We thank V. Lesur, who provided a very valuable review and helped to clarify aspects of the manuscript. I. W. was supported by the French ANR project ANR-13-BS05-0012.

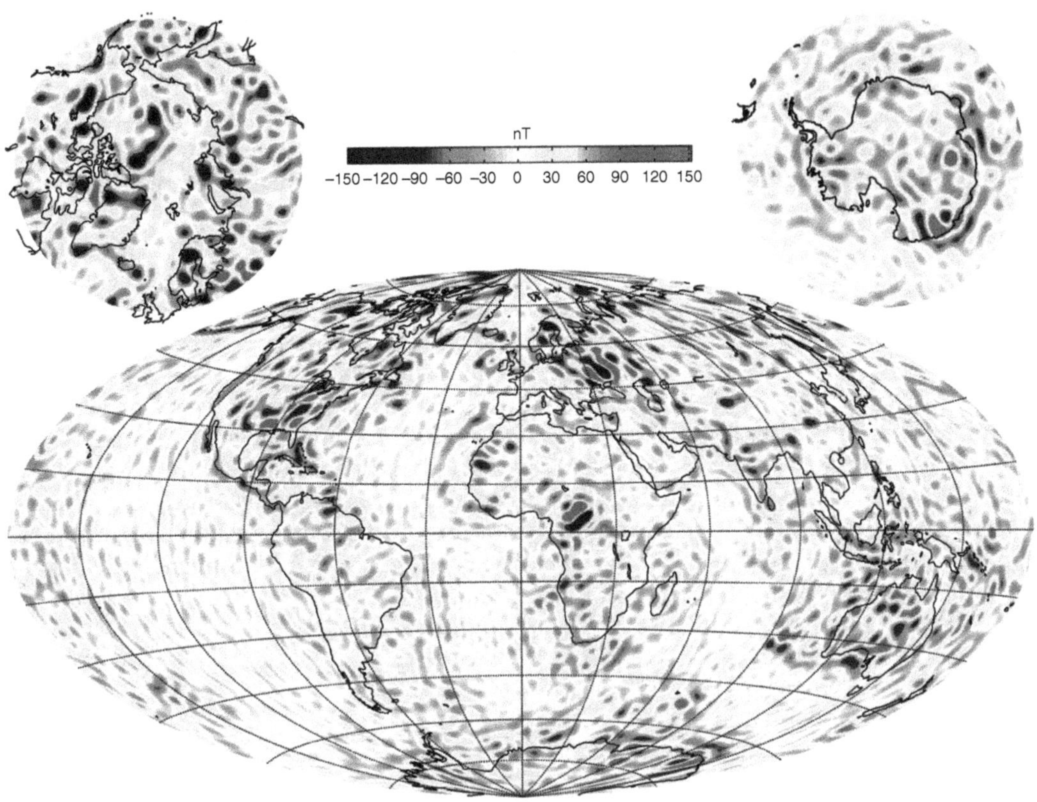

Figure 7.1.1 The vertical component of the lithospheric field at Earth's surface as it derives from MF7 (Maus et al., 2008). Lithospheric anomalies range from ±150 nT. Most prominent are the Bangui anomaly in central Africa, the Kursk anomaly in Russia and the Appalachian anomaly in North America. (A black-and-white version of this figure appears in some formats. For the colour version, please refer to the plate section.)

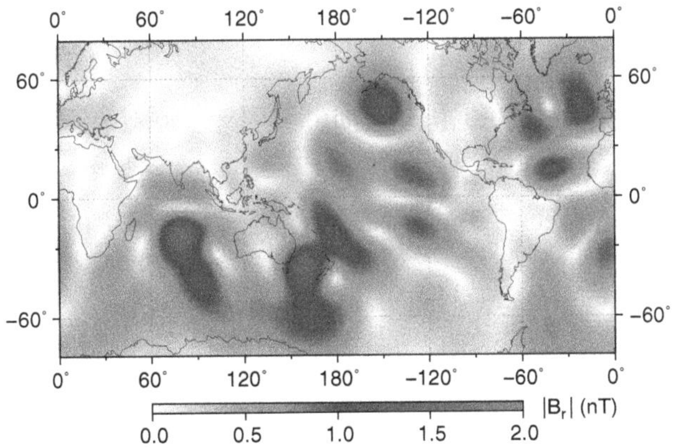

Figure 7.1.2 The amplitude of the vertical tidal magnetic field related to the oceanic M2-tide as it is retrieved from geomagnetic data of the CHAMP mission during the period 2004 to 2009 (Wardinski and Lesur, 2016). (A black-and-white version of this figure appears in some formats. For the colour version, please refer to the plate section.)

to study various aspects of Earth's magnetic field. Further progress in our understanding of these aspects will be facilitated by the combination of geomagnetic satellite missions with ground-based measurements of the Earth magnetic field, allowing a better sampling of the field in space and in time, a condition necessary to disentangle the various magnetic field sources. Such progress will be alongside the refinement of geomagnetic field modelling techniques, their theoretical background and a better statistical description of the different sources.

7.2 Theoretical Background

In this section, we recall the basic formalism of geomagnetic field modelling from measurements made by satellite missions. Generally, it is assumed that measurements of the geomagnetic field, on the ground or at low Earth orbiting satellite altitudes, are made away from magnetic sources lying in the ionosphere and the magnetosphere. When measurements are made within source regions of external magnetic fields, empirical models of the magnetic field generation in these source regions and detailed knowledge of their electric current density are needed to separate the measured field into external and internal parts (see Olsen et al., 2010a, for a detailed description).

7.2.1 Separation of External and Internal Fields Using Spherical Harmonics

In Earth's vicinity, electric current systems are mainly located in the ring current region at ~6 Earth radii, in the magnetopause and in the ionosphere between 80 and 300 km altitude. The majority of geomagnetic satellite measurements are taken in a region at about 450 km altitude. This region can be assumed to be free of magnetic sources. Following the theory of Gauss (1839), the magnetic induction B, which is commonly identified as the magnetic field, is

$$\nabla \cdot B = 0, \tag{7.2.1}$$

which allows us to represent the magnetic field B as gradient of scalar potential V:

$$B = -\nabla V, \tag{7.2.2}$$

where ∇ is the nabla operator and V the potential satisfying the Laplace equation

$$\nabla^2 V = 0. \tag{7.2.3}$$

Equation (7.2.2) is a partial differential equation, which in spherical geometry has a solution that can be expanded in terms of spherical harmonics:

$$V = a\sum_{l=1}^{L}\sum_{m=0}^{l} \left\{(g_l^m \cos(m\phi) + h_l^m \sin(m\phi))\left(\frac{a}{r}\right)^{l+1} P_l^m(\cos\theta) + (q_l^m \cos(m\phi) + s_l^m \sin(m\phi))\left(\frac{r}{a}\right)^{l} P_l^m(\cos\theta)\right\}, \tag{7.2.4}$$

where a is the mean planetary radius, r the radial distance from the planetary centre, θ the colatitude and ϕ the longitude. The $P_l^m(\cos\theta)$ are the Schmidt semi-normalised associated Legendre functions, where l are the degrees and m the order; see Langel (1987) for a detailed derivation. The parameters $\{g_l^m, h_l^m\}$ in the potential equation (7.2.4) are representing the magnetic field that is generated by processes internal to Earth's surface. The parameters scale with $(a/r)^{l+1}$, and their contribution to the potential must vanish at infinity. The terms $\{q_l^m, s_l^m\}$ represent the external magnetic field and remain finite within the region of interest. The two sets of parameters are called the Gauss coefficients of Earth's magnetic field; they are in general varying in time. To this end, the potential can be represented by

$$V = V_{\text{ext}} + V_{\text{int}}, \tag{7.2.5}$$

where it can be demonstrated that external and internal potential fields, $V_{\text{ext}}, V_{\text{int}}$, are mathematically separable.

Different sources contribute to the internal magnetic potential field: the main field, which is generated by dynamo action in Earth's core, and a field due to the magnetisation of the lithosphere. Furthermore, processes lead to internal, but weaker, magnetic signatures, for instance, steady and tidal oceanic currents and currents in the mantle induced by time-varying external fields.

The external potential field is tightly related to processes in the magnetic environment of the Earth. Such processes are driven by solar activity and cause an electrical current flow on the boundary of the magnetosphere across the tail of the magnetosphere, the Chapman–Ferraro current system (Chapman and Ferraro, 1940, 1941). Furthermore, a substantial external magnetic field is generated by an equatorial ring current system inside the magnetosphere. Field-aligned currents and Birkeland currents, which connect the magnetosphere with the ionosphere, generate magnetic fields that are most clearly detectable in polar regions. Current systems in the ionosphere also contribute to an external magnetic field, which is mostly seen as an internal magnetic field, as its sources are below satellite altitudes. Ionospheric currents are highly variable and driven by the solar wind and by interactions of the interplanetary magnetic field and Earth's magnetosphere. Significant magnetic fields are generated by auroral and equatorial electrojets. All these contributions to Earth's magnetic field vary on different temporal scales, but overlap temporally and spatially.

In a spherical coordinate system the magnetic field components are given by

$$B_\phi = \frac{1}{\sin\theta}\sum_{l=1}^{\infty}\sum_{m=0}^{l} m\left\{(g_l^m \sin(m\phi) - h_l^m \cos(m\phi))\left(\frac{a}{r}\right)^{l+2} + (q_l^m \sin(m\phi) - s_l^m \cos(m\phi))\left(\frac{r}{a}\right)^{l-1}\right\} P_l^m(\cos\theta),$$

$$B_\theta = -\sum_{l=1}^{\infty}\sum_{m=0}^{l}\left\{(g_l^m \cos(m\phi) + h_l^m \sin(m\phi))\left(\frac{a}{r}\right)^{l+2} + (q_l^m \cos(m\phi) + s_l^m \sin(m\phi))\left(\frac{r}{a}\right)^{l-1}\right\}\frac{P_l^m(\cos\theta)}{d\theta},$$

$$B_r = \sum_{l=1}^{\infty}\sum_{m=0}^{l}\left\{(l+1)(g_l^m \cos(m\phi) + h_l^m \sin(m\phi))\left(\frac{a}{r}\right)^{l+2} - l(q_l^m \cos(m\phi) + s_l^m \sin(m\phi))\left(\frac{r}{a}\right)^{l-1}\right\}P_l^m(\cos\theta). \tag{7.2.6}$$

For a mathematical accurate separation and modelling of external and internal fields, one needs to measure at least one horizontal intensity B_θ or B_ϕ and the vertical intensity B_r of the field over a closed spherical surface(Langel, 1987).

7.2.2 First-Order Separation of Internal and External Fields

For single satellite mission, the geographical data coverage in space and in time does not allow a sufficient accurate separation of internal and external fields on short temporal scales, i.e. on times scales shorter than a few months in case of the CHAMP satellite mission. This is because the CHAMP satellite mission needed about three months to cover all local times. The geographical data coverage offered by the European Swarm satellite mission, with the availability of measurements along three different orbits, allows a better determination of large spatial scale and short term external and internal field components. However, simplifications and approximations are often needed to capture even the large spatial scales. As pointed out by Langel (1987), it is possible to perform a first-order separation between the large-scale magnetospheric external field (generated by the ring current) and large-scale internal fields by measurements acquired along a single pole to pole orbit. In such case, the maximum temporal resolution is typically constrained by the duration of a satellite half orbit. Assuming that the external magnetospheric field is large scale and symmetric with respect to Earth's dipole axis, then on a sphere $r = a$, the field components in spherical harmonics are simply determined by

$$\begin{aligned}
B_\phi &= \frac{1}{\sin\theta}\sum_{l=1}^{\infty}\sum_{m=0}^{l} m\Big\{(g_l^m \sin(m\phi) - h_l^m \cos(m\phi)) \\
&\quad + (q_l^m \sin(m\phi) - s_l^m \cos(m\phi))\Big\} P_l^m(\cos\theta), \\
B_\theta &= -\sum_{l=1}^{\infty}\sum_{m=0}^{l}\Big\{(g_l^m \cos(m\phi) + h_l^m \sin(m\phi)) \\
&\quad + (q_l^m \cos(m\phi) + s_l^m \sin(m\phi))\Big\} \frac{P_l^m(\cos\theta)}{d\theta}, \\
B_r &= \sum_{l=1}^{\infty}\sum_{m=0}^{l}\Big\{(l+1)(g_l^m \cos(m\phi) + h_l^m \sin(m\phi)) \\
&\quad - l(q_l^m \cos(m\phi) + s_l^m \sin(m\phi))\Big\} P_l^m(\cos\theta),
\end{aligned} \tag{7.2.7}$$

where any radial dependence of the field components vanishes. This idea was followed by Schwarte et al. (2003) in their attempt to model external and internal degree 1 fields and to provide an improved index for the magnetospheric ring current. In this approach, measurements along a single satellite track are used to derive a low-degree spherical harmonic model of degree and order 1. They set

$$\begin{aligned}
B_\theta &= (g_1^0 + q_1^0)\sin\theta \\
&\quad + (g_1^1 + q_1^1)\cos\theta\cos\phi \\
&\quad + (h_1^1 + s_1^1)\cos\theta\sin\phi, \\
B_r &= (2g_1^0 + q_1^0)\cos(\theta) \\
&\quad + (2g_1^1 + q_1^1)\sin\theta\cos\phi \\
&\quad + (2h_1^1 + s_1^1)\sin\theta\sin\phi,
\end{aligned} \tag{7.2.8}$$

where these equations can be simplified to

$$\begin{aligned}
B_\theta &= d_1\sin\theta + d_2\cos\theta\cos\phi + d_3\cos\theta\sin\phi, \\
B_r &= c_1\cos\theta + c_2\sin\theta\cos\phi + c_3\sin\theta\sin\phi,
\end{aligned} \tag{7.2.9}$$

and further provide

$$\begin{aligned}
g_n^m &= \frac{c_n^m + nd_n^m}{2n+1} \\
q_n^m &= \frac{(n+1)d_n^m - c_n^m}{2n+1}.
\end{aligned} \tag{7.2.10}$$

The spherical harmonic coefficients can be found by a simple inversion and provide a good estimate of the magnetic amplitude of the ring current, and compare well with the observatory-based RC-Index (Olsen et al., 2005). However, it is only suitable for low-degree problems, i.e. $n,m < 2$. The reason comes from the orthogonal properties of the spherical harmonics. Spherical harmonic terms of higher degree are linearly dependent on each other for a single satellite track.

7.2.3 Alternative Approaches

Alternative approaches make use of spatially localised basis functions (Chambodut et al., 2005; Stockmann et al., 2009; Beggan et al., 2013), which may be limited in separating external and internal sources, but with the advantage of being less affected by uneven data distributions. This is of particular interest when ground-based data from geomagnetic observatories and repeat stations or airborne data are used to derive magnetic field models. A review of spatially localised functions used in the context of lithospheric magnetic field modelling can be found in Schott and Thébault (2011).

A new approach was developed recently by Holschneider et al. (2016) to separate external and internal geomagnetic fields, where a Bayesian modelling approach has been proposed to distinguish geomagnetic sources based on their spatial correlation structure. The correlation of internal and external magnetic fields can be determined by making assumptions about the spectral distribution of the field energy (McLeod, 1986, 1996). While Baratchart and Gerhards (2017) raised the criticism that this approach may not uniquely separate crustal and core magnetic field contributions for some given data distribution, Lesur et al. (2017) was able to show that a robust separation of inter-annually magnetospheric and core magnetic field contribution from geomagnetic observatory records from 1957 to 2014 can be obtained. Recently, this approach has also been applied to geomagnetic satellite data (Baerenzung, 2018).

Since the launch of the Swarm satellite mission in 2013 gradient estimates of magnetic field components as data are also considered in magnetic field modelling schemes (Kotsiaros et al., 2015; Olsen et al., 2015). The usage of gradient data has been suggested by Khokhlov et al. (1999). The advantage of invoking gradient estimates is the reduction of short temporal and large spatial scale external field contribution, as the gradients are usually computed over short distances between the co-orbiting Swarm satellites A and C flying side by side. This approach is useful to filter out some of the rapid and short variations, but it does not allow extracting the external field variations.

7.2.4 Data Selection and Treatment of External Magnetic Field Signals

The approaches described above all rely on some assumptions regarding the shape and strength of the magnetic fields, particularly of external origin that is assumed to be large scale. However, there is a large variety of contributions that are more complex in space and/or localised. For example, the diurnal solar-quiet (Sq) variation and polar ionospheric fields or the magnetospheric field under perturbed conditions do not satisfy these assumptions. Therefore, data selection plays a significant role in disentangling signals of external and internal magnetic field sources and follows the assumption that selecting data under specific external field conditions may reduce the contribution of external field sources to the magnetic measurements. Data selection schemes are all applied in recent satellite magnetic field models (see GRIMM and CHAOS model series). There is no unique protocol, as it depends on the parameterisation of the models and the used data set. There is, however, a sort of consensus view that

- selecting night-side measurements significantly reduces the contamination of internal field models by the ionospheric Sq field,
- CHAOS model series are considering vector measurements at mid to low latitudes and scalar data otherwise, whereas in the GRIMM model series vector measurements are also considered at high latitude to minimise the contamination by external ionospheric fields in polar regions,
- applying a selection of the measurements during less perturbed magnetic times allows a better separation between the large-scale external fields and the internal field contributions.

Data from quiet times are selected using a maximum threshold criterion for geomagnetic indices, such as the *Dst* index (Sugiura and Kamei, 1991), the vector magnetic disturbance (VMD) index (Thomson and Lesur, 2007), the ring current (RC) index (Olsen et al., 2005, 2014) or the *Kp* index. Additionally to data selection schemes, external magnetic field variations are often parameterised as functions of magnetic activity indexes (Langel and Estes, 1985; Sabaka et al., 2002, 2004). Most of the modelling techniques also apply some geographical weighting, where for instance data in polar regions receive less weight, as data from these regions are notoriously interfered by magnetic signals of field-aligned currents and electrical discharges in the plasma environment of cusp regions.

Both data selection and disturbance field parameterisation ensure external field signatures due to magnetic storms and magnetic daylight activity to be reduced in geomagnetic satellite measurements. As a consequence of these data treatments, external fields are considered to be the main contribution of the model residuals. Interestingly, when the data selection differs from one model to the other, it can be shown that this has a strong influence on the final model, even when the model parameterisation are similar. This was illustrated during the construction of the International Geomagnetic Reference Field after the evaluation and the analysis of the candidate models built by independent teams models (Thébault et al., 2015). Individual candidate models differ on average by about 5 nT and sometimes more than 10 nT at Earth's surface. This shows that even if data selection allows minimising the contributions of small-scale external fields, significant contributions of external fields still contaminate internal field models. This is because one side effect of data selection is to stack systematic external field contributions. Quiet and nighttime data selection is very efficient to reduce the data noise. However, quiet-time external fields can be considered to average out of some typical periods (daily, seasonal, yearly, etc.) and data selection over systematic local times and/or geomagnetic conditions produces a bias difficult to estimate and to quantify (Sabaka et al., 2002, 2015, 2016), and may raise the uncertainty of core field estimates. Other sources completely overlap with each other, and no data selection approaches can address this issue. For example, the large-scale induced lithospheric and oceanic magnetic

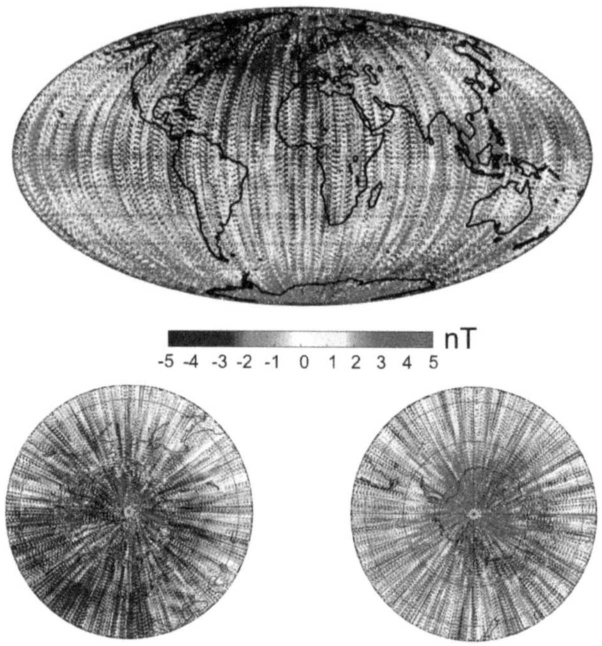

Figure 7.2.1 Residuals of the vertical down component of Swarm A and C measurements. Residuals are computed at satellite altitude. Signatures of the ionospheric field of ±4 nT are faintly visible along the equator in a 20° wide belt. (A black-and-white version of this figure appears in some formats. For the colour version, please refer to the plate section.)

fields range up to 10 nT at Earth's surface and overlap with the contributions of the Earth's core internal field.

Figure 7.2.1 illustrates the success of field separation by a state-of-the-art field modelling approach. The figure shows measurement residuals of Swarm A and C after the removal of a magnetic field model that describes the internal core field, the magnetospheric magnetic field and the static lithospheric field up to spherical harmonic degree 40. The residual field should basically consists of the ionospheric magnetic field and small-scale lithospheric signatures above spherical harmonic degree 40, with an expected rms value of ~3 nT at satellite altitude (460 km). Likely, the ionospheric magnetic field contributes to features along the equator. But, the residuals are dominated by intense and large-scale features in the Northern and Southern Hemispheres with a rms value of ~5 nT. Thébault et al. (2017) argued this may indicate errors in correcting for the large-scale internal and/or external field models, which are caused by variations of the magnetospheric field at timescales shorter that those represented by the magnetospheric and core field models. These variations lead to offsets of (sometimes) 10 nT between the measurements taken by the satellite at adjacent passes. Moreover, these offsets introduce an anisotropic noise particularly strong in horizontal components of lithospheric magnetic field models (Cain et al., 1989; Lesur et al., 2013; Maus et al., 2008).

7.3 Derivation of Magnetic Field Models

Data selected in this manner are used to derive models of the geomagnetic field. The model parameters (the Gauss coefficients) are estimated by a least-squares fit to data selected for a given time interval and at satellite altitude, where the least-squares fit is formulated as an inverse problem. Details of the model inversion are presented in this section.

The theory of solving linear inverse problems has been outlined by Menke (1989) and Gubbins (2004). Solutions of a least-squares fit are generally sought by solving a linear inverse problem, which can be stated as

$$\mathrm{y} = \mathbf{A}\mathrm{m} + e, \tag{7.3.1}$$

where $\mathbf{A}$ is a design matrix mapping the model vector m to the data vector y and e is the error vector. $\mathbf{A}$ is constructed for the model parameters until the truncation level L. The least-squares solution is given by

$$\mathrm{m} = (\mathbf{A}^T\mathbf{A})^{-1}\mathbf{A}^T\mathrm{y}. \tag{7.3.2}$$

The model solution m exists since the matrix $\mathbf{A}$ has full column rank; m provides the best fit to the data, where the differences between the data vector y and the recalculated data $\mathbf{A}$m, measured by the L2-norm,

$$S = \mathrm{Py} - \mathbf{A}\mathrm{mP}^2, \tag{7.3.3}$$

are minimised (e.g. Strang, 1988). When using Equation (7.3.2), the error in Equation (7.3.1) is assumed to be additive, independent and normally distributed with its expectation $E(e) = 0$ and standard deviation $\sigma^2 = 1$. If the errors have a non-uniform variance, then Equation (7.3.2) reads

$$\mathrm{m} = (\mathbf{A}^T C_e^{-1}\mathbf{A})^{-1}\mathbf{A}^T C_e^{-1}\mathrm{y}, \tag{7.3.4}$$

where C_e^{-1} is the inverse of the error covariance matrix. Least-squares problems can be separated into two classes. First, the problem can be qualified as over-determined, where we have sufficient information to uniquely estimate all model parameters. In this case, $\mathbf{A}^T\mathbf{A}$ is regular, and its inverse exists.

Secondly, the problem is under-determined. In this case, we do not have enough information to evaluate all model parameters, and $\mathbf{A}\mathbf{A}^T$ is regular. Most common are cases with contradictory information and inaccessible model parameters due to the lack of information. For these cases, $\mathbf{A}\mathbf{A}^T$ and $\mathbf{A}^T\mathbf{A}$ cannot be numerically inverted even if their inverses formally may exist. Such cases are termed ill-posed and are characterised by a large condition number $k(\mathbf{A})$ of the inverse problem

$$k(\mathbf{A}) = \left|\frac{\lambda_{\max}}{\lambda_{\min}}\right|, \tag{7.3.5}$$

with $\lambda_{\max}$ and $\lambda_{\min}$ being the largest and smallest eigenvalue, respectively, of the matrix $\mathbf{A}$. The condition number quantifies how much the model vector changes with respect to the data vector. For ill-posed problems, small changes of the data vector lead to large changes in the model estimation; the problem becomes sensitive to data error, where data errors will be magnified in the model estimation. In case of an ill-posed inverse problem, evaluations of the least-squares solution Equation (7.3.4) by any numerical technique, such as Gaussian elimination, single value or Cholesky decomposition gives solutions m and covariances of the model parameter, which may be oscillating and large. To circumvent this, Levenberg (1944) introduced the derivation of damped least-squares solution of ill-posed problems. By applying such damping, the solution becomes

$$\mathrm{m} = (\mathbf{A}^T C_e^{-1}\mathbf{A} + C_D^{-1})^{-1}\mathbf{A}^T C_e^{-1}\mathrm{y}, \tag{7.3.6}$$

where the matrix C_D is associated with the damping.

Generally, the derivation of geomagnetic field models by a least-squares fit belongs to the second category, and the algebraic structure of the inverse problem implies that without prior information, the model vector m is not limited or determined by the observation Backus (1988). In the next section we will discuss regularisation schemes, the application of prior information and their effect on the solutions.

Table 7.3.1 *Smoothing Norms for Magnetic Field Inversion after Holme Bloxham (1996)*

	Integral to minimise	Norm function $f(c,l)$	Comments
(a)	$\oint \mathbf{B}^2 dS \mid r = c$	$(l+1)\left(\frac{a}{c}\right)^{(2l+4)}$	Mean square field intensity
(b)	$\oint \mathrm{B_r}^2 dS \mid r = c$	$\frac{(l+1)^2}{2l+1}\left(\frac{a}{c}\right)^{(2l+4)}$	Mean square radial field
(c)	$\oint (\nabla_h \mathrm{B_r})^2 dS \mid r = c$	$\frac{l(l+1)^3}{2l+1}\left(\frac{a}{c}\right)^{(2l+6)}$	Mean square of horizontal derivative of radial field

7.4 The Need for Prior Information

As mentioned, without prior information, the model vector m is not usefully limited by the data, unless the model is a linear combination of the data. Therefore, a model which is not a linear combination of the data vector requires prior information about the true model itself. In spite of infinitely many solutions, a good approximation to the true model can be obtained if some prior knowledge about the nature of the true solution is accessible and may come from other geophysical disciplines, for instance seismic observations. These additional constraints are usually concerned with the degree of sparsity or smoothness of the true solution. Jackson (1979) and Backus (1988b) discuss two forms of prior constraints to be applied in geophysical inverse problems: hard and soft prior bounds.

One way to incorporate prior information is to assume the unknown model parameter to be Gaussian random variables with a mean m_0 and a (prior) covariance matrix. This method is known as stochastic inversion (Franklin, 1970; Jackson, 1979), and other methods, such as Bayesian inferences, rely on similar assumptions. Stochastic inversion has been applied to geomagnetic field modelling first by Gubbins (1983) and is still widely used in this branch of geophysical inverse problems. In the framework of stochastic inversion, the model solution reads

$$\mathrm{m} = (\mathbf{A}^T C_e^{-1} A + C_m^{-1})^{-1} \mathbf{A}^T C_e^{-1} \mathrm{y}. \tag{7.4.1}$$

This is the same as Equation (7.3.6), but now with C_m the prior model covariance matrix. A Gaussian probability distribution with non-zero mean and variance is assigned to the model parameter, m, where the probability distribution is regarded as a soft bound on the model parameters. A non-zero probability exists that m violates these bounds.

So-called hard bounds can be deduced from our understanding of physical processes within the Earth and their observations and are provided in the form

$$F(\mathrm{m}) \leq c, \tag{7.4.2}$$

where $F(\mathrm{m})$ is a functional of the model and c a constant. One such source of prior information that may provide a hard bound on $\mathrm{B_r}$ at the CMB is the heat flux bound. It is assumed that the observed heat flux at Earth's surface can provide a direct constraint on the ohmic heating due to poloidal currents related to the toroidal magnetic field at the core surface. This assumption can be justified by neglecting heat sources in the mantle and crust and that within the core, the toroidal magnetic field is significantly larger than the poloidal magnetic field. Therefore, the ohmic heating inside the core due to toroidal currents related the poloidal magnetic field may be minor. Gubbins' (1975) analysis of the heat flux bound leads to the condition

$$F(\mathbf{B}_r) = \sum_{l=1}^{\infty} \frac{(l+1)(2l+1)(2l+3)}{l} \sum_{m=0}^{l} [(g_l^m)^2 + (h_l^m)^2] \leq Q \tag{7.4.3}$$

for $r = c$, where Q is the observed heat flow out of the Earth surface, i.e. $Q = 3 \times 10^{13}$W.

The condition (7.4.3) assigns to the Gauss coefficients in the model space the probability distribution with a zero mean and the variance

$$\langle |g_l^m(c)|^2 \rangle = Ql(l+1)^{-1}(2l+1)^{-1}(2l+3)^{-1}. \tag{7.4.4}$$

This limits each Gauss coefficient only by requiring it not to produce by itself more ohmic heat in the core than is observed in the surface heat flow. Based on this, Gubbins and Bloxham (1985) derived that near-surface observations of the geomagnetic field cannot resolve the spatial structure of the core field above degree and order 20, whereas Backus (1988a) showed that data uncertainties of 1 nT can resolve Gauss coefficients at the core surface with degree $l \leq 32$ unless the ohmic heating rate exceeds the geothermal flow at Earth's surface. These findings suggest some kind of insufficiency of the applied prior. As pointed out by Backus (1988b), the application of the heat flux norm may lead to suspect results, mainly because of two reasons: (1) as (in his terminology) softening of a hard prior bound, i.e. the derivation of a probability distribution from a hard prior bound adds considerable information to the true model that is not implied by the norm; (2) the application of this norm explicitly assumes, the model coefficients to be independent Gaussian random variables, but this may not be correct. There might be non-linear dependencies between model parameters, which ab initio have been disregarded by the Gaussian assumption and are not mapped by linear correlation measures such as the mean and covariance matrices. Moreover, a further problem may arise from the fact that Ohmic dissipation acts strongly on intermediate to small spatial scales (Calderwood et al., 2003) and therefore may not provide a practical constraint for the large-scale magnetic field at the core surface. Instead, prior information should be supplied in form of a probability distribution, based on objective evidence, a theoretical basis or a hypothesis to be tested (Backus, 1988b).

Holme and Bloxham (1996) discussed a suite of prior constraints and possible regularisation conditions are listed in Table 7.3.1 and can be considered as soft prior bounds. These prior constraints result from hypothesis testing. They control the spatial complexity of the field morphology at the core surface ($r = c$), by setting the diagonal elements of C_m in Equation (7.4.1). Also, a similar formalism can be set up to control the temporal complexity of magnetic field models. These are employed in temporal continuous models like archaeomagnetic models (Korte and Constable, 2005), historical and decadal models (Jackson et al., 2000; Wardinski and Lesur, 2012) and satellite field models (Lesur et al., 2010; Finlay et al., 2016). A further development is facilitated by the use of data assimilation techniques. Data assimilation, central to numerical weather prediction, became of particular interest in geomagnetism over the last decade (Fournier et al., 2010; Aubert and Fournier, 2011). Data assimilation techniques can be seen as strong regularisation techniques in the sense that the derived models are supposed to follow a trajectory in space and in time that belongs to an ensemble of Earth-like dynamo simulations. This approach and its applications in the framework of geomagnetism are described in Fournier et al. (2011, 2015).

7.5 Model Diagnostics

A fundamental difficulty in inverse problems is that an infinite set of models adequately describe our finite set of observations (Backus, 1970; Stark, 1992). Furthermore, most techniques for solving geophysical inverse problems, such as stochastic inversion, Bayesian inference and maximum entropy methods, solve the problem of construction (fit to the data) but do not characterise the non-uniqueness of the inverse problem. There are several possibilities to assess the solutions of linear inverse problems. In the following, we present two analyses to diagnose results of the inversion. Furthermore, we discuss effects of model biases due to model truncation.

7.5.1 Model Comparison

At first, the robustness of magnetic field model features can be assessed by comparing results of different modelling techniques. Here, we provide a brief comparison between CHAOS-6 (Finlay et al., 2016) and GRIMM-3 (Lesur et al., 2011). These two models are based on similar data sets for the period 2001 to 2010 but differ in their data selection criteria (see Section 7.2.4). Another important difference is the prior information that is applied to derive these models. In the derivation of CHAOS-6, temporal and spatial regularisation are omitted, whereas GRIMM-3 put a constraint on the third time derivative of the magnetic field.

Figure 7.5.1 shows the difference of the radial component of the internal main field at Earth's surface derived from CHAOS-6 and GRIMM-3. Significant differences appear close to the auroral oval in the north polar region, and a similar feature may be visible in the Southern Hemisphere. Clearly, this indicates an imprint of external fields into the internal field descriptions, possibly due to field-aligned electrical currents in the cusp regions.

The secular acceleration (SA) differences of these models are shown in Figure 7.5.2. Large discrepancies in the radial secular acceleration appear to be aligned along the equator.

Perhaps this is an indication of the Backus effect and may arise from using different data selection criteria in polar regions (Stern et al., 1980; Holme and Bloxham, 1995). The Backus effect is found to be most significant in sectorial

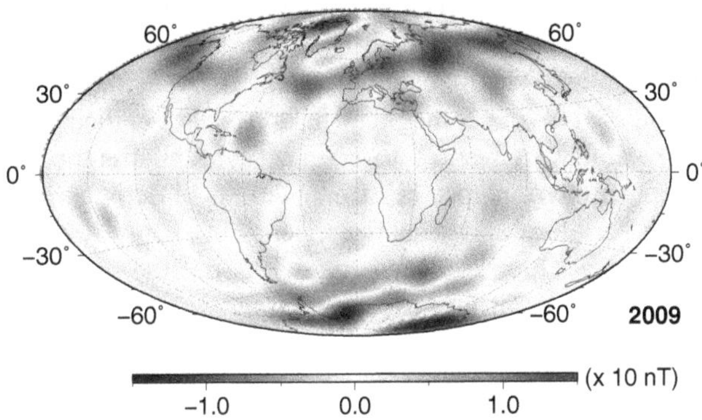

Figure 7.5.1 Differences of the vertical component of the main field between CHAOS-6 and GRIMM-3 at Earth's surface for the epoch 2009. The range of the differences is ±16nT. (A black-and-white version of this figure appears in some formats. For the colour version, please refer to the plate section.)

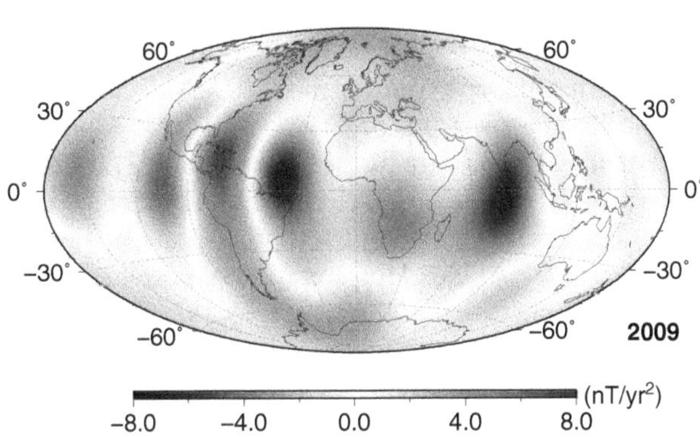

Figure 7.5.2 Differences of the vertical component of the secular acceleration between CHAOS-6 and GRIMM-3 at Earth's surface for the epoch 2009. (A black-and-white version of this figure appears in some formats. For the colour version, please refer to the plate section.)

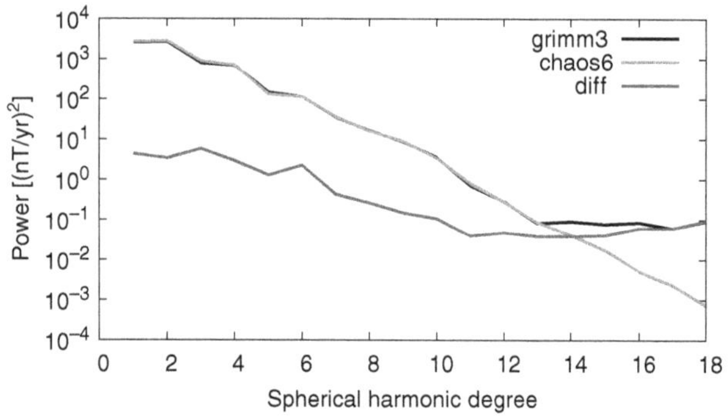

Figure 7.5.3 Power spectra of the two magnetic field models and their difference at Earth's surface.

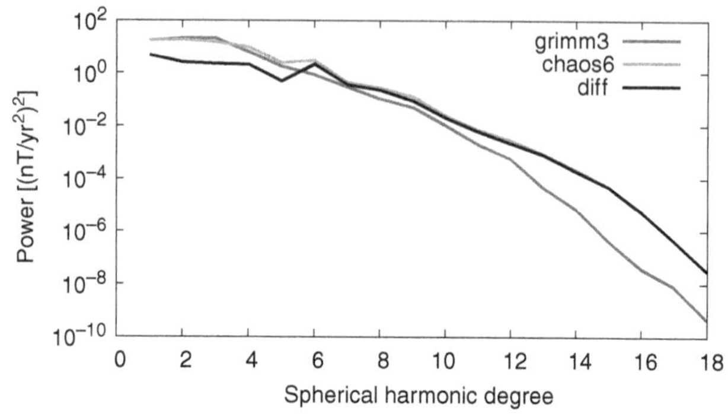

Figure 7.5.4 Power spectra of the two magnetic field models and their difference at Earth's surface.

terms of a spherical harmonic expansion, i.e. g_l^m, h_l^m with l = m (Stern et al., 1980), which leads to enhanced uncertainties of the radial field in the equatorial region. However, there are numerous other effects that may lead to such noise patterns, e.g. Maus Haak (2003) finds similar features caused by magnetic field annihilators.

Further assessment can be obtained by evaluating the spectral content of geomagnetic field models, i.e. their energy distribution with respect to spatial scales. Differences are most prominent in spectra of magnetic field derivatives, i.e. secular variation and secular acceleration. The power spectrum (Lowes, 1966; Mauersberger, 1956) of the secular variation at Earth's surface is computed by

$$W(l) = \left(\frac{a}{r}\right)^{2l+4} \sum_{m=0}^{l} \left((\dot{g}_l^m)^2 + (\dot{h}_l^m)^2 \right), \tag{7.5.1}$$

with $r = a$ the planetary radius. The SA power spectra are computed likewise. Figures 7.5.3 and 7.5.4 provide those power spectra computed for the epoch 2009.

Figure 7.5.3 shows the energy spectra of the secular variation of CHAOS-6 and GRIMM-3 and the difference of their secular variations (SV). The spectra of the SV of GRIMM-3 and CHAOS-6 match each other up to spherical harmonic degree 13. Thereafter, the spectrum of the GRIMM-3 SV becomes flat, whereas the spectrum of CHAOS-6 SV does not show such flattening but shows a further decrease of SV power. This may be related to chosen priors for the model and may suggest that secular variation features on very small spatial scales could be resolved in CHAOS-6. However, the SV power reaches 0.1 $(\mathrm{nT/yr})^2$ at spherical harmonic degree 14, which may be a lower bound of the detectable SV power based on satellite data, because the accuracy of the satellite instrument is ~0.2 nT. Furthermore, other un-modelled slowly varying fields, e.g. oceanic magnetic fields may contribute to uncertainties at these scales. Therefore, values of secular variation power below 0.1 $(\mathrm{nT/yr})^2$ and accordingly resolved secular variation beyond spherical harmonic degree 15 may not be demonstrable by CHAMP satellite measurements.

In Figure 7.5.4 secular acceleration power spectra of the satellite-based field models are shown. The SA power of CHAOS-6 and GRIMM-3 start to differ significantly at spherical harmonic degree 6, i.e. where the power of model differences becomes larger than the SA power of GRIMM-3. At higher spherical harmonic degrees, curves of CHAOS-6 SA power and the differences matches. The different spectral behaviour of the SA powers is certainly due to the different priors in the model derivation, but it may also indicate that SA beyond spherical harmonic degree 6 is not resolved by geomagnetic data from the CHAMP satellite and that different modelling strategies will lead to different estimates of the secular acceleration. Moreover, a similar reasoning based on uncertainties due to un-modelled magnetic fields and the accuracy of the instrument applies for the resolution of the secular acceleration derived from satellite data.

Such comparison highlights the sensitivity of models to priors and to the data selection, in particular the data selection in polar regions. Furthermore, sectorial acceleration coefficients in CHAOS-6 and/or GRIMM-3 are less well determined. There are correlation between coefficients at the same order, but different degrees. Likely, these patterns are rather linked to satellite data selection and external fields handling, than to a lack of directional information in the data (Lesur, personal communication, 2018).

7.5.2 Resolution Analysis

The resolution analysis addresses the question of how much of the model parameters are resolved by observations and determined by prior constraints. The analysis is based on the derivation of the resolution matrix:

$$\mathbf{R} = (\mathbf{A}^T \mathbf{C_e}^{-1} \mathbf{A} + \mathbf{C_m}^{-1})^{-1} \mathbf{A}^T \mathbf{C_e}^{-1} \mathbf{A}, \tag{7.5.2}$$

where $\mathbf{C_m}$ is the model covariance matrix given by the applied prior constraint for the inversion. The derivation of the resolution matrix of the inversion allows to estimate the significance of model parameters. Tarantola (1987) suggested the trace of the resolution matrix could provide an

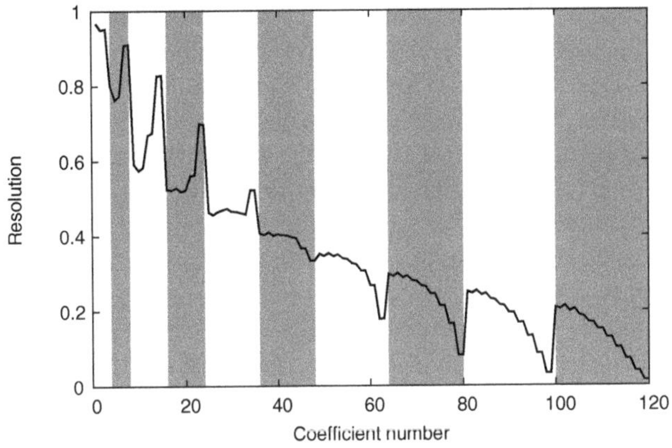

Figure 7.5.5 Diagonal elements of the resolution matrix for a model of Mercury's magnetic field. Grey shaded areas mark coefficients of even spherical harmonic degrees (2–10).

approximate measure of the number of model parameters that are resolved by the data set. Ideally, the resolution matrix would be an identity matrix. Due to inadequacy of the data, a regularisation scheme must be applied in the solving process to obtain a stable solution this is reflected in the form of the resolution matrix. A resolution near 1 signifies that a model parameter is wholly determined by the data, whereas a low resolution means that the model is mostly constrained by the prior information. Low resolution and the inability to satisfy the data entirely are both sources of uncertainty in model estimates.

Figure 7.5.5 shows the diagonal elements of the resolution matrix for an inversion of MESSENGER magnetic field data at Mercury for a spherical harmonic expansion of maximum internal degree $L_{int} = 10$. The value of these elements decreases with the number of coefficients, i.e. the resolution decreases with spherical harmonic degree. The resolution varies noticeably within spherical harmonic degree. The behaviour of the first five spherical harmonic degrees differs from the rest of the model, which may indicate that spherical harmonic degrees $L_{int} > 5$ are insufficiently resolved by the data. The trace of the resolution matrix, which provides an estimate of the degree of freedom of the system, can be computed by the sum of diagonal elements. In the shown case the trace is ~42 and suggests that 42 Gauss coefficients are resolved by the data. This corresponds roughly to a spherical harmonic degree 5. At this degree, the resolution becomes less than 0.5, hinting that the prior dominates the solution.

7.5.3 Covariance Analysis

One possible way to study the results of an inversion is given by the analysis of the covariance matrix of the inverse problem. This matrix is given by

$$\mathbf{C} = \hat{\sigma}^2(\mathbf{A}^T\mathbf{C_e}^{-1}\mathbf{A} + \mathbf{C_m}^{-1})^{-1} \tag{7.5.3}$$

where $\hat{\sigma}^2$ is the misfit between model and observation. The covariance matrix quantifies the uncertainties in the model estimates due to linear dependence of model parameters. Ideally, one would expect this matrix to be purely diagonal, but in fact some non-diagonal elements are not zero, which indicates a dependency between coefficients of the same degree but different order. The correlation between different model parameters may have many causes, likely, due to an insufficient external and internal field separation. But also, un-modelled magnetic fields may leak into the model and can introduce spurious behaviour of the model parameter (see also Section 7.5.4). Such behaviour was revealed by Silva et al. (2012), who found a 2.5 year period of the second time derivative of g_1^0 in a geomagnetic core field model, i.e. CHAOS-3 (Olsen et al., 2010b). The cause for this periodicity, as suggested by Ou et al. (2017), may be an imprint of external field variations and related to the quasi-biennial oscillation of ionospheric and magnetospheric current systems. Also the orbital configuration of satellite missions influences the robustness of the estimated model parameters. Geomagnetic satellite missions show a polar gap, because of their orbital inclination ≠ 90 degrees. This may have an effect on the determination of the g_1^0 term and its temporal derivative, because Earth's external field does not align fully with the main axis of the internal field. The misalignment and local-time effects on timescales shorter than a few years may cause staggering motions of internal and external field axes with respect to each other and, eventually, may allocate external field variation to g_1^0. Another example may be the MESSENGER mission around Mercury. The space probe flew within Mercury's magnetosphere and at low altitude

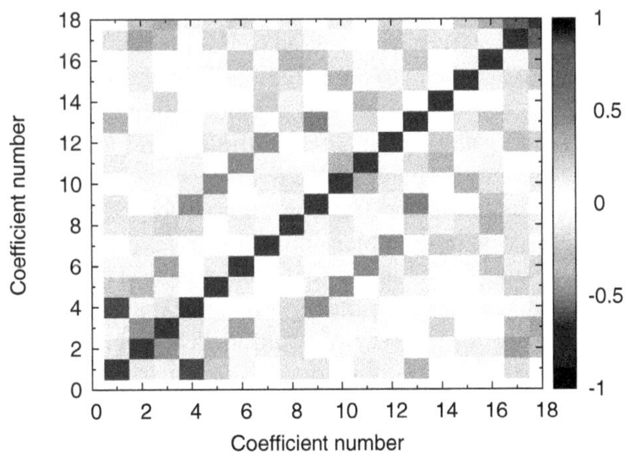

Figure 7.5.6 Covariance matrix of an inversion of MESSENGER magnetic field data. The symmetric plot is organised by coefficient numbers, where the first three coefficients are the Gauss coefficients of the first internal spherical degree, g_2^0 is coefficient 4, g_3^0 is coefficient 9 and the external magnetic field coefficients are coefficients 16 to 18, respectively. (A black-and-white version of this figure appears in some formats. For the colour version, please refer to the plate section.)

(below 1000 km) mainly over the Northern Hemisphere. This led to an uneven distribution, which alienates estimations of spherical harmonic coefficients with even spherical harmonic degrees (Thébault et al., 2018).

Figure 7.5.6 shows an example covariance matrix for an inversion of MESSENGER magnetic field data at Mercury for a spherical harmonic expansion of maximum internal degree $L_{\text{int}} = 3$ and maximum external degree $L_{\text{ext}} = 1$. Diagonal elements display the autocorrelation of coefficients, whereas off-diagonal elements display the cross-correlation between different Gauss coefficients. Significant negative cross-correlation $|c| \sim 0.5$ can be identified between coefficients g_2^0, g_1^n and $g_2^0, g_3^0; g_2^1, g_3^1; \ldots$. These cross-correlations are caused by the uneven hemispherical distribution of MESSENGER data and indicate that these coefficients cannot be fully resolved independently from each other.

Effects due to the leakage of external field variation into model parameters of the internal field and an insufficient separation of external and internal fields due to uneven data distribution could possibly be indicated by significant structures in the covariance matrix of the inverse problem. However, Figure 7.5.6 does not show such correlation structures.

In the previous two section we illustrated resolution and covariance analyses by particular inversions of MESSENGER data for modelling the magnetic field of Mercury. However, these kinds of analyses are applied in many inverse problems and are not limited to magnetic field modelling (e.g. Zhang and Thurber, 2007; Chiao et al., 2014; Wiese et al., 2016).

7.5.4 Spectral Leakage in Global Models

In this section we discuss the effect of spectral leakage, where the expression is adopted from seismology and geodesy. The term may differ from what is used in Fourier analysis, where spectral leakage refers to the effect that changes the spectrum of a product of time series and a window function of finite length (windowing). In this sense, the spectral leakage discussed here may be comparable to an aliasing.

The discussion of spectral leakage is closely related to the treatment of a finite-dimensional linear inverse problem, when the true model is infinite-dimensional. The cause of spectral leakage is the truncation of the expansion (7.2.4), where the spectral energy of un-modelled magnetic fields, i.e. $l > L$ are indefinitely mapped onto spherical harmonic degrees $l \leq L$. Possible sources of such un-modelled magnetic fields could be Earth's crust and core, but also its ocean. This effect should be distinguished from another kind of spectral leakage that may occur between external and internal magnetic field model coefficients, where external magnetic field energy is mapped onto internal magnetic field model coefficients, and vice versa. Lowes and Olsen (2004) estimated the leakage of the ionospheric field into near-zonal internal field coefficients roughly to be 6 nT and the leakage of other fields to be minor. As pointed out by Thébault et al. (2012), this cannot be avoided, but the effects should be detectable by a covariance analysis of the inversion. Furthermore, Whaler and Gubbins (1981) discussed the effect by the inadequate truncation of the spherical harmonic expansion, i.e. when the spatial data distribution differs from the spatial sampling resolution. This may lead to an aliasing between spherical harmonic degrees.

The effects of spectral leakage have also been studied in geodesy (i.e. Jekeli, 1996), where contributions of spectral leakage to gravity field models derived from evenly distributed data are discussed. Here, we focus on spectral leakage that occur when a truncated geomagnetic field model is derived from an uneven data distribution. Though, geomagnetic satellites may provide an uniform data distribution, data selection schemes will turn this into an uneven distribution. Trampert and Snieder (1996) developed a formalism to account and to suppress spectral leakage in global seismic tomography models based on an uneven data distribution, that should apply to a wide range of linear inverse problems.

The observations in Equation (7.3.1) can be related to a set of two model vectors,

$$\mathrm{y} = \mathbf{A}_L \mathrm{m}_L + \mathbf{A}_> \mathrm{m}_> + e, \tag{7.5.4}$$

where $\mathbf{A}_L$ denotes design matrix for a truncated model vector m_L and $\mathbf{A}_>$ the matrix for the model vector $\mathrm{m}_>$ with infinite dimension. Inserting this into Equation (7.3.6) and neglecting the error vector yields

$$\begin{aligned} \mathrm{m} = {} & (\mathbf{A}_L^T C_e^{-1} \mathbf{A}_L + C_{m,L}^{-1})^{-1} \mathbf{A}_L^T Ce^{-1} \mathbf{A}_L \mathrm{m}_L \\ & + (\mathbf{A}_L^T C_e^{-1} \mathbf{A}_L + C_{m,L}^{-1})^{-1} \mathbf{A}_L^T Ce^{-1} \mathbf{A}_> \mathrm{m}_>. \end{aligned} \tag{7.5.5}$$

The first term is similar to the resolution operator (Tarantola, 1987), which relates the true model and its estimate. The second term is assigned to the spectral leakage, and it depends on the data and their spatial sampling, but also on prior constraints, i.e. $C_{m,L}$, for the model estimation.

Trampert and Snieder (1996) provided the formalism for a simplified general case, where the model covariance matrix, C_m, and the error covariance matrix, C_e are assumed to be diagonal. The new model estimate is

$$\mathrm{m} = (\mathbf{A}_L^T W^{-1} \mathbf{A}_L + (\alpha^2/\beta^2) I)^{-1} \mathbf{A}_L^T W \mathrm{y}, \tag{7.5.6}$$

with

$$W = (\mathbf{A}_> \mathbf{A}_>^T + \beta^2 I)^{-1}, \tag{7.5.7}$$

where $\alpha^2 = \sigma_e^2/\sigma_{m,L}^2$ and $\beta^2 = \sigma_e^2/\sigma_{m,>}^2$; $\sigma_e^2, \sigma_{m,L}^2$ and $\sigma_{m,>}^2$ are the variances of the error, the model parameters of the truncated model and the model parameter of the infinite model, respectively, and β is the relative importance of data errors against the variance of neglected model parameters. Evaluations of Equation (7.5.7) tend to be tedious as they invoke derivations of Gram matrices that for some inverse problems do not exist. Values of α^2 and β^2 can be found by an eigenvalue analysis of W. The optimisation of leakage reduction is an iterative process, and Trampert and Snieder (1996) suggested $\alpha^2 < \beta^2$. So far, correction for spectral leakage has only been implemented for global surface wave tomography (Spetzler and Trampert, 2003) and satellite geodesy (Goossens, 2010). However, spectral leakage correction methods should also be applicable to geomagnetic field modelling.

7.6 Concluding Remarks

In this chapter, we provided a review of methods that are applied to construct mathematical models of Earth's magnetic field. These methods have to take into account the different source regions and processes that contribute to the geomagnetic field as it is sampled by satellite missions. Most important is the separation of magnetic signals generated exterior and interior to Earth's surface.

Current problems of an accurate separation of external and internal sources fall into two categories. Problems of the first category can be characterised by an insufficient parameterisation of ionospheric and magnetospheric fields in geomagnetic field modelling, which is caused by a lack of knowledge of magnetospheric and ionospheric dynamo processes. External disturbance fields are largely ignored during internal magnetic field modelling, and therefore these fields overlay the small-scale lithospheric field and rapid time variations of the core field. However, small-scale lithospheric field and rapid time variations of the core field are of particular interest, as they may provide insights on sources and processes that cannot be deduced from other geophysical observations and therefore are important for other scientific disciplines, such as seismology and plate tectonics. Presently, ideas are formulated which may improve our understanding of the processes that are generating these disturbance fields. Joint observations from ground and multi-satellite missions that densely sample the magnetosphere and ionosphere at different distances to Earth's surface and by different geophysical experiments may be one key ingredient. This may lead to precise empirical models, and perhaps the goal could be to model the magnetic field generation in these regions by numerical dynamo simulations of the magneto-hydrodynamic interactions. So far, these ideas are just formulated.

The second category of problems or challenges incorporates the development of the algebraic apparatus. Future progress may be expected from the invocation of hard prior bounds during the inversion of geomagnetic field models, which requires the application of alternative techniques, such as confidence set inferences (Backus, 1989), or methods developed by Donoho (1988) and Stark (1992). Furthermore, joint analyses of geomagnetic observations and dynamo simulations in the framework of data assimilation may allow a better separation of external and internal fields and may provide a substantial improvement of our understanding of outer core processes.

We believe further progress in the understanding of different aspects of Earth's magnetic field will be facilitated by the combination of current and future geomagnetic satellite mission with ground-based measurements of Earth's magnetic field. And, as concluded by Finlay et al. (2017), solutions to this challenge will likely only be feasible as part of a renewed collaboration between the internal and external geomagnetic field communities.

References

Aubert, J. & Fournier, A., 2011. Inferring internal properties of Earth's core dynamics and their evolution from surface observations and a numerical geodynamo model, *Nonlinear Process. Geophys.*, **18**, 657–74.

Backus, G., 1970. Inference from Inadequate and Inaccurate Data, I, *Proc. Natl. Acad. Sci. USA*, **65**, 1–7.

Backus, G. E., 1988a. Bayesian inference in geomagnetism, *Geophys. J. R. Astron. Soc.*, **92**, 125–42.

Backus, G. E., 1988b. Comparing hard and soft prior bounds in geophysical inverse problems, *Geophys. J. R. Astron. Soc.*, **94**, 249–61.

Backus, G. E., 1989. Confidence set interference with a prior quadratic bound, *Geophys. J. R. Astron. Soc.*, **97**, 119–50.

Baerenzung, J., 2018. Sequential assimilation of the Earths magnetic field, 16. Symposium on Study of the Earth's Deep Interior.

Baratchart, L. & Gerhards, C., 2017. On the recovery of core and crustal components of geomagnetic potential fields. Unpublished manuscript.

Beggan, C. D., Saarimäki, J., Whaler, K. A. & Simons, F. J., 2013. Spectral and spatial decomposition of lithospheric magnetic field models using spherical Slepian functions, *Geophys. J. Int.*, **193**, 136–48.

Cain, J. C., Wang, Z., Kluth, C. & Schmitz, D. R., 1989. Derivation of a geomagnetic model to n = 63, *Geophys. J. R. Astron. Soc.*, **97**, 431–41.

Calderwood, A., Roberts, P. & Jones, C., 2003. Energy fluxes and ohmic dissipation in the Earth's core, in *Earth's Core and Lower Mantle*, pp. 100–129, ed. Zhang, K., Soward, A. & Jones, C., CRC Press, Boca Raton, FL.

Chambodut, A., Panet, I., Mandea, M., Diament, M., Holschneider, M. & Jamet, O., 2005. Wavelet frames: An alternative to spherical harmonic representation of potential fields, *Geophys. J. Int.*, **163**, 875–99.

Chapman, S. & Ferraro, V. C. A., 1940. The theory of the first phase of a geomagnetic storm, *J. Geophys. Res.*, **45**, 245.

Chapman, S. & Ferraro, V. C. A., 1941. The geomagnetic ring current: I. Its radial stability, *J. Geophys. Res.*, **46**, 1.

Chiao, L.-Y., Chen, Y.-N. & Gung, Y., 2014. Constructing empirical resolution diagnostics for kriging and minimum curvature gridding, *J. Geophys. Res.*, **119**, 3939–54.

Donoho, D. L., 1988. One-sided inference about functionals of a density, *Ann. Stat.*, **16**, 1390–1420.

Finlay, C. C., Lesur, V., Thébault, E., Vervelidou, F., Morschhauser, A. & Shore, R., 2017. Challenges handling magnetospheric and ionospheric signals in internal geomagnetic field modelling, *Space Sci. Rev.*, **206**, 157–89.

Finlay, C. C., Olsen, N., Kotsiaros, S., Gillet, N. & Tøffner-Clausen, L., 2016. Recent geomagnetic secular variation from Swarm and ground observatories as estimated in the CHAOS-6 geomagnetic field model, *Earth Planets Space*, **68**, 112.

Fournier, A., Aubert, J. & Thebault, E., 2011. Inference on core surface flow from observations and 3-D dynamo modelling, *Geophys. J. Int.*, **186**, 118–36.

Fournier, A., Aubert, J. & Thébault, E., 2015. A candidate secular variation model for IGRF-12 based on Swarm data and inverse geodynamo modelling, *Earth Planets Space*, **67**, 81.

Fournier, A., Hulot, G., Jault, D., Kuang, W., Tangborn, A., Gillet, N., Canet, E., Aubert, J. & Lhuillier, F., 2010. An introduction to data assimilation and predictability in geomagnetism, *Space Sci. Rev.*, **155**, 247–91.

Franklin, J. N., 1970. Well-posed stochastic extension of ill-posed problems, *Geophys. J. R. Astron. Soc.*, **23**, 125–8.

Gauss, C. F., 1839. Allgemeine Theorie des Erdmagnetismus, in *Resultate aus den Beobachtungen des magnetischen Vereins im Jahre 1838*, pp. 1–57, ed. Gauss, C. F. & Weber, W., Leipzig.

Goossens, S., 2010. Applying spectral leakage corrections to gravity field determination from satellite tracking data, *Geophys. J. Int.*, **181**, 1459–72.

Gubbins, D., 1975. Can the Earth's magnetic field be sustained by core oscillations?, *Geophys. Res. Lett.*, **2**, 409–12.

Gubbins, D., 1983. Geomagnetic field analysis – I. Stochastic inversion, *Geophys. J. R. Astron. Soc.*, **73**, 641–52.

Gubbins, D., 2004. *Time Series Analysis and Inverse Theory for Geophysicists*, Cambridge University Press, Cambridge.

Gubbins, D. & Bloxham, J., 1985. Geomagnetic field analysis – III. Magnetic fields on the core-mantle boundary, *Geophys. J. R. Astron. Soc.*, **80**, 695–713.

Helffrich, G. & Kaneshima, S., 2010. Outer-core compositional stratification from observed core wave speed profiles, *Nature*, **468**, 807–10.

Holme, R. & Bloxham, J., 1995. Alleviation of the Backus effect in geomagnetic field modelling, *Geophys. Res. Lett.*, **22**, 1641–4.

Holme, R. & Bloxham, J., 1996. The magnetic fields of Uranus and Neptune: Methods and models, *J. Geophys. Res.*, **101**, 2177–2200.

Holschneider, M., Lesur, V., Mauerberger, S. & Baerenzung, J., 2016. Correlation-based modeling and separation of geomagnetic field components, *J. Geophys. Res.*, **121**, 3142–60.

Jackson, A., Jonkers, A. R. T. & Walker, M. R., 2000. Four centuries of geomagnetic secular variation from historical records, *Philos. Trans. R. Soc. London, Ser. A*, **358**, 957–90.

Jackson, D. D., 1979. The use of a priori data to resolve nonuniqueness in linear inversion, *Geophys. J. R. Astron. Soc.*, **57**, 137–57.

Jekeli, C., 1996. Spherical harmonic analysis, aliasing, and filtering, *J. Geod.*, **70**, 214–23.

Kaneshima, S. & Matsuzawa, T., 2015. Stratification of Earth's outermost core inferred from SmKS array data, *Prog. Earth Planet. Sci.*, **2**, 15.

Khokhlov, A., Hulot, G. & Le Mouel, J.-L., 1999. On the Backus effect – II, *Geophys. J. Int.*, **137**, 816–20.

Korte, M. & Constable, C. G., 2005. Continuous geomagnetic field models for the past 7 millennia: 2. CALS7 K, *Geochem. Geophys. Geosyst.*, **6**, Q02H16, doi: 10.1029/2004GC000801.

Kotsiaros, S., Finlay, C. C. & Olsen, N., 2015. Use of along-track magnetic field differences in lithospheric field modelling, *Geophys. J. Int.*, **200**, 878–87.

Langel, R. A., 1987. The main geomagnetic field, in *Geomagnetism*, vol. 1, ed. Jacobs, J. A., chapter 4, Academic, San Diego, CA.

Langel, R. A. & Estes, R. H., 1985. Large-scale, near-field magnetic fields from external sources and the corresponding induced internal field, *J. Geophys. Res.*, **90**, 2487–94.

Lesur, V., Hamoudi, M., Choi, Y., Dyment, J. & Th′ebault, E., 2016. Building the second version of the World Digital Magnetic Anomaly Map (WDMAM), *Earth Planets Space*, **68**, 27.

Lesur, V., Rother, M., Vervelidou, F., Hamoudi, M. & Th ′ebault, E., 2013. Post-processing scheme for modelling the lithospheric magnetic field, *Solid Earth*, **4**, 105–18.

Lesur, V., Wardinski, I., Baerenzung, J. & Holschneider, M., 2017. On the frequency spectra of the core magnetic field Gauss coefficients, *Phys. Earth Planet. Inter.*, **2**, 164–8.

Lesur, V., Wardinski, I. & Hamoudi, M., 2011. *Third Version of the GFZ Reference Internal Magnetic Model: GRIMM-3*, 25th IUGG General Assembly, Melbourne, Australia.

Lesur, V., Wardinski, I., Hamoudi, M. & Rother, M., 2010. The second generation of the GFZ Reference Internal Magnetic Model: GRIMM-2, *Earth Planets Space*, **62**, 765–73.

Lesur, V., Whaler, K. A. & Wardinski, I., 2015. Are geomagnetic data consistent with stably stratified flow at the core-mantle boundary?, *Geophys. J. Int.*, **201**, 929–46.

Levenberg, K., 1944. A method for the solution of certain non-linear problems in least squares, *Q. Appl. Math.*, **2**, 164–8.

Livermore, P. W., Hollerbach, R. & Finlay, C. C., 2017. An accelerating high-latitude jet in Earth's core, *Nat. Geosci.*, **10**, 62–8.

Lowes, F. J., 1966. Mean-square values on sphere of spherical harmonic vector fields, *J. Geophys. Res.*, **71**, 2179.

Lowes, F. J. & Olsen, N., 2004. A more realistic estimate of the variances and systematic errors in spherical harmonic geomagnetic field models, *Geophys. J. Int.*, **157**, 1027–44.

Mauersberger, P., 1956. Das Mittel der Energiedichte des geomagnetischen Hauptfeldes an der Erdoberfläche und seine säkulare Änderung, *Gerlands Beiträge Geophys.*, **65**, 207–15.

Maus, S. & Haak, V., 2003. Magnetic field annihilators: Invisible magnetization at the magnetic equator, *Geophys. J. Int.*, **155**, 509–13.

Maus, S. & Kuvshinov, A., 2004. Ocean tidal signals in observatory and satellite magnetic measurements, *Geophys. Res. Lett.*, **31**, 15313.

Maus, S., Rother, M., Hemant, K., Stolle, C., Lühr, H., Kuvshinov, A. & Olsen, N., 2006. Earth's lithospheric magnetic field determined to spherical harmonic degree 90 from CHAMP satellite measurements, *Geophys. J. Int.*, **164**, 319–30.

Maus, S., Yin, F., Lühr, H., Manoj, C., Rother, M., Rauberg, J., Michaelis, I., Stolle, C. & Müller, R. D., 2008. Resolution of direction of oceanic magnetic lineations by the sixth-generation lithospheric magnetic field model from CHAMP satellite magnetic measurements, *Geochem. Geophys. Geosyst.*, **9**, Q07021.

McLeod, M. G., 1986. Stochastic processes on a sphere, *Phys. Earth Planet. Inter.*, **43**, 283–99.

McLeod, M. G., 1996. Spatial and temporal power spectra of the geomagnetic field, *J. Geophys. Res.*, **101**, 2745–64.

Menke, W., 1989. *Geophysical Data Analysis: Discrete Inverse Theory*, Academic Press, New York.

Olsen, N., Glassmeier, K.-H. & Jia, X., 2010a. Separation of the magnetic field into external and internal parts, *Space Sci. Rev.*, **152**, 135–57.

Olsen, N., Hulot, G., Lesur, V., Finlay, C. C., Beggan, C., Chulliat, A., Sabaka, T. J., Floberghagen, R., Friis-Christensen, E., Haagmans, R., Kotsiaros, S., Lühr, H., Tøffner-Clausen, L. & Vigneron, P., 2015. The swarm initial field model for the 2014 geomagnetic field, *Geophys. Res. Lett.*, **42**, 1092–8.

Olsen, N., Lühr, H., Finlay, C. C., Sabaka, T. J., Michaelis, I., Rauberg, J. & Tøffner-Clausen, L., 2014. The CHAOS-4 geomagnetic field model, *Geophys. J. Int.*, **197**, 815–27.

Olsen, N., Mandea, M., Sabaka, T. J. & Tøffner-Clausen, L., 2010b. The CHAOS-3 geomagnetic field model and candidates for the 11th generation IGRF, *Earth Planets Space*, **62**, 719–27.

Olsen, N., Sabaka, T. J. & Lowes, F., 2005. New parameterization of external and induced fields in geomagnetic field modeling, and a candidate model for IGRF 2005, *Earth Planets Space*, **57**, 1141–9.

Ou, J., Du, A. & Finlay, C. C., 2017. Quasi-biennial oscillations in the geomagnetic field: Their global characteristics and origin, *J. Geophys. Res.*, **122**, 5043–58.

Sabaka, T. J., Olsen, N. & Langel, R. A., 2002. A comprehensive model of the quiet-time, near-Earth magnetic field: Phase 3, *Geophys. J. Int.*, **151**, 32–68.

Sabaka, T. J., Olsen, N. & Purucker, M. E., 2004. Extending comprehensive models of the Earth's magnetic field with Ørsted and CHAMP data, *Geophys. J. Int.*, **159**, 521–47.

Sabaka, T. J., Olsen, N., Tyler, R. H. & Kuvshinov, A., 2015. CM5, a pre-Swarm comprehensive geomagnetic field model derived from over 12 yr of CHAMP, Ørsted, SAC-C and observatory data, *Geophys. J. Int.*, **200**, 1596–1626.

Sabaka, T. J., Tyler, R. H. & Olsen, N., 2016. Extracting ocean-generated tidal magnetic signals from Swarm data through satellite gradiometry, *Geophys. Res. Lett.*, **43**, 3237–45.

Schott, J. J. & Thébault, E., 2011. Modelling the Earths magnetic field from global to regional scales, in *Geomagnetic Observations and Models*, vol. 1, pp. 229–64, Springer, Netherlands.

Schwarte, J., Lühr, H. & Holme, R., 2003. Improved parameterization of external magnetic field from CHAMP measurements, in *First CHAMP Mission Results for Gravity, Magnetic and Atmospheric Studies*, pp. 240–44, Springer, Berlin.

Silva, L., Jackson, L. & Mound, J., 2012. Assessing the importance and expression of the 6 year geomagnetic oscillation, *J. Geophys. Res.*, **117**, 10101.

Spetzler, J. & Trampert, J., 2003. Implementing spectral leakage corrections in global surface wave tomography, *Geophys. J. Int.*, **155**, 532–8.

Stark, P. B., 1992. Inference in infinite-dimensional inverse problems: Discretization and duality, *J. Geophys. Res.*, **97**, 14055–82.

Stern, D. P., Langel, R. A. & Mead, G. D., 1980. Backus effect observed by Magsat, *Geophys. Res. Lett.*, **7**, 941–4.

Stockmann, R., Finlay, C. C. & Jackson, A., 2009. Imaging Earth's crustal magnetic field with satellite data: A regularized spherical triangle tessellation approach, *Geophys. J. Int.*, **179**, 929–44.

Strang, G., 1988. *Linear Algebra and Its Applications*, Harbourt Brace Jovanovisch, San Diego.

Sugiura, M. & Kamei, T., 1991. Equatorial DST – index 1957–1986, IAGA Bull., 40.

Tarantola, A., 1987. *Inverse Problem Theory: Methods for Data Fitting and Model Parameter Estimation*, Elsevier, Amsterdam.

Thébault, E., Finlay, C. C., Alken, P., Beggan, C. D., Canet, E., Chulliat, A., Langlais, B., Lesur, V., Lowes, F. J., Manoj, C., Rother, M. & Schachtschneider, R., 2015. Evaluation of candidate geomagnetic field models for IGRF-12, *Earth Planets Space*, **67**, 112.

Thébault, E., Langlais, B., Oliveira, J. S., Amit, H. & Leclercq, L., 2018. A time-averaged regional model of the Hermean magnetic field, *Phys. Earth Planet. Inter.*, **276**, 93–105.

Thébault, E., Lesur, V., Kauristie, K. & Shore, R., 2017. Magnetic field data correction in space for modelling the lithospheric magnetic field, *Space Sci. Rev.*, **206**, 191–223.

Thébault, E., Vervelidou, F., Lesur, V. & Hamoudi, M., 2012. The satellite along-track analysis in planetary magnetism, *Geophys. J. Int.*, **188**, 891–907.

Thomson, A. W. P. & Lesur, V., 2007. An improved geomagnetic data selection algorithm for global geomagnetic field modelling, *Geophys. J. Int.*, **169**, 951–63.

Trampert, J. & Snieder, R., 1996. Model estimations biased by truncated expansions: Possible artifacts in seismic tomography, *Science*, **271**, 1257–60.

Tyler, R. H., Maus, S. & Lühr, H., 2003. Satellite observations of magnetic fields due to ocean tidal flow, *Science*, **299**, 239–41.

Wardinski, I. & Lesur, V., 2012. An extended version of the C3 FM geomagnetic field model: Application of a continuous frozen-flux constraint, *Geophys. J. Int.*, **189**, 1409–29.

Whaler, K. A., 1980. Does the whole of the Earth's core convect?, *Nature*, **287**, 528–30.

Whaler, K. A. & Gubbins, D., 1981. Spherical harmonic analysis of the geomagnetic field: An example of a linear inverse problem, *Geophys. J.*, **65**, 645–93.

Wiese, D. N., Killett, B., Watkins, M. M. & Yuan, D.-N., 2016. Antarctic tides from GRACE satellite accelerations, *J. Geophys. Res.*, **121**, 2874–86.

Zhang, H. & Thurber, C. H., 2007. Estimating the model resolution matrix for large seismic tomography problems based on Lanczos bidiagonalization with partial reorthogonalization, *Geophys. J. Int.*, **170**, 337–45.

8

New Insights in Far-Space Measurements

Large-Scale Structures and Processes in the Solar Wind and Terrestrial Magnetosphere

Aurélie Marchaudon

8.1 Introduction

Since the birth of Space Era exactly 60 years ago, considerable knowledge has been gathered on the Earth's space environment as well as on the interplanetary medium and the Sun, and finally on the other planets of the solar system. Space plasmas are ubiquitous in the solar system from the Sun to the planets' closest environment, and they share important common characteristics despite different properties (e.g. density, temperature, magnetic field). They are largely structured by the magnetic field, and large-scale currents are very often found either in thin layers supporting a large-scale magnetic field polarity inversion or at the interface between different plasma regions. The former is found in the solar wind on the form of the heliospheric current sheet (HCS) or in the nightside region of the magnetised planets in the form of the neutral current sheet (NCS). These currents sheets separate the northern and southern magnetic field lines, which exhibit an almost radial opposite direction close to the magnetic equator at relatively large distance either from the Sun or from the planets. The latter is found at the interface between different plasma regions such as between the solar wind and the magnetised environment of the planets where different plasmas cannot mix due to infinite conductivity along magnetic field lines (so called frozen-in condition).

A majority of these plasmas are collisionless apart from very dense media, such as the planetary upper's atmospheres, called ionospheres. In such collisionless media, energy transfer and dissipation cannot result from collisions, and other processes such as shock formation, reconnection and particle–wave interactions play significant roles in plasma dynamics. Shocks are for example found ahead of rapid solar eruptions, where large quantities of mass and energy are released from the Sun's atmosphere into the interplanetary medium. These events, called coronal mass ejections (CMEs), often travel very fast in the interplanetary medium, and a shock has to form ahead of them. More permanent shocks also exist ahead of each planet, where the solar wind flow has to be decelerated by a shock before streaming around the planetary obstacles. These shocks are due to the fact that the solar wind is supersonic and super-Alfvénic, with a velocity higher than the sound and Alfvén waves speeds in the medium (Alfvén waves being low-frequency magnetohydrodynamic (MHD) waves propagating generally along the magnetic field direction with oscillations of the magnetic field and of the ions). Reconnection is also a very important phenomenon in space plasma since it allows large magnetic reconfigurations between magnetic fields of different plasmas, allowing the violation of the frozen-in conditions, the conversion of magnetic to kinetic energy and the transfer of plasma, momentum and energy from one plasma region to the other. Reconnection can be found in the solar eruptions, in the solar wind and in the planetary magnetospheres – the spheres of influence of the magnetised planets.

The task to review in just a few pages, the past 10 years' results obtained by the main space missions dedicated to the Sun–Earth and Sun–planets relationship, is just tremendous. Consequently, this chapter focuses only on a few results obtained regarding the solar wind and the terrestrial magnetosphere. In the following, Section 8.2 will be dedicated to the presentation of a relatively complete set of space scientific missions dedicated to the Sun, the Earth and the other planets of the solar system, highlighting briefly their specificities. Section 8.3 will present recent results on the solar wind, favouring a particular set of events known to have strong consequences on the Earth's magnetosphere, the interplanetary coronal mass ejections (ICMEs). Section 8.4 will give some new insights related to the magnetosheath, known to be the interface region between the solar wind and the Earth's magnetosphere and the large-scale properties of the magnetosphere. Finally, Section 8.5 will very briefly introduce some instrumental perspectives.

8.2 Missions and Instrumentation

Over the past 20 years, numerous space missions have been launched and have led to new discoveries in all the visited regions of the solar system. To test these results, spatial observations are often confronted with numerical simulations (these include MHD, hybrid or kinetic models).

8.2.1 Sun and Solar Wind Missions

Solar missions are mainly dedicated to the study of the solar atmosphere and corona. A non-exhaustive list of solar missions of the past 20 years consists of SOHO (SOlar and Heliospheric Observatory), RHESSI (Reuven Ramaty High Energy Solar Spectroscopic Imager), STEREO (Solar TErrestrial RElations Observatory), HINODE and SDO (Solar Dynamics Observatory), and they are all still functioning (see Figure 8.2.1 for a timeline and duration of these missions). They consequently carry remote instruments, such as full-disc imagers or spectrometers, at different wavelengths (extreme ultra-violet, ultra-violet EUV/UV and visible light) to probe different altitudes inside the solar atmosphere and corona and to follow the evolution of energetic events such as solar eruptions. High-energetic photons detectors or telescopes (X and γ rays) are sometimes present, giving insight into particle accelerations following solar flares. Some missions also carry coronagraph imagers (visible and UV light) and heliospheric imagers (visible light), allowing them to follow events in the corona and the heliosphere. In order to track CMEs from the solar surface up to 1 AU (astronomical unit), the STEREO mission is composed of two identical satellites orbiting at two slightly different distances from the Sun, with one satellite just ahead of the Earth and one just behind it and with angular separation increasing over time. Thanks to these strategic positions on each side of the Earth, triangulation of CMEs was made possible at the beginning of the mission (Rouillard et al., 2010a), before both spacecraft drift apart. Unfortunately, communications were lost with STEREO B in 2014. Although several attempts to contact the spacecraft were made these past 3 years, none of them has yet allowed a recovery.

Solar wind missions such as Ulysses, Wind and ACE (Advanced Composition Explorer) are dedicated to the study of the solar wind and interplanetary magnetic field (IMF) properties. They carry essentially in situ plasma

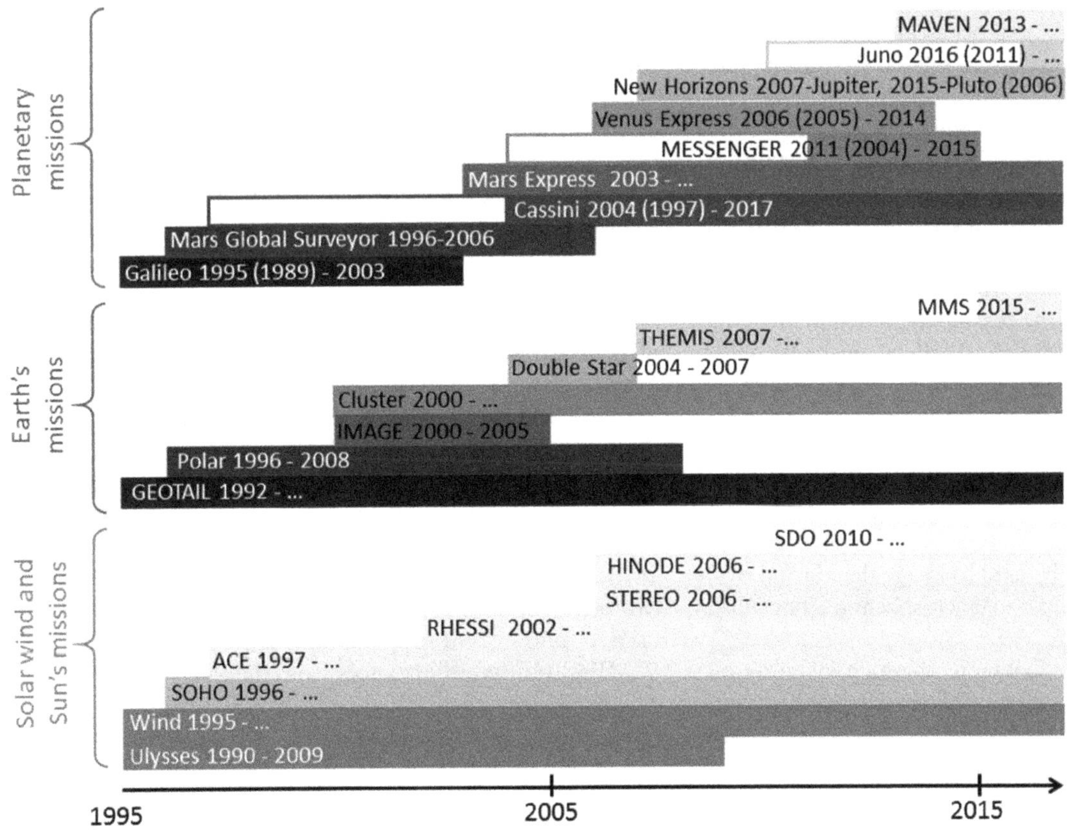

Figure 8.2.1 Timeline of the past 20 years of missions dedicated to the study of the Sun and the solar wind (top row), the terrestrial space environment (middle rows) and the other planets of the solar system for which at least one plasma or field experiment is available on board (lower rows). The start and end (if relevant) dates of operation are indicated for each mission, as well as the launch dates in brackets for cruising missions. (A black-and-white version of this figure appears in some formats. For the colour version, please refer to the plate section.)

measurements, with instrumentation relatively similar to terrestrial missions (see Section 8.2.2 for details). However, they sometimes also carry an isotopic composition sensor, allowing us to study precisely the composition of the solar wind. ACE and Wind are dedicated to the study of solar wind properties in the vicinity of the Earth's magnetosphere. On the other hand, the solar magnetic poles and the solar wind emanating from these regions were uniquely studied over more than one solar cycle (11 years) with the Ulysses mission (McComas et al., 2003), which was placed on a polar orbit around the Sun thanks to a gravity assist by Jupiter.

Finally, several of these missions have spent long duration in space: about 22 years for Wind, 21 years for SOHO and 20 years for ACE. These long-term data sets are crucial to study the properties of the solar surface, the ambient solar wind and the transient structures over almost two solar cycles.

8.2.2 Earth's Magnetospheric Missions

By its proximity, the terrestrial spatial environment has been the focus of many space missions, almost exclusively dedicated to in situ plasma measurements. The main high-altitude ($R > 5$ R_E) missions of the past 20 years are GEOTAIL, Double Star-1, THEMIS (Time History of Events and Macroscale Interactions during Substorms) and MMS (Magnetospheric Multiscale Mission) on equatorial orbits and Polar, IMAGE (Imager for Magnetopause-to-Aurora Global Expansion) and Cluster on polar orbits (see Figure 8.2.1 for timeline and duration of these missions). These missions generally carry on board a comprehensive set of instruments, with particle (electrons and ions) detectors, low-frequency electric and magnetic field experiments, wave experiments and sometimes cameras. In general, spectrometers are used to reconstruct flux-energy distributions of particles, with mass discrimination for the ions (H^+, He^+, O^+) and the pitch-angle distribution being retrieved by the spin of the spacecraft. Highly energetic particles can also be measured by particle telescopes. Magnetic fields are measured by fluxgate magnetometers; electric fields are obtained from double probe electric antennas or electron drift instruments. Wave experiments consist of search-coil magnetometers for electromagnetic waves and sometimes of relaxation sounders for electrostatic waves. Finally, sensitive EUV/UV cameras can also be used to image the auroral oval in the ionosphere and the inner magnetosphere region, called the plasmasphere.

In order to discriminate between spatial and temporal structures, the last generation of missions is generally composed of a constellation of satellites. The Cluster and MMS four-spacecraft missions form a tetrahedron configuration of variable size, and comprehensive data analysis tools have been developed in order to reconstruct, for example, current densities (e.g. Dunlop et al., 2002) or magnetic field line curvature (Shen et al., 2003) or to determine boundaries' normal direction and velocity (e.g. Marchaudon et al., 2004). The five-spacecraft THEMIS mission orbiting at different distances from Earth allowed periodical alignment of the spacecraft either in the magnetotail or in the dayside magnetosheath/magnetosphere. With such alignments, spatiotemporal propagation of magnetospheric structures, such as plasmoids formed by tail reconnection, could be followed by successive impacts at different spacecraft. In 2010, two spacecraft of this constellation were moved to lunar orbits to study the plasma environment around the Moon. Moreover, several missions have spent long durations in space – about 25 years for GEOTAIL, 18 years for Cluster and 10 years for THEMIS – allowing us to construct extremely complete statistical databases and to study effects of seasonal or solar cycle variations on the magnetospheric dynamics. Finally, the fantastic spatiotemporal resolution of the recently launched MMS mission gives new insight on the microphysics of magnetospheric processes.

8.2.3 Planetary Missions

The past 20 years have been a formidable period, during which numerous planets of the solar system have been visited: MESSENGER (MErcury Surface, Space ENvironment, GEochemistry, and Ranging) for Mercury, Venus Express for Venus, Mars Global Surveyor, Mars Express and MAVEN (Mars Atmosphere and Volatile Evolution mission) for Mars, Galileo, one New Horizons flyby and Juno for Jupiter, Cassini for Saturn and one New Horizons flyby again for Pluto (see Figure 8.2.1 for timeline and duration of these missions). These missions have in general multiple objectives, including the study of the surface and/or the atmosphere. Studying the space environment is not necessarily the priority; nevertheless, these missions almost always carry a fluxgate magnetometer (apart from Mars Express and New Horizons), giving access at a minimum to the large-scale magnetic structures surrounding the planet. A number of them carry also particle detectors, electric field measurements and sometimes wave experiments.

Again, several missions have gathered data sets for a long duration: 14 years for Mars Express, 13 years for Cassini before its spectacular end by diving into Saturn's atmosphere in September 2017 and 8 years for Venus Express. These data sets allow precise reconstruction of the spatial environment of Mars, Saturn (and its moons) and Venus. Moreover, several of these planetary missions are regularly used to follow event propagations in the solar wind, such as ICMEs, at different distances from the Sun.

8.3 Solar Wind

8.3.1 Introduction

The solar wind is formed by the expansion of the solar corona into the interplanetary medium and is composed of electrons and ions (essentially protons). From the frozen-in condition where plasma and magnetic field are tied to move together, the Sun's magnetic field is dragged outward by the solar wind plasma while the footprints of the field lines remain globally anchored at the solar surface. With the 27 day solar rotation and the magnetic field surface's anchoring, magnetic field and plasma follow a spiral path in the interplanetary medium, called Parker's spiral (Parker, 1958). Close to the ecliptic, the solar wind is essentially structured by the HCS separating the IMF lines of opposite polarity. The solar wind plasma has two main components, the slow and dense solar wind, whose solar origin is still not fully understood (see Section 8.3.2), and the fast and more tenuous solar wind emanating from the coronal holes, the region of solar open field lines. Coronal holes are located at high latitudes during the quiet part of the 11 year solar cycle, while they are more randomly located when the Sun is active (McComas et al., 2003). The deflection of the rotating solar wind from the high latitudes toward the equator and Parker's spiral rotation contribute to large flapping of the HCS, and the Earth's orbit at 1 AU continuously encounters a highly variable solar wind plasma over time, with a succession of fast and slow streams, as well as variable IMF magnitude and direction. These variable solar wind and IMF conditions strongly affect the coupling efficiency with the magnetosphere of the Earth (see Section 8.4.1) and of the other planets. Some transient large-scale phenomena can largely enhance this coupling and drive magnetic storms in the Earth's environment, such as for example the stream interaction regions (SIRs) and the ICMEs.

SIRs form when the fast solar wind overtakes the slow wind. The encounter between these two different plasmas causes a smooth increase of the solar wind velocity accompanied by strong plasma and magnetic field compressions at the interface. Co-rotating interaction regions (CIRs) correspond to SIRs recurring over several solar rotations. Some of these structures are even capable of driving shocks.

ICMEs are the interplanetary counterpart of strong eruptions (CMEs) occurring at the solar surface and travelling in the interplanetary medium. CMEs are the most powerful structures released by the Sun. In the solar wind, they have relatively well-defined characteristics, although all of them are not necessarily present in each observed ICME, which can complicate their precise identification. The main properties are an enhanced magnitude of the magnetic field with respect to the background IMF, accompanied with a smooth change of the magnetic field direction, relatively low temperature and low β ratio (ratio between kinetic and magnetic pressures of the plasma) and bidirectional streaming of electrons and low-energy protons along magnetic field lines (see e.g. Wu and Lepping, 2011, for other typical properties). A subset of ICMEs show in addition a smooth and coherent rotation of the magnetic field direction, characteristics of a magnetic flux rope with a helicoidal magnetic field more and more twisted toward the edges of the structure. They are called magnetic clouds (MCs). The remaining ICMEs are sometimes called ejecta (EJs). MCs represent 30% to 50% of all observed ICMEs (see e.g. Wu and Lepping, 2007). A large subset of ICMEs (MCs or non-MCs) compress the ambient slower solar wind ahead, driving a strong shock and forming a sheath of compressed and heated solar wind just in front of the ICME.

After a few general results about slow solar wind generation and SIRs/CIRs (see Section 8.3.2), ICMEs will be the essential focus of Section 8.3.3.

8.3.2 Solar Wind Results

Although the origin of the fast solar wind from coronal holes is clear, the emergence of slow solar wind remains more problematic and has been extensively studied these past few years by remote sensing observations from SOHO and STEREO missions (see the review by Abbo et al., 2016). In particular, using an extensive SOHO data set, the source region of the slow wind has been found to possibly lie on open field lines along the boundary of helmet streamers, which are closed magnetic loops formed above active regions (e.g. Abbo et al., 2010). In parallel, remote sensing and in situ observations by STEREO and Wind (Kilpua et al., 2009; Rouillard et al., 2010a, 2010b) have shown that even the slow solar wind is naturally variable and contains small-scale transients continuously but intermittently released and entrained along the slow/fast stream interface of CIRs. Finally, using an older data set from Helios, the properties of the very slow solar wind observed around 0.7 AU, which varies between 200 and 300 km.s^{-1}, has been studied (Sanchez-Diaz et al., 2016). In particular by comparing with MHD modelling results, the interactions with the faster solar wind have been proposed to explain that the very slow solar wind is accelerated in the interplanetary medium and is thus never observed at 1 AU.

Finally, the characterisation of solar structures responsible for moderate magnetic storms in the terrestrial magnetosphere has been extensively studied over one solar cycle, showing that SIRs and CIRs are the main storm drivers during the declining and minimum phases of the solar cycle (e.g. Echer et al., 2013).

8.3.3 ICME Results

Due to their capability to drive strong magnetic storms, ICMEs have been the focus of a very large number of studies during the past 10 years, through comprehensive statistical studies, sometimes over more than 20 years, and through multi-satellite studies, allowing us to characterise the ICME properties at different distances from the Sun and/or at different longitudinal separations (see e.g. Winchester et al., 2017, for a comprehensive review).

8.3.3.1 Speed, Shape and Orientation of ICME

It is well known that ICME properties, such as speed, density or magnetic field, evolve when propagating through the ambient solar wind, and they often largely differ from their original structure as seen in coronagraph images. CMEs' initial speed is also extremely variable at the solar surface. When propagating in the solar wind, ICMEs' speeds tend to reach the background solar wind speed, with the fastest ICMEs being decelerated and the slowest being accelerated (e.g. Iju et al., 2013), the solar distance at which this background velocity is reached being still under debate. However, by a multi-instrument study involving MESSENGER observations close to Mercury and complementary data sets at 1 AU, Winslow et al. (2015) showed that acceleration of ICMEs can still be present far beyond Mercury's orbit and even continuous up to 1 AU.

Moreover, it is useful to fit the shape of the ICME (and more specifically the well-defined ICME-MC) in the interplanetary medium (see Figure 8.3.1 for a schematic of the structure of an ICME). For example, assuming ICME-MC are planar structures and share common characteristics, Janvier et al. (2013) used a statistical data set from the Wind mission to reconstruct analytical models of MC in the interplanetary medium. They found that ICME-MCs can be correctly reproduced by an analytical model with an ellipsoidal shape for the front part linked to two legs radially connecting the ellipse to the solar surface. Fitting this model to STEREO solar coronagraph images of ICMEs worked pretty well. This model was further used to compute magnetic helicity (amount of twisting inside the MC) convected away by MCs in the interplanetary medium (Démoulin et al., 2016). Statistical studies show that a wide range of inclinations of MCs with respect to the ecliptic plane are found (Janvier et al., 2013). However, some studies suggest that MCs' axis could be deflected during propagation (Lynch et al., 2009; Kliem et al., 2012) and even preferentially aligned with the local direction of the HCS close to the solar ecliptic plane (Yurchyshyn, 2008; Isavnin et al., 2014). Using a multipoint data set from MESSENGER, Venus Express, STEREO and ACE, Good and Forsyth (2016) also found that longitudinal extension of MC (separation between both legs) is largely between 15° and 30°.

8.3.3.2 Propagation of ICME

Repeated studies combining remote imaging of the solar corona with in situ measurements of the ambient solar wind have led to a relatively general consensus that almost all ICMEs share the same common magnetic flux rope (MC) configuration at the time of ejection (e.g. Owens et al., 2005). The reasons for the different signatures observed in the solar wind close to 1 AU can be explained either by ICME deformation or erosion during its propagation (see Section 8.3.3.3) or by the fact that the ICME is either crossed close to its apex (almost perpendicular to its central axis), leading to a well-defined magnetic flux rope signature (i.e. an MC), or crossed in the leg region (almost parallel to its central axis), leading to a more complex and unclear signature (such events look more like EJs) (see e.g. Möstl et al., 2010; Owens et al., 2012). These observations are closely related to the propagation direction of the ICME in the solar wind. Indeed, if the ICME is ejected towards the Earth and follows a radial expansion, as is

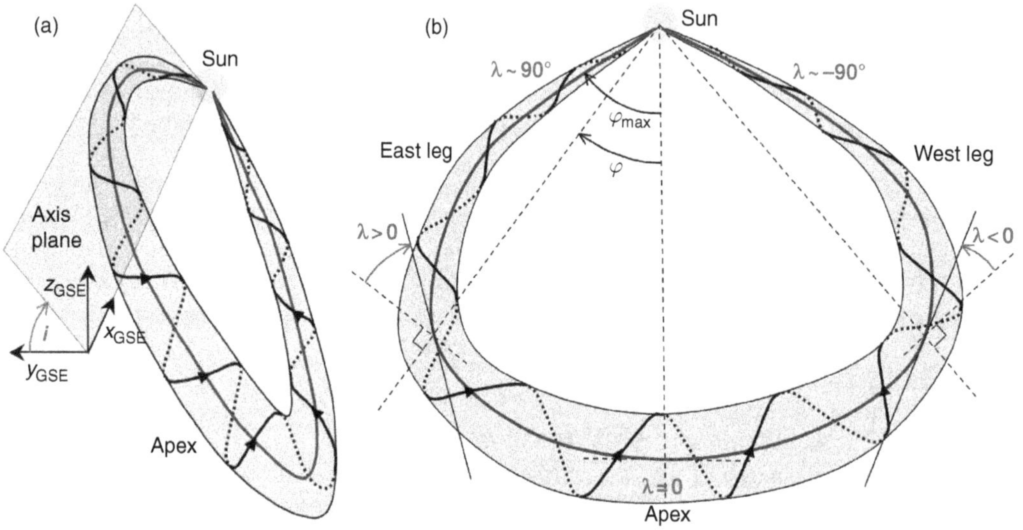

Figure 8.3.1 Schematics of a large-scale planar MC (light grey shape). (a) Schematic in 3-D showing an MC with a certain inclination i with respect to the ecliptic plane (x_{GSE},y_{GSE}). (b) Schematic in 2-D in the plane of the MC, with the angle ϕ being the heliospheric longitude and λ being a characteristic angle evolving monotonously along the MC axis and being 0° at the apex of the MC. The grey line inside the MC grey shape represents the MC axis; the black twisted line represents the magnetic field line at the boundary of the MC. Adapted from Figure 1 of Janvier et al. (2013). (A black-and-white version of this figure appears in some formats. For the colour version, please refer to the plate section.)

commonly assumed, the ICME is crossed close to its apex at 1 AU, and a corresponding MC signature is more often observed. On the other hand, if the ICME propagation deviates from the radial direction, the ICME is more likely to be observed far from its apex (legs region) and takes the form of an EJ at 1 AU. The non-radial expansion of ICMEs has been recently confirmed and found to be due to interaction with large-scale magnetic features close to the solar surface, such as coronal holes or active regions (AR), implying that the deviation occurs very close to the Sun (see e.g. Gopalswamy et al., 2009; Kim et al., 2013). Moreover, by a comprehensive analysis of EUV, coronagraph and heliospheric observations from STEREO and SOHO together with in situ signatures at Earth and Mars (Wind and Mars Express), Möstl et al. (2015) found that a CME was deviated very close to the Sun (below about 2 R_S) by more than 30° from the source region. Possible additional deviation of ICMEs in the interplanetary medium is still an open and crucial question in order to predict the direction and time of arrival of an ICME at Earth (Möstl et al., 2014). Initial results tend to show that the ICME direction remains relatively stable between the corona and the interplanetary medium (within about ±15° of heliocentric longitude of the solar source) (Liu et al., 2013; Davies et al., 2013), although interactions between successive CMEs during their propagation in the interplanetary medium could also cause a deviation from the radial direction (Lugaz et al., 2012).

8.3.3.3 *Interaction between ICMEs and the Erosion of ICMEs*

MC properties can be strongly affected during propagation by the MC interaction with other large-scale solar wind structures, like the HCS, the fast solar wind stream (Gulisano et al., 2012) or other slower CMEs (Liu et al., 2014). Significant erosion of azimuthal flux of the MC due to reconnection (~40%) has been also found to be a regular process occurring in the solar wind (Dasso et al., 2006; Ruffenach et al., 2012) (see Figure 8.3.2). With a comprehensive set of measurements (STEREO, ACE, Wind, and THEMIS), Ruffenach et al. (2012) showed simultaneously magnetic flux erosion of the MC and clear reconnection signatures seen in particle measurements. MCs are found to be eroded essentially within Mercury's orbit and in similar proportions ahead and at rear of the MC. In the former case, erosion is caused by interaction with the sheath magnetic field ahead of the MC. In the latter case, erosion is due to fast solar wind stream overtaking the MC (Ruffenach et al., 2015). Such an erosion process can strongly modify the magnetic polarity of the MC during its propagation and change its coupling efficiency with the Earth's magnetosphere (Lavraud et al., 2014).

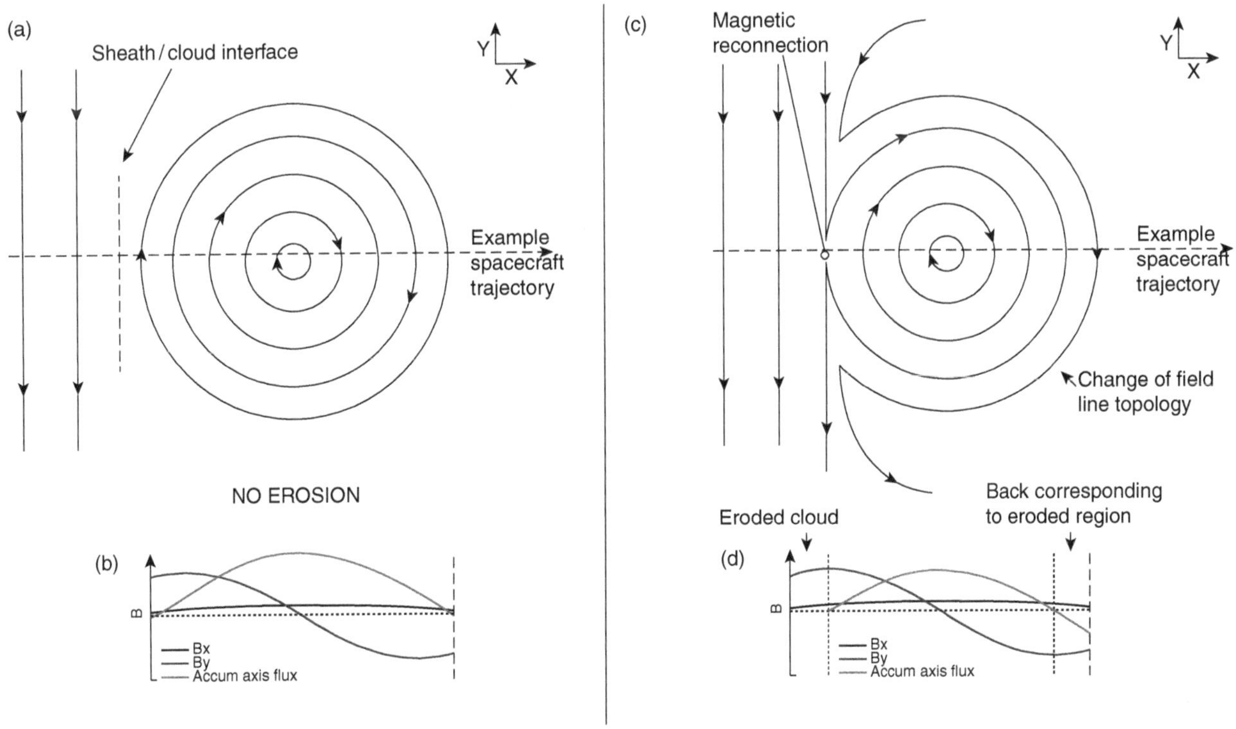

Figure 8.3.2 Schematic representing the magnetic structure of (a) a non-eroded and (c) an eroded MCs by reconnection, seen in the plane perpendicular to the MC axis, together with (b and d) the expected variations in the magnetic field components and accumulated azimuthal flux. Adapted from Figure 1 of Ruffenach et al. (2012).

8.3.3.4 Solar Cycle Variations of ICME Properties

Using observations from long-duration solar wind missions such as Wind or ACE, the evolution of ICME properties (occurrence, magnetic polarity, size, velocity, expansion rate), sheath and eventually driven-shock have also been followed for almost two solar cycles (23 and 24), with a special focus on the deep solar minimum (2007–9). ICMEs move faster and carry a stronger magnetic field (Chi et al., 2016) during the maximum phase of the solar cycle, leading to stronger magnetic storms (Kilpua et al., 2015), with the highest percentage of ICME-driven shocks (Mitsakou and Moussas, 2014). Moreover, the general solar wind properties show large discrepancies between solar cycles 23 and 24 with lower proton density, proton temperature, magnetic field strength, total pressure and Alfvén speed during cycle 24 (McComas et al., 2013; Gopalswamy et al., 2014). ICMEs are more likely to show an MC signature during solar minimum than solar maximum (Lepping et al., 2011; Chi et al., 2016), ICMEs being more likely to propagate close to the solar equator during solar minimum. Moreover, different properties of ICMEs-MCs are also observed in both cycles, with smaller size, slower speed and lower magnetic field strength during solar cycle 24, leading to lower magnetic storm geoeffectivity (Gopalswamy et al., 2015). These differences are even observed for ICMEs-MCs during the two last minima, confirming the peculiarity of the last solar cycle (Lepping et al., 2011).

8.4 Magnetosheath and Magnetosphere

8.4.1 Introduction

As already stated, the supersonic and super-Alfvénic solar wind flow is strongly decelerated and heated through the bow shock formed upstream of the Earth's magnetosphere. In this region, called the magnetosheath, the flow around the obstacle compresses the Earth's magnetic cavity on the dayside and extends it on the nightside, forming the magnetotail. Solar wind properties can face strong modifications during the bow shock crossing, making it sometimes difficult to extrapolate magnetosheath properties close to the magnetopause from solar wind measurements taken ahead of the Earth by ACE for example. From the frozen-in condition, magnetosheath and magnetospheric plasmas cannot mix and stay separated by the magnetosphere's boundary, called the magnetopause.

However, when the IMF and magnetospheric magnetic field lines are anti-parallel, magnetic reconnection (or merging) between these field lines can occur at the magnetopause. This process allows efficient magnetic reconfiguration where terrestrial magnetic field lines can become open and connected to the interplanetary medium (Dungey, 1961) and is recognised as the major mechanism by which energy, plasma and momentum can be transferred from the solar wind to the magnetosphere. The location where IMF and magnetospheric field lines are anti-parallel and reconnection can occur strongly depends on the orientation of the IMF projected onto the plane perpendicular to the Sun–Earth line. The clock angle (CA) defines the IMF direction in this plane defined by the south–north axis generally perpendicular to the Earth's orbit plane (Z-component, directed northward) and the dawn–dusk axis (Y-component, directed duskward) perpendicular to the south–north and Earth–Sun axes (X-component directed sunward). The CA ranges from 0° when the IMF direction is purely northward to 90° (270°) when the IMF direction is duskward (dawnward) to 180° when the IMF is southward. Due to the northward terrestrial magnetic field, magnetopause reconnection occurs close to the subsolar region when the CA is close to 180° (southward IMF) and nightside of the polar cusps (neutral points of the magnetosphere) when the CA is close to 0° (northward IMF) (Dungey, 1963). If the CA is close to 90° or 270° (duskward or dawnward IMF), the reconnection site will be split and displaced toward the flanks, the favoured hemisphere will be opposite on each flank and will depend on the sign of the IMF dawn–dusk component (Crooker, 1979). Complexity also arises from the fact that reconnection can occur when magnetic field lines are not anti-parallel. It is called component reconnection and makes the precise prediction of the magnetopause reconnection location extremely complex (Gonzalez and Mozer, 1974). A current popular model states that the magnetopause reconnection line follows the ridge of maximum magnetic shear where magnetosheath and magnetospheric field lines are the closest to anti-parallel (see Figure 8.4.1). With such a model, called the maximum magnetic shear model, a subsolar tilted reconnection line can appear even if the CA is close to 90° (270°) (Trattner et al., 2007; Fuselier et al., 2011). Finally, other processes at the magnetopause allow momentum and plasma transfer from the solar wind to the magnetosphere, such as diffusion, impulsive penetration and Kelvin–Helmholtz instabilities.

Magnetopause field lines opened through reconnection are dragged by the magnetic tension at the reconnection site and by the solar wind flow mainly in the anti-sunward direction where they are stored in the magnetotail. The convection of these opened field lines is mapped all along the magnetic field down to the ionosphere. The opened magnetic flux added to the magnetotail contributes to its destabilisation through field line stretching. The magnetotail will explosively reconfigure itself through magnetic reconnection and local current system generation and come back to a more stable dipolar configuration during sporadic events called substorms. Nightside newly reconnected magnetic field lines will then be dragged sunward by the magnetic tension. This cycle produce complex plasma circulation in the magnetosphere coupled to large-scale system of currents.

The large coverage of the GEOTAIL, THEMIS and Double Star-1 missions at low latitudes and of the Cluster mission at high latitudes allowed us to obtain a very

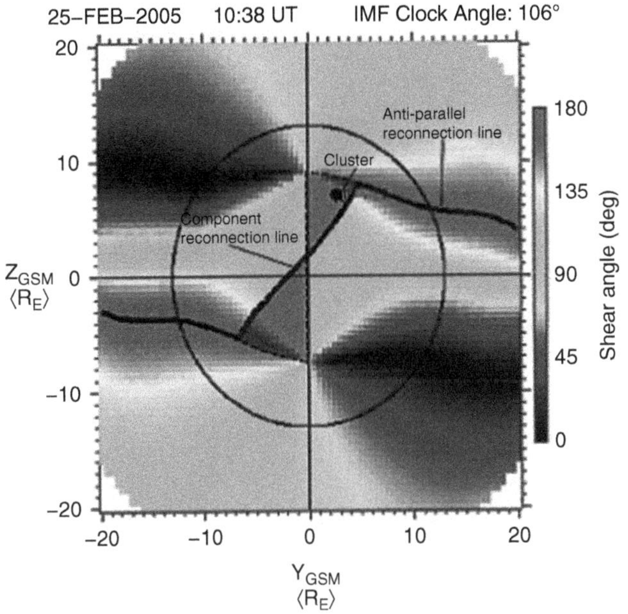

Figure 8.4.1 A 2-D map of shear angles between magnetospheric and magnetosheath magnetic fields at the magnetopause. Dark grey zones on top right and bottom left are near 180° (anti-parallel fields) and dark grey zones on top left and bottom right are near 0° (parallel fields). The shear angles are projected onto the Y-Z GSM plane, and the view is from the Sun. The circle is the terminator projected onto the plane. The maximum shear model predicts that anti-parallel reconnection occurs on the flanks and that there is a component reconnection line that crosses the dayside and connects the two anti-parallel reconnection regions on the flanks. Spacecraft crossing the magnetopause within the grey shaded regions (such as Cluster in this figure) would see different flow directions for reconnection jets from an anti-parallel reconnection line at higher latitudes or a component reconnection line at lower latitudes. Adapted from Figure 1 of Fuselier et al. (2011). (A black-and-white version of this figure appears in some formats. For the colour version, please refer to the plate section.)

comprehensive view of the large-scale properties of the magnetosheath (see Section 8.4.2), of the magnetopause from the sub-solar point to far down along the flanks (Section 8.4.3) and of the magnetotail (see Section 8.4.4).

8.4.2 Magnetosheath Results

Among other interesting new results, the changes of solar wind and IMF parameters across the bow shock have been investigated. One important parameter is the angle formed by the IMF direction with respect to the local normal of the bow shock. If this angle is above 45°, the shock is said quasi-perpendicular, while if it is below 45°, the shock is said quasi-parallel. In a Parker spiral orientation, quasi-parallel shocks will be located on the dawnside of the bow shock and quasi-perpendicular shocks will be located on the duskside. Bow shock crossing of ICME-MCs have been thoroughly investigated with simulations and an extensive set of observations (ACE, Cluster, GEOTAIL, THEMIS) (Turc et al., 2014, 2015). Magnetic field configuration of MCs shows strong modification during the traversal of quasi-parallel shocks while it remains close to identical at the traversal of quasi-perpendicular shocks. If the MC's magnetic field shows a dominant radial component (i.e. along the Sun–Earth direction), the shock will be dominantly quasi-parallel, and the bow shock traversal will modify the MC's magnetic configuration (Turc et al., 2016), potentially modifying its coupling with the magnetosphere.

Magnetosheath properties have also been found to be highly asymmetric. Although higher densities on the dawnside had already been observed in the past (e.g. Paularena et al., 2001), even more enhanced densities during period of low Alfvénic Mach number M_A (ratio between the speed of the medium and the speed of the local Alfvén waves) has been found by Walsh et al. (2012), along with higher temperatures on the dawnside. On the contrary, bulk flow and magnetic field strength have been found to be enhanced on the duskside. These asymmetries have been used to suggest that reconnection is favoured on the duskside magnetopause but with no experimental evidence yet (Walsh et al., 2012). Finally, during periods of low M_A and low plasma β (e.g. during ICME-MC crossing), the magnetosheath flow has been observed to be enhanced only in regions where the flow and the IMF are orthogonal due to the strong asymmetry of the magnetic forces during these periods (Lavraud et al., 2013).

8.4.3 Dayside Magnetosphere Results

Recently, the effect of other solar wind parameters, such as the cone angle, which is the angle between the Earth–Sun line (X-component) and the IMF direction; the plasma β; and the M_A, on reconnection and magnetosheath properties has also been studied. In particular, ICMEs are known to often show lower β and M_A than the ambient solar wind plasma, which in return affects the coupling processes between the solar wind and the magnetosphere.

8.4.3.1 Magnetopause Properties and Asymmetries

Extensive data sets have allowed us to disentangle how the different solar wind parameters affect the magnetopause location. For example, the magnetopause expands outward during a dominant radial IMF (cone angle close to 0° or 180°) by 1 R_E in average (Dušík et al., 2010) and by even more than 4 R_E in the dayside and 7 R_E in the nightside for several atypical events recorded by GEOTAIL and THEMIS (Suvorova et al., 2010). During times of a radial IMF, the total pressure exerted on the subsolar magnetopause is supposed to be lower than for any other IMF direction explaining the outward expansion (Samsonov et al., 2012). However, this explanation has been found to be only partially satisfying and needs more investigations (e.g. Suvorova and Dmitriev, 2015; Pi et al., 2017).

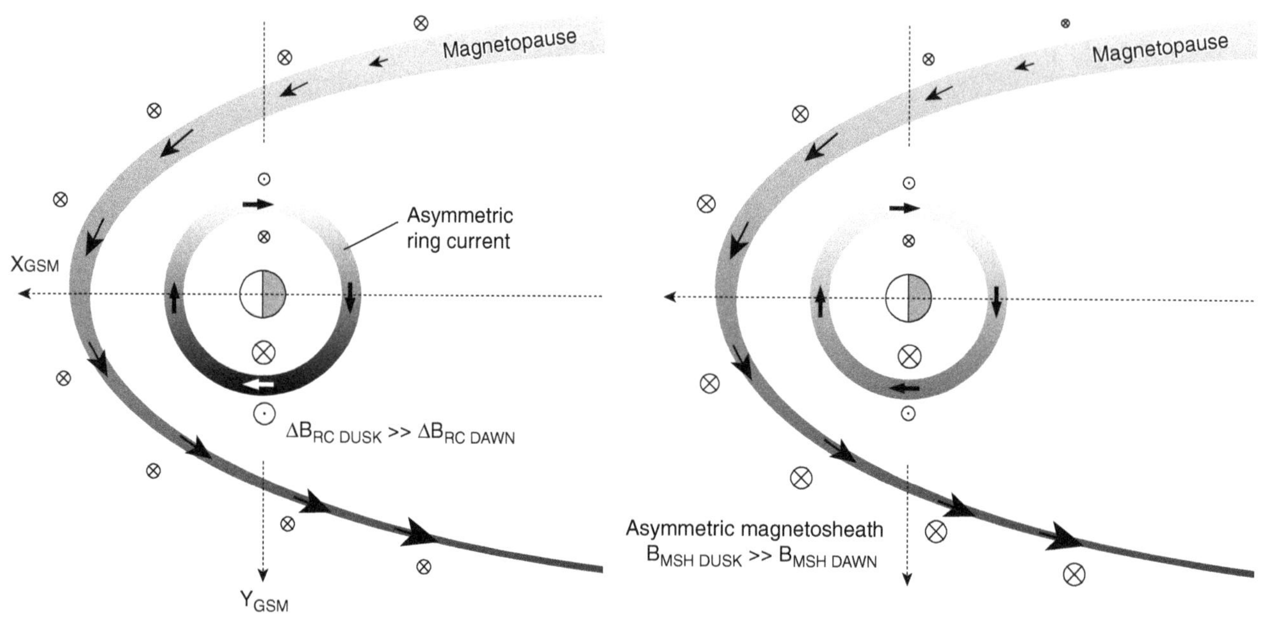

Figure 8.4.2 Illustration of possible explanations for the observed dawn–dusk magnetopause asymmetry. Light shading indicates low current density and darker shadings indicate higher current densities. Arrows indicate current direction and strength. (left) Asymmetries caused by ring current. (right) Asymmetries caused by asymmetric magnetosheath conditions. Adapted from Figure 8 of Haaland and Gjerloev (2013).

With the Cluster data set and its four-spacecraft capability to calculate current and boundary thickness, the magnetopause current has also been observed to show an asymmetric behaviour with more intense current in the duskside magnetopause than in the dawnside; this asymmetry being even enhanced during disturbed conditions (Haaland and Gjerloev, 2013; see Figure 8.4.2). This asymmetry could be caused either by the asymmetric magnetosheath properties (see Section 8.4.2 and Walsh et al., 2012) or by a possible connection with the internal ring current flowing a few R_E around the Earth. Indeed, magnetic indices characterising ring current intensity in different local time sectors show also such a dawn–dusk current asymmetry. Finally, the width of the magnetopause current has also been investigated, showing a thicker magnetopause on the dawnside than on the duskside, leading to a higher current intensity on the duskside than on the dawnside, while a similar total current flows on both flanks of the magnetopause (Haaland et al., 2014).

Lavraud et al. (2013) have also shown that the magnetopause can take on an elongated shape following the IMF direction during low M_A period, with difference reaching more than 5 R_E in radial distance from the Earth. This effect postulated by Lavraud and Borovsky (2008) with MHD simulations is due again to the enhanced magnetosheath flow in certain specific regions (see Section 8.4.2), causing an increasing force normal to the magnetopause and leading to its deformation.

Finally, even density asymmetries have been observed inside the dayside magnetosphere close to the magnetopause. Indeed, the region of dense plasma originating from the Earth's ionosphere and diffusing along trapped dipolar magnetic field lines, called the plasmasphere, can be dynamic. The plasmasphere can elongate through enhanced convection in the inner magnetosphere, forming a plume which can reach the magnetopause, more likely on the dayside post-noon sector. Dense and cold plasmaspheric plasma has been found in the outer magnetosphere and at the magnetopause between 25% and 50% of the time (André and Cully, 2012; Walsh et al., 2013).

8.4.3.2 *Dayside Magnetopause Reconnection and Energy Input*

Intrinsic magnetospheric effects have been found to affect the reconnection location and efficiency. For example, seasonal variations have been found in the location of the subsolar reconnection line during mainly southward IMF through simulations (Cnossen et al., 2012; Hoilijoki et al., 2014; Komar et al., 2015); the reconnection line is systematically displaced in the winter hemisphere during solstice. These simulation results have been confirmed by a Cluster and THEMIS statistical study (Zhu et al., 2015) and multi-spacecraft studies by GEOTAIL and MMS (Kitamura et al., 2016), and by THEMIS and Double Star-1 (Hoilijoki et al., 2014). Causes of this shift need more investigation but could lie in the seasonal shift of the line of maximum magnetic shear between magnetosheath and magnetospheric field lines, caused by the variation of the tilt angle of the terrestrial magnetic dipole (Trattner et al., 2007). Seasonal effect has also been investigated in the coupling efficiency of the solar wind–magnetosphere–ionosphere system. With MHD simulation and Cluster data sets,

Palmroth et al. (2012) have found higher energy inputs during equinox than during solstice, and during solstice, the transfer of energy from the solar wind is mainly directed toward the summer hemisphere (Palmroth et al., 2012; Cnossen et al., 2012; Hoilijoki et al., 2014).

Presence of cold plasma from plasmaspheric plumes on the magnetopause shows a decrease in the velocity of the jets caused by magnetopause reconnection. It has been suggested that mass loading of the magnetosphere by cold ions could have an important effect by changing the properties of the reconnection process and the structure of the reconnection site and by decreasing the reconnection efficiency (Walsh et al., 2014; Toledo-Redondo et al., 2015, 2016). However, other recent studies using the multi-satellite capability of the Cluster mission to calculate the reconnection rate show minimised or no effects when the cold plasma density remains low (Wang et al., 2014, 2015). Consequently, more investigation is necessary of the cold ion contribution to the magnetopause reconnection process.

Finally, external parameters may also have strong impact on the reconnection location efficiency. For example, data from Cluster or Polar missions have allowed us to study parameters seldom studied in the past, like the IMF Earth–Sun X-component (IMF Bx). The maximum magnetic shear model (Trattner et al., 2007) predicts that if the CA is close to 90° or 270° and if the IMF Bx component is dominant, anti-parallel reconnection will dominate with two reconnection sites confined where magnetosheath and magnetospheric field lines are anti-parallel. On the contrary, if the IMF Bx is small, reconnection will occur preferentially along the tilted reconnection line close to the subsolar point (see Figure 8.4.1 for the position of the reconnection line). This model has been tested against Cluster and Polar ion observations in the polar cusp, using the travel time of these particles along magnetic field lines to infer the distance of the reconnection site. However, the results are not fully consistent with the model, since not all events with strong Bx show only anti-parallel reconnection sites (Trattner et al., 2012). In conclusion, the effect of the Bx component on the reconnection site location will need further investigation.

8.4.3.3 Development of Kelvin–Helmholtz Instabilities at the Flank of the Magnetopause

New important results have also been gathered on the Kelvin–Helmholtz instabilities (KHIs). These instabilities are known to develop in the presence of strong flow shear between the high velocity, with a predominantly anti-sunward flow in the magnetosheath and the slightly sunward magnetospheric flow. KHIs happen on the low-latitude flank of the magnetopause during northward IMF when no reconnection is able to develop on the dayside magnetopause and are characterised by undulations travelling anti-sunward. If the non-linear phase of the instability is reached, the undulations can steepen enough to form rolled-up vortices, where reconnection is thought to occur (Nykyri and Otto, 2001). Thanks to a statistical study with THEMIS data, Lin et al. (2014) have shown that the Kelvin–Helmholtz (KH) wave period is shorter during high solar wind speed and longer with larger IMF clock angle, but the solar wind as controlling factor of the KHI needs to be further investigated. Moreover, recent results have shown that KHI could develop also during southward IMF (Hwang et al., 2011; Yan et al., 2014) or at high latitudes behind the cusp (Hwang et al., 2012). The physical reason for the unexpected development of KHI during these events deserves more investigation. With GEOTAIL and Double Star-1 data, Taylor et al. (2012) have also shown that KHIs are more likely to develop on the duskside flank. This asymmetric behaviour could be either due to the asymmetric properties of the magnetosheath and the magnetopause (see Sections 8.4.3.1 and 8.4.3.2) or due to the contact of plasmaspheric plume on the post-noon sector magnetopause with KHIs appearing more easily and closer to the subsolar magnetopause in the presence of cold ions (Walsh et al., 2015).

8.4.4 Magnetotail Results

The dynamics of the nightside magnetosphere is extremely complex and would need a full chapter in itself to describe the complex structure of the NCS at large and small scales and the surrounding plasma sheet (see e.g. Petrukovich et al., 2015, for a review) or to give all the latest insights regarding the substorm processes from the generation mechanism and the controlling (external or/and internal) parameters to the energy dissipation in the ionosphere (see e.g. Sergeev et al., 2012, and references therein).

The NCS at near-Earth distance located at the interface between the two magnetotail lobes shows also highly asymmetric properties. Using the four-spacecraft capability of the Cluster mission to calculate the intensity and the thickness of the NCS, statistical study shows strong dawn–dusk asymmetries between 15 and 20 R_E from Earth, with lower current intensity, thicker current sheet and lower radius of curvature of magnetic field lines in the dawnside than in the duskside (Artemyev et al., 2011). These asymmetries could be explained by the existence of a transient proton population with higher density in the duskward and carrying a large part of the current. The magnetotail current has also been found to be thinner in a large pre-midnight sector (Rong et al., 2011). As thinner current sheets are more likely to disrupt and be diverted toward the ionosphere as part of the substorm process, these observations are consistent with the occurrence of substorm onset in the same region. The occurrence of fast earthward flow in the plasma sheet surrounding the NCS and related to magnetotail reconnection is also higher in the pre-midnight sector (Lee et al., 2012). Using an extensive GEOTAIL data set, the NCS is also studied at a larger distance from Earth (20–100 R_E) and shows similar extreme asymmetric properties (Vasko et al.,

2015). The occurrence of thin current sheets is also similar, implying that the distant tail could be as important as the near-current sheet for magnetotail dynamics. Finally, since the discovery with Cluster of a highly inclined NCS (Zhang et al., 2002) and the oscillations of the NCS (Sergeev et al., 2004) propagating toward the flanks, new results show that such a wavy NCS keeps the same current characteristics as opposed to a horizontal NCS (intensity, thickness), but a part of the current is now aligned with the magnetic field (Vasko et al., 2014).

8.5 Conclusion

This chapter was an attempt to present new insights on far spacecraft measurements. However, the choice of the results presented here is extremely subjective, as important results published in the literature are not described here. Understanding the physics of the Sun–Earth and Sun–planets systems is an extremely exciting task. Results presented here show that although progress is made in this matter, there are still numerous open questions which will need to be addressed in the future in order to fully comprehend these complex relationships and be able to predict them for space weather purposes. In the near future, new space missions are planned and will allow new breakthroughs. The Sun and solar wind will benefit because of the recent launch of the Parker Solar Probe mission (August 2018) and the upcoming launch of the Solar Orbiter mission (expected in February 2020), which will approach closer to the Sun than any earlier missions (about 0.25 AU for the latter and even about 10 R_S for the former). Some planets in the solar system will be visited by new missions in the next two decades: Bepi Colombo for Mercury (launched in October 2018 with an orbital insertion planned in 2025) and JUICE for Jupiter and its moons (expected launch in 2022 and arrival in 2029). Finally, the Earth's magnetosphere, while less well served than during these past two decades, will benefit from the small mission SMILE (expected launch in 2023), which will mainly carry X-ray and UV imaging cameras to combine images of the auroral ovals, and of the magnetosheath and polar cusps.

References

Abbo, L., E. Antonucci, Z. Mikić, J. A. Linker, P. Riley, R. Lonello (2010), Characterization of the slow wind in the outer corona, Adv. Space Res., 46, 1400, doi: 10.1016/j.asr.2010.08.008.

Abbo L., L. Ofman, S. K. Antiochos, V. H. Hansteen, L. Harra, Y.-K. Ko, G. Lapenta, B. Li, P. Riley, L. Strachan, R. von Steiger and Y.-M. Wang (2016), Slow solar wind: Observations and modeling, Space Science Reviews, 201, 55, doi: 10.1007/s11214-016-0264-1.

André, M., and C. M. Cully (2012), Low-energy ions: A previously hidden solar system particle population, Geophys. Res. Lett., 39, L03101, doi: 10.1029/2011GL050242.

Artemyev, A. V., A. A. Petrukovich, R. Nakamura and L. M. Zelenyi (2011), Cluster statistics of thin current sheets in the Earth magnetotail: Specifics of the dawn flank, proton temperature profiles and electrostatic effects, J. Geophys. Res., 116, A09233, doi: 10.1029/2011JA016801.

Chi, Y., et al. (2016), Statistical study of the interplanetary coronal mass ejections from 1995 to 2015, Sol. Phys., 291, 2419, doi: 10.1007/s11207-016-0971-5.

Cnossen, I., M. Wiltberger and J. E. Ouellette (2012), The effects of seasonal and diurnal variations in the Earth's magnetic dipole orientation on solar wind–magnetosphere–ionosphere coupling, J. Geophys. Res., 117, A11211, doi: 10.1029/2012JA017825.

Crooker, N. U. (1979), Dayside merging and cusp geometry, J. Geophys. Res., 84(A3), 951–9, doi: 10.1029/JA084iA03p00951.

Dasso, S., C. H. Mandrini, P. Démoulin and M. L. Luoni (2006), A new model-independent method to compute magnetic helicity in magnetic clouds, A&A, 455, 349–59, doi: 10.1051/0004-6361:20064806.

Davies, J. A., C. H. Perry, R. M. G. M. Trines, R. A. Harrison, N. Lugaz, C. Möstl, Y. D. Liu and K. Steed (2013), Establishing a stereoscopic technique for determining the kinematic properties of solar wind transients based on a generalized self-similarly expanding circular geometry, Astrophys. J., 777(2), doi: 10.1088/0004-637X/777/2/167.

Démoulin, P., M. Janvier and S. Dasso (2016), Magnetic flux and helicity of magnetic clouds, Sol. Phys., 291, 531, doi: 10.1007/s11207-015-0836-3.

Dungey, J. W. (1961), Interplanetary magnetic field and the auroral zones, Phys. Rev. Lett., 6(2), 7–48, doi: 10.1103/PhysRevLett.6.47.

Dungey, J. W. (1963), The structure of the exosphere or adventures in velocity space, in *Geophysics: The Earth's Environment*, edited by C. Dewitt, J. Hieblot, and A. Lebeau, pp. 505–50, Gordon and Breach, New York.

Dunlop, M. W., A. Balogh, K.-H. Glassmeier and P. Robert (2002), Four-point Cluster application of magnetic field analysis tools: The Curlometer, J. Geophys. Res., 107, 1384, doi: 10.1029/2001JA005088.

Dušík, Š., G. Granko, J. Šafránková, Z. Němeček and K. Jelínek (2010), IMF cone angle control of the magnetopause location: Statistical study, Geophys. Res. Lett., 37, L19103, doi: 10.1029/2010GL044965.

Echer E., B. T. Tsurutani and W. D. Gonzalez (2013), Interplanetary origins of moderate (-100 nT < Dst ≤ -50 nT) geomagnetic storms during solar cycle 23 (1996–2008), J. Geophys. Res., 118, 385–92, doi: 10.1029/2012JA018086.

Fuselier, S. A., K. J. Trattner and S. M. Petrinec (2011), Antiparallel and component reconnection at the dayside magnetopause, J. Geophys. Res., 116, A10227, doi: 10.1029/2011JA016888.

Gonzalez, W.D., and F. S. Mozer (1974). A quantitative model for the potential resulting from reconnection with an arbitrary interplanetary magnetic field. J. Geophys. Res., 79, doi: 10.1029/JA079i028p04186.

Good S. W., and R. J. Forsyth (2016), Interplanetary coronal mass ejections observed by MESSENGER and Venus Express, Sol. Phys., 291, 239, doi: 10.1007/s11207-015-0828-3.

Gopalswamy, N., S. Akiyama, S. Yashiro, H. Xie, P. Mäkelä and G. Michalek (2014), Anomalous expansion of coronal mass ejections during solar cycle 24 and its space weather implications, Geophys. Res. Lett., 41, 2673–80, doi: 10.1002/2014GL059858.

Gopalswamy, N., P. Mäkelä, H. Xie, S. Akiyama and S. Yashiro (2009), CME interactions with coronal holes and their interplanetary consequences, J. Geophys. Res., 114, A00A22, doi: 10.1029/2008JA013686.

Gopalswamy, N., S. Yashiro, H. Xie, S. Akiyama and P. Mäkelä (2015), Properties and geoeffectiveness of magnetic clouds during solar cycles 23 and 24, J. Geophys. Res., 120, 9221–45, doi: 10.1002/2015JA021446.

Gulisano A. M., P. Démoulin, S. Dasso and L. Rodriguez (2012), Expansion of magnetic clouds in the outer heliosphere, A&A, 543, A107, doi: 10.1051/0004-6361/201118748.

Haaland, S., and J. Gjerloev (2013), On the relation between asymmetries in the ring current and magnetopause current, J. Geophys. Res., 118, 7593–7604, doi: 10.1002/2013JA019345.

Haaland, S., J. Reistad, P. Tenfjord, J. Gjerloev, L. Maes, J. DeKeyser, R. Maggiolo, C. Anekallu and N. Dorville (2014), Characteristics of the flank magnetopause: Cluster observations, J. Geophys. Res., 119, 9019–37, doi: 10.1002/2014JA020539.

Hoilijoki, S., V. M. Souza, B. M. Walsh, P. Janhunen and M. Palmroth (2014), Magnetopause reconnection and energy conversion as influenced by the dipole tilt and the IMF Bx, J. Geophys. Res., 119, 4484–94, doi: 10.1002/2013JA019693.

Hwang, K.-J., M. L. Goldstein, M. M. Kuznetsova, Y. Wang, A. F. Viñas and D. G. Sibeck (2012), The first in situ observation of Kelvin–Helmholtz waves at high-latitude magnetopause during strongly dawnward interplanetary magnetic field conditions, J. Geophys. Res., 117, A08233, doi: 10.1029/2011JA017256.

Hwang, K.-J., M. M. Kuznetsova, F. Sahraoui, M. L. Goldstein, E. Lee and G. K. Parks (2011), Kelvin–Helmholtz waves under southward interplanetary magnetic field, J. Geophys. Res., 116, A08210, doi: 10.1029/2011JA016596.

Iju, T., M. Tokumaru and K. Fujiki (2013), Radial speed evolution of interplanetary coronal mass ejections during solar cycle 23, Sol. Phys., 288, 331, doi: 10.1007/s11207-013-0297-5.

Isavnin, A., A. Vourlidas and E. K. J. Kilpua (2014), Three-dimensional evolution of flux-rope CMEs and its relation to the local orientation of the heliospheric current sheet, Sol. Phys., 289, 2141, doi: 10.1007/s11207-013-0468-4.

Janvier, M., P. Démoulin and S. Dasso (2013), Global axis shape of magnetic clouds deduced from the distribution of their local axis orientation, A&A, 556, A50, doi: 10.1051/0004-6361/201321442.

Kilpua, E. K. J., J. G. Luhmann, J. Gosling, Y. Li, H. Elliott, C. T. Russell, L. Jian, A. B. Galvin, D. Larson, P. Schroeder, K. Simunac and G. Petrie (2009), Small solar wind transients and their connection to the large-scale coronal structure, Sol. Phys., 256, 327, doi: 10.1007/s11207-009-9366-1.

Kilpua, E. K. J., N. Olspert, A. Grigorievskiy, M. J. Käpylä, E. I. Tanskanen, H. Miyahara, R. Kataoka, J. Pelt and Y. D. Liu (2015), Statistical study of strong and extreme geomagnetic disturbances and solar cycle characteristics, Astrophys. J., 806(2), doi: 10.1088/0004-637X/806/2/272.

Kim R.-S., N. Gopalswamy, K.-S. Cho, Y.-J. Moon and S. Yashiro (2013), Propagation characteristics of CMEs associated with magnetic clouds and ejecta, Sol. Phys., 284, 77–88, doi: 10.1007/s11207-013-0230-y.

Kitamura, N., et al. (2016), Shift of the magnetopause reconnection line to the winter hemisphere under southward IMF conditions: Geotail and MMS observations, Geophys. Res. Lett., 43, 5581–8, doi: 10.1002/2016GL069095.

Kliem, B., T. Török and W. T. Thompson (2012), A parametric study of erupting flux rope rotation: Modeling the 'cartwheel CME' on 9 April 2008, Sol. Phys., 281, 137, doi: 10.1007/s11207-012-9990-z.

Komar, C. M., R. L. Fermo and P. A. Cassak (2015), Comparative analysis of dayside magnetic reconnection models in global magnetosphere simulations, J. Geophys. Res., 120, 276–94, doi: 10.1002/2014JA020587.

Lavraud, B. and J. E. Borovsky (2008), Altered solar wind–magnetosphere interaction at low Mach numbers: Coronal mass ejections, J. Geophys. Res., 113, A00B08, doi: 10.1029/2008JA013192.

Lavraud, B., et al. (2013), Asymmetry of magnetosheath flows and magnetopause shape during low Alfvén Mach number solar wind, J. Geophys. Res., 118, 1089–1100, doi: 10.1002/jgra.50145.

Lavraud, B., A. Ruffenach, A. P. Rouillard, P. Kajdic, W. B. Manchester and N. Lugaz (2014), Geo-effectiveness and radial dependence of magnetic cloud erosion by magnetic reconnection, J. Geophys. Res., 119, 26–35, doi: 10.1002/2013JA019154.

Lee, D.-Y., H.-S. Kim, S. Ohtani and M. Y. Park (2012), Statistical characteristics of plasma flows associated with magnetic dipolarizations in the near-tail region of r < 12 RE, J. Geophys. Res., 117, A01207, doi: 10.1029/2011JA017246.

Lepping, R. C., C.-C. Wu, D.B. Berdichevsky and A. Szabo (2011), Magnetic clouds at/near the 2007–2009 solar minimum: Frequency of occurrence and some unusual properties, Sol. Phys., 274, 345, doi: 10.1007/s11207-010-9646-9.

Lin, D., C. Wang, W. Li, B. Tang, X. Guo and Z. Peng (2014), Properties of Kelvin–Helmholtz waves at the magnetopause under northward interplanetary magnetic field: Statistical study, J. Geophys. Res., 119, 7485–94, doi: 10.1002/2014JA020379.

Liu, Y. D., et al. (2013), On Sun-to-Earth propagation of coronal mass ejections, Astrophys. J., 769(1), doi: 10.1088/0004-637X/769/1/45.

Liu, Y. D., Z. Yang, R. Wang, J. G. Luhmann, J. D. Richardson and N. Lugaz (2014), Sun-to-Earth characteristics of two coronal mass ejections interacting near 1 AU: Formation of a complex ejecta and generation of a two-step geomagnetic storm, Astrophys. J. Lett., 793(2), doi: 10.1088/2041-8205/793/2/L41.

Lugaz, N., C. J. Farrugia, J. A. Davies, C. Möstl, C. J. Davis, I. I. Roussev and M. Temmer (2012), The deflection of the two interacting coronal mass ejections of 2010 May 23–24 as revealed by combined in situ measurements and heliospheric imaging, Astrophys. J., 759(1), doi: 10.1088/0004-637X/759/1/68.

Lynch, B. J., S. K. Antiochos, Y. Li, J. G. Luhmann and C. R. DeVore (2009), Rotation of coronal mass ejections during eruption, Astrophys. J., 697(2), doi: 10.1088/0004-637X/697/2/1918.

McComas, D. J., H. A. Elliott, N. A. Schwadron, J. T. Gosling, R. M. Skoug and B. E. Goldstein (2003), The three-dimensional solar wind around solar maximum, Geophys. Res. Lett., 30, 1517, doi: 10.1029/2003GL017136.

McComas, D. J., N. Angold, H. A. Elliott, G. Livadiotis, N. A. Schwadron, R. M. Skoug and C. W. Smith, Weakest solar wind of the space age and the current 'mini' solar maximum (2013), Astrophys. J., 779(1), doi: 10.1088/0004-637X/779/1/2.

Marchaudon, A., J.-C. Cerisier, J.-M. Bosqued, M. W. Dunlop, J. A. Wild, P. M. Décréau, E. Förster, D. Fontaine and H. Laakso (2004), Transient plasma injections in the dayside magnetosphere: One-to-one correlated observations by Cluster and SuperDARN, Ann. Geophys., 22, 141–58, doi: 10.5194/angeo-22-141-2004.

Mitsakou, E., and X. Moussas (2014), Statistical study of ICMEs and their sheaths during solar cycle 23 (1996–2008), Sol. Phys., 289, 3137, doi: 10.1007/s11207-014-0505-y.

Möstl, C., et al. (2014), Connecting speeds, directions and arrival times of 22 coronal mass ejections from the Sun to 1 AU, Astrophys. J., 787(2), doi: 10.1088/0004-637X/787/2/119.

Möstl, C., et al. (2015), Strong coronal channelling and interplanetary evolution of a solar storm up to Earth and Mars, Nat. Commun., 6(7135), doi: 10.1038/ncomms8135.

Möstl, C., M. Temmer, T. Rollett, C. J. Farrugia, Y. Liu, A. M. Veronig, M. Leitner, A. B. Galvin and H. K. Biernat (2010), STEREO and Wind observations of a fast ICME flank triggering a prolonged geomagnetic storm on 5–7 April 2010, Geophys. Res. Lett., 37, L24103, doi: 10.1029/2010GL045175.

Nykyri, K., and A. Otto (2001), Plasma transport at the magnetospheric boundary due to reconnection in Kelvin-Helmholtz vortices, Geophys. Res. Lett., 28, 3565, doi: 10.1029/2001GL013239.

Owens, M. J., P. J. Cargill, C. Pagel, G. L. Siscoe and N. U. Crooker (2005), Characteristic magnetic field and speed properties of interplanetary coronal mass ejections and their sheath regions. J. Geophys. Res., 110, A01105, doi: 10.1029/2004JA010814.

Owens, M. J., P. Démoulin, N. P. Savani, B. Lavraud and A. Ruffenach (2012), Implications of non-cylindrical flux ropes for magnetic cloud reconstruction techniques and the interpretation of double flux rope events, Sol. Phys., 278, 435, doi: 10.1007/s11207-012-9939-2.

Palmroth, M., R. C. Fear and I. Honkonen (2012), Magnetopause energy transfer dependence on the interplanetary magnetic field and the Earth's magnetic dipole axis orientation, Ann. Geophys., 30, 515–26, doi: 10.5194/angeo-30-515-2012.

Parker, E. N. (1958), Dynamics of the interplanetary gas and magnetic fields, Astrophys. J., 128, 664.

Paularena, K. I., J. D. Richardson, M. A. Kolpak, C. R. Jackson and G. L. Siscoe (2001), A dawn–dusk density asymmetry in Earth's magnetosheath, J. Geophys. Res., 106(A11), 25377–25394, doi: 10.1029/2000JA000177.

Petrukovich A., A. Artemyev, I. Vasko, R. Nakamura and L. Zelenyi (2015), Current sheets in the Earth magnetotail: Plasma and magnetic field structure with Cluster project observations, Space Sci. Rev., 188, 311–37, doi: 10.1007/s11214-014-0126-7.

Pi, G., J.-H. Shue, K. Grygorov, H.-M. Li, Z. Němeček, J. Šafránková, Y.-H. Yang and K. Wang (2017), Evolution of the magnetic field structure outside the magnetopause under radial IMF conditions, J. Geophys. Res., 122, 4051–63, doi: 10.1002/2015JA021809.

Rong, Z. J., W. X. Wan, C. Shen, X. Li, M. W. Dunlop, A. A. Petrukovich, T. L. Zhang and E. Lucek (2011), Statistical survey on the magnetic structure in magnetotail current sheets, J. Geophys. Res., 116, A09218, doi: 10.1029/2011JA016489.

Rouillard, A. P., et al. (2010a), Intermittent release of transients in the slow solar wind: 1. Remote sensing observations, J. Geophys. Res., 115, A04103, doi: 10.1029/2009JA014471.

Rouillard, A. P., et al. (2010b), Intermittent release of transients in the slow solar wind: 2. In situ evidence, J. Geophys. Res., 115, A04104, doi: 10.1029/2009JA014472.

Ruffenach, A., et al. (2015), Statistical study of magnetic cloud erosion by magnetic reconnection, J. Geophys. Res., 120, 43–60, doi: 10.1002/2014JA020628.

Ruffenach, A., et al. (2012), Multispacecraft observation of magnetic cloud erosion by magnetic reconnection during propagation, J. Geophys. Res., 117, A09101, doi: 10.1029/2012JA017624.

Samsonov, A. A., Z. Němeček, J. Šafránková and K. Jelínek (2012), Why does the subsolar magnetopause move sunward for radial interplanetary magnetic field?, J. Geophys. Res., 117, A05221, doi: 10.1029/2011JA017429.

Sanchez-Diaz, E., A. P. Rouillard, B. Lavraud, K. Segura, C. Tao, R. Pinto, N. R. Sheeley Jr and I. Plotnikov (2016), The very slow solar wind: Properties, origin and variability, J. Geophys. Res., 121, 2830–41, doi: 10.1002/2016JA022433.

Shen, C., X. Li, M. Dunlop, Z. X. Liu, A. Balogh, D. N. Baker, M. Hapgood and X. Wang (2003), Analyses on the geometrical structure of magnetic field in the current sheet based on Cluster measurements, J. Geophys. Res., 108(A5), 1168, doi: 10.1029/2002JA009612.

Sergeev, V. A., V. Angelopoulos and R. Nakamura (2012), Recent advances in understanding substorm dynamics, Geophys. Res. Lett., 39, L05101, doi: 10.1029/2012GL050859.

Sergeev, V., A. Runov, W. Baumjohann, R. Nakamura, T. L. Zhang, A. Balogh, P. Louarn, J. Sauvaud and H. Reme (2004), Orientation and propagation of current sheet oscillations, Geophys. Res. Lett., 31, 5807, doi: 10.1029/2003GL019346.

Suvorova, A. V., and A. V. Dmitriev (2015), Magnetopause inflation under radial IMF: Comparison of models, Earth Space Sci., 2, 107–14, doi: 10.1002/2014EA000084.

Suvorova, A. V., J.-H. Shue, A. V. Dmitriev, D. G. Sibeck, J. P. McFadden, H. Hasegawa, K. Ackerson, K. Jelínek, J. Šafránková and Z. Němeček (2010), Magnetopause expansions for quasi-radial interplanetary magnetic field: THEMIS and Geotail observations, J. Geophys. Res., 115, A10216, doi: 10.1029/2010JA015404.

Taylor M. G., et al. (2012), Spatial distribution of rolled up Kelvin–Helmholtz vortices at Earth's dayside and flank magnetopause, Ann. Geophys., 30, 1025–35, doi: 10.5194/angeo-30-1025-2012.

Toledo-Redondo, S., M. André, A. Vaivads, Yu. V. Khotyaintsev, B. Lavraud, D. B. Graham, A. Divin and N. Aunai (2016), Cold ion heating at the dayside magnetopause during magnetic reconnection, Geophys. Res. Lett., 43, 58–66, doi: 10.1002/2015GL067187.

Toledo-Redondo, S., A. Vaivads, M. André and Y. V. Khotyaintsev (2015), Modification of the Hall physics in magnetic reconnection due to cold ions at the Earth's magnetopause, Geophys. Res. Lett., 42, 6146–54, doi: 10.1002/2015GL065129.

Trattner, K. J., J. S. Mulcock, S. M. Petrinec and S. A. Fuselier (2007), Probing the boundary between anti-parallel and component reconnection during southwards interplanetary magnetic field conditions, J. Geophys. Res., 112, A08210, doi: 10.1029/2007JA012270.

Trattner, K. J., S. M. Petrinec, S. A. Fuselier and T. D. Phan (2012), The location of reconnection at the magnetopause: Testing the maximum magnetic shear model with THEMIS observations, J. Geophys. Res., 117, A01201, doi: 10.1029/2011JA016959.

Turc, L., C. P. Escoubet, D. Fontaine, E. K. J. Kilpua and S. Enestam (2016), Cone angle control of the interaction of magnetic clouds with the Earth's bow shock, Geophys. Res. Lett., 43, 4781–9, doi: 10.1002/2016GL068818.

Turc, L., D. Fontaine, P. Savoini and E. K. J. Kilpua (2014), Magnetic clouds' structure in the magnetosheath as observed by Cluster and Geotail: four case studies, Ann. Geophys., 32, 1247–61, doi: 10.5194/angeo-32-1247-2014.

Turc, L., D. Fontaine, P. Savoini and R. Modolo (2015), 3D hybrid simulations of the interaction of a magnetic cloud with a bow shock, J. Geophys. Res., 120, 6133–51, doi: 10.1002/2015JA021318.

Vasko, I. Y., A. V. Artemyev, A. A. Petrukovich, R. Nakamura and L. M. Zelenyi (2014), The structure of strongly tilted current sheets in the Earth magnetotail, Ann. Geophys., 32, 133, doi: 10.5194/angeo-32-133-2014.

Vasko, I. Y., A. A. Petrukovich, A. V. Artemyev, R. Nakamura and L. M. Zelenyi (2015), Earth's distant magnetotail current sheet near and beyond lunar orbit, J. Geophys. Res., 120, 8663–80, doi: 10.1002/2015JA021633.

Walsh, B. M., T. D. Phan, D. G. Sibeck and V. M. Souza (2014), The plasmaspheric plume and magnetopause reconnection, Geophys. Res. Lett., 41, 223–8, doi: 10.1002/2013GL058802.

Walsh, B. M., D. G. Sibeck, Y. Nishimura and V. Angelopoulos (2013), Statistical analysis of the plasmaspheric plume at the magnetopause, *J. Geophys. Res.*, 118, 4844–51, doi: 10.1002/jgra.50458.

Walsh, B. M., D. G. Sibeck, Y. Wang and D. H. Fairfield (2012), Dawn–dusk asymmetries in the Earth's magnetosheath, J. Geophys. Res., 117, A12211, doi: 10.1029/2012JA018240.

Walsh, B. M., E. G. Thomas, K.-J. Hwang, J. B. H. Baker, J. M. Ruohoniemi and J. W. Bonnell (2015), Dense plasma and Kelvin–Helmholtz waves at Earth's dayside magnetopause, J. Geophys. Res., 120, 5560–73, doi: 10.1002/2015JA021014.

Wang, S., L. M. Kistler, C. G. Mouikis, Y. Liu and K. J. Genestreti (2014), Hot magnetospheric O+ and cold ion behavior in magnetopause reconnection: Cluster observations, J. Geophys. Res., 119, 9601–23, doi: 10.1002/2014JA020402.

Wang, S., L. M. Kistler, C. G. Mouikis and S. M. Petrinec (2015), Dependence of the dayside magnetopause reconnection rate on local conditions, J. Geophys. Res., 120, 6386–6408, doi: 10.1002/2015JA021524.

Winchester, W., E. K. J. Kilpua, Y. D. Liu, N. Lugaz, P. Riley, T. Török and B. Vršnak (2017), The physical processes of CME/ICME evolution, Space Sci. Rev., 212, 1159, doi: 10.1007/s11214-017-0394-0.

Winslow, R. M., N. Lugaz, L. C. Philpott, N. A. Schwadron, C. J. Farrugia, B. J. Anderson and C. W. Smith (2015), Interplanetary coronal mass ejections from MESSENGER orbital observations at Mercury, J. Geophys. Res., 120, 6101–18, doi: 10.1002/2015JA021200.

Wu, C.-C., and R. P. Lepping (2007), Comparison of the characteristics of magnetic clouds and magnetic cloud-like structures for the events of 1995–2003, Sol. Phys., 242, 159, doi: 10.1007/s11207-007-0323-6.

Wu C.-C., and R. P. Lepping (2011), Statistical comparison of magnetic clouds with interplanetary coronal mass ejections for solar cycle 23, Sol. Phys., 269, 141, doi: 10.1007/s11207-010-9684-3.

Yan, G. Q., F. S. Mozer, C. Shen, T. Chen, G. K. Parks, C. L. Cai and J. P. McFadden (2014), Kelvin–Helmholtz vortices observed by THEMIS at the duskside of the magnetopause under southward interplanetary magnetic field, Geophys. Res. Lett., 41, 4427–34, doi: 10.1002/2014GL060589.

Yurchyshyn, V. (2008), Relationship between EIT posteruption arcades, coronal mass ejections, the coronal neutral line, and magnetic clouds, Astrophys. J. Lett., 675(1), doi: 10.1086/533413.

Zhang, T. L., W. Baumjohann, R. Nakamura, A. Balogh and K. Glassmeier (2002), A wavy twisted neutral sheet observed by CLUSTER, Geophys. Res. Lett., 29, 1899, doi: 10.1029/2002GL015544.

Zhu, C. B., H. Zhang, Y. S. Ge, Z. Y. Pu, W. L. Liu, W. X. Wan, L. B. Liu, Y. D. Chen, H. J. Le and Y. F. Wang (2015), Dipole tilt angle effect on magnetic reconnection locations on the magnetopause, J. Geophys. Res., 120, 5344–54, doi: 10.1002/2015JA020989.

PART III

Spatial and Temporal Variations of the Geomagnetic Field

9

Spatial and Temporal Changes of the Geomagnetic Field

Insights from Forward and Inverse Core Field Models

*Nicolas Gillet**

9.1 Introduction

The past two decades have seen our knowledge on the geomagnetic field and the dynamics within the Earth's core strongly modified by the new possibilities offered by modern computers, and the inflow of continuous satellite observations, in complement of ground-based records. We are able to run geodynamo simulations that resemble the Earth: some are dipole dominated, and under some conditions, they show polarity reversals (Jones, 2015). At the same time, as we push the parameters towards Earth-like values, numerical geodynamos no longer show reversals (Schaeffer et al., 2017). We indeed face several severe obstacles: no simulation is currently able to mimic all timescales together, and turbulence is not fully rendered. We also suffer from limited geophysical constraints. Direct geomagnetic records (20 years of continuous satellite data, less than 200 years of observatory series, four centuries of historical measurements; see Jackson and Finlay, 2015) cover a short era in comparison with key timescales of the core dynamics, while we know from palaeomagnetic studies that changes occur at all periods up to some 100 kyr (Constable and Johnson, 2005). The available window is also cut towards high frequencies, since periods shorter than $O(1)$ years are hidden by signals from external (ionospheric and magnetospheric) sources (e.g. Finlay et al., 2017).

In this chapter I review the main recent findings concerning time changes of the geomagnetic field on periods ranging from a couple of years to millennia, covering observational constraints, dynamical modelling and the combination of the two through data assimilation studies. It is on purpose that I exclude mechanical forcings (e.g. precession, tides). These are undoubtedly a source of core motions and an alternative scenario to planetary dynamos (Le Bars et al., 2015). However, up to now, they have failed at producing Earth-like dynamos. This may just be a question of time, as much more effort has been put on the convective side.

I first recall the main dynamical features captured with geomagnetic data: from interannual to decadal changes with modern direct records, to centennial and longer fluctuations as seen with historical data, archaeological artefacts or sediment series (Sections 9.2.3–9.2.5). I highlight methodological issues associated with the complex separation of magnetic sources and the difficult estimation of model uncertainties (Sections 9.2.1–9.2.2). Then I present some milestones in terms of timescale separation if one wishes to reach Earth-like dynamics, and some current limitations of geodynamo simulations (Section 9.3.1). Several reduced models are next illustrated (Sections 9.3.2–9.3.3), developed on purpose to analyse core field changes from interannual to centennial periods, based on magneto–Coriolis (MC) or magneto–Archimedes–Coriolis (MAC) waves. In Section 9.3.4 I recall important findings from numerical simulations that integrate primitive equations. I report in Section 9.4 efforts carried out to merge the information contained into geophysical observations with that coming from dynamical models, a process known as data assimilation: attempts at using geodynamo simulations for either static or dynamical recoveries of the core state (Sections 9.4.1–9.4.2), and applications to the construction of field models (Section 9.4.3). I finally recall crucial issues in Section 9.5.

9.2 Geomagnetic Field Modelling

9.2.1 Separation of Sources

We shall ignore electrical currents in the mantle (a condition valid for periods longer than a few months; see Jault, 2015),

* I am grateful to Nils Olsen and Chris Finlay for providing the revised observatory monthly series and to Sabrina Sanchez for her ensemble of AmR field models. Discussions with colleagues helped clarify my mind on several points: archaeomagnetic field modelling (Erwan Thébault), stratified layers (Jérémie Vidal, Nathanael Schaeffer and Dominique Jault), geodynamo simulations (the last two plus Julien Aubert). I thank David Cébron for his comments on the manuscript before submission. N. G. is partially supported by the French Centre National d'Etudes Spatiales (CNES) for the study of Earth's core dynamics in the context of the Swarm mission of ESA. ISTerre is part of Labex OSUG@2020 (ANR10 LABX56).

so that the magnetic field B derives from a potential: $B = -\nabla V$. From the solenoidal condition $\nabla \cdot B = 0$, V satisfies Laplace's equation, $\nabla^2 V = 0$. In the absence of electrical currents, internal and external solutions arise, which in spherical coordinates (r,θ,ϕ) are naturally expressed by means of spherical harmonics separation (Olsen et al., 2010) and Gauss coefficients (denoted by g_n^m and h_n^m, with n and m the degree and order).

A proper separation of internal and external signals requires the knowledge of several components of the field on a closed surface. This is unfortunately never completely achieved. Ignoring some components of the field leads to non-uniqueness issues (Sabaka et al., 2010). The coverage is obviously uneven for archaeomagnetic, observatory and historical records, in favour of lands (resp. seas) for ground-based stations (resp. ship logs; see Jonkers et al., 2003). Such disparities in the spatial sampling imply ambiguities between Gauss coefficients (see Gillet et al., 2013, for an application to the observatory era). The major North–South discrepancy in the coverage by archaeological artefacts and lavas translates into sensitivity kernels much biased towards the North and the Middle East (e.g. Korte and Constable, 2011).

Magnetic records from space are available continuously from low-orbiting satellite data since 1999, with the Ørsted, SAC-C, CHAMP and Swarm missions (Finlay et al., 2017). With their quasi-polar trajectory, they cover an almost entire spherical surface, but the remaining polar gaps nevertheless induce larger uncertainties on zonal coefficients (Olsen and Kotsiaros, 2012). Furthermore, the sampling of external signals towards low frequencies (longer than a few months) is hindered by the slow drift of the satellite orbit in local time. The associated difficulty in describing slow evolutions of external sources is a major limit to the recovery of rapid (shorter than a couple of years) internal field changes. Another important issue when considering satellite data is the ionosphere being internal to the spacecraft trajectory: one thus cannot properly separate core and ionospheric field changes by solving Laplace's equation. Finally, the current-free separation leading to magnetic potential fields breaks down in places (such as auroral regions) where the satellite trajectory crosses electrical currents (Finlay et al., 2017).

Separating the signature of the several sources in satellite and ground-based stations is carried out by means of field modelling (i.e. space-time interpolation). Complete descriptions often rely on subtle weighting of the many parameters entering this inverse problem (e.g. the comprehensive approach by Sabaka et al., 2015), although attempts at objectifying the prior information now appear (Holschneider et al., 2016). The absence of global deterministic physical models able to isolate the external activity also leads us to consider several magnetic indices (e.g. Olsen et al., 2014; Thomson and Lesur, 2007), used to select quiet-time data, and onto which rapid external variations may be anchored (see Lesur et al., 2010; Finlay et al., 2016).

9.2.2 Field Model Uncertainties

Solutions for internal model coefficients are obtained by minimising a cost function, a sum of a measure of the misfit to magnetic data and a measure of the model complexity. In the case of quadratic norms, the penalty function takes the form

$$J(x) = \|y - H(x)\|_R^2 + \|x - \overline{x}\|_P^2, \qquad (9.2.1)$$

with the notation $\| x \|_M^2 = x^T M^{-1} x$. Vectors x and y respectively store model parameters and observations; $\overline{x}$ is a background model, H is the observation operator and matrices P and R store the a priori cross-covariances of respectively the model anomaly to $\overline{x}$ and the observation error.

Models of the main field (MF) originating from the core shall be used as input data in reanalyses of the core dynamics (see Section 9.4), a reason why it is necessary to estimate uncertainties on Gauss coefficients. With quadratic measures as in Equation (9.2.1), posterior uncertainties may be described by the inverse of the Hessian matrix (which measures the curvature of J in the neighbourhood of the solution $\mathbf{x}^*$),

$$C = \left(\nabla H(x^*)^T R^{-1} \nabla H(x^*) + P^{-1}\right)^{-1}. \qquad (9.2.2)$$

Estimating a posteriori model uncertainties thus requires reasonable estimates of both the observation errors and the prior information on the unknown parameters (through respectively R and P). Unfortunately, none of these is easy to handle.

A first concern is the numerical possibility of calculating C in cases where millions of data and/or parameters are to be considered. If the model size is too large (e.g. when co-estimating rapidly changing external sources; Sabaka et al., 2015), storing C will not be possible. In the case of a large number of data, considering cross-covariant data errors will be out of reach, possibly leading to biases in the resulting field model. Alternatively, one may consider a data set reduced at given frequencies, for instance with observatory series (Macmillan and Olsen, 2013), or spatially averaged, as done with virtual observatories (Mandea & Olsen, 2006; Barrois et al, 2018).

In any case, an a priori model norm is often penalised in the inversion process, in order to ensure the spectral convergence of the model in space and time and avoid generating too many short-length or short-timescale oscillations. This is achieved by applying regularisations which for instance damp the second or third time derivative, on the top of a spatial norm (following Bloxham and Jackson, 1992). By doing so, one a priori assumes that short wavelengths are tiny, which is not based on any physical insight. As a consequence, the a posteriori uncertainties are severely underestimated at short lengths/timescales, and the obtained model should be considered as a time-weighted estimate towards high harmonic degrees. This drawback is partly alleviated by the use of more realistic second-order statistics when constructing the prior matrix P (Gillet et al., 2013) or correlation based modelling (Holschneider et al., 2016). In such cases it becomes meaningful to consider the formal posterior covariance matrix (9.2.2) as a measure of model uncertainties, if data errors are decently estimated.

However, assessing observation errors also is a tricky issue. One should consider both measurement errors (the accuracy of the instruments) and errors associated with unmodelled sources. Estimating the former is not straightforward: see the protocols for providing standard deviations in observatory series (Lesur et al., 2017a) or uncertainties on data from the Swarm mission (Tøffner-Clausen et al., 2016). But more importantly, the signatures of unmodelled processes often dominate the error budget. For instance, even the most up-to-date satellite field models do not capture entirely the ionospheric processes, in particular in the auroral region (Finlay et al., 2017). Another example concerns archaeomagnetic records that are influenced by many processes, posterior to the age of the target magnetisation, and which may alter the minerals within the sample or the orientation of the recorded field (e.g. Constable and Korte, 2015). In most cases, such unmodelled sources result in biases and/or errors correlated in space and/or time. This prevents us from considering C a perfect measure of a posteriori model uncertainties. For instance, by increasing the amount N of data, formal posterior errors will decrease as $1/\sqrt{N}$ (consequence of the central limit theorem). Illusory small uncertainties will result if such records are influenced by biases. Alternatively, Monte Carlo (ensemble) strategies to sample observations may be used to estimate model errors, an approach widely used in archaeomagnetic field modelling (Panovska et al., 2015) but which does not remove the drawbacks induced by regularisations, as illustrated below.

9.2.3 Centennial to Millennial Geomagnetic Changes

Despite these numerous difficulties (non-exhaustive, even not speaking about dating uncertainties associated with archaeomagnetic artefacts and sediment cores), interesting features are seen in Gauss coefficient series of archaeomagnetic field models deduced from global databases (e.g. GEOMAGIA; Brown et al., 2015). Such models bring out some debated, intriguing non-dipolar features. Oscillations of period ≈ 250 years are observed for instance in Western Europe intensity records (Genevey et al., 2016). These suggest either complex combinations of Gauss coefficient series or oscillations at twice this period in some specific coefficients. Interestingly, even though global model predictions partly filter out centennial changes seen in regional records, fluctuations of period ≈ 500 years are isolated in Gauss coefficient series (see Figure 9.2.1). These are particularly clear for models inverted from archaeological and lava records, while considering sediment series makes some of the centennial fluctuation disappear (see h_2^2 prior to 500 AD). According to the uncertainties provided with the field models, such oscillations are resolved. The discrepancy between models containing or not containing sediment data thus calls for some explanation.

Let us consider the normalised, time-averaged model uncertainty per degree n, as measured by

$$\chi(n) = \left[\left(\int_{t_s}^{t_e}\sum_{i=1}^{N_e}\sum_{m=0}^{n} {g_n^m}_i(t)^2 + {h_n^m}_i(t)^2 \mathrm{d}t\right)^{-1}\right.$$

$$\left.\int_{t_s}^{t_e}\sum_{i=1}^{N_e}\sum_{m=0}^{n}\left({g_n^m}_i(t) - \langle g_n^m\rangle\right)^2 + \left({h_n^m}_i(t) - \langle h_n^m\rangle\right)^2 \mathrm{d}t\right]^{1/2}, \quad (9.2.3)$$

for an ensemble of $N_e = 20$ realisations. Brackets denote the ensemble average. We logically witness (Figure 9.2.2) a decrease in χ when incorporating sediment cores on the top of archaeological data. However, $\chi(n)$ saturates towards high degrees, a consequence of the damping used in the model construction (see Section 9.2.2): uncertainties tend to be underestimated towards small length scales. Furthermore, the model uncertainty is lower than the difference between models built with and without sediment data: in at least one of these two models, posterior error bars must be too low. Alternatives to this issue may be found (see Section 9.4.2) with the use of stochastic priors (Hellio and Gillet, 2018) or dynamo norms (Sanchez, 2016). These latter spatial constraints provide larger posterior uncertainties in comparison with regularised models, as shown in Figure 9.2.2, reaching about 100% error at degree 5.

But the prior information implicitly introduced through the damping is not the only thing responsible for the too large discrepancy between models incorporating or not incorporating sediment data. As highlighted by Licht et al. (2013), assessing the relative weights of archaeological and sediment data is not straightforward. The sedimentation rate associated with each core actually tends to smooth in time the magnetic signal recorded in sediments (Pavón-Carrasco et al., 2014). Added to the large dating errors, this process acts as a low-pass filter. As a consequence, sediment records should be considered as time-weighted data. Accounting for such a process in the construction of the field models is feasible as far as algorithms are concerned. However, this would involve estimating, for each sediment core, a time-dependent filter translating sedimentation rates into time-weighting functions.

Sediment and archaeological data also indicate signals on longer periods, although the existence of clearly isolated periodicities is debated (Nilsson et al., 2011). Given the short era of observatory data, historical databases (Jonkers et al., 2003; Arneitz et al., 2017) are the only remaining source to study centennial field changes to a higher accuracy. The most striking time-dependent pattern, extracted from the *gufm1* field model (Jackson et al.,

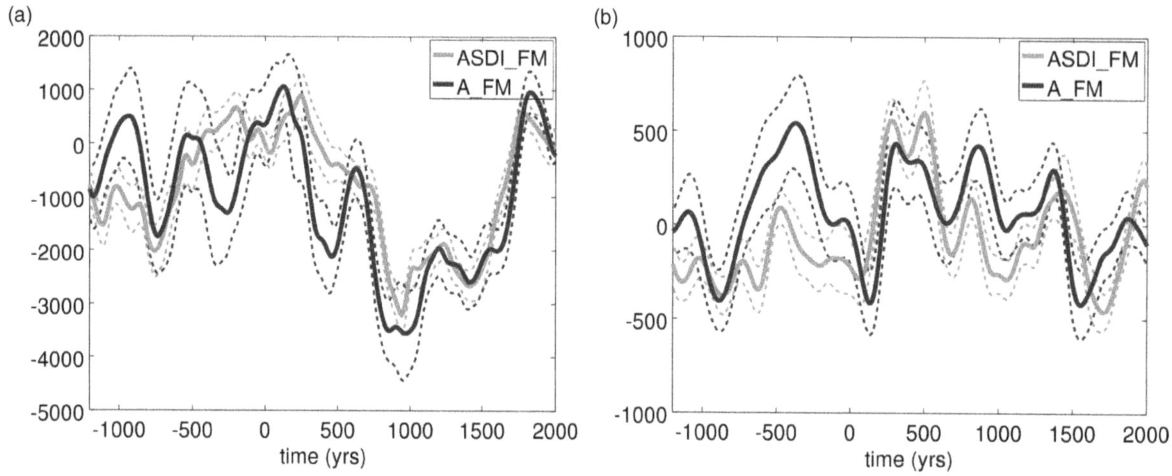

Figure 9.2.1 Time series of Gauss coefficients h_2^2 (top) and h_4^2 (bottom) for the field models A_FM (black) and ASDI_FM (grey) by Licht et al. (2013). The former is built from archaeological and lava data only; the latter incorporates sediment data. Dotted lines represent $\pm 1\sigma$ uncertainties.

2000), is the westward drift of intense flux patches in the equatorial belt. Mainly symmetric with respect to the equator, and the most obvious at azimuthal wave number $m = 5$, they travel at a speed of about 17 km/yr, which translates into a period around 275 years (Finlay and Jackson, 2003). Physical mechanism candidates for explaining such features are presented in Section 9.3.

9.2.4 Getting Rid of Geomagnetic Jerks?

The era of modern magnetic measurements (observatory and satellites) has brought several surprises. Short magnetic series may appear boring at first sight. However, what actually matters, as far as its main source (the dynamics within the core) is concerned, is actually the secular variation $\partial \boldsymbol{B}/\partial t$ (SV). Indeed, this latter is directly related to the fluid velocity $\boldsymbol{u}$ within the core, through the induction equation

$$\frac{\partial \boldsymbol{B}}{\partial t} = \nabla \times (\boldsymbol{u} \times \boldsymbol{B}) + \eta \nabla^2 \boldsymbol{B}, \qquad (9.2.4)$$

where $\eta \sim 1\ \mathrm{m}^2/\mathrm{s}$ is the magnetic diffusivity of the metallic core. For this reason, SV series are scrutinised, which put the focus on the so-called geomagnetic jerks (see Mandea et al., 2010, for a review), or sudden changes in the rate of change of the field.

More and more of these features show up as more accurate records accumulate – uncertainties in observatory series have been significantly reduced with the advent of proton magnetometers in the 1960s and of digital acquisition later on (Turner et al., 2015). Figure 9.2.3 shows some examples of SV changes with monthly series at middle, low and high latitudes. By removing some estimates of external contributions, the scatter at middle to low latitudes is much reduced, and some of the sudden changes in the SV trend clearly appear only in the cleaned series. The reduction of the scatter

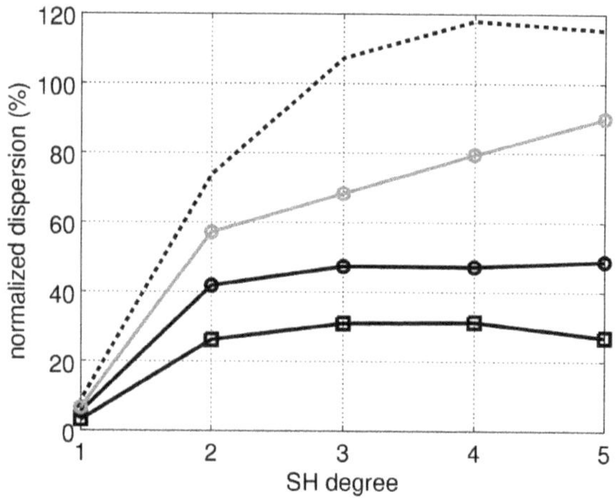

Figure 9.2.2 Normalised dispersion $\chi(n)$, as measured with Equation (9.2.3), for the models A_FM and ASDI_FM (resp. black circles and squares; Licht et al., 2013) and the model AmR (grey circles; Sanchez et al., 2016). The ensemble results from either bootstrap (A FM and ASDI_FM) or ensemble Kalman filter (AmR) methods. The black dotted line shows the norm, per degree n, of the difference between A_FM and ASDI_FM, normalised by the norm of the models.

is much less impressive at high latitudes, because there, several important external contributions are not satisfactorily modelled. The difficult characterisation of jerks comes down to the harsh separation of internal and external sources at the very period of interest, from months to a few years. This point is highlighted with the power spectral density (PSD) of revised monthly series shown in Figure 9.2.4. Even if the dispersion associated with external fields is much reduced at mid-latitudes, remaining external contributions dominate over core signals towards short periods. Important spectral lines show up in particular at periods of 1 year, 6 months etc.,

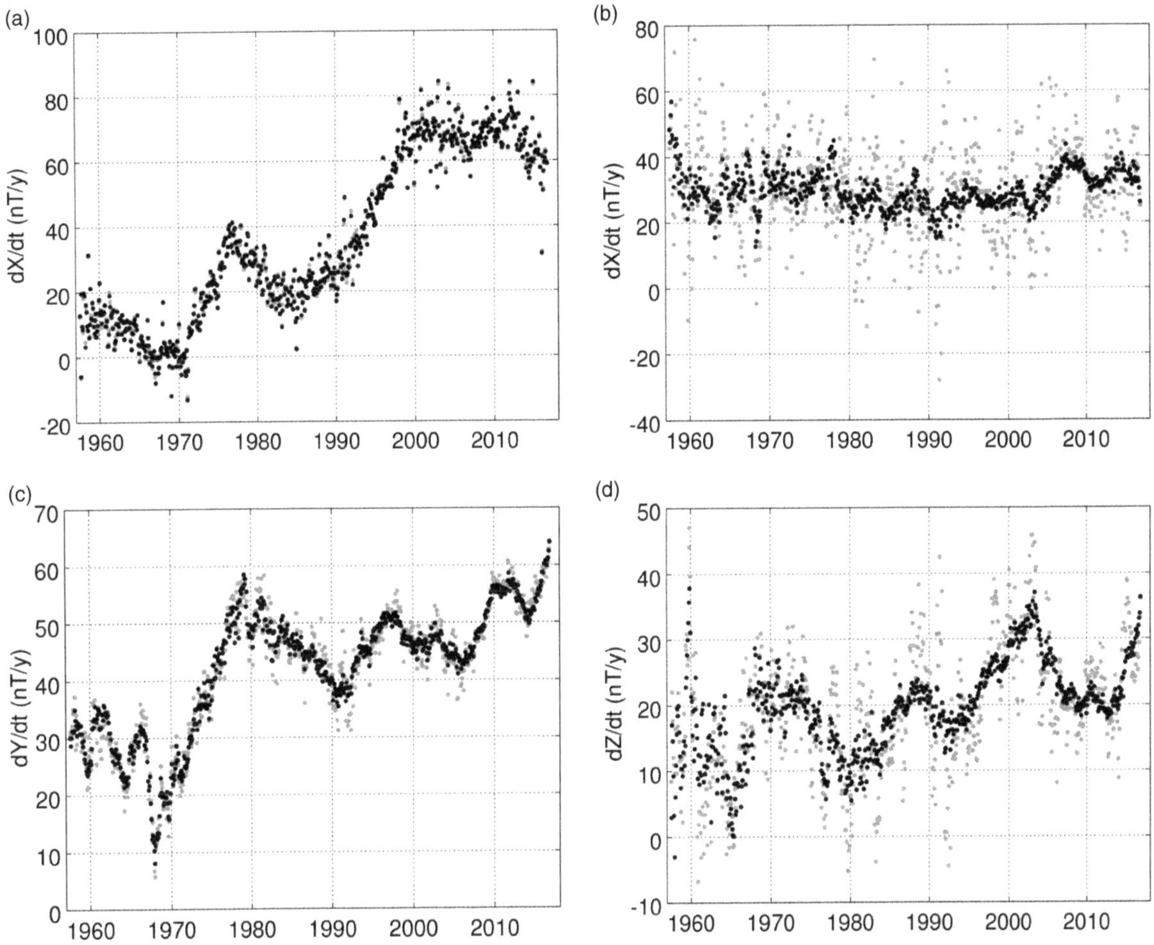

Figure: 9.2.3 SV series (annual differences of monthly means) from ground-based data. (top) dX/dt at Thule (77°N, 69°W, left) and M'Bour (14°N, 17°W, right). (bottom) dY/dt (left) and dZ/dt (right) at Chambon-la-Forêt (48°N, 2°E). (X, Y, Z) are local Cartesian coordinates, respectively, northward, eastward and downward. In grey, ordinary monthly values calculated from hourly means. In black, the revised monthly means (Olsen et al., 2014): CHAOS magnetospheric field model predictions (using as input a preliminary extended version of the RC index) and CM4 ionospheric field predictions (Sabaka et al., 2004) (using as input the F10.7 index) are first removed from hourly means, before a Huber-weighted robust monthly mean is computed. Hourly mean data (ftp://ftp.nerc-murchison.ac.uk/geomag/Swarm/AUX_OBS/hour/) have been produced by the British Geological Survey for ESA within the framework of the Swarm mission (Macmillan and Olsen, 2013).

which are filtered out when considering annual differences of monthly series, as in Figure 9.2.3.

Long-standing debates have concerned the community: whether jerks were global or not, delayed or not from one place to the other, etc. Alternatively, one may wonder whether we should at all localise jerks in time. The interpretation of SV series as resulting from processes such that the magnetic field is not twice time differentiable (in a range of frequencies) suggests that jerks should naturally emerge at any time (Gillet et al., 2013). This translates from the −4 slope found for the PSD of ground-based series (De Santis et al., 2003; see also Figure 9.2.4) and from a similar slope in the Gauss coefficients PSD (Lesur et al., 2017b). In this framework, one should not consider jerks as being characteristic of internal field processes temporally isolated. Instead, it calls for a physical interpretation of the −4 slope, which is also found in Gauss coefficient series from geodynamo simulations (see Section 9.3.1).

9.2.5 On Secular Acceleration Pulses

The ubiquitous existence of jerks might seem at odds with one focus put forward by the continuous era of satellite measurements: the occurrence of pulses in the norm of the secular acceleration $\partial^2 \boldsymbol{B}/\partial t^2$ (SA). In satellite field models, SA pulses show up every 3 years or so (Chulliat and Maus, 2014; Finlay et al., 2016). They seem to occur in between epochs where geomagnetic jerks most likely happen, in which case jerks might not arrive (completely) at any time. At the core surface (of radius $c = 3485$ km), they are carried by the largest length scales (degrees $n \leq 8$), although the SA is less and less constrained towards shorter

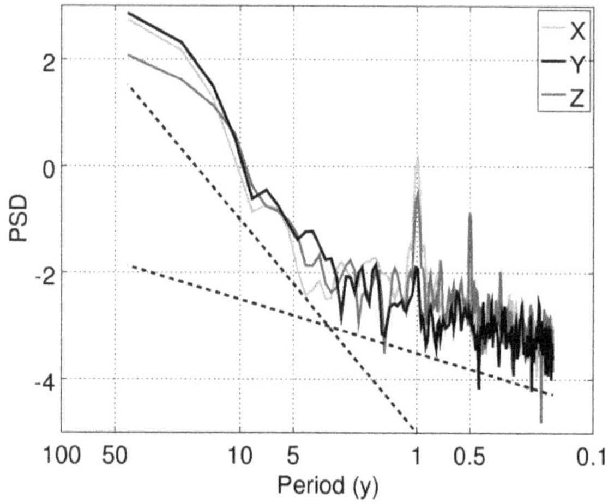

Figure 9.2.4 PSD of revised monthly means for the *X, Y* and *Z* components at Chambon-la-Forêt (48°N, 2°E), obtained by a multi-tapper analysis after removing the end-to-end line of the series. Dotted lines recall the −4 and −1 slopes.

wavelengths (Lesur et al., 2010). SA pulses were first highlighted in the equatorial belt (within ±30° latitude) in the Atlantic Hemisphere, though the more accurate, more recent ones also display important features at high latitudes in the Northern Hemisphere. A 3 year period in the SA norm implies 6 year oscillations in some Gauss coefficients: four SA pulses thus only correspond to two periods, so that longer series are required to assess the exact periodicity of SA pulses, if any, from spectral analyses. Similar SA events nevertheless seem to have occurred earlier on, as suggested by Soloviev et al. (2017) from the analysis of observatory series.

Their interpretation is currently giving rise to many conjectures: MAC waves in stratified layers are candidates (Chulliat et al., 2015), although SA pulses can entirely be analysed as the signature of quasi-geostrophic (QG) core flows (Finlay et al., 2016). In all cases, fluid motions responsible for such events must contain a significant non-axisymmetric component. As a consequence, it cannot be the signature of only torsional motions (geostrophic, or axisymmetric and invariant along the rotation axis; see Section 9.3.2), which also occur at approximately a 6 year period (Gillet et al., 2010). It is curious that magnetic data require intriguing non-zonal motions at the same period where geostrophic motions are seen to oscillate. The possibility of a mechanism coupling these two dynamics is an open question.

There is nevertheless another possibility for this 3 year apparent periodicity in SA pulses, in link with (1) the blue SA spatial power spectrum at the CMB and (2) the way magnetic models are constructed. Let us define the SA norm per harmonic degree as a function of time t for a truncation degree N (the blue SA power spectrum means Ψ_N does not converge as N increases),

$$\Phi(n,t) = \sqrt{(n+1)\sum_{m=0}^{n}\left(\partial_t^2 g_n^m(t)^2 + \partial_t^2 h_n^m(t)^2\right)}, \tag{9.2.5}$$

and the SA norm per harmonic degree as a function of frequency f,

$$\Phi^\dagger(n,f) = \sqrt{(n+1)\sum_{m=0}^{n}\left(|\alpha_n^m|^2(f) + |\beta_n^m|^2(f)\right)}, \tag{9.2.6}$$

where $\alpha_n^m(f)$ and $\beta_n^m(f)$ stand for the Fourier transform of respectively $\partial_t^2 g_n^m(t)$ and $\partial_t^2 h_n^m(t)$. From Equation (9.2.5) one defines the total SA norm:

$$\Psi_N(t) = \sqrt{\sum_{n=1}^{N}\Phi^2(n,t)}, \tag{9.2.7}$$

Figure 9.2.5 shows $\Phi^\dagger(n,f)$ for the CHAOS-6 field model (Finlay et al., 2016). Lower degrees clearly present more power on short timescales than high degrees, which are more affected by the temporal damping imposed in the model construction. At the same time, since at the CMB smaller structures are more energetic, time changes in the total SA norm are dominated by the resolved SV changes at the smaller length scales, in practice at degrees n from 5 to 8 (see Figure 9.2.6). This translates into 3 year SA pulses, coherent with a cut-off period at about 6 years for the resolution of degrees 5 to 8 in Figure 9.2.4. For $n \geq 9$, interannual SV changes are not recovered, and these smaller length scales do not imprint SA pulses, while at large length scales (in particular $n \leq 2$), the SA norm oscillates at periods shorter than 3 years.

As an illustration, let's consider a synthetic one-dimensional 'SV' series $d\varphi/dt$ generated as an order 1 auto-regressive process (AR-1). Such a process naturally shows jerklike features (Figure 9.2.7, top). By construction, it is not differentiable: the spectrum for its time derivative

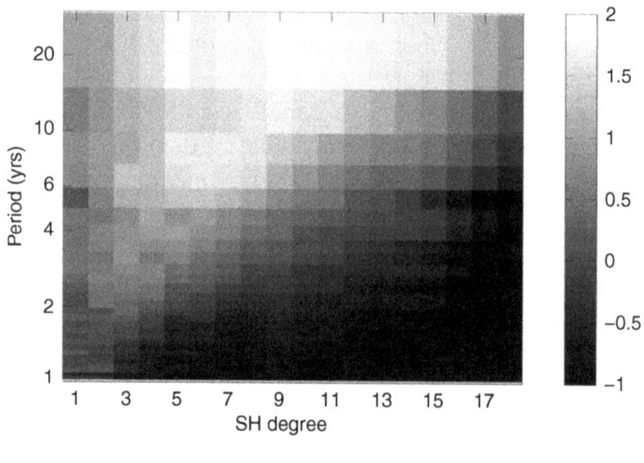

Figure 9.2.5 SA norm $\Phi^\dagger$ as a function of spherical harmonic degree n and period $T = 1/f$, as defined in Equation (9.2.6), for the model CHAOS-6-x4 ($\log_{10}$ scale, in nT/yr^2).

$d^2\varphi/dt^2$ (or 'SA') would be white, i.e. pulses in the SA norm $|d^2\varphi/dt^2|$ occur all the time, with an amplitude depending on the sampling rate. Considering now the same series but low-pass filtered, we see oscillations in the 'SA norm' of period half the shortest non-filtered timescale (Figure 9.2.7, bottom). This test does not mean there is no signal associated with reported SA pulses. These sign interannual SV changes at intermediate length scales, whose phase appears coherent within several degrees at a period about 6 years (Figure 9.2.6), in link with the localised patches isolated in

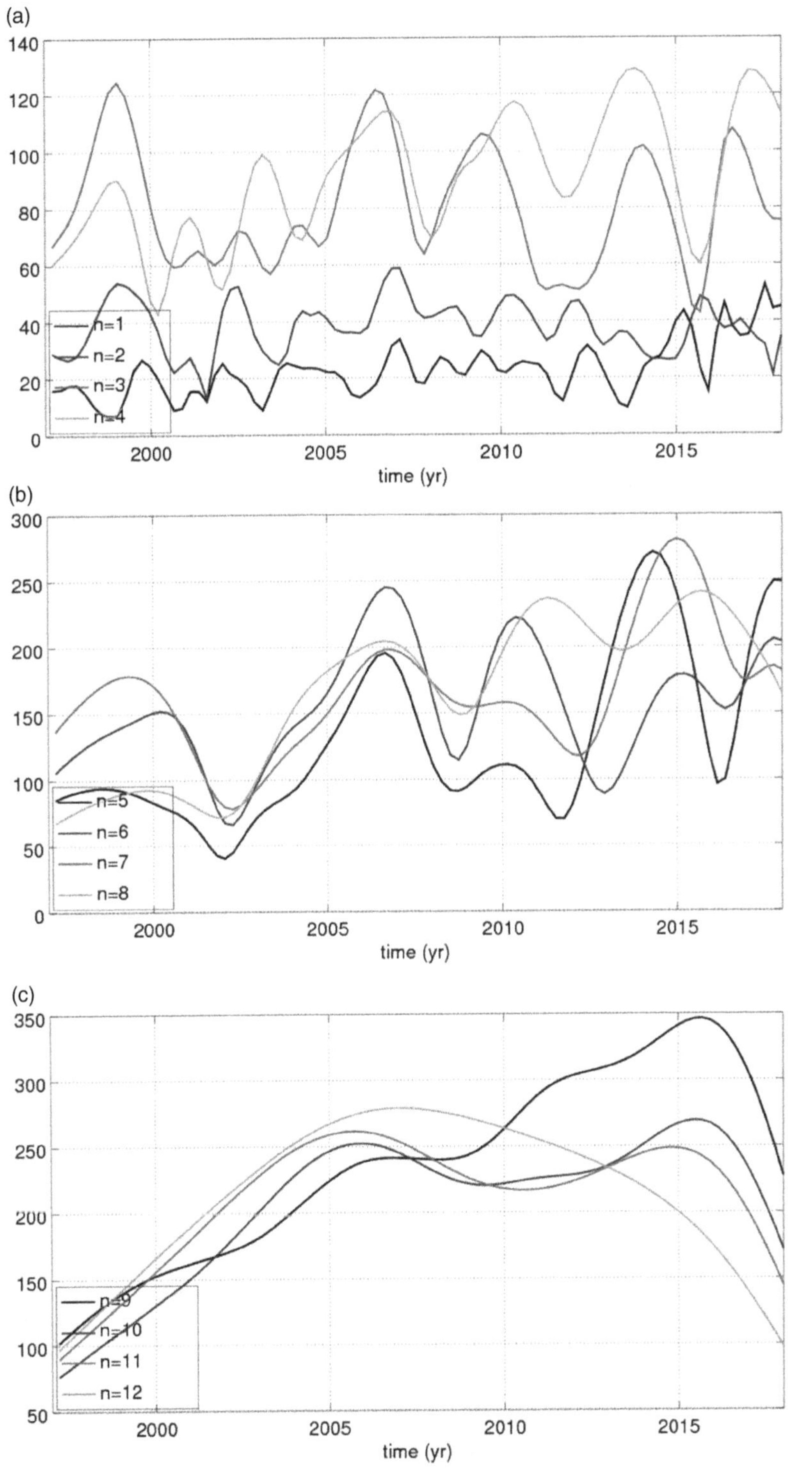

Figure 9.2.6 SA norm Φ as a function of spherical harmonic degree n and time (in nT/yr^2), for the model CHAOS-6-x4: degrees (top) $n \in [1, 4]$, (middle) $n \in [5, 8]$ and (bottom) $n \in [9, 12]$.

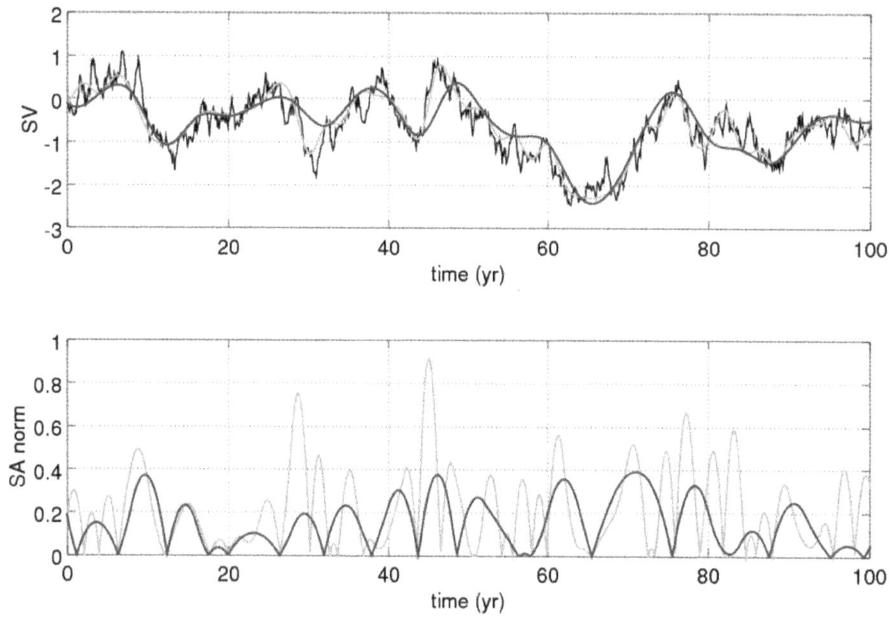

Figure 9.2.7 (top) 1D synthetic 'SV' time series $d\varphi/dt$ (black) generated as an AR-1 process, with a correlation time of 10 yr and a unit variance: the series are interpollated using splines with a sampling of respectively 2 yr (light grey) and 4 yr (dark grey). (bottom) 'SA' norm series $|d^2\varphi/dt^2|$ evaluated for the two interpollated series.

field models. However, one should keep in mind that SA pulses should not necessarily be interpreted as resulting from periodic processes, acknowledging our inability to recover rapid fluctuations at small length scales.

Finally, SV and SA highlight the core dynamics at different frequencies (time derivatives enhance short periods). Given the red temporal spectrum observed in magnetic series at decadal to centennial periods (of slope about respectively −4 and −2; see De Santis et al., 2003; Panovska et al., 2013), the SV is dominated by decadal or longer fluctuations, while the SA emphasises the shorter accessible time changes (a couple of years). The question of the continuation of the −4 temporal spectrum towards shorter periods, and a possible cut-off frequency in link with the mantle conductivity or the magnetic dissipation in the fluid core, remains open.

9.3 Dynamical Models of Earth's Core Dynamics

9.3.1 Relevant Timescales

Minimalist models of the dynamics within the Earth's core must include, on top of the induction Equation (9.2.4), the momentum equation in the rotating frame of our planet,

$$\frac{\mathrm{D}\mathbf{u}}{\mathrm{D}t} + 2\Omega \mathbf{1}_z \times \mathbf{u} = -\nabla P + \frac{1}{\rho\mu}(\nabla \times \mathbf{B}) \times \mathbf{B} + \mathbf{F} + \nu\nabla^2\mathbf{u}, \tag{9.3.1}$$

where $\mathbf{1}_z$ is the unit vector along the rotation axis, $\Omega = 7.2710^{-5}$ rad/s is the rotation rate, P is the modified pressure (including the centrifugal force), $\nu \sim 10^{-5}\mathrm{m}^2/\mathrm{s}$ is the kinematic viscosity, $\mu = 4\pi\ 10^{-7}$ H/m is the magnetic permeability of free space, $\rho \simeq 10^4\ \mathrm{kg/m}^3$ is the core density and $\mathbf{F}$ contains body forces (such as buoyancy). Several characteristic timescales can be extracted from Equations (9.2.4) and (9.3.1):

- the vortex turn-over time $\tau_u = L/U$ for a circulation of size L at a flow speed U
- the magnetic diffusion time $\tau_\eta = L^2/\eta$
- the time $\tau_a = D/V_a$ taken for a perturbation to cross the core, at the Alfvén speed $V_a = B/\sqrt{\rho\mu}$, with B the magnetic field magnitude and $D = 2265$ km the gap between the inner and outer cores
- the time $\tau_i = D/(\Omega L)$ (characteristic of inertial waves) taken for a perturbation of size L to form a Taylor column of height $H \sim D$ (Cardin and Olson, 2015)

I refer to Nataf and Schaeffer (2015) for a discussion of timescales and spatiotemporal spectra in planetary cores. For the sake of simplicity, let us consider planetary size structures ($L \sim D$), $U(D) \simeq 10$ km/yr, characteristic of the westward drift at the core surface (Finlay and Jackson, 2003), and $B \sim 3$ mT for the field strength within the core (Gillet et al., 2010). From these we obtain

$$\tau_\eta \sim 100{,}000\ \mathrm{yr} \gg \tau_u \sim 200\ \mathrm{yr} \gg \tau_a \sim 2.5\ \mathrm{yr} \gg \tau_i \sim 4\ 10^{-4}\ \mathrm{yr}. \tag{9.3.2}$$

I consider below convective dynamos that couple Equations (9.2.4–9.3.1), where $\mathbf{F}$ stands for buoyancy, with a conservation equation for heat (often in a codensity formulation that mixes thermal and chemical density anomalies with a single diffusivity). From the above timescales one can build several dimensionless numbers of interest, summarised in Table 9.3.1 for the

Table 9.3.1 *Several Dimensionless Numbers, Estimated for the Earth's Core and for Numerical Simulations of the Geodynamo*

	P_m	E	R_m	A	S	λ
Name	magnetic Prandtl	Ekman	magnetic Reynolds	Alfvén	Lundquist	Lehnert
Definition	$\frac{\nu}{\eta}$	$\frac{\nu}{\Omega D^2}$	$\frac{\tau_\eta}{\tau_u} = \frac{UD}{\eta}$	$\frac{\tau_a}{\tau_u} = \frac{U}{V_a}$	$\frac{\tau_\eta}{\tau_a} = \frac{V_a D}{\eta}$	$\frac{\tau_i}{\tau_a} = \frac{V_a}{\Omega D}$
Earth	10^{-5}	10^{-14}	10^3	10^{-2}	4×10^4	10^{-4}
simulations	$> 10^{-1}$	$> 10^{-7}$	10^3	$> 3 \times 10^{-1}$	$< 10^3$	$> 10^{-4}$

Note. Adapted from Schaeffer et al. (2017).

Earth and numerical geodynamos. I focus on dynamos targeting Earth-like parameters (low Ekman and magnetic Prandtl numbers). For these two input parameters, simulations are still far away from geophysical values. Interestingly, ratios of magnetic diffusion and induction (the magnetic Reynolds number) are nevertheless correctly recovered by numerical studies. The Earth's core is luckily close to dynamo onset, as three-dimensional computations would struggle to reach significantly larger values of R_m. The ratio of inertial to Alfvén timescales also is well recovered (as measured by a Lehnert number $\lambda \ll 1$). This transpires through a rotation-dominated dynamics on short periods (quasi-geostrophy): motions organise primarily in columns aligned with the rotation axis (Jault, 2008; Soderlund et al., 2012).

To be compared with geophysical observations, simulations must be dimensionalised, a procedure that involves a priori choices. One often resorts to the magnetic diffusion time when long sequences ($>O(10)$ kyr) are to be considered, while focusing on rapid changes, a scaling based on the advective turn-over time appears efficient (Olson et al., 2012). Interestingly, τ_u in simulations is connected to the timescale τ_n ratio of the MF to SV spatial power spectra. Indeed, Lhuillier et al. (2011) approximate $\tau_n \simeq 3\tau_u/n$. Measurable from geomagnetic field models, τ_n can thus be used to compare geophysical and numerical outputs.

Focusing on periods from decades up to some 10 kyr, dynamo simulations share some spectral properties with observations. A −4 slope of Gauss coefficient temporal spectra is found at decadal to centennial periods (see Section 9.2.4), and a transition from a −4 to a −2 slope towards long periods is observed for the axial dipole (Olson et al., 2012; Bouligand et al., 2016; Buffett and Matsui, 2015) as well as in geophysical records of the virtual axial dipole moment (Constable and Johnson, 2005). However, what is missed by computations is actually low values of the Alfvén number, which means all timescales between τ_a and τ_u (a few years to a few hundreds of years) are at least partly shrunk: while the magnetic energy is 10^4 times larger than the kinetic energy at large length scales in the Earth's core, both share similar magnitudes in simulations. The good correspondence in PSD highlighted above should not hide the too narrow separation of timescales, a consequence of enhanced diffusive processes (Bouligand et al., 2016).

Reaching weaker values of A is actually computationally very expensive, because it implies increasing both the rotation rate and the input thermal forcing (Schaeffer et al., 2017). The alternative is two-fold. One avenue consists of parameterising turbulent magnetohydrodynamic (MHD) processes, as performed by Aubert et al. (2017). Enhancing dissipation for large wave numbers, the authors manage to follow a path in the direction of Earth-like parameters. Alternatively, one may consider dynamos for $P_m \geq 1$ (Kageyama et al., 2008), so that the strong magnetic energy branch is captured (Dormy et al., 2017). Such dynamos nevertheless show much less axial invariance, as they reach significantly larger values of λ. Furthermore, what they bring is at the expense of losing wave-dominated rapid dynamics, by enhancing the role played by viscous processes.

These limits call today for considering alternative reduced equation systems. I discuss below two families of dynamical models relevant for interpreting magnetic data: magnetostrophic waves (Section 9.3.2), where Coriolis and Lorentz forces play the dominant roles, and MAC waves (Section 9.3.3) that occur in the presence of a stratified layer at the top of the core.

9.3.2 Magnetostrophic Waves and Taylor's State

Major developments have been carried out ignoring at first sight the source term in the momentum equation. Viscosity too shall be neglected in this section, on the argument $E \ll 1$. There remain two coupled equations (induction and momentum) giving birth to slow and fast MC modes (Malkus, 1967), with restoring mechanisms from Coriolis and Lorentz forces. Noting $\omega_i = O(2\Omega)$ the frequency of inertial waves (anisotropic and dispersive) and $\omega_a \sim V_a/L$ that of Alfvén waves (non-dispersive), the frequencies of fast and slow MC waves evolve respectively as (Finlay, 2008)

$$\omega_{MC}^f \sim \omega_i, \quad \omega_{MC}^s \sim \omega_a^2/\omega_i. \tag{9.3.3}$$

If rapid MC modes are essentially inertial waves, too fast to be detected today from geomagnetic data, the period of the gravest slow ones is $\tau^s_{MC} \sim 2\Omega L^2/V_a = O(500)$ years for $L \sim$ 2000 km. They are candidates to interpret major features, such as westward drift of equatorial patches or fluctuations found in archaeomagnetic series (see Section 9.2.3).

Slow MC waves are magnetostrophic, in the sense that inertia is negligible. They thus satisfy Taylor's (1963) condition,

$$\forall s, \mathbf{1}_\phi \cdot \iint_\Sigma \nabla \times \mathbf{B} \times \mathbf{B} d\Sigma = 0, \tag{9.3.4}$$

obtained by integrating Equation (3.1) along geostrophic cylinders $\Sigma(s)$ (encapsulated in the outer core, and which axis coincides with the rotation axis), with s the cylindrical radius and $\mathbf{1}_\phi$ the unit vector in the azimuthal direction. Said differently, in the absence of inertia and viscosity, since the projections of the Coriolis force and buoyancy on geostrophic cylinders vanish, the only remaining term comes from the Lorentz force. Numerical attempts at reaching magnetostrophic dynamos have been carried out (Livermore et al., 2011; Roberts and Wu, 2014), involving challenging computational issues (Walker et al., 1998).

Magnetostrophic waves contain a subfamily, QG-MC modes (Hide, 1966). The strong axial invariance of flows found in rotating MHD simulations gave birth to a renewed interest in such waves (Canet et al., 2014). Such modal analyses constitute a first step towards reduced models advecting quadratic quantities of the magnetic field in the equatorial plane (Canet et al., 2009; Jault and Finlay, 2015), where the main sources of non-linearities arise from the electromotive force in Equation (9.2.4). The time stepping of such equations is in its early phase (Maffei and Jackson, 2017).

Interestingly, non-linear interactions of (QG-)MC modes are able to fill the entire frequency spectrum from interannual to centennial periods, where the most accurate magnetic observations are available. However, similar intermediate timescales also emerge from the description of hybrid inertial–Alfvén waves by Bardsley and Davidson (2016): perturbations propagate along $\mathbf{1}_z$ to form columns that travel along field lines at the (local) Alfvén speed.

The validity of the QG assumption is nevertheless debated. Kinematic core flow inversions, applied to the most recent era covered by satellites, require a core surface kinetic energy dominated by equatorially symmetric structures (Gillet et al., 2011; Aubert, 2014). This observation is nevertheless tempered by the aspiration for the flow to locally cross the equator, in particular under Indonesia (Bloxham, 1989; Baerenzung et al., 2016; Barrois et al., 2017), where some of the most intense SV activity is detected. Dynamo simulations too require some breaking of axial invariance for Earth-like dipolarity to be obtained (Garcia et al., 2017), and deviations from non-magnetic QG solutions are also found in linear magnetoconvection solutions with an imposed dipole field (Sreenivasan and Jones, 2011). This motivates the derivation of baroclinic QG models, where motions, still strongly anisotropic, are not axially invariant (Calkins et al., 2013).

A specific family of QG motions has been the focus of many studies since their theoretical description by Braginsky (1970): torsional (or azimuthal) motions of geostrophic cylinders. These arise when re-instating inertia in Equation (9.3.4): any perturbation to Taylor's constraint shall give rise to a wavelike response. Because inertial waves are much faster than Alfvén waves at large length scales, torsional waves are actually the only Alfvén waves detectable in the Earth's core (as the Coriolis force averaged over geostrophic cylinders vanishes). These waves have been detected from magnetic observations, propagating outward from the inner core (Gillet et al., 2010). Their detection offered the possibility to probe the magnetic field in the bulk of the core. Their excitation mechanism is a source of debate: Teed et al. (2015) argue from magnetoconvection simulations for instabilities related to large Lorentz torques on the edge of the tangent cylinder (the geostrophic cylinder tangent to the inner core), while Gillet et al. (2017) consider the outward propagation as the signature of propagating normal modes in the presence of a conductive layer at the base of the mantle, triggered either on the tangent cylinder or in the bulk of the fluid core. Band-pass-filtered geostrophic motions inferred from magnetic data (Figure 9.3.1) show waves modulated over decadal periods, though it is difficult to accurately isolate spectral lines (and thus the wave forms), given (1) the limited time span covered by accurate observations, (2) the blurred SV signal towards high frequencies (see Section 9.2.5) and (3) the red spectrum of core motions, which are less energetic towards short periods.

Their apparent partial absorption at the equator is a potential source of information on the deep mantle conductivity (hardly detectable from above) in a case where an electromagnetic torque couples the core to the mantle (Schaeffer and Jault, 2016). In this context, mantle conductances deduced from torsional waves (3×10^7–10^8 S; Gillet et al., 2017) appear relatively low in comparison with previous estimates from either core notations (Buffett et al., 2002) or kinematic core flows (Holme, 1998). Spread over a thick ($\sim$1000 km) shell, such values remain nevertheless higher than what is currently accepted from induction studies (typically between 0 and 10 S/m; Velímský, 2010). There remain several possibilities: (1) materials of electrical conductivity intermediate between $O(10^2)$ S/m and that of the core (10^5 S/m) may be shrunk into a thin layer close to the CMB (to which induction studies are insensitive), (2) the dissipation in models of torsional waves may be underestimated, or (3) the core–mantle coupling could arise from another mechanism than electromagnetic torques (e.g. topographic).

Another motivation for the study of geostrophic motions is their link to changes in the length of day (LOD; Jault and Finlay, 2015). These show modulated fluctuations of

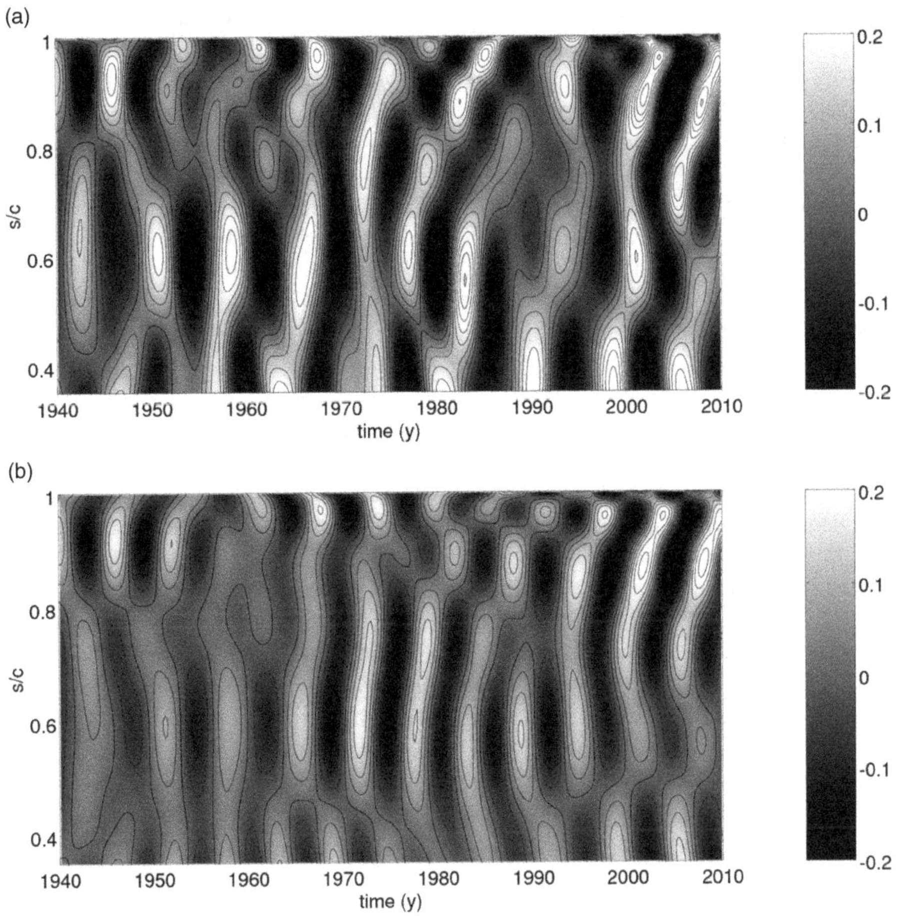

Figure 9.3.1 Geostrophic velocity as a function of time and cylindrical radius (in km/yr), between the inner core radius ($s/c = 0.35$) and the core equator ($s/c = 1$). From kinematic QG inversion (Gillet et al., 2015b) band-pass filtered for $T \in [4, 9.5]$ years (top) and $T \in [5, 8]$ years (bottom). (A black-and-white version of this figure appears in some formats. For the colour version, please refer to the plate section.)

typically 0.2 ms around 6 year periods (Chao et al., 2014) convincingly predicted by geostrophic flow models inverted from magnetic data (Gillet et al., 2015b). In such a scenario, the larger LOD changes at decadal periods (of amplitude a few ms) would result from the interaction of non-geostrophic motions with the background magnetic field (as derived from Taylor's condition), though an alternative scenario is reported below. At periods decadal and longer, the dynamics of both non-geostrophic and geostrophic motions could be associated with MC waves, as these potentially couple zonal to non-zonal motions in the presence of non-axisymmetric background magnetic fields (Labbé et al., 2015).

9.3.3 MAC Waves in a Stratified Layer

Motivated by the detection of a low seismic velocity layer at the top of the core (some 100 km thick; Helffrich and Kaneshima, 2010), an extra complexity may be introduced. As the inner core crystallises, the fluid phase is enriched in light elements that potentially accumulate at the top of the core (a point that nevertheless requires elaborate scenarios; see Brodholt and Badro, 2017). Low seismic velocities can be translated into density gradient. The re-evaluation to larger values of the outer core thermal conductivity from high temperature, high pressure ab initio calculations and laboratory experiments (e.g. Pozzo et al., 2012; Ohta et al., 2016) also calls for a subadiabatic heat flux and a stable layer at the top of the core (but see Konôpková et al., 2016).

A resulting stable stratification would carry gravity waves of frequency ω_g characterised by the Brünt–Vaïsala frequency $N = \sqrt{-(g/\rho)(d\rho/dr)}$, where g is the gravity acceleration. The penetration of geostrophic columns through such a layer depends on the relative periods of inertial and gravity waves (Takehiro and Lister, 2001; Vidal and Schaeffer, 2015), which much vary with the considered length scales. In the Earth's core we do not know accurately how N compares with Ω. In the case of QG inertial (i.e. Rossby) waves, the shorter the length scale, the longer the period, so that large-length-scale modes (the ones we may see) are less likely to be affected by stratification. Interestingly, most nodes of such waves concentrate near the equator, where intense SA pulses have first been highlighted (see Section 9.2.5).

Vidal and Schaeffer (2015) found a weak effect of a simple geometry imposed magnetic field on the penetration depth for Rossby modes. On the other end of the time spectrum (towards zero frequency), Takehiro (2015) and Takehiro and Sasaki (2017) show that the presence a magnetic field couples the stratified layer to the deeper buoyant layer, if the radial propagation of Alfvén waves along poloidal field lines is faster than their damping by diffusive processes. Applied to the Earth, the strong value of the Lundquist number S implies deep convective motions to penetrate into the stable layer. If dynamo simulations with a top stratified layer show a lower level of SV at the CMB (Christensen and Wicht, 2008), it is because of relatively low values of S, which may apply to planets like Mercury (possibly Saturn). An alternative scenario for reconciling magnetic and seismic observations involves lateral heat flux heterogeneities: locally buoyant areas may lie within a globally stable density gradient below the CMB (Olson et al., 2017). We still lack definitive arguments on the influence of stratification at decadal periods: the penetration depth of slow QG-MC waves into a stratified layer, or their coupling with MAC modes in the top layer, requires dedicated study.

Meanwhile, in the absence of thermochemical diffusion, and under the Boussinesq approximation, the system may be reduced following Braginsky's (1993) hidden ocean model. In the presence of a magnetic field, the linearised system of equations leads to MAC waves within the stably stratified layer. Their period evolves as (Finlay, 2008)

$$\omega_{\text{MAC}} = \pm\omega^{s}_{\text{MC}}\left(1+\frac{\omega_g^2}{\omega_a^2}\right)^{1/2}, \tag{9.3.5}$$

which applied to Earth-like values ($\omega_a \ll \omega_g$) gives $\omega_{\text{MAC}} \simeq \pm\,\omega_a\omega_g/\omega_i$. Associated periods span interannual to decadal timescales, the reason why they have been considered to explain features such as SA pulses (Chulliat et al., 2015) or fluctuations in the axial dipole SV (Buffett, 2014).

Interestingly, if zonal motions are not directly affected by stratification (they do not cut iso-density surfaces), their coupling to non-zonal motion may render them sensitive to the density gradient. As a consequence, there exist axisymmetric MAC waves (Braginsky, 1999) which carry angular momentum and thus a signature in the LOD. Recent numerical estimates, however, suggest a too small angular momentum budget in the upper layer (Buffett et al., 2016), which would call for a magnetic coupling with geostrophic motions in the deeper layer.

9.3.4 Insights from Numerical Geodynamo Simulations

There is no such dynamo computation suited to cover the entire frequency range (Meduri and Wicht, 2016). Dynamos whose parameters E and P_m are the closer to Earth-like values do not show polarity reversals – though it is not possible to integrate primitive equations for long enough to likely see a reversal with extreme parameters. Reversals are reserved to simulations with high E, for which forcings strong enough to produce reversals are easier to compute. On the contrary, torsional waves excited by Lorentz torques require fast enough rotation rates (Teed et al., 2015), and they are continuously excited for the lowermost values of P_m and E (Schaeffer et al., 2017). As another example, MC waves can be detected in simulations, provided a strong enough field is generated (on the strong branch of Dormy et al., 2017), which calls for using either large P_m or low E (Hori et al., 2017).

Dynamo simulations show today the first signs for the physics thought to occur in the Earth's core and probed by direct magnetic records. It is also an extensive tool to tackle issues such as core–mantle coupling. As an example, the westward drift and the asymmetry between Atlantic and Pacific Hemispheres found in modern SV maps is interpreted by Aubert et al. (2013) as the by-product of a gravitational torque between the inner core and the mantle, in link with tomographic seismic anomaly maps. Alternatively to the thermally driven asymmetries and persistent flux lobes (e.g. Gubbins et al., 2007; Aubert et al., 2008), the rich dynamics operating in the most up-to-date dynamo simulations with isotropic forcing starts to show some breaking of the longitudinal symmetry naturally emerging. Schaeffer et al. (2017) indeed extracted almost axially invariant eccentric circulations persistent over centuries, similar to that inverted by Pais and Jault (2008) from satellite field models. A comparison of eccentric circulations found from magnetic data and the two above computations is shown in Figure 9.3.2. The origin of large-length-scale circulations in computations is subject to discussion: Sakuraba and Roberts (2009) advocate for an enlarging effect in link with the fixed-flux thermal boundary condition, while Yadav et al. (2016) relate the energy cascade towards wider structures to highly supercritical convection allowing to reach a MAC equilibrium of forces (and a stronger effect of the magnetic field). Schaeffer et al. (2017) also find wider patterns where the field is stronger.

9.4 The Advent of Geomagnetic Data Assimilation

The past decade has seen rising geomagnetic data assimilation studies. These aim at mixing information from observations and from a dynamical model to constrain the state of the Earth's core – a discrete representation of unknown fields, for instance **u, B** (for a review in the context of geomagnetism, see Fournier et al., 2010). As such, they depart from kinematic core flow inversions. The entry to the fluid core is made via the radial component of the magnetic field at the core surface. This latter, continuous through the CMB when the mantle is insulating, is obtained

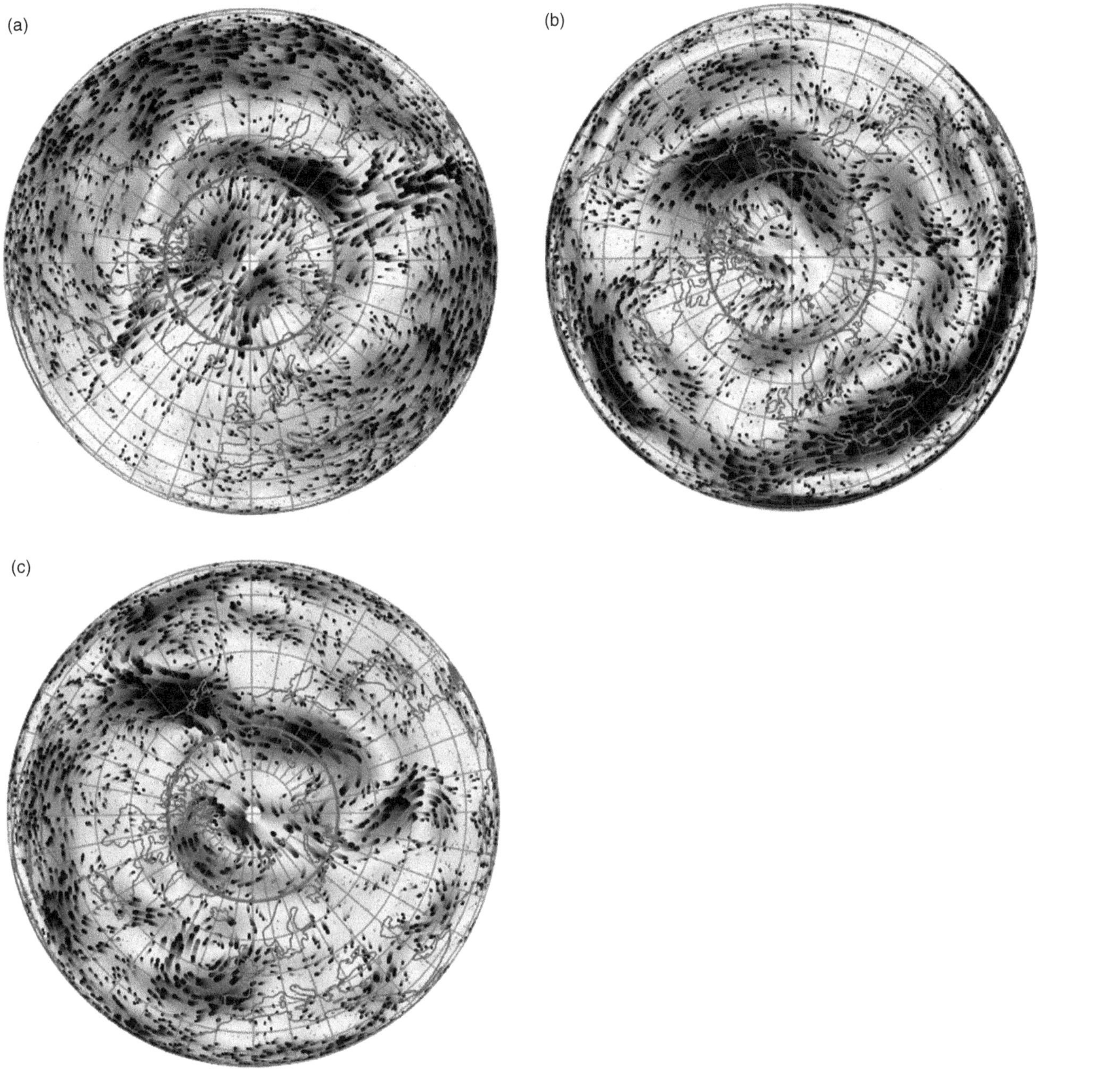

Figure 9.3.2 Equatorial cross sections of core motions, viewed from the north pole: passive tracer trajectories (black) superimposed with u_ϕ (colour scale within ± 30 km/yr, red westward). (top) Geodynamo simulation with isotropic forcing ($E = 10^{-6}$, $P_m = 0.2$; Schaeffer et al., 2017), low-pass filtered for $T \geq 1000$ years. (middle) Kinematic QG inversion in 2008 (Gillet et al., 2015b). (bottom) Geodynamo simulation with non isotropic forcing favouring a westward drift in the Atlantic hemisphere ($E = 1.2\ 10^{-5}$, $P_m = 2.5$; Aubert et al., 2013). (A black-and-white version of this figure appears in some formats. For the colour version, please refer to the plate section.)

by downward continuation of a potential field from magnetic records above the Earth's surface (see Section 9.2.1).

The main assimilation techniques essentially split between variational and probabilistic approaches. With the former, one adjusts the initial condition to minimise the misfit to the data over the whole considered time window, subject to the numerical constraint imposed by means of adjoint equations. The latter avenue derives from Bayes' rule; it has been more widely used, especially when the forward model consists of geodynamo equations, as initiated by Liu et al. (2007). It consists in operating a series of analyses (a linear regression given the propagated model error covariance matrix, $\mathrm{P}^f(t)$) and forecasts (advection of the core state by the model) each time observations are available. I recall below the several attempts that have been carried out so far, with either self-consistent geodynamo simulations or reduced models. The pertinence of the considered model will depend on the timescale of interest.

9.4.1 Snapshot Inference Using Geodynamo Norms

The most up-to-date simulations (with the lower values of A and E) are not accessible yet for assimilation studies, as ensemble strategies followed to approximate P^f require to compute simultaneously several tens (or more) of geodynamo runs (prohibitive using cutting-edge computations). As a consequence, progress is still needed for directly adjusting the trajectory of dynamo simulations using modern observations and their interannual to decadal fluctuations (see Section 9.2.5).

Such computations nevertheless provide Earth-like snapshot images of the core surface field (Christensen et al., 2010). This encouraged the combination of spatial cross-covariances, obtained from long geodynamo free runs, with MF and SV observations, into snapshot images of the core state (Fournier et al., 2011). In this framework, observing SV data (on the top of MF data) helps constrain unobserved quantities, such as core motions, via the radial component of Equation (9.2.4), when the dynamics of the system is either ignored on purpose (Aubert, 2014) or biased due to input parameters that are far from realistic (Kuang and Tangborn, 2015). This comes down to inferring the core state in a kinematic manner, using a dynamo norm as a spatial prior.

Core flow maps obtained with such dynamo norms recover the planetary-scale eccentric gyre first revealed with kinematic QG inversion (Pais and Jault, 2008). They also give partial access to the underlying magnetic field and co-density organisation (Aubert et al., 2013; Aubert, 2014). These latter two fields are indirectly related to surface MF and SV observations, through their statistical cross-correlation with the large-length-scale magnetic and velocity fields at the core surface.

9.4.2 Geodynamo-Driven Dynamical Reconstructions

Geodynamo simulations are arguably suitable for dynamically inferring changes in the core state on centennial and longer periods. This paves the way for assimilation of historical and aechaeomagnetic data, implementing time constraints from primitive equations. The first attempts (Kuang et al., 2009) have been performed with 'optimal interpolation' (OI) sequential techniques (i.e. with a static P^f), and univariate statistics (ignoring most cross-covariances), in order to reduce the computational cost. It is, however, known that space and time cross-correlations are important (Fournier et al., 2011; Tangborn and Kuang, 2015).

In this context, Fournier et al. (2013) used an ensemble Kalman filter (EnKF). Updating the forecast statistics was possible thanks to the limited amount of data, in practice $n_b(n_b+2)$, with n_b the spherical harmonic truncation degree when one considers field models as observations ($n_b \simeq 5$ and 13 for respectively aechaeomagnetic and historical models). Using synthetic experiments at relatively high values of E, Fournier et al. show it is possible to retrieve some structures of the unobserved quantities (such as the field and the flow in the bulk of the core) from surface observations of B_r. They advocate for ensemble sizes significantly larger than the size of the data vector, in order to reduce the warm-up time to about 1000 years, and thus significantly supplement OI techniques that do not update P^f.

Sanchez (2016) recently extended this approach to the reanalysis of pointwise (synthetic) aechaeomagnetic observations at the Earth's surface. Similarly to Fournier et al. (2013), she finds it possible to recover part of the unobserved state even when data are mostly concentrated in a single hemisphere. This is encouraging: by directly inverting for ancient data, one avoids ad hoc assumptions made to recover Gauss coefficients and their associated uncertainties (see Section 9.2.2). It nevertheless requires the migration in time of data at the precise epochs where analyses are performed. Alternatively, one may use uncertainties as provided by stochastic inversions (Hellio and Gillet, 2018). These works pave the way for applications to geophysical data, though it certainly requires a closer look at errors associated with the dynamical model itself.

This last issue is two-fold. First, rapid changes as captured with modern data are not yet reproduced with primitive equation-based models: the model is imperfect and should be complemented. Second, the observation is very sparse: we have access only to the large length scales of B_r at the top of the fluid core. This induces subgrid errors (because the electromotive force in Equation (9.2.4) is non-linear), which dominate measurement errors and must be accounted for (Pais and Jault, 2008; Baerenzung et al., 2016). One could add model errors upon measurement errors with no other modification to the algorithms. However, errors mentioned above are correlated in time (Gillet et al., 2015b) – the reason why they have been considered by means of stochastic differential equations by Barrois et al, 2017, 2018). The authors show with an augmented-state OI filter how mandatory it is to account for model uncertainties to correctly recover transient surface core motions. Baerenzung et al. (2017) follow a similar stochastic avenue using instead an EnKF, involving some 10,000 ensemble members to obtain well-conditioned P^f. They emphasise two different timescales for the dynamics: a millennial planetary gyre superimposed with more rapid and spatially localised features. These two geophysical applications of data assimilation tools illustrate how mixing constraints from dynamical models and observations allows us to extract information about the core physics.

9.4.3 Dynamically Constrained Geomagnetic Field Models

Early assimilation strategies have been introduced with operational perspectives (production of field models, SV predictions). Sanchez et al. (2016) proposed series of snapshot aechaeomagnetic field models under a dynamo norm, showing that some information may be retrieved up to spherical harmonic degrees $n = 5$, depending on the considered epoch. Applications to space weather were anticipated by Aubert (2015), who generated a decadal forecast of the South Atlantic anomaly with geodynamo equations, starting from an initial regression of MF and SV data under a dynamo norm. Beggan and Whaler (2010) tested the improvement of SV predictions with piecewise constant flows advecting the core surface field within an EnKF framework.

The production of 5 year SV predictions involving assimilation tools has been proposed in the context of the International Geomagnetic Reference Field. Kuang et al. (2010) plugged geodynamo equations into univariate OI schemes; Fournier et al. (2015) performed a multivariate snapshot inference implying geodynamo cross-covariances; and Gillet et al. (2015a) integrated a reduced stochastic QG model into an augmented-state Kalman filter. Ultimately, producing unbiased estimates will require mixing dynamo equations in the bad parameter regime or involving parameterised subgrid processes, together with (stochastic) parameterisation of unresolved quantities.

9.5 Concluding Remarks

To close this chapter, I remind of some of the main difficulties that we must face in order to take a step forward. On long periods, on the top of new records, magnetic field modellers need accurate error estimates (for both measurements and unmodelled processes) associated with archeomagnetic and sediment records, to be incorporated with realistic prior information into the inverse problem. On shorter periods, the main barrier is related to external signals that hide the information coming from the core, preventing from resolving rapid changes towards short wavelengths. Overcoming this difficulty will possibly require considering dynamical models of the ionosphere when inverting for satellite and observatory data.

Numerical models of the core hardly reach the condition $A < O(1)$. To significantly lower A, one must resort to subgrid-scale parameterisation. These have to be invented, since no such model is currently suitable for turbulent rotating MHD, where reverse energy cascades make large-length-scale features very sensitive to unresolved circulations. In front of such complex issues, there is a need for reduced models, for two reasons. First, they help better understand the underlying physics, possibly allowing for short-cuts in the expensive modelling from primitive equations. Second, as they rely on a smaller number of parameters, they are ideal for integration into data assimilation algorithms. These indeed represent heavily underdetermined inverse problems: having access only to large-length-scale, low-frequency B_r at the CMB, only about 200 observations each year are available to constrain the entire core state.

The implementation of geomagnetic assimilation tools requires realistic measures of the prognostic model imperfections. If early stochastic models are developed, their integration into three-dimensional models remains untrodden ground. Similarly, localisation methods widely used for reanalysing surface envelopes (ocean and atmosphere) with moderate ensemble sizes (Oke et al., 2007) will be important to implement, to propagate at best the information through the succession of analyses. This nevertheless requires specific thoughts, as in the Earth's core, the thick spherical shell geometry, together with a momentum balance dominated by Coriolis and Lorentz forces, makes non-local interactions ubiquitous.

References

P. Arneitz, R. Leonhardt, E. Schnepp, B. Heilig, F. Mayrhofer, P. Kovacs, P. Hejda, F. Valach, G. Vadasz, C. Hammerl et al. The histmag database: combining historical, archaeomagnetic and volcanic data. *Geophys. J. Int.*, 210(3):1347–59, 2017.

J. Aubert. Earth's core internal dynamics 1840–2010 imaged by inverse geodynamo modelling. *Geophys. J. Int.*, 197(3):1321–34, 2014.

J. Aubert. Geomagnetic forecasts driven by thermal wind dynamics in the Earth's core. *Geophys. J. Int.*, 203(3): 1738–51, 2015.

J. Aubert, H. Amit, G. Hulot and P. Olson. Thermochemical flows couple the Earth's inner core growth to mantle heterogeneity. *Nature*, 454(7205):758, 2008.

J. Aubert, C. C. Finlay and A. Fournier. Bottom-up control of geomagnetic secular variation by the Earth's inner core. *Nature*, 502(7470):219, 2013.

J. Aubert, T. Gastine and A. Fournier. Spherical convective dynamos in the rapidly rotating asymptotic regime. *J. Fluid Mech.*, 813:558–93, 2017.

J. Baerenzung, M. Holschneider and V. Lesur. The flow at the Earth's core-mantle boundary under weak prior constraints. *J. Geophys. Res.*, 121(3): 1343–64, 2016.

J. Baerenzung, M. Holschneider, J. Wicht, S. Sanchez and V. Lesur. Modeling and predicting the short term evolution of the geomagnetic field. *arXiv preprint arXiv:1710.06188*, 2017.

O. Bardsley and P. A. Davidson. Inertial–Alfvén waves as columnar helices in planetary cores. *J. Fluid Mech.*, 805, 2016.

O. Barrois, N. Gillet and J. Aubert. Contributions to the geomagnetic secular variation from a reanalysis of core surface dynamics. *Geophys. J. Int.*, 211(1):50–68, 2017.

Barrois, O., Hammer, M., Finlay, C., Martin, Y., & Gillet, N., 2018. Assimilation of ground and satellite magnetic measurements: inference of core surface magnetic and velocity field changes, *Geophys. J. Int.*, 215(1), 695–712.

C. Beggan and K. Whaler. Forecasting secular variation using core flows. *Earth Planets Space*, 62(10):11, 2010.

J. Bloxham. Simple models of fluid flow at the core surface derived from geomagnetic field models. *Geophys. J. Int.*, 99(1):173–82, 1989.

J. Bloxham and A. Jackson. Time-dependent mapping of the magnetic field at the core-mantle boundary. *J. Geophys. Res.*, 97(B13):19537–63, 1992.

C. Bouligand, N. Gillet, D. Jault, N. Schaeffer, A. Fournier and J. Aubert. Frequency spectrum of the geomagnetic field harmonic coefficients from dynamo simulations. *Geophys. J. Int.*, 2016.

S. I. Braginsky. Torsional magnetohydrodynamic vibrations in the Earth's core and variations in day length. *Geomag. Aeron.*, 10:1–8, 1970.

S. I. Braginsky. MAC-oscillations of the hidden ocean of the core. *J. Geomag. Geoelec.*, 45(11–12):1517–38, 1993.

S. I. Braginsky. Dynamics of the stably stratified ocean at the top of the core. *Phys. Earth Planet. Int.*, 111(1): 21–34, 1999.

J. Brodholt and J. Badro. Composition of the low seismic velocity E' layer at the top of Earth's core. *Geophys. Res. Lett.*, 2017.

M. C. Brown, F. Donadini, M. Korte, A. Nilsson, K. Korhonen, A. Lodge, S. N. Lengyel and C. G. Constable. GEOMAGIA50. v3: 1. general structure and modifications to the archeological and volcanic database. *Earth Planets Space*, 67(1):83, 2015.

B. Buffett. Geomagnetic fluctuations reveal stable stratification at the top of the Earth's core. *Nature*, 507(7493): 484, 2014.

B. Buffett and H. Matsui. A power spectrum for the geomagnetic dipole moment. *Earth Planet. Sci. Lett.*, 411: 20–26, 2015.

B. Buffett, P. Mathews and T. Herring. Modeling of nutation and precession: Effects of electromagnetic coupling. *J. Geophys. Res.: Solid Earth*, 107(B4), 2002.

B. Buffett, N. Knezek and R. Holme. Evidence for MAC waves at the top of Earth's core and implications for variations in length of day. *Geophys. J. Int.*, 204(3): 1789–1800, 2016.

M. A. Calkins, K. Julien and P. Marti. Three-dimensional quasi-geostrophic convection in the rotating cylindrical annulus with steeply sloping endwalls. *J. Fluid Mech.*, 732:214–44, 2013.

E. Canet, A. Fournier and D. Jault. Forward and adjoint quasi-geostrophic models of the geomagnetic secular variation. *J. Geophys. Res.*, 114(B11), 2009.

E. Canet, C. Finlay and A. Fournier. Hydromagnetic quasi-geostrophic modes in rapidly rotating planetary cores. *Phys. Earth Planet. Inter.*, 229:1–15, 2014.

P. Cardin and P. Olson. Experiments on core dynamics. In *Treatise on Geophysics*, 2nd edn. Elsevier BV, Amsterdam, pages 317–39, 2015.

B. F. Chao, W. Chung, Z. Shih and Y. Hsieh. Earth's rotation variations: A wavelet analysis. *Terra Nova*, 26 (4):260–64, 2014.

U. R. Christensen and J. Wicht. Models of magnetic field generation in partly stable planetary cores: Applications to Mercury and Saturn. *Icarus*, 196(1):16–34, 2008.

U. R. Christensen, J. Aubert and G. Hulot. Conditions for Earth-like geodynamo models. *Earth Planet. Sci. Lett.*, 296 (3):487–96, 2010.

A. Chulliat and S. Maus. Geomagnetic secular acceleration, jerks, and a localized standing wave at the core surface from 2000 to 2010. *J. Geophys. Res.*, 119(3):1531–43, 2014.

A. Chulliat, P. Alken and S. Maus. Fast equatorial waves propagating at the top of the Earth's core. *Geophys. Res. Lett.*, 42 (9):3321–9, 2015.

C. G. Constable and C. Johnson. A paleomagnetic power spectrum. *Phys. Earth Planet. Inter.*, 153:61–73, 2005.

C. G. Constable and M. Korte. Centennial to millennialscale geomagnetic field variations. In *Treatise on Geophysics*, 2nd edn., Elsevier BV, Amsterdam, pages 309–41. 2015.

A. De Santis, D. Barraclough and R. Tozzi. Spatial and temporal spectra of the geomagnetic field and their scaling properties. *Phys. Earth Planet. Inter.*, 135:125–34, 2003.

E. Dormy, L. Oruba and L. Petitdemange. Three branches of dynamo action. *Fluid Dyn. Res.*, 2017.

C. Finlay, V. Lesur, E. Thèbault, F. Vervelidou, A. Morschhauser and R. Shore. Challenges handling magnetospheric and ionospheric signals in internal geomagnetic field modelling. *Space Sci. Rev.*, 206(1–4):157–89, 2017.

C. C. Finlay. Course 8 waves in the presence of magnetic fields, rotation and convection. *Les Houches*, 88:403–50, 2008.

C. C. Finlay and A. Jackson. Equatorially dominated magnetic field change at the surface of Earth's core. *Science*, 300 (5628):2084–6, 2003.

C. C. Finlay, N. Olsen, S. Kotsiaros, N. Gillet and L. Tøffner-Clausen. Recent geomagnetic secular variation from Swarm. *Earth Planets Space*, 68(1):1–18, 2016.

A. Fournier, G. Hulot, D. Jault, W. Kuang, A. Tangborn, N. Gillet, E. Canet, J. Aubert and F. Lhuillier. An introduction to data assimilation and predictability in geomagnetism. *Space Sci. Rev.*, 155(1–4):247–91, 2010.

A. Fournier, J. Aubert and E. Thèbault. Inference on core surface flow from observations and 3-D dynamo modelling. *Geophys. J. Int.*, 186(1):118–36, 2011.

A. Fournier, L. Nerger and J. Aubert. An ensemble Kalman filter for the time-dependent analysis of the geomagnetic field. *Geophys. Geochem. Geosyst.*, 14(10): 4035–43, 2013.

A. Fournier, J. Aubert and E. Thèbault. A candidate secular variation model for IGRF-12 based on swarm data and inverse geodynamo modelling. *Earth Planets Space*, 67 (1):1–17, 2015.

F. Garcia, L. Oruba and E. Dormy. Equatorial symmetry breaking and the loss of dipolarity in rapidly rotating dynamos. *Geophys. Astrophys. Fluid Dyn.*, 1–14, 2017.

A. Genevey, Y. Gallet, S. Jesset, E. Thèbault, J. Bouillon, A. Lefèvre and M. Le Goff. New archeointensity data from french early medieval pottery production (6th–10th century ad). Tracing 1500 years of geomagnetic field intensity variations in Western Europe. *Phys. Earth Planet. Inter.*, 257:205–19, 2016.

N. Gillet, D. Jault, E. Canet and A. Fournier. Fast torsional waves and strong magnetic field within the Earth's core. *Nature*, 465 (7294):74, 2010.

N. Gillet, N. Schaeffer and D. Jault. Rationale and geophysical evidence for quasi-geostrophic rapid dynamics within the Earth's outer core. *Phys. Earth Planet. Inter.*, 187:380–90, 2011.

N. Gillet, D. Jault, C. C. Finlay and N. Olsen. Stochastic modelling of the Earth's magnetic field: inversion for covariances over the observatory era. *Geophys. Geochem. Geosyst.*, 14:766–86, 2013.

N. Gillet, O. Barrois and C. C. Finlay. Stochastic forecasting of the geomagnetic field from the COV-OBS.x1 geomagnetic field model, and candidate models for IGRF-12. *Earth Planets Space*, 67(1):71, 2015a.

N. Gillet, D. Jault and C. Finlay. Planetary gyre, time-dependent eddies, torsional waves, and equatorial jets at the Earth's core surface. *J. Geophys. Res.*, 120(6):3991–4013, 2015b.

N. Gillet, D. Jault and E. Canet. Excitation of traveling torsional normal modes in an Earth's core model. *Geophys. J. Int.*, 2017.

D. Gubbins, A. P. Willis and B. Sreenivasan. Correlation of Earth's magnetic field with lower mantle thermal and seismic structure. *Phys. Earth Planet. Inter.*, 162(3):256–60, 2007.

G. Helffrich and S. Kaneshima. Outer-core compositional stratification from observed core wave speed profiles. *Nature*, 468 (7325):807, 2010.

G. Hellio and N. Gillet. Geomagnetic field changes over the past three millennia, and their uncertainties. *Geophys. J. Int.*, submitted, 2018.

R. Hide. Free hydromagnetic oscillations of the Earth's core and the theory of the geomagnetic secular variation. *Philos. Trans. R. Soc. London A*, 259(1107):615–47, 1966.

R. Holme. Electromagnetic core-mantle coupling II: Probing deep mantle conductance. In *The Core-Mantle Boundary Region*, American Geophysical Union, Washington, DC, pages 139–51, 1998.

M. Holschneider, V. Lesur, S. Mauerberger and J. Baerenzung. Correlation-based modeling and separation of geomagnetic field components. *J. Geophys. Res.*, 121(5):3142–60, 2016.

K. Hori, R. Teed and C. Jones. The dynamics of magnetic Rossby waves in spherical dynamo simulations: A signature of strong-field dynamos? *Phys. Earth Planet. Inter.*, 2017.

A. Jackson and C. Finlay. Geomagnetic secular variation and its applications to the core. In *Treatise on Geophysics*, 2nd edn., Elsevier BV, Amsterdam, pages 137–84, 2015.

A. Jackson, A. R. T. Jonkers and M. R. Walker. Four centuries of geomagnetic secular variation from historical records. *Philos. Trans. R. Soc. London A*, 358:957–90, 2000.

D. Jault. Axial invariance of rapidly varying diffusionless motions in the Earth's core interior. *Phys. Earth Planet. Inter.*, 166 (1):67–76, 2008.

D. Jault. Illuminating the electrical conductivity of the lowermost mantle from below. *Geophys. J. Inter.*, 202:482–96, 2015.

D. Jault and C. Finlay. Waves in the core and mechanical core-mantle interactions. In *Treatise on Geophysics*, 2nd edn., Elsevier BV, Amsterdam, pages 225–45, 2015.

C. Jones. Thermal and compositional convection in the outer core. In *Treatise on Geophysics*, 2nd edn., Elsevier BV, Amsterdam, pages 115–59, 2015.

A. R. Jonkers, A. Jackson and A. Murray. Four centuries of geomagnetic data from historical records. *Rev. Geophys.*, 41 (2), 2003.

A. Kageyama, T. Miyagoshi and T. Sato. Formation of current coils in geodynamo simulations. *Nature*, 454(7208):1106, 2008.

Z. Konôpková, R. S. McWilliams, N. Gómez-Pèrez and A. F. Goncharov. Direct measurement of thermal conductivity in solid iron at planetary core conditions. *Nature*, 534:99–101, 2016.

M. Korte and C. Constable. Improving geomagnetic field reconstructions for 0–3 ka. *Phys. Earth Planet. Inter.*, 188:247–59, 2011.

W. Kuang and A. Tangborn. Dynamic responses of the Earth's outer core to assimilation of observed geomagnetic secular variation. *Prog. Earth Planet. Sci.*, 2(1):40, 2015.

W. Kuang, A. Tangborn, Z. Wei and T. Sabaka. Constraining a numerical geodynamo model with 100 years of surface observations. *Geophys. J. Int.*, 179(3):1458–68, 2009.

W. Kuang, Z. Wei, R. Holme and A. Tangborn. Prediction of geomagnetic field with data assimilation: a candidate secular variation model for IGRF-11. *Earth Planets Space*, 62(10):7, 2010.

F. Labbé, D. Jault and N. Gillet. On magnetostrophic inertialess waves in quasi-geostrophic models of planetary cores. *Geophys. Astrophys. Fluid Dyn.*, 109(6):587–610, 2015.

M. Le Bars, D. Cèbron and P. Le Gal. Flows driven by libration, precession, and tides. *Ann. Rev. Fluid Mech.*, 47:163–93, 2015.

V. Lesur, I. Wardinski, M. Hamoudi and M. Rother. The second generation of the GFZ reference internal magnetic model: GRIMM-2. *Earth Planets Space*, 62(10):6, 2010.

V. Lesur, B. Heumez, A. Telali, X. Lalanne and A. Soloviev. Estimating error statistics for Chambonla-Forêt observatory definitive data. *Ann. Geophys.*, 35:939, 2017a.

V. Lesur, I. Wardinski, J. Baerenzung and M. Holschneider. On the frequency spectra of the core magnetic field gauss coefficients. *Phys. Earth Planet. Inter.*, 2017b.

F. Lhuillier, J. Aubert and G. Hulot. Earth's dynamo limit of predictability controlled by magnetic dissipation. *Geophys. J. Int.*, 186(2):492–508, 2011.

A. Licht, G. Hulot, Y. Gallet and E. Thèbault. Ensembles of low degree archeomagnetic field models for the past three millennia. *Phys. Earth Planet. Inter.*, 224:38–67, 2013.

D. Liu, A. Tangborn and W. Kuang. Observing system simulation experiments in geomagnetic data assimilation. *J. Geophys. Res.*, 112(B8), 2007.

P. W. Livermore, G. Ierley and A. Jackson. The evolution of a magnetic field subject to taylor s constraint using a projection operator. *Geophys. J. Int.*, 187(2):690–704, 2011.

S. Macmillan and N. Olsen. Observatory data and the Swarm mission. *Earth Planets Space*, 65(11):15, 2013.

S. Maffei and A. Jackson. Kinematic validation of a quasigeostrophic model for the fast dynamics in the Earth's outer core. *Geophys. J. Int.*, 2017.

W. V. Malkus. Hydromagnetic planetary waves. *J. Fluid Mech.*, 28 (4):793–802, 1967.

M. Mandea and N. Olsen. A new approach to directly determine the secular variation from magnetic satellite observations. *Geophys. Res. Lett.*, 33(15), 2006.

M. Mandea, R. Holme, A. Pais, K. Pinheiro, A. Jackson and G. Verbanac. Geomagnetic jerks: Rapid core field variations and core dynamics. *Space Sci. Rev.*, 155(1):147–75, 2010.

D. G. Meduri and J. Wicht. A simple stochastic model for dipole moment fluctuations in numerical dynamo simulations. *Frontiers Earth Sci.*, 4:38, 2016.

H.-C. Nataf and N. Schaeffer. Turbulence in the core. In *Treatise on Geophysics*, 2nd edn., Elsevier BV, Amsterdam, pages 161–81. 2015.

A. Nilsson, R. Muscheler and I. Snowball. Millennial scale cyclicity in the geodynamo inferred from a dipole tilt reconstruction. *Earth Planet. Sci. Lett.*, 311(3):299–305, 2011.

K. Ohta, Y. Kuwayama, K. Hirose, K. Shimizu and Y. Ohishi. Experimental determination of the electrical resistivity of iron at Earth's core conditions. *Nature*, 534(7605):95–8, 2016.

P. R. Oke, P. Sakov and S. P. Corney. Impacts of localisation in the enkf and enoi: experiments with a small model. *Ocean Dyn.*, 57(1):32–45, 2007.

N. Olsen and S. Kotsiaros. The geomagnetic field gradient tensor, properties and parametrization in terms of spherical harmonics. *Int. J. Geomath.*, 2012.

N. Olsen, K.-H. Glassmeier and X. Jia. Separation of the magnetic field into external and internal parts. *Space Sci. Rev.*, 152 (1–4):135–57, 2010.

N. Olsen, H. Lühr, C. C. Finlay, T. J. Sabaka, I. Michaelis, J. Rauberg and L. Tøffner-Clausen. The CHAOS-4 geomagnetic field model. *Geophys. J. Int.*, 197(2):815–27, 2014.

P. Olson, U. Christensen and P. Driscoll. From superchrons to secular variation: A broadband dynamo frequency spectrum for the geomagnetic dipole. *Earth Planet. Sci. Lett.*, 319–20:75–82, 2012.

P. Olson, M. Landeau and E. Reynolds. Dynamo tests for stratification below the core-mantle boundary. *Phys. Earth Planet. Inter.*, 271:1–18, 2017.

M. Pais and D. Jault. Quasi-geostrophic flows responsible for the secular variation of the Earth's magnetic field. *Geophys. J. Int.*, 173(2):421–43, 2008.

S. Panovska, C. Finlay and A. Hirt. Observed periodicities and the spectrum of field variations in Holocene magnetic records. *Earth Planet. Sci. Lett.*, 379:88–94, 2013.

S. Panovska, M. Korte, C. Finlay and C. Constable. Limitations in paleomagnetic data and modelling techniques and their impact on holocene geomagnetic field models. *Geophys. J. Int.*, 202(1):402–18, 2015.

F. J. Pavón-Carrasco, M. L. Osete, J. M. Torta and A. De Santis. A geomagnetic field model for the holocene based on archaeomagnetic and lava flow data. *Earth Planet. Sci. Lett.*, 388:98–109, 2014.

M. Pozzo, C. Davies, D. Gubbins and D. Alfe`. Thermal and electrical conductivity of iron at Earth's core conditions. *Nature*, 485(7398):355–58, 2012.

P. H. Roberts and C.-C. Wu. On the modified taylor constraint. *Geophys. Astrophys. Fluid Dyn.*, 108(6):696–715, 2014.

T. J. Sabaka, N. Olsen and M. E. Purucker. Extending comprehensive models of the earth's magnetic field with Ørsted and CHAMP data. *Geophys. J. Int.*, 159(2): 521–47, 2004.

T. J. Sabaka, G. Hulot and N. Olsen. Mathematical properties relevant to geomagnetic field modeling. In *Handbook of Geomathematics*, Springer, pages 503–38, 2010.

T. J. Sabaka, N. Olsen, R. H. Tyler and A. Kuvshinov. Cm5, a pre-swarm comprehensive geomagnetic field model derived from over 12 yr of CHAMP, Ørsted, SAC-C and observatory data. *Geophys. J. Int.*, 200(3):1596–1626, 2015.

A. Sakuraba and P. H. Roberts. Generation of a strong magnetic field using uniform heat flux at the surface of the core. *Nat. Geosci.*, 2(11):802–5, 2009.

S. Sanchez. *Assimilation of geomagnetic data into dynamo models, an archeomagnetic study*. PhD thesis, Institut de Physique du Globe de Paris, 2016.

S. Sanchez, A. Fournier, J. Aubert, E. Cosme and Y. Gallet. Modelling the archaeomagnetic field under spatial constraints from dynamo simulations: a resolution analysis. *Geophys. J. Int.*, 207:983–1002, 2016.

N. Schaeffer and D. Jault. Electrical conductivity of the lowermost mantle explains absorption of core torsional waves at the equator. *Geophys. Res. Lett.*, 43(10):4922–8, 2016.

N. Schaeffer, D. Jault, H.-C. Nataf and A. Fournier. Turbulent geodynamo simulations: A leap towards Earth's core. *Geophys. J. Int.*, 211(1):1–29, 2017.

K. M. Soderlund, E. M. King and J. M. Aurnou. The influence of magnetic fields in planetary dynamo models. *Earth Planet. Sci. Lett.*, 333:9–20, 2012.

A. Soloviev, A. Chulliat and S. Bogoutdinov. Detection of secular acceleration pulses from magnetic observatory data. *Phys. Earth Planet. Inter.*, 270:128–42, 2017.

B. Sreenivasan and C. A. Jones. Helicity generation and subcritical behaviour in rapidly rotating dynamos. *J. Fluid Mech.*, 688:5, 2011.

S.-I. Takehiro. Penetration of Alfvèn waves into an upper stably-stratified layer excited by magnetoconvection in rotating spherical shells. *Phys. Earth Planet. Inter.*, 241:37–43, 2015.

S.-I. Takehiro and J. R. Lister. Penetration of columnar convection into an outer stably stratified layer in rapidly rotating spherical fluid shells. *Earth Planet. Sci. Lett.*, 187(3):357–66, 2001.

S.-I. Takehiro and Y. Sasaki. Penetration of steady fluid motions into an outer stable layer excited by MHD thermal convection in rotating spherical shells. *Phys. Earth Planet. Inter.*, 2017.

A. Tangborn and W. Kuang. Geodynamo model and error parameter estimation using geomagnetic data assimilation. *Geophys. J. Int.*, 200(1):664–75, 2015.

J. Taylor. The magneto-hydrodynamics of a rotating fluid and the Earth's dynamo problem. *Proc. R. Soc. London A*, 274 (1357):274–83, 1963.

R. J. Teed, C. A. Jones and S. M. Tobias. The transition to Earth-like torsional oscillations in magnetoconvection simulations. *Earth Planet. Sci. Lett.*, 419:22–31, 2015.

A. W. Thomson and V. Lesur. An improved geomagnetic data selection algorithm for global geomagnetic field modelling. *Geophys. J. Int.*, 169(3):951–63, 2007.

L. Tøffner-Clausen, V. Lesur, N. Olsen and C. C. Finlay. In-flight scalar calibration and characterisation of the Swarm magnetometry package. *Earth Planets Space*, 68(1):129, 2016.

G. Turner, J. Rasson and C. Reeves. Observation and measurement techniques. In *Treatise on Geophysics*, 2nd edn., Elsevier BV, Amsterdam, pages 91–135, 2015.

J. Velímskỳ. Electrical conductivity in the lower mantle: Constraints from CHAMP satellite data by time-domain em induction modelling. *Phys. Earth Planet. Inter.*, 180(3):111–17, 2010.

J. Vidal and N. Schaeffer. Quasi-geostrophic modes in the Earth's fluid core with an outer stably stratified layer. *Geophys. J. Int.*, 202(3):2182–93, 2015.

M. Walker, C. Barenghi and C. Jones. A note on dynamo action at asymptotically small ekman number. *Geophys. Astrophys. Fluid Dyn.*, 88(1–4):261–75, 1998.

R. K. Yadav, T. Gastine, U. R. Christensen, S. J. Wolk and K. Poppenhaeger. Approaching a realistic force balance in geodynamo simulations. *Proc. Natl. Acad. Sci. USA*, 113 (43):12065–70, 2016.

10
The Global Lithospheric Magnetic Field
World Magnetic Anomaly Maps and Models

*Vincent Lesur and Erwan Thébault**

10.1 Introduction

Some of the processes that shaped the Earth's surface, in particular the slow drift of continents associated with plate tectonics, are recorded in the rock magnetisation and therefore have a signature that can be clearly identified in magnetic survey data. It is therefore a worthwhile task to gather available data to have a global view of the magnetic signals generated in the lithosphere at scales ranging from few kilometres to several thousands of kilometres. This task is the primary objective of the World Digital Magnetic Anomaly Map (WDMAM): an international scientific project supported by the International Association of Geomagnetism and Aeronomy (IAGA) and the UNESCO Commission for the Geological Map of the World (CGMW).

The first version of the WDMAM was released in 2007 (Korhonen et al., 2007) after years of effort from a dedicated IAGA task force, and more specifically from Juha Korhonen, who led this project. This version has been the first truly global compilation of available total intensity anomaly data (i.e. spatial variations of the magnetic field total intensity around a reference). It resulted from the cooperation of several groups, each having proposed a candidate to the map. Three are described in Hamoudi et al. (2007), Hemant et al. (2007) and Maus et al. (2007), respectively, but other candidates were not associated with a publication in a scientific journal. These candidates were evaluated and eventually merged to produce the final version. This version included all data available to the project at that time, but soon new data sets, or reprocessed data sets, were released. Furthermore, the model used to fill oceanic areas not covered by data was not sufficiently accurate. So, a new version of the map was planned. For this second version, only two groups prepared a candidate, an Institut de Physique du Globe de Paris (IPGP) candidate, focusing on marine areas, and a GeoForshungsZentrum Potsdam (GFZ) candidate, dedicated to continental areas. The two groups combined their efforts to produce a new grid that was evaluated in the same way as the first version. This second version was released during the International Union of Geodesy and Geophysics General (IUGG) assembly, held in Prague in June–July 2015 (Catalán et al., 2016). Technical details on the way the map was built can be found in Dyment et al. (2015) and Lesur et al. (2016).

All this work and publications would not have been possible without countries, institutions, scientists and private companies agreeing to release their survey magnetic data for this project. These survey data include marine and aeromagnetic data and more recently, data obtained from satellite missions as CHAMP and Swarm. The former types of near-surface surveys are expensive to run but allow the rapid acquisition of information on the geological structure underlying relatively large areas. Therefore they have been heavily used for oil and gas exploration as well as for finding areas with potential economic value for mining activities. Typically the wavelengths covered are as small as few hundred metres for recent surveys and extend sometimes up to a few hundred kilometres. Satellite data also provide information on the geological structure of the Earth's lithosphere, but only at large scale. Whatever is the origin of these data, whether they have been acquired with airborne survey, during marine cruises or as part of satellite missions, they all have to be processed to extract signals generated in the lithosphere (e.g. Ravat et al., 2003; Maus et al., 2007; Ravat et al., 2009; Quesnel et al., 2009; Lesur et al., 2013; Thébault et al., 2016). In particular the raw data have to be corrected from external geomagnetic fields such as the fields generated in the ionosphere and magnetosphere, and from fields of internal origin as the core and induced fields. Since these contributions are only partially known, the survey data, that are often only total intensity anomaly values, are contaminated by biases and noise. However, the main difficulties are linked to the lack of metadata. Often the

* The authors thank the Centre National d'Etudes Spatiales (CNES) for its support.

reference datum of a given data set is unknown leading to erroneous positioning of the data. Even more often, the main field referenced model is not given, or its reference time has errors, introducing offsets between adjacent surveys and slightly distorted total intensity anomaly fields.

In this chapter we shortly recall the processes that are used to compile the WDMAM. A detailed description is not necessary since full details are given in the original papers, however, we give some general steps in Section 10.2. Then, in Section 10.3, we describe and compare the choices made to build the successive versions of the WDMAM. In Section 10.4, we discuss field models that have been derived from worldwide magnetic data compilations, and in Section 10.5, we discuss possible steps to push further the analysis of these magnetic data.

10.2 Building the WDMAM

Most of the magnetic data provided by countries, institutes or scientists for building the WDMAM are compilations of processed and gridded values. Grids have different cell size, orientation, altitude, datum and reference main field models.

The cell sizes range from 1 × 1 km to 5 × 5 km depending on the grid, but the true resolution can be much less than that, depending on the original data collection and gridding specifications. Most of the grids are a UTM projection grid, and their orientations give the local true geographic North and East at the reference longitude of the grid. Unfortunately, this latter longitude is not always given and has to be found by comparison with adjacent data or other geographic information. Some other grids are built on a regular spacing in latitude and longitude. The altitudes are generally either 1 km or 5 km above the reference ellipsoid WGS 84. For relatively old or strictly national surveys, other reference ellipsoids may be used. Gridded data provided without information on the reference ellipsoid are generally not included in the WDMAM. Overall, the main difficulty is the lack of information on the used reference main field model. It is often the International Geomagnetic Reference Field (IGRF). Of course the magnetic main field varies with time, but successive versions of the IGRF do not give always the same model for a given epoch. So to have all the necessary information, not only the reference epoch for the data has to be provided but also the IGRF version. Again, this information is rarely given in the metadata. In the worst scenario, no reference field is used, but a constant offset is subtracted, sometimes even linear trends.

To use these data, the reference field model has to be known, or to be estimated, usually by trial and error, where the choice is made by calculating magnetic anomaly statistics; absolute mean and anomaly variance should be as small as possible. Then, data are processed to have the CM4 (Sabaka et al., 2004) as the reference field model. They are upward continued to 5 km altitude and generally re-gridded to get finally a regular 0.05° grid in latitude, longitude. This processing is a heavy task; fortunately, a significant amount of data are provided as large compilations, with all the required metadata. This is the case for grids for Australia, Antarctica, the US and Canada and for the Russia, South Africa and Antarctica compilations.

Marine data are generally provided partially processed, or simply as raw data. All these were handled by T. Ishihara and others, following a technique described in Quesnel et al. (2009). Once processed, these data have to be gridded, while making sure that isolated tracks are not lost or significantly modified. We note that the two versions of the WDMAM differ greatly for the treatment of the marine data. In the first version, marine data were upward continued to 5 km altitude, as over continental areas, and preference was given to ready-to-use compilations that included oceanic areas (e.g. the North America compilation). In the second version of the WDMAM, the marine data were kept at sea level and preferred over existing compilations, such that all oceanic areas were treated in a consistent way.

For areas of the globe where no marine or aeromagnetic data were available, synthetic data derived from lithospheric field models based on satellite data were used. Of course, these data do not have a resolution comparable to aeromagnetic or marine data. This difference of resolution is apparent in areas where small-scale anomalies were to be expected from knowledge of the oceanic crust age. Since reliable, but not perfect, oceanic crust age models exist, it was decided to build synthetic anomaly data from them. These synthetic data were processed in the same way as measured data. There are significant differences in the way these synthetic data were derived in the two successive versions of the WDMAM (Dyment et al., 2015).

In the next step the different data sets have to be merged to have a global view of the lithospheric field anomalies. This is done in a consistent way between independent data sets, but of course, inside a data set, the way sub-grids have been merged depends on the data set providers. The merging process often requires the adoption of a preference order. It is relatively simple to set this order, which is important only when there is an overlap between data sets. The main point is that true measured data are always preferred to synthetic data derived from models. Also, it is clear that a smooth interpolation between data sets is not always the best way to proceed. For example, the information content along an isolated marine track is totally different and much higher quality than synthetic data. In that case no interpolation is possible, and data points associated with marine tracks data are left unchanged, regardless of the information provided through synthetic data. In the same way, no interpolation is possible between marine grids that are at sea level and continental grids that are at 5 km altitude. The differences in the way two adjacent areas have been treated require setting an index for each point of the WDMAM grid. The index given indicates the data set from which the point value originates. In case of interpolation, the index is set to zero. The way data sets are interpolated is described for the second version of the WDMAM in Lesur et al. (2016).

In the final step of the processing, the long wavelengths of the data compilation are modified such that they fit the observed magnetic signals at satellite altitudes. Several models of the lithospheric field, based on satellite data, exist, and a choice has to be made. In the second version of the WDMAM, the field model used is described in Lesur et al. (2013). It would have been pertinent to use the CM4 model lithospheric field (Sabaka et al., 2004), since its core field part is already used as a reference model, but it has a limited spatial resolution. However, in this second version of the WDMAM, the relevant information is provided such that the user can modify the reference model if necessary. In Lesur et al. (2016), it is pointed out that this substitution of the long wavelengths is not as trivial a task as one may think.

Overall, it is very important to remember that the WDMAM is not a homogeneous compilation of magnetic anomaly data, and that

1 large parts of the maps do not have a good resolution because no near-surface data are available in these areas
2 significant parts of the marine areas are not covered by data, and the high-resolution information provided by the WDMAM, when available, is based on models, not data
3 the reference altitude of the WDMAM over marine and continental areas is not the same
4 although authors of the WDMAM paid attention to possible discontinuities between adjacent surveys, some spurious anomalies may still be present, that are not linked to the underlying geology, but to characteristics of different surveys

Therefore, before applying any analysis or deriving any new results from the map, it is essential to refer to the index associated with each data point of the WDMAM compilation. This index indicates the origin data set of the given value, hence if the value provided is a true data value, is a synthetic datum derived from a model or stems from an interpolation process.

Figure 10.4.2 gives, as an example, an image of the WDMAM over the western Atlantic Ocean and part of eastern North America.

10.3 Versions of the WDMAM

The two versions of the WDMAM have been built nearly 10 years apart, and during that time, two new data sets have been provided to improve the map: Algeria and Morocco. These two data sets are of the best quality and are the only detailed surveys, with Nigeria, to cover the northwest of Africa. However, there are other important differences between the map versions.

In the first version a compilation of European aeromagnetic data due to Wonik et al. (2001) was used. Unfortunately, the reference ellipsoid for this data set was not given. Several have been tried, but none of them reconciled known positions of the magnetic anomalies in the UK with anomalies in the data set when those of Central Europe were at their expected positions. In the second version it was therefore attempted to rebuild this European compilation. Data sets for Portugal, Spain, France, Italy, Austria, Germany, Romania and the UK and a Fennoscandian compilation were used. Other countries were often covered with data of lesser quality. Only a few areas were not covered with low-altitude data, as for example Switzerland or the Netherlands. The anomaly field over Europe is therefore much better described in the second version of the map compared to the first.

For the rest of the world, some data sets have been updated, as Russia, but the main difference between the two versions is in the way data have been interpolated and merged. In the first version of the map the data sets over continental areas, except Australia, were interpolated using a collocation least squares method (Maus et al., 2007). This gave visually very good results but generated significant deviations from original aeromagnetic data sets. Feedback from the scientists who processed the original data indicated that a closer fit to the grids they obtained was recommended. Therefore, in the second version of the map, the data sets were cropped to limit as much as possible overlaps, with grid values estimated through a weighted averaging scheme in these overlapping areas (Lesur et al., 2016). The different data sets were clearly identified through indices and as little processed as possible. Otherwise, the final grid has the same resolution over continental areas and the same altitude for the two versions of the map.

This is not true for oceanic regions. As described in the previous section, there are some major differences in this case between the two versions of the map. In the second WDMAM version:

1 All areas covered by marine data have grids at sea level, as opposed to 5 km altitude in the first version.
2 All data have been processed in a consistent way for all oceanic areas, using everywhere the same interpolation technique (Lesur et al., 2016).
3 The magnetisation model of the oceanic floor has been fully revised and adjusted (see Dyment et al., 2015; Lesur et al., 2016).

The first point clearly increases the resolution of the data but introduces a discontinuity between marine and continental areas. This was nonetheless preferred because marine data are usually not too close to coasts and correspond to relatively deep water. Shallow water areas are, when surveyed, covered by aeromagnetic data associated with continental surveys. For marine areas, as for continental areas, the user should systematically check the index associated with a data point to know its origin and therefore the gridded data altitude.

10.4 Models of the Earth Magnetic Field Generated in the Lithosphere

The WDMAM compilation can be further analysed by generating a lithospheric field model. Sampling theorems on a sphere (e.g. Lesur, 2006) indicate that the maximum spherical harmonic degree that could be reached with a regular 0.05° × 0.05° grid is 1799, although a higher resolution is possible along longitudes. Estimating Gauss coefficients at such high spherical harmonic degrees is, however, pointless, because there are large areas in the WDMAM where there is no near-surface data and therefore no information on the short-wavelength lithospheric magnetic field. The model derived together with the WDMAM (Lesur et al., 2016) goes up to degree 800. We point out that it is not necessary to work with an oblate spheroidal coordinate system because the spherical harmonic model does not go to infinity. The model is therefore not singular, even if its reference sphere is below the sources of the lithospheric field. One just needs to be careful in the interpretation of global statistical properties such as power spectra. The latter may be slightly divergent at high spherical harmonic degrees.

Global lithospheric total field models cannot be fully defined from magnetic anomaly data. This is because a projection of the lithospheric field on the main field direction is generally a very good approximation of anomaly data. It follows that the lithospheric field model is poorly constrained in the direction perpendicular to the main field. The data inversion process that leads to the lithospheric field model needs therefore to be regularised. The usual way of handling this is to minimise together with the weighted squared differences between model and data the squared norm of the lithospheric field component perpendicular to the main field direction. Typically, the balance between these two terms is controlled by a 'regularising parameter' that is chosen rather arbitrarily by the scientist building the model.

The choice of regularising parameter made by modellers is clearly visible in the power spectra of the derived models. As an example, the power spectra of the WDMAM-MOD (a model derived from the WDMAM and presented in Lesur et al., 2016), the EMM 2015 and the EMM 2017 (see www.ngdc.noaa.gov/geomag/EMM/index.html) are shown in Figure 10.4.1. It is obvious that a much stronger regularisation has been applied to the EMM 2017 model. The two other models have roughly the same behaviour, although the EMM 2015 spectrum flattens at lower spherical harmonic degrees. There is less energy in the EMM models at degrees around 130, possibly due to the way marine data are handled. We note also that the WDMAM-MOD lacks power at low degrees, but this is likely due to the way long-wavelength satellite data have been merged into the compilation rather than due to the effect of the regularisation. Given our current knowledge, the only criterion to estimate the small wavelengths of the lithospheric field is the fit to the data. A very strong regularisation, as applied to the EMM 2017, necessarily deteriorates the fit to the data. The scientists who have derived this model clearly assumed a high level of noise in their aeromagnetic and marine data compilation. Indeed, estimating this level of noise is difficult; however, a possible way forward is to assume that the lithospheric field is only generated by induced magnetisation. This, combined with prior knowledge of the geology of the lithosphere, may allow us to constrain locally the direction of the lithospheric field and to reduce the non-uniqueness. Recent works (e.g. Gubbins et al., 2011) show how to get information on the rock magnetisation from magnetic field models. This will be described in the next section.

Independently of the regularisation applied, spherical harmonic models cannot fit sharp discontinuities in the magnetic anomaly data set. These are smoothed, and spurious oscillations are generated. Typically, this happens when an area covered by high-resolution aeromagnetic data neighbours an area without near-surface data. These oscillations are, however, small, and despite them, models remain extremely good representations of the lithospheric magnetic field. The models are often preferred to data compilations because they allow us to describe the field at any altitude and position. Furthermore, they allow the study of important properties of the lithosphere, as described in the next section.

Figure 10.4.2 gives an image of the radial down-component of the lithospheric field model WDMAM-MOD. The model can be compared with the original WDMAM anomaly data. The lack of resolution of the model compared to the anomaly map is obvious.

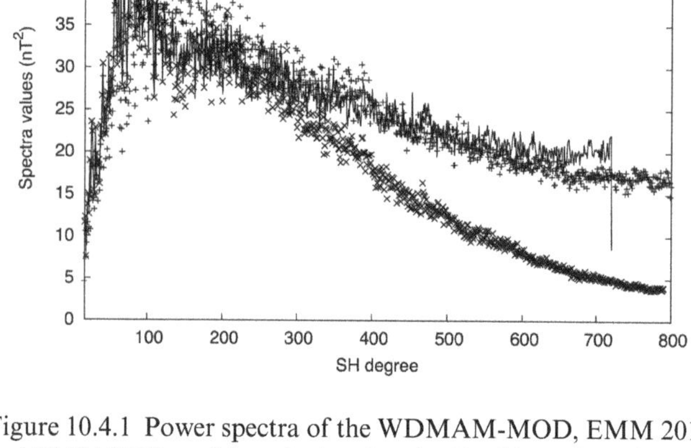

Figure 10.4.1 Power spectra of the WDMAM-MOD, EMM 2015 and EMM 2017 models, calculated at the Earth's surface, for spherical harmonic degrees ranging from 16 to 800.

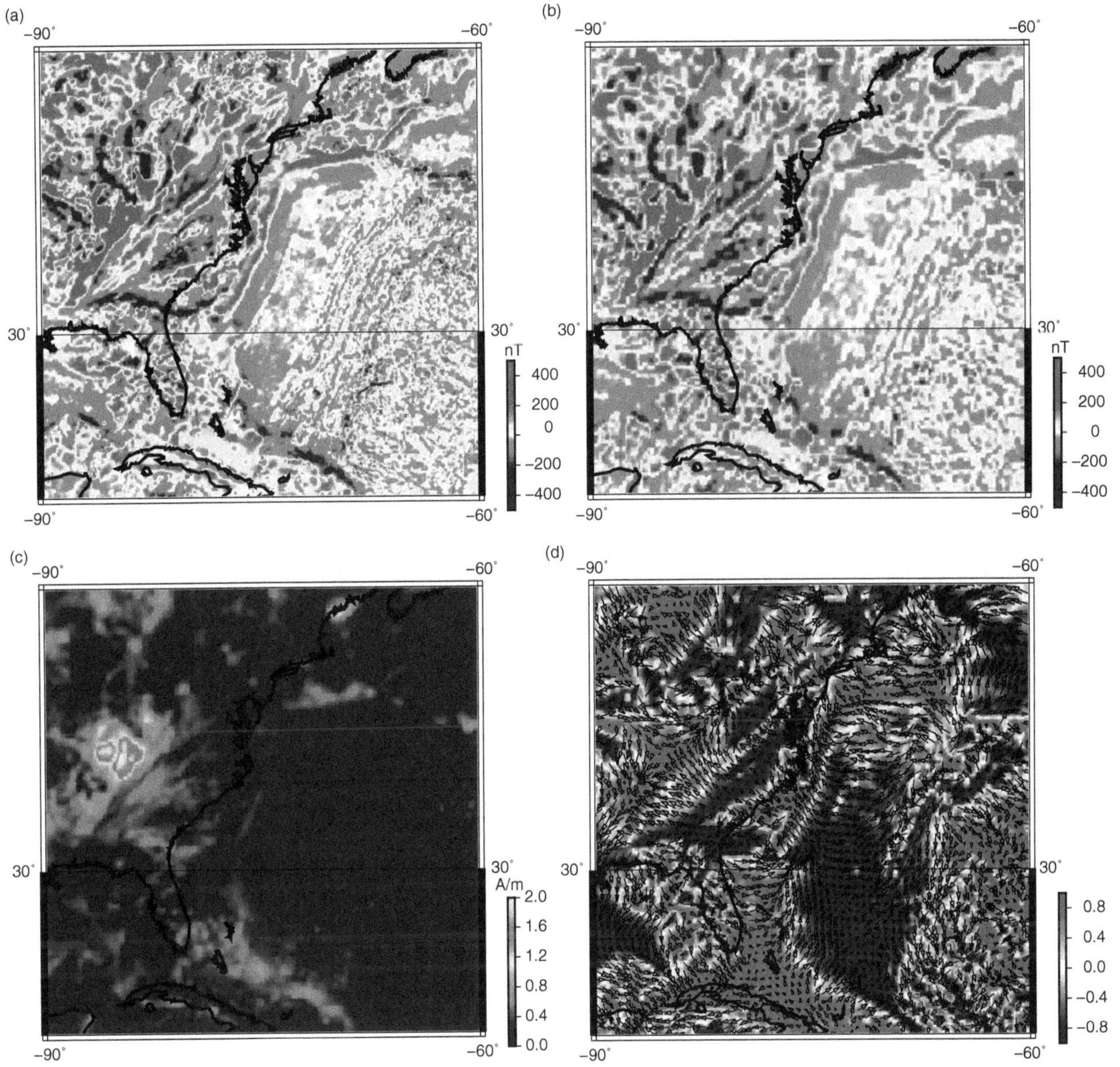

Figure 10.4.2 (top left) Total intensity anomaly data from the WDMAM, mainly over the western Atlantic and continental eastern North America. Marine data are at sea level altitude, whereas over continents, data are at 5 km altitude above the WGS84. Some data gaps are present along the US shoreline north of the Gulf of Mexico and also over parts of Cuba. (top right) Vertical down-component of the lithospheric magnetic field model wdmam-mod, truncated at spherical harmonic 720 and plotted at the Earth reference radius 6371.2 km. (bottom left) Estimated intensity of the averaged visible magnetisation over a layer of 50 km for a spherical Earth of radius 6371.2 km. (bottom right) Same map, but for the magnetisation direction. The normalised vertical down-component value is shown in colour scale; the arrow indicates the horizontal magnetisation direction. When the vertical component absolute value is 1, the horizontal component is necessarily null. (A black-and-white version of this figure appears in some formats. For the colour version, please refer to the plate section.)

10.5 On the Interpretation of Global Anomaly Data

Magnetic anomaly data are not easy to analyse because there is no simple relation between the anomaly data and the geological structure they originate from. However, the magnetic crust and its thickness can be related qualitatively to tectonic and thermal processes. Magnetic anomaly variations are associated with rifting, compositions, crustal age variations, crustal thickness, heat flow anomalies and variation of the Earth's magnetic field along time (e.g. Fox Maule et al., 2005; Besse and Courtillot, 2002). At the near-surface, volcanism and related igneous processes, such as dike emplacement, rifting and faulting, act to modify pre-existing magnetic signatures, thus providing insights on the processes involved. Studying magnetic anomaly data is therefore a very effective way of investigating the underlying geology, in particular in remote regions where rock outcrops are absent. In exploration geophysics, when dealing with wavelengths shorter than 100 km, methods have been developed to interpret these data, e.g. the Spector and Grant spectral method (Spector and Grant, 1970), pole reduction, spatial derivatives, upward and downward continuation and Euler and Werner deconvolution (Blakely, 1995). At global or regional scales the first techniques reduce generally to building magnetic field models, whereas at large scales, methods equivalent to Euler and Werner deconvolution have been seldom used (see e.g. Ravat et al., 2002, for an example application).

Large-scale quantitative estimates about the magnetisation and the magnetic crustal thickness can be provided further by forward modelling methods such as the equivalent source dipole representation (Mayhew, 1979), assuming that induced magnetisations dominate over remanent magnetisations in continental crust and that vertical thickness variations dominate over lateral susceptibility variations (Purucker et al., 2002). These forward modelling approaches have shown that the bare thickness and susceptibility contrasts between continental and oceanic crusts can explain large-scale magnetic anomalies (Counil et al., 1991). Also, large magnetic signatures over superchrons in the Indian and Atlantic Oceans are well predicted by models based on remanent magnetism and oceanic crustal thickness values (Dyment and Arkani-Hamed, 1998; Dyment et al., 2015).

The equivalent source representation can also be used to estimate the depth to the bottom of magnetic sources, which is assumed to correspond to the Curie temperature at which rocks lose their spontaneous magnetisation. Large-scale magnetic anomalies detected from satellites were used to estimate the average heat flow over Antarctica (Fox Maule et al., 2005). On more intermediate length scales, the stratification of the Earth's crust and the spatial correlation of the crustal magnetisation can no longer be neglected. Therefore hypotheses about the depths to the top and bottom of magnetic sources and parameters related to the geology are often required (Bouligand et al., 2009). However, these forward modelling approaches are not sufficient to determine the physical parameters of the sources, which may depend locally on many different processes.

Other issues arise because the lithospheric field length scales larger than 2500 km remain unknown due to the masking core field and because of magnetic annihilators that are magnetisation distributions within the crust that produce no external magnetic field signatures. The direction and strength of rock magnetisation are therefore quantities particularly difficult to extract from magnetic field measurements (Parker et al., 1987). The reconstruction of the missing wavelengths and statistical descriptions of the distribution of magnetisation can be obtained by comparing anomaly power spectra against statistical spectra derived analytically either globally (Jackson, 1994; Voorhies et al., 2002; Thébault and Vervelidou, 2015) or by successive regional analyses (Vervelidou and Thébault, 2015).

Independently of these statistical approaches, recent, very interesting developments may allow us to further progress in the analysis of magnetic data. They rely on the remarkable work presented in Gubbins et al. (2011) at global scale, which has been recently extended to flat geometries (Gubbins et al., 2017). In this work, the vector field of the magnetisation $\mathcal{M}$ at a radius $r = a$ is decomposed into three components:

$$\mathcal{M}(a,\theta,\phi) = \mathcal{M}_i(a,\theta,\phi) + \mathcal{M}_e(a,\theta,\phi) + \mathcal{M}_t(a,\theta,\phi), \quad (10.5.1)$$

with

$$\mathcal{M}_i = \sum_{\ell,m} i_\ell^m \hat{Y}_{\ell,\ell+1}^m(\theta,\phi), \quad \mathcal{M}_e = \sum_{\ell,m} e_\ell^m \hat{Y}_{\ell,\ell-1}^m(\theta,\phi),$$

$$\mathcal{M}_t = \sum_{\ell,m} t_\ell^m \hat{Y}_{\ell,\ell}^m(\theta,\phi), \quad (10.5.2)$$

where the $\hat{Y}_{\ell,\ell+1}^m, \hat{Y}_{\ell,\ell-1}^m$ and $\hat{Y}_{\ell,\ell}^m$ are vector spherical harmonics defined by

$$\hat{Y}_{\ell,\ell+1}^m(\theta,\phi) = -a\ (r/a)^{\ell+2}\ \nabla\left((a/r)^{\ell+1}\ Y_\ell^m(\theta,\phi)\right) \quad (10.5.3)$$

$$\hat{Y}_{\ell,\ell-1}^m(\theta,\phi) = -a\ (a/r)^{\ell-1}\ \nabla\left((r/a)^{\ell}\ Y_\ell^m(\theta,\phi)\right) \quad (10.5.4)$$

$$\hat{Y}_{\ell,\ell}^m(\theta,\phi) = -\hat{\mathbf{r}} \times \nabla_h Y_\ell^m(\theta,\phi) \quad (10.5.5)$$

The normalisation we use is not the same as in Gubbins et al. (2011). Here $Y_\ell^m(\theta,\phi)$ are the usual Schmidt semi-normalised real spherical harmonics used in geomagnetism, where negative orders, $m < 0$, are associated with $\sin(|m|\phi)$ terms, whereas null or positive orders, $m \geq 0$, are associated with $\cos(|m|\phi)$ terms. We have

$$\frac{1}{4\pi}\int_0^{2\pi}\int_0^{\pi} \hat{Y}_{\ell,\ell'}^m(\theta,\phi)\hat{Y}_{\ell,\ell''}^m(\theta,\phi)\sin\theta\, d\theta\, d\phi$$

$$= \begin{cases} 0 & \text{if } \ell' \neq \ell'', \\ \ell+1 & \text{if } \ell' = \ell'' = \ell+1, \\ \ell & \text{if } \ell' = \ell'' = \ell-1 \\ \ell(\ell+1)/(2\ell+1) & \text{if } \ell' = \ell'' = \ell. \end{cases} \quad (10.5.6)$$

Such a decomposition of the magnetisation is able to describe any possible magnetisation distribution physically acceptable. Similarly, with this normalisation, the magnetic field generated in the lithosphere can be described above the Earth surface as

$$\mathbf{B}_i = \sum_{\ell,m} g_\ell^m\ (a/r)^{\ell+2}\ \hat{Y}_{\ell,\ell+1}^m(\theta,\phi), \quad (10.5.7)$$

with the usual Gauss coefficients. In Gubbins et al. (2011), it is shown that only the magnetisation $\mathcal{M}_e$ contributes to the field generated in the lithosphere. The other parts of $\mathcal{M}$ do not generate signals. It is also shown that the coefficients e_ℓ^m can be estimated from the g_ℓ^m of the lithospheric field. With the normalisation used here, and following Vervelidou et al. (2017b, Equation A.6), we obtain

$$g_\ell^m = -\mu_0 \frac{\ell}{\ell+2}\left(1-(1-d/a)^{\ell+2}\right) e_\ell^m, \qquad (10.5.8)$$

where a is the radius of the magnetised layer of thickness d. This radius is also the reference radius for the definition of the lithospheric field Gauss coefficients. This part of the magnetisation that generates signals is called the *visible* magnetisation in Vervelidou et al. (2017a) and Vervelidou et al. (2017b). From this result and under a few hypotheses (e.g. pure induced magnetisation), it is possible to recover the full magnetisation of the rocks, i.e. to estimate $\mathcal{M}_i$ and $\mathcal{M}_t$, the *hidden* parts of the magnetisation. One of the advantages of working with the visible magnetisation is that it is a property of the rocks, and therefore there is no need for pole reduction or other deconvolution to know the position of the bodies generating the magnetic field. As an example, Figure 10.4.2 shows the original magnetic anomaly field of the WDMAM, the vertical down-component of the associated magnetic field model at 6371.2 km in kilometres truncated at spherical harmonic degree 720 and the intensity and direction of the calculated *visible* magnetisation.

It is clear through this example that the WDMAM has a higher resolution than the corresponding model. This is obvious not only over oceanic areas but also over the continental eastern North America. The *visible* magnetisation has the same resolution as the model but seems smoother. This is particularly obvious for the map of magnetisation intensity and comes from the fact that a small uniformly magnetised body gives rise to a total intensity anomaly field that generally changes sign above, or close to, the body location. The anomaly field seems therefore rougher than the estimate magnetisation.

The *visible* magnetisation directions shown in Figure 10.4.2 are obtained without making any assumptions regarding the magnetisation direction. They are clearly not aligned with the main field direction, but it does not mean that the magnetisation is not induced. It is possible to define a distribution of susceptibility that for a dipolar-inducing field generates the *visible* magnetisation obtained here and for which the full magnetisation, i.e. the combination of the *visible* and *hidden* magnetisation, is aligned with the inducing field. The directions obtained here can be understood as perturbations of the true magnetisation direction because, first, only the lithospheric field wavelengths corresponding to spherical harmonics of degree larger than 16 are known, and second, the Runcorn theorem (Runcorn, 1975) tells us that a constant susceptibility in a spherical shell does not generate visible magnetisation for an inducing field internal to the shell. We note that the obtained magnetisation directions tend to be vertical, either up or down, and that when they have a significant horizontal component, it is often on the edge of strongly magnetised areas. In these areas of strong gradients of magnetisation intensity, the direction of visible magnetisation is not always aligned with the gradient direction. From this example, it is clear that the direction of visible magnetisation is intuitively particularly difficult to predict and does not reflect the true magnetisation direction.

10.6 Conclusion

Magnetic anomaly maps give the essence of the crustal magnetic field that provide useful qualitative and quantitative information about the magnetised rocks in the subsurface. Currently there is little overlap between the wavelengths measured by satellite and airborne or marine surveys, and both data sets have to be merged by ad hoc procedures. Characterising the anomaly field at all spatial dimensions is a continuous effort involving significant data processing and data management as well as reprocessing of old aeromagnetic and marine data, new data acquisition in remote areas and at satellite altitudes and standardisation of data reduction. This should ultimately provide a more complete and extensive view of the magnetic crust at all scales and over its entire thickness. Combined with novel joint analysis and forward modelling techniques, it will certainly offer a different understanding of the geophysical processes responsible for the visible magnetic crustal field.

References

Besse, J. & Courtillot, V. (2002), 'Apparent and true polar wander and the geometry of the geomagnetic field over the last 200 myr', *J. Geophys. Res.* **B11**(107).

Blakely, R. (1995), *Potentiel Theory in Gravity and Magnetic Applications*, Cambridge University Press, Cambridge.

Bouligand, C., Glen, J. & Blakely, R. (2009), 'Mapping curie temperature depth in the western united states with a fractal model for crustal magnetization', *J. Geophys. Res.* **114**, B11104.

Catalán, M., Dyment, J., Lesur, V., Thébault, E., Hamoudi, M., Choi, Y., Santis, A. D., Takemi, I., Korhonen, J., Litvinova, T., Luís, J., Meyer, B., Milligan, P., Masao, N., Okuma, S., Pilkington, M., Purucker, M., Ravat, D., Gaina, C., Maus, S., Quesnel, Y., Saltus, R. & Taylor, P. (2016), 'Making a better magnetic map', *EOS* **97**.

Counil, J.-L., Cohen, Y. & Achache, J. (1991), 'A global continent-ocean magnetization contrast: spherical harmonic analysis', *Earth Planet. Sci. Lett.* **103**, 354–64.

Dyment, J. & Arkani-Hamed, J. (1998), 'Contribution of lithospheric remanent magnetization to satellite magnetic anomalies over the world's oceans', *J. Geophys. Res.* **103**, 15423–42.

Dyment, J., Choi, Y., Hamoudi, M., Lesur, V. & Thébault, E. (2015), 'Global equivalent magnetization of the oceanic lithosphere', *Earth Planet. Sci. Lett.* **430**, 54–65.

Fox Maule, C., Purucker, M. E., Olsen, N. & Mosegaard, K. (2005), 'Heat flux anomalies in Antarctica revealed by satellite magnetic data', *Science* **309**, 464–67.

Fox Maule, C., Purucker, M. & Olsen, N. (2005), 'The magnetic crustal thickness of Greenland', in C. Reigber, H. Lhr, P. Schwintzer & J. Wickert, eds., *Earth Observation with CHAMP, Results from Three Years in Orbit*, Springer.

Gubbins, D., Ivers, D., Masterton, S. M. & Winch, D. E. (2011), 'Analysis of lithospheric magnetization in vector spherical harmonics', *Geophys. J. Int.* **187**, 99–117.

Gubbins, D., Ivers, D. & Williams, S. (2017), 'Analysis of regional crustal magnetization in vector cartesian harmonics', *Geophys. J. Int.* **211**, 1285–95.

Hamoudi, M., Thébault, E., Lesur, V. & Mandea, M. (2007), 'Geoforschungszentrum anomaly magnetic map (gamma): A candidate model for the world digital magnetic anomaly map', *Geochem Geophys. Geosyst.* **8**.

Hemant, K., Thébault, E., Mandea, M., Ravat, D. & Maus, S. (2007), 'Magnetic anomaly map of the world: merging satellite, airborne, marine and ground-based magnetic data sets', *Earth Planet Sci. Lett.* **260**, 56–71.

Jackson, A. (1994), 'Statistical treatment of crustal magnetization', *Geophys. J. Int.* **119**, 991–8.

Korhonen, J., Fairhead, J., Hamoudi, M., Hemant, K., Lesur, V., Mandea, M., Maus, S., Purucker, M., Ravat, D., Sazonova, T. & Th_ebault, E. (2007), *Magnetic Anomaly Map of the World/ Carte des anomalies magnetiques du monde*, 1st edn., Commission for the Geological Map of the World, Paris.

Lesur, V. (2006), 'Introducing localized constraints in global geomagnetic field modelling', *Earth Planets Space* **58**, 477–83.

Lesur, V., Hamoudi, M., Choi, Y., Dyment, J. & Thébault, E. (2016), 'Building the second version of the world digital magnetic anomaly map (WDMAM)', *Earth Planets Space* 68, 27.

Lesur, V., Rother, M., Vervelidou, F., Hamoudi, M. & Thébault, E. (2013), 'Post-processing scheme for modeling the lithospheric magnetic field', *Solid Earth* **4**, 105–18.

Maus, S., Lühr, H., Hemant, K., Balasis, G., Ritter, P. & Stolle, C. (2007), 'Fifth generation lithospheric magnetic field model from CHAMP satellite measurements', *Geochem. Geophys. Geosys.* **8**, Q05013, doi: 10.1029/2006GC001521.

Maus, S., Sazonova, T., Hemant, K., Fairhead, J. & Ravat, D. (2007), 'National geophysical data center candidate for the world digital magnetic anomaly map', *Geochem. Geophys. Geosyst.* **8**, Q06017.

Mayhew, M. (1979), 'Inversion of satellite magnetic anomaly data', *J. Geophys.* **45**, 119–28.

Parker, R. L., Shure, L. & Hildebrand, J. A. (1987), 'The application of inverse theory to seamount magnetism', *Rev. Geophys.* 25, 17–40.

Purucker, M., Langlais, B., Olsen, N., Hulot, G. & Mandea, M. (2002), 'The southern edge of cratonic north america: Evidence from new stallite magnetometer observations', *Geophys. Res. Lett.* **29**(15), 8000.

Quesnel, Y., Catalåan, M. & Ishihara, T. (2009), 'A new global marine magnetic anomaly data set', *J. Geophys. Res.* **114**, B04106.

Ravat, D., Finn, C., Hill, P., Kucks, R., Phillips, J., Blakely, R., Bouligand, C., Sabaka, T., Elshayat, A., Aref, A. & Elawadi, E. (2009), A preliminary, full spectrum, magnetic anomaly grid of the united states with improved long wavelengths for studying continental dynamics, Open-File Report 2009 1258, US Geological Survey.

Ravat, D., Hildenbrand, T. & Roest, W. (2003), 'New way of processing near-surface magnetic data: the utility of the comprehensive model of the magnetic field', *Leadind Edge* **22**, 784–5.

Ravat, D., Wang, B., Wildermuth, E. & Taylor, P. (2002), 'Gradients in the interpretation of satellite-altitude magnetic data: An example from central Africa', *J. Geodyn.* **33**(1–2), 131–42.

Runcorn, S. K. (1975), 'On the interpretation of lunar magnetism', *Phys. Earth Planet. Inter.* **10**, 327–35.

Sabaka, T. J., Olsen, N. & Purucker, M. E. (2004), 'Extending comprehensive models of the Earth's magnetic field with Ørsted and CHAMP data', *Geophys. J. Int.* **159**, 521–47.

Spector, A. & Grant, F. (1970), 'Statistical models for interpreting aeromagnetic data', *Geophysics* **35**, 293–302.

Thébault, E. & Vervelidou, F. (2015), 'A statistical spatial power spectrum of the Earth's lithospheric magnetic field', *Geophys. J. Int.* **201**(2), 605–20.

Thébault, E., Vigneron, P., Langlais, B. & Hulot, G. (2016), 'A swarm lithospheric magnetic field model to SH degree 80', *Earth Planets Space* **68**(1), 126.

Vervelidou, F., Lesur, V., Grott, M., Morschhauser, A. & Lillis, R. J. (2017b), 'Constraining the date of the martian dynamo shutdown by means of craters' magnetization signatures', *J. Geophys. Res. Planets* **122**.

Vervelidou, F., Lesur, V., Morschhauser, A. & Grott, M. (2017a), 'On the accuracy of paleopole estimations from magnetic measurements', *Geophys. J. Int.* **211**, 1669–78.

Vervelidou, F. & Thébault, E. (2015), 'Global maps of the magnetic thickness and magnetization of the earth's lithosphere', *Earth Planets Space* **67**(1), 1–19.

Voorhies, C. V., Sabaka, T. & Purucker, M. (2002), 'On magnetic spectra of Earth and Mars', *J. Geophy. Res.* **107(E6)**, 5034.

Wonik, T., Trippler, K., Geipel, H., Greinwald, S. & Pashkevitch, I. (2001), 'Magnetic anomaly map for northern, western, and eastern Europe', *Terra Nova* **13**(3), 203–13.

11
The Ionospheric Field

The ionosphere boundary between the magnetosphere and atmosphere is often considered thin in the magnetosphere–ionosphere–thermosphere system. This approximation is not valid at the inner boundary, where height variation is important in ionosphere–thermosphere (I-T) coupling, particularly with respect to momentum/energy transfer. Here the Cowling channel and energy coupling between regions are better modeled including altitude variations.

In the equatorial region the equatorial plasma fountain results from a field perpendicular ExB drift and field aligned plasma diffusion, while the equatorial ionization anomaly is formed by removal of equatorial plasma by upward ExB drift. Under magnetic storm conditions an eastward prompt penetration electric field and neutral winds contribute. The polar cap ionosphere and auroral zones transfer solar wind energy into the magnetosphere. In the polar cap key indicators for energy/momentum transfer to the solar wind I-T system are the cross-polar cap potential/electric field, and the relationship to the interplanetary magnetic field where linear and non-linear relationships may occur.

Models have been produced to describe various aspects of the coupled system. In the auroral zones aurora are associated with different regions and processes; substorm-associated aurora, shock associated aurora, pulsation aurora, cusp aurora and mid-latitude aurora. These categories and recent models are referenced.

11.1 Ionosphere–Thermosphere Coupling

Robert J. Strangeway

11.1.1 Ionosphere–Thermosphere Coupling

While this section covers the topic of ionosphere–thermosphere coupling, the overall context will be the role of ionosphere–thermosphere (I-T) coupling in the context of the coupled magnetosphere–ionosphere–thermosphere system. One theme that will be present throughout this section is the idea that the ionosphere cannot always be treated as a thin layer that is simply an inner boundary to the magnetosphere. As an example, it is the usual practice to use the height-integrated Pedersen and Hall conductivities when treating the ionosphere as an inner boundary. We will see in the subsequent discussion that the height variation within the ionosphere is an important aspect of ionosphere–thermosphere coupling. We will do this by addressing two topics: momentum transfer and energy transfer.

11.1.2 Momentum Transfer

Momentum transfer is involved in two primary aspects of the coupled I-T system, wave-mediated coupling and ionospheric currents. In order to investigate these we will first provide a simplified set of the momentum equations in the I-T system, after Strangeway and Raeder (2001) and Strangeway (2012).

First, for each species, we will define a generic force term $\mathbf{F}_s$:

$$\mathbf{F}_s = -\nabla P_s - n_s m_s \frac{d\mathbf{U}_s}{dt_s}, \tag{11.1.1}$$

where the subscript s indicates species, which can be electrons, ions or neutral particles. P_s is the pressure for the species, which we have assumed to be isotropic for simplicity, n_s is the species number density, m_s is the species mass, $\mathbf{U}_s$ is the species bulk velocity and $d/dt_s = \partial/\partial t_s + \mathbf{U}_s \cdot \nabla$ is the total time derivative for that species. The primary simplification here is to include the forces and species inertia in $\mathbf{F}_s$. In particular, additional forces such as gravity, non-isotropic pressure or viscosity-related forces could be included in (11.1.1).

We have not included the electric field $\mathbf{E}$ in $\mathbf{F}_s$. Instead we will define a modified electric field

$$\widetilde{\mathbf{E}} = \mathbf{E} - \frac{\mathbf{F}_e}{ne} - \frac{m_e \nu_{ei}}{ne^2}\mathbf{j}. \tag{11.1.2}$$

In (11.1.2) we have assumed quasi-neutrality, i.e. $n_e \approx n_i \approx n$, where the subscripts i and e correspond to ions and electrons, respectively, e is the magnitude of the electron charge, $\mathbf{F}_e$ is (11.1.1) as specified for the electron fluid, ν_{ei} is the Coulomb collision frequency between electrons and ions and $\mathbf{j}$ is the current density, $\mathbf{j} = ne(\mathbf{U}_i - \mathbf{U}_e)$.

With these definitions, the electron and ion fluid momentum equations become

$$m_e\nu_{en}(\mathbf{U}_e - \mathbf{U}_n) = -e\left(\widetilde{\mathbf{E}} + \mathbf{U}_e \times \mathbf{B}\right), \tag{11.1.3}$$

where ν_{en} is the electron-neutral collision frequency, and

$$m_i\nu_{in}(\mathbf{U}_i - \mathbf{U}_n) = e\left(\widetilde{\mathbf{E}} + \frac{\mathbf{F}}{ne} + \mathbf{U}_i \times \mathbf{B}\right), \tag{11.1.4}$$

with ν_{in} being the ion-neutral collision frequency, $\mathbf{F} = \mathbf{F}_i + \mathbf{F}_e$, and $\mathbf{B}$ is the ambient magnetic field.

As shown by Strangeway (2012), the advantage of the momentum equations as given by (11.1.3) and (11.1.4) is that they can be recast in a form that relates the current density to the modified electric field in the neutral atmosphere frame of reference ($\widetilde{\mathbf{E}} + \mathbf{U}_n \times \mathbf{B}$) with terms that include the standard Pedersen and Hall conductivities plus terms related to the additional forces acting on the plasma as given by $\mathbf{F}$, most notably the ion inertia term. The ion inertia must be included if we are considering Alfvén wave–mediated coupling, as discussed, for example, by Lysak (1999), Song et al. (2005), Tu et al. (2011, 2014), Lysak et al. (2013) and Yoshikawa et al. (2013).

We note that Parker (1996), Song et al. (2005) and Tu et al. (2011, 2014) also emphasized that the neutral atmosphere should be included in the set of momentum equations. For the neutral particles their momentum equation can be expressed as

$$n_n m_n \nu_{ni}(\mathbf{U}_n - \mathbf{U}_i) = \mathbf{F}_n, \tag{11.1.5}$$

where, as with the other species, $\mathbf{F}_n$ includes both the inertia of the neutral particles and other forces to the degree of approximation desired.

In order to couple the plasma to the neutral particles we make use of momentum conservation for the collisions, i.e.

$$n_n m_n \nu_{ni} = n_i m_i \nu_{in}, \tag{11.1.6}$$

and the left-hand side of (11.1.4) can be replaced by $-\mathbf{F}_n/n_i$. For very-low-frequency waves ($\omega \lesssim \nu_{ni}$), $\mathbf{U}_n \approx \mathbf{U}_i$, and the neutral particles provide the mass density in the Alfvén speed.

In order to evaluate the relative importance of collisions we plot a sketch of the altitude profile of the collision frequencies in Figure 11.1.1. This figure shows the collision frequencies for the dayside ionosphere (based on Richmond, 1995). We have also included the electron and ion gyrofrequencies ($\Omega_e = eB/m_e$ and $\Omega_i = eB/m_i$, respectively) and the ion plasma frequency ω_{pi} as a proxy for the plasma density. For simplicity we have assumed only singly charged oxygen ions. The dashed line shows $\nu_{en} + \nu_{ei}$. We have not shown the neutral-ion collision frequency ν_{ni}, but this depends on the plasma density and roughly follows the altitude profile of ν_{ei}, albeit several orders of magnitude smaller, with neutral-ion characteristic collision times typically being of the order hours.

Figure 11.1.1 is primarily meant as a guide in determining where collisions become important in the species momentum equations. Inspection of (11.1.3) and (11.1.4) shows that for the momentum perpendicular to the local magnetic field the transition from weakly collisional to strongly collisional occurs when $\nu_s \approx \Omega_s$ for a particular species. Electrons are largely collisionless, even though their collision frequency is higher than the ion collision frequency, only becoming collisional in the bottomside ionosphere. The ions, on the other hand, become strongly collisional in the upper part of the E-region.

Combining (11.1.3) and (11.1.4), we find that, for motion perpendicular to the magnetic field,

$$nm_i\nu_{in}(\mathbf{U}_{\perp i} - \mathbf{U}_{\perp n}) \approx \mathbf{j} \times \mathbf{B} + nm_i\, d\mathbf{U}_{\perp i}/dt, \tag{11.1.7}$$

where we have assumed the plasma is cold. (Note that pressure should be retained for motion parallel to the magnetic field, especially at high latitudes, as that includes the effect of the ambipolar electric field associated with electron heating, $E_{\|} = -\nabla_{\|}P_e/ne$.) For sufficiently low frequencies the last term on the right-hand side of (11.1.7) can be

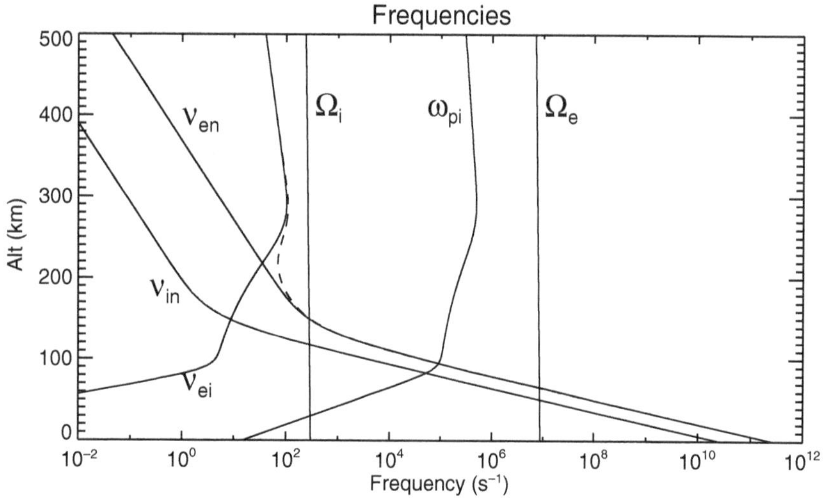

Figure 11.1.1 Sketch of the dayside ionosphere collision frequencies near local noon and ~45°N, after Strangeway (2009), based on Richmond (1995).

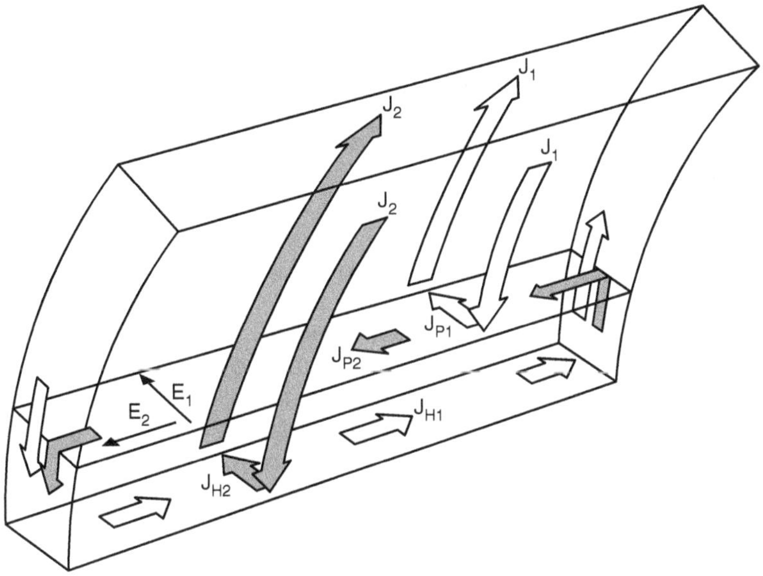

Figure 11.1.2 Sketch of a Cowling channel where a secondary current system (gray arrows) is set up to cancel the Hall current of a primary current system (white arrows), from Russell et al. (2016).

ignored, and the direction of the plasma motion is given by $\mathbf{j} \times \mathbf{B}$.

One other consequence of Figure 11.1.1 is that for the perpendicular component of (11.1.3) we can ignore both collisions and the electron inertia because of the low electron mass. If we can further assume $\mathbf{F}_{\perp e} \approx 0$, then (11.1.3) reduces to the electron frozen-in condition $\mathbf{E}_{\perp} + \mathbf{U}_e \times \mathbf{B} = 0$. In that case the Pedersen and Hall conductivities become

$$\sigma_P \approx \frac{ne}{B}\left(\frac{\nu_{in}/\Omega_i}{1+\nu_{in}^2/\Omega_i^2}\right) \tag{11.1.8a}$$

and

$$\sigma_H \approx \frac{ne}{B}\left(\frac{\nu_{in}^2/\Omega_i^2}{1+\nu_{in}^2/\Omega_i^2}\right). \tag{11.1.8b}$$

For altitudes for which $\nu_{in} < \Omega_i$ the Pedersen conductivity is larger, with the Hall conductivity being larger for $\nu_{in} > \Omega_i$. Thus Hall currents dominate at lower altitudes (see Figure 11.1.1). This has consequences for the concept of what is known as the Cowling conductivity as applied to the auroral ionosphere.

In its simplest form, Cowling conductivity arises from a current system where a secondary current system is established to cancel the Hall current from a primary current system. This is shown in Figure 11.1.2. Here we show a longitudinally extended primary current system and associated flow channel in the northern auroral ionosphere. The curved lines are meant to indicate magnetic field lines, and the long horizontal direction is assumed to be nominally east–west aligned. The upward current, labeled J_1, is assumed to extend throughout the sheet defined by the magnetic field. This current would be carried by precipitating electrons, and the current would therefore correspond to a longitudinally extended auroral arc. This current closes with the corresponding downward current through the Pedersen current J_{P1}. The flow channel is assumed to be narrow in the north–south direction. For simplicity, while the Pedersen current would flow over an extended altitude range, we have followed the approach of Fujii et al. (2011, 2012), where we assume the Pedersen current flows in a layer at a different altitude than the Hall current. Clearly, in actuality, these are volume currents (as are the field-aligned currents), but the main point here is that the Pedersen and Hall currents are not co-located. Within the Pedersen current layer the plasma flows from left to right (or equivalently from east to west). At lower altitudes we see the primary Hall current J_{H1}, flowing in the opposite direction to the plasma flow in the Pedersen layer. This is to be expected since in steady state the electric field cannot have a curl, and the electrons, being frozen-in, flow from east to west in the Hall current layer, in the $\mathbf{E}_1 \times \mathbf{B}$ direction, while the ions tend to flow along the direction of E_1 in this layer.

The Cowling channel arises out of the assumption that a secondary electric field, E_2, frequently referred to as a polarization electric field, is set up so that the secondary Pedersen current, J_{P2}, opposes the primary Hall current, J_{H1}. A complete Cowling channel is one in which the two currents cancel, and we define $\alpha = J_{P2}/J_{H1}$ as a parameter that specifies the completeness of the channel. For $\alpha = 0$, there is no polarization electric field, while $J_{P2} = J_{H1}$ for $\alpha = 1$. An aside may be in order here: Yoshikawa et al. (2013) also discuss the consequences of Cowling conductivity, but with a geometry that is orthogonal to that presented in Figure 11.1.2, with the primary Pedersen current flowing in the longitudinal direction, rather than northward as shown in Figure 11.1.2. Here we are assuming that the upward current of the primary current system is carried by precipitating electrons in an auroral arc. Yoshikawa et al. (2013) primarily address Alfvén wave–mediated coupling, rather than coupling within an auroral arc system. In the Yoshikawa et al. (2013) geometry there would be no longitudinally extended auroral field-aligned currents for a complete Cowling channel.

The secondary current system is shown in gray in Figure 11.1.2, and there is a secondary Hall current, J_{H2}, that also closes within the magnetosphere through the field-aligned currents J_2. The total field-aligned current that closes the northward current is given by

$$J_N = J_{P1} + J_{H2} = \left(1 + \alpha \frac{\Sigma_H^2}{\Sigma_P^2}\right) \Sigma_P E_1 = \Sigma_C E_1 \tag{11.1.9}$$

where Σ_p and Σ_H are the height-integrated Pedersen and Hall conductivities, respectively. Usually $\Sigma_H > \Sigma_P$ and the Cowling conductivity, Σ_C, for a full Cowling channel is about an order of magnitude larger than Σ_P for typical Σ_H/Σ_P ratios.

There are several aspects of the Cowling channel that raise questions when considered within the context of the coupled system. First, as already noted, in steady state the electric field is curl-free. This means that the secondary, polarization electric field must map into the magnetosphere. This secondary electric field is given by

$$E_2 = \alpha \frac{\Sigma_H}{\Sigma_P} E_1 \tag{11.1.10}$$

and the secondary electric field can be larger than the primary electric field. This implies that the plasma flow within the magnetosphere is mainly normal to the initially assumed flow channel.

Second, the two layer model shown in Figure 11.1.2 emphasizes that the secondary Pedersen current flows at a different altitude than the primary Hall current. This suggests that there should a large magnetic field perturbation between the two current layers, with a maximum at the altitude where $\nu_{in}/\Omega_i = \alpha \Sigma_H/\Sigma_P$, as this is where the eastward current changes sign. Unfortunately, this is an altitude that is difficult to probe with spacecraft because of the large atmospheric drag, but observations of such a perturbation would be a strong indication of the presence of a Cowling channel. There is also a question as to how a Cowling channel could occur within coupled magnetosphere–ionosphere–thermosphere models, as the Cowling channel may be a manifestation of field-aligned current closure entirely within the ionosphere. In Figure 11.1.2 one consequence of a complete Cowling channel is that no field-aligned currents flow into the magnetosphere at the east and west ends of the flow channel, where the primary closure currents (in white, at the ends of the channel) become zero. In the absence of the Cowling channel, under the assumption that the Hall conductivity is small outside of the flow channel, then a global model that uses a thin ionosphere boundary condition would of necessity have field-aligned currents that flow at the ends of the flow channel.

One other aspect of momentum coupling in the thermosphere concerns cross-field transport of momentum, and hence flows. In particular, Kikuchi (2014, and references therein) note that because of the high conductivity of the E-region ionosphere, waves may more easily propagate from high latitudes to lower latitudes as the transverse-magnetic TM mode in the surface–ionosphere waveguide. On the other hand, Lysak (2004) has modeled the waveguide above the ionosphere, associated with a local maximum in Alfvén speed, and shown that waves can propagate rapidly across the magnetic field in the ionospheric waveguide. Wave propagation for both waveguides has been explored by Lysak and Yoshikawa (2006) and Lysak et al. (2013). The latter paper uses specific rather than height-integrated conductivities. This adds some fine-scale structure at low altitudes, but the ionospheric waveguide itself is essentially unaltered.

11.1.3 Energetics in the Coupled System

As noted by Strangeway (2012), following the classical work of Banks and Kockarts (1973), the heating rate for a species r through collisions with the other species s is given by

$$\left.\frac{\delta Q_r}{\delta t}\right|_s = \frac{2m_r n_r \nu_{rs}}{m_r + m_s}\left[\frac{3}{2}k_b(T_s - T_r) + \frac{1}{2}m_s|\mathbf{U}_s - \mathbf{U}_r|^2\right]. \tag{11.1.11}$$

Here the species can again include ions, electrons or neutrals. The heating rate can be divided into two parts. The first is associated with the temperature difference, $T_s - T_r$, and this heating (or cooling, depending on the sign) acts to try to bring the species into thermal equilibrium. The second term on the right-hand side depends on the difference in streaming velocities and is frequently referred to as frictional heating. It should be noted that for simplicity we have assumed the collision frequency associated with both types of heating is given by the collision frequency used in the momentum equations (11.1.2), (11.1.3) and (11.1.4). This is reasonable for the frictional heating, but may be less so for the heating through temperature differences, since here the collisions are related to the thermal velocity of the underlying distributions. Nevertheless, when comparing heating rates, this simplification is probably adequate.

It should be noted that (11.1.11) gives a volume heating rate, and the corresponding change in temperature is given by

$$\frac{3}{2} n_r \left.\frac{dk_b T_r}{dt_r}\right|_s = \left.\frac{\delta Q_r}{\delta t}\right|_s, \tag{11.1.12}$$

where d/dt_r is the total time derivative following the species r and we have neglected the effects of species compression and heat flux. As noted by Strangeway (2012), the neutral species dominate the number (and mass) density in the ionosphere and the minority species (ions and electrons) are heated much more rapidly than the neutrals. Furthermore, in terms of frictional heating, even though the electron-neutral collision frequency is larger than the ion-neutral frequency (cf. Figure 11.1.1), the lower mass of the electrons means that there is less energy available as streaming energy, and the ion temperature increases more rapidly than the electrons. Thus the effect of enhanced flows in terms of heating is to first raise the ion temperature, and then through temperature equilibration slowly raise the temperature of the neutrals on a timescale of the order $1/\nu_{ni}$, which, as noted above, is of the order of hours.

We can relate the frictional heating rate for the ions and neutrals to the Joule dissipation in the ionosphere and hence to Poynting flux into the ionosphere. This provides a link between magnetospheric processes and ionospheric heating. From (11.1.7), ignoring the inertia term on the right-hand side,

$$\begin{aligned} nm_i\nu_{in}|\mathbf{U}_{\perp i} - \mathbf{U}_{\perp n}|^2 &\approx (\mathbf{U}_{\perp i} - \mathbf{U}_{\perp n}) \cdot (\mathbf{j} \times \mathbf{B}) \\ &= \mathbf{j} \cdot (\mathbf{E}_\perp + \mathbf{U}_n \times \mathbf{B}), \end{aligned} \tag{11.1.13}$$

where we have assumed the frozen-in electron condition $\mathbf{E}_\perp + \mathbf{U}_i \times \mathbf{B} - \mathbf{j} \times \mathbf{B}/ne = 0$. The left-hand side of (11.1.13) equals the sum of the frictional heating rate for ions and neutrals. Thus, from Poynting's theorem, in steady state the frictional heating requires Poynting flux into the ionosphere.

The left-hand side of (11.1.13) corresponds to the total heating rate for the ions and neutrals, as can be seen by adding the corresponding heating rates for each species given by (11.1.11), noting the heating associated with the temperature differences cancel, and again making use of (11.1.6). Because $\nu_{in} \gg \nu_{ni}$, the effect of frictional heating is to first raise the ion temperature so that (11.1.11) ≈ 0 for the ions. In that case the neutral species heating rate from the temperature difference between ions and neutrals also equals the ion frictional heating rate, and the net effect is that after the initial interval of ion heating all the ion and neutral frictional heating acts to heat the neutrals, albeit on the neutral-ion collision timescale.

There are some issues with the simple picture outlined here, however. There have been reports of significant thermospheric heating in the cusp region (Lühr et al., 2004), where there appears to be upwelling of the atmosphere on relatively short timescales. This may have major effects on low-altitude spacecraft, increasing the drag and possibly shortening mission lifetimes. Cosgrove et al. (2014) have generated maps of the Poynting flux into the topside ionosphere using data from the Fast Auroral Snapshot small Explorer. These maps clearly show enhanced Poynting flux in the cusp region, presumably associated with reconnection at the dayside magnetopause. We would therefore expect upwelling of the ions, and this has been observed with radars (e.g. Wahlund et al., 1992), where the regions of enhanced ion heating result in type 1 ionospheric outflows.

Knipp et al. (2011) presented examples of large Poynting fluxes observed in the dayside cusp, and Crowley et al. (2010) modeled an event included in the database generated by Knipp et al. (2011), where the CHAMP spacecraft showed enhanced thermospheric densities, inferred from atmospheric drag measured by the spacecraft. Crowley et al. (2010) attributed the thermospheric upwelling to increased Joule dissipation observed in the cusp, as was also concluded by Lühr et al. (2004). Crowley et al. (2010) used the TIME-GCM model (Roble and Ridley, 1994) that includes full thermospheric dynamics. This model showed a relatively prompt thermospheric response to the imposed Joule heating. How is this reconciled with the statement above that the timescale for temperature changes in the neutral atmosphere is given by the neutral-ion collision time?

The answer to this seeming contradiction is how changes in scale height affect the altitude profile. For oxygen atoms that have 1000 K temperature, the scale height is ~53 km. If, for discussion purposes, we fix the oxygen density at 200 km altitude, the oxygen density reduces by a factor of ~0.058 at 350 km altitude, around the perigee of the CHAMP spacecraft. If we assume the temperature increases to 1500 K (a factor of 1.5) due to heating at lower altitudes, where ν_{ni} is larger, with thermal conduction increasing the temperature at higher altitudes, then the density ratio at 350 km becomes ~0.150, and the oxygen atom density is almost a factor of 3 larger. The point here is that while the thermospheric density changes do require heating, the change in density at a particular altitude is not linear with temperature. For increasing altitude the change in density becomes relatively larger as the scale height increases.

Another factor that may come into play is the presences of large-amplitude small-scale field-aligned current structures, as discussed by Lühr et al. (2004). These can be difficult to resolve on low-altitude spacecraft, but they may have significant impact on the Joule heating. Such small-scale structures could be a signature of Alfvén waves. But it is not clear how well small-scale Alfvén waves couple to the ionosphere, because of the impedance mismatch between the Alfvén wave conductance ($\Sigma_{VA} = 1/\mu_0 V_A$, where V_A is the Alfvén speed) and the height-integrated Pedersen conductivity (Σ_P). Typically, Σ_{VA} is of the order 1 S, compared to $\Sigma_P \sim 10$ S in the auroral zone. In addition, some of the Alfvén wave energy may be converted to precipitating electron energy flux (e.g. Chaston et al., 2003).

Precipitating 'soft' electrons, i.e. with energies around 100 eV, are a second source of ion outflows, and Wahlund et al. (1992) referred to these as type 2 outflows. In this case the precipitating electrons enhance the topside electron temperature and the resultant ambipolar electric field increases the ion scale height. Indeed, Strangeway et al. (2005) derived outflow scaling laws based on cusp observations by the FAST spacecraft that related the outflows to both large-scale (DC) Poynting flux and electron precipitation. These scaling laws were extended to include Alfvénic Poynting flux by Brambles et al. (2011), who found that the Alfvénic Poynting flux seemed to best characterize the outflow response when used within global simulations. It should be noted, however, that the scaling laws do not explicitly address the underlying physics, and there is much effort on the part of the community to generate better, more physics-based approaches to understanding how the magnetosphere–ionosphere–thermosphere interaction results in ions escaping to the magnetosphere.

11.1.4 Summary

In this section we have emphasized the role of momentum and energy coupling in the magnetosphere–ionosphere–thermosphere system. One aspect of this is that the ionosphere cannot be treated as a thin layer in every instance. Because of the strong altitude dependence of the collision frequencies the Alfvén wave characteristics are strongly altitude dependent. How such waves travel in and through the ionosphere is better addressed using an ionosphere that allows for altitude variation. This is also the case when considering current flows within the ionosphere. The Cowling channel, in particular, would be better modeled using height-dependent conductivities, as this may affect how the secondary currents that constitute the Cowling channel are set up within the high-latitude ionosphere. In terms of energy coupling, the heating rates for ions and neutrals are strongly altitude dependent. Most thermospheric models, by their nature, include altitude variation for the thermosphere. How the ions are treated in such models also needs care. The ions may follow the $\mathbf{E} \times \mathbf{B}$ flow direction at higher altitudes, but the flows are distorted by collisions at lower altitudes.

11.2 Equatorial Ionization Anomaly

Balan Nanan

11.2.1 Equatorial Ionospheric Anomaly

A brief review of the recent developments in the understanding of the equatorial plasma fountain (EPF) and equatorial ionization anomaly (EIA) is presented. It is clarified that (1) the EPF is not upward $\mathbf{E} \times \mathbf{B}$ plasma drift at the equator followed by downward plasma diffusion, but it is field perpendicular $\mathbf{E} \times \mathbf{B}$ plasma drift and field-aligned plasma diffusion acting together along the field lines at all altitudes and plasma flowing in the direction of the resultant. (2) The EIA is formed not from the accumulation of plasma at the crests but mainly from the removal of plasma from around the equator by the upward $\mathbf{E} \times \mathbf{B}$ drift with small accumulations when the crests are within ~ ± 20° magnetic latitude and no accumulation beyond ~ ± 25°. (3) The strong EIA (over 100%) under magnetically active conditions arises from the simultaneous action of eastward prompt penetration electric field and equatorward neutral wind.

11.2.2 Introduction

Because the Sun shines over the equator, the ionospheric density was expected to vary from a maximum at the equator to a minimum at high latitudes. But when the density variation was measured (Namba and Maeda, 1939; Appleton, 1946), it was found to exhibit an unexpected large structure with a trough around the equator, crests near ±15° magnetic latitudes and crest-to-trough ratio of about 1.6 in daytime peak electron density (Nmax). This large structure known as the equatorial ionization anomaly (EIA) covers about half the global area in 24 hours, and the position of the crests and the crest-to-trough ratio vary with various geophysical conditions.

Following the discovery, several theories like the diffusion theory (Mitra, 1947) and electrodynamic drift theory (Martyn, 1955) were suggested to explain the anomaly (EIA). According to diffusion theory (Mitra, 1946), plasma diffuses along the geomagnetic field lines from the (otherwise) high-density region at the equator to the low density region at higher latitudes under the influence of gravity and pressure gradient forces, which was suggested to produce the anomaly. Martyn (1955) proposed that, in addition to diffusion, upward $\mathbf{E} \times \mathbf{B}$ drift of plasma should also be important in causing the anomaly. Later when computers became powerful enough, Rishbeth (1963) showed that diffusion is important but not sufficient. The combination of the drift and diffusion theories which generate the equatorial plasma fountain (EPF), on the other hand, was successful in explaining the observations (Moffett and Hanson, 1965). Hanson and Moffett (1966) presented the first pictures of EPF and EIA in the Northern Hemisphere from steady state model calculations using assumed $\mathbf{E} \times \mathbf{B}$ drift velocity (the drift velocity had yet to be measured). Anderson (1973) showed the neutral wind modulation of EIA.

The Sheffield University Plasmasphere Ionosphere Model (SUPIM) (Bailey and Balan, 1996; Bailey et al., 1997), which incorporates measured values of $\mathbf{E} \times \mathbf{B}$ drift velocity (Fejer et al., 1991) and neutral wind velocity (from horizontal wind model HWM) (Hedin et al., 1995), showed detailed pictures of the plasma fountain with and without neutral wind and their effects in producing and modulating EIA (Balan and Bailey, 1995; Balan et al., 1997). SUPIM also showed the impulsive strengthening of the fountain during pre-reversal enhancement (PRE) of the upward drift and reverse plasma fountain during the following downward drift. There are a number of other physics-based models (not referred due to page limitation), which have also reproduced the EIA. Review articles on EIA were presented by Rajaram (1977), Moffett (1979), Anderson (1981) and Stening (1992).

During the early stages of major geomagnetic storms especially superstorms, the daytime plasma fountain becomes a super fountain (Balan et al., 2009) and the EIA becomes strong with over 1000% increase in density at the crests that shift up to ±30° latitudes (e.g. Mannucci et al., 2005; Balan et al., 2011). Such strong EIA was suggested to occur due to strong eastward prompt penetration electric field PPEF (Kelley et al., 2004) though modeling studies later showed that PPEF alone is unlikely to produce the strong EIA (Balan et al., 2009). The strong EIA was explained in terms of the combined effects of strong eastward PPEF and fast storm-time equatorward winds (e.g. Lin et al., 2005; Balan et al., 2010, 2011; Lu et al., 2012).

As described, theoretical models have quite successfully reproduced the observed EIA and its variations under quiet and active conditions. However, easy-to-understand misinterpretations of EPF and EIA have caused misunderstandings which have led unlikely mechanisms to be proposed for the strong EIA under active conditions. The development of EIA in the morning by ~1030 LT, though **E × B** drift turns upward at ~0630 LT, also has indirectly supported the misinterpretations. This brief review is intended to clarify the misinterpretations using SUPIM.

SUPIM (Bailey and Balan, 1996) solves the coupled time-dependent equations of continuity, momentum and energy for the electrons and ions (O^+, H^+, He^+, $N_2{}^+$, NO^+ and $O_2{}^+$) along closed eccentric-dipole geomagnetic field lines. Recently, the N^+ ion has also been included and the lower altitude boundary is extended down from 110 km to 80 km (Souza et al., 2013). The model inputs are the measured **E** × **B** drift velocity (Fejer et al., 1991), neutral wind velocities from HWM93 (Hedin et al., 1995) and neutral densities from NRLMSISE (Picone et al., 2002). HWM and NRLMSISE give the wind velocities and densities as function of altitude, latitude, local time, season and level of solar and magnetic activities. The **E** × **B** drift velocities used for quiet conditions are measured at the equatorial station Jicamarca and low-latitude station Arecibo (Fejer et al., 1991). These drifts and zero drift are applied at apex altitudes of 90–700 km and 1500–5000 km and above 55,000 km, with interpolations for the apex altitudes in between. The apex interval varies with apex as 2.5 km for 90–120 km, 5 km for 125–300 km, 10 km for 310–500 and 20 km for 520–1000 km; and above 1000 km the apex varies at intervals corresponding to 1° latitude. The shorter intervals at lower apex are meant for conductivity calculations. The drift velocity varies with altitude and latitude according to the field line geometry.

11.2.3 EIA under Quiet Conditions

Though diffusion of plasma from the equator to higher latitudes was suggested to explain EIA (Mitra, 1946), it alone was found insufficient (Rishbeth, 1963) as illustrated in Figure 11.2.1, which shows the electron density Ne (in units of 10^5 cm^{-3}) variation obtained from the model calculations carried out with no **E** × **B** drift. Diffusion alone produces only a weak anomaly in the morning with crests at around ±10° magnetic latitudes (Figure 11.2.1a) due to the large plasma pressure gradient between the equator and higher latitudes, and the anomaly becomes weaker with time and disappears before noon when the pressure gradient is small. The Ne variation at noon (13 LT; Figure 11.2.1b) is centered at the equator at ~330 km height with a density of ~19 units. The density in the Northern (winter) Hemisphere is slightly greater than that in the Southern Hemisphere due to winter anomaly (larger O/N_2 ratio compared to summer); the modeling day 30 October is at the beginning of winter in the north. It may be noted the density at the latitudes (~±15°) of the crests is ~17 units.

As mentioned above, the equatorial plasma fountain (EPF) has often been viewed as vertically upward **E** × **B** plasma drift at the equator followed by downward diffusion (e.g. Balan et al., 1995; Lu et al., 2012). This easy-to-understand picture has two incorrect interpretations: (1) the drift is interpreted to act only at the equator and (2) diffusion is interpreted to follow the drift. In reality and in models, the field perpendicular **E** × **B** plasma drift and field-aligned plasma diffusion present along the field lines at all altitudes act together, and plasma flows in

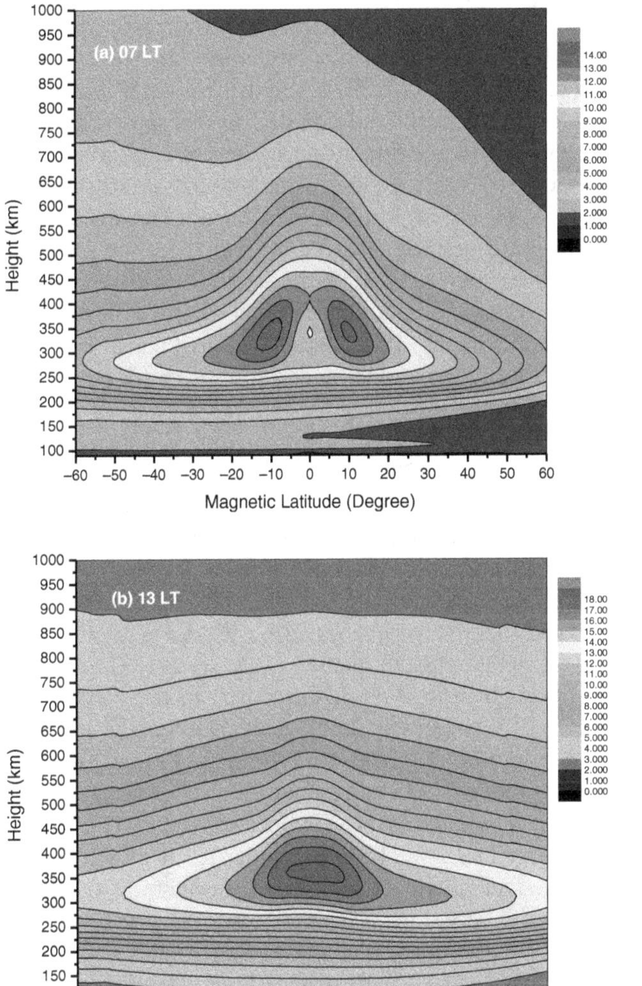

Figure 11.2.1 Altitude–latitude map of electron density Ne at (a) 0700 LT and (b) 1300 LT modeled by SUPIM for magnetically quiet (Ap = 4) conditions at high solar activity (day 303, F10.7 = 268) in Jicamarca longitude (283°E) with no **E** × **B** drift and no neutral wind. Unit of Ne is 10^5 cm^{-3}. Positive latitude is north. (A black-and-white version of this figure appears in some formats. For the colour version, please refer to the plate section.)

the direction of the resultant. The fountain can rise up to ~1000 km in height over the equator and cover up to ~±30° in latitudes (at 300 km) at high solar activity (Balan et al., 1997). The altitude–latitude coverage varies with the time of the day, season and level of solar activity and has large day-to-day variability due to the variability of the **E** × **B** drift.

The EIA crests have often been viewed as forming from the accumulation of the downward diffusing plasma (e.g. Kelley et al., 2004; Lu et al., 2012). This also is not true because it is not clear how downward diffusing plasma could accumulate at high-altitude-latitude locations (e.g. ~300 km at ±16° latitudes) where field lines are inclined, or what stops the downward diffusion at these locations against gravity. In fact, even in this scenario, plasma diffuses to the top of the crests and at the same time it also diffuses down from the bottom of the crests to lower altitudes where it gets lost by chemical recombination. In reality and in models, the EIA is formed mainly from the removal of plasma from around the equator by the upward **E** × **B** drift with small accumulation at the crests as long as the crests are at lower latitudes. The amount of accumulation reduces as the crests shift to higher latitudes where field lines become more inclined and accumulation becomes zero when the crests are beyond ±25° latitudes (Balan et al., 2009).

The process of formation of the EIA can be illustrated by comparing Figures 11.2.1 and 11.2.2. Figure 11.2.2a is obtained from the model calculations carried out for the same conditions as for Figure 11.2.1 but using an assumed upward **E** × **B** drift (Figure 11.2.2b) applied for an hour just before noon (1100–1200 LT) and no drift at other local times. The drift is applied only just before noon so that the EIA formation (around noon of photo-chemical equilibrium) is not dominated either by production or chemical loss of ionization. The Ne variation at 1300 LT (Figure 11.2.2a) exhibits a well-developed (nearly) symmetric EIA with crests at ~±15° latitudes at ~400 km height; density at the crests is ~18 units and crest-to-tough ratio is ~1.5. Compared to the Ne variation with no EIA (Figure 11.2.2b), the density at the trough of EIA is reduced by ~40% (from 19 units to 12 units) while the density at the crests (18 units; Figure11.2.2a) is nearly equal to the density (17 units) at the same location (~±15°) but with no EIA (Figure 11.2.1b). This clearly indicates that the EIA crests are formed mainly by the removal of plasma from around the equator with very small accumulation at the crests.

However, the density in the topside ionosphere around the equator (~±10°) with drift (Figure 11.2.2a) is greater than that without the drift (Figure 11.2.1a) due to the (nearly) horizontal field lines supporting the upward drifted plasma. This results in smaller trough-to-crest ratio in TEC than in Nmax (Balan and Bailey, 1995). Figure 11.2.2a also shows clear F_3 layer developed around the equator.

The model results also show a well-developed EIA shortly after the application of the drift (Figure 11.2.2a), within about an hour in this case. However, observations (and modeling using measured **E** × **B** drift) suggest long duration for the development. In the morning at high solar activity the EIA develops by ~1030 LT though the **E** × **B** drift turns upward by 0630 LT, taking about 240 min for the development of EIA. It takes such a long time because in the morning the actual drift is weak and slow varying and ionization production is rapid, so that the slow removal of plasma from around the equator is filled by production. The misinterpretations of EPF and EIA also require a long time for the plasma at the equator to drift up in altitude and then diffuse down along the field lines to form EIA.

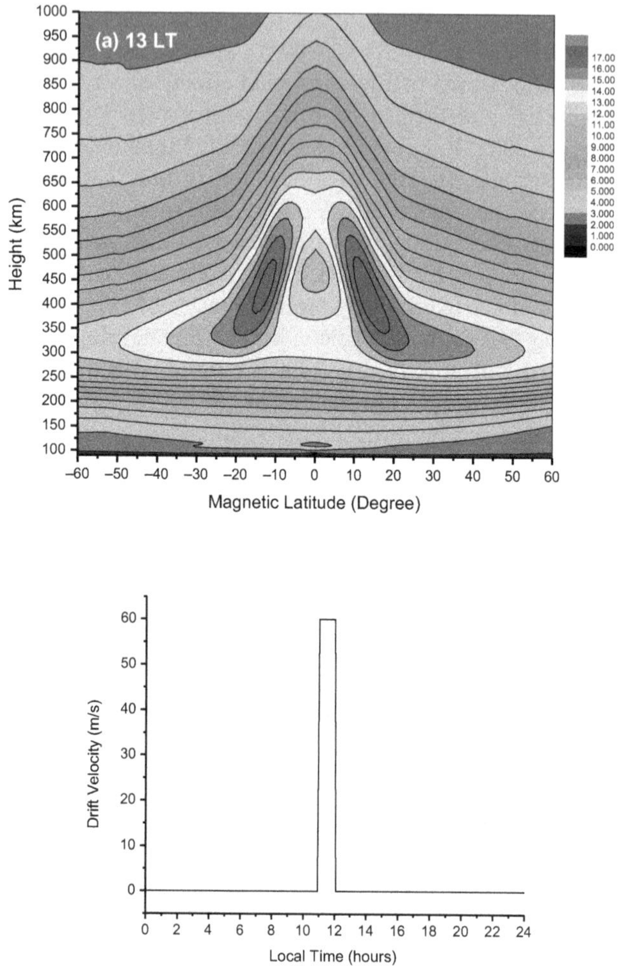

Figure 11.2.2 (a) Altitude–latitude map of Ne at 1300 LT modeled by SUPIM for magnetically quiet (Ap = 4) conditions at high solar activity (day 303, F10.7 = 268) in Jicamarca longitude (283°E) using (b) **E** × **B** drift only around noon. Positive latitude is north. (A black-and-white version of this figure appears in some formats. For the colour version, please refer to the plate section.)

11.2.4 EIA under Active Conditions

As mentioned in Section 11.2.3, the EIA becomes strong during early (around noon) stages of major geomagnetic storms. The strongest recorded EIA occurred during the extreme storm on 30 October 2003 (e.g. Mannucci et al., 2005), which has been studied in a number of papers (e.g. Lin et al., 2005; Liu et al., 2015; Balan et al., 2011). The strong EIA in electron density Ne measured by the CHAMP satellite at ~400 km height during this event (Figure 11.2.3a) has a very wide trough, sharp crests at ~±30° latitudes and crest-to-trough ratio ~10; the density at the crests compared to previous quiet-time values is over 1000%. CHAMP also measured a simultaneous fast equatorward neutral surge (Figure 11.2.3b) that crossed over the equator (Balan et al., 2011). Though the electric field was not measured during this event, the **E** × **B** drift inferred from the RTI (range time intensity) map of the VHF radar in Brazil (Figure 11.2.3c) at nearly the same longitude gives the highest recorded upward drift of ~1200 m s^{-1} (Abdu et al., 2008). The impulsive action of the two drivers together produced the strongest EIA (Figure 11.2.3a).

The strong EIA, however, was suggested to be caused by strong eastward PPEF alone. In the fully sunlit equatorial ionosphere the **E** × **B** drift due to eastward PPEF drives plasma upward so fast that it cannot recombine. This plasma spills over into the anomaly, which was suggested to cause the strong EIA (Kelley et al., 2004). This explanation linked to the easy-to-understand pictures of the plasma fountain and EIA was found inadequate because strong eastward PPEF alone cannot strengthen the EIA crests though it can produce a super plasma fountain and shift the crests to higher latitudes (Balan et al., 2009, 2013). Detailed explanations of how the impulsive action of the two drivers (strong eastward PPEF and fast equatorward wind) produce the strong EIA have been reported through basic principles and model calculations (Balan et al., 2011, 2013).

However, model calculations by SUPIM using eastward PPEF and quiet-time neutral wind during evening hours produced strong EIA, which was interpreted as due to eastward PPEF (Lin et al., 2005). This interpretation is misleading because the quiet-time wind velocity has become equatorward well before the PPEF. Model results by SAMI-2 using eastward PPEF alone showed small strengthening of EIA by about 50% (Huba et al., 2005). Caution is to be taken here not to interpret that strong EIA (over 100% increases in Nmax and TEC) is also due to eastward PPEF.

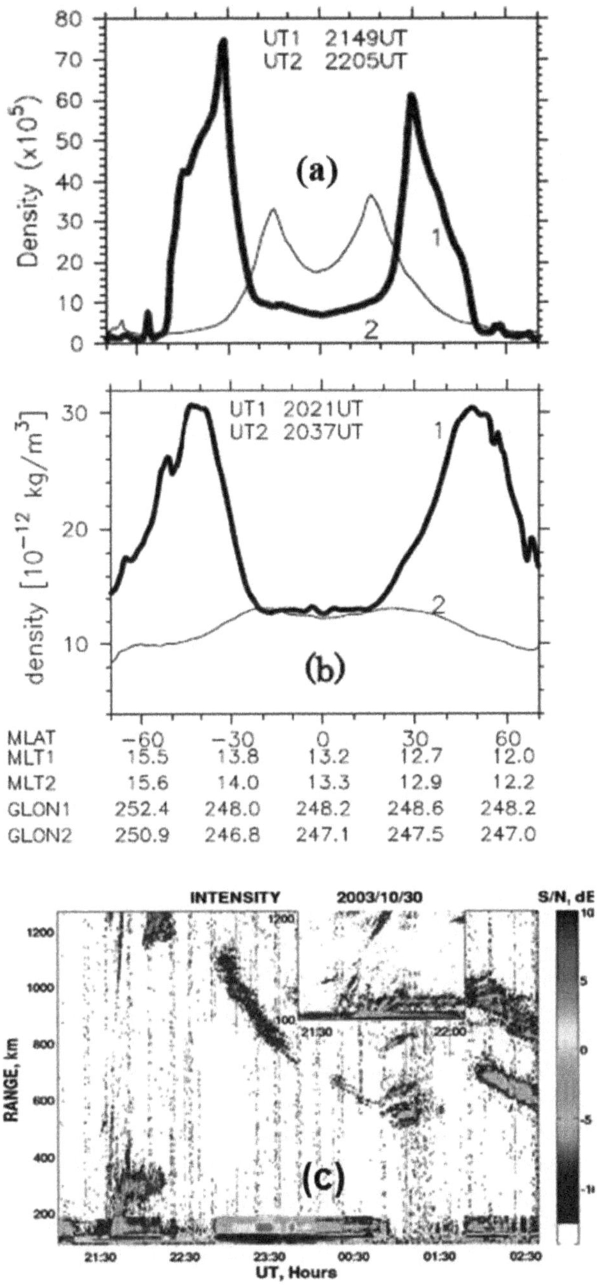

Figure 11.2.3 Latitude variations of (a) electron density Ne and (b) neutral density N measured by the CHAMP satellite at equatorial crossing time 2149 UT on 30 October 2003 (curves 1) compared with previous quiet day variations at a nearly same crossing time 2205 UT (curves 2). Latitude, longitude and UT of the satellite passes are listed at the bottom. Positive latitude is north. (c) RTI map of the VHF radar in Brazil on 30 October 2003 (after Abdu et al., 2008). (A black-and-white version of this figure appears in some formats. For the colour version, please refer to the plate section.)

11.2.5 Summary

Though theoretical models have reproduced the EIA and its variations long ago, the easy-to-understand misinterpretations of the EPF and EIA have caused misunderstandings. This brief review provides clarifications.

- It is clarified that the EPF is not upward **E × B** plasma drift at the equator followed by downward plasma diffusion along the field lines, but it is field perpendicular **E × B** plasma drift and field-aligned plasma diffusion acting together along the field lines at all altitudes and plasma flowing in the direction of the resultant.
- The EIA is formed not from the accumulation of plasma at the crests but mainly from the removal of plasma from

around the equator by the upward **E × B** drift with small accumulation when the crests are within ~±20° magnetic latitudes; the amount of accumulation reduces with latitude and becomes zero by ~±25°.

- The time required for the development of EIA is less than the time expected from the easy-to-understand misinterpretations.
- During early stages of the daytime (around noon) main phase of major geomagnetic storms the plasma fountain becomes a super fountain and EIA becomes strong not due to eastward prompt penetration electric field (PPEF) alone but due to the combined effect of eastward PPEF and storm-time equatorward winds (SEW).
- SEW (with normal electric field) alone can produce strong EIA while PPEF alone is unlikely.

11.3 High-Latitude and Polar Cap Ionosphere

*C. Robert Clauer**

This section reviews the advances in research that have focused on the high-latitude ionosphere, including the polar cap during the years 2011–17. The phenomena observed in the high-latitude ionosphere, including the polar cleft, the polar cap and the auroral zone are related to the transfer of solar wind energy into the magnetosphere and the transport of that energy through the magnetosphere. The phenomena in these regions can be investigated to understand the process of energy coupling, the efficiency of the coupling, and the evolution of the energy as it is transported through the magnetosphere in the form of magnetic energy as field lines are stretched, or in the form of kinetic energy as plasma is accelerated. The entire outer magnetosphere maps to a small area in the polar region, thus making the use of arrays of ground-based instruments important and unique to the exploration and understanding of these phenomena. Methodologies that utilize observational data from high-latitude arrays together with in situ measurements by spacecraft are particularly powerful. Utilizing improved observational arrays on the ground and multi-satellite arrays in space, together with advanced numerical computer simulations, and improvements in physics theory, we see considerable advancement in our understanding of the dynamic phenomena that occur in the high-latitude and polar cap ionosphere in the published literature during the past 6 years.

Using the astrophysics data system (http://ui.adsabs.harvard.edu) to search the scientific literature yields 1111 refereed papers that contain the term 'high-latitude ionosphere', 222 refereed papers for the term 'polar cap ionosphere' and 345 papers for the term 'cross-polar cap potential', where these terms exist somewhere within the abstract or full text of the paper.

Within the allotted space it is not possible to consider all of the possible research areas within the high-latitude ionosphere, so I will focus the discussion on advances to the understanding of the apparent polar cap electric potential saturation phenomena, the ionosphere under extreme solar wind driving, and advances in understanding high-latitude conductance and I will mention some papers that relate peripherally to these topics. For a more complete selection of citations relevant to other topics related to the high-latitude ionosphere, I refer the reader to the astrophysics data system specified above or the various other scientific publication search engines.

A key measurement of the interaction that couples energy and momentum from the solar wind to the magnetosphere–ionosphere system is the electric cross-polar cap potential (CPCP). Early investigations showed a linear relationship between solar wind driving measured by the southward component of the interplanetary magnetic field (IMF) and the CPCP (Reiff et al., 1981; Reiff and Luhmann, 1986). Later investigations, however, that examined a larger range of solar wind driving associated with magnetic storms showed that at extreme levels of southward IMF, the relationship becomes non-linear and this has been described as polar potential saturation. It has never been shown, however, that the potential develops to some asymptotic level, only that the rate of potential development decreases. In some apparently contradictory evidence, Weimer et al. (2017) report that the region 1 field-aligned current shows a linear response to the interplanetary electric field and does not show signs of saturation, though conductance changes may play a role in the resolution of this.

Theories to account for the non-linear behavior of the CPCP focus on limiting the coupling between the solar wind and magnetosphere during periods of stronger driving. Various mechanisms are proposed to accomplish this (Siscoe et al., 2002a, 2002b; Kivelson and Ridley, 2008; Lopez et al., 2010).

The impact of strong solar wind driving on the cross-polar cap potential (CPCP) and polar cap electric field measured in the high-latitude ionosphere has been a topic that has been advanced by several papers (Wilder et al., 2011; Maimaiti et al., 2017; Burke et al., 2017; Myllys et al., 2016). Kubota et al. (2017) investigate the polar cap potential saturation using a global MHD simulation. An investigation of solar wind parameters and mechanisms that impact the CPCP is reported by Myllys et al. (2017).

Wilder et al. (2010) showed that the reverse polar convection potential shows the same nonlinear behavior during periods of extreme northward IMF as the CPCP does for periods of extreme southward IMF. An important and unique insight has been obtained through observations of an event in which the reverse polar convection potential does not saturate (Clauer et al., 2016). The event is

* This work at Virginia Tech has been supported by National Science Foundation grant PLR-1543364 and by support from the Commonwealth of Virginia.

associated with extremely rare solar wind conditions that make a strong case for the importance of magnetosheath plasma beta in controlling the saturation phenomena. As beta increases, the reconnection efficiency increases and the CPCP increases. Small beta decreases the reconnection efficiency. This is a strong endorsement of the Lopez et al. (2010) magnetosheath force balance model for regulating the dayside reconnection potential.

In the Lopez et al. (2010) Force Balance theory, current generated at the bow shock closes through the magnetosheath to the magnetopause and couples with Birkeland currents. Under moderate solar wind driving conditions, the gradient in the plasma pressure controls the magnetosheath flow. At large IMF magnitudes, however, the bow shock/magnetosheath current system can become sufficiently strong that the **J**×**B** force becomes dominant. As driving increases, the **J**×**B** force diverts flow away from the merging line, thus decreasing the rate of energy coupling. This is responsible for limiting the development of the CPCP during times of strong driving. Thus, in this model, the transition from magnetosheath flow dominated by plasma pressure to one dominated by magnetic pressure determines of the CPCP development is linear or becomes nonlinear.

Wilder et al. (2015) utilize MHD global simulations to examine the role of plasma beta as well as magnetic flux tube deformation in the saturation problem. Lin et al. (2017) examine the role of solar wind density using MHD simulations to examine the results presented by Clauer et al. (2016). Both confirm the importance of plasma beta as a controlling parameter for dayside magnetic reconnection with the IMF imbedded in the magnetosheath plasma.

Coupling between the solar wind–magnetosphere–ionosphere–thermosphere was the topic of many papers. Storm-time coupling was investigated by Conner et al. (2014), Sandholt and Farrugia (2014) and Sandholt et al. (2015). High-latitude electrodynamic modeling using an assimilative procedure has been reported by Cousins et al. (2015). A statistical investigation of high-latitude Birkeland currents using AMPERE data is reported by Coxon et al. (2016). AMPERE provides a global estimate of the region 1 and region 2 field-aligned currents using magnetic measurements from the global array of polar orbiting Iridium satellites (Anderson et al., 2014). Empirical models to provide global estimates of the region 1 and region 2 field-aligned currents have also been presented by Weimer et al. (2017).

The examination of extreme events and strong magnetospheric driving conditions on the high-latitude ionosphere has appeared prominently in the recent literature. Storm-time impacts to the high-latitude ionosphere have been reported in Astafyeva et al. (2015), Durgonics et al. (2017), Chen et al. (2016a, 2016b) and Prikryl et al. (2014, 2016). Electron energization and precipitation at high latitudes during solar wind high-speed streams has been investigated by Grandin et al. (2017). Strong solar wind driving can also affect the dayside high latitudes during strong northward IMF to produce intense heating (Wilder et al., 2012).

Characteristics of the high-latitude ionosphere – density, density irregularities and conductivity – was also an area of extensive publication. High-latitude ionospheric irregularities was a new and particularly active topic both in terms of observations and models (Datta-Barua et al., 2015; Cherniak and Zakharenkova, 2016, 2017; Deshpande et al., 2014, 2016; Loucks et al., 2017; Park et al., 2017). Advances to high-latitude ionospheric conductivity models have been reported by McGranaghan et al. (2015, 2016a, 2016b). Further development of ionospheric models has been reported by Chisham (2017). TEC variations observed in the polar cap ionosphere have been reported by Watson et al. (2016). Examination of local conductivity variations has been shown to be very important to the investigation of conjugate magnetic variations (Hartinger et al., 2017) and to the determination of current systems as produced by a current generator or voltage generator mechanism. This problem results because high-latitude ground magnetometer measurements are very local in nature. It is not correct to assume the summer and winter conductivities to be large and small respectively in the analysis of conjugate ground magnetic data, since local conductivity enhancements in the winter hemisphere have a strong influence on the magnetic perturbations recorded on the ground.

The coupling between the high-latitude Northern and Southern Hemispheres and the various electrodynamic and topological asymmetries was another important topic advance during the past 6 years (Laundal et al., 2017; Förster and Haaland, 2015; Cnossen and Förster, 2016). Conjugate observations of traveling convection vortices initiated by transient solar wind pressure pulses was a topic for several papers (Kim et al., 2013, 2014, 2015, 2017). Northern and southern conjugate responses of TEC to solar flares have been reported by Bahair et al. (2011).

A review of the literature during the past decade has shown considerable advancement in our understanding of the high-latitude ionosphere, the possible resolution of some controversies and the development of new problems to resolve. Identifying the importance of plasma beta to regulate the solar wind–magnetosphere coupling process and resulting CPCP has the possibility to resolve many of the different theoretical explanations for the saturation of the CPCP. However, a better understanding of the high-latitude ionospheric conductivity may be required to resolve the linear response of the region 1 coupling currents and the non-linear response of the high-latitude ionospheric electric field and CPCP. Certainly the ionosphere must play a role, since the region 1 currents are important to determine the shape of the magnetosphere flaring angle, and this, in turn, affects the magnetosheath pressure. The potential influence of magnetospheric processes on the ionospheric processes that produce density irregularities at high latitudes is another emerging area of important investigation. Finally, investigation of the whole geospace system, including the symmetries and asymmetries found in both hemispheres as the system couples with the dynamic solar wind, is another area of growth in research and of growing importance for space weather models.

11.4 Auroral Processes

Yusuke Ebihara

11.4.1 Substorm-Associated Evolution of Aurora

A substorm cycle is divided into growth, expansion and recovery phases in terms of morphology of auroral forms. A noticeable auroral form in the growth phase is a north–south (N-S) or east–west (E-W) arc. Origin of the auroral arcs is still problematic. One idea is that the N-S and E-W auroral arcs are a projection of the processes taking place in the equatorial plane in the magnetotail, and they are a prominent precursor of the substorm expansion (Nishimura et al., 2011, and references therein). According to the global magnetohydrodynamics (MHD) simulation called REPPU (Tanaka, 2015), the N-S and E-W auroral structures are suggested to originate in structured plasma pressure distribution in the lobe. The auroral structures move equatorward during the growth phase because the structured pressure distribution moves toward the equatorial plane due to the magnetospheric convection (Ebihara and Tanaka, 2016). The substorm expansion onset is closely related to a sudden intensification of upward field-aligned currents (FACs) on the nightside. The FACs are believed to be a part of the current wedge that is a diversion of the cross-tail current (Kepko et al., 2015, and references therein). The REPPU code provides a different point of view. The current line (an integral curve of current density vector) extending from the onset position is not connected with the cross-tail current because of the intensification of the electric current flowing perpendicular to the magnetic field. Instead, the current line is connected with the cusp/mantle dynamo region by way of the newly developed dynamo region in the inner magnetosphere (Ebihara and Tanaka, 2015a). A westward traveling surge is a remarkable feature in the expansion phase, but its magnetospheric counterpart is a matter of debate. According to multi-spacecraft measurements of Cluster, the surge is associated with a U-shaped potential structure, and a major part of these parallel potential drops were concentrated below ~2.6 Re altitude. Convergent electric fields were observed over the surge (Marklund et al., 2012). An unsettled problem is the polarity of the electric field associated with the surge: convergent electric field or divergent electric field. The REPPU code, which can reproduce the evolution of auroral forms as shown in Figure 11.4.1, shows that both the convergent and divergent electric fields may coexist along the same field line. The simulation result shows that a broad, convergent electric field is situated at high altitude, while a narrow, divergent electric field is situated at low altitude and in the ionosphere. The convergent electric field is associated with the dynamo action in the near-Earth dynamo region, while the divergent one is associated with the dynamo action at the leading edge of the surge where the gradient of the ionospheric Hall current is large (Ebihara and Tanaka, 2015b). The surge is developed in corporation with the magnetospheric and ionospheric processes. Further studies are needed to resolve these problems.

11.4.2 Shock-Associated Aurora

Interplanetary shocks are known to intensify the brightness of aurora. This is called shock aurora. At the least, there are four mechanisms for the shock aurora: pitch angle scattering of particles interacting with excited waves, field-aligned currents associated with compressional waves, parallel electric fields associated with magnetic reconnection and parallel electric fields associated with a shear at the magnetopause. Recent

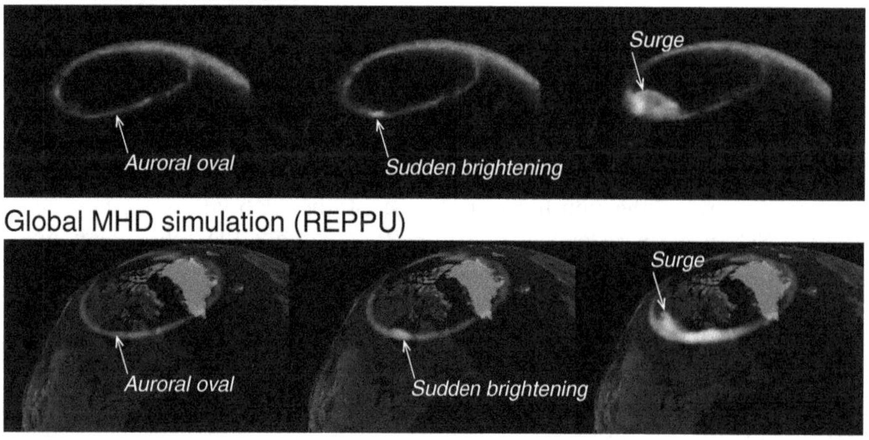

Figure 11.4.1 Evolution of auroral substorm (top) observed by the Polar satellite and (bottom) simulated by the REPPU code, a global magnetohydrodynamics (MHD) simulation.

observations show that auroral streaks moved anti-sunward along the poleward boundary of the auroral oval on the morning side. A field-aligned potential drop associated with shears at the magnetopause is suggested (Zhou et al., 2017). Fine-scale, transient auroral arcs also appeared on a closed field line in the post-noon sector within an interval of 1–2 min (Motoba et al., 2014). The dayside diffuse aurora intensified within 10 s, suggesting immediate pitch angle scattering of electrons due to interaction with the whistler-mode chorus waves excited by the compression of the inner magnetosphere (Liu et al., 2015).

11.4.3 Pulsating Aurora

Pulsating aurora refers to a class of aurora that exhibits quasi-periodic or irregular variations in time with periods of a few seconds to a few tens of seconds. In addition to well-known 3±1 Hz modulation, a faster modulation, a series of 20–30 ms pulses, was found to coexist (Kataoka et al., 2012). Whistler-mode chorus waves, or electron cyclotron harmonics waves, are thought to modulate the degree of scattering of electrons in the magnetosphere. Time-of-flight analysis for precipitating electrons suggests that the scattering region is confined within ±15° magnetic latitude when field-aligned propagation of the chorus waves is assumed (Nishiyama et al., 2011). The pulsating aurora was found to be well correlated with tens of keV electron fluxes at geosynchronous orbit (Jaynes et al., 2013) and the amplitude of the chorus waves near the equatorial plane (Nishimura et al., 2011). An immediate question is the mechanism that directly determines the growth of the chorus waves quasi-periodically. The amplitude of the chorus waves is correlated with compressional Pc4–5 pulsations (Li et al., 2011a) and plasma density in the magnetosphere (Li et al., 2011b). A field line resonance in the Pc4–5 range as a result of substorm injection has been suggested (Jaynes et al., 2015). The following features are also reported. The first feature is pulsating aurora at 630.0 nm [OI]. The altitude of the 630.0 nm [OI] pulsating aurora is systematically lower than that of the steady state 630.0 nm [OI] aurora (Liang et al., 2016). The second one is the auroral brightness caused by reflected electrons bouncing between the two hemispheres (Samara et al., 2017). Thirdly, substantial ionization was observed at >68 km altitude in association with pulsating aurora, suggesting precipitation of electrons with energy up to ~200 keV (Miyoshi et al., 2015).

11.4.4 Flickering Aurora

The flickering aurora is probably the fastest type of aurora, in which luminosity changes rapidly with a frequency range of 3–15 Hz. Very rapid flickering aurora in a frequency range of 50–80 Hz was found to coexist with a slowly varying one having a frequency range of 10 Hz. This is consistent with the hypothesis that the flickering aurora is caused by resonant interaction with electromagnetic ion cyclotron (EMIC) waves in multi-ionic species (Fukuda et al., 2017).

11.4.5 Cusp (Midday) Aurora

The cusp aurora is dominated by red emission at 630.0 nm [OI], primarily caused by precipitation of magnetosheath-like soft electrons. The cusp aurora contains highly structured emission at 427.8 nm (N_2^+). This may imply that the magnetosheath electrons had to be accelerated, probably by dispersive Alfvén waves (Mende et al., 2016). Auroral forms sometimes extend equatorward from a persistent midday aurora when $Kp \leq 3$ and the X-component of interplanetary magnetic field (IMF) is dominant (Rodriguez et al., 2012). The other type of equatorward extension is an N-S-aligned discrete arc. This may be associated with equatorward motion of the open-closed boundary (Han et al., 2015). Thermally excited red emission is occasionally important in determining the emission altitude in the cusp when the electron temperature is high (Carlson et al., 2013). A spot of proton aurora is also prominent in the cusp. The protons are believed to come from the magnetosheath by way of the dayside reconnection region. In the high-altitude cusp region, there is a local minimum of the strength of the magnetic field, so that some protons are locally trapped. Resonant interaction with EMIC waves has been suggested to scatter the protons, leading to precipitation into the ionosphere (Xiao et al., 2013).

11.4.6 Mid-latitude Aurora

A spot of the proton aurora frequently appears in the mid-latitude (subauroral) region. The isolated proton aurora was found to undergo modulations in conjunction with rising tone emissions (about 1 Hz per several tens of seconds) with subpacket structures of Pc1 geomagnetic pulsations (Nomura et al., 2016; Ozaki et al., 2016). This observation strongly suggests that the isolated proton aurora results from pitch angle scattering due to non-linear interaction with triggered EMIC emissions. Interaction with fast magnetosonic waves is also suggested to generate the subauroral proton aurora (Xiao et al., 2015).

11.4.7 Global Auroral Models

An empirical model that specifies the global distribution of the aurora is valuable for understanding dynamics of the magnetosphere, energy input into the atmosphere, natural hazards related to geomagnetically induced current and historical documents describing aurora. The OVATION Prime-2013 model predicts the locations of diffuse aurora, monoenergetic, broadband and ion precipitations as a function of solar wind parameters together with an optimized solar wind–magnetosphere coupling functions (Newell et al., 2014). The Auroral Precipitation Model (APM) specifies diffuse auroral zone, auroral oval precipitation, and soft diffuse precipitation for given AL and Dst indices (Vorobjev et al., 2013). By incorporating secular variation of the geomagnetic field, the AUREST (AURora ESTimation) model predicts high- and low-latitude boundaries of the region where aurora is visible for the past 10,000 years (Korte and Stolze, 2016).

References

Abdu, M. A., et al. (2008), 'Abnormal evening vertical plasma drift and effects on ESF and EIA over Brazil-South Atlantic sector during the 30 October 2003 superstorm', J. Geophys. Res., 113, A07313, doi: 10.1029/2007JA012844.

Anderson, B. J., H. Korth, C. L. Waters, D. L. Green, V. G. Merkin, R. J. Barnes and L. P. Dyrud (2014), 'Development of large-scale Birkeland currents determined from the Active Magnetosphere and Planetary Electrodynamics Response Experiment', Geophys Res. Lett., 41, 3017–25, doi: 10.1002/2014GL059941.

Anderson, D. N. (1973), 'A theoretical study of ionospheric F region equatorial anomaly – I. Theory', Planet. Space Sci., 21, 409.

Anderson, D. N. (1981), 'Modelling the ambient low latitude F region ionosphere – A review', J. Atmos. Terr. Phys., 43, 753.

Appleton, E. V. (1946), 'Two anomalies in the ionosphere', Nature, 157, 691.

Astafyeva, Elvira, Irina Zakharenkova and Forster Matthias (2015), 'Ionospheric response to the 2015 St. Patrick's Day storm: A global multi-instrument overview', J. Geophys. Res., 120, 9023–37, doi: 10.1002/2015JA021629.

Bahair, Siti Aminah, Mardina Abdullah and Baharuding Yatim (2011), 'The response of TEC at quasi-conjugate point GPS stations during solar flares', Acta Geophys., 59, 407–27, doi: 10.2478/s11600-010-0054-1.

Bailey, G. J. and N. Balan (1996), 'A low latitude Ionosphere-plasmasphere model', in STEP Hand Book of Ionospheric Models, ed. R. W. Schunk, Utah State University, Logan.

Bailey, G. J., N. Balan and Y. Z. Su (1997), 'The Sheffield University plasmasphere-ionosphere model – a review', J. Atmos. Terr. Phys., 59, 1541.

Balan, N, G. J. Bailey, M. A. Abdu, K. I. Oyama, P. G. Richards, J. MacDougall and I. S. Batista (1997), 'Equatorial plasma fountain and its effects over three locations: Evidence for an additional layer, the F3 layer', J. Geophys. Res., 102, 2047–56.

Balan, N., K. Shiokawa, Y. Otsuka, S. Watanabe and G. J. Bailey (2009), 'Super plasma fountain and equatorial ionization anomaly during penetration electric field', J. Geophys. Res., 114, A03310, doi: 10.1029/2008JA013768.

Balan, N. and G. J. Bailey (1995), 'Equatorial plasma fountain and its effects – possibility of an additional layer', J. Geophys. Res., 100, 21421.

Balan, N., K. Shiokawa, Y. Otsuka, T. Kikuchi, D. Vijaya Lekshmi, S. Kawamura, M. Yamamoto and G. J. Bailey (2010), 'A physical mechanism of positive ionospheric storms at low and mid latitudes through observations and modeling', J. Geophys. Res., 115, A02304, doi: 10.1029/2009JA014515.

Balan, N., M. Yamamoto, J. Y. Liu, Y. Otsuak, H. Liu and H. Lühr (2011), 'New aspects of thermospheric and ionospheric storms revealed by CHAMP', J. Geophys. Res., 116, A07305, doi: 10.1029/2010JA0160399.

Balan, N., Y. Otsuka, M. Nishioka, J. Y. Liu and G. Bailey (2013), 'Physical mechanisms of the ionospheric storms at equatorial and higher latitudes during MP and RP of geomagnetic storms', J. Geophys. Res., 118, 2660–69, 2012JA018557.

Banks, P. M. and G. Kockarts (1973), *Aeronomy, Part B*, Academic Press, New York.

Brambles, O. J., W. Lotko, B. Zhang, M. Wiltberger, J. Lyon and R. J. Strangeway (2011), 'Magnetosphere sawtooth oscillations induced by ionospheric outflow', Science, 332, 1183–6, doi: 10.1126/science.1202869.

Burke, W. J., L. M. Kilcommons and M. R. Hairston (2017), 'Storm time coupling between the magnetosheath and the polar ionosphere', J. Geophys. Res., 122, doi: 10.1002/2017JA024101.

Carlson, H. C., K. Oksavik and J. I. Moen (2013), 'Thermally excited 630.0 nm O(1D) emission in the cusp: A frequent high-altitude transient signature', J. Geopphys. Res., 118(9), 5842–52.

Chaston, C. C., J. W. Bonnell, C. W. Carlson, J. P. McFadden, R. E. Ergun and R. J. Strangeway (2003), 'Properties of small-scale Alfvén waves and accelerated electrons from FAST', J. Geophys. Res., 108, 8003, doi: 10.1029/2002JA009420.

Chen, C. H, C. H. Lin, T. Matsuo and W. H. Chen (2016b), 'Ionospheric data assimilation modeling of the 2015 St. Patrick's Day geomagnetic storm', J. Geophys. Res., 121, 11549–59, doi: 10.1002/2016JA023346.

Chen, C. H., C. H. Lin, T. Matsuo, W. H. Chen, I. T. Lee, J. Y. Liu, J. T. Lin and C. T. Hsu (2016a), 'Ionospheric data assimilation with thermosphere-ionosphere-electrodynamics general circulation model and GPS-TEC during geomagnetic storm conditions', J. Geophys. Res., 121, 5708–22, doi: 10.1002/2015JA021787.

Cherniak, Lurii and Irina Zakharenkova (2017), 'New advantages of the combined GPS and GLONASS observations for high-latitude ionospheric irregularities monitoring: case study of June 2015 geomagnetic storm', Earth Planets Space, 69, doi: 10.1186/s40623-017-0652-0.

Cherniak, Luril and Irina Zakharenkova (2016), 'High-latitude ionospheric irregularities: differences between ground- and space-based GPS measurements during the 2015 St. Patrick's Day storm', Earth Planets Space, 68, doi: 10.1186/s40623-016-0506-1.

Chisham, G. (2017), 'A new methodology for the development of high-latitude ionospheric climatologies and empirical models', J. Geophys. Res., 122, 932–47, doi: 10.1002/2016JA023235.

Clauer, C. Robert, Zhonghua Xu, M. Maimaiti, J. Michael Ruohoneimi, Wayne Scales, Michael D. Hartinger, Michael J. Nicolls, Stephen Kaeppler, Frederick D. Wilder and Ramon E. Lopez (2016), 'Investigation of a rare event where the polar ionospheric reverse convection potential does not saturate during a period of extreme northward IMF solar wind driving', J. Geophys. Res., 121, 5422–35, doi: 10.1002/2016JA022557.

Cnossen, Ingrid and Matthias Förster (2016), 'North-south asymmetries in the polar thermosphere-ionosphere system: Solar cycle and seasonal influences', J. Geophys. Res., 121, 612–27, doi: 10.1002/2015JA021750.

Connor, H. K., E. Zesta, D. M. Ober and J. Raeder (2014), 'The relation between transpolar potential and reconnection rates during sudden enhancement of solar wind dynamic pressure: OpenGGCM-CTIM results', J. Geophys. Res., 119, 3411–29, doi: 10.1002/2013JA019728.

Cosgrove, R. B., H. Bahcivan, S. Chen, R. J. Strangeway, J. Ortega, M. Alhassan, Y. Xu, M. Van Welie, J. Rehberger, S. Musielak and N. Cahill (2014), 'Empirical model of Poynting flux derived from FAST data and a cusp signature', J. Geophys. Res., 119, 411–30, doi: 10.1002/2013JA019105.

Cousins, E. D. P., Tomoko Matsuo and A. D. Richmond (2015), 'Mapping high-latitude ionospheric electrodynamics with SuperDARN and AMPERE', J. Geophys. Res., 120, 5854–70, doi: 10.1002/2014JA020463.

Coxon, J. C., S. E. Milan, J. A. Carter, L. B. N. Clausen, B. J. Anderson and H. Korth (2016), 'Seasonal and diurnal variations in AMPERE observations of the Birkeland currents compared to modeled results', J. Geophys. Res., 121, 4027–40, doi: 10.1002/2015JA022050.

Crowley, G., D. J. Knipp, K. A. Drake, J. Lei, E. Sutton and H. Lühr (2010), 'Thermospheric density enhancements in the dayside cusp region during strong BY conditions', Geophys. Res. Lett., 37, L07110, doi: 10.1029/2009GL042143.

Datta-Barua, S., Y. Su, K. Deshpande, D. Maladinovich, G. S. Bust, D. Hampton and G. Crowley (2015), 'First light from a kilometer-baseline scintillation auroral GPS array', Geophys. Res. Lett., 42, 3639–46, doi: 10.1002/2015GL063556.

Deshpande, K., G. S. Bust, C. R. Clauer, C. L. Rino and C. S. Carrano (2014), 'Satellite-beacon Ionospheric scintillation Global Model of the upper Atmosphere (SIGMA) I: High latitude sensitivity study of the model parameters', J. Geophys. Res., 119, 4026–43, doi: 10.1002/2013JA019699.

Deshpande, K., G. S. Bust, C. R. Clauer, W. A. Scales, N. A. Frissell, J. M. Ruohoniemi, L. Spogli, C. Mitchell and A. T. Weatherwax (2016), 'Satellite-beacon Ionospheric-scintillation Global Model of the upper Atmosphere (SIGMA)II: Inverse modeling with high-latitude observations to deduce irregularity physics', J. Geophys. Res., 121, 9188–9203, doi: 10.1002/2016JA022943.

Durgonics, Tibor, Attila Komjathy, Olga Verkhoglyadova, Esayas B. Shume, Hans-Henrik Benzon, Anthony J. Mannucci, Mark D. Butala, Per Høeg and Richard B. Langley (2017), 'Multi-instrument observations of a geomagnetic storm and its effects on the Arctic ionosphere: A case study of the 19 February 2014 storm', Radio Sci., 52, 146–65, doi: 10.1002/2016RS006106.

Ebihara, Y. and T. Tanaka (2015a), 'Substorm simulation: Insight into the mechanisms of initial brightening', J. Geophys. Res., 120(9), 7270–88, doi: 10.1002/2015JA021516.

Ebihara, Y. and T. Tanaka (2015b), 'Substorm simulation: Formation of westward traveling surge', J. Geophys. Res., 120(12), 10,466–84, doi: 10.1002/2015JA021697.

Ebihara, Y. and T. Tanaka (2016), 'Substorm simulation: Quiet and N-S arcs preceding auroral breakup', J. Geophys. Res., 121(2), 1201–18, doi: 10.1002/2015JA021831.

Fejer, B. G., E. R. Depaula, S. A. Gonzales and R. F. Woodman (1991), 'Average vertical and zonal F region plasma drifts over Jicamarca', J. Geophys. Res., 96, 13901.

Förster, M. and S. Haaland (2015), 'Interhemispheric differences in ionospheric convection: Cluster EDI observations revisited', J. Geophys. Res., 120, 5805–23, doi: 10.1002/2014JA020774.

Fujii, R., O. Amm, H. Vanhamäki, A. Yoshikawa and A. Ieda (2012), 'An application of the finite length Cowling channel model to auroral arcs with longitudinal variations', J. Geophys. Res., 117, A11217, doi: 10.1029/2012JA017953.

Fujii, R., O. Amm, A. Yoshikawa, A. Ieda and H. Vanhamäki (2011), 'Reformulation and energy flow of the Cowling channel', J. Geophys. Res., 116, A02305, doi: 10.1029/2010JA015989.

Fukuda, Y., R. Kataoka, H. A. Uchida, Y. Miyoshi, D. Hampton, K. Shiokawa, Y. Ebihara, D. Whiter, N. Iwagami and K. Seki (2017), 'First evidence of patchy flickering aurora modulated by multi-ion electromagnetic ion cyclotron waves', Geophys. Res. Lett., 44(9), 3963–70, doi: 10.1002/2017GL072956.

Grandin, M., A. T. Aikio, A. Kozlovsky, T. Ulich and T. Raita (2017), 'Cosmic radio noise absorption in the high latitude ionosphere during solar wind high-speed streams', J. Geophys. Res., 122, 5203–23, doi: 10.1002/2017JA023923.

Han, D. S., X. C. Chen, J. J. Liu, Q. Qiu, K. Keika, Z. J. Hu, J. M. Liu, H. Q. Hu and H. G. Yang (2015), 'An extensive survey of dayside diffuse aurora based on optical observations at Yellow River Station', J. Geophys. Res., 120(9), 7447–65, doi: 10.1002/2015JA021699.

Hanson, W. B. and R. J. Moffett (1966), 'Ionization transport effects in the equatorial F region', J. Geophys. Res., 71, 5559.

Hartinger, M. D., Z. Xu, C. R. Clauer, Y. Yu, D. Weimer, H. Kim, V. Pilipenko, D. T. Welling, R. Behlke and A. N. Willer (2017), 'Associating ground magnetometer observations with current or voltage generators', J. Geophys. Res., doi: 10.1002/2017JA0241402016.

Hedin, A. E., E. L. Fleming, A. H. Manson, F. L. Schmidlin, S. K. Avery, R. R. Clark, G. J. Fraser, T. Tsuda, F. Vial and R. Vincent (1995), 'Empirical wind model for the upper, middle and lower atmosphere', J. Atmos. Terr. Phys., 58, 1421–47.

Huba, J. D., G. Joice, S. Sazykin, R. Wolf and R. Spiro (2005), 'Simulation study of penetration electric field effects on the low- to mid-latitude ionosphere', Geophys. Res. Lett., 32, 123101, doi: 10.1029/2005GL024162.

Jaynes, A. N., M. R. Lessard, J. V. Rodriguez, E. Donovan, T. M. Loto'Aniu and K. Rychert (2013), 'Pulsating auroral electron flux modulations in the equatorial magnetosphere', J. Geophys. Res., 118(8), 4884–94, doi: 10.1002/jgra.50434.

Jaynes, A. N., M. R. Lessard, K. Takahashi, A. F. Ali, D. M. Malaspina, R. G. Michell, E. L. Spanswick et al. (2015), 'Correlated Pc4-5 ULF waves, whistler-mode chorus, and pulsating aurora observed by the Van Allen Probes and ground-based systems', J. Geophys. Res., 120(10), 8749–61, doi: 10.1002/2015JA021380.

Kataoka, R., Y. Miyoshi, D. Hampton, T. Ishii and Y. Kozako (2012), 'Pulsating aurora beyond the ultra-low-frequency range', J. Geophys. Res., 117(A8), A08336, doi: 10.1029/2012JA017987.

Kelley, M. C., M. N. Vlasov, J. C. Foster and A. J. Coster (2004), 'A quantitative explanation for the phenomenon known as storm-enhanced density', Geophys. Res. Lett., 31, L19809, doi: 10.1029/2004GL020875.

Kepko, L., R. L. McPherron, O. Amm, S. Apatenkov, W. Baumjohann, W. Birn, M. Lester, R. Nakamura, T. I. Pulkkinen and V. Sergeev (2015), 'Substorm current wedge revisited', Space Sci. Rev., 190(1–4), 1–46, doi: 10.1007/s11214-014-0124-9.

Kikuchi, T. (2014), 'Transmission line model for the near-instantaneous transmission of the ionospheric electric field and currents to the equator', J. Geophys. Res., 119, 1131–56, doi: 10.1002/2013JA019515.

Kim, H., C. R. Clauer, A. J. Gerrard, M. J. Engebretson, M. D. Hartinger, M. R. Lessard, J. Matzka, D. G. Sibeck, H. J. Singer, C. Stolle, D. R. Weimer and Z. Xu (2017), 'Conjugate observations of electromagnetic iono-cyclotron waves associated with traveling convection vortex events', J. Geophys. Res., 122, 7336–52, doi: 10.1002/2017JA024108.

Kim, H., C. R. Clauer, K. Deshpande, M. R. Lessard, A. T. Weatherwax, G. S. Bust, G. Crowley and T. E. Humphreys (2014), 'Ionospheric irregularities during a substorm event: Observations of ULF pulsations and GPS scintillations', J. Atmos. Sol. Terr. Phys., 114, 1–8, doi: 10.1016/j.jastp.2014.03.006.

Kim, H., C. R. Clauer, M. J. Engebretson, J. Matzka, D. G. Sibeck, H. J. Singer, C. Stolle, D. R. Weimer and Z. Xu (2015), 'Conjugate observations of traveling convection vortices associated with transient events at the magnetopause', J. Geophys. Res., 120, 2015–35, doi: 10.1002/2014JA020743.

Kim, H., X. Cai, C. R. Clauer, B. S. R. Kunduri, J. Matzka, C. Stolle and D. R. Weimer (2013), 'Geomagnetic response to solar wind dynamic pressure impulse events at high-latitude conjugate points', J. Geophys. Res., 118, 6055–71, doi: 10.1002/jgra50555.

Kivelson, M. G. and A. J. Ridley (2008), 'Saturation of the polar cap potential: Inference from Alfvén wing arguments', J. Geophys. Res., 113, doi: 10.1029/2007JA012,302.

Knipp, D., S. Eriksson, L. Kilcommons, G. Crowley, J. Lei, M. Hairston and K. Drake (2011), 'Extreme Poynting flux in the dayside thermosphere: Examples and statistics', Geophys. Res. Lett., 38, L16102, doi: 10.1029/2011GL048302.

Korte, M. and S. Stolze (2016), 'Variations in mid-latitude auroral activity during the Holocene', Archaeometry, 58 (1), 159–76, doi: 10.1111/arcm.12152.

Kubota, Y., T. Nagatsuma, M. Den, T. Tanaka and S. Fujita (2017), 'Polar cap potential saturation during the Bastille Day storm event using global MHD simulation', J. Geophys. Res., 122, 4398–4409, doi: 10.1002/2016JA023851.

Laundal, K. M., I. Cnossen, S. E. Milan, S. E. Haaland, J. Coxon, N. M. Pedatella, M. Förster and J. P. Reistad (2017), 'North-South asymmetries in Earth's magnetic field. Effects on high-latitude geospace', Space Sci. Rev., 206, 225–57, doi: 10.1007/s11214-016-0273-0.

Li, W., J. Bortnik, R. M. Thorne, Y. Nishimura, V. Angelopoulos and L. Chen (2011b), 'Modulation of whistler mode chorus waves: 2. Role of density variations', J. Geophys. Res., 116 (A6), A06206, doi: 10.1029/2010JA016313.

Li, W., R. M. Thorne, J. Bortnik, Y. Nishimura and V. Angelopoulos (2011a), 'Modulation of whistler mode chorus waves: 1. Role of compressional Pc4-5 pulsations', J. Geophys. Res., 116(A6), A06205, doi: 10.1029/2010JA016312.

Liang, J., E. Donovan, B. Jackel, E. Spanswick and M. Gillies (2016), 'On the 630 nm red-line pulsating aurora: Red-line emission geospace observatory observations and model simulations', J. Geophys. Res., 121(8), 7988–8012, doi: 10.1002/2016JA022901.

Lin, C. H., A. D. Richmond, R. A. Heelis, G. J. Bailey, G. Lu, J. Y. Liu, H. C. Yeh and S. Y. Su (2005), 'Theoretical study of the low and mid latitude ionospheric electron density enhancement during the October 2003 storm: Relative importance of the neutral wind and the electric field', J. Geophys. Res., 110, A12312, doi: 10.1029/2005JA011304.

Lin, D., B. Zhang, W. A. Scales, M. Wiltberger, C. R. Clauer and Z. Xu (2017), 'The role of solar wind density in cross polar cap potential saturation under northward interplanetary magnetic field', Geophys. Res. Lett., doi: 10.1002/2017GL075275.

Liu, J., H. Hu, D. Han, H. Yang and M. Lester (2015), 'Simultaneous ground-based optical and SuperDARN observations of the shock aurora at MLT noon', Earth Planet. Space, 67, 120, doi: 10.1186/s40623-015-0291-2.

Lopez, R. E., R. Bruntz, E. J. Mitchell, M. Wiltberger, J. G Lyon and V. G. Merkin (2010), 'Role of magnetosheath force balance in regulating the dayside reconnection potential', J. Geophys. Res., 115, doi: 10.1029/2009JA014597.

Loucks, D., S. Palo, M. Pilinski, G. Crowley, I. Azeem and D. Hampton (2017), 'High-latitude GPS phase scintillation from E region electron density gradients during the 20–21 December 2015 geomagnetic storm', J. Geophys. Res. Space Phys., 122(7), 7473–90, doi: 10.1002/2016JA023839.

Lu, G., L. P. Goncharenko, M. J. Nicolls, A. I. Maute, A. J. Coster and L. J. Paxton (2012), 'Ionospheric and thermospheric variations associated with prompt penetration electric fields', J. Geophys. Res., 117, A08312, doi: 10.1029/2012JA017769.

Lühr, H., M. Rother, W. Köhler, P. Ritter and L. Grunwaldt (2004), 'Thermospheric up-welling in the cusp region: Evidence from CHAMP observations', Geophys. Res. Lett., 31, L06805, doi: 10.1029/2003GL019314.

Lysak, R. L. (1999), 'Propagation of Alfvén waves through the ionosphere: Dependence on ionospheric parameters', J. Geophys. Res., 104(10), 10017–30.

Lysak, R. L. (2004), 'Magnetosphere–ionosphere coupling by Alfvén waves at midlatitudes', J. Geophys. Res., 109, A07201, doi: 10.1029/2004JA010454.

Lysak, R. L. and A. Yoshikawa (2006), 'Resonant cavities and waveguides in the ionosphere and atmosphere', in *Magnetospheric ULF Waves: Synthesis and New Directions*, ed. K. Takahashi, P. J. Chi, R. E. Denton and R. L. Lysak, American Geophysical Union, Washington, DC, doi: 10.1029/169GM19.

Lysak, R. L., C. L. Waters and M. D. Sciffer (2013), 'Modeling of the ionospheric Alfvén resonator in dipolar geometry', J. Geophys. Res., 118, 1514–28, doi: 10.1002/jgra.50090.

Maimaiti, Maimaitirebike, John Michael Ruohoniemi, J. B H. Baker, Robert Clauer, Michael J. Nicolls and Marc R. Hairston (2017), 'RISR-N observations of the IMF By influence on reverse convection during extreme northward IMF', J. Geophys. Res., 122, doi: 10.1002/2016JA023612.

Mannucci, A. J., B. T. Tsurutani, B. A. Iijima, A. Komjathy, A. Saito, W. D. Gonzalez, F. L. Guarnieri, J. U. Kozyra and R. Skoug (2005), 'Dayside global ionospheric response to the major interplanetary events of October 29–30, 2003 Halloween Storms', Geophys. Res. Lett., 32, L12S02, doi: 10.1029/2004GL021467.

Marklund, G. T., S. Sadeghi, B. Li, O. Amm, J. A. Cumnock, Y. Zhang, H. Nilsson et al. (2012), 'Cluster multipoint study of the acceleration potential pattern and electrodynamics of an auroral surge and its associated horn arc', J. Geophys. Res., 117(10), A10223, doi: 10.1029/2012JA018046.

Martyn, D. F. (1955), 'Theory of height and ionization density changes at the maximum of a Chapman-like region, taking account of ion production, decay, diffusion and total drift', in *Proceedings, Cambridge Conference*, p. 254, Physical Society, London.

McGranaghan, Ryan, Delores J. Knipp and Tomoko Matsuo (2016a), 'High-latitude ionospheric conductivity variability in three dimensions', Geophys. Res. Lett., 43, 7867–77, doi: 10.1002/2016GL070253.

McGranaghan, Ryan, Delores J. Knipp, Matsuo Tomoko and Ellen Cousins (2016b), 'Optimal interpolation analysis of high-latitude ionospheric Hall and Pedersen conductivities: Application to assimilative ionospheric electrodynamics reconstruction', J. Geophys. Res., 121, 4898–4923, doi: 10.1002/2016JA022486.

McGranaghan, Ryan, Delores J. Knipp, Matsuo Tomoko, Humberto Godinez, Robert J. Redmon, Stanley C. Solomon and Steven K. Morley (2015), 'Modes of high-latitude auroral conductance variability derived from DMSP energetic electron precipitation observations: Empirical orthogonal function analysis', J. Geophys. Res., 120, 11013–31, doi: 10.1002/2015JA021828.

Mende, S. B., H. U. Frey and V. Angelopoulos (2016), 'Source of the dayside cusp aurora', J. Geophys. Res., 121(8), 7728–38, doi: 10.1002/2016JA022657.

Mitra, S. K. (1946), 'Geomagnetic control of region F2 of the ionosphere', Nature, 158, 668.

Miyoshi, Y., S. Oyama, S. Saito, S. Kurita, H. Fujiwara, R. Kataoka, Y. Ebihara et al. (2015), 'Energetic electron precipitation associated with pulsating aurora: EISCAT and Van Allen Probe observations', J. Geophys. Res., 120(4), 2754–66, doi: 10.1002/2014JA020690.

Moffett, R. J. (1979), 'The equatorial anomaly in the electron distribution of the terrestrial F region', Fund. Cosmic Phys., 4, 313.

Moffett, R. J. and W. B. Hanson (1965), 'Effect of ionization transport on the equatorial F region', Nature, 206, 705.

Motoba, T., Y. Ebihara, A. Kadokura and A. T. Weatherwax (2014), 'Fine-scale transient arcs seen in a shock aurora', J. Geophys. Res., 119(8), 6249–55, doi: 10.1002/2014JA020229.

Myllys, M., E. K. J. Kilpua, B. Lavraud and J. I. Pulkkinen (2016), 'Solar wind–magnetosphere coupling efficiency during ejecta and sheath-driven geomagnetic storms', J. Geophys. Res., 121, 4378–96, doi: 10.1002/2016JA022407.

Myllys, M., E. K. J. Kipua and B. Lavraud (2017), 'Interplay of solar wind parameters and physical mechanisms producing the saturation of the cross polar cap potential', Geophys. Res. Lett., 44, 3019–27, doi: 10.1002/2017GL072676.

Namba, S. and K.-I. Maeda (1939), *Radio Wave Propagation*, report, Corona, Tokyo.

Newell, P. T., K. Liou, Y. Zhang, T. Sotirelis, L. J. Paxton and E. J. Mitchell (2014), 'OVATION Prime-2013: Extension of auroral precipitation model to higher disturbance levels', Space Weather, 12(6), 368–79, doi: 10.1002/2014SW001056.

Nishimura, Y., J. Bortnik, W. Li, R. M. Thorne, L. Chen, L. R. Lyons, V. Angelopoulos et al. (2011), 'Multievent study of the correlation between pulsating aurora and whistler mode chorus emissions', J. Geophys. Res, 116(A11), A11221, doi: 10.1029/2011JA016876.

Nishimura, Y., L. R. Lyons, V. Angelopoulos, T. Kikuchi, S. Zou and S. B. Mende (2011), 'Relations between multiple auroral streamers, pre-onset thin arc formation, and substorm auroral onset', J. Geophys. Res., 116(A9), A09214, doi: 10.1029/2011JA016768.

Nishiyama, T., T. Sakanoi, Y. Miyoshi, Y. Katoh, K. Asamura, S. Okano and M. Hirahara (2011), 'The source region and its characteristic of pulsating aurora based on the Reimei observations', J. Geophys. Res., 116(3), A03226, doi: 10.1029/2010JA015507.

Nomura, R., K. Shiokawa, Y. Omura, Y. Ebihara, Y. Miyoshi, K. Sakaguchi, Y. Otsuka and M. Connors (2016), 'Pulsating proton aurora caused by rising tone Pc1 waves', J. Geophys. Res., 121(2), 1608–18, doi: 10.1002/2015JA021681.

Ozaki, M., K. Shiokawa, Y. Miyoshi, R. Kataoka, S. Yagitani, T. Inoue, Y. Ebihara et al. (2016), 'Fast modulations of pulsating proton aurora related to subpacket structures of Pc1 geomagnetic pulsations at subauroral latitudes', Geophys. Res. Lett., 43(15), 7859–66, doi: 10.1002/2016GL070008.

Park, Jaeheung, Hermann Lühr, Gurarn Kervalishvili, Jan Rauberg, Claudia Stolle, Young-Sil Kwak, and Woo Kyoung Lee (2017), 'Morphology of high-latitude plasma density perturbations as deduced from the total electron content measurements onboard the Swarm constellation', J. Geophys. Res., 122, 1338–59, doi: 10.1002/2016JA023086.

Parker, E. N. (1996), 'The alternative paradigm for magnetospheric physics', J. Geophys. Res., 101(10), 10587–625.

Picone, J. M., A. E. Hedin, D. Drob and A. C. Aikin (2002), 'NRLMSISE-00 empirical model of the atmosphere: Statistical comparisons and scientific issues', J. Geophys. Res., 107(A12), A1468, doi: 10.1029/2002JA009430.

Prikryl, P., R. Ghoddousi-Fard, J. M. Weygand, A. Viljanen, M. Connors, D. W. Danskin, P. T. Jayachandran, K. S. Jacobsen, Y. L. Andalsvik, E. G. Thomas, J. M. ruohoniemi, T. Durgonics, K. Oksavik, Y. Zhang, E. Spanswick, M. Aquino and V. Sreeja (2016), 'GPS phase scintillation at high latitudes during the geomagnetic storm of 17–18 March 2015', J. Geophys. Res., 121, 10448–65, doi: 10.1002/2016JA023171.

Prikryl, Paul, P. Thayyil Jayachandran, Sajan C. Mushini and Ian G. Richardson (2014), 'High-latitude GPS phase scintillation and cycle slips during high-speed solar wind streams and interplanetary coronal mass ejections: A superposed epoch analysis', Earth Planets Space, 66, doi: 10.1186/1880-5981-66-62.

Rajaram, G. (1977), 'Structure of the equatorial F region, topside and bottomside – A review', J. Atmos. Terr. Phys., 39, 1125.

Reiff, P. H. and J. G. Luhmann (1986), 'Solar wind control of the polar cap voltage', in *Solar Wind–Magnetosphere Coupling*, ed. Y. Kamide and J. A. Slavin, p. 507, Terra Scientific, Tokyo.

Reiff, P., R. Spiro and T. Hill (1981), 'Dependence of polar cap potential on interplanetary parameters', J. Geophys. Res., 86 (7), 639.

Richmond, A. D. (1995), 'Ionospheric electrodynamics', in *Handbook of Atmospheric Electrodynamics*, vol. II, ed. H. Volland, pp. 249–90, CRC Press, Boca Raton, FL.

Rishbeth, H., A. J. Lyon and M. Peart (1963), 'Diffusion in the equatorial F layer', J. Geophys. Res., 68, 2559.

Roble, R. G. and E. C. Ridley (1994), 'Thermosphere-ionosphere-mesosphere-electrodynamics general circulation model (time-GCM): Equinox solar cycle minimum simulations (300–500 km)', Geophys. Res. Lett., 21, 417–20, doi: 10.1029/93GL03391.

Rodriguez, J. V., H. C. Carlson and R. A. Heelis (2012), 'Auroral forms that extend equatorward from the persistent midday aurora during geomagnetically quiet periods', J. Atmos. Sol. Terr. Phys., 86, 6–24, doi: 10.1016/j.jastp.2012.06.001.

Russell, C. T., J. G. Luhmann and R. J. Strangeway (2016), *Space Physics*, Cambridge University Press, Cambridge.

Samara, M., R. G. Michell and G. V. Khazanov (2017), 'First optical observations of interhemispheric electron reflections within pulsating aurora', Geophys. Res. Lett., 44(6), 2618–23, doi: 10.1002/2017GL072794.

Sandholt, P. E. and C. J. Farrugia (2014), 'Aspects of magnetosphere-ionosphere coupling in sawtooth substorms: A case study', Ann. Geophys., 32, 1277–91, doi: 10.5194/angeo-32-1277-2014.

Sandholt, P. E., C. J. Farrugia and W. F. Denig (2015), 'Transitions between states of magnetotail-ionosphere coupling and the role of solar wind dynamic pressure: the 25 July 2004 interplanetary CME case', Ann. Geophys., 33, 427–36, doi: 10.5194/angeo-33-427-2015.

Siscoe, G. L., G. M. Erickson, B. U. O. Sonnerup, N. C. Maynard, J. A. Schoendorf, K. D. Siebert, D. R. Weimer, W. W. White and G. R. Wilson (2002a), 'Hill model of transpolar potential saturation: Comparisons with MHD simulations', J. Geophys. Res., 107(A6), doi: 10.1029/2001JA000109.

Siscoe, G. L., N. U. Crooker and K. D. Siebert (2002b), 'Transpolar potential saturation: Roles of region 1 current system and solar wind ram pressure', J. Geophys. Res., 107 (A10), doi: 10.1029/2001JA009176.

Song, P., V. M. Vasyliūnas and L. Ma (2005), 'A three-fluid model of solar wind–magnetosphere–ionosphere–thermosphere coupling', in *Multiscale Coupling of Sun-Earth Processes*, ed. A. T. Y. Lui, Y. Kamide and G. Consolini, pp. 447–56, Elsevier, Amsterdam, doi: 10.1016/B978-044451881-1/50033-2.

Souza, J. R., W. D. Asevedo Jr, P. C. P. dos Santos, A. Petry, G. J. Bailey, I. S. Batista and M. A. Abdu (2013), 'Longitudinal variation of the equatorial ionosphere: Modeling and experimental results', Adv. Space Res., 51, 654–60, doi: 10.1016/j.asr.2012.01.023.

Stening R. J. (1992), 'Modeling the low-latitude F region', J. Atmos. Terr. Phys., 54, 1387.

Strangeway, R. J. (2009), 'Space environment and scientific missions: Magnetic fields in space', IEEE T. Magn., 45(10), 4486–92.

Strangeway, R. J. (2012), 'The equivalence of Joule dissipation and frictional heating in the collisional ionosphere', J. Geophys. Res., 117, A02310, doi: 10.1029/2011JA017302.

Strangeway, R. J. and J. Raeder (2001), 'On the transition from collisionless to collisional magnetohydrodynamics', J. Geophys. Res., 106(A2), 1955–60, doi: 10.1029/2000JA900116.

Strangeway, R. J., R. E. Ergun, Y.-J. Su, C. W. Carlson and R. C. Elphic (2005), 'Factors controlling ionospheric outflows as observed at intermediate altitudes', J. Geophys. Res., 110, A03221, doi: 10.1029/2004JA010829.

Tanaka, T. (2015), 'Substorm auroral dynamics reproduced by advanced global magnetosphere–ionosphere (M-I) coupling simulation', in *Auroral Dynamics and Space Weather*, pp. 177–90, John Wiley, Hoboken, NJ, doi: 10.1002/9781118978719.ch13.

Tu, J., P. Song and V. M. Vasyliūnas (2011), 'Ionosphere/thermosphere heating determined from dynamic magnetosphere-ionosphere/thermosphere coupling', J. Geophys. Res., 116, A09311, doi: 10.1029/2011JA016620.

Tu, J., P. Song and V. M. Vasyliūnas (2014), 'Inductive-dynamic magnetosphere-ionosphere coupling via MHD waves', J. Geophys. Res., 119, 530–47, doi: 10.1002/2013JA018982.

Vorobjev, V. G., O. I. Yagodkina and Yu. V. Katkalov (2013), 'Auroral precipitation model and its applications to ionospheric and magnetospheric studies', J. Atmos. Sol. Terr. Phys., 102, 157–71, doi: 10.1016/j.jastp.2013.05.007.

Wahlund, J. E., H. J. Opgenoorth, I. Haggstrom, K. J. Winser and G. O. L. Jones (1992), 'EISCAT observations of topside ionospheric ion outflows during auroral activity: revisited', J. Geophys. Res., 97, 3019–37, doi: 10.1029/91JA02438.

Watson, Chris, P. T. Jayachandran and John W. MacDougall (2016), 'Characteristics of GPS TEC variations in the polar cap ionosphere', J. Geophys. Res., 121, 4748–68, doi: 10.1002/2015JA022275.

Weimer, D. R., T. R. Edwards and Nils Olsen (2017), 'Linear response of field-aligned currents to the interplanetary electric field', J. Geophys. Res., 122, 8502–15, doi: 10.1002/2017JA024372.

Wilder, F. D., C. R. Clauer, J. B. H. Baker, E. P. Cousins and M. R. Hairston (2011), 'Inter-hemispheric observations of dayside convection under northward IMF', J. Geophys. Res., 116, doi: 10.1029/2011JA016748.

Wilder, F. D., G. Crowley, B. J. Anderson and A. D. Richmond (2012), 'Intense dayside Joule heating during the 5 April 2010 geomagnetic storm recovery phase observed by AMIE and AMPERE', J. Geophys. Res., 117, doi: 10.1029/2011JA017262.

Wilder, F. D., S. Eriksson and M. Wiltberger (2015), 'The role of magnetic flux tube deformation and magnetosheath plasma beta in the saturation of the Region 1 field-aligned current system', J. Geophys. Res., 120, 2036–51, doi: 10.1002/2014JA020533.

Wilder, Frederick, Robert Clauer and Joseph Baker (2010), 'Polar cap electric field saturation during IMF Bz north and south conditions', J. Geophys. Res., 115, A10230, doi: 10.1029/2010JA015487.

Xiao, F., Q. Zong, Z. Su, C. Yang, Z. He, Y. Wang and Z. Gao (2013), 'Determining the mechanism of cusp proton aurora', Sci. Rep., 3, 1654, doi: 10.1038/srep01654.

Xiao, F., Q. Zong, Y. Wang, Z. He, Z. Su, C. Yang and Q. Zhou (2015), 'Generation of proton aurora by magnetosonic waves', Sci. Rep., 4, 5190, doi: 10.1038/srep05190.

Yoshikawa, A., O. Amm, H. Vanhamäki, A. Nakamizo and R. Fujii (2013), 'Theory of Cowling channel formation by reflection of shear Alfven waves from the auroral ionosphere', J. Geophys. Res., 118, 6416–25, doi: 10.1002/jgra.50514.

Zhou, X., G. Haerendel, J. I. Moen, E. Trondsen, L. Clausen, R. J. Strangeway, B. Lybekk and D. A. Lorentzen (2017), 'Shock aurora: Field-aligned discrete structures moving along the dawnside oval', J. Geophys. Res., 122(3), 3145–62, doi: 10.1002/2016JA022666.

12 The Magnetosphere

Since the discovery of the magnetosphere-magnetotail system in the1950s-1960s), and the associated beginning of the satellite era, we have gained a well-informed understanding of this space plasma region permeated by the geomagnetic field and home to a variety of charged particle populations and plasma waves. Over the last six decades, IAGA has played an important role in supporting international magnetospheric research. Here we provide an overview of recent developments in energy transport from the solar wind into the Earth's environment. The topics include magnetosphere energy input and the role of the boundary layer.

Solar wind interaction with the magnetosphere, which creates geomagnetic activity, and the response of the region leading to substorms and steady magnetospheric convection are discussed. The charged particle energy (eV to MeV) inherent/contained in the magnetospheric ring current and Van Allen radiation belts establish many properties of the region, giving rise to boundary regions and waves. Results from recent state of the art and currently operating Earth orbiting satellites (Cluster, THEMIS, Van Allen Probes, Magnetospheric Multiscale), are providing exciting new results on waves from ultra-low frequency (ULF) from a few milliHertz, up to upper hybrid waves and continuum radiation in the 1–2 MHz band. Finally, current understanding of the plasmasphere and associated boundary the plasmapause, are considered.

12.1 Magnetosphere Energy Input

Stephen A. Fuselier

12.1.1 The Magnetopause

The magnetopause is the boundary of the Earth's magnetosphere where the internal magnetic pressure of the Earth's magnetic field balances the external dynamic pressure of the shocked solar wind in the magnetosheath. Along the Earth–Sun line, this boundary is located about 64,000 km from the center of the Earth, or about 10 Earth Radii (where 1 Earth radius = 6378 km) and the magnetopause is approximately a paraboloid of revolution around the Earth–Sun line. Normally, pressure balance in two separate plasmas includes magnetic and plasma dynamic and thermal pressures. However, for the solar wind–magnetosphere interaction, the plasma in the magnetosphere is much less dense than the magnetosheath plasma (typically by 3 orders of magnitude), and therefore the Earth's magnetic field dominates the internal, magnetospheric pressure. Similarly, the solar wind contains a magnetic field from the Sun, but its contribution to the total pressure is much less than the plasma dynamic pressure. The thermal pressure in the solar wind is also much less than the dynamic pressure.

12.1.2 Energy Transfer across the Magnetopause

Energy is transferred from the solar wind across the magnetopause and into the magnetosphere. The Earth's magnetosphere is externally driven, with the solar wind as the energy source for internal magnetospheric dynamics. However, the external driving does not mean that the Earth's magnetosphere has no internal plasma source or internal energy source. The high-latitude ionosphere is an internal plasma source and the importance of this source compared to the solar wind source is still not clear. However, recent results from the Magnetospheric Multiscale mission indicate that this internal plasma source has limited impact on the energy transfer processes at the dayside magnetopause (Fuselier et al., 2017). The Earth's rotation adds energy to the plasma in the inner magnetosphere. However, the rotation rate is relatively slow (compared to, for example, the rates at Jupiter or Saturn) and is not sufficient to produce an internal energy source that rivals that in the solar wind. In contrast, the magnetospheres of Jupiter and Saturn are largely internally driven. Both planets have moons that provide copious amounts of plasma to their large magnetospheres and the rapid rotation of the two planets provides significant internal energy.

Because the Earth's magnetosphere is externally driven by the solar wind, it is important to understand the energy transfer across the magnetopause. In the 1960s and 1970s, several mechanisms were proposed to account for energy transfer across this boundary. These included finite gyroradii effects, non-linear wave–particle interactions – including diffusion and the Kelvin–Helmholtz instability – so-called impulsive penetration and magnetic reconnection. As the time resolution of space plasma instrumentation became commensurate with the crossing time of a spacecraft across the magnetopause current layer, it became apparent that magnetic reconnection played an important role in the energy transfer across the magnetopause.

Through the 1980s, it became increasingly apparent that magnetic reconnection was the dominant mechanism for energy and mass transfer across the magnetopause. Other mechanisms could not explain the wide variety of observational evidence linked to the reconnection process. For example, a limited test of wave intensities at the magnetopause indicates that these wave intensities are too small to account for the mass transfer across the magnetopause (see e.g. Hultqvist et al., 1999, chapter 5). Also, the Kelvin–Helmholtz instability, while active at the flanks of the magnetopause under a limited range of solar wind interplanetary magnetic field (IMF) orientations, occurs between the magnetosheath and an existing boundary layer inside the magnetopause that is produced by magnetic reconnection at the dayside magnetopause. The relative importance of plasma and energy transfer mechanisms and the observational evidence for magnetic reconnection are summarized in the book *Magnetospheric Plasma Sources and Losses* (Hultqvist et al., 1999, chapter 5). It is now well established that reconnection is the dominant mechanism for transfer of solar wind plasma and energy across the magnetopause.

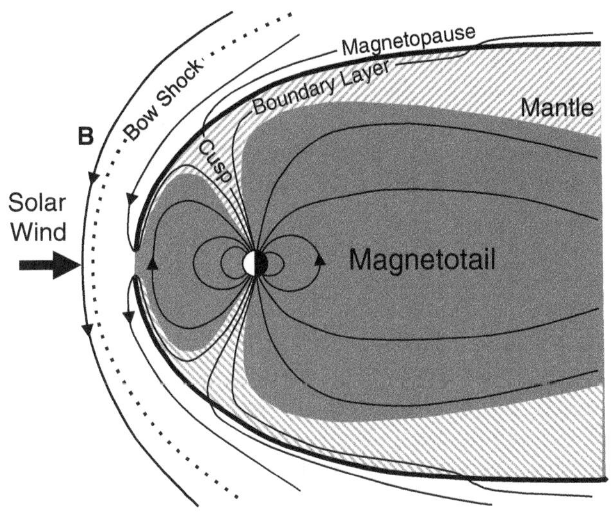

Figure 12.2.1 Schematic of the Earth's magnetosphere showing the boundary layers adjacent to the magnetopause. The noon–midnight cut through the magnetosphere shows the relatively thin boundary layer on the dayside, the magnetospheric cusps, and the mantle near the tail magnetopause. These boundary layers are all threaded by reconnected magnetic fields when the IMF is southward. Thus, these boundary layers are all interconnected by the reconnection process.

12.2 Magnetopause Boundary Layer

Stephen A. Fuselier

12.2.1 Early Concepts

As the time resolution of space plasma instrumentation improved, the existence of a thick boundary layer at the magnetopause became apparent (for a more complete historical review of magnetospheric boundary layers, see Eastman, 2003). A relatively thin boundary layer/current layer was predicted very early on by Chapman and Ferraro (1931) and later quantified by Ferraro (1952). Assuming the solar wind is unmagnetized, Ferraro (1952) showed that the gyromotion of solar wind ions and electrons in opposite directions produces a current on the magnetopause surface. More importantly, it predicts a very thin current layer of the order of the gyroradius of a solar wind ion in the Earth's magnetic field at the magnetopause. The gyroradius of a 1 keV solar wind proton in a 40 nT magnetopause magnetic field is only about 100 km. Thus, the presence of a boundary layer with thickness of hundreds to tens of thousands of kilometers at the magnetopause is clearly inconsistent with this early prediction. The inconsistency stems from the fact that the solar wind is not unmagnetized. The plasma beta (the ratio of the thermal to magnetic energy) in the solar wind and shocked magnetosheath is approximately 1, indicating that magnetic and plasma forces are equally important in the plasma external to the Earth's magnetopause.

The boundary layer at the dayside magnetopause is one of what seemed to be several magnetopause boundary layers in different locations with seemingly different properties. The relatively thin (but still up to thousands of kilometers thick) boundary layer observed at the dayside magnetopause was originally called the low-latitude boundary layer (Haerendel et al., 1978). At higher latitudes, the high-altitude cusp and high-latitude boundary or entry layer were identified (Haerendel et al., 1978). Finally, further tailward, the plasma mantle was identified (Rosenbauer et al., 1975).

Figure 12.2.1 shows a simplified noon–midnight cross section of the Earth's magnetosphere that depicts the large-scale boundary layer or layers adjacent to, but inside the magnetopause. Not shown in this noon–midnight cross section is the boundary layer at low latitudes on the flanks of the magnetopause, extending well past the terminator. In some ways, this flank boundary layer is considered an extension of the low-latitude boundary layer.

12.2.2 Connecting the Boundary Layers through Magnetic Reconnection

Figure 12.2.1 shows magnetic reconnection at the subsolar point on the magnetopause along the Earth–Sun line. Reconnected magnetic field lines have one 'foot' in the ionosphere and the other in the solar wind. Under the convective motion of the solar wind, these field lines are dragged over the Earth's polar regions and into the Earth's magnetotail. Entry of plasma all along the open magnetic field produces a single, interconnected boundary layer that extends from the subsolar point along the magnetopause to the distant tail. In this picture, the low-latitude boundary layer, the cusp and entry layer and the mantle are all interconnected and threaded by reconnected field lines. The properties of the plasma along these field lines changes as the plasma flows into the magnetosphere, travels down to the high-latitude ionosphere in the cusp and reflects and returns along field lines in the mantle. Thus, the seemingly disconnected boundary layers identified in the 1970s are all interconnected through magnetic reconnection, and the seemingly different properties are just the result of plasma energization along the magnetopause, convection into and through the magnetosphere and mixing with the ionospheric plasma from the high-latitude ionosphere. This interconnection is strong evidence that reconnection dominates the entry of plasma and energy into the magnetosphere.

In 2015, NASA launched the Magnetospheric Multiscale mission. This four-spacecraft mission uses the Earth's magnetosphere as a laboratory to study magnetic reconnection. In particular, the mission focuses on magnetic reconnection in the near-Earth environment at the dayside magnetopause and in the nightside magnetotail. MMS has two unique capabilities. First, it has plasma instrumentation with much higher time resolution than any other magnetospheric mission. Second, the spacecraft are capable of operations very close together, with separations as small as 7 km achieved during the mission. Using these new capabilities, MMS is uncovering the physics of magnetic reconnection. In particular, it focuses on the dynamics of electrons in the region where they become demagnetized and the magnetosheath and magnetosphere magnetic fields interconnect. The high time resolution has proven to be important for understanding the physics in this small region, and MMS has confirmed many of the theoretical predictions about reconnection and raised some important new questions about the process (Burch and Phan, 2016).

Figure 12.2.2 shows a schematic of this very small region where reconnection is initiated. At the dayside magnetopause, magnetic reconnection is highly asymmetric. What that means is that the plasmas in the magnetosheath and in the magnetosphere have very different densities and magnetic field magnitudes (see Cassak and Fuselier, 2016). Typically, the mass density in the magnetosheath (ρ_1 in Figure 12.2.2) is dominated by solar wind protons and is about 30 cm^{-3}. The mass density in the magnetosphere (ρ_2 in Figure 12.2.2) is also typically dominated by protons, but is typically between 0.1 and 1 cm^{-3}. Thus, there is a nearly 3 order of magnitude difference in the densities across the magnetopause. The magnetosheath magnetic field magnitude (B_1 in Figure 12.2.2) is typically about a factor of 2 less

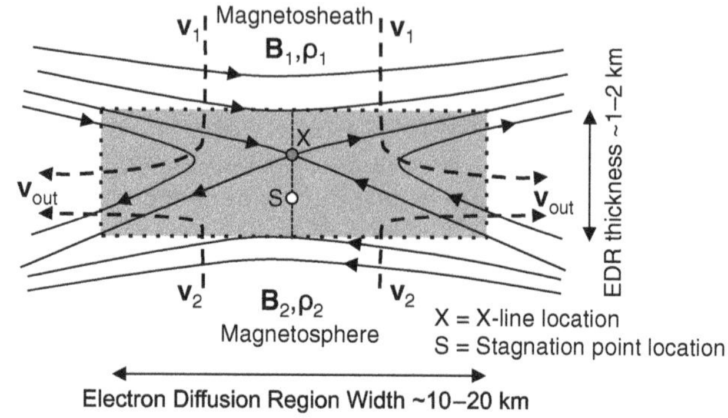

Figure 12.2.2 Adapted from Cassak and Shay (2007). The electron diffusion region at the magnetopause is modified considerably from symmetric reconnection because the magnetosheath density is much larger than the magnetospheric density. As a result, the X-line and stagnation point are separated in the region. Overall, the electron diffusion region is very small and requires very high time resolution measurements from the MMS mission to elucidate its properties. The outflow from the diffusion region ultimately becomes the boundary layer at the magnetopause.

than the magnetospheric magnetic field magnitude (B_2 in Figure 12.2.2).

These huge differences have a profound effect on the structure of the reconnection electron diffusion region (Cassak and Shay, 2007) and the dynamics of the electrons within the region. The MMS mission is now confirming many of these predicted effects and identifying new kinetic processes occurring in the diffusion region as a result of the asymmetric nature of reconnection at the magnetopause. As Figure 12.2.2 shows, one of the important consequences of this asymmetry is the separation of the X-line, where the magnetic field magnitude goes to zero (in anti-parallel reconnection), and the stagnation line, where the plasma bulk flow goes to zero. Another consequence of the asymmetry is the motion of ions through the diffusion region. As a result of much higher proton density on the magnetosheath side, the magnetosheath ions penetrate deeper into the magnetosphere than the electrons. This sets up a normal electric field (Malakit et al., 2013) and a charge imbalance that ultimately results in ion-scale physics as the magnetospheric electrons rush in to balance the excess ion charge. In addition, magnetosheath electrons crossing the stagnation point execute cycloidal, or meandering orbits and appear as "crescents" in the MMS electron distribution functions (Burch et al., 2016). The MMS mission is currently elucidating the details of this complex region. The coupling of the electron diffusion region with the much larger scale boundary layer remains one of the important challenges for the future.

12.3 Substorms and Steady Magnetospheric Convection

*Robert L. McPherron**

12.3.1 Introduction

The field of solar–terrestrial physics is concerned with the effects of the Sun on the Earth and the region of space around it. It began with the study of magnetic storms and their association with solar flares (Stewart, 1861). In the study of storms it was noted that there were frequent perturbations in the Earth's magnetic field of about 3 hour duration during the main phase. These were called *polar substorms* (Chapman, 1962). Today these disturbances are called *magnetospheric substorms*. More recently, magnetic indices have shown that there are other forms of magnetic activity and that their behavior depends on the strength and steadiness of the solar wind driver (McPherron, 2015; McPherron et al., 2008). These other forms have come to be called *modes of response*. The modes include the substorm, steady magnetospheric convection and sawtooth events. In addition, there are pseudo-breakups and poleward boundary intensifications which play significant roles in structuring magnetic activity. All of these phenomena have been shown to depend on magnetic reconnection (Hubert et al., 2017).

In this brief report we review substorms and steady magnetospheric convection. For each we use electrojet indices to summarize the behavior of the magnetic field. For our study we use lists of onset times to do superposed epoch analysis revealing the systematic response of the surface magnetic field. Pseudo-breakups and sawtooth injection events appear to be undeveloped substorms and quasi-periodic repetition of substorms, respectively. Poleward boundary intensifications appear to be reconnection at a retreating x-line produced by an earlier substorm.

12.3.2 The Magnetospheric Substorm

The substorm has been a subject of study since 1962, when Chapman (1962) described the short-term fluctuations in magnetic field during the main phase of storms and called them polar substorms. The concept was extended by Akasofu (1964) and Akasofu et al. (1965), who described the behavior of aurora during polar substorms. The concept was generalized by Jelly and Brice (1967) and Coroniti et al. (1968) to include a variety of related phenomena throughout the magnetosphere and is now known as the magnetospheric substorm. It was quickly shown that substorms generally occur when the interplanetary magnetic field (IMF) is southward (Fairfield and Cahill, 1966; Hirshberg and Colburn, 1969), supporting the idea, suggested by Dungey (1961), that magnetic and auroral activity are driven by magnetic reconnection. These ideas were incorporated into what became known as the near-Earth neutral line model (NENL) of substorms. An essential element of this model was the diversion of tail current through the ionosphere during the expansion phase as the substorm current wedge (McPherron et al., 1973).

The NENL model is based on the idea of unbalanced reconnection. Strong reconnection on the dayside and none or weak reconnection on the night side results in a systematic sequence of changes in the configuration of the magnetosphere that inevitably lead to the onset of fast nightside reconnection close to the Earth. The substorm growth phase is the consequence of these changes (McPherron, 1970). Fast nightside reconnection begins near Earth when the plasma sheet has thinned sufficiently. It creates the expansion phase of the substorm. When the reconnection site moves tailward, the recovery phase begins attempting to re-create the initial configuration. If the solar wind driver is fluctuating, the sequence repeats with another substorm. However, if the solar wind driver remains steady at moderate levels, the magnetosphere enters a state of balanced reconnection (DeJong et al., 2007) called *steady magnetospheric convection*. If the solar wind driver is strong, the balance breaks down, and a quasi-periodic sequence of substorms called *sawtooth events* occurs (Cai and Clauer, 2009).

There were immediate objections to the neutral line model that persist to this day (Frank et al., 1976; Lui et al., 1976). Numerous alternative explanations were suggested, as reviewed in McPherron (1991). New observations by the AMPTE spacecraft led Lui et al. (1988) and Lui et al. (1991) to suggest a current disruption model. This model postulated that the substorm begins at the inner edge of the plasma sheet by plasma instability that leads to diversion of tail current into the ionosphere. Others have adopted this concept, if not the same instability. These models argue that the current wedge and auroral expansion are independent of reconnection and are eventually responsible for its onset. When it was discovered that the auroral breakup does not map along field lines to the neutral line (Baker et al., 1996) the neutral line model was revised by Shiokawa et al. (1997) and Shiokawa et al. (1998), who suggested that breakup is caused by the pileup of plasma flows created by reconnection. The sparsity of spacecraft in the tail has made it impossible to prove that all substorm

* The author's time in preparing this chapter has been supported by two grants from the National Science Foundation, NSF-GEO Baseline AGS-1606014 and NSF-GEM 1602588, and one grant from NASA, HGI NNX17AB83 G. Data used in preparation of the figures were provided by the NASA National Space Science Data Center (https://spdf.gsfc.nasa.gov/), the World Data Center for Geomagnetism in Kyoto, Japan (http://wdc.kugi.kyoto-u.ac.jp/), the SuperMag Project at Johns Hopkins Applied Physics Laboratory (http://supermag.jhuapl.edu/) and the THEMIS Data Base in Berkeley, California (http://themis.ssl.berkeley.edu/). To demonstrate the systematic features of the various response modes, we have developed a new magnetic index called the mid-latitude positive bay index (MPB) and created lists of substorm onsets from the SuperMag SML index and MPB indices, a list of steady magnetospheric convection events from the SML index and a list of sawtooth onsets. The SML and MPB index series and the substorm and SMC lists cover the interval 1980–2015. These assets are available from the author on request.

expansions are associated with flows (Ohtani, 2004), so some researchers maintain that expansion starts close to the Earth and later triggers reconnection. This controversy between the outside-in and inside-out models has come to be known as the 2 minute problem, as this is the time for signals to propagate either way between the two locations (Friedrich et al., 2001).

The Geotail and THEMIS spacecraft missions were launched to resolve this controversy (Angelopoulos, 2008; Nishida, 1994). Geotail clearly established that the near-Earth neutral line is typically located between 22 and 30 R_E and forms close to expansion onset. THEMIS observations with five spacecraft supposedly demonstrated that reconnection precedes expansion onset (Angelopoulos et al., 2008; 2009), but Lui (2009) immediately argued that timing was incorrectly done, and the controversy continues. More recently a variant of both models has been developed by Lyons et al. (2010), Nishimura et al. (2010), Nishimura et al. (2011) and Nishimura et al. (2014). In this new model, dayside reconnection creates a polar cap flow channel that eventually impinges on the plasma sheet boundary beyond the Moon, causing a burst of reconnection and a flow channel that eventually reaches the Earth, triggering current disruption. They note that tailward flows are seen at the Moon during the expansion phase but do not discuss how the current disruption might trigger the formation of the near-Earth neutral line.

To illustrate the effects of a substorm seen in magnetic indices, we next present results of a superposed epoch analysis of the SML and MPB indices relative to expansion onset. SML is defined as the SuperMag Lower index, which is AL constructed from as many as 200 stations. MPB is the mid-latitude positive bay index, which is the power in magnetic perturbations caused by the field-aligned currents of the current wedge. To accomplish this, we have developed an algorithm for detecting the onset of the expansion phase in the SuperMag Lower (SML) index. For the 36 years of available data, this program finds 109,405 onsets. If we limit the events to decreases larger than −80 nT and half width greater than 14 min, the list contains 52,937, or 4 substorms per day. The characteristics of isolated substorms are illustrated by selecting substorms from this list that satisfy three constraints. The average SML index in the interval [−2, −1] hours before onset, and [2, 3] hours after, is greater than −80 nT, and the minimum of the negative bay is less than −200 nT. There are 7100 events in this subset. We have used this list of isolated moderate to strong substorms to perform superposed epoch analysis. The results are presented in Figure 12.3.1. Vertical dotted lines are start of growth phase and end of expansion phase. Figure 12.3.1a displays the optimum coupling function for the AL index (McPherron et al., 2015). The median behavior is a gradual increase in coupling starting more than 2.5 hours before the SML onset that plateaus about 40 min before onset. At onset, coupling begins to decrease, reaching the median level in about 2 hours. Figure 12.3.1b presents the mid-latitude positive bay index (MPB) (McPherron and Chu, 2017). There is a gradual increase for 2 hours before SML onset, followed in several minutes by a rapid rise. The index peaks in about 30 min and then decays to background in the next 2.5 hours. In Figure 12.3.1c, the ensemble median of the SML index is plotted. It indicates about 65 min growth phase followed by a rapid drop, reaching a minimum in 30 min. The median drop in SML is only −125 nT. Figure 12.3.1d presents the SML steadiness parameters defined as the ratio of the 30 min running standard deviation to the running average. The averages are over the 30 min before the current point. The largest fluctuations occur during the expansion phase. Similar results for the AL index have been published by Weimer (1992) using 55 substorms.

12.3.3 Steady Magnetospheric Convection

In the Dungey (1961) model the rates of magnetic reconnection on the day and night sides were balanced. In such cases, there are no changes in the magnetospheric configuration with time, and the classic substorm may not occur. A possible example was noted by Caan et al. (1973) and was the inspiration that led Pytte et al. (1978) to describe the convection bay phenomenon. Early studies of this phenomenon based on studies of a few events are reviewed by Sergeev et al. (1996). An important result of this work is that SMC are associated with a double auroral oval and that transient auroral activations and magnetic spikes occur at its poleward edge. Today these are called poleward boundary intensifications (PBI). O'Brien et al. (2002) and McPherron et al. (2005) used properties of the AU and AL indices to expand the number of SMC events to more than 2000 and showed that they tend to occur during low solar wind speed and moderately negative B_z. They also found that the rate of change of speed and B_z tend to be lower during SMC than at other times. Their analysis demonstrated that nearly all SMC start during the recovery phase of an earlier substorm and about half end with a substorm.

An alternative approach to the identification of SMC events was developed by DeJong and Clauer (2005) and DeJong et al. (2007). If reconnection is balanced, the rate at which open flux is added to the polar cap should equal the rate at which it is removed. Under these conditions there will be no change in the size of the polar cap. The authors used auroral images from the Polar spacecraft to estimate polar cap flux finding intervals when it was relatively constant. DeJong et al. (2009) used 51 events to obtain probability distributions for different solar wind variables. In doing this, they introduced a quantitative measure of steadiness defined as the ratio of a 30 min running standard deviation to the running mean. They found that IMF B_z is much steadier during SMC than other times. A list of nearly 3000 SMC events of duration longer than 90 min was selected by Kissinger et al. (2011b) and used in statistical studies (Kissinger et al., 2012a; Kissinger et al., 2012b; Kissinger et al., 2013). This work showed that solar wind conditions most conducive to the occurrence of SMC occur in the 2 days surrounding the stream interface between high- and low-speed streams in the declining phase of the solar cycle. For

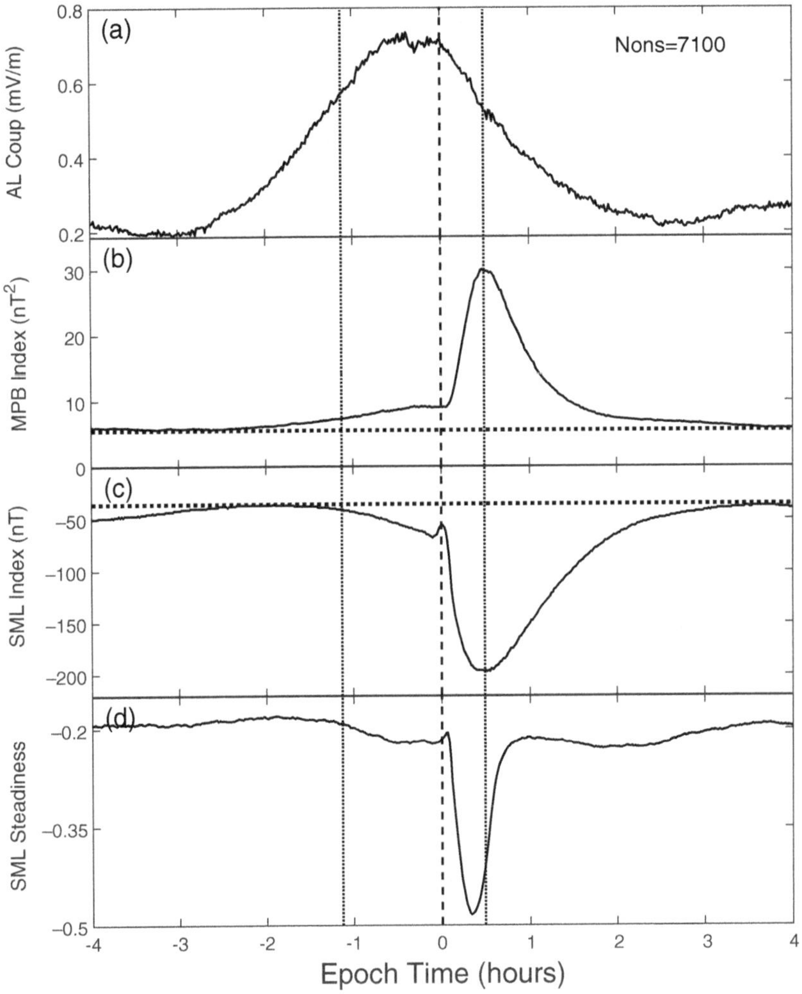

Figure 12.3.1 The systematic response of several magnetic indices relative to the onset of a magnetospheric substorm is displayed. (a) Ensemble median of the optimum AL coupling function for 7100 isolated substorms. (b) Mid-latitude positive bay (MPB) index for these events. (c) SuperMag Lower (SML) index. (d) SML steadiness parameter. The thick vertical dashed line is the time of negative bay onsets in SML. The thin dotted lines indicate the start of substorm growth phase and the end of the expansion phase.

a subset of these events with THEMIS plasma flow data, McPherron et al. (2011) determined that the number of flows reaching 12 R_E during SMC was much smaller than at other times. Kissinger et al. (2012b) constructed maps of the average plasma density, temperature, pressure and flow speed during different types of geomagnetic activity. They found that during SMC, the total pressure in the plasma sheet is higher and extends radially farther from the Earth than at other times. Also the plasma flow is more organized, and it does not penetrate to the Earth as it does during substorms. Instead, it is deflected toward the flanks of the magnetotail and presumably flows around the Earth to the dayside.

We have developed a computer algorithm to identify SMC events using the Kissinger et al. (2011a) selection criteria. For the years 1980 through 2015 (13,149 days) we find 13,363 events or 1 per day. We have performed a superposed epoch analysis of the SML index using this list with the results summarized in Figure 12.3.2. Each panel has at least two traces. The dashed trace shows ensemble averages using the start time of the SMC. A thick line shows averages relative to substorm onset in a 3 hour interval prior to the start of the SMC. In Figure 12.3.2b the third dotted trace shows the averages for a subset of SMC that emerged from quiet conditions.

Figure 12.3.2a presents the number of SMC in progress as a function of time relative to the reference time. There are no SMC in progress prior to epoch zero. The dashed trace represents the average relative to the start of the SMC (vertical dashed line). Every event in the list has a duration of 90 min or longer. As duration increases, the number of events decreases and the curve falls. The thick line shows the average number of SMC in progress relative to the substorm onset prior to the SMC. Since the SMC may start anywhere from 1 to 180 min after the SML onset, there are no SMC present until after SML onset. As more SMC start, the curve gradually rises until 90 min, when some SMC have already ended, and others have not yet started. To avoid contamination of the averages by other types of activity beyond the end of an SMC, all data from an event's end to the right edge of the ensembles were flagged. We have similarly flagged intervals from substorm onset to the start of the SMC in the ensemble of SML index traces.

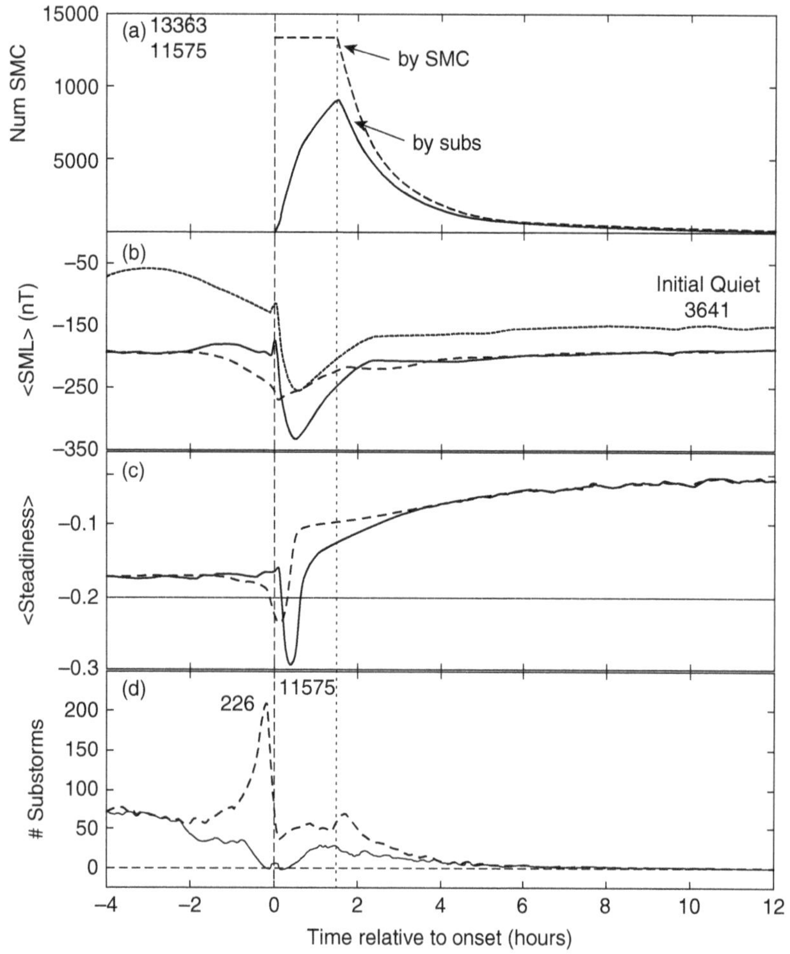

Figure 12.3.2 Ensemble averages of SML index during steady magnetospheric convection events. The vertical dashed line is either the start of the SMC or the onset of a substorm preceding the SMC. Thick lines show the average relative to substorm onset, while dashed lines are relative to start SMC. (a) Number of SMC in progress at a given time relative to the reference time.
(b) Ensemble averages of SML index relative to the two reference times. The dotted line is the average for SMC preceded by a quiet interval.
(c) Steadiness parameter for the SML index.
(d) Total number of substorm onsets counted at each minute relative to the start of the SMC. This number should be zero after zero epoch for the dashed curve using SMC start as reference. There is a spike of height 11,575 at epoch zero in the solid trace (SML onsets).

Figure 12.3.2b shows the average of the SML index relative to the two reference times. The dashed line indicates that there was a significant disturbance in the SML index prior to the SMC events presumably a substorm. The start of the SMC appears to correspond to the beginning of recovery from this disturbance. The thick trace was obtained by averaging relative to the onset of a substorm. The pattern is identical to that of a substorm described in the previous section. Following the substorm, the SML index recovers to the level present before the substorm. We have identified a subset of SMC that began from quiet conditions (SML > −100 nT). The dotted trace demonstrates that for these events SML does not recover to pre-substorm levels.

Figure 12.3.2c presents the ensemble average of the steadiness of the SML index. This parameter is defined as the ratio of the 30 min running standard deviation to the running mean. The time at the end of the window is taken as the time of the steadiness. One of the SMC selection criterion requires that this be greater than −0.2 in 90% of a 90 min interval. The thin line indicates that this condition is not satisfied before the start of the SMC but is afterwards. The thick trace using substorm onset for reference clearly demonstrates that it is the substorm expansion that causes the steadiness to fall below −0.2 and that the criterion begins to be satisfied at the end of the expansion phase. It should be noted that both steadiness traces rise continuously toward the end of the averaging interval. This implies that the longer SMC events are associated with steadier SML index.

Figure 12.3.2d displays the total number of SML onsets at each minute relative to the reference time. The upper trace with peak at 226 shows the tendency for SML onsets to occur prior to the start of the SMC. However, it also shows that substorm onsets were detected during the SMC. Clearly future work is required to create lists where substorms and SMC are mutually exclusive.

12.3.4 Other Response Modes

It is very common for the aurora to become activated for a short interval, causing a minor poleward bulge and weak magnetic disturbance. Such events were described in an early study of auroral morphology and were called *pseudo-breakups* (Elvey, 1957). These have many of the characteristics of the auroral breakup during a substorm but are weaker and rarely expand beyond the range of a single all-sky camera. Pseudo-breakups often occur in the late growth phase at the equatorward edge of the auroral oval and appear to be precursors to a full breakup. They are often mentioned in the literature, but no adequate explanation for their behavior has been proposed.

Another phenomenon that is common during strong steady driving is sawtooth injection events. These events are best seen in the flux of ions at synchronous orbit which has the form of a slow decrease followed by a sharp rise giving the appearance of a sawtooth (Belian et al., 1995). These events frequently consist of a sequence of 3–10 quasi-periodic teeth with average separation of 2.5 hours (Cai and Clauer, 2009). Initially, these were thought to be a unique phenomenon, but numerous studies have shown that each sharp rise is correlated with a substorm expansion that is unusual only in its azimuthal extent often expanding into the afternoon sector. Sawtooth events appear to simply be quasi-periodic large substorms.

A third phenomenon is often seen in the recovery phase of substorms, during steady magnetospheric convection and during sawtooth events. This is manifested as sudden intensifications of the aurora at the poleward edge of the auroral oval. These are called *poleward boundary intensifications* or PBI (de la Beaujardiere et al., 1994). Frequently, a narrow stream of aurora will emerge from the bright region, moving equatorward. It has been demonstrated that these are associated with a burst of reconnection. During isolated substorms, they tend to occur in the recovery phase, but in complex activity, it is difficult to determine whether the PBI and streamers are precursors to the next substorm or remnants of the last one.

12.3.5 Discussion

There is abundant evidence that geomagnetic activity is primarily caused by magnetic reconnection when the IMF is southward. The dimensions of the magnetosphere, the speed of the solar wind and the conductivity of the ionosphere constrain the rates at which different processes take place. Consequently, characteristic timescales appear and geomagnetic activity is organized into specific modes of response. These include quiet times, the magnetospheric substorm and steady magnetospheric convection. Not discussed in this section are magnetic storms. In addition, there are three other distinct types of magnetic activity that appear related to these modes. These include sawtooth injection events, poleward boundary intensifications and pseudo-breakups. The magnetospheric substorm appears to be the most fundamental response. As yet there is no universally accepted explanation for the cause of the substorm onset and expansion.

There are several issues that lead to controversy. First is the timing issue. Most signatures of substorm onset are imprecise, many with errors of the order of several minutes. Second, the signatures are frequently ambiguous, such as the occurrence of pseudo-breakups before main breakups. Third, the existing network of auroral imagers is sparse with many gaps due to geography and weather. Ground magnetometers are often not in appropriate locations or their response is ambiguous. In space the situation is worse. Even with five spacecraft, it is highly likely that the onset of reconnection and the fast flows it generates will often be missed. Another serious issue is field line mapping. Above about 65° latitude, small changes in the tail current alter the equatorial crossing point of a field line by very large amounts. Usually PBI are attributed to reconnection at a distant x-line, but this is not necessarily the case. A reconnection event at any distance that begins to close lobe field lines in the lobes will project to a high-latitude edge of the oval.

12.4 Radiation Belts and Ring Current

Daniel N. Baker

12.4.1 Introduction

The first major scientific discovery of the Space Age (Van Allen et al., 1958) was that the Earth is shrouded in toroids, or 'belts', of very high energy, magnetically trapped charged particles. Early observations of the radiation environment suggested that the Van Allen belts could be delineated into an inner zone dominated by high-energy protons and an outer zone dominated by high-energy electrons. Subsequent studies showed that electrons in the energy range 100 keV $< E <$ 1 MeV often populated both the inner and outer zones with a pronounced 'slot' region relatively devoid of energetic electrons existing between them. The energy distribution, spatial extent and particle species makeup of the Van Allen belts has been subsequently explored by several space missions (Baker, 2014).

12.4.2 Radiation Belt Structure and Dynamics

Figure 12.4.1 is a schematic that includes the Earth's radiation belts. Closest to the Earth's surface is the inner Van Allen zone. This belt extends from just above the dense atmosphere out to an equatorial altitude of about 10,000 km above the Earth's surface. The inner Van Allen belt comprises dominantly very energetic protons (ranging up to multiple GeV energies). Recent work using instruments on NASA's Van Allen Probes Mission has shown fascinating temporal and spatial behavior for the inner zone proton population (see Selesnick et al., 2014). These results demonstrate that protons with energies from ~10 MeV to ~100 MeV are quite stable in time near the geocentric radial distance of $r \sim 1.5\ R_E$ (Earth radius = 6372 km), at which the inner zone proton fluxes peak. However, an outer 'shoulder' of the radial distribution

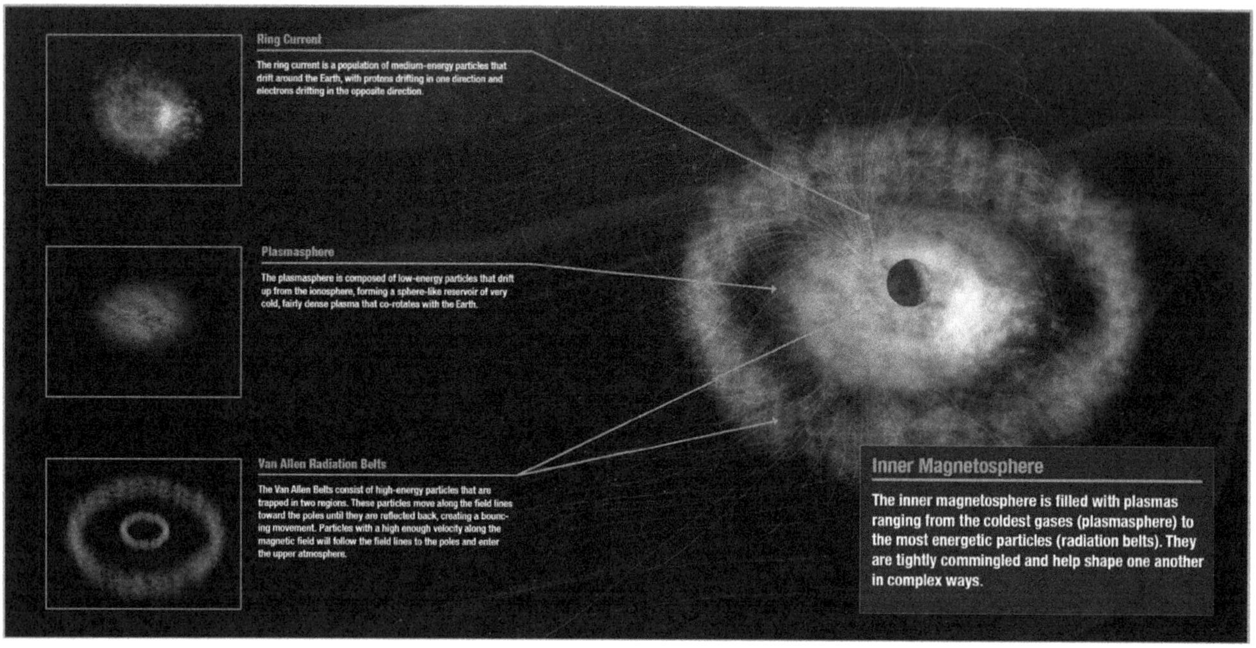

Figure 12.4.1 A diagram of the inner magnetosphere showing the commingled nature of the plasmasphere, ring current and Van Allen radiation belts within the same volume of space near Earth (NASA/Troy Benesch). (A black-and-white version of this figure appears in some formats. For the colour version, please refer to the plate section.)

from $1.7 \leq r \leq 2.5$ R_E shows tremendous temporal variability for protons with $E > 60$ MeV. These variable proton fluxes are probably due primarily to evolution of trapped solar energetic protons.

The inner Van Allen zone also has copious fluxes of low- and medium-energy electrons (Fennell et al., 2015; Li et al., 2015) as revealed by Van Allen Probes and other spacecraft data sets. However, the Van Allen Probes era (September 2012 to present) has provided many new discoveries about inner magnetospheric ultra-relativistic electrons with energies $E > 2.5$ MeV. In particular, initial results after major storm intervals surprisingly showed essentially no detectable prompt ultra-relativistic electron fluxes in the region $L < 2.8$ (Baker et al., 2014), but over longer timescales these electrons did eventually move inward from locations beyond this radial distance to $L \sim 2.5$. The paucity of very energetic electrons in the inner magnetosphere immediately following major magnetic storms is quite striking.

The slot region of the radiation belts extends from roughly $L \sim 2.0$ to $L \sim 3.0$ depending on energy and particle species. (L is the geocentric distance in Earth radii at which a dipole magnetic field line crosses the magnetic equatorial plane.) The slot is a region often relatively devoid of energetic electrons. However, during strong geomagnetic storm periods, the gap between the inner and outer zone can be filled to a large degree by moderate-energy (and even high-energy) electrons. For example, in the intense 'Halloween' storm period of late October and November 2003, the slot region was filled with multi-MeV electrons for several weeks (Baker et al., 2004).

Finally, the outer Van Allen radiation belt represents in many ways the most variable part of the Earth's radiation environment. The outer zone is broad in spatial extent (from $r \sim 3$ R_E to $r \sim 6.5$ R_E). It comprises mildly to highly relativistic electrons (~100 keV to 10 MeV) and varies widely in particle intensity. Commercial, military and scientific satellites operating in medium-Earth orbit (MEO) and geostationary Earth orbit (GEO) number in the multiple hundreds worldwide. All of these operating spacecraft are subject to outer Van Allen belt space weather impacts.

Recent observations by the NASA dual-spacecraft Van Allen Probes mission have revealed several unexpected properties of the radiation belts, especially at highly relativistic ($E > 2$ MeV) and ultra-relativistic ($E > 5$ MeV) kinetic energies. Using high-spatial-resolution and high-temporal-resolution data from the Van Allen Probes, it is observed that multiple belts can exist concurrently and that an exceedingly sharp inner boundary exists for ultra-relativistic electrons (Baker et al., 2013, 2014). Using additionally available Van Allen Probes data, it is demonstrated that these remarkable features of energetic electrons are not due to a physical boundary within Earth's intrinsic magnetic field. Rather, it is likely that human-generated electromagnetic transmitter wave fields help produce such effects. These unique measurements show that human-made wave–particle scattering effects deep inside the Earth's magnetosphere can contribute to an almost 'impenetrable' barrier through which the most energetic Van Allen belt electrons often cannot migrate.

12.4.3 Ring Current, Plasmasphere and Radiation Belt Coupling

As described in Chapter 4 of this volume, a key current system in the magnetic equatorial plane is the extraterrestrial ring current. The ring current plays a major role by shaping the overall magnetosphere. During geomagnetic storms, ring current energetic particle populations are increased greatly, with peak fluxes occurring in the inner magnetosphere. The ring current distorts the basic dipolar magnetic field configuration and thus changes the radiation belts in often dramatic ways as well.

The ring current is a westward-directed electric current carried by equatorially confined populations of ions and electrons surrounding Earth. The ring current often overlaps with both the outer radiation belt and the plasmasphere (see Figure 12.4.1). The ring current varies in density, composition and structure, but normally extends between 2 and 9 Earth radii. Changes in ring current composition can occur suddenly during times of geomagnetic storms and other increased geomagnetic activity. The source of ring current plasma can vary during these times, with ionospheric outflow dominating over protons of solar wind origin during storm times. For extensive reviews on the ring current, see Daglis et al. (1999) and Daglis and Kozyra (2002). A more recent paper by Keika et al. (2013) summarizes work on inner magnetospheric O^+ energization and associated ring current injection/pressure effects.

The storm-time ring current is seen to arise from (1) the injection of energetic particles into the inner magnetosphere during magnetospheric 'substorms' and (2) fast transport of energetic charged particles from the nightside plasma sheet deep into the inner magnetosphere ($r < 4R_E$) due to intensification of the Earth's dawn–dusk convection electric field during extended periods of strong southward IMF. Present understanding favors the enhanced convection model over the substorm plasma injection model. The storm-time growth of the ring current typically lasts less than a day and constitutes the main phase of a magnetic storm.

The *Dst* index is an indicator of the ring current strength as measured from a network of ground-based magnetometer stations. This index is central in space weather investigations. The so-called *Dst* effect (Li et al., 1997a, 1997b; Kim and Chan, 1997) appears as adiabatic changes in radiation belt electron populations during times of dramatic geomagnetic field changes that are associated with changes in the ring current. The result can be a sudden apparent loss or enhancement in radiation belt fluxes. However, the effect is, in fact, an entirely adiabatic change that eventually reduces as the ring current subsides. The real loss and acceleration during the main phase of storms can therefore be very difficult to separate from this *Dst* effect.

After the main phase, the ring current decays, returning to its pre-storm state in a period of several days (Jacobs, 1987). During the storm recovery phase, particle transport slows, which allows loss processes to reduce ring current particle fluxes back to their normal levels. The main loss processes during both the storm main and recovery phases involve ring current particle charge exchange with neutral (low-energy) hydrogen atoms in the geocorona and Coulomb collisions with the thermal plasma of the plasmasphere. Another process contributing to ring current decay is the precipitation loss of ring current particles into the atmosphere as a result of wave–particle interactions.

As noted above, the magnetic field in the magnetosphere very much controls the location and extent of the radiation belt particle populations. As the ring current waxes and wanes, the overall magnetic field configuration changes in significant ways. This, in turn, causes the radiation belts to expand in radial extent or to shrink depending on the ring current strength. Thus, it is clear that the inner part of Earth's magnetosphere is a complex, highly interconnected system of cold, warm and very hot particles all interacting with one another through the vehicle of the geomagnetic field.

Enhancements in the storm-time ring current can have direct impacts on radiation belt losses in two ways (1) through wave–particle interactions or (2) through increased magnetopause shadowing due to diamagnetic effects. Electromagnetic ion cyclotron (EMIC) waves are driven by unstable ring current proton distributions (e.g. Anderson and Hamilton, 1993) and are able to produce losses of the very highest-energy radiation belt electrons and protons (Meredith et al., 2003). The other mechanism involves a build-up of the partial ring current by enhanced convection and repeated substorm injections (Gkioulidou et al., 2014). This expands electron drift orbits to the point where magnetopause shadowing can have a greater effect on the core population (Ukhorskiy et al., 2006). This can lead to dramatic dropout features in the measured fluxes (Shprits et al., 2006). Such a process irreversibly reduces the content of the outer radiation belt region and can cause almost total drainage of the pre-existing radiation belt populations.

Acceleration events of relativistic electrons can also be influenced by ring current dynamics. Extended periods of electron substorm injections can build up and maintain the 'source' and 'seed' populations that are required to produce relativistic and ultra-relativistic electrons. Chorus wave growth, which plays a central role in relativistic electron enhancements, is driven by temperature anisotropies in ring current energy electrons. The plasmasphere is the cold, dense toroidal plasma population that typically extends from the ionosphere out to a variable outer boundary known as the plasmapause (e.g. Carpenter and Lemaire, 2004). The plasmapause varies in local time, and can be as low as 2 R_E or as far out as beyond 6 R_E, depending on geomagnetic activity (Chappell, 1972). There is often much structure observed in the plasmapause/plasmasphere boundary layer, including persistent features such as the duskside bulge and plume region (Darrouzet et al., 2009).

Plasmaspheric densities often are substantially altered during large geomagnetic storm events due to interplanetary shocks or coronal mass ejection impacts. Afterwards, several processes act to refill the pre-storm population from the ionosphere through e.g. field-aligned diffusion, charge exchange and cross-field transport. During storm recovery phases, the rate of field-aligned refilling is quite slow (⊠days). Thus, variability in local plasmaspheric densities can be accordingly rather high during dynamic conditions associated with repetitive storms. Even after more than five decades of study (see review by Singh and Horwitz, 1992), plasmasphere refilling dynamics and their impacts on overall plasmasphere structuring are still a subject of intense study with many unsolved problems (Gallagher and Comfort, 2016).

The dynamic plasmasphere has long been recognized as a key agent in radiation belt configurations due to multiple interaction pathways between the cold plasma and more energetic particles. Earlier work has shown a clear correlation between the plasmapause location and innermost extent of the multi-MeV electron outer radiation belt (Li et al., 2006). However, more recent detailed observations across a wide range of energies have revealed that the relationship between the plasmapause and relativistic electrons is complex and highly energy and activity dependent. The plasmapause can provide a shielding effect as well as cause greater loss rates. This depends on the electron energy and the location within the geomagnetic field. The primary loss mechanism for radiation belt electrons within the plasmasphere is by cyclotron resonance interactions with very-low-frequency hiss waves. This wave mode is efficient at pitch angle scattering relativistic electrons into the loss cone (Thorne et al., 1973). At lower L-shells, the resonance is more favorable for higher energies (Summers et al., 2007) and thus the outer radiation belt can be quickly sculpted by the location of the plasmasphere and the hiss therein.

The lifetimes for relativistic electrons due to hiss scattering vary with magnetic field strength, magnetic local time, electron energy and geomagnetic activity. In addition to this constant loss mechanism, radial diffusion is at work in this region, although it acts much more slowly here than in the outer belt region. Therefore, the dynamics inside the plasmasphere are a complicated balance of inward diffusion and slow scattering losses and are highly dependent on the location and portion of the population being observed. There are times when the plasmasphere can retreat quickly, allowing for acceleration at very high energies to take place just outside the plasmapause boundary at low L-shells. After a time, the plasmasphere can refill to higher radial distances, effectively enveloping the newly accelerated electrons in a protective shell (Baker et al., 2013). Deep inside the plasmasphere, ultra-relativistic electrons can have lifetimes of tens to hundreds of days.

12.5 Recent Advances in Satellite Missions

Craig A. Kletzing

12.5.1 Introduction

The Van Allen Probes mission (launched in August 2012), the Magnetospheric Multiscale (MMS) mission (launched in March 2015), the THEMIS mission (launched in 2007) and the Cluster mission (launched in August 2000) are the four most fully instrumented, multi-spacecraft missions developed over the past two decades to investigate magnetospheric processes.

12.5.2 Van Allen Probes Mission

The Van Allen Probes comprises two identically instrumented spacecraft orbiting close to the Earth's equatorial plane (10° inclination) with field and particle measurements covering the full range of plasma parameters essential for understanding radiation belt physics. Several important advances have resulted in much improved understanding of the radiation belts.

One of the earliest discoveries of the Van Allen Probes mission was the existence of a third radiation belt of relativistic electrons. This occurred shortly after the launch of the spacecraft and was created when the outer belt electrons (from $L = \sim 4\text{–}6\ R_E$) were eroded to give a much thinner belt at lower L values. The outer belt then began to repopulate from larger L values with a gap between the eroded belt and the newly reforming belt (Baker et al., 2013).

Another set of important advances was the firm establishment of chorus waves as a local accelerator of energetic radiation belt electrons. Observations of a clear peak in phase space density within the outer belt demonstrated that local acceleration was taking place. With gradients in phase space density toward both higher and lower L values, only a local acceleration process provided a viable explanation. At the same time, wave measurements indicated chorus activity suggesting the link between waves and particles (Reeves et al., 2013).

Following this result, numerical simulations using a proxy for chorus activity derived from multiple spacecraft showed that the chorus waves could explain the rapid acceleration of electrons to relativistic energies matching both the observed timescale and observed energies (Thorne et al., 2013). Figure 12.5.1 shows this result, comparing modeled particle fluxes and those measured on the Van Allen Probes.

The Van Allen Probes mission has also uncovered many new details of the particles in the inner radiation belts. The improved particle instrumentation on the Van Allen Probes has allowed for characterization and removal of energetic ion contamination that has obscured the electron measurements on prior missions. This has revealed that relativistic electrons above 1 MeV only very rarely penetrate to L values less than 2.5. Indeed, over the first 5 years of the Van Allen Probes mission, no inward radial diffusion of particles at these energies beyond this has been observed and prior mission have seen such

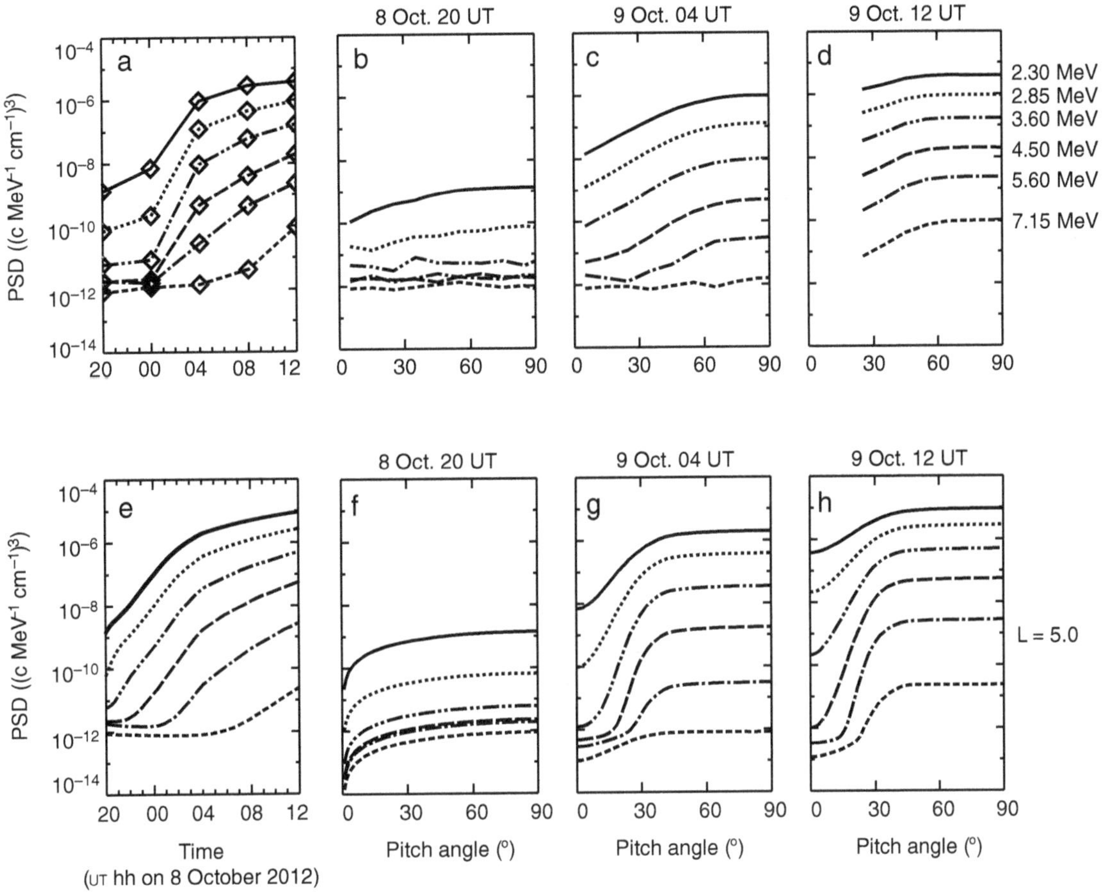

Figure 12.5.1 (a–d) Comparison between Van Allen Probes observations and (e–h) model results showing good agreement in both energy, timescale and pitch angle evolutions for an event in October 2012.

diffusion only very rarely (Fennel et al., 2015). However, lower-energy particles in the few hundreds of keV have been observed to move inward past this location with no apparent restriction.

Wave observations from the Van Allen Probes have shown local generation of plasmaspheric hiss, revealed structure in hiss, differentiated between parallel propagating and oblique chorus, with the largest-amplitude chorus being mainly parallel propagating.

Current and future work is now focusing on statistical studies using the more than 6 years of data collected by the mission. Such data have never before existed, and these studies are beginning to yield a clear picture of both wave and particle features. Additionally, the launch of the Japanese Arase satellite in 2016 is enabling satellite-to-satellite comparisons of data. Because Arase is more highly inclined than the Van Allen Probes, conjunction data between the three satellites will allow wave propagation and particle evolution studies never before achieved.

12.5.3 The Magnetospheric Multiscale (MMS) Mission

The MMS mission is composed of four identically instrumented spacecraft with some of the highest time resolution particle and wave measurements ever flown in the magnetosphere. The mission was designed with a primary focus of uncovering the detailed physics of magnetic reconnection. Magnetic reconnection, which connects the Sun's magnetic field to that of the Earth and converts magnetic energy into particle kinetic energy and heat, occurs on very small spatial scales, below the ion and electron inertial scale. In particular, the 'electron diffusion' region is at a scale small enough that electrons become demagnetized as they evolve through the reconnecting magnetic field.

The first report of observations by MMS from within the electron diffusion region showed that, as models had predicted, electrons were demagnetized and accelerated. This event was highly sheared with very nearly oppositely directed solar and terrestrial magnetic fields. The conversion of magnetic energy to particle energy was confirmed and a full set of electric field measurements in conjunction with direct measurement of currents demonstrated how this energy conversion takes place. Additionally, the high cadence measurements allowed direct measurement of the electron population, which carries the energy delivered to the particles by the reconnecting magnetic field (Burch et al., 2016).

While idealized models of magnetic reconnection have zero magnetic field at the heart of the reconnection region, reconnection can take place with a non-zero background magnetic field termed 'guide field' reconnection. Such an event was reported by Burch and Phan (2016) in which a significant guide field was present. They reported two regions of current which formed a bifurcated current system. One region had strong currents and electric fields, while the other was much more quiescent. Modeling with particle-in-cell simulations has shown that this type of guide field reconnection is indeed reproduced in this type of simulation.

Details of the energy conversion process have been examined using the MMS data set to estimate the various terms in the generalized Ohm's law (Torbert et al., 2016). This showed that electron inertial effects and electron pressure gradients were most important but were not the only terms contributing to Ohm's law. Comparison with simulation showed reasonable agreement.

Although MMS has produced the most detailed look at reconnection ever made, it has also sampled other processes of the dayside magnetosphere, exploring ion microphysics, waves and turbulence, flux transfer events and Kelvin–Helmholz waves. With MMS's detailed field and particle measurements, the structure of the magnetopause, its dynamics and details of the magnetosheath have been reported.

The first phase of the mission was dedicated to dayside events, although some work was done on nightside physics. However, as the orbit has precessed, the apogee has been raised, and MMS is now sampling the tail reconnection region. These results are still being analyzed but will provide the same highly detailed measurements of tail reconnection.

12.5.4 The THEMIS Mission

The THEMIS mission was launched as five identically instrumented spacecraft. In 2011, two of the THEMIS spacecraft were put into orbit around the Moon and were rechristened as ARTEMIS. In 2014, similar to Cluster, THEMIS reported formation of a plume which extended from the plasmaspheric to the magnetopause under active geomagnetic conditions. Using combined ground and spacecraft measurements, they showed that the plume rose from the ionosphere and then extended outward, persisting for hours (Walsh et al., 2014).

In the ion foreshock, a region just outside the Earth's bow shock, THEMIS found that electrons can be accelerated to relativistic energies. While such energetic electrons are commonly found in the radiation belts, this report showed they are present in other regions as well. This study adds important insights into how particle acceleration can occur (Wilson et al., 2016). In a later THEMIS result, a statistical study showed that there are different mechanisms for ion and electron energization in the foreshock. Both show a positive correlation with increased solar wind speed.

THEMIS has also produced studies of magnetospheric Kelvin–Helmholtz waves, comparing observations with simulation models. Among the results, they have reported that solar wind fluctuations influence the velocity and magnitude of the waves, with stronger fluctuations driving larger waves.

The ARTEMIS part of the mission has reported that they observe molecular ion outflow from the Earth at lunar distances during geomagnetically active times. A model–data comparison suggests that they escape by drifting around the magnetosphere and entering the boundary layer somewhere at the dayside magnetopause, then escaping tailward. This is important both in terms of atmospheric escape but also is a source of volatiles to the lunar regolith.

12.5.5 The Cluster Mission

The Cluster mission has four identically instrumented satellites. Although the spacecraft have been in orbit for more than 18 years, they are still returning useful scientific results. Cluster has connected the presences of trans-polar arcs with high-energy plasma from the lobe which has become trapped on high-latitude field lines by reconnection in the tail (Fear et al., 2014). This addressed a long-standing problem of the source of the plasma creating these auroral features.

Another problem in auroral physics, first identified by the Freja mission, is that of the 'black' aurora. These are regions in the auroral zone in which there is no visible light at all but which are embedded in regions of active aurora. Cluster observations have shown that, as predicted, these occur in regions of down-going current collocated with intense electric fields which prevent electron precipitation and can evacuate ionospheric plasma into the outer magnetosphere.

As discussed below, 'equatorial noise' is a fast magnetosonic wave that has been observed for decades in the inner magnetosphere. Cluster observations of these waves have shown that the frequency structure of these waves, which shows multiple harmonics at integer multiples of the ion cyclotron frequency, is driven by ion ring distributions (Balikhin et al., 2015). They compared observed distributions of the ions with theory, deriving growth rates that agreed with the wave observations.

Cluster has also observed signatures of plasmaspheric plumes and ion outflow from the ionosphere. The plumes generally occurred under IMF B_z southward conditions, while the outflows were found near the Earth's magnetopause during B_z northward conditions. A statistical study showed that these ionospheric outflows were observed about 10% of the time when the spacecraft were near the magnetopause. The plumes were found to be primarily duskside phenomena, in agreement with prior IMAGE observations.

The Cluster spacecraft are nearing their end of life, but as these examples (and many before them) have shown, the mission has had an incredible return on investment.

12.6 Plasma Wave Interactions

Craig A. Kletzing

12.6.1 Wave Measurement Techniques

Plasma waves are a ubiquitous presence throughout the magnetosphere. They range in frequency from the ultra low frequency (but not ULF in the radio spectrum sense) of a few mHz associated with large-scale oscillations of the geomagnetic field up to waves in the 1–2 MHz regime and above characterized by upper-hybrid waves and continuum radiation.

Historically, most observations of magnetospheric waves above a few Hz were made along a single axis of the electric or magnetic field and provided spectral information derived from filter banks. This operational mode was driven by both telemetry constraints and hardware limitations, but provided a good picture of the overall wave environment in the various regions of the magnetosphere.

Modern wave measurements have progressed significantly due to increases in computation capability on board spacecraft coupled with higher telemetry rates. This has enabled fully three-dimensional measurements of both the electric and magnetic field that permit computation of many wave properties such as Poynting flux, wave normal direction, polarization and ellipticity (Kletzing et al., 2013, and references therein). With this more complete set of wave properties, the detailed physics of wave propagation can be studied. Another recent innovation has been the development of wave–particle correlators which directly compare particle arrival times to the phase of a wave (Kletzing, 2017, and references therein).

12.6.2 Low-Frequency Waves

Waves below 0.1–1 Hz are typically described as ULF waves. These waves are predominantly observed in the inner magnetosphere because they represent global modes of oscillation of the geomagnetic field. Recent measurements have reported resonance of these wave with both ions and electrons with clear peaks when the drift motion of the particles match the wave frequencies.

Alfvén waves are observed in a variety of regions from near the reconnection sites on the dayside to the auroral region and the inner magnetosphere. In the inner magnetosphere, Alfvén waves have been identified as potential source of acceleration of ring current ions. The waves are observed to be excited by injections of plasma from the tail and are excited at the plasmapause (Chaston et al., 2015). In the auroral zone these waves have been identified for several decades as a key mechanism of auroral acceleration. Recent laboratory experiments have begun to directly identify the response of electrons in the wave field and have confirmed basic theoretical understanding of the physics of the waves (Schroeder, 2016, and references therein).

At frequencies between the gyro frequencies of various ions, primarily H+, He+ and O+, electromagnetic ion cyclotron waves (EMIC) are observed. These waves are thought to be a mechanism for loss of radiation belt electrons, causing them to precipitate into the auroral ionosphere. However, recent studies remain ambiguous as to how strong the connection to particle loss is.

12.6.3 Intermediate-Frequency Waves

This category of waves lies between the ion cyclotron frequency and the electron cyclotron frequency, consisting of magnetosonic waves and whistler-mode waves. Perhaps most well-known are whistlers themselves, which are launched by lightning and then propagate through the ionosphere into the inner magnetosphere. They are commonly observed on the Van Allen Probes and were well known before the Space Age from listening to radio waves as they reached the opposite hemisphere and penetrated through the ionosphere.

Magnetosonic waves have been reported for several decades and were termed *equatorial noise*. These waves, which predominantly occur on the dayside, are linearly polarized and are very localized near the magnetic equator. Theory suggests that they should be capable of energizing electrons to high energies, but this has not yet been confirmed by observation. One very interesting new observation from the Van Allen Probes has shown that these waves can have frequency structure in which frequency rises slowly over a few minutes up to the lower hybrid cut-off (Boardsen, 2014). No model has yet reproduced these features.

One of the most important whistler-mode waves is that of chorus waves. These waves occur in intense, short duration (<0.5 s) structures which rise in frequency and are observed to split into two bands separated by one-half of the electron cyclotron frequency. The lower band is typically 0.1–0.5 f_{ce} and the upper band 0.5–0.8 f_{ce}. Models of wave generation show that they are very non-linear and that they can take electrons from tens to a few hundred keV up to relativistic energies. As discussed in the previous section, these waves are thought to be a major contributor to outer radiation belt electron energization.

Also in the whistler mode is plasmaspheric hiss, which is observed inside the Earth's plasmasphere and is a key mechanism for scattering energetic electrons into the loss cone so that they are lost to the atmosphere. In the outer magnetosphere, whistler-mode waves have been observed by MMS near the electron diffusion region and are thought to play a role in the reconnection process.

12.6.4 Higher-Frequency Waves

At frequencies above the electron cyclotron frequency, a variety of waves are observed. In the inner magnetosphere,

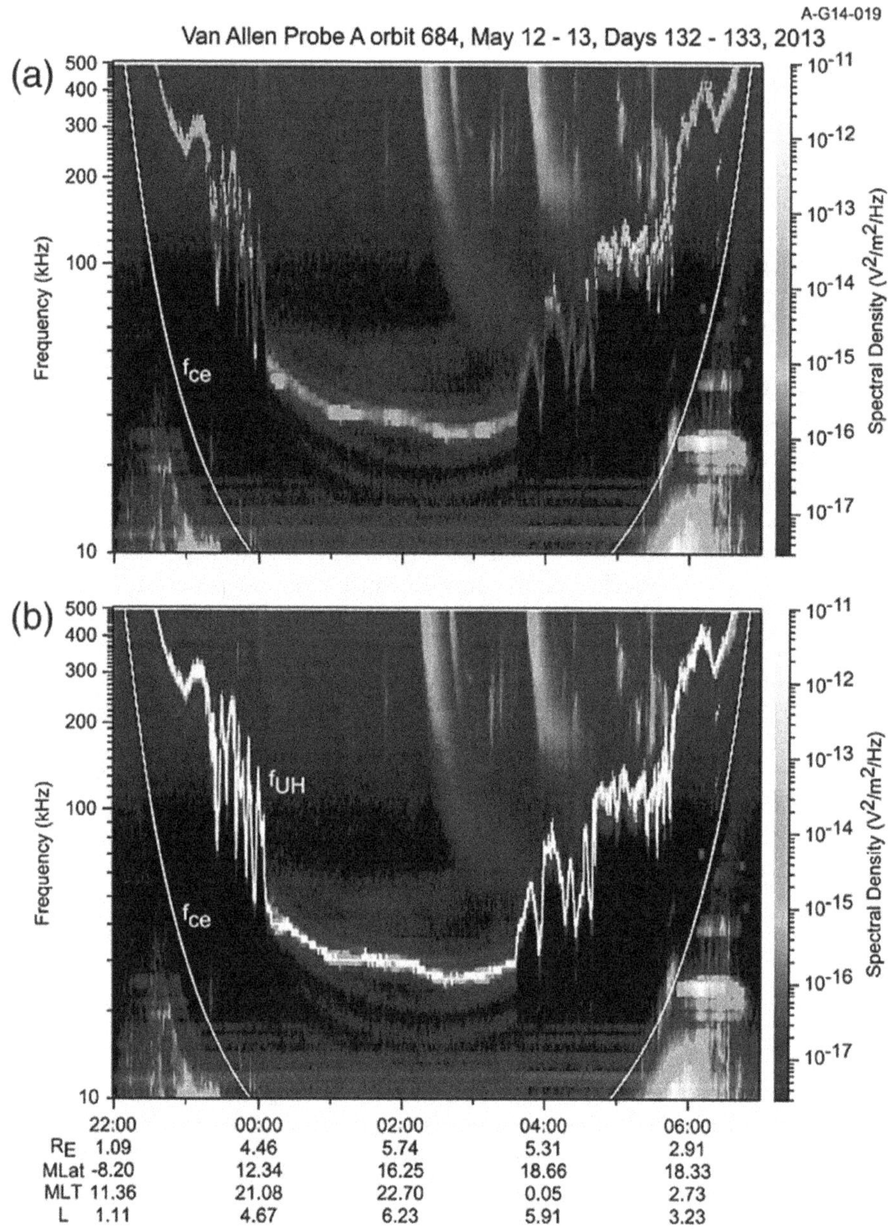

Figure 12.6.1 Data from the HFR receiver on the Van Allen Probes showing (a) the upper-hybrid line and (b) the trace of this line, which is used to determine plasma density.

electron cyclotron harmonics are occasionally observed, particularly during active times. Two of the most common wave types in this range are Langmuir waves and upper-hybrid waves. Both are observed throughout the magnetosphere, and often seen as a weak spectral feature excited by thermal motion of the plasma. At ionospheric altitudes, these waves have been observed to be driven by instabilities derived from the precipitation of auroral particles.

In the inner magnetosphere, upper-hybrid waves range in frequency from hundreds of kHz at values of L below 2 to a few tens of kHz at $L = 6$. This range of frequencies is determined by the magnetic field and the background electron density These emissions are so ubiquitous that the Van Allen Probes wave receivers have exploited this dependence on density and magnetic field to produce one of the most complete records of plasma density at $L \leq 6$ (Kurth et al., 2015). Having such a complete record of density has proved to be invaluable for detailed theory–model comparisons with data. An example of the upper-hybrid line and its digitization to derive density is shown in Figure 12.6.1.

Above the plasma frequency, so-called 'continuum radiation' is observed. This is wave power which, because of its high frequency, propagates as a free-space mode but is cut off at the plasma frequency. Its lower cut-off can also be used as a measure of plasma density.

12.7 Plasmasphere and Plasmapause

*Viviane Pierrard**

12.7.1 Introduction

The plasmasphere is the extension of the ionosphere at higher altitudes. This roughly toroidal region of the magnetosphere surrounds the Earth at low and middle latitudes. It contains a relatively dense plasma of low-energy (typically 1 eV) electrons and protons, with also a small amount of helium ions. It was discovered in 1963 from the analysis of VLF whistler wave data. The outer surface of the plasmasphere is generally characterized by a sharp decrease of the plasma density: the plasmapause. Whistlers and in situ satellite observations show that the plasmapause forms closer to Earth when the geomagnetic activity level is enhanced. Depending on the strength of the level of geomagnetic activity, the equatorial position of the observed plasmapause knee changes from 7 R_E to 2.5 R_E. The position of the plasmapause is due to the interplay between the electric field associated with the Earth's rotation (that dominates at small radial distances) and the convection electric field associated with geomagnetic activity (that dominates at larger distance) (Lemaire and Pierrard, 2008). The depleted region beyond the plasmapause is called the plasma trough.

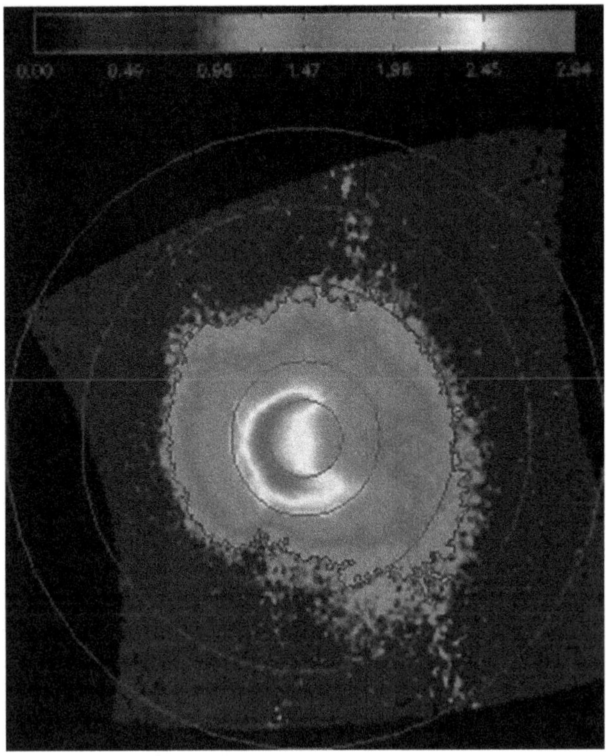

Figure 12.7.1 The global plasmasphere seen by EUV/IMAGE on 22 June 2001 at 0519 UT (when the satellite was above the North Pole and projected in the geomagnetic equatorial plane). The irregular line corresponds to the threshold of 40% of the maximum luminosity and illustrates the plasmapause position. The circles correspond to the Earth's surface and L = 2, 4, 6 and 8. (A black-and-white version of this figure appears in some formats. For the colour version, please refer to the plate section.)

12.7.2 Recent Progress with Observations

Recent spacecraft observations have revolutionized our understanding of the plasmasphere. Especially, IMAGE/EUV provided from 2000 to 2006 the first global images of the plasmasphere in the equatorial plane when the spacecraft was above the North Pole (Goldstein et al., 2005; Darrouzet et al., 2009). These images allowed magnetic local time (MLT) analyses of the plasmapause evolution with time (Pierrard and Cabrera, 2006). With sensitive EUV cameras, IMAGE provided global views of the plasmasphere by imaging the distribution of He^+ ions in its 30.4 nm resonance line integrated along the line of sight. An example is illustrated in Figure 12.7.1. New structures like plumes, notches and shoulders have been detected. More recently, the first global meridian images of the plasmasphere were obtained with TEX instrument on board KAGUYA and provided useful information on physical processes of plasmapause formation (Murakami et al., 2016).

Moreover, many spacecraft determine the plasma density along their orbit and show in situ spatial and temporal structures inside and outside the plasmasphere and the evolution of the plasmapause positions. The four CLUSTER spacecraft make multipoint measurements and the observations of the instrument WHISPER (that determines the low-energy plasmaspheric density) could be complemented by measurements of higher energy instruments to analyze the links between the plasmapause and the radiation belts boundaries (Darrouzet et al., 2013). Data from Van Allen Probes also showed the influence of the plasmasphere to limit the ultra-relativistic electrons of cosmic rays and solar origin from reaching low altitudes (Baker et al., 2014).

Simultaneous THEMIS observations are also available to complement the plasmaspheric observations (Bandic et al., 2017). Statistical studies have been made to determine the dependence of the plasmapause positions on solar and geomagnetic indices (Verbanac et al., 2015, with Cluster; Bandic et al., 2016, with CRRES for instance) and the thickness of the plasmapause boundary layer (Kotova et al., 2018, with MAGION). Empirical relations between the plasmapause equatorial distance and a variety of geomagnetic indices like *Kp*, *Dst* or AE have been deduced as well as their dependence and propagation in MLT.

* V. Pierrard thanks the Solar-Terrestrial Center of Excellence and the Royal Belgian Institute for Space Aeronomy.

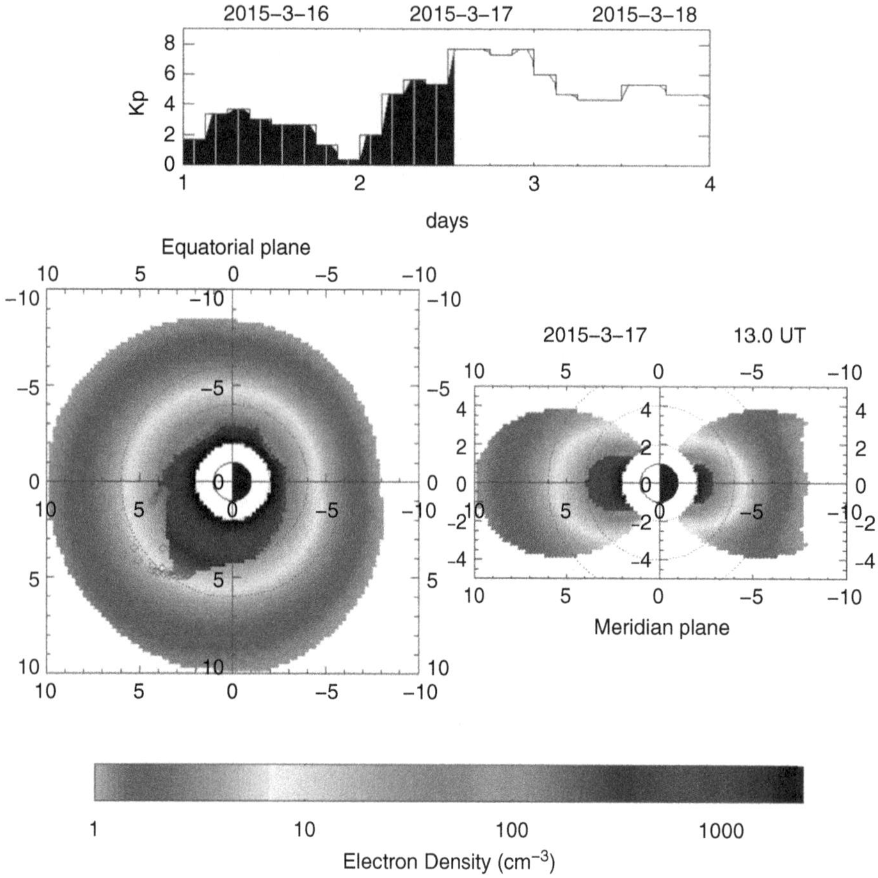

Figure 12.7.2 Density of the electrons in the equatorial (left) and meridian plane (right) obtained with the SPM plasmaspheric model (Pierrard and Voiculescu, 2011) for the date of 17 March 2015 at 1300 UT, during a geomagnetic storm. The plasmasphere corresponds to the dark region with higher density, the plasmapause is illustrated by the diamonds and the low-density plasma beyond the plasmapause corresponds to the plasma trough. The upper panel shows the planetary geomagnetic activity index of Bartels *Kp* observed from 16 to 18 March 2015. (A black-and-white version of this figure appears in some formats. For the colour version, please refer to the plate section.)

12.7.3 Recent Progress in Modeling

In addition to empirical models, more physics-based models have been developed (Pierrard et al., 2009, for a review). Recent progress in plasmasphere modeling was obtained by the development of three-dimensional dynamic models taking into account the influence of the geomagnetic activity. The highly dynamic region of the plasmasphere is disturbed during geomagnetic storms and substorms, with formation of a sharp plasmapause closer to the Earth and the generation of a plume in the afternoon MLT sector rotating with the Earth. The position of the plasmapause, the limit of the plasmasphere, is highly dynamic and depends on the level of geomagnetic activity. During quiet times, the plasmapause is located further from the Earth and the ionosphere refills the plasmasphere in a few days typically.

Figure 12.7.2 illustrates an example of the plasmasphere electron density as obtained from the SWIFF (Space Weather Integrated Forecasting Framework) Plasmasphere Model (SPM) (Pierrard and Voiculescu, 2011) for the date of 17 March 2015 during a geomagnetic storm. During magnetic storms, the plasmasphere is eroded and structures like plasma plumes and channels can appear. Such a plume is clearly visible on the figure. This model is available for free run at any chosen date on the space weather portal (www.spaceweather.eu) and is now provided also in real time on ESA website of SSA (Space Situational Awareness) at http://swe .ssa.esa.int/space-radiation. It provides the time variations of plasmapause position in the geomagnetic and meridian planes, the density of electrons and ions inside and outside the plasmasphere and the temperature of the particles and is coupled with the ionosphere to be able to simulate refilling processes after storms (Pierrard and Voiculescu, 2011). The density and temperature in the ionosphere are modulated by day/night activity. Different waves circulate inside and outside the plasmasphere and also along the plasmapause, so that such a model of the plasmasphere is useful also to determine its influence on the radiation belts dynamics.

The results of the simulations have been compared with various satellite observations to determine the most relevant physical mechanisms of plasmapause formation and the most appropriate electric field configurations depending on geomagnetic activity (Lemaire and Pierrard, 2008; Pierrard et al., 2008, Verbanac et al., 2018). As predicted by quasi-interchange mechanism of plasmapause formation used in SPM model, the plasmapause positions at the post-midnight observed from the meridian perspective by KAGUYA clearly show a plasmapause formation first near the equatorial region and in the post-midnight sector.

References

Akasofu, S. I. (1964), 'The development of the auroral substorm', Planet. Space Sci., 12(4), 273–82. doi: 10.1016/0032-0633(64)90151-5

Akasofu, S. I., S. Chapman and C. I. Meng (1965), 'Polar electrojet', J. Atmos. Terr. Phys., 27(11–1), 1275. doi: 10.1016/0021-9169(65)90087-5

Anderson, B. J., and D. C. Hamilton (1993), 'Electromagnetic ion cyclotron waves stimulated by modest magnetospheric compressions', J. Geophys. Res., 98(A7). doi: 10.1029/93JA00605.

Angelopoulos, V. (2008), 'The THEMIS mission', Space Sci. Rev., 141(1–4), 5–34. doi: 10.1007/s11214-008-9336-1

Angelopoulos, V., et al. (2008), 'Tail reconnection triggering substorm onset', Science, 321(5891), 931–5. doi: 10.1126/science.1160495

Angelopoulos, V., et al. (2009), 'Response to comment on "Tail reconnection triggering substorm onset"', Science, 324(5933). doi: 10.1126/science.1168045

Baker, D. N., et al. (2013), 'A long-lived relativistic electron storage ring embedded in Earth's outer Van Allen belt', Science, 6129, 186–90. doi: 10.1126/science.1233518

Baker, D. N., et al. (2014). 'An impenetrable barrier to ultrarelativistic electrons in the Van Allen radiation belts', Nat. Lett., 515, 531–4. doi: 10.1038/nature13956

Baker, D., T. Pulkkinen, V. Angelopoulos, W. Baumjohann and R. L. McPherron (1996), 'Neutral line model of substorms: Past results and present view', J. Geophys. Res., 101(A6), 12975–13010. doi: 10.1029/95JA03753

Baker, D. N., et al. (2014), 'An impenetrable barrier to ultra-relativistic electrons in the Van Allen radiation belts', Nature, 515, 531–4.

Baker, D. N. (2014). 'New twists in Earth's radiation belts', Am. Sci., 102(5), 374–381.

Baker, D.N., S. G. Kanekal, X. Li, S. P. Monk, J. Goldstein and J. L. Burch (2004), 'An extreme distortion of the Van Allen belt arising from the "Hallowe'en" solar storm in 2003', Nature, 432, 878–81. doi: 10.1038/nature03116

Baker, D. N., S. G. Kanekal, V. C. Hoxie, M. G. Henderson, X. Li, H. E. Spence, S. R. Elkington, R. H. Friedel, J. Goldstein, M. K. Hudson, G. D. Reeves, R. M. Thorne, C. A. Kletzing and S. G. Claudepierre (2013), 'A long-lived relativistic electron storage ring embedded within the Earth's outer Van Allen Radiation Zone', Science, 340, 186–90. doi: 10.1126/science.1233518

Balikhin, M. A., et al. (2015), 'Observations of discrete harmonics emerging from equatorial noise', Nat. Comm., 6, 7703. doi: 10.1038/ncomms8703

Bandic, M., G. Verbanac, M. Moldwin, V. Pierrard and G. Piredda (2016), 'MLT dependence in the relationship between plasmapause, solar wind and geomagnetic activity based on CRRES: 1990–1991', J. Geophys. Res., 121, 4397–4408. doi: 10.1002/2015JA022278

Bandic, M., G. Verbanac, V. Pierrard and J. Cho (2017), 'Evidence of MLT propagation of the plasmapause inferred from THEMIS data', J. Atmos. Sol. Terr. Phys., 161, 55–63. doi: 10.1016/j.jastp.2017.05.005

Belian, R. D., T. E. Cayton and G. D. Reeves (1995), 'Quasi-periodic, substorm associated, global flux variations observed at geosynchronous orbit', in *Space Plasmas: Coupling between Small and Medium Scale Processes*, ed. M. Ashour-Abdalla, T. Chang and P. Dusenbery, p. 143, American Geophysical Union, Washington, DC. doi: 10.1029/GM086p0143

Boardsen, S. A., et al. (2014), 'Van Allen Probe observations of periodic rising frequencies of the fast magnetosonic mode', Geophys. Res.Lett., 41, 8161–8. doi: 10.1002/2014GL062 020

Burch, J. L., and T. D. Phan (2016), 'Magnetic reconnection at the dayside magnetopause: Advances with MMS', Geophys. Res. Lett., 43, 16, 8327. doi: 10.1002/2016GL069787

Burch, J. L., et al. (2016), 'Electron-scale measurements of magnetic reconnection in space', Science, 352, aaf2939. doi: 10.1126/science.aaf2939

Caan, M. N., R. L. McPherron and C. T. Russell (1973), 'Solar wind and substorm-related changes in the lobes of the geomagnetic tail', J. Geophys. Res., 78(34), 8087–96. doi: 10.1029/JA078i034p08087

Cai, X., and C. R. Clauer (2009), 'Investigation of the period of sawtooth events', J. Geophys. Res., 114, 9. doi: 10.1029/2008ja013764

Carpenter, D. L., and J. Lemaire (2004). 'The plasmasphere boundary layer', Ann. Geophys., 22, 4291–8. doi: 10.5194/angeo-22-4291

Cassak, P. A., and M. A. Shay (2007), 'Scaling of asymmetric magnetic reconnection: General theory and collisional simulations', Phys. Plasmas, 14, 102114.

Cassak, P. A., and S. A. Fuselier (2016), 'Reconnection at Earth's dayside magnetopause', in *Magnetic Reconnection*, ed. W. Gonzalez and E. Parker, pp. 213–76, Springer, Switzerland. doi: 10.1007/978-3-319-26432-5

Chapman, S. (1962), 'Earth storms: Retrospect and prospect', J. Phys. Soc. Jpn., 17(A-I), 6.

Chapman, S., and V. C. A. Ferraro (1931), 'A new theory of magnetic storms, Part I, The initial phase', Terr. Mag. Atmos. Elect., 36, 7.

Chappell, C. R. (1972), 'Recent satellite measurements of the morphology and dynamics of the plasmasphere', Rev. Geophys., 10(4), 951–79. doi: 10.1029/RG010i004p00951

Chaston, C. C., et al. (2015), 'Extreme ionospheric ion energization and electron heating in Alfvén waves in the storm time inner magnetosphere', Geophys. Res. Lett., 42, 10531–40. doi: 10.1002/2015GL066674

Coroniti, F. V., R. L. McPherron and G. K. Parks (1968), 'Studies of magnetospheric substorm. 3. Concept of magnetospheric substorm and its relation to electron precipitation and micropulsations', J. Geophys. Res., 73(5), 1715–22. doi: 10.1029/JA073i005p01715

Daglis, I. A., and J. U. Kozyra (2002), 'Outstanding issues of ring current dynamics', J. Atmos. Sol. Terr. Phys., 64(2), 253–64. doi: 10.1016/S1364-6826(01)00087-6.

Daglis, I. A., R. M. Thorne, W. Baumjohann and S. Orsini (1999), 'The terrestrial ring current: Origin, formation, and decay', Rev. Geophys., 37(4), 407–38. doi: 10.1029/1999RG900009

Darrouzet, F., J. De Keyser and V. Pierrard (2009), *The Earth's Plasmasphere: Cluster and IMAGE – A Modern Perspective*, Springer, New York.

Darrouzet, F., et al. (2009), 'Plasmaspheric density structures and dynamics: Properties observed by the CLUSTER and IMAGE missions', Space Sci. Rev., 145(55). doi: 10.1007/s11214-008-9438-9

Darrouzet, F., V. Pierrard, S. Benck, G. Lointier, J. Cabrera, K. Borremans, N. Ganushkina, and J. De Keyser (2013), 'Links between the plasmapause and the radiation belts boundaries as observed by the instruments CIS, RAPID and WHISPER on CLUSTER', J. Geophys. Res., 118, 4176–88. doi: 10.1002/jgra.50239

De La Beaujardiere, O., L. R. Lyons, J. M. Ruohomemi, E. Friss-Christensen, C. Danielsen, F. Rich and P. Newell (1994), 'Quiet-time intensifications along the poleward auroral boundary near midnight', J. Geophys. Res., 99(A1), 287–98. doi: 10.1029/93JA01947

DeJong, A. D., A. J. Ridley, X. Cai and C. R. Clauer (2009), 'A statistical study of BRIs (SMCs), isolated substorms, and individual sawtooth injections', J. Geophys. Res., 114. doi: 10.1029/2008ja013870

DeJong, A. D., and C. R. Clauer (2005), 'Polar UVI images to study steady magnetospheric convection events: Initial results', Geophys. Res. Lett, 32(24), 4. doi: 10.1029/2005gl024498

DeJong, A. D., X. Cai, R. C. Clauer and J. F. Spann (2007), 'Aurora and open magnetic flux during isolated substorms, sawteeth, and SMC events', Ann. Geophys., 25(8), 1865–76. doi: 10.5194/angeo-25-1865-2007

Dungey, J. W. (1961), 'Interplanetary magnetic field and the auroral zones', Phys. Res. Lett., 6, 47–8.

Eastman, T. E. (2003), 'Historical review (pre-1980) of magnetospheric boundary layers and the low-latitude boundary layer', in *Earth's Low-Latitude Boundary Layer*, Geophys. Monograph 133, ed. P. T. Newell and T. Onsager, pp. 1–12, American Geophysical Union, Washington, DC.

Elvey, C. T. (1957), 'Problems of auroral morphology', Proc. Natl. Acad. Sci. USA, 43(1), 63–75.

Fairfield, D. H., and L. J. Cahill Jr (1966), 'Transition region magnetic field and polar magnetic disturbances', J. Geophys. Res., 71(1), 155–69.

Fear, R. C., et al. (2014), 'Direct observation of closed magnetic flux trapped in the high latitude magnetosphere', Science, 346, 6216. doi: 10.1126/science.1257377

Fennell, J. F., et al. (2015), 'Van Allen Probes show the inner radiation zone contains no MeV electrons: ECT/MagEIS data', Geophys. Res. Lett., 42, 1283–9. doi: 10.1002/2014GL062874

Fennell, J. F., S. G. Claudepierre, J. B. Blake, T. P. O'Brien, J. H. Clemmons, D. N. Baker, H. E. Spence and G. D. Reeves (2015), 'Van Allen Probes show that the inner radiation zone contains no MeV electrons: ECT/MagEIS data', Geophys. Res. Lett., 31(5), 1283–9. doi: 10.1002/2014GL062874

Ferrero, V. C. A. (1952), 'On the theory of the first phase of a geomagnetic storm I the new illustrative calculation based on an idealized (plane not cylindrical) model field distribution', J. Geophys. Res., 57(15), 1952.

Frank, L., K. Ackerson and R. Lepping (1976), 'On hot tenuous plasmas, fireballs, and boundary layers in the Earth's magnetotail', J. Geophys. Res., 81(34), 5859–81. doi: 10.1029/JA081i034p05859

Friedrich, E., J. Samson, I. Voronkov and G. Rostoker (2001), 'Dynamics of the substorm expansive phase', J. Geophys. Res., 106(A7), 13145–63. doi: 10.1029/2000JA000292

Fuselier, S. A., J. L. Burch, J. Mukherjee, K. J. Genestreti, S. K. Vines, R. Gomez, J. Goldstein, K. J. Trattner, S. M. Petrinec, B. Lavraud and R. J. Strangeway (2017), 'Magnetospheric ion influence at the dayside magnetopause', J. Geophys. Res., 122, 8617–31. doi: 10.1002/2017JA02415

Gallagher, D. L., and R. H. Comfort (2016), 'Unsolved problems in plasmasphere refilling', J. Geophys. Res., 121, 1447–51. doi: 10.1002/2015JA022279

Gkioulidou, M., A. Y. Ukhorskiy, D. G. Mitchell, T. Sotirelis, B. H. Mauk and L. J. Lanzerotti (2014), 'The role of small-scale ion injections in the buildup of Earth's ring current pressure: Van Allen Probes observations of the 17 March 2013 storm', J. Geophys. Res., 119, 7327–42. doi: 10.1002/2014JA020096

Goldstein, J., B. R. Sandel, W. T. Forrester, M. F. Thomsen and M. R. Hairston (2005), 'Global plasmasphere evolution 22–23 April 2001', J. Geophys. Res., 110, A12218. doi: 10.1029/2005JA011282

Haerendel, G., G. Paschmann, N. Sckopke, H. Rosenbauer and P. C. Hedgecock (1978), 'The frontside boundary layer of the magnetosphere and the problem of reconnection', J. Geophys. Res., 83, 3195.

Hirshberg, J., and D. S. Colburn (1969), 'Interplanetary field and geomagnetic variations-A unified view', Planet. Space Sci., 17, 1183–1206.

Hubert, B., J. C. Gerard, S. E. Milan and S. W. H. Cowley (2017), 'Magnetic reconnection during steady magnetospheric convection and other magnetospheric modes', Ann. Geophys., 35 (3), 505–24. doi: 10.5194/angeo-35-505-2017

Hultqvist, B., M. Øieroset, G. Paschmann and R. Treumann (Eds.) (1999), *Magnetospheric Plasma Source and Losses*, Kluwer, Dordrecht.

Jacobs, J. A. (Ed.) (1987), *Geomagnetism*. 2 vols. Academic Press, New York.

Jelly, D., and N. Brice (1967), 'Changes In Van Allen radiation associated with polar substorms', J. Geophys. Res., 72(23), 5919–31.

Keika, K., L. M. Kistler and P. C. Brandt (2013), 'Energization of O+ ions in the Earth's inner magnetosphere and the effects on ring current buildup: A review of previous observations and possible mechanisms', J. Geophys. Res., 118, 4441–64. doi: 10.1002/jgra.50371

Kim, H.-J., and A. A. Chan (1997), 'Fully adiabatic changes in storm time relativistic electron fluxes', J. Geophys. Res., 102 (A10), 22107–16. doi: 10.1029/97JA01814

Kissinger, J., F. D. Wilder, R. L. McPherron, T. S. Hsu, J. B. H. Baker and L. Kepko (2013), 'Statistical occurrence and dynamics of the Harang discontinuity during steady magnetospheric convection', J. Geophys. Res., 118(8), 5127–35. doi: 10.1002/jgra.50503

Kissinger, J., R. L. McPherron, T. S. Hsu and V. Angelopoulos (2011b), 'Steady magnetospheric convection and stream interfaces: Relationship over a solar cycle', J. Geophys. Res., 116. doi: 10.1029/2010ja015763

Kissinger, J., R. L. McPherron, T. S. Hsu and V. Angelopoulos (2012b), 'Diversion of plasma due to high pressure in the inner magnetosphere during steady magnetospheric convection', J. Geophys. Res., 117(A5), A05206. doi: 10.1029/2012ja017579

Kissinger, J., R. L. McPherron, T.-S. Hsu and V. Angelopoulos (2011a), 'Steady magnetospheric convection and stream interfaces: Relationship over a solar cycle', J. Geophys. Res., 116 (13), 1–11. doi: 10.1029/2010JA015763

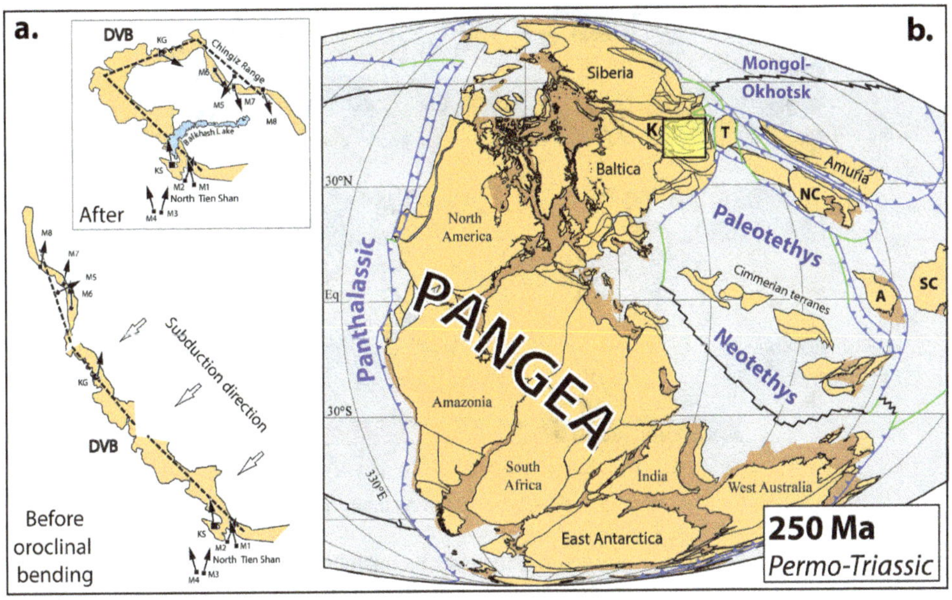

Figure 2.2.2 (a) The relationship between primary Silurian and Devonian declinations (solid arrows) and structural trends in the so-called Devonian Volcanic Belt (DVB) of Kazakhstan in an inferred Middle Devonian configuration before oroclinal bending started and a Permo-Triassic configuration after oroclinal bending (see Abrajevitch et al., 2007). (b) Pangea geography at around 250 Ma, including the opening Neotethys Ocean, the still large but diminishing Paleotethys Ocean, Mongol-Okhotsk and parts of the Panthalassic Ocean. Cimmerian includes terranes from Alborz to Sibumasu, which were parts of the Gondwanan margin in the Early Permian. Solid blue lines are subduction zones with teeth on the upper plate, black lines are spreading centers, and green lines are transform plate margins (Domeier and Torsvik, 2014). Paleomagnetic reconstruction with longitude calibrated with the plume generation zone method (Torsvik et al., 2014a). A, Annamia; K, Kazakhstan (black box approximately the box in panel a, top); NC, North China; SC, South China; T, Tarim.

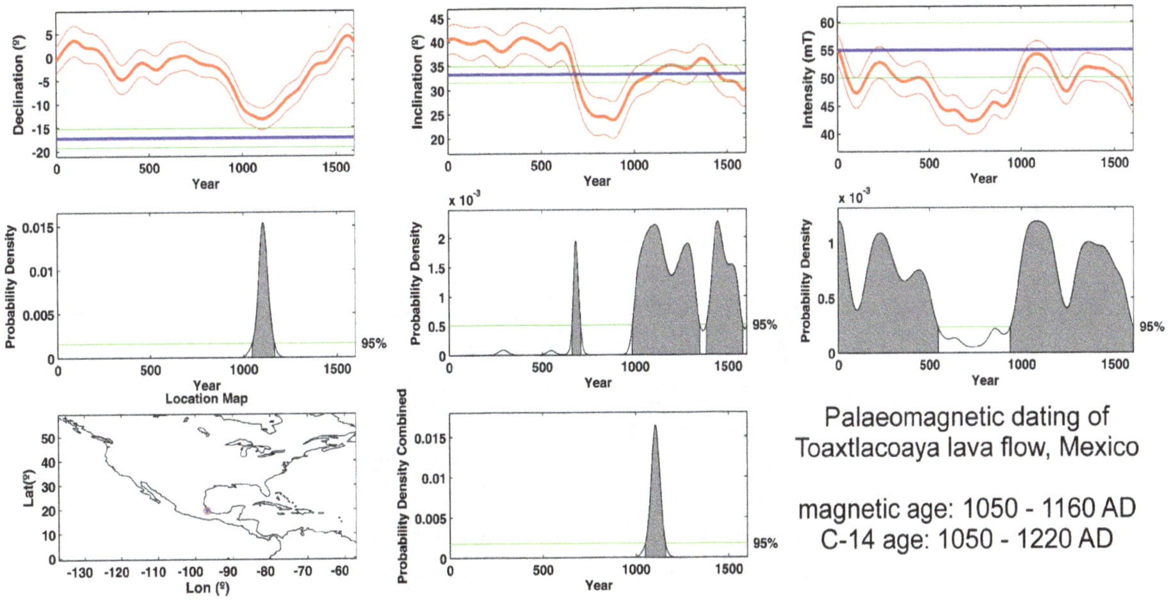

Figure 2.3.1 Paleomagnetic dating of a Holocene lava flow using the archaeo_dating Matlab tool (Pavón-Carrasco et al., 2011). Thick red curves are SV model curves with their 95% uncertainty shown as thin red lines; blue lines are measured paleomagnetic declination, inclination and paleointensity with their 95% uncertainties. Probability density curves indicate the possible ages, with 95% confidence limits. Note how the different field elements provide different possible age ranges but only one combined probability density. Also note the good coincidence between paleomagnetic and C-14 ages.

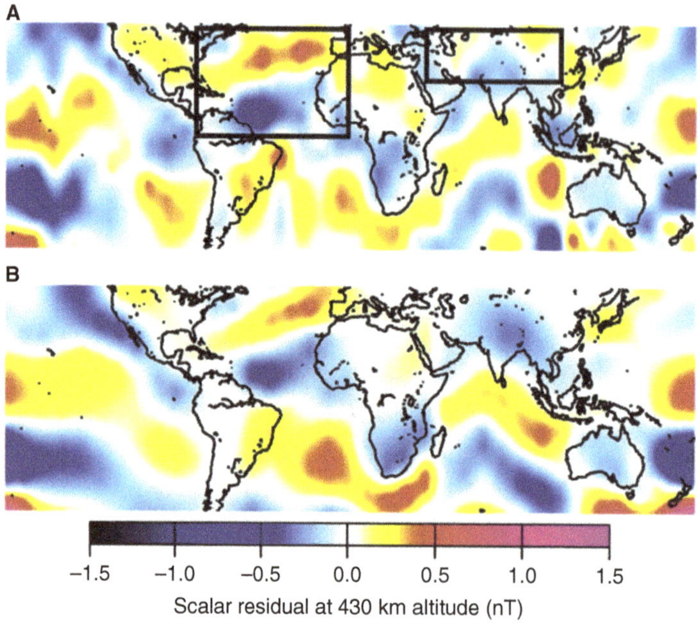

Figure 2.4.1 Earth's magnetic field in June 2014 as observed by the Swarm constellation. Image from ESA/DTU Space, 'Swarm reveals Earth's changing magnetism', ESA, 19 June 2014, www.esa.int/Our_Activities/Observing_the_Earth/Swarm/Swarm_reveals_Earth_s_changing_magnetism.

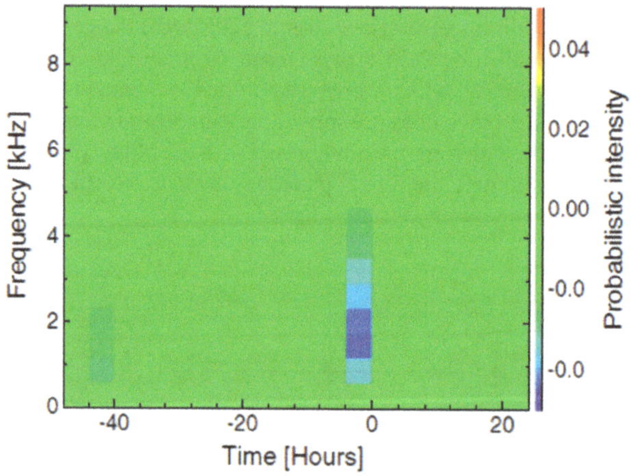

Figure 2.4.4 Frequency-time dependence of the probabilistic intensity obtained from the night-time electric field data measured within 440 km of the epicenters of earthquakes with magnitudes >5.0 during DEMETER mission.After Pisa et al. (2013).

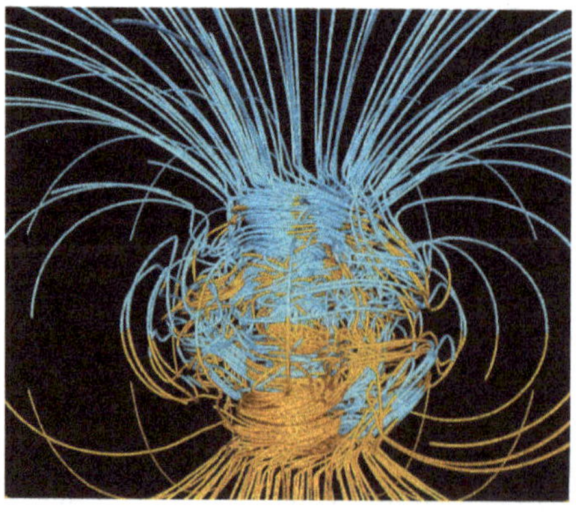

Figure 4.0.1 Computer simulation of the Earth's magnetic field in a period of normal polarity between magnetic reversals. The lines represent magnetic field lines: blue when the field points toward the centre of the Earth and yellow when it points away. The rotation axis of the Earth is centred and vertical in the diagram.From Glatzmaier and Roberts (1995).

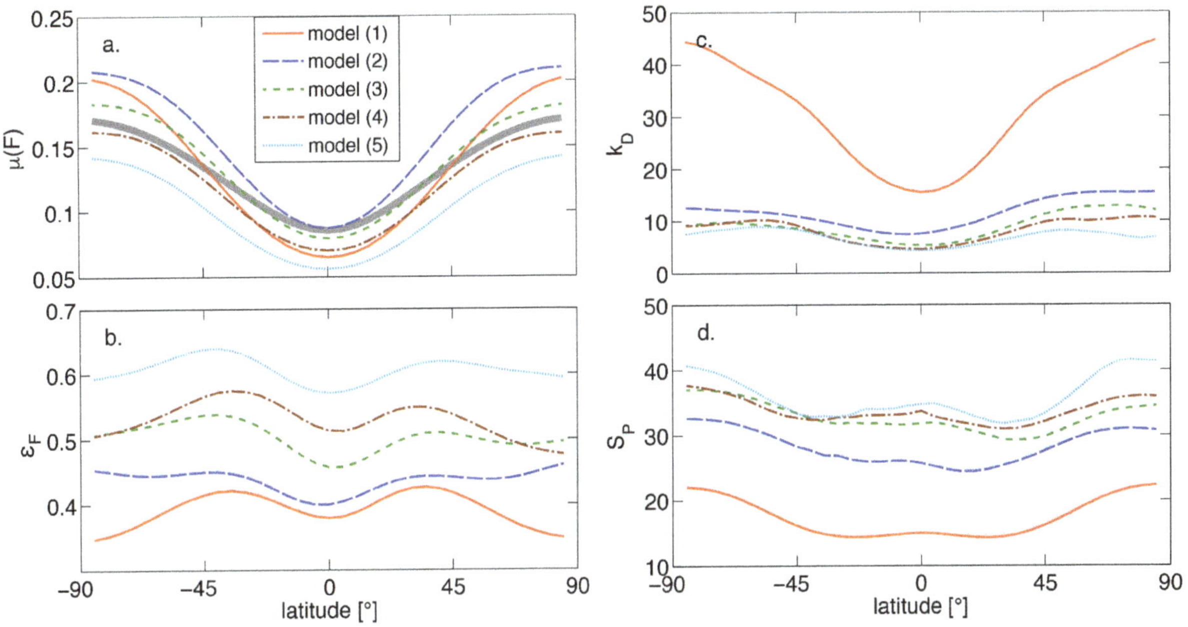

Figure 5.2.1 Latitudinal variation of (a) the average intensity μ_F, (b) the relative variability in intensity ε_F, (c) the precision parameter of directions k_D and (d) the angular dispersion of poles S_P for the five investigated models. The thick gray curve in panel (a) is the dipole formula fit to the mean of the five models (after Lhuillier and Gilder, 2013).

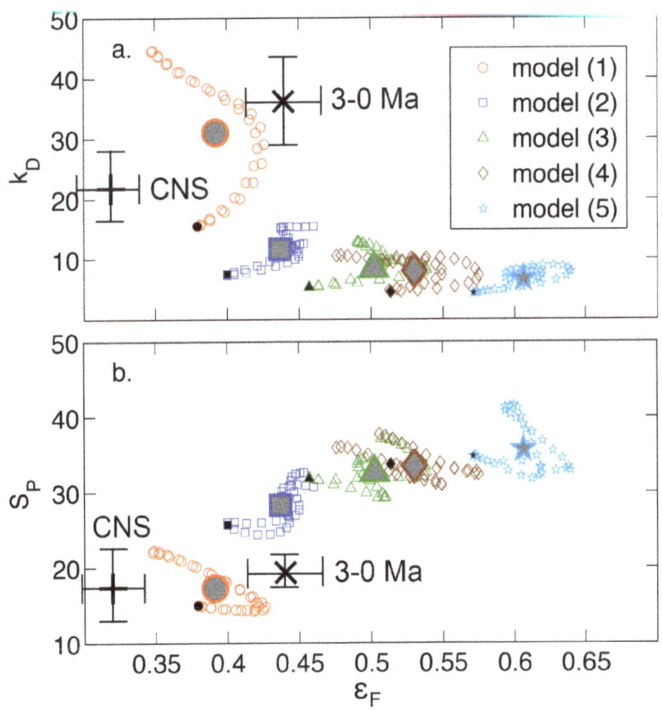

Figure 5.2.2 The precision parameter for directions (k_D) and the angular dispersion of poles (S_P) as a function of the relative variability in intensity (ε_F) for each of the 35 latitudes for each of the five dynamo models. Equatorial values are in black; large symbols represent the overall averages (after Lhuillier and Gilder, 2013). Model 1 has the lowest reversal rate (2/Myr) and 5 the highest (13/Myr). These models predict that dynamos with lower reversal frequencies will be more stable with slower, less variable secular variation rates than dynamos with higher reversal frequencies. The plot includes experimental results from Tuoyun (Xinjiang, China) for the Cretaceous normal superchron (CNS) and from Boring (Oregon, USA) for the last 3 Ma (3–0 Ma). Uncertainties on ε_F (1σ) were derived from bootstrapping the data; 95% confidence limits for k_D and S_P were calculated according to Cox (1969).

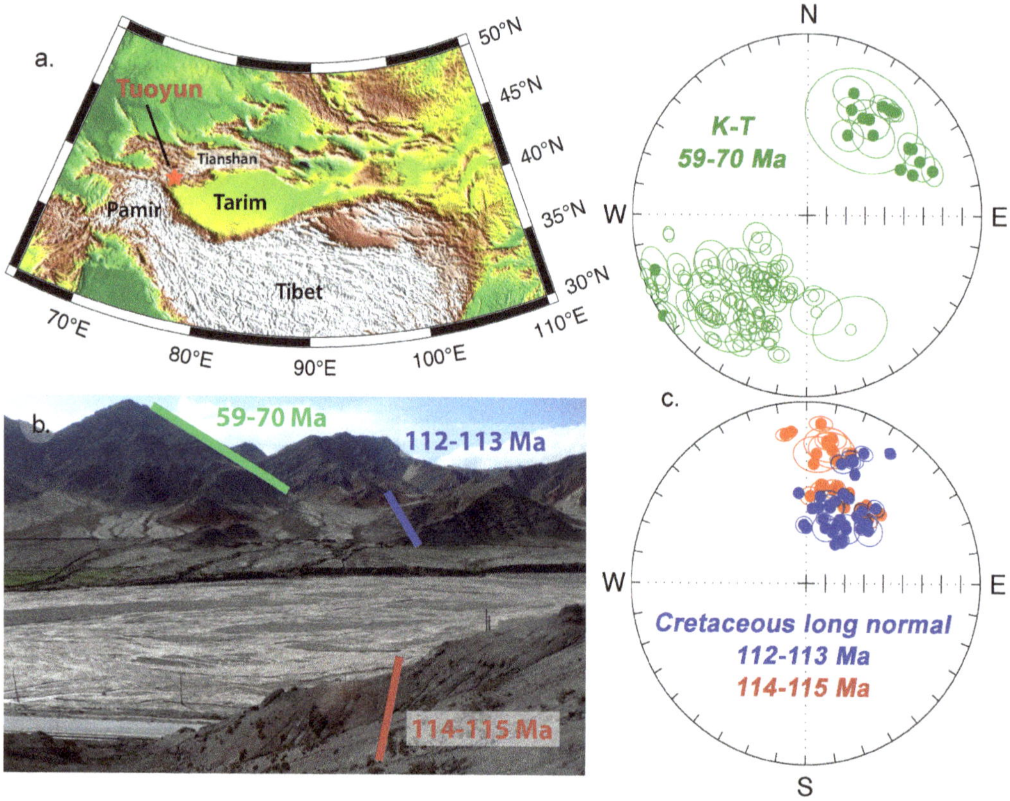

Figure 5.3.1 (a) Topographic locality map and (b) photographic overview of the two Cretaceous long normal (112–115) sections and the Cretaceous–Tertiary (K-T) boundary section investigated by Lhuillier et al. (2016). Stereonet plots on the right are the 75 and 56 site mean directions from the K-T and superchron sections, respectively (two sites were omitted from the original data set).

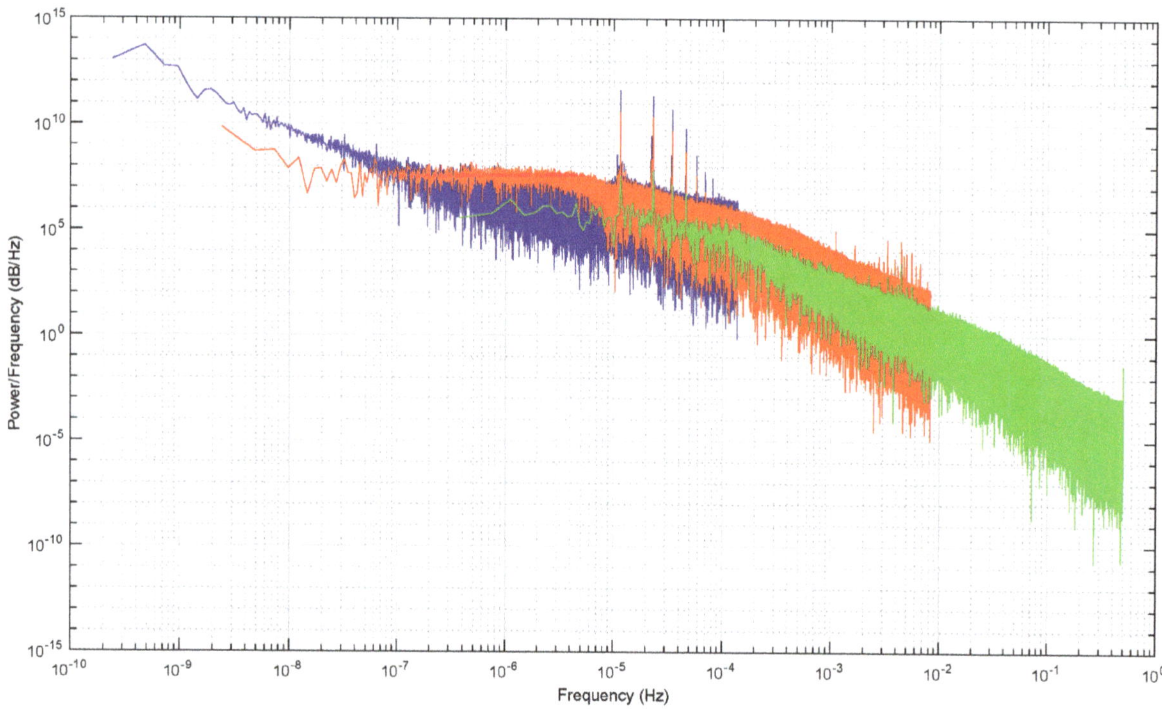

Figure 6.1.1 Periodogram of data recorded at the Chambon-la-Forêt magnetic observatory between 1883 and 2014 (hourly data, in blue), 1996 and 2008 (minute data, in red) and in January 2014 (1 s data, in green). The power generally decreases with frequency, as longer periods such as the secular variation have larger amplitudes than smaller periods. Peaks at the 24 hour period (about 1.16×10^{-5} Hz) and harmonics are caused by the ionospheric Sq field. Peaks at higher frequencies are caused by geomagnetic pulsations. Minute data cover only one solar cycle (cycle 23), while hourly data cover 12 solar cycles, hence the level differences below 10^{-5} Hz.

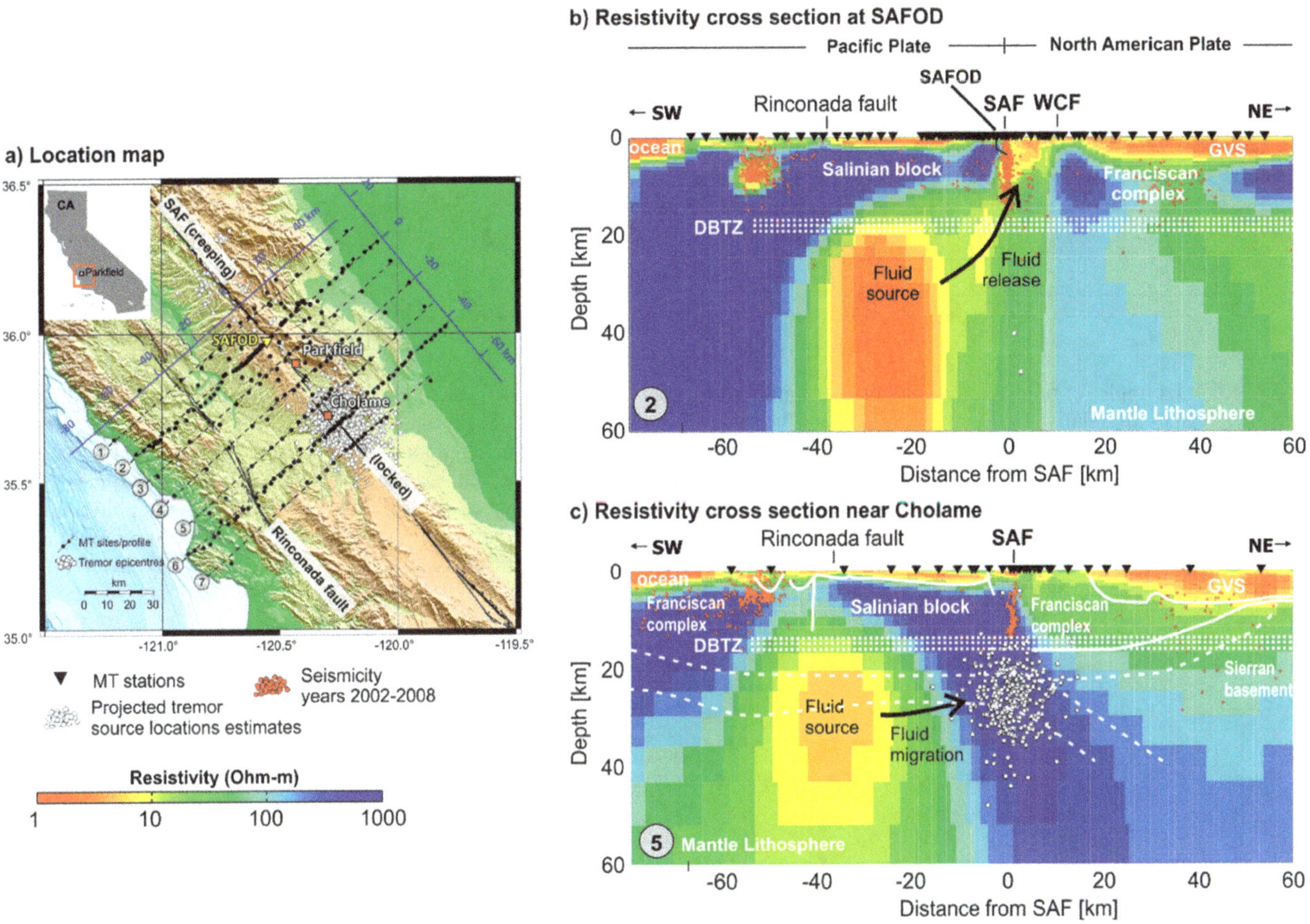

Figure 6.2.1 Map view. Black dots mark MT stations along seven ~130 km long profiles. Black lines indicate the San Andreas and Rinconada faults. The northwestern part of the SAF is described as creeping, while the southeastern section is locked. Regional-scale resistivity models across the SAF (a) at SAFOD near Parkfield and (b) at the source region of tremor near Cholame. WCF = Waltham canyon fault. BDTZ = brittle-ductile transition zone. Superimposed are boundaries of geological units (white lines). Modified after Becken and Ritter (2012).

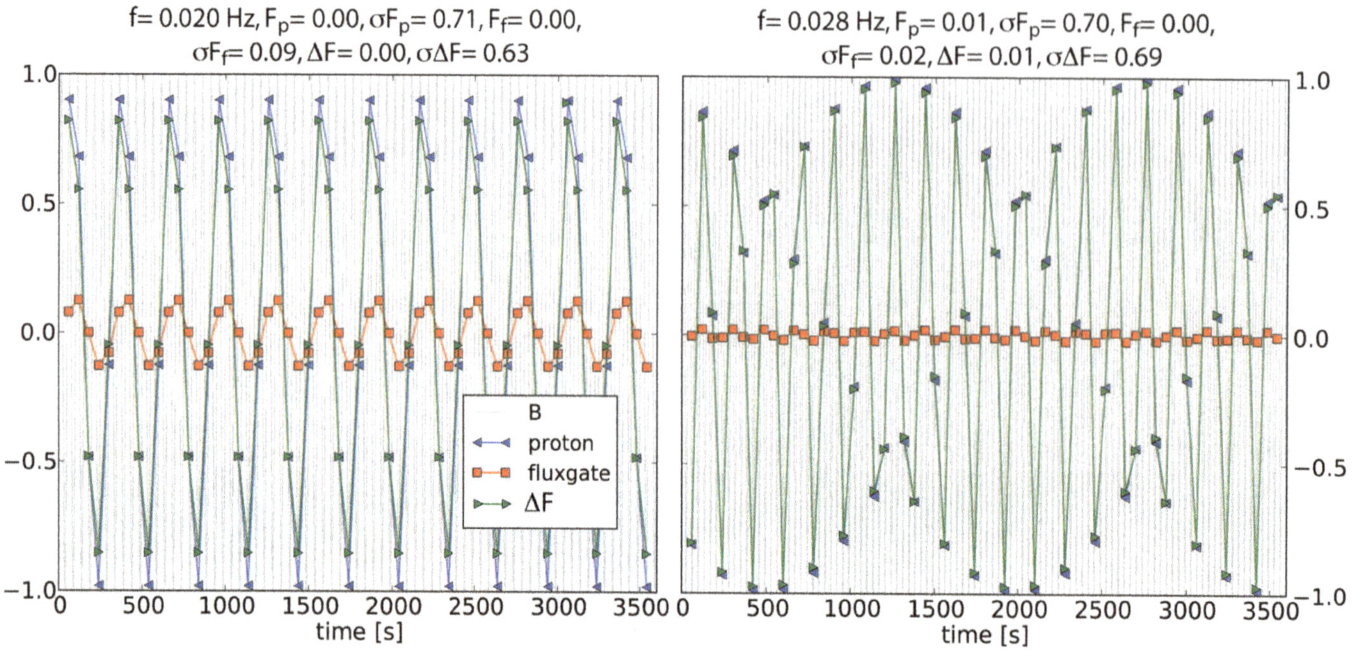

Figure 6.3.4 Examples of numerical simulations for input frequencies of 0.020 Hz (left) and 0.028 Hz (right) over a 1 hour period. All numerical values are relative to the amplitude (±1) of the artificial input magnetic field (B, light gray), which is a sinusoidal waveform consisting of 1000 data points per second over a 1 hour duration. One minute instrument readings are calculated based on the Intermagnet protocol for non-filtered, non-Overhauser effect proton (blue triangles) and Gaussian-filtered fluxgate (red squares) magnetometers. Observed measurement oscillations are due to aliasing effects and are smaller for the fluxgate readings due to the Gaussian filter. Green triangles show the scalar differences of the measurements for each minute (ΔF). Listed above each plot are the hourly mean values of F_p, F_f and ΔF as well as their standard deviations (σF_p, σF_f, $\sigma \Delta F$) (subscripts p = proton, f = fluxgate). This protocol was performed for 10^5 discrete frequencies to produce Figure 6.3.5.

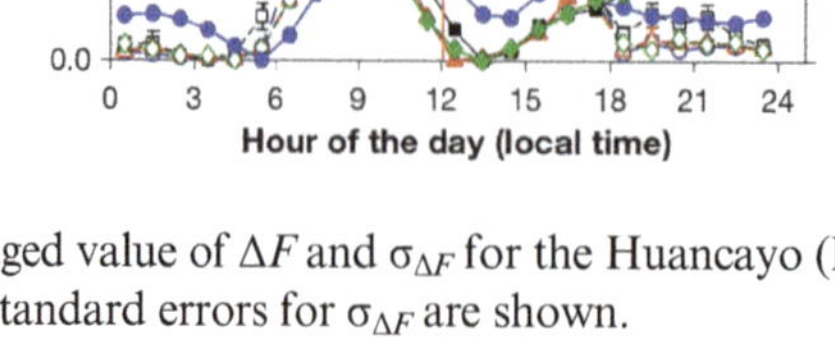

Figure 6.3.8 Daily variation of the hourly averaged value of ΔF and $\sigma_{\Delta F}$ for the Huancayo (HUA, Peru) observatory from 2002 to 2005. The data are scaled so that the minimum value = 0; standard errors for $\sigma_{\Delta F}$ are shown.

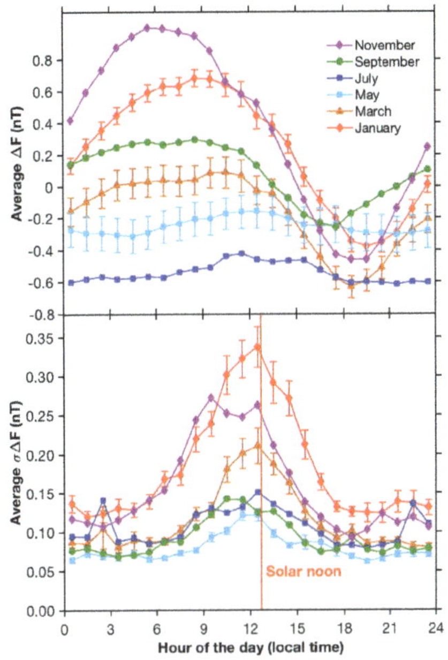

Figure 6.3.9 Daily variation of the hourly averaged value of (a) ΔF and (b) $\sigma_{\Delta F}$ for selected months at the Dumont d'Urville (DRV, Antarctica) observatory in 2004. Standard errors for select months are shown.

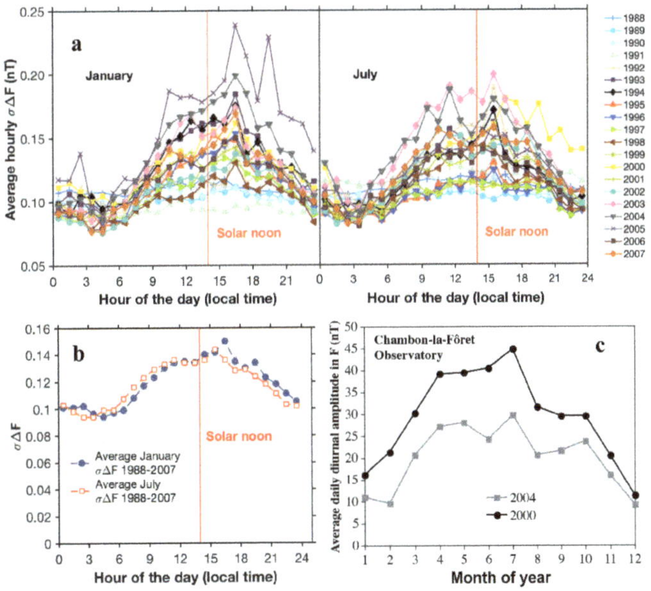

Figure 6.3.10 (a) Daily variation of the hourly averaged value of $\sigma_{\Delta F}$ for January and July for the years 1988 to 2007 for the Chambon-la-Forêt (CLF, France) observatory (January 1999 data not included). (b) Average daily variation of the hourly averaged value of $\sigma_{\Delta F}$ for January and July for the years 1988 to 2007 for the Chambon-la-Forêt (CLF, France) observatory. (c) Average daily diurnal amplitude in total field (F) per month in 2000 and 2004 from the Chambon-la-Forêt (CLF, France) observatory.

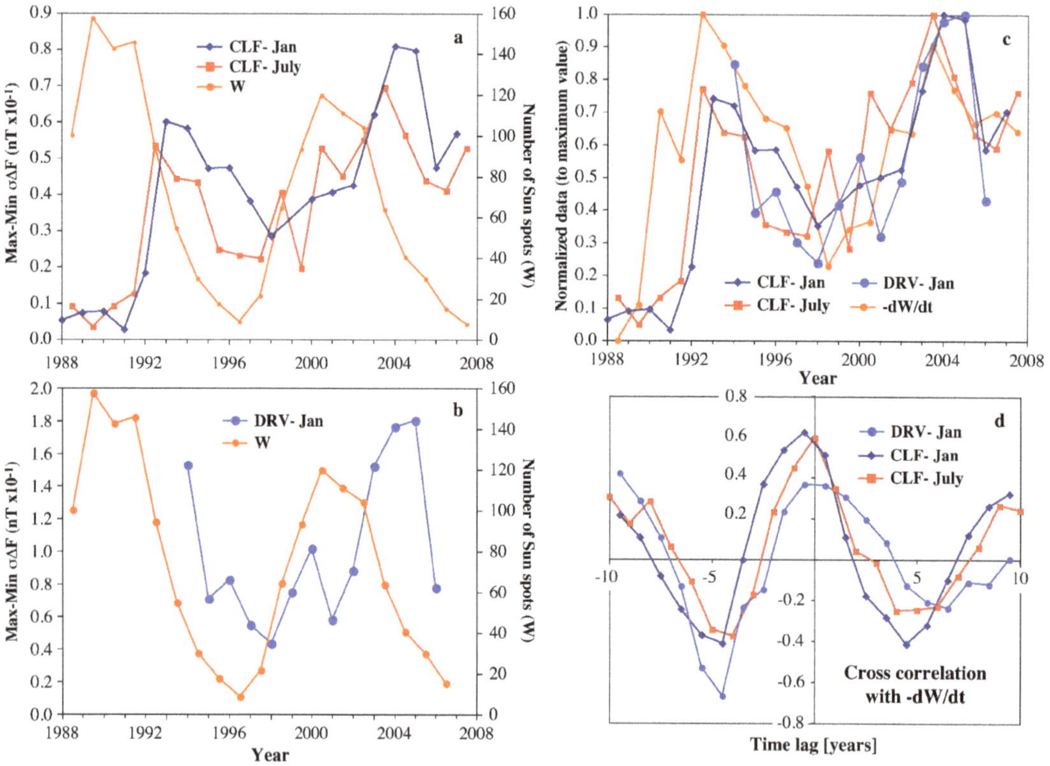

Figure 6.3.11 Difference between the average maximum and minimum variation of $\sigma_{\Delta F}$ compared against sunspot (Wolf, W) number. (a) January and July from 1988 to 2007 for the Chambon-la-Forêt (CLF) observatory. (b) January from 1994 to 2006 for the Dumont d'Urville (DRV) observatory. (c) Same observatory data as in (a) and (b), normalized by the maximum value, and plotted with the negative rate of change of the yearly averaged W ($-dW/dt$), also normalized to the maximum value. (d) Cross-correlation functions of the average maximum-minimum variation of $\sigma_{\Delta F}$ with the negative rate of change of the sunspot (Wolf, W) number. Cross-correlation functions were calculated with Python's xcorr algorithm using time series shifted to zero mean and normalized to a maximum value of one. That the cross correlations maximize for all three time series around zero time lag demonstrates that the observed signals are in phase with the negative rate of change of the sunspot number.

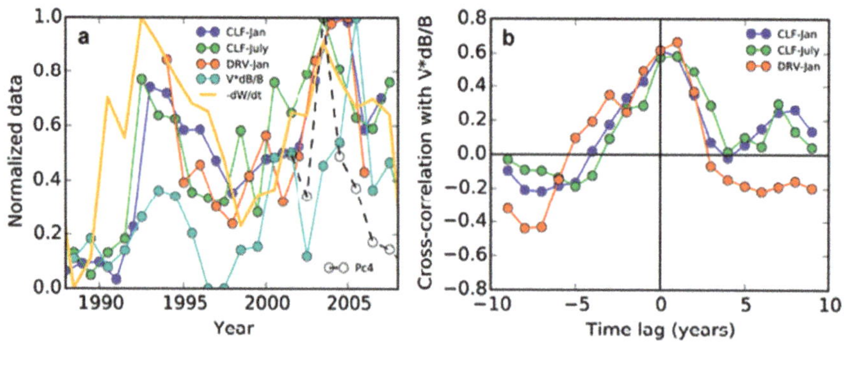

Figure 6.3.12 (a) Normalized proxy of magnetic field fluctuations at the ground ($\sigma_{\Delta F}$, Figure 6.3.11) compared to the turbulence penetration parameter $V_{ay}{\bullet}\sigma_B/B$ (light blue, 1 min data from the OMNI database: http://omniweb.gsfc.nasa.gov) and wave power of the Pc4 magnetic field fluctuations in the inner magnetosphere (black dashed line, from the MAARBLE project, www.maarble.eu/, with data obtainable at the Cluster Science Archive, www.cosmos.esa.int/web/csa). Scaling uses the normalization $X' = \frac{X - X_{\min}}{X_{max} - X_{\min}}$. (b) Cross-correlation functions between $\sigma_{\Delta F}$ and the turbulence penetration parameter $V{\bullet}\sigma_B/B$. Cross-correlation is done to a time series shifted by subtracting the mean values of the variables and normalizing to a maximum value of one.

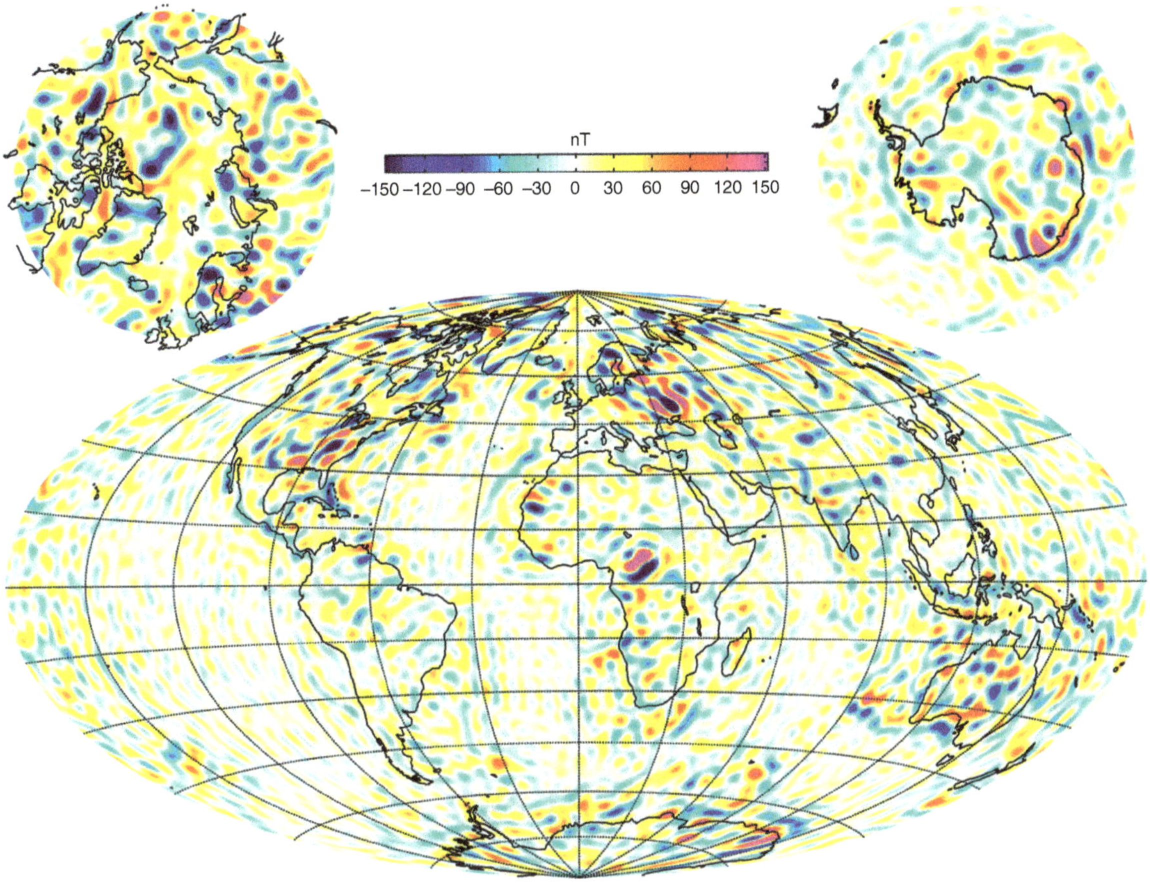

Figure 7.1.1 The vertical component of the lithospheric field at Earth's surface as it derives from MF7 (Maus et al., 2008). Lithospheric anomalies range from ±150 nT. Most prominent are the Bangui anomaly in central Africa, the Kursk anomaly in Russia and the Appalachian anomaly in North America.

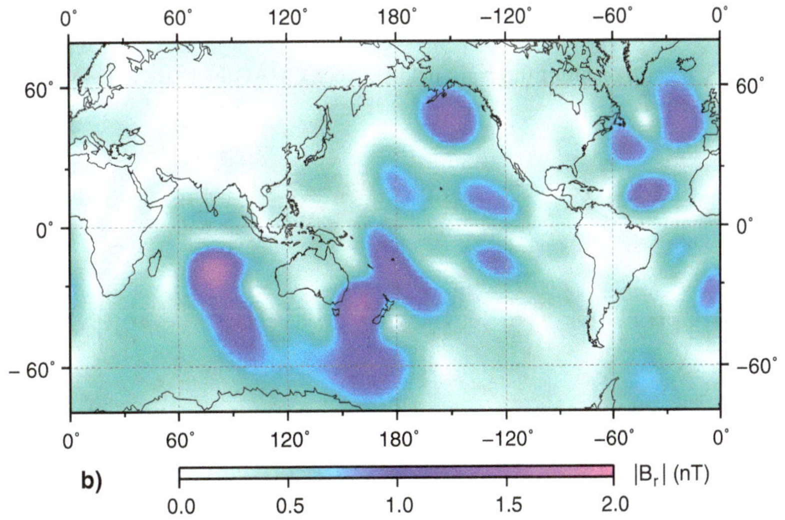

Figure 7.1.2 The amplitude of the vertical tidal magnetic field related to the oceanic M2-tide as it is retrieved from geomagnetic data of the CHAMP mission during the period 2004 to 2009 (Wardinski and Lesur, 2016).

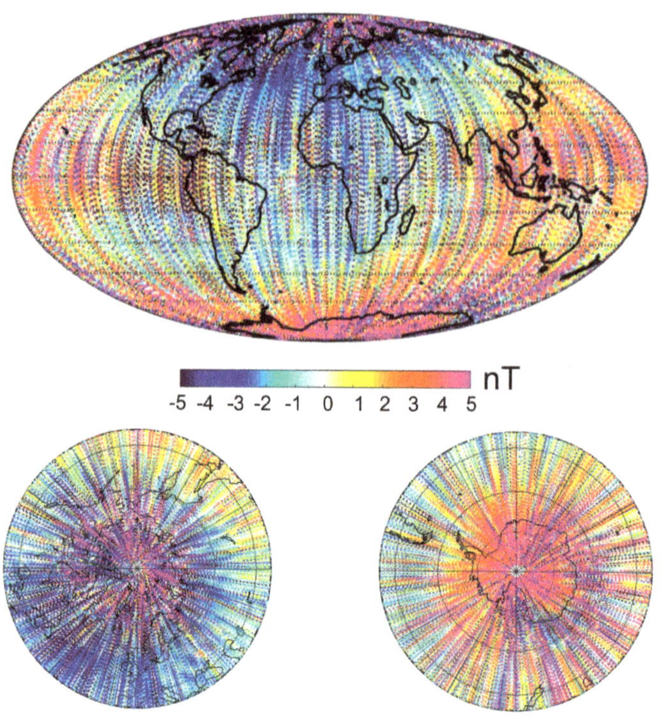

Figure 7.2.1 Residuals of the vertical down component of Swarm A and C measurements. Residuals are computed at satellite altitude. Signatures of the ionospheric field of ±4 nT are faintly visible along the equator in a 20° wide belt.

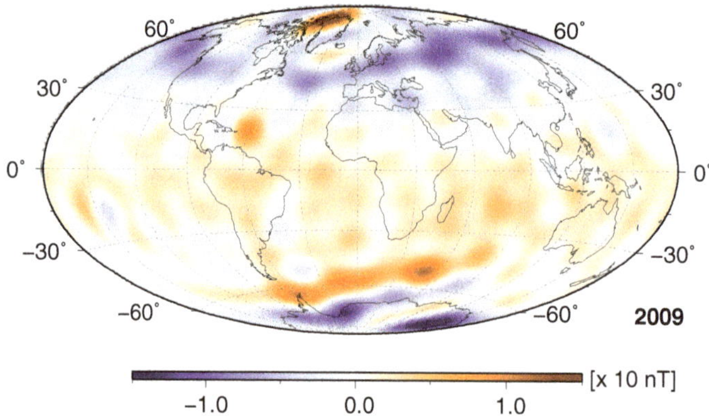

Figure 7.5.1 Differences of the vertical component of the main field between CHAOS-6 and GRIMM-3 at Earth's surface for the epoch 2009. The range of the differences is ±16nT.

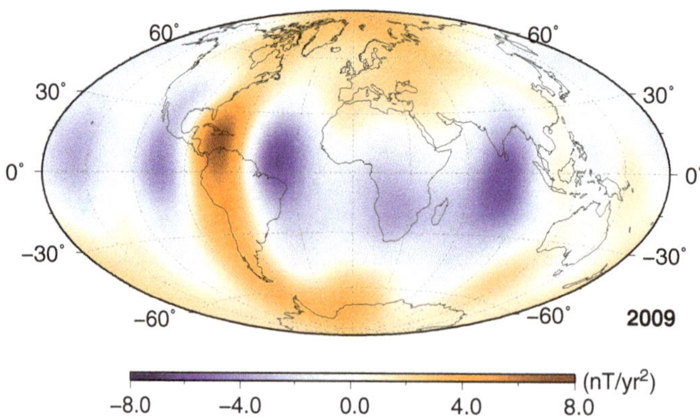

Figure 7.5.2 Differences of the vertical component of the secular acceleration between CHAOS-6 and GRIMM-3 at Earth's surface for the epoch 2009.

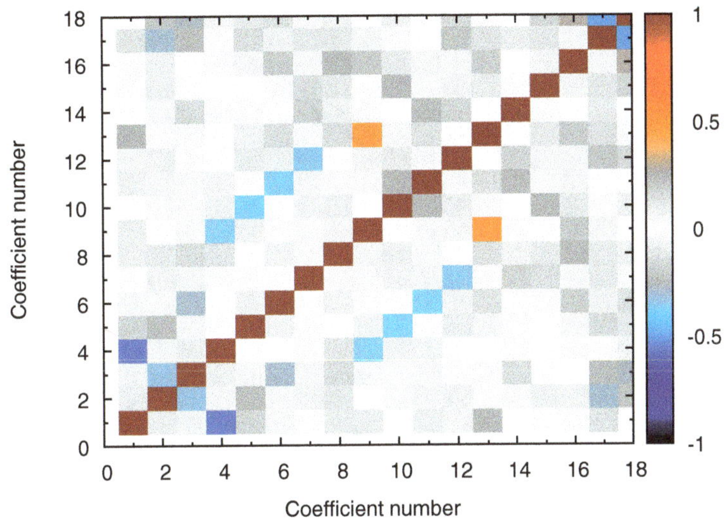

Figure 7.5.6 Covariance matrix of an inversion of MESSENGER magnetic field data. The symmetric plot is organised by coefficient numbers, where the first three coefficients are the Gauss coefficients of the first internal spherical degree, g_2^0 is coefficient 4, g_3^0 is coefficient 9 and the external magnetic field coefficients are coefficients 16 to 18, respectively.

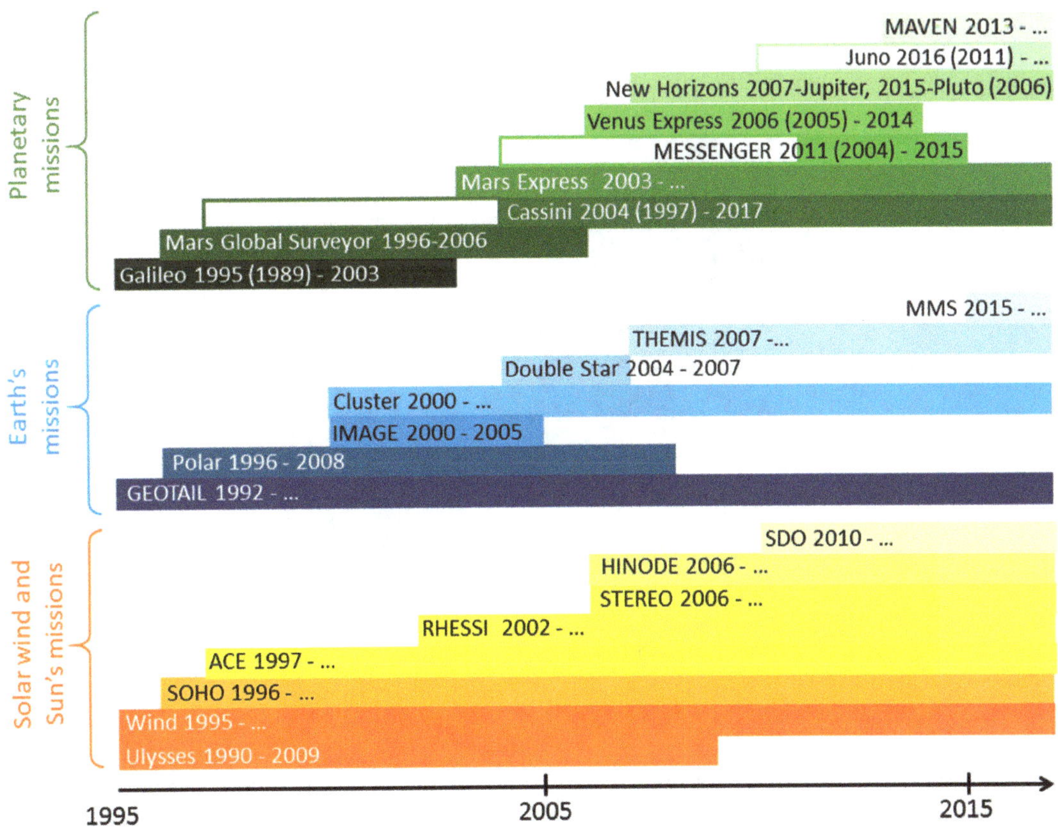

Figure 8.2.1 Timeline of the past 20 years of missions dedicated to the study of the Sun and the solar wind (orange/yellow), the terrestrial space environment (blue) and the other planets of the solar system for which at least one plasma or field experiment is available on board (green). The start and end (if relevant) dates of operation are indicated for each mission, as well as the launch dates in brackets for cruising missions.

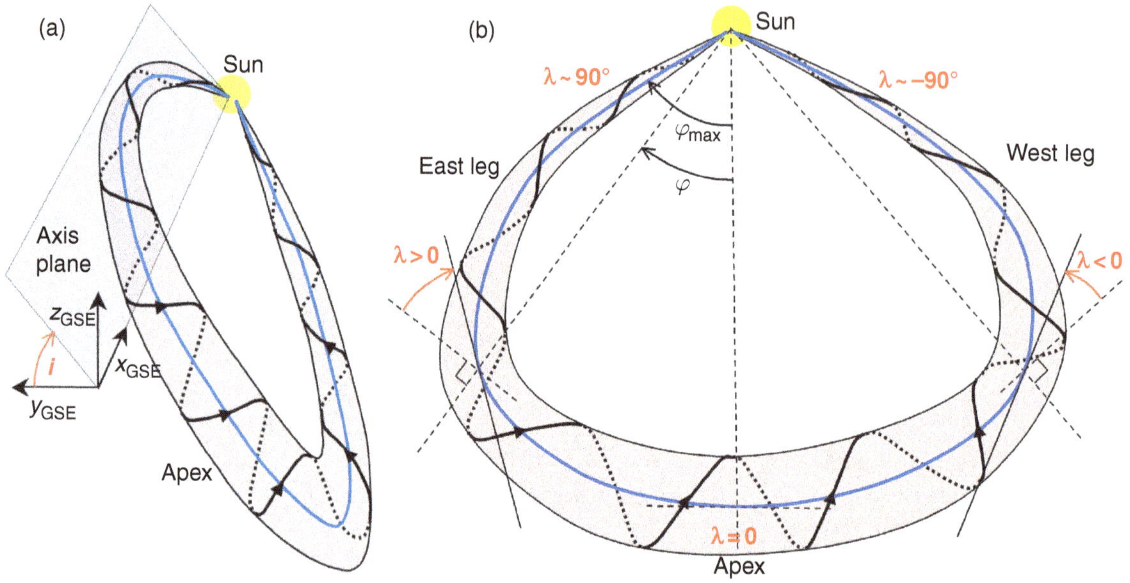

Figure 8.3.1 Schematics of a large-scale planar MC (pink shape). (a) Schematic in 3-D showing a MC with a certain inclination i with respect to the ecliptic plane (x_{GSE},y_{GSE}). (b) Schematic in 2-D in the plane of the MC, with the angle ϕ being the heliospheric longitude and λ being a characteristic angle evolving monotonously along the MC axis and being 0° at the apex of the MC. The blue line represents the MC axis; the black twisted line represents the magnetic field line at the boundary of the MC.Adapted from Figure 1 of Janvier et al. (2013).

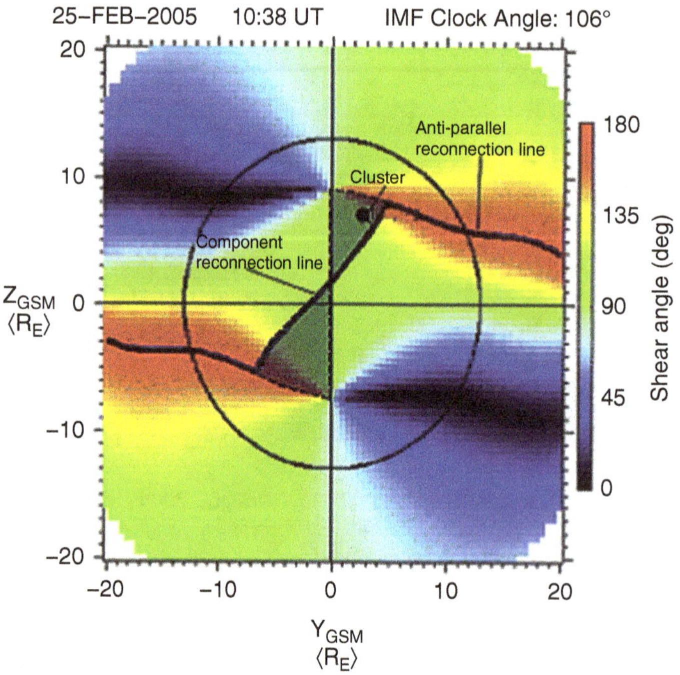

Figure 8.4.1 A 2-D map of shear angles between magnetospheric and magnetosheath magnetic fields at the magnetopause. Red is near 180° (anti-parallel fields) and dark purple is near 0°. The shear angles are projected onto the Y-Z GSM plane, and the view is from the Sun. The circle is the terminator projected onto the plane. The maximum shear model predicts that anti-parallel reconnection occurs on the flanks and that there is a component reconnection line that crosses the dayside and connects the two anti-parallel reconnection regions on the flanks. Spacecraft crossing the magnetopause within the grey shaded regions (such as Cluster in this figure) would see different flow directions for reconnection jets from an anti-parallel reconnection line at higher latitudes or a component reconnection line at lower latitudes. Adapted from Figure 1 of Fuselier et al. (2011).

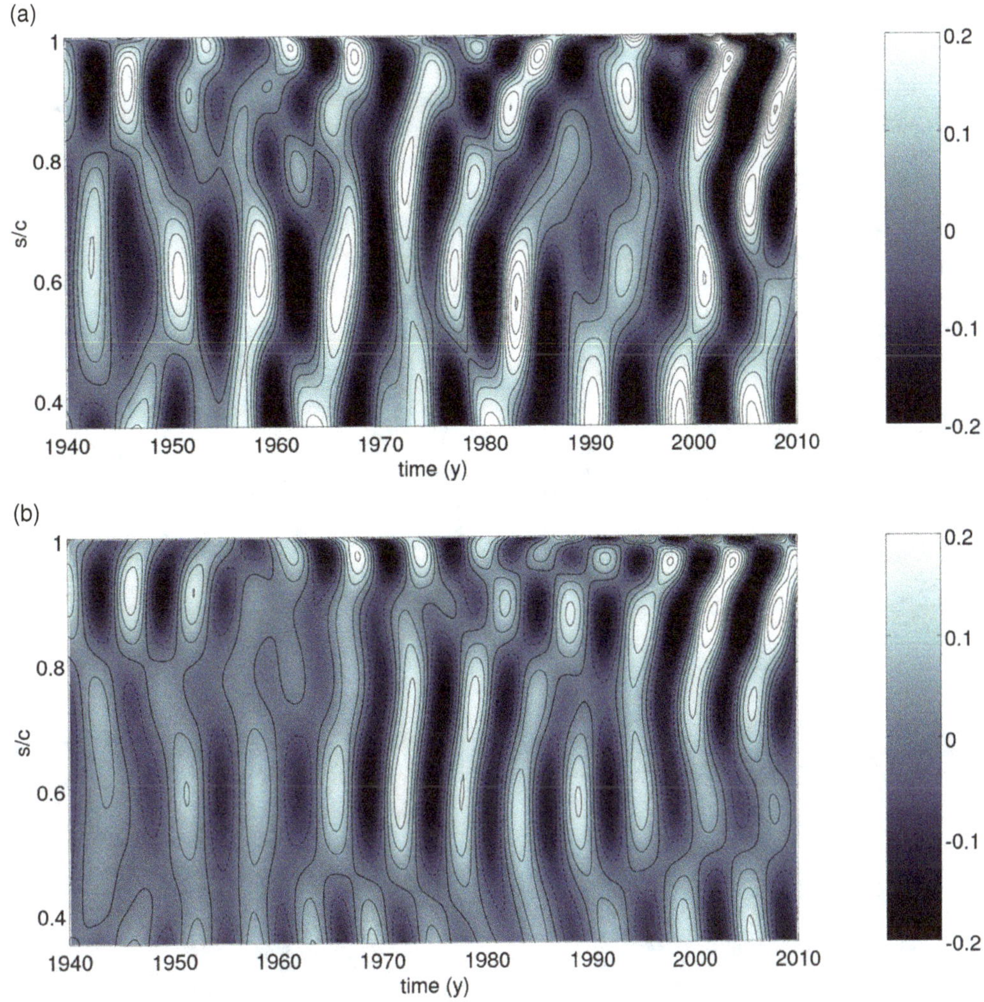

Figure 9.3.1 Geostrophic velocity as a function of time and cylindrical radius (in km/yr), between the inner core radius ($s/c = 0.35$) and the core equator ($s/c = 1$). From kinematic QG inversion (Gillet et al., 2015b) band-pass filtered for $T \in [4, 9.5]$ years (top) and $T \in [5, 8]$ years (bottom).

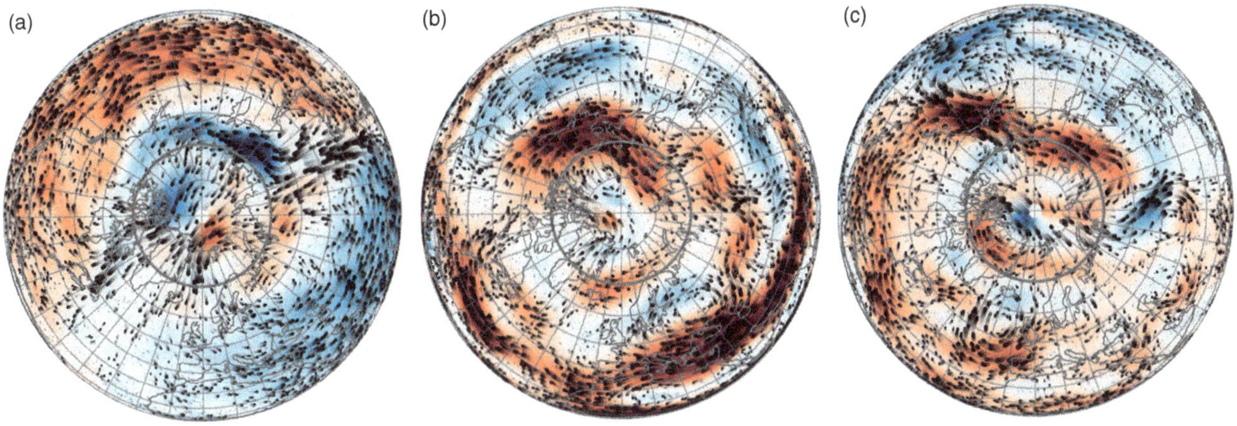

Figure 9.3.2 Equatorial cross sections of core motions, viewed from the north pole: passive tracer trajectories (black) superimposed with u_ϕ (colour scale within ± 30 km/yr, red westward). (top) Geodynamo simulation with isotropic forcing ($E = 10^{-6}$, $P_m = 0.2$; Schaeffer et al., 2017), low-pass filtered for $T \geq 1000$ years. (middle) Kinematic QG inversion in 2008 (Gillet et al., 2015b). (bottom) Geodynamo simulation with non-isotropic forcing favouring a westward drift in the Atlantic hemisphere ($E = 1.2\ 10^{-5}$, $P_m = 2.5$; Aubert et al., 2013).

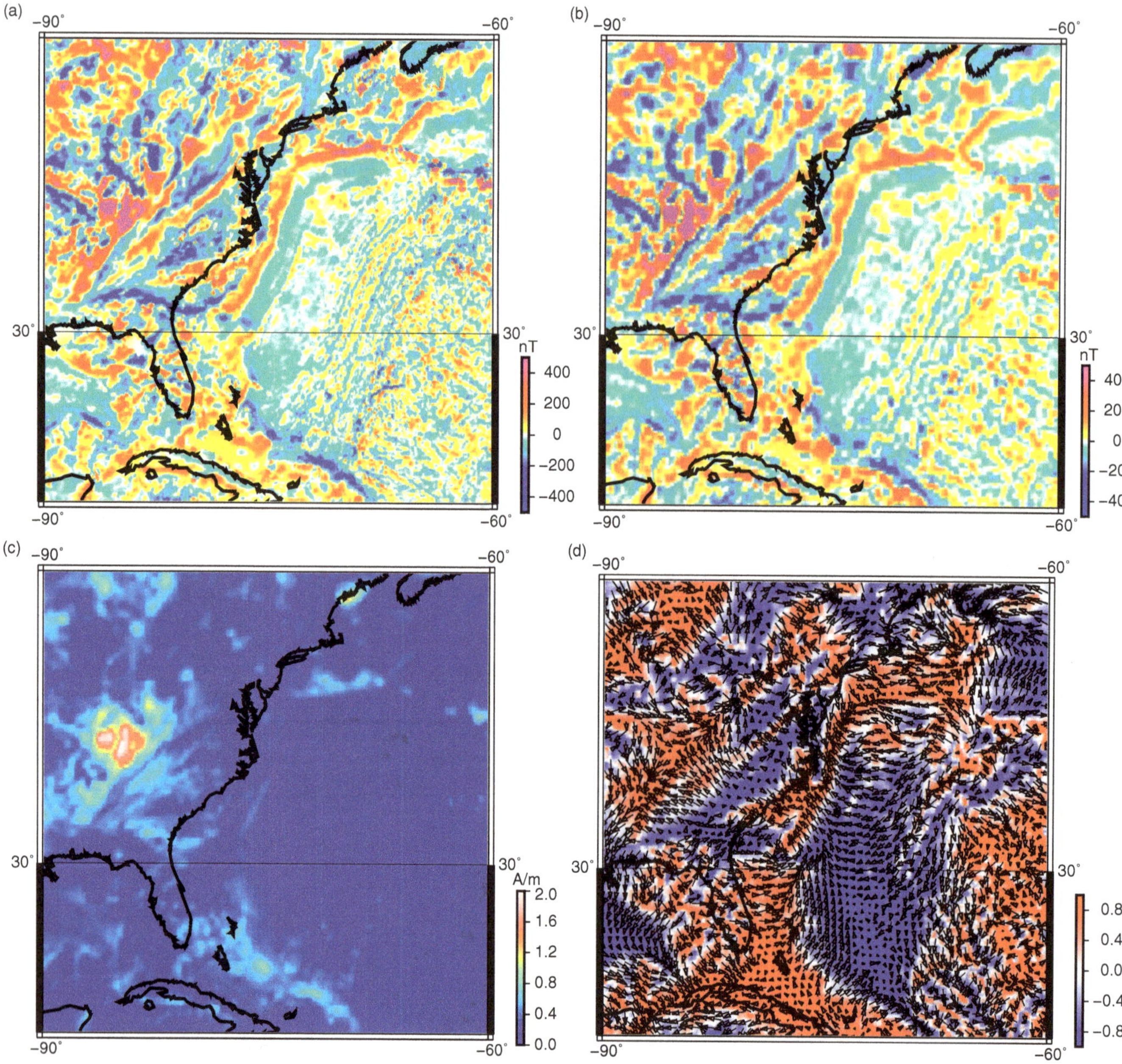

Figure 10.4.2 (top left) Total intensity anomaly data from the WDMAM, mainly over the western Atlantic and continental eastern North America. Marine data are at sea level altitude, whereas over continents, data are at 5 km altitude above the WGS84. Some data gaps are present along the US shoreline north of the Gulf of Mexico and also over parts of Cuba. (top right) Vertical down-component of the lithospheric magnetic field model wdmam-mod, truncated at spherical harmonic 720 and plotted at the Earth reference radius 6371.2 km. (bottom left) Estimated intensity of the averaged visible magnetisation over a layer of 50 km for a spherical Earth of radius 6371.2 km. (bottom right) Same map, but for the magnetisation direction. The normalised vertical down-component value is shown in colour scale; the arrow indicates the horizontal magnetisation direction. When the vertical component absolute value is 1, the horizontal component is necessarily null.

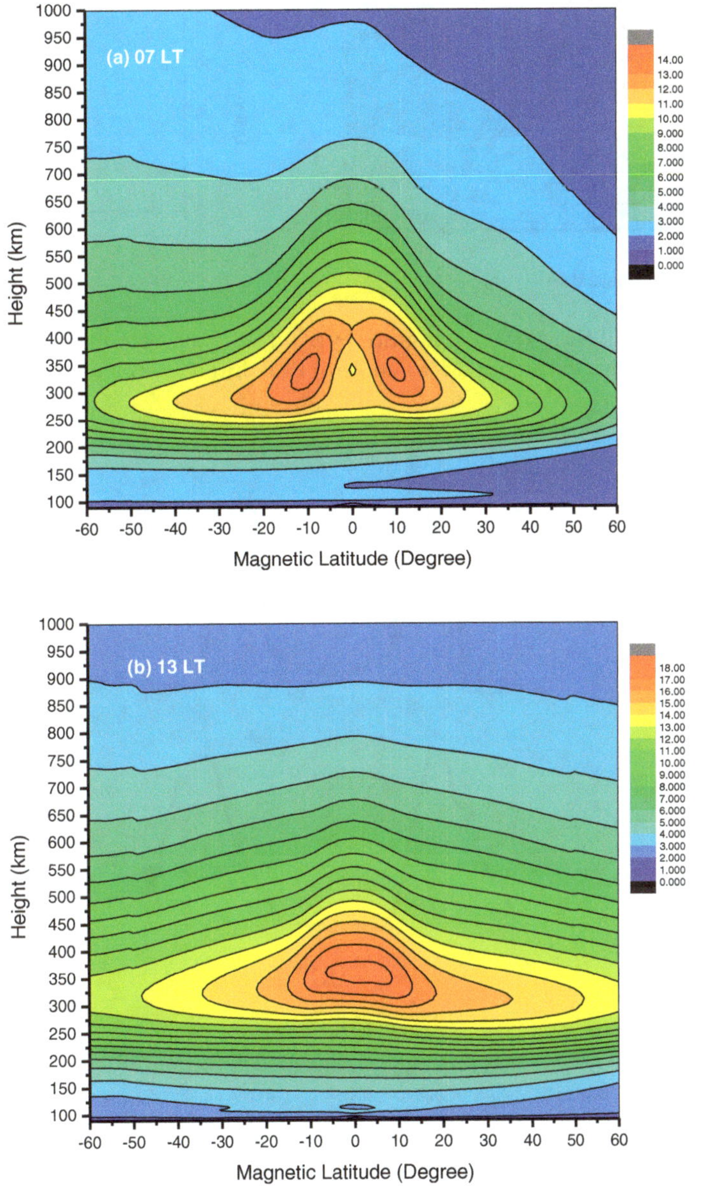

Figure 11.2.1 Altitude–latitude map of electron density Ne at (a) 0700 LT and (b) 1300 LT modeled by SUPIM for magnetically quiet (Ap = 4) conditions at high solar activity (day 303, F10.7 = 268) in Jicamarca longitude (283°E) with no **E×B** drift and no neutral wind. Unit of Ne is 10^5 cm^{-3}. Positive latitude is north.

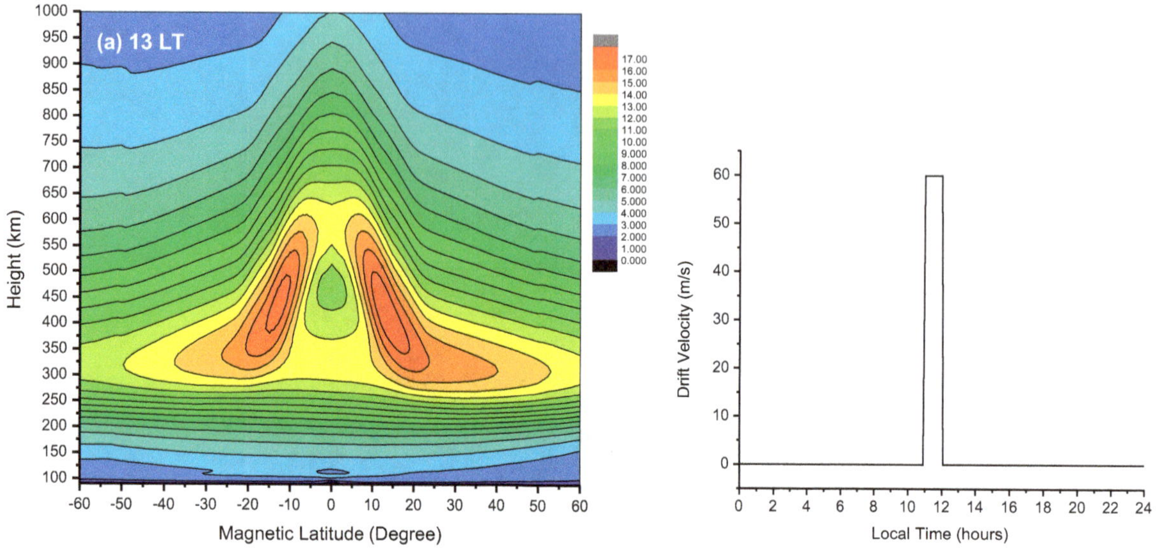

Figure 11.2.2 (a) Altitude–latitude map of Ne at 1300 LT modeled by SUPIM for magnetically quiet (Ap = 4) conditions at high solar activity (day 303, F10.7 = 268) in Jicamarca longitude (283°E) using (b) **E×B** drift only around noon. Positive latitude is north.

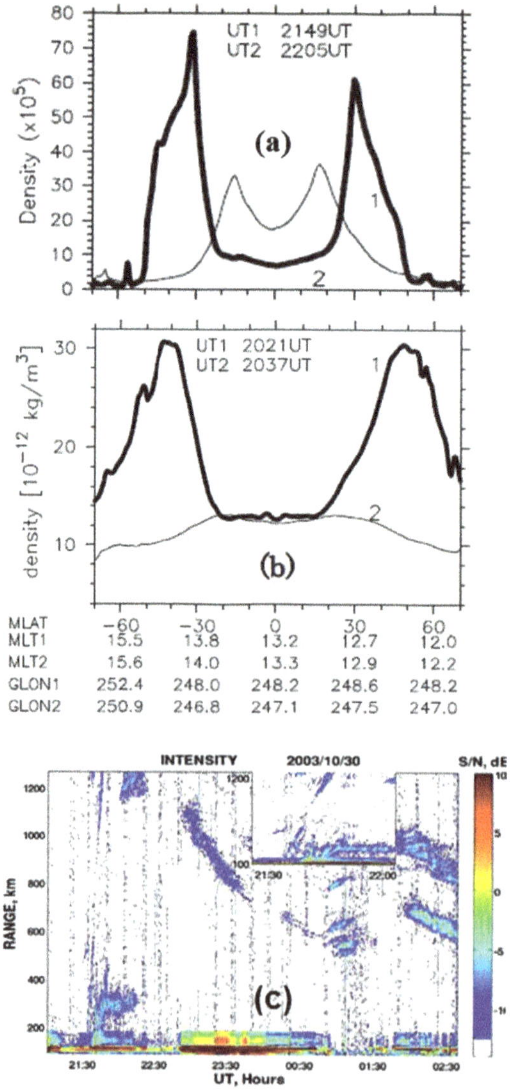

Figure 11.2.3 Latitude variations of (a) electron density Ne and (b) neutral density N measured by the CHAMP satellite at equatorial crossing time 2149 UT on 30 October 2003 (curves 1) compared with previous quiet day variations at a nearly same crossing time 2205 UT (curves 2). Latitude, longitude and UT of the satellite passes are listed at the bottom. Positive latitude is north. (c) RTI map of the VHF radar in Brazil on 30 October 2003 (after Abdu et al., 2008).

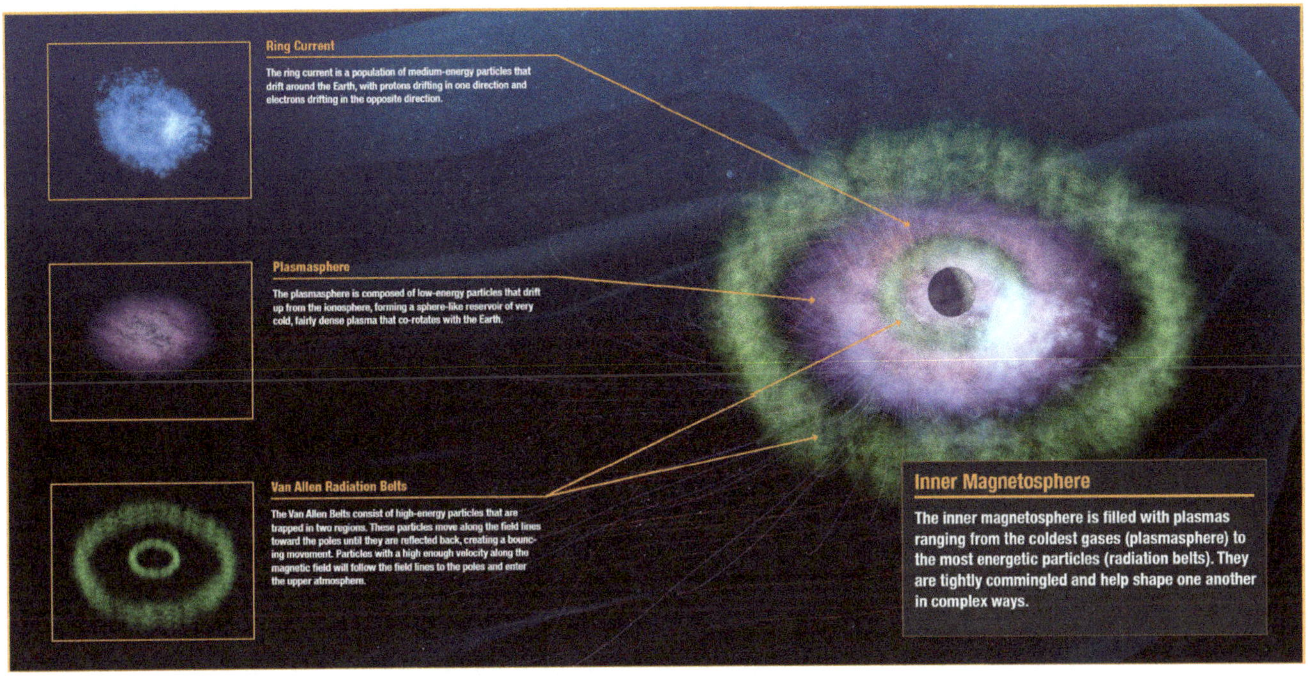

Figure 12.4.1 A diagram of the inner magnetosphere showing the commingled nature of the plasmasphere, ring current and Van Allen radiation belts within the same volume of space near Earth (NASA/Troy Benesch).

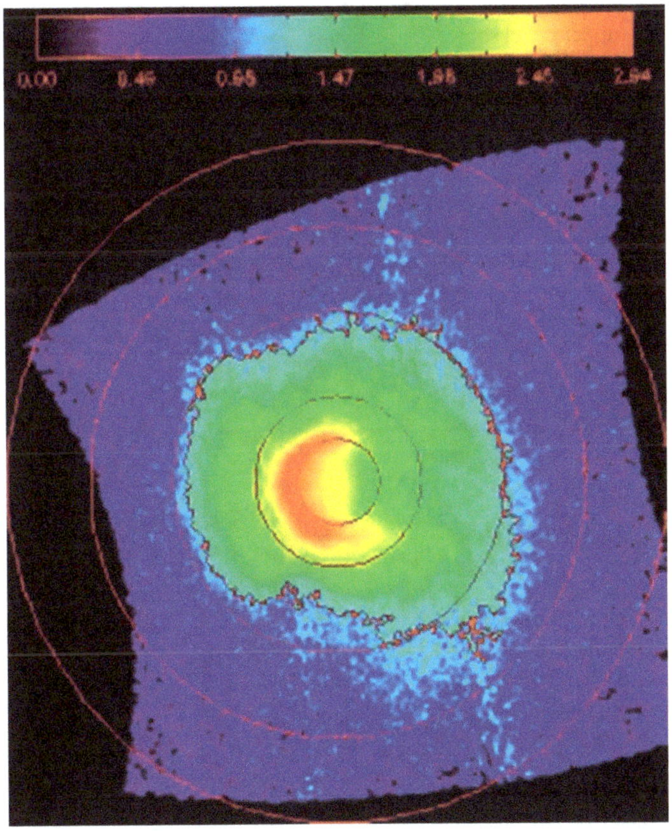

Figure 12.7.1 The global plasmasphere seen by EUV/IMAGE on 22 June 2001 at 0519 UT (when the satellite was above the North Pole and projected in the geomagnetic equatorial plane). The irregular line corresponds to the threshold of 40% of the maximum luminosity and illustrates the plasmapause position. The circles correspond to the Earth's surface and L = 2, 4, 6 and 8.

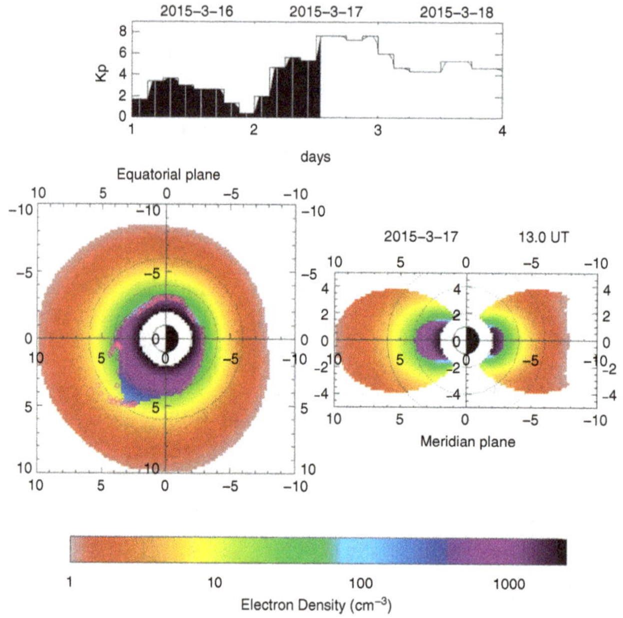

Figure 12.7.2 Density of the electrons in the equatorial (left) and meridian plane (right) obtained with the SPM plasmaspheric model (Pierrard and Voiculescu, 2011) for the date of 17 March 2015 at 1300 UT, during a geomagnetic storm. The plasmasphere corresponds to the dark region with higher density, the plasmapause is illustrated by the diamonds and the low-density plasma beyond the plasmapause corresponds to the plasma trough. The upper panel shows the planetary geomagnetic activity index of Bartels *Kp* observed from 16 to 18 March 2015.

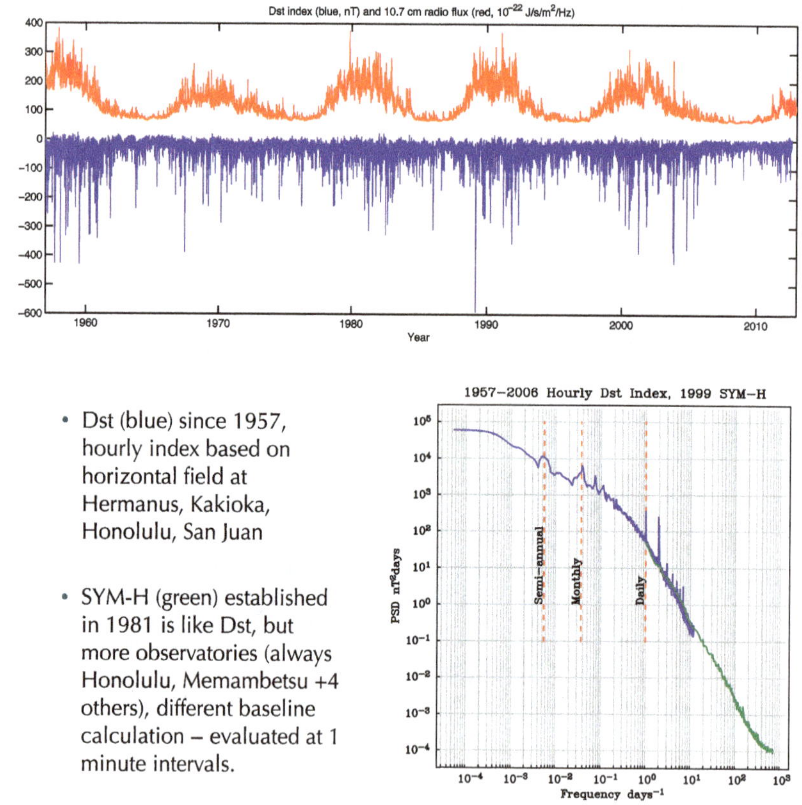

Figure 13.1.2 (a) Hourly *Dst* index variations from 1957 on compared with 10.7 cm radio flux. (b) the power spectrum of *Dst* from 1957 to 2006 and of *SYM-H* from 1999. *SYM-H* has 1 min sampling.

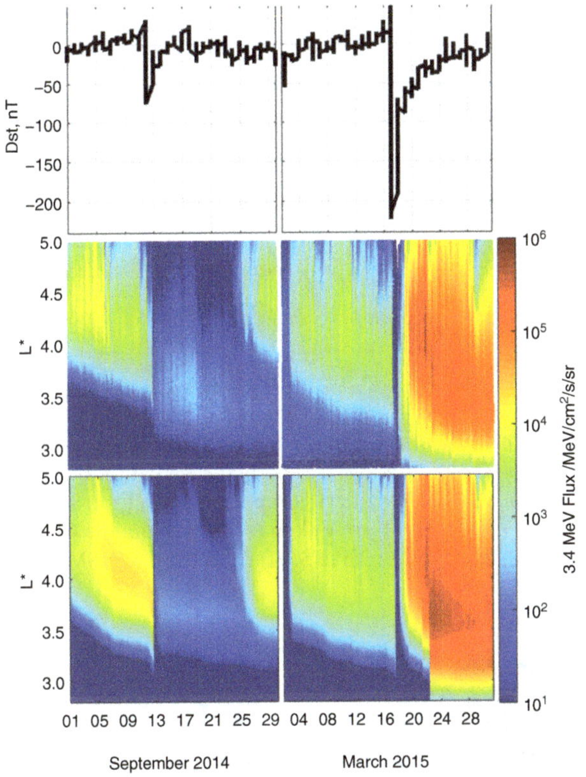

Figure 13.3.1 Comparison of ULF wave radial diffusion model results for the magnetic storms in September 2014 (left column; cf. Ozeke et al., 2017) and March 2015 (right column). (top) Storm-time disturbance index, *Dst*; (middle) 3.4 MeV electron flux observed by the Van Allen Probes; and (bottom) ULF wave-driven radial diffusion model results of electron flux at a fixed energy at 3.4 MeV using ULF wave power observed by ground-based magnetometers and mapped to the equatorial plane. The simulation results in the bottom row combine multiple first adiabatic invariant conserving radial diffusion simulations, each driven by the flux variations at the outer boundary of the simulation domain at $L^* = 5$ and combined to generate flux at fixed energy for direct comparison to the Van Allen Probe observations (cf. Mann et al., 2016, 2018; Ozeke et al., 2017). See text for more details.

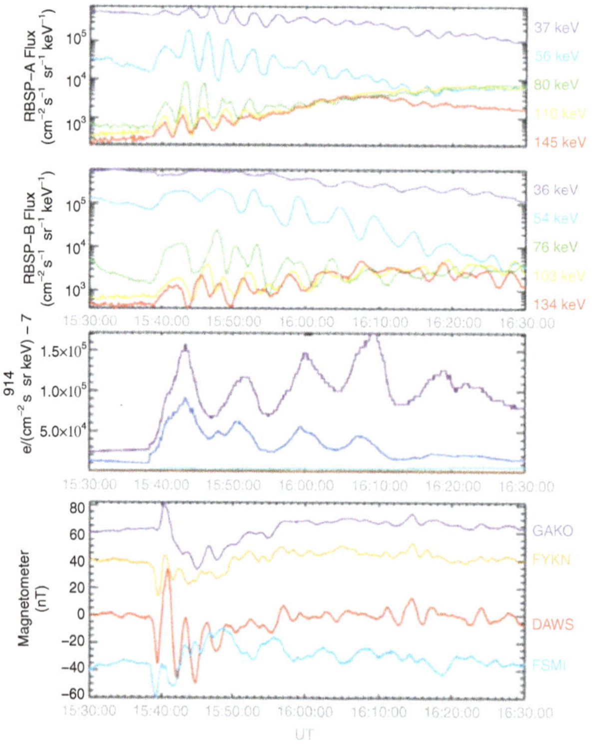

Figure 13.3.2 Observations of coherent ULF modulation of ~50–80 keV energetic electron flux on 31 October 2012 (cf. Claudepierre et al., 2013). Energetic electrons from Van Allen Probes A and B (formerly the Radiation Belt Storm Probes (RBSP)) are shown in the top two panels and from GOES 14 (g14; third panel), as well as the H-component fields from selected ground-based magnetometers (bottom panel). From west to east: Gakona (GAKO) and Fort Yukon (FYKN) from the Geophysical Institute Magnetometer Array (GIMA) in Alaska, and Dawson City (DAWS) and Fort Smith (FSMI) from the CARISMA array (Mann et al., 2008).

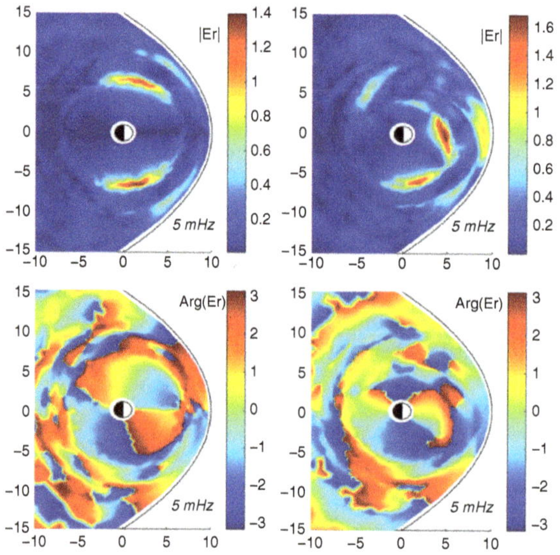

Figure 13.3.3 Simulations of 5 mHz ULF wave propagation driven inside the magnetosphere by a magnetopause driver using the model of Degeling et al. (2018). (left) Profiles of the magnitude of the radial electric field Er (top), and its phase in radians (bottom), in the equatorial plane for a symmetric plasmasphere. (right) Equatorial profiles of 5 mHz ULF wave Er magnitude (top) and phase (bottom) arising from the same magnetopause driver, but where 24 hours of Volland-Stern cold plasmasphere evolution have produced a storm-time plasmaspheric drainage plume, which significantly distorts the ULF wave power distributions in MLT, as well as its penetration to low L in the local morning sector. See text for more details.

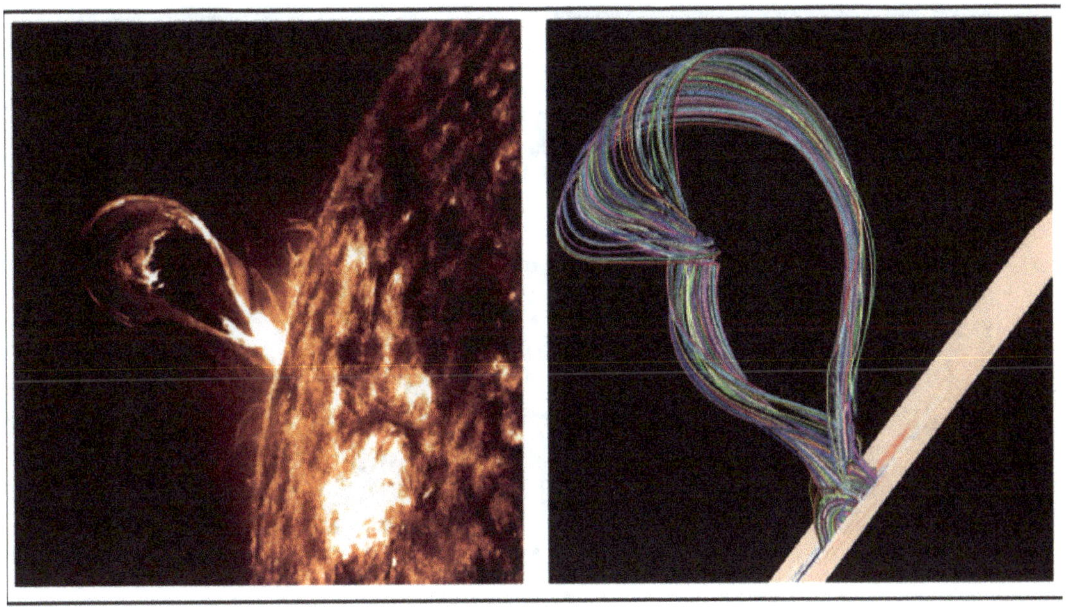

Figure 14.2.1 (left) Typical shape of the erupting flux rope observed by SDO/AIA on 14 October 2012, at 2:32 UT. (right) Late stage of numerical model evolution of the flux rope for the active region AR10930.Image adapted from Amari et al. (2014, Figures 4c and 4d).

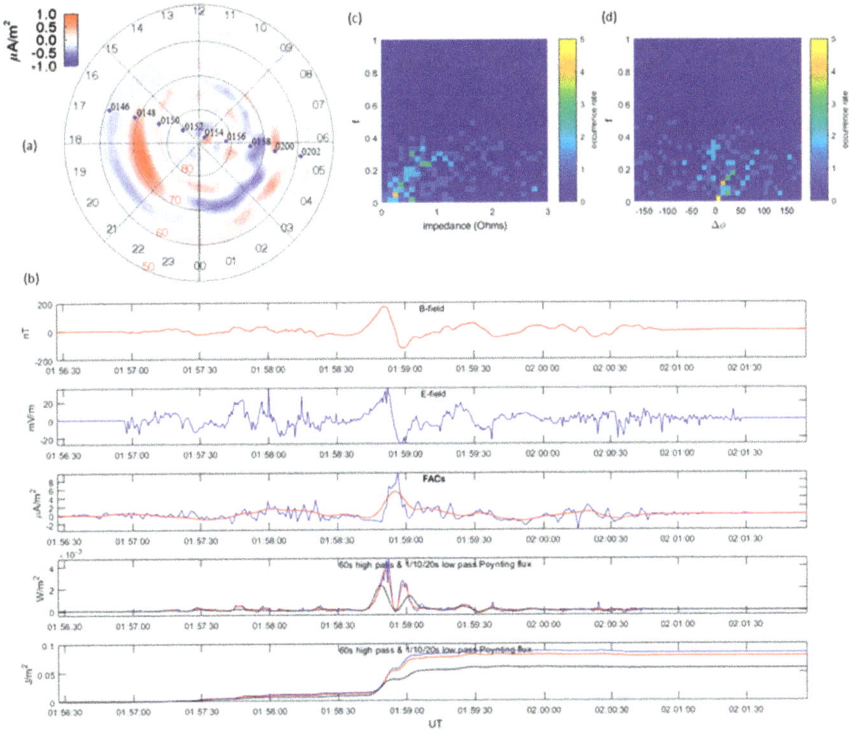

Figure 15.1.1 Field-aligned currents (FACs) and Alfvén waves in the coupled magnetosphere–ionosphere system on 7 January 2016. (a) FAC maps as estimated using the AMPERE experiment (Anderson et al., 2000) superposed with the orbit of the Swarm A satellite from 0146 UT to 0200 UT. (b) Swarm A measurements of detrended magnetic (top panel) and electric (second panel) fields, the Swarm FAC (third panel) derived using the single (blue) and dual (red) spacecraft techniques (cf. Ritter et al., 2013), as well as the instantaneous Poynting flux (fourth panel) for the cases where the electric and magnetic fields used to compute it have been time domain filtered at 1 s (blue), 10 s (red) and 20 s (black), and (bottom panel) the Poynting flux integrated along this section of the Swarm A orbit computed using the same filtering. (c, d) Occurrence distributions of the Alfvén wave impedance magnitude and phase, respectively, as a function of frequency for the polar pass shown in panel (b). See text for details.

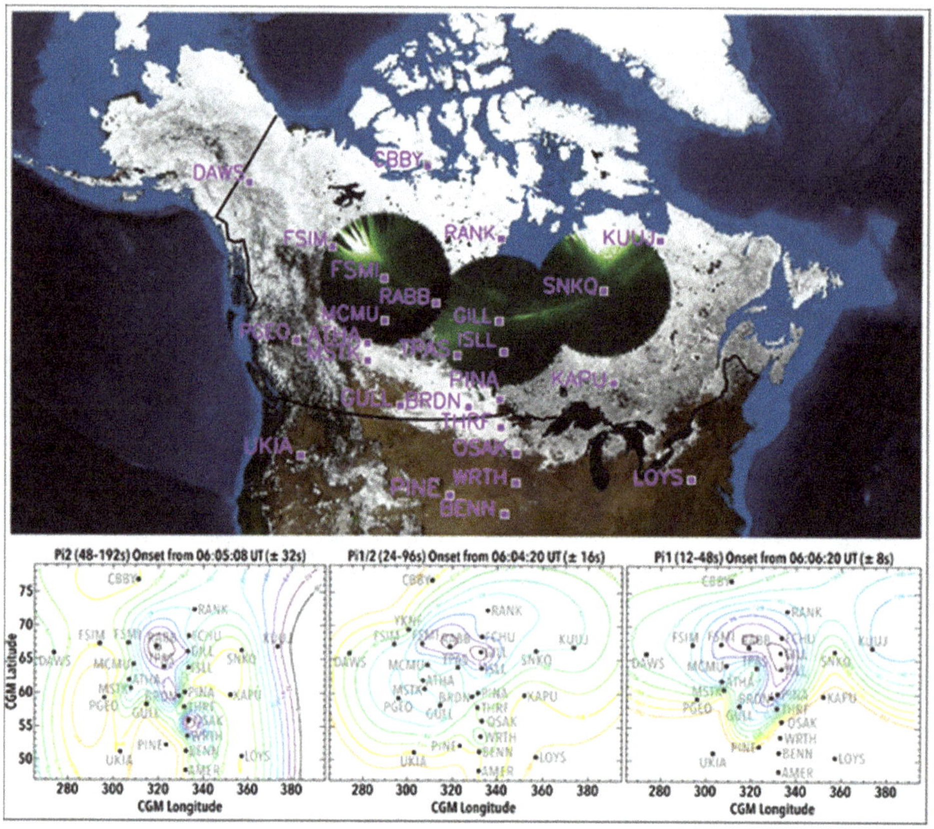

Figure 15.1.2 (top) Selected ground magnetometer stations from the CARISMA, THEMIS, CANMOS and McMAC magnetometer chains. Over-plotted are false colour white light all-sky imager (ASI) data from the GILL, FSMI and SNKQ THEMIS ASIs at 06:04:20 UT. This corresponds to the time of auroral brightening in this small substorm event on 5 March 2008. (bottom) AWESOME timing contours for the Pi2 (48–192 s; left), Pi1/2 (24–96 s; middle) and Pi1 (12–48 s; right) wavelet bands.Adapted from Rae et al. (2017, Figures 1 and 4).

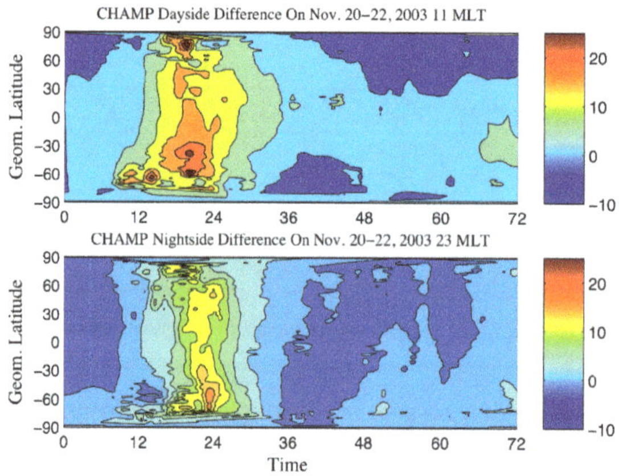

Figure 15.2.1 Storm-time deviation of the thermospheric density during the 20 November 2003 magnetic storm. Values are in units of 10^{-12} kg/m^3 and relative to the pre-storm quiet-time level.Modified after Liu and Lühr (2005, Figure 4).

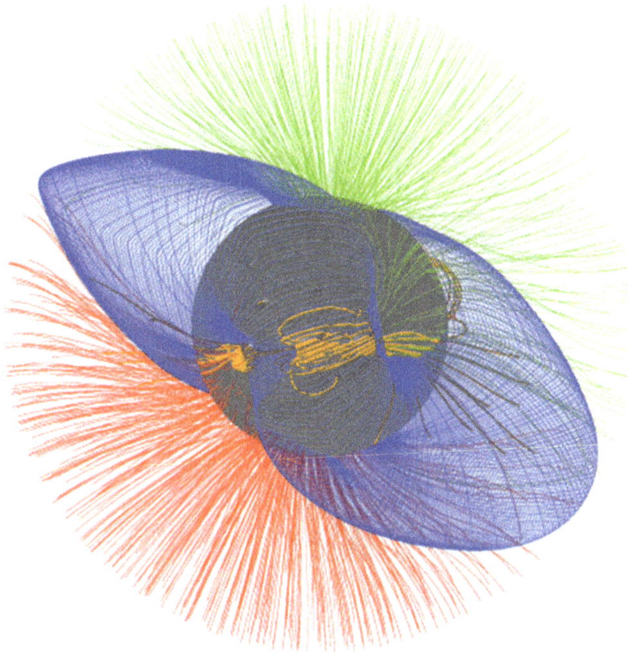

Figure 18.8.1 Potential field source surface model of the corona. Open magnetic fields that extend into the heliosphere are shown in green (outward or positive fields) and red (inward or negative fields). Closed magnetic field lines are shown in blue. The outer limit of the model is the spherical source surface at 2.5 solar radii. The open magnetic field lines are orthogonal to the source surface. (From http://gong.nso.edu/science/gongpfss/gongpfss.html, courtesy Gordon Petrie, National Solar Observatory.)

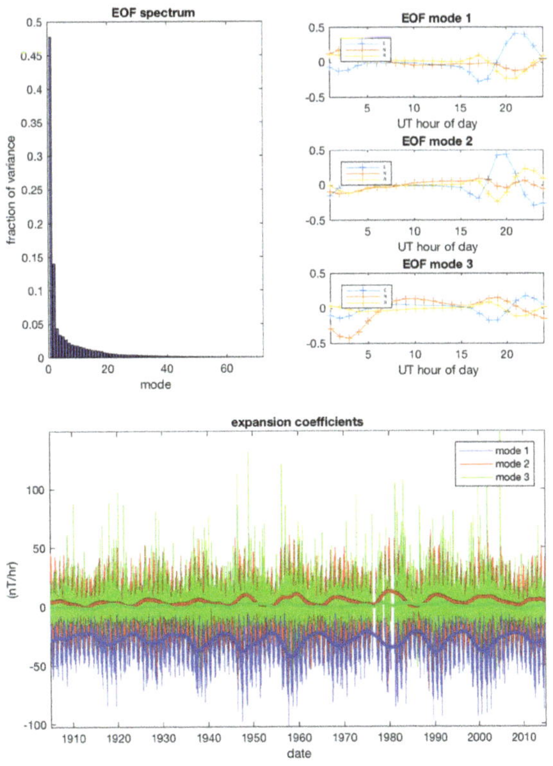

Figure 19.3.2 EOF analyses of the day vs hour data matrix of geomagnetic data from Honolulu. The EOF spectrum (upper left panel) shows that more than half of the variance is explained by the first two EOF modes; the EOFs (daily structures of *E, N, R*) for the first three modes are shown in the upper right panel. The associated expansion coefficients (lower panel) of the first two modes primarily show a seasonal cycle modulated by the solar cycle (as indicated by the thicker curves obtained by low-pass filtering the series with a three-year moving window). Because seasons as well as the solar cycle can be predicted, around half of the variance *E, N, R* is expected also to be predictable using only the first two EOF modes.

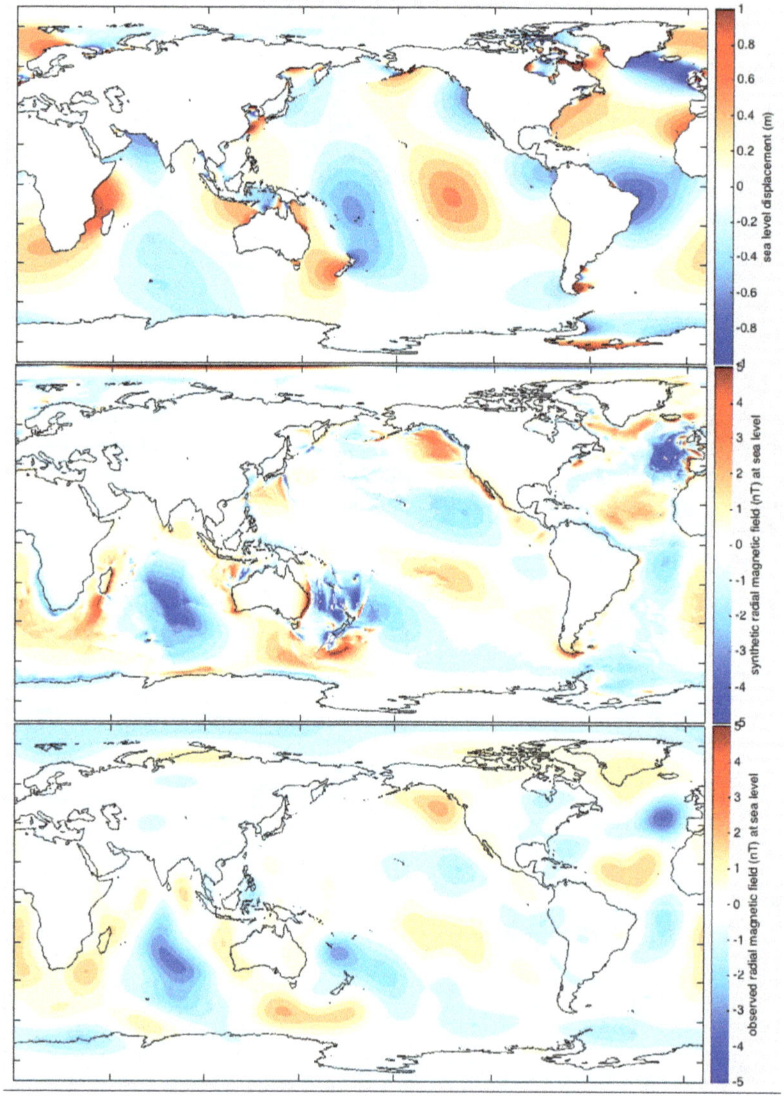

Figure 19.3.3 Movie frame of M_2 tidal surface displacement (top frame) and the radial component of magnetic field generated by the tidal flow as calculated in a numerical model (middle frame) and as seen in observations (bottom frame). The agreement between the bottom two frames strongly supports the assumption that the observed M_2 magnetic fluctuations are primarily due to ocean tides.

Kissinger, J., R. L. McPherron, T.-S. Hsu, V. Angelopoulos and X. Chu (2012a), 'Necessity of substorm expansions in the initiation of steady magnetospheric convection', Geophys. Res. Lett., 39(L15105), 1–5. doi: 10.1029/2012GL052599

Kletzing, C. A., et al. (2017), 'Phase sorting wave-particle correlator', J. Geophys. Res., 122, 2069–78. doi: 10.1002/2016JA023334

Kletzing, C. A., et al. (2013), 'The Electric and Magnetic Field Instrument Suite and Integrated Science (EMFISIS) on RBSP', Space Sci. Rev. doi: 10.1007/s11214-013-9993-6

Kotova, G., M. Verigin, J. Lemaire, V. Pierrard, V. Bezrukikh and J. Smilauer (2018), 'Experimental study of the plasmasphere boundary layer using MAGION 5 data', J. Geophys. Res., 123, 1251–9. doi: 10.1002/2017JA024590

Kurth, W. S., et al. (2015), 'Electron densities inferred from plasma wave spectra obtained by the Waves instrument on Van Allen Probes', J. Geophys. Res., 120, 904–14, doi: 10.1002/2014JA020857

Lemaire, J., and V. Pierrard (2008), 'Comparison between two theoretical mechanisms for the formation of the plasmapause and relevant observations', Geomagn. Aeron., 48(5), 553–70. doi: 10.1134/S0016793208050010

Li, X., D. N. Baker, M. Temerin, D. Larson, R. P. Lin, G. D. Reeves, M. Looper, S. G. Kanekal and R. A. Mewaldt (1997a), 'Are energetic electrons in the solar wind the source of the outer radiation belt?', Geophys. Res. Lett., 24(8), 923–6. doi: 10.1029/97GL00543

Li, X., D. N. Baker, M. Temerin, T. D. Cayton, E. G. D. Reeves, R. A. Christensen, J. B. Blake, M. D. Looper, R. Nakamura and S. G. Kanekal (1997b), 'Multisatellite observations of the outer zone electron variation during the November 3–4, 1993, magnetic storm', J. Geophys. Res., 102(A7), 14123–40. doi: 10.1029/97JA01101

Li, X., D. N. Baker, T. P. O'Brien, L. Xie and Q. G. Zong (2006), 'Correlation between the inner edge of outer radiation belt electrons and the innermost plasmapause location', Geophys. Res. Lett., 33(14). doi: 10.1029/2006GL026294

Li, X., R. S. Selesnick, D. N. Baker, A. N. Jaynes, S. G. Kanekal, Q. Schiller, L. Blum, J. Fennell and J. B. Blake (2015), 'Upper limit on the inner radiation belt MeV electron intensity', J. Geophys. Res., 120(2), 1215–28. doi: 10.1002/2014JA020777

Lui, A. T. Y. (2009), 'Comment on "Tail reconnection triggering substorm onset"', Science, 324(5933), 3. doi: 10.1126/science.1167726

Lui, A. T. Y., C.-I. Meng and S.-I. Akasofu (1976), 'Search for the magnetic neutral line in the near-Earth plasma sheet, 1. Critical reexamination of earlier studies on magnetic field observations', J. Geophys. Res., 81(34), 5934–40. doi: 10.1029/JA081i034p05934

Lui, A. T. Y., C.-L. Chang, A. Mankofsky, H.-K. Wong and D. Winske (1991), 'A cross-field current instability for substorm expansions', J. Geophys. Res., 96(A7), 11389–401. doi: 10.1029/91JA00892

Lui, A. T. Y., R. E. Lopez, S. M. Krimigis, R. W. McEntire, L. J. Zanetti and T. A. Potemra (1988), 'A case study of magnetotail current sheet disruption and diversion', Geophys. Res. Lett, 15 (7), 721–4. doi: 10.1029/GL015i007p00721

Lyons, L. R., Y. Nishimura, Y. Shi, S. Zou, H. J. Kim, V. Angelopoulos, C. Heinselman, M. J. Nicolls and K. H. Fornacon (2010), 'Substorm triggering by new plasma intrusion: Incoherent-scatter radar observations', J. Geophys. Res., 115(13). doi: 10.1029/2009ja015168

Malakit, K., M. A. Shay, P. A. Cassak and D. Ruffolo (2013), 'New electric field in asymmetric magnetic reconnection', Phys. Rev. Lett., 111, 135001.

McPherron, R. L. (1970), 'Growth phase of magnetospheric substorms', J. Geophys. Res., 75(28), 5592–9. doi: 10.1029/JA075i028p05592

McPherron, R. L. (1991), 'Physical processes producing magnetospheric substorms and magnetic storms', in *Geomagnetism*, vol. 4, ed. J. Jacobs, pp. 593 739, Academic Press, London.

McPherron, R. L. (2015), 'Earth's Magnetotail', in *Magnetotails in the Solar System*, ed. A. Keiling, C. M. Jackman and P. A. Delamere, pp. 61–84, Blackwell Science, Oxford. doi: 10.1002/9781118842324.ch4

McPherron, R. L., and X. Chu (2017), 'The mid-latitude positive bay and the MPB Index of substorm activity', Space Sci. Rev., 206(1–4), 91–122. doi: 10.1007/s11214-016-0316-6

McPherron, R. L., C. T. Russell and M. Aubry (1973), 'Satellite studies of magnetospheric substorms on August 15, 1968, 9. Phenomenological model for substorms', J. Geophys. Res., 78 (16), 3131–49.

McPherron, R. L., J. M. Weygand and T. S. Hsu (2008), 'Response of the Earth's magnetosphere to changes in the solar wind', J. Atmos. Sol. Terr. Phys., 70(2–4), 303–15. doi: 10.1016/j.jastp.2007.08.040

McPherron, R. L., T. P. O'Brien and S. M. Thompson (2005), 'Solar wind drivers for steady magnetospheric convection', in *Multiscale Coupling of Sun-Earth Processes*, ed. A. T. Y. Lui, Y. Kamide and G. Consolini, pp. 113–24, Elsevier, Amsterdam. doi: 10.1016/B978-044451881-1/50009-5

McPherron, R. L., T. S. Hsu, J. Kissinger, X. Chu and V. Angelopoulos (2011), 'Characteristics of plasma flows at the inner edge of the plasma sheet', J. Geophys. Res., 116. doi: 10.1029/2010ja015923

McPherron, R. L., T.-S. Hsu and X. Chu (2015), 'An optimum solar wind coupling function for the AL index', J. Geophys. Res., 120(4), 2494–2515. doi: 10.1002/2014ja020619

Meredith, N. P., et al. (2003), 'Statistical analysis of relativistic electron energies for cyclotron resonance with EMIC waves observed on CRRES', J. Geophys. Res., 208, 1250.

Murakami, G., K. Yoshioka, A. Yamazaki, Y. Nishimura, I. Yoshikawa and M. Fujimoto (2016), 'The plasmapause formation seen from meridian perspective by KAGUYA', J. Geophys. Res., 121(11), 973–11984. doi: 10.1002/2016JA023377

Nishida, A. (1994), 'The Geotail mission', Geophys. Res. Lett, 21 (25), 2871–3. doi: 10.1029/94gl01223

Nishimura, Y., et al. (2014), 'Day-night coupling by a localized flow channel visualized by polar cap patch propagation', Geophys. Res. Lett, 41(11), 3701–9. doi: 10.1002/2014gl060301

Nishimura, Y., L. Lyons, S. Zou, V. Angelopoulos and S. Mende (2010), 'Substorm triggering by new plasma intrusion: THEMIS all-sky imager observations', J. Geophys. Res., 115(A7), A07222. doi: 10.1029/2009ja015166

Nishimura, Y., L. R. Lyons, V. Angelopoulos, T. Kikuchi, S. Zou and S. B. Mende (2011), 'Relations between multiple auroral streamers, pre-onset thin arc formation, and substorm auroral onset', J. Geophys. Res., 116(10). doi: 10.1029/2011ja016768

O'Brien, T. P., S. M. Thompson and R. L. McPherron (2002), 'Steady magnetospheric convection: Statistical signatures in the solar wind and AE', Geophys. Res. Lett., 29(7). doi: 10.1029/2001GL014641

Ohtani, S. (2004), 'Flow bursts in the plasma sheet and auroral substorm onset: observational constraints on connection between midtail and near-Earth substorm processes', Space Sci. Rev., 113(1–2), 77–96. doi: 10.1023/B:SPAC.0000042940.59358.2 f

Pierrard, V., and J. Cabrera (2006), 'Dynamical simulations of plasmapause deformations', Space Sci. Rev., 122(1–4), 119–26. doi: 10.1007/s11214-005-5670-8

Pierrard, V., J. Goldstein, N. André, V. K. Jordanova, G. A. Kotova, J. F. Lemaire, M. W. Liemohn and H. Matsui (2009), 'Recent progress in physics-based models of the plasmasphere', Space Sci. Rev., 145, 193–229. doi: 10.1007/s11214-008-9480-7

Pierrard, V., G. Khazanov, J. Cabrera and J. Lemaire (2008), 'Influence of the convection electric field models on predicted plasmapause positions during the magnetic storms', J. Geophys. Res., 113, A08212, 1–21. doi: 10.1029/2007JA012612

Pierrard, V., and M. Voiculescu (2011), 'The 3D model of the plasmasphere coupled to the ionosphere', Geophys. Res. Lett., 38, L12104. doi: 10.1029/2011GL047767

Poppe, A. R., et al. (2016), 'ARTEMIS observations of terrestrial ionospheric molecular ion outflow at the Moon', Geophys. Res. Lett., 43, 6749–58, doi: 10.1002/2016GL069715

Pytte, T., R. L. McPherron, E. W. Hones and H. L. West (1978), 'Multiple satellite studies of magnetospheric substorms: Distinction between polar magnetic substorms and convection-driven bays', J. Geophys. Res., 83(Na2), 663–79. doi: 10.1029/JA083iA02p00663

Reeves, G. D., et al. (2013), 'Electron acceleration in the heart of the Van Allen radiation belts', Science, 1237743. doi: 10.1126/science.1237743

Rosenbauer, H., H. Grünwaldt, M. D. Montgomery, G. Paschmann and N. Sckopke (1975), 'Heos 2 plasma obserations in the distant polar magnetosphere: The plasma mantle', J. Geophys. Res., 80, 2723–37.

Russell, A. J. B., T. Karlsson and A. N. Wright (2015), 'Magnetospheric signatures of ionospheric density cavities observed by Cluster', J. Geophys. Res., 120, 1876–87.

Schroeder, J. W. R., et al. (2016), 'Direct measurement of electron sloshing of an inertial Alfvén wave', Geophys. Res. Lett., 43, 4701–7. doi: 10.1002/2016GL068865

Selesnick, R. S., D. N. Baker, A. N. Jaynes, X. Li, S. G. Kanekal, M. K. Hudson and B. T. Kress (2014), 'Observations of the inner radiation belt: CRAND and trapped solar protons', J. Geophys. Res., 119. doi: 10.1002/2014JA020188

Sergeev, V. A., R. J. Pellinen and T. I. Pulkkinen (1996), 'Steady magnetospheric convection: a review of recent results', Space Sci. Rev., 75(3–4), 551–604.

Shiokawa, K., et al. (1998), 'High-speed ion flow, substorm current wedge, and multiple Pi 2 pulsations', J. Geophys. Res., 103(A3), 4491–4507. doi: 10.1029/97ja01680

Shiokawa, K., W. Baumjohann and G. Haerendel (1997), 'Braking of high-speed flows in the near-Earth tail', Geophys. Res. Lett, 24(10), 1179–82. doi: 10.1029/97gl01062

Shprits, Y. Y., R. M. Thorne, R. Friedel, G. D. Reeves, J. Fennell, D. N. Baker and S. G. Kanekal (2006), 'Outward radial diffusion driven by losses at magnetopause', J. Geophys. Res., 111, A11214. doi: 10.1029/2006JA011657

Singh, S., and J. L. Horwitz (1992), 'Plasmasphere refilling: Recent observations and modeling', J. Geophys. Res., 97, 1049–79. doi: 10.1029/91JA02602

Stewart, B. (1861), 'On the great magnetic disturbance of 28 Aug. to 7 Sep. 1859', Philos. Trans. London, 151, 423–30.

Summers, D., B. Ni and N. P. Meredith (2007), 'Timescales for radiation belt electron acceleration and loss due to resonant wave-particle interactions: 2. Evaluation for VLF chorus, ELF hiss, and electromagnetic ion cyclotron waves', J. Geophys. Res., 112, A04207. doi: 10.1029/2006JA011993

Thorne, R. L., et. al. (2013), 'Rapid local acceleration of relativistic radiation-belt electrons by magnetospheric chorus', Nature, 504(7480), 411. doi: 10.1038/nature12889

Thorne, R. M., E. J. Smith, R. K. Burton and R. E. Holzer (1973), 'Plasmaspheric hiss', J. Geophys. Res., 78(10), 1581–96. doi: 10.1029/JA078i010p01581

Torbert, R. B., et al. (2016), 'Estimates of terms in Ohm's law during an encounter with an electron diffusion region', Geophys. Res. Lett., 43, 5918. doi: 10.1002/2016GL069553

Ukhorskiy, A. Y., B. J. Anderson, P. C. Brandt and N. A. Tsyganenko (2006), 'Storm time evolution of the outer radiation belt: Transport and losses', J. Geophys. Res., 111, A11S03. doi: 10.1029/2006JA011690

Van Allen, J. A., G. H. Ludwig, E. C. Ray and C. E. McIlwain (1958), 'Observations of high intensity radiation by satellites 1958 Alpha and Gamma', Jet Propul., 28, 588–92.

Verbanac, G., M. Bandic and V. Pierrard (2018), 'MLT plasmapause characteristics: comparison between THEMIS observations and numerical simulations', J. Geophys. Res., 123, 2000–2017. doi: 10.1002/2017JA024573

Verbanac, G., V. Pierrard, M. Bandic, F. Darrouzet, J.-L. Rauch and P. Décréau (2015), 'Relationship between plasmapause, solar wind and geomagnetic activity between 2007 and 2011 using Cluster data', Ann. Geophys., 33, 1271–83. doi: 10.5194/angeo-33-1271-2015

Walsh, B. M., et al. (1974), 'Simultaneous ground- and space-based observations of the plasmaspheric plume and reconnection', Science, 343(6175), 1122–5. doi: 10.1126/science.1247212

Walsh, B. M., T. D. Phan, D. G. Siebeck and V. M. Souza (2014), 'The plasmaspheric plume and magnetopause reconnection', Geophys. Res. Lett., 41(2), 223–8. doi: 10.1002/2013GL058802

Weimer, D. R. (1992), 'Characteristic time scale of substorm expansion and recovery', Proc. Int. Conf. Substorms, ICS-1, 581–6.

Wilson, L. B., III (2016), 'Relativistic electrons produced by foreshock disturbances observed upstream of the Earth's bow shock', Phys. Rev. Lett., 117(21). doi: 10.1103/PhysRevLett.117.215101

13
Temporal Field Variations

13.1 From Reversals to Very-Low-Frequency Waves

Catherine Constable

The geomagnetic field on Earth's surface and in surrounding regions results from diverse sources, ranging from Earth's core, through its mantle and lithosphere and across the ionosphere, magnetosphere and heliosphere. The individual contributions exhibit a broad range of magnitudes, spatial scales and variability, but the instantaneous state of Earth's magnetic environment largely reflects dynamic interactions between two important sources, a self-sustaining dynamo in the liquid outer core producing the dominantly dipolar internal field, and the solar wind, the fully ionized plasma emanating from the Sun. The pressure balance between them defines the magnetopause, inside which lies the magnetosphere, a region extending to about 10 Earth radii and dominated by the geomagnetic field. Much closer to Earth, the ionosphere is a partially ionized gas lying between the neutral density atmosphere and the larger magnetosphere. Enhanced currents flow in the dayside ionosphere as a result of the diurnal cycle of solar heating. Below the ionosphere, lightning activity (sferics) in the neutral atmosphere is responsible for much of the electromagnetic activity at frequencies above 1 Hz. Kinetic and electromagnetic energy are provided to the magnetosphere, ionosphere and upper atmosphere by the solar wind depositing charged particles into Earth's magnetosphere, mainly in the polar regions. The solar wind controls space weather and climate, and the internal magnetic field is a factor contributing to the retention of Earth's atmosphere. Earth's lithosphere also contributes to the overall field via both induced and remanent magnetization of iron bearing minerals in the crust, which has an averaged thickness of about 40 km. At greater depths in Earth's mantle, temperatures increase above the Curie points of magnetic minerals and electromagnetic induction by time-varying external fields is the primary source of magnetic field variations.

The mathematical and physical foundations for describing various geomagnetic phenomena are given in some detail by Backus et al. (1996). The natural spectrum of electromagnetic variations surrounding the Earth extends across an enormous frequency range (Figure 13.1.1) and is controlled by a variety of internal and external physical processes. Table 13.1.1 outlines the contributions from the various internal sources, along with their typical magnitudes and scales for temporal and spatial variations. Details of internal field observations and their interpretations including EM induction studies are covered elsewhere in this volume and in the *Treatise of Geophysics*, vol. 5 (Kono, 2015).

Temporal variations in just the internal geomagnetic field occur on a huge range of timescales, and are presumed to have occurred since the inception of Earth's dynamo shortly after planetary formation. The longest-term changes are investigated using paleomagnetism (Tauxe, 2010) combined with modern geochronological methods and use geological materials as proxy magnetometers. The magnetic field may have been present as early as 3–4 billion years ago (Tarduno et al., 2015), but we know little about actual field strength and structure then, or of any significant changes since the field began. Field intensity is the key to balancing the solar wind, which solar analog models suggest would have been stronger in the very distant past (Tarduno et al., 2010). There is speculation that the normal standoff distance for the magnetopause may have been smaller then, more like 5 *RE* than the current value of around 10 *RE*. This diminished value is not dissimilar to the standoff distance observed during large magnetic storms associated with solar coronal mass ejections such as the 1859 Carrington event (Carrington, 1859) or the more recent 2003, so-called Halloween storm (Webb and Allen, 2004).

The most dramatic changes in the paleomagnetic record are associated with geomagnetic polarity reversals and excursions during which field strength at Earth's surface may decrease by factors of 10–20 and large deviations from the dominant dipole structure are detected. The timing of geomagnetic reversals is well known (geologically speaking) for the 0–160 Ma interval (Gee and Kent, 2007), from spatial variations in magnetic anomalies in oceanic crust erupted at mid-ocean ridge spreading centers and subsequently cooled through the Curie temperatures of the relevant magnetic minerals. The average reversal rate has varied over Earth's history, and is around 2 Myr^{-1} for the past 160

Table 13.1.1 *Typical Properties of Various Internal Geomagnetic Variations*

Type of contribution	Period or frequency	Typical amplitude
Internal core field	Annual to billion year	20−100 μT
Changes in reversal rate geomagnetic dipole reversal	0–~10/My	100−200 μT
rate geomagnetic excursion/cryptochron rate	~2/My	~30−100 μT
	0–8/My	
Time taken for reversal or excursion	10^2−10^4 years	
Paleosecular variations	1−100 ky	
Secular variation	Annual	± ~100 nT/yr
Lithospheric field	Static	± 100 nT, up to 1μT
Induced magnetic fields	11 years to kHz	± 30 nT

My, but this interval also includes well-documented very long stable polarity intervals like the 40 My Cretaceous Normal Superchron. The most recent reversal occurred at about 780 ka, and the well-documented Laschamp excursion at about 40 ka. Estimates of the time taken for a magnetic reversal range from a few hundred years for purely directional changes to some tens of thousands of years for global changes in axial dipole moment. Details of field amplitude and morphological changes during excursions and polarity reversals are difficult to obtain because of limitations in the geological record. This complicates efforts to understand the consequences of magnetic reversals for interactions with external magnetic field variations, although there have been attempts to simulate the effects (e.g. Siscoe and Chen, 1975; Glassmeier et al., 2004; Stadelmann et al., 2010).

The most reliable and detailed information about prehistoric magnetic temporal variations comes from geological and archeological materials providing paleomagnetic records on centennial to million year timescales (Tauxe, 2010). Spatial and temporal sampling are limited, but global estimates of axial dipole moment variations exist for the past 2 million years (e.g. Ziegler et al., 2011), with temporal resolution of about 10,000 years. Beyond the most basic dipole description, several time-varying, low-degree spherical harmonic models of the field are available for the past 10,000 years with temporal resolution of a few hundred years (e.g. Constable and Korte, 2015; Constable et al., 2016).

Turning to direct geomagnetic field observations we enter a frequency range where there is significant overlap between internal and external magnetic field variations. Numerous reviews (e.g. Sabaka et al., 2010; Olsen et al., 2010) outline the mathematical and physical basis for geomagnetic field modeling and discuss the modern data from geomagnetic observatories, repeat stations and low-Earth-orbiting satellites that are used to infer the field structure and its rate of change. Large-scale compilations of direct surface observations were used to produce the *gufm1* field model, a time-varying spherical harmonic representation to degree and order 14, that spans 400 years from 1590 to 1990 CE (Jackson et al., 2000) and forms an excellent standard for interpreting decadal to centennial scale secular internal field variations. The advent of satellite observations has led to a truly global view of the field reflected in the modern International Geomagnetic Reference Field and the CHAOS series (Finlay et al., 2016). CHAOS-6X, using combined observations from the Swarm satellite mission and ground observatories, now covers the interval 1999–2017 with full vector satellite observations, and the highest available global resolution of secular variation and secular acceleration. The 10 kyr, *gufm1,* and CHAOS models clearly show the evolution on centennial to annual timescales of the South Atlantic Magnetic Anomaly, a rapidly changing low field region produced by reverse flux patches at the core surface beneath Southern Africa, that indicates the importance of non-dipole field structure and is important in the context of space weather. The 10 kyr models put the current decay in magnetic field strength into a broader temporal context and indicate that it is not unusual.

On decadal timescales both internal and external field variations contribute, and the separation of physical sources is to a large extent based on understanding of their spatial and temporal structures. Figure 13.1.1 presents a broad amplitude spectrum of the geomagnetic field as a function of frequency, ranging across 10^{-15} Hz to 10^4 Hz with spatial origin of the sources roughly divided according to the characteristic timescales of their variations.There are some significant overlaps. Table 13.1.2 lists the primary external sources. Lithospheric contributions to the field are essentially static (see, e.g., the MF7 model to degree and order 110 at http://geomag.us/models/MF7.html) and make no contribution to the frequency spectrum, but separating them from the large-scale magnetospheric and ionospheric field variations can be troublesome because of temporal-spatial aliasing in the satellite observations that provide the best global coverage.

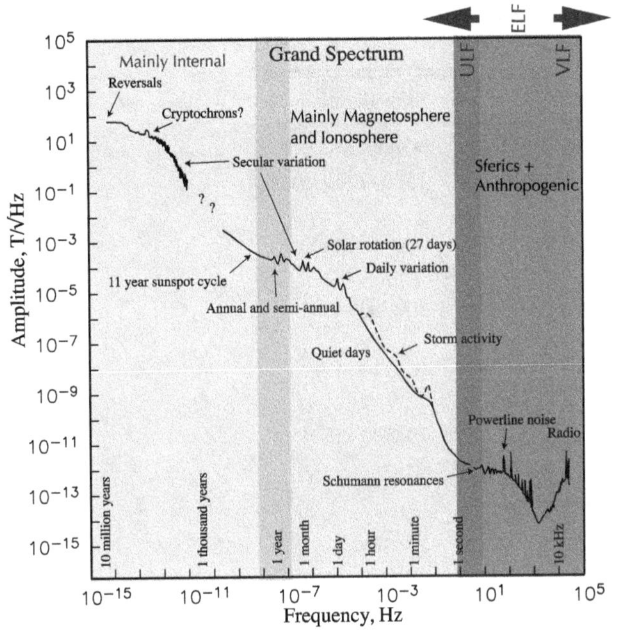

Figure 13.1.1 Amplitude spectrum of geomagnetic variations, modified from Constable and Constable (2004). Shading denotes separation of physical sources as a function of frequency.

The geomagnetic spectrum generally falls off with increasing frequency with some apparent periodic signals superimposed. Fluid flow in the highly electrically conductive outer core produces both very long- and short-term secular variations in the magnetic field, which propagate upward through the much lower conductivity mantle and lithosphere. The largest and longest-term changes are associated with geomagnetic reversals and excursions of the dipole part of the internal field. Finite electrical conductivity of the mantle effectively filters variations in the core field on timescales much less than a year (Backus, 1983). Thus the internal part of the spectrum is perforce greatly diminished in the frequency range dominated by the solar cycle, and magnetospheric processes.

A proxy for magnetospheric time variations is provided by hourly variations of the *Dst* index derived from a subset of magnetic observatories and the more recent *SYM-H* index which has 1 min sampling (see Figure 13.1.2). The indices are intended to represent the strength of the external mainly dipolar ring current magnetospheric activity, which varies with strength of the solar wind, and is correlated with the 10.7 cm solar radio flux which clearly shows solar cycle variations. A power spectrum of *Dst* and *SYM-H* variations shows peaks at roughly semi-annual, monthly (27 day solar rotation) and daily plus daily harmonics (likely reflecting ionospheric contamination) periods, superimposed on the generally red spectrum.

Above the insulating atmosphere is the relatively electrically conductive ionosphere, which supports *Sq* currents as a result of dayside solar heating. Outside the solid Earth the magnetosphere is deformed and modulated by the solar wind, compressed on the Sunward side and elongated on the nightside. The level of asymmetric variations in structure of the magnetosphere represents departures from the dominantly dipolar external ring current structure and is described by an additional proxy index known as *ASY* (see http://wdc.kugi.kyoto-u.ac.jp/aeasy/index.html for detailed explanation). The volume within the magnetosphere contains the highly dynamic Van Allen radiation belts, which are layers of energetic charged particles: usually there are two (sometimes three) main belts depending on solar activity and typically ranging in altitude from about 1000 to 60,000 km. In the outer belt, at a distance of about 3 Earth radii, the magnetospheric ring current which acts to oppose the main field is modulated by geomagnetic activity represented by *Dst*. Particles in the innermost belt can reach atmospheric altitudes, mostly in the South Atlantic region. Magnetic fields generated in the magnetosphere and ionosphere propagate by induction into the conductive Earth, providing information on electrical conductivity variations in the crust and mantle. Low-Earth-orbiting satellites that measure the magnetic field fly above the lower boundary of the ionosphere, but below the magnetospheric induction sources.

The various categories of external field signals are described in Table 13.1.2. Note that even the largest external contributions are small in comparison with the strength of the internal part, which currently ranges from about 22 μT over Brazil (near the South Atlantic Anomaly) to more than 65 μT near the poles.

Electromagnetic sources at ~1 Hz and higher include parts of the ultra-low-frequency (ULF) range, $f < 3$ Hz, the extremely low-frequency (ELF) band, $3 < f < 3 \times 10^3$ Hz, and the VLF range from 3 to 30 kHz. These sources provide information about the upper atmosphere (ionosphere and magnetosphere) as well as the lower atmosphere and electrical structure of Earth's lithosphere. Knowledge of ionospheric processes is thus informed by developments in both magnetospheric and atmospheric science.

At frequencies >1 Hz, the spectrum is dominated by sferics and anthropogenic noise. The Schumann resonances at about 7–8 Hz and higher harmonics originate from upward propagation of lightning into the global ionospheric waveguide. Locally, lightning energy peaks at about 10 kHz but rapid attenuation with distance in the Earth/ionsophere waveguide at this frequency impedes long-range detection (Shvets and Hayakawa, 2011). At frequencies higher than 10^4 Hz the global EM spectrum starts to rise in response to (presumably) man-made signals. This dominance obscures any contributions from whistlers and dawn chorus, whose interactions with the dipole field and radiation belt particles are nonetheless important. In Figure 13.1.1, there is a clear frequency

Table 13.1.2 *Typical Properties of Various External Geomagnetic Variations (modified from Olsen, 2007).*

Type of Variation	Symbol	Period or Frequency	Amplitude	Skin Depth
Solar Cycle		11 years	10–20 nT	>2000 km
Annual		12 months	5 nT	1500–2000 km
Semiannual		6 months	5 nT	
Storm-time	Dst	hours to weeks	50–500 nT	300–1000 km
Regular Daily		12 months	5 nT	
at mid-latitudes	Sq	24 h and harmonics	20–50 nT	300–600 km
at low latitudes	EEJ	12 months	50–100 nT	
Substorms	DP	10 min to 2 h	100 nT (1000 nT at p.l.)	100–300 km
Pulsations	ULF	0.3-600s		20–100 km
regular	Pc	150-600s (pc5)	10 nT (100nT at p.l.)	
continuous		45-150s (pc4)	2 nT	
pulsations		5-45s (pc2,3)	0.5 nT	
		0.2-5s (pc1)	0.1nT	
Irregular Pulsations	Pi	1-150s	1nT	
Ionospheric Alfvén Resonances	IAR	0.5 – 10 Hz		
Extreme low frequency	ELF Sferics	3 Hz- 3kHz	<0.1 nT	10 m to Kms
Schumann Resonances		8 Hz and harmonics	<0.1 nT	
Plasmaspheric Hiss		200 Hz – 2kHz		
Chorus, Whistlers, Lightning	VLF	3-30kHz		m to 10s of m

Note. Signals with significant latitude dependence are given values at polar latitudes (p.l. > 65°).

separation in the various natural sources, with signals from magnetic storms, and daily variation in the ionosphere dying away around 1 Hz and the sferics losing energy around 1–3 kHz. These are well known as 'dead bands' with very low signal in the magnetotelluric (MT) source field used for induction studies which aim to recover subsurface geoelectrical structure (Chave and Jones, 2012).

There are important applications of EM signals at frequencies >1 Hz. Whistlers and related ionospheric phenomena in the VLF range (Helliwell, 2006) were first identified in the late nineteenth and early twentieth century and subsequent research alerted the community to the importance of these signals in the study of the outer ionosphere and magnetosphere, satellite communications, nuclear explosion detection and general solar terrestrial relationships. The cold war generated considerable interest in using the ionospheric waveguide for global communications with submarines. More recently there has been a surge in activity related to understanding the global atmospheric electrical circuit and especially changes in thunderstorm and lightning activity in studies of global warming and climate variations. High-frequency ground observations of electric and magnetic fields can be used to study the Schumann resonances, improving knowledge of the global electric circuit and the role of sferics in the origin and structure of ELF fields. The subsurface conductivity structure completes the global electric circuit (GEC), so global induction studies can be expected to play an increasingly important role in understanding Earth's electric field.

More detail could be provided about external magnetic field variations. Progress since the advent of satellite observations has been enhanced by the integration of a theoretical framework based on plasma physics and fluid dynamics. A general introduction to Earth's external EM environment is given by Kivelson and Russell (1995). For specifics of ionospheric studies, see for example Kelley (2009). Nickolaenko and Hayakawa (2014) discuss resonant EM fields produced by lightning in the Earth-ionosphere cavity. Their monograph contains a comprehensive review of efforts to use the Schumann resonances for global sensing of thunderstorm activity and of the lower ionosphere at a planetary scale. A further volume by Surkov and Hayakawa (2014a) is devoted to ULF and ELF EM fields, while Helliwell (2006) relates the history of discovery and subsequent development of whistlers and related ionospheric phenomena.

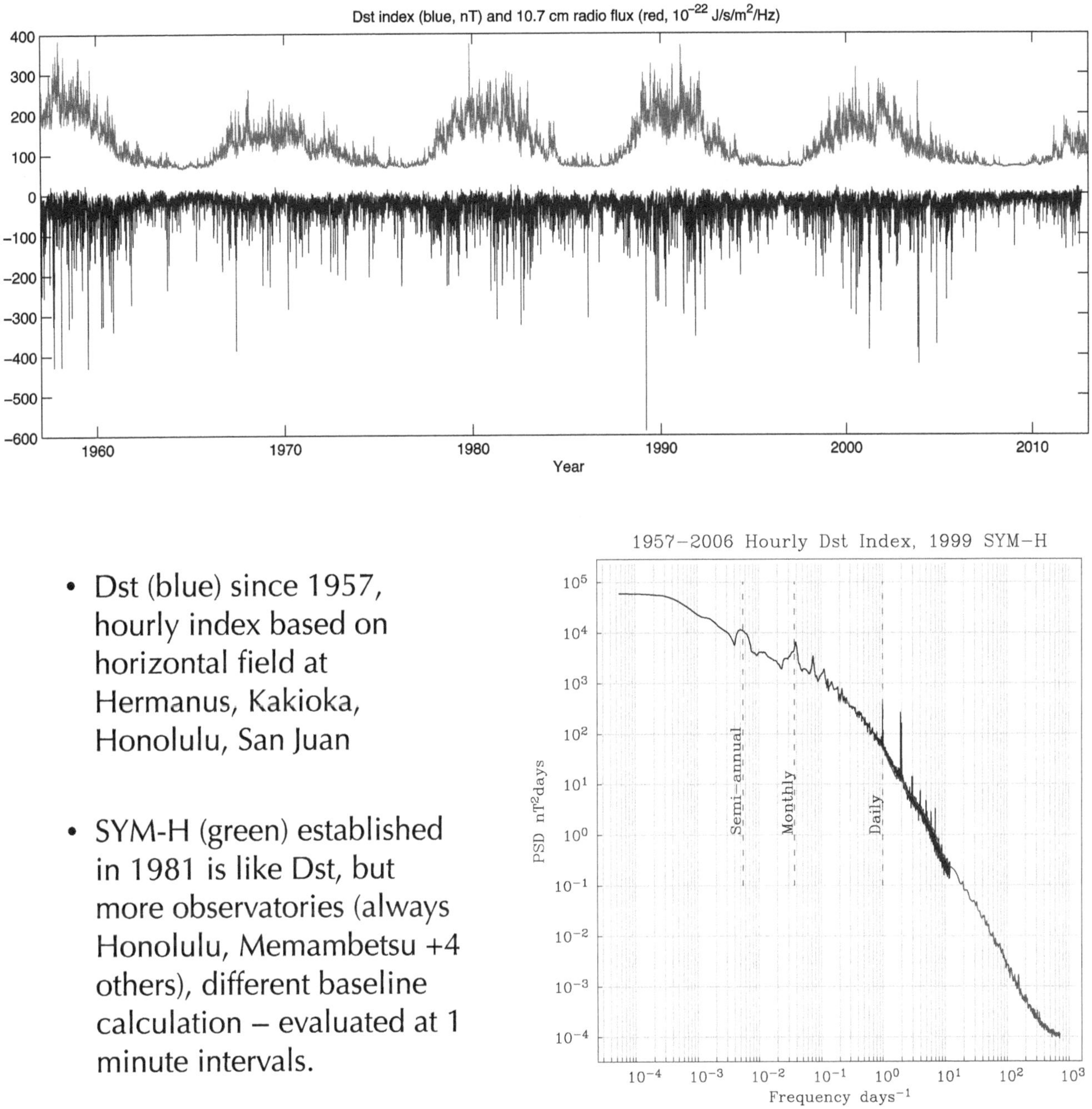

Figure 13.1.2 (a) Hourly *Dst* index variations from 1957 on compared with 10.7 cm radio flux. (b) the power spectrum of *Dst* from 1957 to 2006 and of *SYM-H* from 1999. *SYM-H* has 1 min sampling. (A black-and-white version of this figure appears in some formats. For the colour version, please refer to the plate section.)

13.2 Ultra-Low Frequency Wave Dynamics

*Ian R. Mann**

13.2.1 Introduction

In response to non-steady driving from the solar wind, the geospace environment is filled with plasma waves. Plasma waves in the ultra-low frequency (ULF) band, with scale sizes comparable to the scale size of the magnetosphere, have frequencies from around 1 mHz to over 1 Hz, and play a fundamental role in transporting energy throughout the coupled and dynamical geospace system. For example, this can occur as a result of solar wind impulses (e.g. Kepko et al., 2002; Claudepierre et al., 2009), and/or the action of the Kelvin–Helmholtz instability at the magnetopause (e.g. Chen and Hasegawa, 1974; Mann et al., 1999; Rae et al., 2005). ULF

* I. R. M. thanks Andy Kale for the artwork in Figure 13.2.1, and Canadian Natural Sciences and Engineering Research Council (NSERC) for financial support.

waves also play a role as an agent for the transfer of momentum from the solar wind into the ionosphere, and in establishing the patterns of ionospheric convection (e.g. Dungey, 1961; see also Chapter 15 of this volume). ULF waves can also interact with the energetic particle populations in the magnetosphere, causing the acceleration, transport and loss of ring current and radiation belt particles (see Section 13.3 for more details; see also the book edited by Balasis et al., 2016). ULF waves are also excited by the processes of energy release and dynamics in the nightside magnetosphere and magnetotail, including during substorms. This is discussed in more detail in Sections 12.3 and 15.1 of this volume.

In addition to energy transfer, the dynamical processes active in the coupled solar wind–magnetosophere–ionosphere system drive waves which can be used to remote sense and diagnose such processes non-locally through magnetoseismology (see e. g. the book by Menk and Waters, 2013) and ULF wave ray tracing (e.g. Tamao, 1964; Chi et al., 2009). Since the characteristics of the observed waves are affected by the medium through which they propagate, they can also be used to infer properties of the plasma, and especially the plasma mass density (e.g. Menk et al., 2014, and references therein; see also this volume Chapter 15 and Section 13.4). At magnetospheric boundaries such as at the plasmapause, there can be different gradients in ion number and mass densities due to spatial gradients in heavy ion populations, and these can also be investigated using ULF wave techniques (see e.g. Fraser et al., 2005). For example, Kale et al. (2009) investigated the dynamics of the plasmapause and heavy ions during the Halloween 2003 superstorm revealing a prevalence of O+ ions and an inward motion of the plasmapause to at least $L \sim 2.4$.

13.2.2 ULF Wave Characterization and Mode Coupling

In early studies of ULF waves, the frequency of the wave disturbance and the characteristic of the observed waveform were used to classify magnetic pulsations. Such a characterization scheme was proposed by Jacobs et al. (1964) and recommended to IAGA in 1963 (see Table 13.2.1). In the modern era, it has become well-established that a variety of either external and internal sources can drive waves in each category, perhaps with the exception of Pi2, a category which has become synonymous with the short wavetrain irregular pulsations seen at substorm onset – but even there it is now understood that many different processes active at substorm onset can produce waves in the Pi2 band (see e.g. the review by Keiling and Takahashi, 2011). The role of ULF waves in magnetosphere–ionosphere coupling, including during substorms, is discussed in Section 15.1.

In this chapter, we use a magnetohydrodynamic (MHD) theory description for wave propagation, highlighting additional kinetic effects where appropriate. Under such an approach the magnetic restoring force arises from the Lorentz $\boldsymbol{j} \times \boldsymbol{B}$ force per unit volume, which can be split into a force characterized by an isotropic magnetic pressure

Table 13.2.1 *Classification Scheme of Pulsations of Continuous (Pc) or Irregular (Pi) Nature According to the Classification Scheme of Jacobs et al. (1964)*

Continuous pulsations		Irregular pulsations	
Type	Period range (s)	Type	Period range (s)
Pc1	0.2–5.0	Pi1	1–40
Pc2	5–10	Pi2	40–150
Pc3	10–45		
Pc4	45–150		
Pc5	150–600		

gradient and a field-aligned magnetic tension force (see e.g. Kivelson and Russell, 1995). In a warm plasma there is an additional thermal pressure gradient force. The field-guided MHD mode of Alfvén (1942) and the more isotropic fast magnetoacoustic mode (hereafter the fast mode) (see e.g. Alfvén and Fälthammar, 1963) can both propagate through the magnetosphere providing pathways for the transport of energy throughout the system. In warm plasma MHD, there is also a slow magnetoacoustic mode which is usually assumed to be rapidly Landau damped by thermal particles in the Earth's magnetosphere; in a collisionless plasma such a slow phase speed mode also cannot be described using a simple compressibility approximation (e.g. Southwood and Hughes, 1983).

Dungey (1954) was the first to propose that two field-guided Alfvén waves, propagating in opposite directions after reflection from the ionosphere, could generate standing Alfvén wave harmonics with discrete frequencies on closed field lines. These frequencies vary from field line to field line, due to variation in both magnetic field and density, and are damped due to the dissipation of wave energy by Ohmic heating in the ionosphere. Early models used dipole fields to estimate the expected Alfvén eigenfrequencies as a function of McIlwain dipole L-value (e.g. Orr and Matthew, 1971, and references therein), finding good agreement with observations from ground-based magnetometers (e.g. Samson et al., 1971). More detailed computations of the damping rates of Alfvén waves in a background dipole magnetic field are provided for example by Allan and Knox (1979a, 1979b; see also Ozeke and Mann, 2005; Ozeke et al., 2005), and in a more realistic compressed and non-axisymmetric magnetosphere by, for example, Kabin et al. (2007).

In the inhomogeneous magnetosphere, fast modes propagating across field lines can become coupled to standing Alfvén modes through the process known as field line resonance (FLR) where the local standing field-aligned Alfvén wave frequency matches that of a driving fast mode (Tamao, 1966; Chen and Hasegawa, 1974; Southwood, 1974). In addition to the characteristics of the background plasma and its inhomogeneity, the nature of the wave propagation and mode coupling between fast and Alfvén waves is also determined by the wave azimuthal mode number, m. When $m = 0$, fast and Alfvén waves are decoupled; for other m-values the fast and Alfvén waves are coupled by the FLR processes, with the coupling rate determined by both the m-value as well as the local plasma gradient. The m-value also allows the fast and Alfvén wave modes to be distinguished by mode polarization. For low-m waves, Alfvén waves in an axisymmetric dipole magnetic field adopt a toroidal polarization, with magnetic field and velocity perturbations polarized predominantly in the azimuthal direction, and electric fields in the radial direction, resulting in a field-aligned Poynting flux (see Figure 13.2.1a). By contrast, low-m fast modes have a poloidal polarization and behave like 'breathing modes' with magnetic perturbations predominantly both parallel to the background field and in the radial direction, and electric fields predominantly in the azimuthal direction. Such polarizations can be used to identify these modes in the in-situ fields observed by satellites (see e.g. the discussion in Ozeke et al., 2005, and references therein; see also Section 13.4).

Waves with poloidal polarization are of particular interest, because their azimuthal electric field is parallel to the direction of the drift of energetic ring current and radiation belt particles. This results not only in the possibility that the waves can do positive (negative) work on the particles to accelerate (decelerate) them, but as discussed in Section 13.2.3, under certain conditions this interaction can be resonant causing a strong coupling between the waves and particles. Significantly, as discussed by Radoski (1967), in the limit of large $m \to \infty$ the Alfvén wave still propagates, remains incompressible, but adopts a poloidal polarization in the form of a guided poloidal mode (Figure 13.2.1b). For intermediate m-values, both fast and Alfvén modes can demonstrate mixed polarizations, with $m = 0$ and $m \to \infty$ being limiting cases (e.g. Mann et al., 1995, 1997; Mann and Wright, 1995). High, but finite, m-value waves may also remain significant for wave-energetic particle interactions.

Interestingly, in the inhomogeneous magnetosphere, where the standing field-aligned Alfvén eigenfrequencies are a function of L-shell, once Alfvén waves have been driven across an L-shell envelope of field lines, phase mixing causes internal radial structure inside the FLRs whose spatial scales decrease with time, t, $\propto 1/t$ (e.g. Mann et al., 1995). In the case of high-m guided poloidal modes, the phase mixing causes a time-dependent polarization rotation from poloidal to toroidal during the waves' lifetime (Mann and Wright, 1995). This can complicate the interpretation of in-situ field data; nonetheless ULF wave polarization remains a powerful tool for diagnosing the MHD wave mode and the nature of mode coupling and other wave characteristics (e.g. Chisham et al., 1997; Ozeke et al., 2005; Rae et al., 2005; Mann et al., 1998; Dai et al., 2015).

13.2.3 Wave Excitation Mechanisms

It is a useful phenomenological distinction to separate the excitation mechanisms of large-scale Pc3-5 ULF waves into external and internal to the magnetosphere.

13.2.3.1 External Drivers of ULF Waves

The external drivers of low-m dayside long-period ULF waves, with scale sizes of the order of the global magnetosphere, can be linked to structures and features in the solar wind. For example, increases in solar wind mass density and/or velocity can cause direct variations in dynamic pressure in the upstream solar wind. These upstream dynamic pressure changes, which survive processing through the bow shock, can cause the impulsive excitation of ULF waves inside the magnetospheric cavity (e.g. Kivelson and Southwood, 1985; Sibeck, 1990; Kepko et al., 2002; Takahashi and Ukhorski, 2007; Claudepierre et al., 2009, 2010). Interestingly, recent research has also shown how rotations in the IMF, when processed by an interaction with the bow shock, can generate downstream pressure pulses often associated with hot flow anomalies (HFAs) which can also impact the magnetopause and drive ULF waves (e.g. Sibeck et al., 1999). More recent work by Hartinger et al. (2013) suggests that ion-foreshock associated processes might be a much larger source of long-period Pc5 ULF wave power in the magnetosphere than previously thought, in addition to their long association with Pc3 waves arising from the specular reflection of solar wind ions at the foreshock (e.g. Le and Russell, 1994, and references therein).

These compressional waves typically propagate through the inhomogeneous magnetosphere until they are reflected at a turning point, and if additionally reflected from the magnetopause can cause the creation of a series of cavity modes with discrete frequencies (e.g. Kivelson and Southwood, 1985, 1986; Mann et al., 1995). As described by Wright (1994), the outer magnetosphere is probably more appropriately modeled as a waveguide which is open downtail (cf. Figure 13.2.1a). However, despite dispersion in the waveguide, the weak-dependence of the waveguide mode eigenfrequency on m-value, and the fact that the lowest m-value waveguide modes have the slowest downtail group speed, mean that such waveguide modes typically have frequencies close to those expected when $m = 0$ (Wright, 1994). These discrete frequency cavity/waveguide modes can then drive discrete frequency standing FLRs (e.g. Rickard and Wright, 1994; Claudepierre et al., 2009), the cavity/waveguide model offering a potential explanation for the long-standing ground-based observations of discrete frequency FLRs (e.g. Samson et al., 1971; Mathie et al., 1999a, 1999b; Rae et al., 2005), with Mann et al. (1998) reporting the observation of the downtail propagation of a dispersing waveguide mode wavepacket. Low-m FLRs demonstrate tailward phase propagation in ground magnetometer data (e.g. Mathie and Mann, 2000) as well as with HF radars (e.g. Ruohoniemi et al., 1991; Samson et al., 1991; Fenrich et al., 1995; Yeoman et al., 1997) consistent with the waveguide driver hypothesis.

Fast solar wind speeds are also thought to cause the excitation of the '*wind-over-water*' Kelvin–Helmholtz instability (KHI) along the magnetopause (see Figure

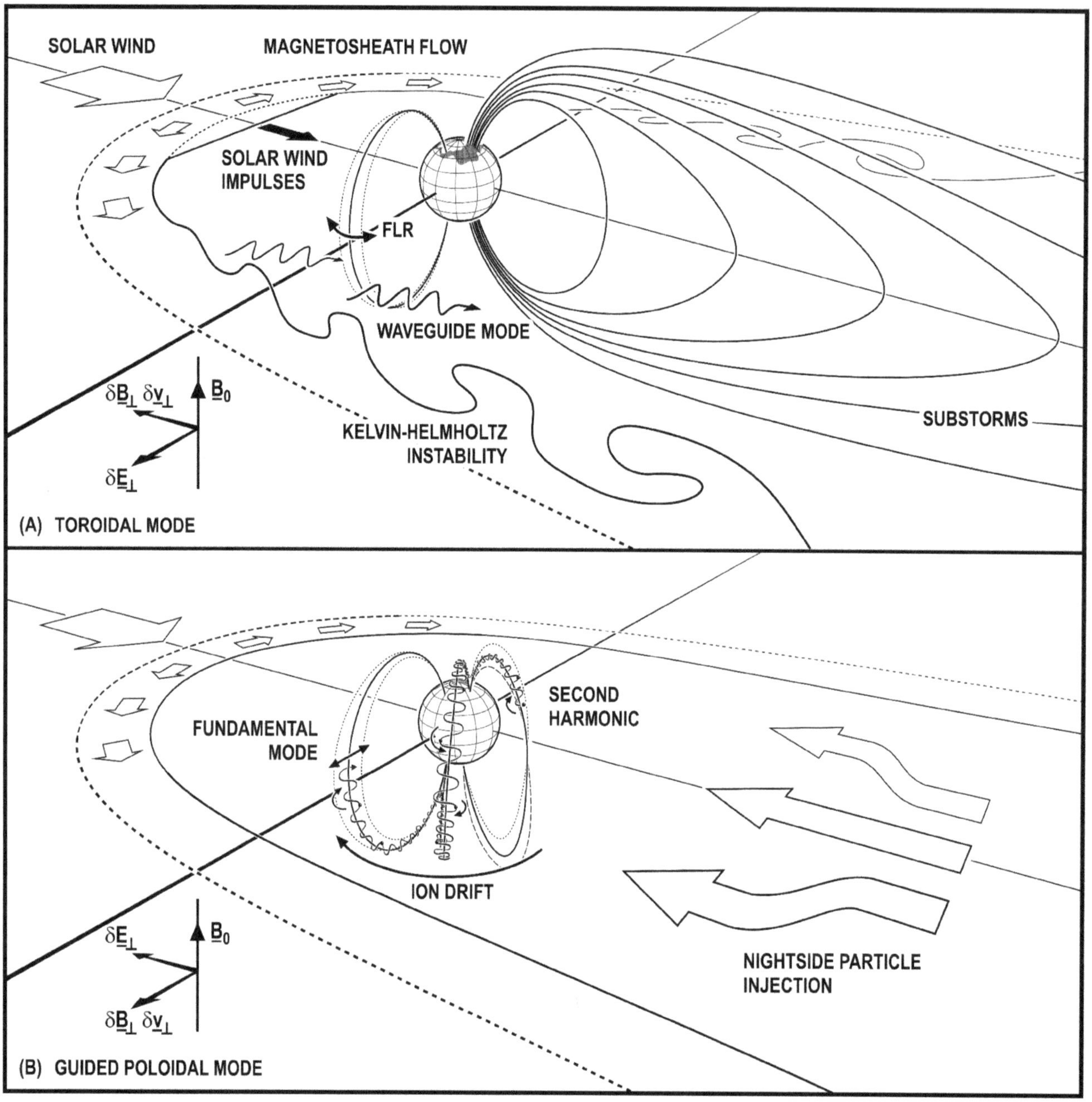

Figure 13.2.1 Schematic of (a) external excitation of waveguide modes in the magnetosphere by solar wind impulses and/or flank Kelvin–Helmholtz instability (KHI) and ULF excitation from nightside disturbances and magnetotail flows from substorms. Resulting field-aligned standing Alfvén wave field line resonances (FLRs) have low azimuthal wavenumber (m) with a transverse toroidal polarization with respect to the background magnetic field (B_0), and (b) internal excitation of high-m guided poloidal Alfvén waves arising from drift- or drift-bounce resonance with energetic particles, for example as injected from the magnetotail. The westward ion drift trajectory of the gyro-centers of ions is shown, and the inset shows the polarization of the guided poloidal Alfvén waves with respect to B_0.

13.2.1a). The KHI is often associated with the development of magnetopause surface waves, and these have long been associated with the driving of ULF waves in the magnetosphere (e.g. Chen and Hasegawa, 1974), with recent studies suggesting that the magnetopause is very often unstable to the KHI (e.g. Kavosi and Raeder, 2015). Especially for low-frequency modes with long perpendicular wavelengths, the KHI can allow compressional fast mode energy from these surface modes to penetrate deep into the magnetosphere and drive FLRs. However, the KHI can also couple to propagating body modes which allows energy extracted from the solar wind to propagate into the magnetosphere and perhaps efficiently excite KHI-energized cavity/waveguide modes through the process of over-reflection (Walker, 2000; Mann et al., 1999;

Mills et al., 1999). The nature of the modes excited in the waveguide affects the characteristics of the resulting FLRs, such that their features can be used to diagnose the excitation mechanism of the waveguide modes, and in particular to distinguish between more impulsive and Kelvin-Helmholtz-like magnetopause running pulse drivers (Mann and Wright, 1999; Mathie and Mann, 2000).

The absence of the L-independent frequency signatures expected for cavity/waveguide modes in in-situ satellite magnetic field spectra has challenged the cavity mode hypothesis (e.g. Kivelson et al., 1997). However, subsequent studies (e.g. Waters et al., 2002, and references therein) and examinations using the trajectories of virtual spacecraft in coupled ULF wave simulation models have demonstrated why such signatures might be expected to be rather weak (Rickard and Wright, 1995). At lower L-shells, however, there is clear evidence in satellite magnetic and electric field data for the existence of cavity modes trapped inside the plasmasphere. Here, a cavity is formed by reflection from a turning point inside the plasmasphere and the plasmapause (e.g. Takahashi et al., 2010), energy being fed into these plasmaspheric cavity modes as a result of mode tunneling from the outer magnetosphere (e.g. Lee and Lysak, 1999).

There remains a challenge to explain the low-frequencies of the lowest harmonics of the cavity/waveguide modes of the outer magnetosphere. Harrold and Samson (1992) suggested that the cavity outer boundary might be located at the bow shock rather than the magnetopause, while Mann et al. (1999) emphasized the possibility that the lowest cavity/waveguide harmonic might be represented as a radial quarter wavelength mode. More recently, Plasche et al. (2009) offered an alternative suggestion for the observed discrete frequency modes, proposing Kruskal-Schwarzschild modes confined to the magnetopause boundary layer as an alternative explanation. The reader is referred to the review by Wright and Mann (2006) for a more detailed discussion of the nature and characteristics of ULF waves in the large-scale waveguides of the dayside and nightside magnetosphere.

13.2.3.2 Internal Drivers of ULF Waves

ULF waves can also be driven internally, inside the magnetosphere, as a result of wave–particle interactions with energetic particle populations which are trapped on closed magnetospheric field lines. Unstable ring current ion distributions with available free energy with $df/dW > 0$, where f is the ion phase space density and W is the ion energy, have been associated with ULF wave excitation in the Pc3–5 range as a result of drift-bounce resonances (see Figure 13.2.1b). As described by Southwood et al. (1969), such free energy can be provided either by positive energy spectral gradients ($\partial f/\partial W > 0$), or by spatial gradients ($df/dL > 0$), or a combination of both, depending upon local time and local plasma conditions (see also e.g. Ozeke and Mann, 2001). Such energy transfer is usually examined in the context of the drift-bounce resonance which occurs when $\omega - m\omega_d = N\omega_b$, where ω and m are wave frequency and azimuthal wavenumber, ω_b is the particle bounce frequency and ω_d is the bounce-average drift frequency of the resonant particles (e.g. Southwood et al., 1969; Southwood, 1976; Chen and Hasegawa, 1991).

Such internal wave–particle excitation usually results in the driving of Alfvén waves with a poloidal polarization, which follows since the particles do work on the waves driving energy transfer at the rate given by $dW/dt = qE.v_D$, and in a dipole field the particle drift velocity v_D is in the azimuthal direction. As discussed for example by Elkington et al. (2003), in a compressed dipole field additional $m \pm 1$ drift resonances can also become important and might potentially allow wave–particle energy transfer involving both poloidal and toroidal modes. The temporal evolution of ULF waves excited by energetic ring current ions can also involve a polarization rotation with time from poloidal to toroidal, especially following the end of the driving phase (Mann and Wright, 1995; Mann et al., 1997). Data from ground-based magnetometers has provided strong evidence that this polarization rotation may develop as predicted by the theory, for example in relation to giant pulsations (e.g. Chisham et al., 1997). Klimushkin (2000, and references therein) have also assessed such polarization rotations including the effects of curved magnetic field geometries, and the transfer of energy from the poloidal to the toroidal resonance surface.

There are numerous papers which provide evidence from satellite data supporting the drift-bounce resonance excitation theory (e.g. Min et al., 2017; Yeoman et al., 2016; Takahashi et al., 2013, and references therein), including not only wave polarizations in space but also evidence for free energy in particle spectra (e.g. Hughes et al., 1978; Takahashi et al., 2018). Measurements of the ionospheric flows which are created by standing Alfvén waves in FLRs can also provide a diagnostic for their excitation mechanisms, with low-m externally driven and high-m internally driven FLRs being separable using the SuperDarn HF radar network (e.g. Fenrich et al., 1995; Fenrich and Samson, 1997), the high-m waves usually demonstrating westward phase propagation (see also e.g. Yeoman and Wright, 2001; Baddeley et al., 2002). Some recent studies have also identified high-m waves with eastward phase propagation, which suggests that energetic ring current electrons could also drive ULF waves via drift-bounce resonance (e.g. James et al., 2013). A number of authors have also identified an association between co-located high- and low-m FLRs, suggesting some high-m waves might be driven via an internal non-linear KHI (e.g. Allan and Wright, 1997; Wright and Yeoman, 1999).

Overall, and as discussed for example by Ozeke and Mann (2001, 2008) and Murphy et al. (2014), it seems likely that waves in the ULF band play an important role in energetic particle dynamics in the magnetosphere. This can span substorm injections, ring current dynamics, and might even include a role for internally excited ULF waves driven by keV energy ring current particles in radiation belt electron dynamics (e.g. Ozeke and Mann, 2008). In such a case moderate azimuthal wavenumber ULF waves excited by drift-bounce resonance with ring current ions might additionally be resonant with radiation belt electrons through the $N = 0$ drift resonance. ULF wave-radiation belt electron interactions are discussed in more detail in Section 13.3.

13.3 Ultra-Low Frequency Wave–Particle Interactions in the Radiation Belts

Ian R. Mann, Louis G. Ozeke, Alex Degeling, Kyle R. Murphy and Robert Rankin

13.3.1 Introduction

Understanding the dynamics of the relativistic electrons in the Earth's Van Allen radiation belts remains one of the most important and fundamental problems in space physics. One of the major challenges in making scientific advances arises because the dynamics of the outer radiation belt are believed to be controlled by a delicate balance between electron acceleration and loss processes (e.g. Reeves et al., 2003). Enhancements in the outer radiation belt flux can be caused by multiple different processes which may occur at the same time during magnetic storms, such as local acceleration of the electrons by chorus waves (e.g. Reeves et al., 2013; Li et al., 2015a, 2016a) and inward ULF wave radial diffusion, (e.g. Ozeke et al., 2014a, 2014b; Mann et al., 2016; Li et al., 2015b, 2016b). Geomagnetic storms are also times when multiple competing loss processes may be present. For example, enhanced outward radial diffusion and magnetopause shadowing can cause the loss of electrons through the magnetopause (e.g. Shprits et al., 2006; Loto'aniu et al., 2010; Turner et al., 2012; Mann et al., 2016), while pitch angle scattering of the electrons into the loss cone can cause electron loss to the atmosphere by wave–particle interactions with waves such as chorus, hiss, or electromagnetic ion cyclotron (EMIC) waves (see e.g. the review by Millan and Thorne, 2007).

In this section we review the role of ULF wave–particle interactions in the radiation belts. In particular, we examine the role of long-period Pc3–5 waves in radiation belt transport, acceleration and loss. We briefly also examine the role of ULF waves in the Pc1 band, in particular EMIC waves, and their possible role in radiation belt loss through the scattering of radiation belt electrons into the atmosphere. Wave–particle interactions of radiation belt electrons with chorus waves and with plasmaspheric hiss are examined elsewhere in this volume in Section 12.6.

13.3.2 Long-Period ULF Wave–Particle Interactions

It is now well-known that there is a strong correlation between long-period Pc5 ULF wave power and MeV electron flux enhancements in the Van Allen belts (e.g. Mathie and Mann, 2000; Mann et al., 2004, 2012). ULF wave interactions with MeV electrons can occur through drift resonance (cf. Section 13.2.3), and electrons with energies up to around 5 MeV or more can interact with ULF wave frequencies ~0.1–100 mHz in the outer zone of the electron radiation belt. The characteristics of the amplitudes, frequency spectra, and even azimuthal wavenumbers of ULF waves can be resolved from the ground using data from magnetometer arrays around the world. Recent work has used such capabilities to demonstrate the value of ground-based ULF wave studies for understanding the critical impact of ULF waves on the dynamics of these relativistic so-called satellite 'killer' electrons in the Van Allen belts (e.g. Loto'aniu et al., 2010; Rae et al., 2012; Ozeke et al., 2009, 2012a, 2012b, 2014a, 2014b; Claudepierre et al., 2013; Mann et al., 2004, 2012, 2016).

Stochastic ULF wave-driven radiation belt electron transport can occur by radial diffusion through violation of the third adiabatic invariant, and conservation of the first and second adiabatic invariants (e.g. Fälthammar, 1965; Schulz and Lanzerotti, 1974; Rostoker et al., 1998; Mathie and Mann, 2001; Mann et al., 2016; Ozeke et al., 2012b, 2014b, 2017). In these radial diffusion models, the dynamics of outer radiation belt electron flux in the magnetic equatorial plane can be described using the radial diffusion equation:

$$\frac{\partial f}{\partial t} = L^2 \frac{\partial}{\partial L}\left[\frac{D_{LL}}{L^2}\frac{\partial f}{\partial L}\right] - \frac{f}{\tau}. \tag{13.3.1}$$

Here, f represents the phase space density of the electrons and it is assumed that the first and second adiabatic invariants, M and J, are conserved (see Schulz and Lanzerotti, 1974). The diffusion coefficient and the electron lifetime are represented by D_{LL} and τ, respectively, and depending on the local gradient of particle phase space density, diffusive transport can be either inwards (e.g. Falthammer, 1965; Schulz and Lanzerotti, 1974; Brautigam and Albert, 2000; Ozeke et al., 2014) or outwards (e.g. Shprits et al., 2006; Loto'aniu et al., 2010; Turner et al., 2012; Mann et al., 2016).

ULF wave radial diffusion can therefore enhance the flux in the belts through inward transport of a new source population from the outer boundary, or cause losses through outwards radial diffusion where the gradients are reversed – for example during periods of magnetopause shadowing (e.g. Turner et al., 2012; Mann et al., 2016). In radial diffusion models, the flux dynamics in the domain inside the outer boundary are controlled by the rate of transport as specified by a combination of the radial diffusion coefficient, D_{LL}, and the local phase space density gradient (e.g. Schulz and Lanzerotti, 1974), and the competing losses as specified by τ. Typically, an inner boundary condition that $f = 0$ is also applied at the inner boundary at $L = 1$. The solutions to Equation (13.3.1) only give the electron phase density space, f. However, Equation (13.3.1) can be solved for multiple different first adiabatic invariants, M, and the simulations combined together for a direct comparison to measurements from spacecraft energetic particle detectors at fixed energy (see e.g. Ozeke et al., 2014a, 2014b, 2017; Mann et al., 2016, 2018).

Prior empirical approaches to specifying D_{LL} have used the planetary magnetic activity index Kp (e.g. Brautigam and Albert, 2000). This has been improved in more recent studies by using Canadian Array for Real-time Investigations of Magnetic Activity (CARISMA; Mann et al., 2008) ground magnetometer observations to characterize the ULF wave power and hence the electric diffusion coefficient D_{LL}^{E} as a function of Kp (Ozeke et al., 2012a; see also Rae et al., 2012; Ozeke et al., 2012b, 2014; Mann et al., 2012). This represents a pragmatic approach for ULF wave-driven radial diffusion simulations since D_{LL}^{E} is typically much stronger than the magnetic diffusion coefficient, D_{LL}^{M} (Ozeke et al., 2014a; Tu et al., 2012; see also Brizard and Chan, 2001; Fei et al., 2006).

These approaches require the mapping of ULF wave magnetic amplitude on the ground through the ionosphere and into the equatorial plane and typically use a model for guided Alfvén wave eigenmodes (e.g. Ozeke et al., 2009; see also e.g. Sciffer and Waters, 2011). During storm times, the Kp-dependence of ULF wave power may also be different from the solar cycle average (e.g. Dimitrakoudis et al., 2015), suggesting additional statistical characterizations using the storm-time disturbance index Dst, with proximity to the magnetopause potentially also being important (Murphy et al., 2015). These modeling approaches also need to specify the loss timescale, τ, and typically use plasma wave characterizations (such as Orlova et al., 2016; Gu et al., 2012), and an empirical Carpenter and Anderson (1992) plasmapause location to separate hiss and chorus losses, as was done for example by Ozeke et al. (2017) and Mann et al. (2016).

Further improvements can be made using observed event-based ULF wave fields to constrain the rates of diffusion, for example as done by Mann et al. (2016, 2018) who used ULF wave diffusion and magnetopause shadowing to explain the formation of a remnant storage ring and subsequent third radiation belt morphology reported by Baker et al. (2013). Figure 13.3.1 uses such a modeling approach applied to the September 2014 storm studied by Ozeke et al. (2017) (left column), and to the March 2015 superstorm (right column) one of the strongest magnetic storms in the era of the operation of NASA Van Allen Probes (e.g. Mauk et al., 2013). In each storm there was a very rapid hour timescale loss in the main phase (e.g. Olifer et al., 2018). In the September 2014 event, this was followed by an extended period of radiation belt depletion which existed for around two weeks following a very fast radiation belt extinction before ULF wave transport of a renewed source of flux at the boundary replenished the belt (Ozeke et al., 2017). In contrast, in the March 2015 event, the fast losses were followed almost immediately by very fast and strong radiation belt acceleration from fast inwards radial diffusion.

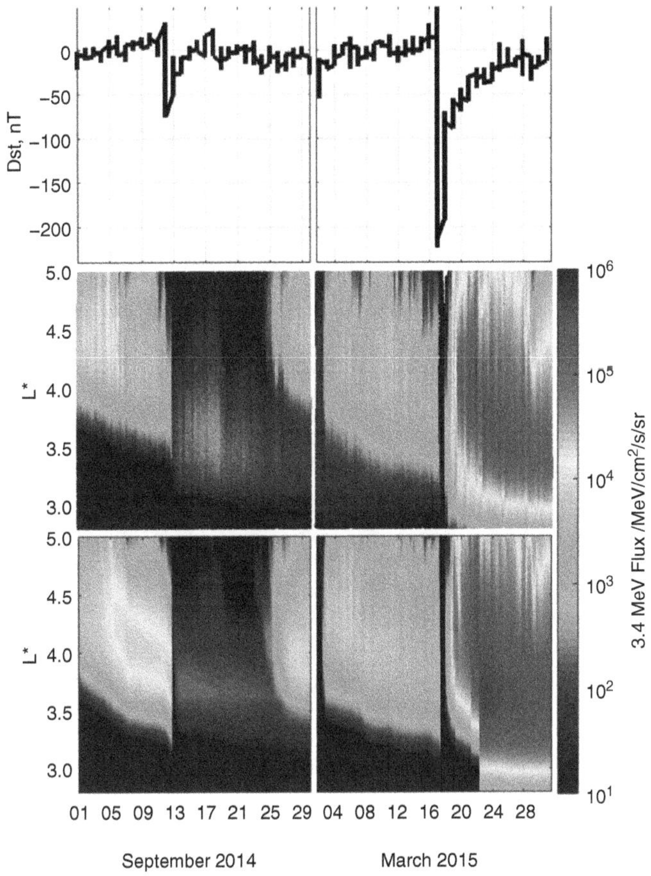

Figure 13.3.1 Comparison of ULF wave radial diffusion model results for the magnetic storms in September 2014 (left column; cf. Ozeke et al., 2017) and March 2015 (right column). (top) Storm-time disturbance index, Dst; (middle) 3.4 MeV electron flux observed by the Van Allen Probes; and (bottom) ULF wave-driven radial diffusion model results of electron flux at a fixed energy at 3.4 MeV using ULF wave power observed by ground-based magnetometers and mapped to the equatorial plane. The simulation results in the bottom row combine multiple first adiabatic invariant conserving radial diffusion simulations, each driven by the flux variations at the outer boundary of the simulation domain at $L^* = 5$ and combined to generate flux at fixed energy for direct comparison to the Van Allen Probe observations (cf. Mann et al., 2016, 2018; Ozeke et al., 2017). See text for more details. (A black-and-white version of this figure appears in some formats. For the colour version, please refer to the plate section.)

In Figure 13.3.1, the rates of radial diffusion were defined using event-based ground-based ULF wave observations, and the agreement between the Van Allen Probes observations (middle row) and the modeling (bottom row) is very good. Note that these radial diffusion simulations do not include any local VLF wave chorus acceleration. Although empirical

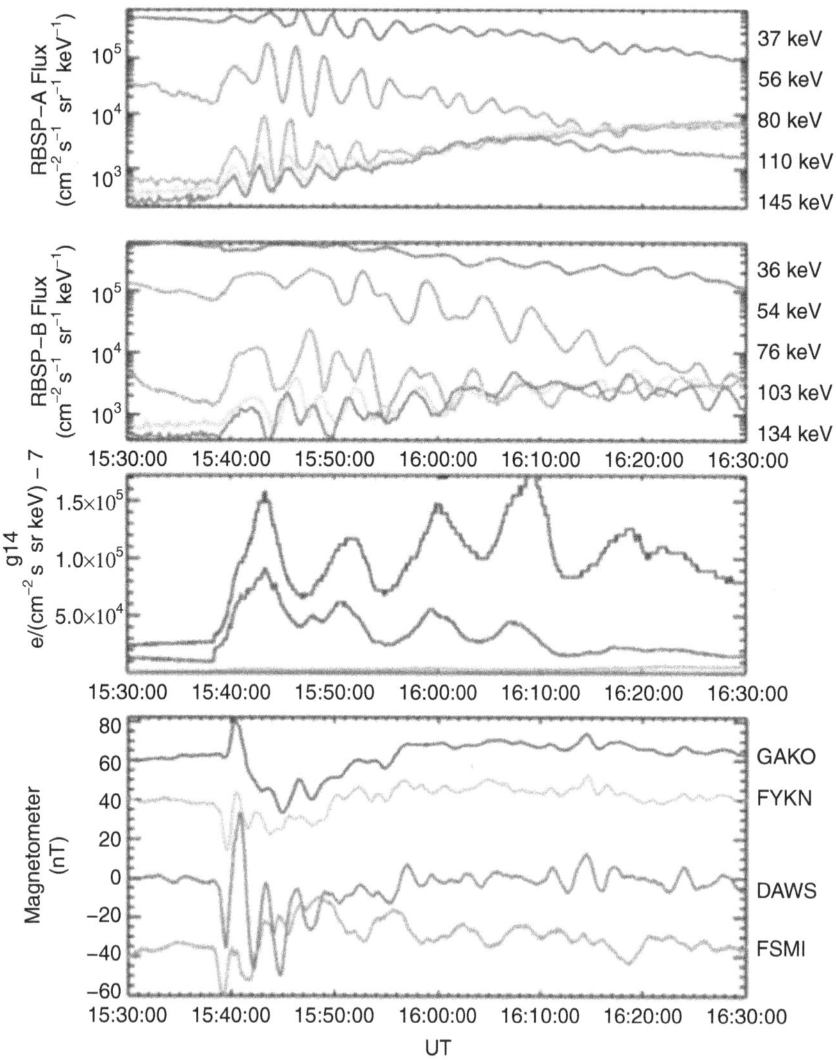

Figure 13.3.2 Observations of coherent ULF modulation of ~50–80 keV energetic electron flux on 31 October 2012 (cf. Claudepierre et al., 2013). Energetic electrons from Van Allen Probes A and B (formerly the Radiation Belt Storm Probes (RBSP)) are shown in the top two panels and from GOES 14 (g14; third panel), as well as the H-component fields from selected ground-based magnetometers (bottom panel). From west to east: Gakona (GAKO) and Fort Yukon (FYKN) from the Geophysical Institute Magnetometer Array (GIMA) in Alaska, and Dawson City (DAWS) and Fort Smith (FSMI) from the CARISMA array (Mann et al., 2008). (A black-and-white version of this figure appears in some formats. For the colour version, please refer to the plate section.)

representations of chorus and hiss loss inside and outside the plasmapause are included through τ (using the method explained in Ozeke et al., 2017, and Mann et al., 2018), there are no additional losses due to specifications of electromagnetic ion cyclotron (EMIC) wave scattering into the atmosphere (cf. Section 13.3.3). Note, however, that in the manner described by Ozeke et al. (2017), fast hour timescale extinction across the belt is included in the main phase at minimum Dst for both storm simulation runs shown in Figure 13.3.1. This extinction rapidly extinguishes the entire belts, effectively decoupling the pre- and post-storm fluxes, and is required in order to get good model agreement with the observations. Whether this loss can be explained by EMIC wave losses, by fast transport to the magnetopause, or requires some other as yet unidentified mechanisms is not known – but represents an important focus for future research.

Typically, ULF wave electron radial transport in the Van Allen belts is assumed to be diffusive. However, at times the response might be more coherent (see Figure 13.3.2) delivering acceleration on a timescale proportional to time rather than the square root of time as in the case of diffusion (e.g. Mann et al., 2013). Evidence for coherent ULF wave acceleration was recently presented by Mann et al. (2013) at energies around 500 keV using data from the Combined Release and Radiation Effects Satellite (CRRES), and by Claudepierre et al. (2013) using data from the NASA Van Allen Probes at energies of ~50–80 keV. These observations used the long-predicted, but rarely observed, signature of 180 degree phase change expected across the resonant energy (e.g. Southwood, 1976) to identify a coherent interaction. As described by Mann et al. (2013), instruments with broad energy channels will integrate over these energy-dependent flux modulations masking the signatures of the coherent ULF wave interaction. However, data from the Energetic Particle Composition and Thermal (ECT; Spence et al., 2013) suite on the NASA Van Allen Probes have sufficient energy resolution that the details of coherent resonances can not only be resolved (Claudepierre et al., 2013), but are also a relatively common phenomena (e.g. Zhou et al., 2016).

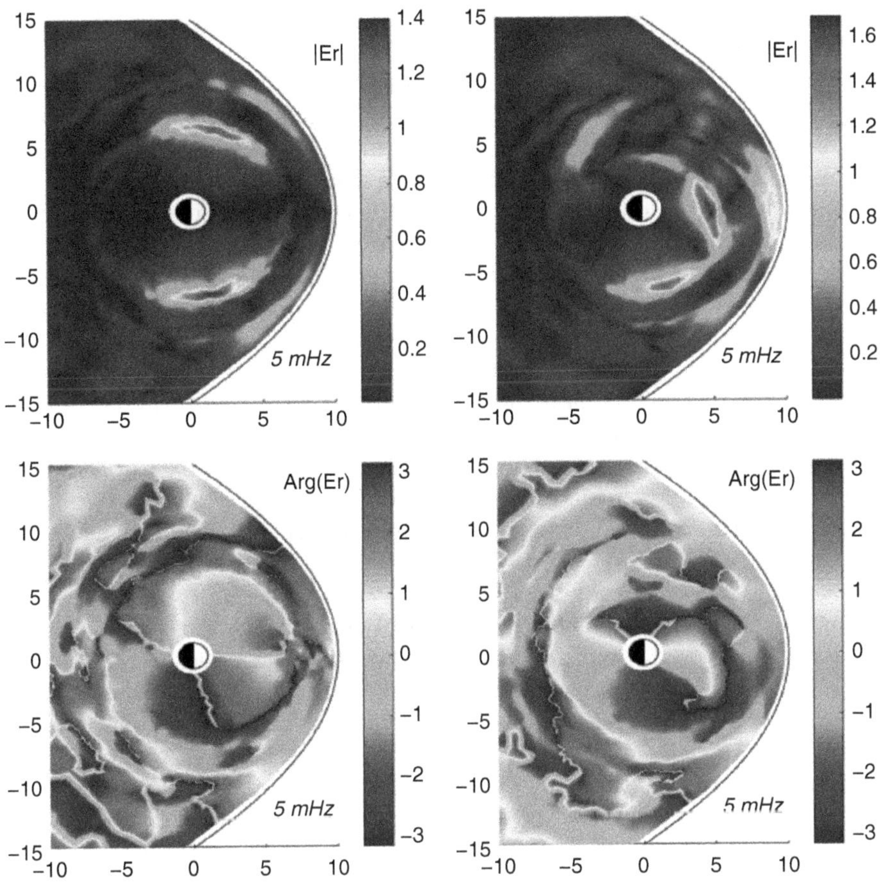

Figure 13.3.3 Simulations of 5 mHz ULF wave propagation driven inside the magnetosphere by a magnetopause driver using the model of Degeling et al. (2018). (left) Profiles of the magnitude of the radial electric field Er (top), and its phase in radians (bottom), in the equatorial plane for a symmetric plasmasphere. (right) Equatorial profiles of 5 mHz ULF wave Er magnitude (top) and phase (bottom) arising from the same magnetopause driver, but where 24 hours of Volland-Stern cold plasmasphere evolution have produced a storm-time plasmaspheric drainage plume, which significantly distorts the ULF wave power distributions in MLT, as well as its penetration to low L in the local morning sector. See text for more details. (A black-and-white version of this figure appears in some formats. For the colour version, please refer to the plate section.)

As shown in Figure 13.3.2 (top two panels), during the 31 October 2012 event studied by Claudepierre et al. (2013), despite their relatively close proximity Van Allen Probes A and B saw very different energetic particle modulation. At higher altitudes, the GOES 14 satellite (Figure 13.3.2, third panel) showed a different response from the Van Allen Probes, suggesting a region of wave–particle interaction localized in both L and MLT. Conjugate ground-based magnetometers (Figure 13.3.2, bottom panel) verified the longitudinally localized nature of the driving ULF wave, with only the Dawson City (DAWS) station observing strong coherence of the observed wave magnetic fields with the particle flux modulation observed on Van Allen Probe A. Observations such as these highlight the importance of multi-point observations at meso- and global scales. Modern dense ground-based magnetometer networks offer the promise for further advances, although the nature of wave propagation to the ground must be carefully considered (e.g. Ozeke et al., 2009; Sciffer and Waters, 2011). Further work is needed to assess the relative importance, prevalence and impact of coherent verses diffusive ULF wave transport for explaining Van Allen belt dynamics.

Since the nature and strength of ULF wave–particle interactions depends on the characteristics of wave propagation into the inner magnetosphere, the 3-D structure of the total plasma mass density, and hence Alfvén speed profiles, is likely critically important. As discussed by Mann et al. (2006), this provides a remarkable cross-energy coupling between, for example, eV and MeV energy particles spanning six orders of magnitude through the intermediary of ULF waves. The storm-time dynamics of the plasmasphere, including the generation of the storm-time plasmaspheric drainage plume (e.g. Goldstein et al., 2004), are expected to significantly affect wave refraction and reflection. Such changes can occur on very rapid hour-type timescales (e.g. Kale et al., 2009), and can significantly impact the penetration of ULF wave power to low L-shells with impacts for electron transport and penetration into the slot region (Loto'aniu et al., 2006).

Figure 13.3.3 illustrates the impact of cold plasma structure on ULF wave propagation. The linear warm plasma MHD waves, launched from perturbations of the dayside magnetopause, propagate in a medium which includes the advection of cold plasma in the plasmasphere and the generation of a plasmaspheric drainage plume in a Volland-Stern convection electric field (Degeling et al., 2018). Wave refraction and field line resonance excitation in the non-uniform and non-axisymmetric magnetosphere significantly change the distribution of ULF wave power. In the left column of Figure 13.3.3, 5 mHz waves are continuously launched from a broad magnetopause source centered at 12 MLT into a magnetosphere with an azimuthally symmetric plasmasphere confined by a plasmapause between 4 and 5 Earth radii. Strong, narrow peaks in the amplitude of the radial electric field, E_r, occur at $L = 6$, accompanied by a phase change of π radians across the peak. This indicates the excitation of field line resonances (FLRs) in the morning and afternoon local times sectors with a high degree of symmetry about the noon meridian, as expected.

In Figure 13.3.3 (right column), the same 5 mHz magnetopause ULF wave source is used. However, the cold plasma density profile evolves under the action of constant 0.25 mV/m dawn-dusk convection electric field for 24 hours creating a plasmaspheric drainage plume in the afternoon sector. While the morning sector FLR location has remained largely unaltered from the symmetric case, the afternoon sector FLR has shifted substantially. This FLR is now centered at approximately 12 MLT and at an L-shell of less than 5, due to an increase in density across the dayside magnetosphere which extends into the afternoon sector. Overall the axis of symmetry in the dayside ULF wave profile has now shifted toward the morning sector, resulting in an eastward ULF wave phase propagation across most of the dayside magnetosphere. The plamaspheric drainage plume structure, which extends in an arc from 18 MLT at low L to approximately 15 MLT at the magnetopause, also has a profound effect on the wave structure, with an FLR-like peak appearing along the duskward edge of the density plume. Standing wave (node/antinode) structures are also visible at higher L-shells in the late afternoon sector, within the density plume. Each of these three factors: deeper penetration of ULF wave power due to the increased density, eastward phase propagation across most of the dayside, and the excitation of standing modes within the plasmaspheric density plume itself, are all potentially beneficial for electron energization by ULF waves. Similar results, and more details of the methodology, are presented in Degeling et al. (2018).

Overall, these simulations indicate that the evolving three-dimensional structure of the cold plasma distribution in the storm-time magnetosphere can significantly affect not only the characteristics of the ULF waves excited by the solar wind, but also impact the nature of their potential acceleration and transport of Van Allen belt electrons. This is expected to be an important topic for future study.

13.3.3 Short-Period Pc1 EMIC Radiation Belt Wave–Particle Interactions

Short-period electromagnetic ion cyclotron (EMIC) plasma waves in the ULF Pc1 band (cf. Table 13.2.1) have long been hypothesized as a possible additional radiation belt electron loss mechanism involving scattering into the atmosphere (e.g. Horne and Thorne, 1998; Albert, 2003; Meredith et al., 2003; Summers et al., 2007; Silin et al., 2011). A Doppler-shifted resonance of the EMIC wave fields into the range of electron gyrofrequencies is the basis of the interaction, and this can occur at relativistic energies as a result of the fast field-aligned speed of the electrons. Cold plasma approximations predict stronger EMIC wave interactions at energies >~2 MeV, but in regions of high plasma density and/or low magnetic field lower energy interactions may still be possible (e.g. Meredith et al., 2003). A warm plasma treatment by Silin et al. (2011) indicates that the inference from cold plasma theory that the region around the ion gyrofrequencies are likely preferable for MeV electron interactions is incorrect, and that the critical parameter for strong interactions is the parallel wavenumber. These authors show that any EMIC wave frequency dependence on the intensity of the interaction is weak, being determined primarily only implicitly from the parallel wavenumber through the dispersion relation. However, even using warm plasma theory Silin et al. (2011) conclude that interactions of EMIC waves with electrons with energies below ~2 MeV are unlikely.

A recent observational study by Usanova et al. (2014) identified a clear connection between periods of EMIC wave occurrence and impacts on ultra-relativistic (>~2 MeV) electrons. Usanova et al. compared EMIC waves observed by ground magnetometer networks at $L = 4$–4.5 with ultra-relativistic electrons monitored close to the equatorial plane by the NASA Van Allen Probes and observed a clear narrowing of the electron pitch angle distributions connected to EMIC wave occurrence. Usanova et al. (2014) also used a model to assess the pitch angle dependence of EMIC wave pitch angle diffusion coefficients, obtaining good agreement with the observations. They were hence able to verify for the first time that EMIC waves can interact with radiation belt electrons, and significantly, can lead to the loss of moderate pitch angle ultra-relativistic radiation belt electrons by EMIC wave scattering into the atmosphere. Radiation belt electron populations are often peaked at 90°, and at high pitch angles EMIC waves alone cannot scatter electrons into the loss cone. While EMIC waves acting in concert with other plasma wave modes such as whistler waves could cause losses which reach the core of the distribution (e.g. Mourenas et al., 2016), whether this results in significant EMIC wave-related losses into the heart of the distribution at high pitch angles remains to be determined.

Recent work by Hendry et al. (2016) using ground-based magnetometer measurements of EMIC waves from Halley station combined with Polar Operational Environmental Satellites (POES) observations of electron precipitation has provided further evidence for EMIC loss. They found that of the precipitation events which were observed proximal to the Halley Bay ground magnetometer station, around 90% could be linked to EMIC waves, which typically demonstrated a rising tone. Hendry et al. (2017) extended this work to further connect EMIC waves to 300 keV electron precipitation. This lower energy precipitation link remains to be explained, although recently Chen et al.

(2016) suggested that non-resonant EMIC wave scattering could allow lower energy scattering into the loss cone.

13.3.4 Future Perspectives

Although establishing the relative importance of different competing acceleration and loss processes remains an active area of research, it is now clear that ULF wave–particle interactions play a key role. Observational characterizations of ULF wave power seems to be important for delivering improved modeling and forecast capabilities using ULF wave radial diffusion models, especially if this is to reach the point of predictability. Even in more complex radiation belt models which additionally include more accurate representations of higher-frequency hiss, chorus and EMIC wave–particle interactions, long-period ULF wave radial transport remains critically important.

However, a number of challenges still remain. For example, high-temporal-resolution specification of the outer boundary condition is critical in the context of both magnetopause shadowing and inward ULF wave transport during belt replenishment (cf. Mann et al., 2016; Ozeke et al., 2017). For example, Olifer et al. (2018) recently showed a very close correspondence between the location of the last closed drift shell (LCDS) and hour timescale relativistic electron losses monitored using measurements from the Global Positioning System (GPS) satellite constellation. Further assessment of the relative impacts from magnetopause shadowing and plasma wave scattering losses to the atmosphere appears to be especially important. Flux recoveries also need to be accurately assessed as a function of energy, not least because this could allow both inward and outward radial diffusion at different first adiabatic invariants to occur the same time. The potential role of coherent ULF wave interactions (e.g. Ukhorskiy et al., 2006; Mann et al., 2013) also remains to be fully quantified.

Nonetheless, the advent of expanded data availability from more extensive modern ground-based magnetometer networks, combined with that from flotillas of multiple well-instrumented spacecraft, means that the future looks very bright for ULF wave-related radiation belt research.

13.4 Observations, Diagnostics and Applications

Colin L. Waters

13.4.1 Introduction

The geomagnetic field varies over a range of temporal scales arising from internal, geological (long-term) and external, space physics (shorter-term) phenomena. These short and long timescales overlap for annual up to solar cycle (11 year) periods. Perhaps one exception is the debated idea that ULF (mHz) signals might be associated with earthquakes. This section focuses on the space physics sources of temporal variations in the geomagnetic field, from 1000 s periods to kHz frequencies.

The first observations of the geomagnetic field appear to involve the 'compass' during the Han Dynasty (China) around 300–200 BC (Lowrie, 2007). However, origins of space physics may be traced to earlier observations of the aurora (Kivelson and Russell, 1995). Therefore, progress in understanding shorter-period geomagnetic field variability has resulted from research in both Earth (geophysics) and space (plasma) science, while developments in sensor technology have allowed the detection and analyses of small amplitude, short-period variations. Since the formation of the IUGG in 1919, improvements in the sensitivity, increases in the number and location of field and plasma sensors, coupled with advances in data analysis methods have provided many advances in understanding geomagnetic field variations.

During the thirteenth General Assembly of the IUGG (August 1963), a subcommittee was tasked with classifying geomagnetic field variations seen in time series records, resulting in the categories shown in Table 13.2.1 (Section 13.2.2). Saito (1969) provided a review of knowledge at the time these notations were introduced. The labels in Table 13.2.1 have persisted despite improved understanding of the various generation and propagation characteristics of these variations as plasma waves in near-Earth space, as documented by subsequent reviews (McPherron et al., 1972; Orr, 1973; Kimura, 1974; Southwood and Hughes, 1983; Odera, 1986; Arnoldy et al., 1988; Samson, 1991; Engebretson et al., 1994; McPherron, 2005; Takahashi et al., 2006). Consequently, research has for the most part moved beyond morphological studies to focus on (1) the generation, resonant and/or propagation physics (Section 13.2), (2) impacts of these variations on their environment (e.g. radiation belt particles (Section 13.3), human technology in space and on the ground) and (3) use of magnetic variation data for diagnostics such as remote sensing the composition of the magnetosphere and the dynamic properties of magnetospheric boundaries. The various sources, diagnostic uses and properties of the geomagnetic variations listed in Table 13.2.1 are described by Menk and Waters (2013). This section highlights developments in observations, diagnostic uses and applications related to ultra-low-frequency (ULF) geomagnetic variations in addition to higher-frequency waves such as chorus and hiss.

13.4.2 Observations from Earth's Surface

The apparatus used to detect minute timescale variations of the geomagnetic field originally (~1750) comprised a suspended compass needle viewed with a microscope (Chapman and Bartels, 1962). The microscope was subsequently replaced by mirror reflected, optical signal amplification for recording onto photosensitive paper but the suspended magnet design was used as late as the 1970s.

For shorter-period variations, a large wire coil is a simple, effective, low-noise sensor. Induction coil designs remain popular in addition to total field, vector sensors such as the fluxgate magnetometer (Serson, 1973). Modern, large spatial magnetometer arrays employ both designs in order to capture magnetic variations from quasi-DC to VLF frequencies.

ULF variations originate from magnetized plasma dynamics (Alfvén and Fälthammar, 1963; Section 13.2) and so ULF wave properties contain information about the structure of near-Earth space, energy transport, plasma wave conversion and interactions with charged particles. Measurements of amplitude and phase of the magnetic and/or electric field components provide a rich source of information. Spatial magnetometer arrays have been, and remain, an essential tool along with suitable data processing (e.g. Samson, 1983). An enormous volume is accessed by mapping from Earth's surface along the geomagnetic field into space. Therefore, collaboration across national and international boundaries remains essential for ULF research. Fortunately, such cooperation has allowed for the development of 'super' instrument arrays such as INTERMAGNET (www.intermagnet.org), and SuperMAG (http://supermag.jhuapl.edu; Gjerloev, 2012). Recent data products from SuperMAG provides global Pc5 frequency, polarization azimuth and ellipticity information, important for diagnostics of energetic particle dynamics (Section 13.3). While the number of magnetometers included in SuperMAG has increased to more than 300, there are still regions of sparse data coverage. A statistical-based regression technique that folds in SuperMAG indices 'fills-in' the data to provide degree spatial resolution over the Northern Hemisphere (Waters et al., 2015).

A detailed understanding of the transition of ULF variations from space to the ground through the ionosphere is important for developing ground-based remote sensing and space diagnostic techniques. A number of features of this process have been discovered including the 'back-emf' inductive feedback of the ionosphere, effects of an oblique field geometry and details of ULF wave mode coupling in the ionosphere affecting wave polarization and amplitude modification (Yoshikawa and Itonaga, 2000; Sciffer et al., 2004, 2005). These are necessary for developing ground-based methods to remote sense ULF electric field amplitudes in space, particularly those involving particle energization in the radiation belts (Ozeke et al., 2009; Sciffer and Waters, 2011; Shah et al., 2016). Alfvén wave mode coupling (Section 13.2) depends on the ionosphere Hall conductivity. Traditionally, the terms *toroidal* and *poloidal* have been used to label cold plasma, ULF wave modes according to azimuthal wavenumber ($m = 0$ and $m \rightarrow \infty$). A 'pure toriodal' mode requires zero ionospheric Hall conductance (Sciffer et al., 2005), non-twist magnetic field geometry (Kabin et al., 2007; Waters et al., 2007) and $m = 0$. The ionospheric Hall current also increases the ring time of field line resonances (FLRs) enhancing methods for remote sensing the magnetospheric plasma (Waters et al., 2013).

Attenuation of ULF wave fields measured at the ground for wavelengths smaller than the ionospheric height (~120 km) has long been recognized. Therefore, high-m azimuthal structured waves (Section 13.2.3) often require detection at ionospheric altitudes or in space using multi-spacecraft platforms. HF radar (e.g. SuperDARN) and imaging riometer data have provided insights into their relationship with drift- and bounce-resonance and energization of radiation belt charged particle dynamics (Baddeley et al., 2002; Beharrell et al., 2010; James et al., 2016; Section 13.3).

Gyro-resonance generates higher-frequency waves which are in the Pc3–4 range for ions in the solar wind and in the Pc1 band within the magnetosphere. Until recently, 0.2–5 Hz waves were thought to originate from ion temperature anisotropy regions in the equatorial region, based on the 'bouncing' wave packet concept developed from ground-based observations. However, recent findings that implicate EMIC waves in energetic radiation belt particle loss and the development of methods to use these waves for diagnostics of magnetosphere plasma boundaries and storm related plasma plumes have questioned this model (Demekhov, 2007; Loto'aniu et al., 2005; Engebretson et al., 2008; Usanova et al., 2008; Halford et al., 2010). Formulating a definitive source mechanism for Pc1 waves remains an active research area.

The deployment of spatial arrays of induction coil sensors has provided spectacular spectra of multi-harmonic waves in the Ionosphere Alfvén Resonator (IAR; Surkov and Hayakawa, 2014b). Observations and modeling have allowed the development of diagnostic capability of the top-side ionosphere using IAR harmonics (Lysak, 1993; Potapova et al., 2017). Modulation of the ionospheric plasma by ULF variations have also been reported in total electron content (TEC), Doppler sounder and GPS measurements (Wright et al., 1999; Waters and Cox, 2009; Belakhovsky et al., 2016). The physics involves the continuity equation

$$\frac{\partial N}{\partial t} = q - l - \boldsymbol{v}.\nabla \mathrm{N} - \mathrm{N}\nabla.\mathbf{v},$$

where q and l are production and loss rates and $\mathbf{v}$ is the plasma velocity resulting from the ULF variations in the remaining advection and divergence terms. Bulk motion of electrons associated with the advection mechanism appears to be the main source of ULF wave induced, Doppler shift in HF signals that propagate via the ionosphere (Norouzi-Sedeh, 2013).

The close association between ULF variations and space weather has motivated the development of ULF indices. A ULF index that captures the wave power in the solar wind, magnetosphere and on the ground over 2–7 mHz has been developed (Pilipenko et al., 2017). This hourly index uses data from Northern Hemisphere ground stations and GOES, Wind, ACE and IMP8 spacecraft and informs studies of solar wind-ionosphere interactions, substorms, ULF power distribution in the magnetosphere, radiation belt electron energization and earthquake precursor research. The Pc3 index (Marshall, 1996) was developed in order to inform the geomagnetic survey industry of possible contamination of mineral survey data from naturally occurring magnetic field variations (e.g. Vallee et al., 2005).

13.4.3 Observations from Space

Magnetic, electric and particle sensors on board spacecraft provide in situ, co-located field and particle information. Resonant structures and wave–particle interactions are more clearly seen in spacecraft data which also provide validation of remote-sensing methods (e.g. Takahashi, 1990, 2015). The THEMIS, Magnetospheric Multiscale Spacecraft (MMS), Cluster and Van Allen spacecraft have focused efforts to understand the physics of Earth's radiation belts via wave–particle behavior from ULF to VLF (Section 13.3). These data, combined with simulation studies, show that ULF wave properties are influenced by ionospheric conditions (Shah et al., 2016). Higher-frequency chorus and hiss waves also participate in particle energization and loss (Section 13.3). Many aspects remain unknown, given that a complete theory of chorus has not been developed, particularly 'chirp' spectral features and non-linear wave growth physics.

13.4.4 Diagnostics and Applications Using ULF to VLF Variations

At Pc3–5 frequencies, wavelengths are comparable with the size of the magnetosphere and wave properties contain information on various boundary structures. The dynamics of the open-closed field line boundary may be monitored using the presence or lack of FLRs at high latitudes (Ables and Fraser, 2005). The persistence of ~0.3 mHz spaced, ULF frequencies in the Pc5 band has prompted several explanations including cavity and waveguide modes both internal and external to the magnetosphere (Samson et al., 1992; Fenrich et al., 1995), inherent solar wind variations and surface modes of the magnetopause (Plaschke and Glassmeier, 2011). A diagnostic capability of the outer magnetosphere may be developed once the physical mechanisms that explain the properties of these variations, including the local time dependence (Norouzi-Sedeh et al., 2015), has been established.

The magnetic field geometry at high latitudes maps many magnetospheric regions to a relatively small area on the ground. Various features of Pc1–2 (0.1–5 Hz) appear to reflect properties of the source regions, including the cusp, plasma mantle, plasma sheet and low latitude boundary layers and plasmatrough (Menk et al., 1992; Neudegg et al., 1995). Understanding generation and propagation of Pc1–2; distilling their role in energetic particle loss from the radiation belts; and mapping the unstructured, pearl and other emission types to magnetosphere boundaries provide ample motivation for studies of these emissions.

Remote-sensing magnetospheric cold plasma dynamics has matured since routine detection of FLR frequencies using the cross-spectral phase was introduced (Waters et al., 1991). Analyses, diagnostic applications and results of FLR and travel-time methods are described by Menk and Waters (2013). FLR properties derived from ground magnetometer data are used to (1) track the plasmapause location; (2) determine plasmasphere refilling rates after storm intervals; (3) remote sense plasma drainage plumes and density bite-outs, locations that appear to be rich in EMIC signals; (4) estimate plasma composition by combining FLR data with VLF whistler measurements and dependence on magnetic activity; (5) reveal solar cycle variations in magnetospheric mass densities; and (6) detect differences in the conductivity of conjugate ionospheres using quarter-mode resonances (Menk et al., 2014; Jorgensen et al., 2017; Kale et al., 2009; Grew et al., 2007; Takahashi et al., 2010; Obana et al., 2015).

In addition to advances in understanding and using ULF variations in space, ULF variations at Earth's surface impact modern technology and infrastructure. Aeromagnetic surveys for mineral prospecting use ULF activity time series and indices and ULF variations have been implicated as possible precursors of earthquakes (Fraser-Smith et al., 1990; Hayakawa, 2016). This is vigorously debated, highlighting difficulties with data analyses, interpretation and understanding of the physical mechanisms (Campbell, 2009; Masci, 2011). Many studies rely on the separation of naturally occurring ULF from possible earthquake signatures using magnetic activity indices, such as *Kp*. Difficulties with this approach were discussed by Currie and Waters (2014), highlighting the need for a closer connection between the space and geophysics communities.

Geomagnetic variations during magnetic active periods induce electric currents which enter electrical energy supply networks (e.g. Boteler, 2011). Research was initially focused on high-latitude regions, but recent studies have shown transformer vulnerability to these geomagnetic induced currents (GICs) at middle and low latitudes (e.g. Marshall et al., 2013). In addition to developing a prediction capability, a key area of GIC research is the development of methods to map the detected geomagnetic field variations to spatial and temporal variations in the ground electric field which are used by industry software to understand the impacts on a network. The progress, challenges and status of GIC research and applications is discussed by Pulkkinen et al. (2017).

13.4.5 Outstanding Research Questions

The increase in the number of ground magnetometer installations and multi-spacecraft missions and developments in data analyses and computer simulations since the 1950s has progressed geomagnetic variation research from descriptive to predictive capability in many areas. However, many outstanding puzzles and tasks remain, and some of the key questions follow, ordered from low- to high-frequency variations:

- What is the explanation for ~0.3 mHz spaced, repeatable, discrete frequencies in the Pc5 band. Why might these exist in the solar wind?
- How does the diurnal variation in the Pc5 spectra reflect magnetosphere dynamics and properties (e.g. Norouzi-Sedeh et al., 2015)?
- What additional Pc4–5 data are required in order to further develop diagnostic capability of magnetosphere boundaries?
- Given our understanding of Pc3–5 variations, what additional information is required in order to develop a predictive capability for radiation belt particle dynamics?
- How important are 'global' and/or plasmasphere cavity modes?
- What is the origin of polar cap Pc3 variations? Do they involve nightside boundaries and instability regions?
- Where do daytime Pi2 (ground) come from? These appear to be absent in spacecraft data.
- What is(are) the source mechanism(s) for EMIC waves?
- What is the role of plasma plumes and magnetosphere compression events in the generation and propagation of EMIC waves?
- How robust is an ionospheric *F*-region diagnostic using IAR harmonics? What properties of the top-side ionosphere can be reliably derived from these data?
- What are the relative contributions of low-*m*, high-*m* Pc4–5, EMIC and chorus/hiss waves to radiation belt particle energization and loss?
- What magnetosphere parameters control the chirp and growth of chorus waves?
- Do ULF geomagnetic field variations occur before earthquakes? If so, what is the physical mechanism?
- How important are ULF variations to the generation of geomagnetic induced currents?
- What are the barriers to developing a predictive capability for GIC events and how can these be addressed?

These and other puzzles ensure ULF wave research remains a key area for understanding near-Earth space.

References

Ables, S. T. and B. J. Fraser (2005). Observing the open-closed boundary using cusp-latitude magnetometers. Geophys. Res. Lett., **32**, L10104, doi: 10.1029/2005GL022824.

Albert, J. M. (2003). Evaluation of quasi-linear diffusion coefficients for EMIC waves in a multispecies plasma. J. Geophys. Res., **108**, 1249, doi: 10.1029/2002JA009792.

Alfvén, H. (1942). Existence of electromagnetic-hydromagnetic waves. Nature, **150**, 405.

Alfvén, H., and C. G. Fälthammar (1963). *Cosmical Electrodynamics*, Oxford University Press, Oxford.

Allan, W., and F. B. Knox (1979a). A dipole field model for axisymmetric Alfvén waves with finite ionosphere conductivities. Planet. Space Sci., **27**(1), 79–85.

Allan, W., and F. B. Knox (1979b). The effect of finite ionosphere conductivities on axisymmetric toroidal Alfvén wave resonances. Planet. Space Sci., **27**(7), 939–50.

Allan, W., and A. N. Wright (1997), Large-m waves generated by small-m field line resonances via the nonlinear Kelvin–Helmholtz instability. J. Geophys. Res., **102**(A9), 19927–33, doi: 10.1029/97JA01489.

Arnoldy, R. L., L. J. Cahill Jr, M. J. Engebretson, L. J. Lanzerotti and A. Wolfe (1988). Review of hydromagnetic wave studies in the Antarctic. Rev. Geophys., **26**, 181–201.

Backus, G., R. Parker and C. Constable (1996). *Foundations of Geomagnetism*. Cambridge University Press, Cambridge.

Backus, G. E. (1983). Application of mantle filter theory to the magnetic jerk of 1969. Geophys. J. Int., **74**(3), 713–46.

Baddeley, L. J., T. K. Yeoman, D. M. Wright et al. (2002). Morning sector drift-bounce resonance driven ULF waves observed in artificially-induced HF radar backscatter. Ann. Geophys., **20**(9), 1487–98.

Baker, D. N., S. G. Kanekal, V. C. Hoxie, M. G. Henderson, X. Li, H. E. Spence, S. R. Elkington, R. H. W. Friedel, J. Goldstein, M. K. Hudson and G. D. Reeves (2013). A long-lived relativistic electron storage ring embedded in Earth's outer Van Allen belt. Science, **340**(6129), 186–90.

Balasis, G., I. A. Daglis and I. R. Mann, eds. (2016). *Waves, Particles, and Storms in Geospace: A Complex Interplay*. Oxford University Press, Oxford.

Beharrell, M., A. J. Kavanagh and F. Honary (2010). On the origin of high m magnetospheric waves. J. Geophys. Res., **115**, A02201, doi: 10.1029/2009JA014709.

Belakhovsky, V., V. Pilipenko, D. Murr, E. Fedorov and A. Kozlovsky (2016). Modulation of the ionosphere by Pc5 waves observed simultaneously by GPS/TEC and EISCAT. Earth Planets Space, **68**, 102, doi: 10.1186/s40623-016–0480-7.

Boteler, D. H. (2011). Space weather effects on power systems, in *Space Weather*, ed. P. Song, H. J. Singer and G. L. Siscoe, American Geophysical Union, Washington, DC, doi: 10.1002/GM125p0347.

Bourdarie, S., R. H. W. Friedel, J. Fennell, S. Kanekal and T. E. Cayton (2005). Radiation belt representation of the energetic electron environment: Model and data synthesis using the Salammbô radiation belt transport code and Los Alamos geosynchronous and GPS energetic particle data. Space Weather, **3**, S04S01, doi: 10.1029/2004SW000065.

Brautigam, D. H. and J. M. Albert (2000). Radial diffusion analysis of outer radiation belt electrons during the October 9, 1990, magnetic storm. J. Geophys. Res., **105** (A1), 291–309.

Brizard, A. J. and A. A. Chan (2001). Relativistic bounce-averaged quasilinear diffusion equation for low-frequency electromagnetic fluctuations. Phys. Plasmas, **8**(11), 4762–71.

Campbell, W. H. (2009). Natural magnetic disturbance fields, not precursors, preceding the Loma Prieta earthquake. J. Geophys. Res., **114**, A05307, doi: 10.1029/2008JA013932.

Carpenter, D. and R. Anderson (1992). An ISEE/Whistler model of equatorial electron density in the magnetosphere. J. Geophys. Res., **97**, 1097–1108.

Carrington, R. C. (1859). Description of a Singular Appearance seen in the Sun on September 1, 1859. Monthly Notices R. Astron. Soc., **20**(1), 13–15.

Chapman, S. and J. Bartels (1962). *Geomagnetism*, vol. 1, Clarendon Press, Oxford.

Chave, A. D., and A. G. Jones, eds. (2012). *The Magnetotelluric Method: Theory and Practice*, Cambridge University Press, Cambridge.

Chen, L., and A. Hasegawa (1974). A theory of long-period magnetic pulsations: 1. Steady state excitation of field line resonance. J. Geophys. Res., **79**(7), 1024–32.

Chen, L., and A. Hasegawa (1991). Kinetic theory of geomagnetic pulsations: 1. Internal excitations by energetic particles. J. Geophys. Res., **96**(A2), 1503–12, doi: 10.1029/90JA02346.

Chen, L., R. M. Thorne, J. Bortnik and X.-J. Zhang (2016). Nonresonant interactions of electromagnetic ion cyclotron waves with relativistic electrons. J. Geophys. Res., **121**, 9913–25, doi: 10.1002/2016JA022813.

Chi, P. J., C. T. Russell, J. C. Foster, M. B. Moldwin, M. J. Engebretson and I. R. Mann (2005). Density enhancement in plasmasphere-ionosphere plasma during the 2003 Halloween Superstorm: Observations along the 330th magnetic meridian in North America. Geophys. Res. Lett., **32**, L03S07, doi: 10.1029/2004GL021722.

Chi, P. J., C. T. Russell and S. Ohtani (2009). Substorm onset timing via traveltime magnetoseismology. Geophys. Res. Lett., **36**(8).

Chisham, G., I. R. Mann and D. Orr (1997). A statistical study of giant pulsation latitudinal polarization and amplitude variation. J. Geophys. Res., **102**(A5), 9619–29.

Claudepierre, S. G., et al. (2013). Van Allen Probes observation of localized drift resonance between poloidal mode ultra-low frequency waves and 60 keV electrons. Geophys. Res. Lett., **40**, 4491–7, doi: 10.1002/grl.50901.

Claudepierre, S. G., M. K. Hudson, W. Lotko, J. G. Lyon and R. E. Denton (2010). Solar wind driving of magnetospheric ULF waves: Field line resonances driven by dynamic pressure fluctuations. J. Geophys. Res., **115**(A11), doi: 10.1029/2010JA015399.

Claudepierre, S. G., M. Wiltberger, S. R. Elkington, W. Lotko and M. K. Hudson (2009). Magnetospheric cavity modes driven by solar wind dynamic pressure fluctuations. Geophys. Res. Lett., **36**(13), doi: 10.1029/2009GL039045.

Constable, C., and M. Korte (2015). Centennial- to millennial-scale geomagnetic field variations, in *Treatise on Geophysics*, vol. 5, pp. 309–41, Elsevier, New York.

Constable, C., M. Korte and S. Panovska (2016). Persistent high paleosecular variation activity in the Southern Hemisphere for at least 10000 years. Earth Planet. Sci. Lett., **453**, 78–86.

Constable, C. G., and S. C. Constable (2004). Satellite magnetic field measurements: Applications in studying the deep earth, in *The State of the Planet: Frontiers and Challenges in Geophysics*, American Geophysical Union, Washington, DC.

Currie, J. L., and C. L. Waters (2014). On the use of geomagnetic indices and ULF waves for earthquake precursor signatures. J. Geophys. Res., **119**, 992–1003, doi: 10.1002/2013JA019530.

Dai, L., K. Takahashi, R. L. Lysak et al. (2015). Storm time occurrence and spatial distribution of Pc4 poloidal ULF waves in the inner magnetosphere: A Van Allen Probes statistical study. *J. Geophys. Res.*, **120**(6), 4748–62.

Degeling, A. W., I. J. Rae, C. E. J. Watt, Q. Q. Shi, R. Rankin and Q.-C. Zong (2018). Control of ULF wave accessibility to the inner magnetosphere by the convection of plasma density. J. Geophys. Res., **123**, doi: 10.1002/2017ja024874.

Demekhov, A. G. (2007). Recent progress in understanding Pc1 pearl formation. J. Atmos. Sol. Terr. Phys., **69**, 1609–22.

Dent, Z. C., I. R. Mann, F. W. Menk, J. Goldstein, C. R. Wilford, M. A. Clilverd, and L. G. Ozeke (2003). A coordinated ground-based and IMAGE satellite study of quiet-time plasmaspheric density profiles. Geophys. Res. Lett., **30**, 1600, doi: 10.1029/2003GL016946.

Dent, Z. C., I. R. Mann, J. Goldstein, F. W. Menk and L. G. Ozeke (2006). Plasmaspheric depletion, refilling, and plasmapause dynamics: A coordinated ground-based and IMAGE satellite study. J. Geophys. Res., **111**, A03205, doi: 10.1029/2005JA011046.

Dimitrakoudis, S., I. R. Mann, G. Balasis, C. Papadimitriou, A. Anastasiadis and I. A. Daglis (2015). Accurately specifying storm-time ULF wave radial diffusion in the radiation belts. Geophys. Res. Lett., **42**, 5711–18, doi: 10.1002/2015GL064707.

Dungey, J. W. (1954). Electrodynamics of the outer atmosphere. Pennsylvania State University lonos. Res. Lab. Sci. Rept. No. 69.

Dungey, J. W. (1961). Interplanetary magnetic field and the auroral zones. Phys. Rev. *Lett.*, **6**(2), 47.

Elkington, S. R., M. K. Hudson and A. A. Chan (2003). Resonant acceleration and diffusion of outer zone electrons in an asymmetric geomagnetic field. J. Geophys. Res., **108**(A3).

Engebretson, M. J., K. Takahashi and M. Scholer, eds. (1994). *Solar Wind Sources of Magnetospheric Ultra-Low Frequency Waves*, AGU Monogr. 81, American Geophysical Union, Washington, DC, doi: 10.1029/GM081p00xi.

Engebretson, M. J., M. R. Lessard, J. Bortnik, J. C. Green, R. B. Thorne, D. L. Detrick, A. T. Weatherwax, J. Mannionen, N. J. Petit, J. L. Posch and M. C. Rose (2008). Pc1-Pc2 waves and energetic particle precipitation during and after magnetic storms: Superposed epoch analysis and case studies. J. Geophys. Res., **113**, A01211, doi: 10.1029/2007JA012362.

Fälthammar, C.-G. (1965). Effects of time-dependent electric fields on geomagnetically trapped radiation. J. Geophys. Res., **70** (11), 2503–16, doi: 10.1029/JZ070i011p02503.

Fei, Y., A. A. Chan, S. R. Elkington and M. J. Wiltberger (2006). Radial diffusion and MHD particle simulations of relativistic electron transport by ULF waves in the September 1998 storm. J. Geophys. Res., **111**, A12209, doi: 10.1029/2005JA011211.

Fenrich, F. R., and J. C. Samson (1997). Growth and decay of field line resonances. J. Geophys. Res., **102**(A9), 20031–9.

Fenrich, F. R., J. C. Samson, G. Sofko and R. A. Greenwald (1995). ULF high-and low-m field line resonances observed with the Super Dual Auroral Radar Network. J. Geophys. Res., **100**(A11), 21535–47.

Finlay, C. C., N. Olsen, S. Kotsiaros, N. Gillet and L. Tøffner-Clausen (2016). Recent geomagnetic secular variation from

Swarm and ground observatories as estimated in the CHAOS-6 geomagnetic field model. Earth Planets Space, **68**(1), 1.

Fraser, B. J., J. L. Horwitz, J. A. Slavin, Z. C. Dent and I. R. Mann (2005). Heavy ion mass loading of the geomagnetic field near the plasmapause and ULF wave implications. Geophys. Res. Lett., **32**, L04102, doi: 10.1029/2004GL021315.

Fraser-Smith, A. C., A. Bernardi, P. R. McGill, M. E. Ladd, R. A. Helliwell and O. G. Villard (1990). Low-frequency magnetic field measurements near the epicenter of the Ms 7.1 Loma Prieta earthquake. Geophys. Res. Lett., **17**, 1465–8, doi: 10.1029/GL017i009p01465.

Gee, J. S., and D. V. Kent (2007). Source of oceanic magnetic anomalies and the geomagnetic polarity timescale, in *Treatise on Geophysics*, vol. 5, pp. 419–60, Elsevier, New York.

Gjerloev, J. W. (2012). The SuperMAG data processing technique. J. Geophys. Res., **117**, A09213, doi: 10.1029/2012JA017683.

Glassmeier, K. H., J. Vogt, A. Stadelmann and S. Buchert (2004). Concerning long-term geomagnetic variations and space climatology. Ann. Geophys., **22**(10), 3669–77.

Goldstein, J., B. R. Sandel, M. F. Thomsen, M. Spasojević and P. H. Reiff (2004). Simultaneous remote sensing and in situ observations of plasmaspheric drainage plumes. J. Geophys. Res., **109**, A03202, doi: 10.1029/2003JA010281.

Grebowsky, J. M. (1970). Model study of plasmapause motion. J. Geophys. Res., **75**(22), 4329–33, doi: 10.1029/JA075i022p04329.

Grew, R. S., F. W. Menk, M. A. Clilverd and B. R. Sandel (2007). Mass and electron densities in the inner magnetosphere during a prolonged disturbed interval. Geophys. Res. Lett., 34, L02108, doi: 10.1029/2006GL028254.

Gu, X., Y. Y. Shprits and B. Ni (2012). Parameterized lifetime of radiation belt electrons interacting with lower-band and upper-band oblique chorus waves. Geophys. Res. Lett., **39**, L15102, doi: 10.1029/2012GL052519.

Halford, A. J., B. J. Fraser and S. K. Morley (2010). EMIC wave activity during geomagnetic storm and nonstorm periods: CRRES results. J. Geophys. Res., **115**, A12248, doi: 10.1029/2010JA015716.

Harrold, B. G., and J. C. Samson (1992). Standing ULF modes of the magnetosphere: A theory. Geophys. Res. Lett., **19**(18), 1811–14.

Hartinger, M. D., D. L. Turner, F. Plaschke, V. Angelopoulos, and H. Singer (2013). The role of transient ion foreshock phenomena in driving Pc5 ULF wave activity. J. Geophys. Res., **118**(1), 299–312.

Hayakawa, M. (2016). Earthquake prediction with electromagnetic phenomena. AIP Conf. Proc., 1709, 020002, doi: 10.1063/1.4941199.

Helliwell, R. A. (2006). *Whistlers and Related Ionospheric Phenonemena*. Dover, Mineola, NY.

Hendry, A. T., C. J. Rodger and M. A. Clilverd (2017). Evidence of sub-MeV EMIC-driven electron precipitation. Geophys. Res. Lett., **44**, 1210–18, doi: 10.1002/2016GL071807.

Hendry, A. T., C. J. Rodger, M. A. Clilverd, M. J. Engebretson, I. R. Mann, M. R. Lessard, T. Raita, and D. K. Milling (2016). Confirmation of EMIC wave driven relativistic electron precipitation. J. Geophys. Res., **121**, doi: 10.1002/2015JA022224

Horne, R. B., and R. M. Thorne (1998). Potential waves for relativistic electron scattering and stochastic acceleration during magnetic storms. Geophys. Res. Lett., **25**(15), 3011–14.

Horne, R. B., S. A. Glauert, N. P. Meredith, H. Koskinen, R. Vainio, A. Afanasiev, N. Y. Ganushkina, O. A. Amariutei, D. Boscher, A. Sicard and V. Maget (2013). Forecasting the Earth's radiation belts and modelling solar energetic particle events: Recent results from SPACECAST. J. Space Weather Space Clim., **3**, A20.

Hughes, W. J., D. J. Southwood, B. Mauk, R. L. McPherron and J. N. Barfield (1978). Alfvén waves generated by an inverted plasma energy distribution. Nature, **275**(5675), 43–5.

Jackson, A., A. R. T. Jonkers and M. R. Walker (2000). Four centuries of geomagnetic secular variation from historical records. Philos. Trans. R. Soc. London A, **358**, 957–90.

Jacobs, J. A., Y. Kato, S. Matsushita and V. A. Troitskaya (1964). Classification of geomagnetic micropulsations. J. Geophys. Res., **69**(1), 180–81.

James, M. K., T. K. Yeoman, P. N. Mager and D. Y. Klimushkin (2013). The spatio-temporal characteristics of ULF waves driven by substorm injected particles. J. Geophys. Res., **118**, 1737–49, doi: 10.1002/jgra.50131.

James, M. K., T. K Yeoman, P. N. Mager and D. Y. Klimushkin (2016). Multiradar observations of substorm-driven ULF waves. J. Geophys. Res., **121**, 5213–32, doi: 10.1002/2015JA022102.

Jorgensen, A. M., B. Heilig, M. Vellante, J. Lichtenberger, J. Reda, F. Valach and I. Mandic (2017). Comparing the dynamic global core plasma model with ground-based plasma desnity observations. J. Geophys. Res., **122**, 7997–8013, doi: 10.1002/2016JA023229.

Kabin, K., R. Rankin, I. R. Mann, A. W. Degeling and R. Marchand (2007). Polarization properties of standing shear Alfvén waves in non-axisymmetric background magnetic fields. Ann. Geophys., **25**(3), 815–22.

Kale, Z. C., I. R. Mann, C. L. Waters, M. Vellante, T. L. Zhang and F. Honary (2009). Plasmaspheric dynamics resulting from the Hallowe'en 2003 geomagnetic storms. J. Geophys. Res., **114**, A08204, doi: 10.1029/2009JA014194.

Kavosi, S. and J. Raeder (2015). Ubiquity of Kelvin–Helmholtz waves at Earth's magnetopause. Nat. Comm., **6**, 7019.

Keiling, A., and K. Takahashi (2011). Review of Pi2 models. Space Sci. Rev., **161**(1–4), 63–148.

Kelley, M. C. (2009). *The Earth's Ionosphere: Plasma Physics and Electrodynamics*, 2nd edn., Elsevier, New York.

Kepko, L., H. E. Spence and H. J. Singer (2002). ULF waves in the solar wind as direct drivers of magnetospheric pulsations. Geophys. Res. Lett., **29**(8), doi: 10.1029/2001GL014405.

Kimura, I. (1974). Interrelation between VLF and ULF emissions. Space Sci. Rev., **16**, 389–411.

Kivelson, M. G., and C. T. Russell, eds. (1995). Introduction to Space Physics. Cambridge University Press, Cambridge.

Kivelson, M. G., and D. J. Southwood (1985). Resonant ULF waves: A new interpretation. Geophys. Res. Lett., **12**(1), 49–52.

Kivelson, M. G., and D. J. Southwood (1986). Coupling of global magnetospheric MHD eigenmodes to field line resonances. J. Geophys. Res., **91**(A4), 4345–51.

Kivelson, M. G., M. Cao, R. L. McPherron and R. J. Walker (1997). A possible signature of magnetic cavity mode oscillations in ISEE spacecraft observations. J. Geomagn. Geoelec., **49**(9), 1079–98.

Kivelson, M. G., and C. T. Russell, eds. (1995). *Introduction to Space Physics*. Cambridge University Press, Cambridge.

Klimushkin, D. Y. (2000). The propagation of high-m Alfvén waves in the Earth's magnetosphere and their interaction with high-energy particles. J. Geophys. Res., **105**(A10), 23303–10, doi: 10.1029/1999JA000396.

Kono, M. (2015). Geomagnetism: An introduction and overview, in *Treatise on Geophysics*, vol. 5, pp. 1–31, Elsevier, New York.

Le, G., and C. T. Russell (1994). The morphology of ULF waves in the Earth's foreshock, in *Solar Wind Sources of Magnetospheric Ultra-Low-Frequency Waves*, ed. M. J. Engebretson, K. Takahashi and M. Scholer, American Geophysical Union, Washington, DC, doi: 10.1029/GM081p0087.

Lee, D.-H., and R. L. Lysak (1999). MHD waves in a three-dimensional dipolar magnetic field: A search for Pi2 pulsations. J. Geophys. Res., **104**(A12), 28691–99, doi: 10.1029/1999JA900377.

Li, W., et al. (2016a). Radiation belt electron acceleration during the 17 March 2015 geomagnetic storm: Observations and simulations. J. Geophys. Res., **121**, 5520–36.

Li, W., R. Thorne, J. Bortnik, D. Baker, G. Reeves, S. Kanekal, H. Spence and J. Green (2015a). Solar wind conditions leading to efficient radiation belt electron acceleration: A superposed epoch analysis. Geophys. Res. Lett., **42**, 6906–15.

Li, Z., M. Hudson, B. Kress and J. Paral (2015b). Three-dimensional test particle simulation of the 17–18 March 2013 CME shock-driven storm. Geophys. Res. Lett., **42**, 5679–85.

Li, Z., M. Hudson, J. Paral, M. Wiltberger and D. Turner (2016b). Global ULF wave analysis of radial diffusion coefficients using a global MHD model for the 17 March 2015 storm. J. Geophys. Res., **121**(7), 6196–6206.

Loto'aniu, T. M., B. J. Fraser and C. L. Waters (2005). Propagation of electromagnetic ion cyclotron wave energy in the magnetosphere. J. Geophys. Res., **110**, A07214, doi: 10.1029/2004JA010816.

Loto'aniu, T. M., I. R. Mann, L. G. Ozeke, A. A. Chan, Z. C. Dent and D. K. Milling (2006). Radial diffusion of relativistic electrons into the radiation belt slot region during the 2003 Halloween geomagnetic storms. J. Geophys. Res., **111**, A04218, doi: 10.1029/2005JA011355.

Loto'Aniu, T. M., H. J. Singer, C. L. Waters, V. Angelopoulos, I. R. Mann, S. R. Elkington and J. W. Bonnell (2010) Relativistic electron loss due to ultralow frequency waves and enhanced outward radial diffusion. J. Geophys. Res., **115**, A12245, doi: 10.1029/2010JA015755.

Lowrie, W. (2007). *Fundamentals of Geophysics*, Cambridge University Press, Cambridge.

Lysak, R. L. (1993). Generalized model of the ionospheric Alfvén resonator, in *Auroral Plasma Dynamics*, ed. R. L. Lysak, Geophys. Monogr. 80, American Geophysical Union, Washington, DC.

Mann, I. R., et al. (2018). Reply to 'The dynamics of Van Allen belts revisited'. Nat. Phys., **14**(2), 103, doi: 10.1038/nphys4351.

Mann, I. R., K. R. Murphy, L. G. Ozeke, I. J. Rae, D. K. Milling, A. A. Kale and F. F. Honary (2012). The role of ultralow frequency waves in radiation belt dynamics, in *Dynamics of the Earth's Radiation Belts and Inner Magnetosphere*, ed. D. Summers, I. R. Mann, D. N. Baker and M. Schulz, American Geophysical Union, Washington, DC, doi: 10.1029/2012GM001349.

Mann, I. R., L. G. Ozeke, K. R. Murphy, S. Claudepierre, D. Turner, D. N. Baker, I. J. Rae, A. Kale, D. K. Milling and A. Boyd (2016). Explaining the dynamics of the ultra-relativistic third Van Allen radiation belt. Nat. Phys., **12**(10), 978, doi: 10.1038/nphys3799.

Mann, I. R., and A. N. Wright (1995). Finite lifetimes of ideal poloidal Alfvén waves. J. Geophys. Res., **100**(A12), 23677–86.

Mann, I. R., and A. N. Wright (1999). Diagnosing the excitation mechanisms of Pc5 magnetospheric flank waveguide modes and FLRs. Geophys. Res. Lett., **26**(16), 2609–12.

Mann, I. R., K. G. Balmain, J. B. Blake, D. Boteler, S. Bourdarie, J. H. Clemmons, Z. C. Dent, A. W. Degeling, R. Fedosejeves, J. F. Fennell and B. J. Fraser (2006). The outer radiation belt injection, transport, acceleration and loss satellite (ORBITALS): A Canadian small satellite mission for ILWS. Adv. Space Res., **38**(8), 1838–60.

Mann, I. R., G. Chisham and S. D. Bale (1998). Multisatellite and ground-based observations of a tailward propagating Pc5 magnetospheric waveguide mode. J. Geophys. Res., **103** (A3), 4657–69.

Mann, I. R., E. A. Lee, S. G. Claudepierre, J. F. Fennell, A. Degeling, I. J. Rae, D. N. Baker, G. D. Reeves, H. E. Spence, L. G. Ozeke and R. Rankin (2013). Discovery of the action of a geophysical synchrotron in the Earth's Van Allen radiation belts. Nat. Comm., **4**, 2795.

Mann, I. R., D. K. Milling, I. J. Rae, L. G. Ozeke, A. Kale et al. (2008). The upgraded CARISMA magnetometer array in the THEMIS era. Space Sci. Rev., **141**(1–4), 413–51.

Mann, I. R., T. P. O'Brien and D. K. Milling (2004). Correlations between ULF wave power, solar wind speed, and relativistic electron flux in the magnetosphere: Solar cycle dependence. J. Atmos. Sol. Terr. Phys., **66**(2), 187–98.

Mann, I. R., A. N. Wright and P. S. Cally (1995). Coupling of magnetospheric cavity modes to field line resonances: A study of resonance widths. J. Geophys. Res., **100**(A10), 19441–56.

Mann, I. R., A. N. Wright and A. W. Hood (1997). Multiple-timescales analysis of ideal poloidal Alfvén waves. J. Geophys. Res., **102**(A2), 2381–90.

Mann, I. R., A. N. Wright, K. J. Mills and V. M. Nakariakov (1999). Excitation of magnetospheric waveguide modes by magnetosheath flows. J. Geophys. Res., **104**(A1), 333–53.

Marshall, R. A., H. Gorniak, T. V. der Walt, C. L. Waters, M. D. Sciffer, M. Miller, M. Dalzell, T. Daly, G. Pouferis, G. Hesse, and P. Wilkinson (2013). Observations of geomagnetically induced currents in the Australian power network. Space Weather, **11**, doi: 10.1029/2012SW000849.

Marshall, R. A. (1996). Geomagnetic pulsation service, IPS Radio and Space Services Internal Report, IPS TR-96-02, Aus. Govt. Dep. Admin. Services.

Masci, F. (2011). On the seismogenic increase of the ratio of the ULF geomagnetic field components. Phys. Earth Planet. Inter., **187**, 19–32, doi: 10.1016/j.pepi.2011.05.001.

Mathie, R. A., and I. R. Mann (2001). On the solar wind control of Pc5 ULF pulsation power at mid-latitudes: Implications for MeV electron acceleration in the outer radiation belt. J. Geophys. Res., **106**(A12), 29783–96, doi: 10.1029/2001JA000002.

Mathie, R. A., F. W. Menk, I. R. Mann and D. Orr (1999a). Discrete field line resonances and the Alfvén continuum in the outer magnetosphere. Geophys. Res. Lett., **26**(6), 659–62.

Mathie, R. A., and I. R. Mann (2000). A correlation between extended intervals of ULF wave power and storm-time geosynchronous relativistic electron flux enhancements. Geophys. Res. Lett., **27**(20), 3261–4.

Mathie, R. A., and I. R. Mann (2000). Observations of Pc5 field line resonance azimuthal phase speeds: A diagnostic of their excitation mechanism. J. Geophys. Res., **105**(A5), 10713–28.

Mathie, R. A., I. R. Mann, F. W. Menk and D. Orr (1999b). Pc5 ULF pulsations associated with waveguide modes observed with the IMAGE magnetometer array. J. Geophys. Res., **104** (A4), 7025–36.

Mauk, B. H., N. J. Fox, S. G. Kanekal, R. L. Kessel, D. G. Sibeck and A. Ukhorskiy (2013). Science objectives and rationale for the radiation belt storm probes mission. Space Sci. Rev., **179** (1–4), 3–27.

McPherron, R. L., C. T. Russell and P. J. Coleman (1972). Fluctuating magnetic fields in the magnetosphere. Space Sci. Rev., **13**, 411–54.

McPherron, R. L. (2005). Magnetic pulsations: Their sources and relation to solar wind and geomagnetic activity. Surv. Geophys., **26**, 545–92.

Menk, F. W., D. Orr, M. A. Clilverd, A. J. Smith, C. L. Waters, D. K. Milling and B. J. Fraser (1999). Monitoring spatial and temporal variations in the dayside plasmasphere using geomagnetic field line resonances. J. Geophys. Res., **104**(A9), 19955–69, doi: 10.1029/1999JA900205.

Menk, F. W., Z. C. Kale, M. Sciffer, P. Robinson, C. L. Waters, R. Grew, M. Clilverd and I. R. Mann (2014). Remote sensing the plasmasphere, plasmapause, plumes and other features using ground-based magnetometers. J. Space Weather Space Clim., **4**, A34.

Menk, F. W., and C. L. Waters (2013). *Magnetoseismology: Ground-Based Remote Sensing of Earth's Magnetosphere*. John Wiley, Hoboken, NJ.

Menk, F. W., B. J. Fraser, H. J. Hansen, P. T. Newell, C.-I. Meng and R. J. Morris (1992). Identification of the magnetospheric cusp and cleft using Pc1-2 ULF pulsations. J. Atmos. Terr. Phys., **54**(7/8), 1021–43.

Meredith, N. P., R. M. Thorne, R. B. Horne, D. Summers, B. J. Fraser and R. R. Anderson (2003). Statistical analysis of relativistic electron energies for cyclotron resonance with EMIC waves observed on CRRES. J. Geophys. Res., **108**, 1250, doi: 10.1029/2002JA009700.

Millan, R. M., and R. M. Thorne (2007). Review of radiation belt relativistic electron losses. J. Atmos. Sol. Terr. Phys., **69**, 362–77.

Mills, K. J., A. N. Wright and I. R. Mann (1999). Kelvin-Helmholtz driven modes of the magnetosphere. Phys. Plasmas, **6**(10), 4070–87.

Min, K., K. Takahashi, A. Y. Ukhorskiy et al. (2017). Second harmonic poloidal waves observed by Van Allen Probes in the dusk-midnight sector. J. Geophys. Res., **122**(3), 3013–39.

Morley, S. K., J. P. Sullivan, M. R. Carver, R. M. Kippen, R. H. W. Friedel, G. D. Reeves and M. G. Henderson (2017). Energetic particle data from the Global Positioning System constellation. Space Weather, **15**, 283–9, doi: 10.1002/2017SW001604.

Mourenas, D., A. V. Artemyev, Q. Ma, O. V. Agapitov and W. Li (2016). Fast dropouts of multi-MeV electrons due to combined effects of EMIC and whistler mode waves. Geophys. Res. Lett., **43**, doi: 10.1002/2016GL068921.

Murphy, K. R., I. R. Mann and D. G. Sibeck (2015). On the dependence of storm time ULF wave power on magnetopause location: Impacts for ULF wave radial diffusion. Geophys. Res. Lett., **42**, 9676–84, doi: 10.1002/2015GL066592.

Murphy, K. R., I. R. Mann and L. G. Ozeke (2014). A ULF wave driver of ring current energization. Geophys. Res. Lett., **41** (19), 6595–6602.

Neudegg, D. A., B. J. Fraser, F. W. Menk, H. J. Hansen, G. B. Burns, R. J. Morris and M. J. Underwood (1995). Sources and velocities of Pc1-2 ULF waves at high latitudes. Geophys. Res. Lett., **22**(21), 2965–8, doi: 10.1029/95GL02939.

Nickolaenko, A., and M. Hayakawa (2014). *Schumann Resonance for Tyros*. Springer Japan, Tokyo.

Norouzi-Sedeh, L. (2013). Doppler clutter in HF radar systems produced by ULF waves, PhD thesis, University of Newcastle, NSW, Australia.

Norouzi-Sedeh, L., C. L. Waters and F. W. Menk (2015). Survey of ULF wave signatures seen in the Tasman International Geospace Environment Radar data. J. Geophys. Res., **120**, doi: 10.1002/2014JA020652.

Obana, Y., C. L. Waters, M. D. Sciffer, F. W. Menk, R. L. Lysak, K. Shiokawa, A. W. Hurst and T. Petersen (2015). Resonance structure and mode transition of quarter-wave ULF pulsations around the dawn terminator. J. Geophys. Res., **120**, 4194–4212, doi: 10.1002/2015JA021096.

Odera, T. J. (1986). Solar wind controlled pulsations: A review. Rev. Geophys., **24**, 55–74.

Olifer, L., I. R. Mann, S. K. Morley, L. G. Ozeke and D. Choi (2018). On the role of last closed drift shell dynamics in driving fast losses and Van Allen radiation belt extinction. J. Geophys. Res., **123**, doi: 10.1029/2018JA025190.

Olsen, N. (2007). Natural sources for electromagnetic induction studies, in *Encyclopedia of Geomagnetism and Paleomagnetism*, ed. D. Gubbins and E. Herrero-Bervera, pp. 696–700, Springer, New York.

Olsen, N., G. Hulot and T. J. Sabaka (2010). Sources of the geomagnetic field and the modern data that enable their investigation, in *Handbook of Geomathematics*, ed. W. Freeden, M. Z. Nashed and T. Sonar, Springer, Berlin.

Orlova, K., Y. Shprits and M. Spasojevic (2016). New global loss model of energetic and relativistic electrons based on Van Allen Probes measurements. J. Geophys. Res., **121**, 1308–14, doi: 10.1002/2015JA021878.

Orr, D., and J. A. Matthew (1971). The variation of geomagnetic micropulsation periods with latitude and the plasmapause. Planet. Space Sci., **19**(8), 897–905.

Orr, D. (1973). Magnetic pulsations within the magnetosphere: A review. J. Atmos. Terr. Phys., **35**, 1–50.

Ozeke, L. G., and I. R. Mann (2008). Energization of radiation belt electrons by ring current ion driven ULF waves. J. Geophys. Res., **113**, A02201, doi: 10.1029/2007JA012468.

Ozeke, L. G., I. R. Mann and I. J. Rae (2009). Mapping guided Alfvén wave magnetic field amplitudes observed on the ground to equatorial electric field amplitudes in space. J. Geophys. Res., **114**, A01214, doi: 10.1029/2008JA013041.

Ozeke, L. G., I. R. Mann, D. L. Turner, K. R. Murphy, A. W. Degeling, I. J. Rae and D. K. Milling (2014b). Modeling cross L shell impacts of magnetopause shadowing and ULF wave radial diffusion in the Van Allen belts. Geophys. Res. Lett., **41**, 6556–62.

Ozeke, L. G., I. R. Mann, K. R. Murphy, D. G. Sibeck and D. N. Baker (2017). Ultra-relativistic radiation belt extinction and ULF wave radial diffusion: Modeling the September 2014 extended dropout event. Geophys. Res. Lett., **44**, 2624–33, doi: 10.1002/2017GL072811.

Ozeke, L. G., I. R. Mann, K. R. Murphy, I. J. Rae, D. K. Milling, S. R. Elkington, A. A. Chan and H. J. Singer (2012a). ULF wave derived radiation belt radial diffusion coefficients. J. Geophys. Res., **117**, A04222.

Ozeke, L. G., I. R. Mann, K. R. Murphy, I. Jonathan Rae and D. K. Milling (2014a). Analytic expressions for ULF wave radiation belt radial diffusion coefficients. J. Geophys. Res., **119**, 1587–1605.

Ozeke, L. G., I. R. Mann, K. R. Murphy, I. J. Rae and A. A. Chan (2012b). ULF wave–driven radial diffusion simulations of the outer radiation belt, in *Dynamics of the Earth's Radiation Belts and Inner Magnetosphere*, ed. D. Summers, I. R. Mann, D. N. Baker and M. Schulz, American Geophysical Union, Washington, DC, doi: 10.1029/2012GM001332.

Ozeke, L. G., and I. R. Mann (2001). Modeling the properties of high-m Alfvén waves driven by the drift-bounce resonance mechanism. J. Geophys. Res., **106**(A8), 15583–97.

Ozeke, L. G., and I. R. Mann (2005). High and low ionospheric conductivity standing guided Alfvén wave eigenfrequencies: A model for plasma density mapping. J. Geophys. Res., **110** (A4), doi: 10.1029/2004JA010719.

Ozeke, L. G., I. R. Mann and J. T. Mathews (2005). The influence of asymmetric ionospheric Pedersen conductances on the field-aligned phase variation of guided toroidal and guided poloidal Alfvén waves. J. Geophys. Res., **110**(A8), doi: 10.1029/2005JA011167.

Pilipenko, V. A., O. V. Kozyreva, M. J. Engebretson and A. A. Soloviev (2017). ULF wave power index for space weather and geophysical applications: A review. Russ. J. Earth Sci., **17**, doi: 10.2205/2017ES000597.

Plaschke, F., K. H. Glassmeier, H. U. Auster et al. (2009). Standing Alfvén waves at the magnetopause. Geophys. Res. Lett., **36**(2), doi: 10.1029/2008GL036411.

Plaschke, F., and K.-H. Glassmeier (2011). Properties of standing Kruskal-Schwarzschild-modes at the magnetopause. Ann. Geophys., **29**, 1793–1807, doi: 10.5194/angeo-29-1793-2011.

Potapova, A. S., T. N. Polyushkina, B. Tsegmed, A. V. Oinats, A. Yu. Pashinin, I. K. Edemskiy, A. A. Mylnikova and K. G. Ratovsky (2017). Considering the potential of IAR emissions for ionospheric sounding. J. Atmos. Sol. Terr. Phys., **164**, 229–34.

Pulkkinen, A., E. Bernabeu, A. Thomson, A. Viljanen, R. Pirjola, D. Boteler, J. Eichner, P. J. Cilliers, D. Welling, N. P. Savani, R. S. Weigel, J. J. Love, C. Balch, C. M. Ngwira, G. Crowley, A. Schultz, R. Kataoka, B. Anderson, D. Fugate, J. J. Simpson and M. MacAlester (2017). Geomagnetically induced currents: Science, engineering and applications readiness. Space Weather, doi: 10.1002/2016SW001501.

Radoski, H. R. (1967). Highly asymmetric MHD resonances: The guided poloidal mode. J. Geophys. Res., **72**(15), 4026–7.

Rae, I. J., I. R. Mann, K. R. Murphy, L. G. Ozeke, D. K. Milling, A. A. Chan, S. R. Elkington and F. Honary (2012). Ground-based magnetometer determination of in situ Pc4–5 ULF electric field wave spectra as a function of solar wind speed. J. Geophys. Res., **117**, A04221, doi: 10.1029/2011JA017335.

Rae, I. J., E. F. Donovan, I. R. Mann et al. (2005). Evolution and characteristics of global Pc5 ULF waves during a high solar wind speed interval. J. Geophys. Res., **110**(A12), doi: 10.1029/2005JA011007.

Reeves, G. D., Y. Chen, G. S. Cunningham, R. W. H. Friedel, M. G. Henderson, V. K. Jordanova, J. Koller, S. K. Morley, M. F. Thomsen and S. Zaharia (2012). Dynamic Radiation Environment Assimilation Model: DREAM. Space Weather, **10**, S03006, doi: 10.1029/2011SW000729.

Reeves, G., K. McAdams, R. Friedel and T. O'Brien (2003). Acceleration and loss of relativistic electrons during small geomagnetic storms. Geophys. Res.Lett., **30**, doi: 10.1002/2015GL066376.

Reeves, G. D., H. E. Spence, M. G. Henderson, S. K. Morley, R. H. W. Friedel, H. O. Funsten, D. N. Baker, S. G. Kanekal, J. B. Blake, J. F. Fennell and S. G. Claudepierre (2013). Electron acceleration in the heart of the Van Allen radiation belts. Science, **341**(6149), 991–4.

Rickard, G. J., and A. N. Wright (1995). ULF pulsations in a magnetospheric waveguide: Comparison of real and simulated satellite data. J. Geophys. Res., **100**(A3), 3531–7.

Rickard, G. J., and A. N. Wright (1994). Alfvén resonance excitation and fast wave propagation in magnetospheric waveguides. J. Geophys. Res., **99**(A7), 13455–64.

Rostoker, G., S. Skone and D. N. Baker (1998). On the origin of relativistic electrons in the magnetosphere associated with some geomagnetic storms. Geophys. Res. Lett., **25**(19), 3701–4.

Ruohoniemi, J. M., R. A. Greenwald, K. B. Baker and J. C. Samson (1991). HF radar observations of Pc 5 field line resonances in the midnight/early morning MLT sector. J. Geophys. Res., **96**(A9), 15697–710, doi: 10.1029/91JA00795.

Sabaka, T. J., G. Hulot and N. Olsen (2010). Mathematical Properties Relevant to Geomagnetic Field Modeling, in *Handbook of Geomathematics*, pp. 503–38, Springer, Berlin.

Saito, T. (1969). Geomagnetic pulsations. Space Sci. Rev., **10**, 319–412.

Samson, J. C., R. A. Greenwald, J. M. Ruohoniemi, T. J. Hughes and D. D. Wallis (1991). Magnetometer and radar observations of magnetohydrodynamic cavity modes in the Earth's magnetosphere. Can. J. Phys., **69**, 929.

Samson, J. C., B. G. Harrold, J. M. Ruohoniemi, R. A. Greenwald and A. D. M. Walker (1992). Field line resonances associated with MHD waveguides in the magnetosphere. Geophys. Res. Lett., **19**, 441–4, doi: 10.1029/92GL00116.

Samson, J. C. (1991). Geomagnetic pulsations and plasma waves in the Earth's magnetosphere, in *Geomagnetism*, vol. 4, ed. J. A. Jacobs, chapter 4, Academic Press, London.

Samson, J. C., J. A. Jacobs and G. Rostoker (1971). Latitude-dependent characteristics of long-period geomagnetic micropulsations. J. Geophys. Res., **76**(16), 3675–83.

Samson, J. C. (1983). Pure states, polarized waves, and principal components in the spectra of multiple, geophysical time series. Geophys. J. R. Astron. Soc., **72**, 647–64.

Schulz, M., and L. J. Lanzerotti (1974). *Particle Diffusion in the Radiation Belts*, Physics and Chemistry in Space 7, Springer, Berlin.

Sciffer, M. D., and C. L. Waters (2011). Relationship between ULF wave mode mix, equatorial electric fields, and ground magnetometer data. J. Geophys. Res., **116**, A06202, doi: 10.1029/2010JA016307.

Sciffer, M. D., C. L. Waters and F. W. Menk (2005). A numerical model to investigate the polarisation azimuth of ULF waves through an ionosphere with oblique magnetic fields. Ann. Geophys., **23**, 3457–71.

Sciffer, M. D., C. L. Waters and F. W. Menk (2004). Propagation of ULF waves through the ionosphere: Inductive effect for oblique magnetic fields. Ann. Geophys., **22**, 1155–69.

Serson, P. H. (1973). Instrumentation for induction studies on land. Phys. Earth Planet. Inter., **7**, 313–22.

Shah, A. S., C. L. Waters, M. D. Sciffer and F. W. Menk (2016). Energization of outer radiation belt electrons during storm recovery phase. J. Geophys. Res., **121**, 10845–60, doi: 10.1002/2016JA023245.

Shprits, Y. Y., R. M. Thorne, R. Friedel, G. D. Reeves, J. Fennell, D. N. Baker and S. G. Kanekal (2006). Outward radial diffusion driven by losses at magnetopause. J. Geophys. Res., **111**, A11214, doi: 10.1029/2006JA011657.

Shprits, Y. Y., D. Subbotin, A. Drozdov, M. E. Usanova, A. Kellerman, K. Orlova, D. N. Baker, D. L. Turner and K. C. Kim (2013). Unusual stable trapping of the ultrarelativistic electrons in the Van Allen radiation belts. Nat. Phys., **9**(11), 699.

Shvets, A., and M. Hayakawa (2011). Global lightning activity on the basis of inversions of natural ELF electromagnetic data observed at multiple stations around the world. Surv. Geophys., **32**(6), 705–32.

Sibeck, D. G. (1990). A model for the transient magnetospheric response to sudden solar wind dynamic pressure variations. J. Geophys. Res., **95**(A4), 3755–71.

Sibeck, D. G., N. L. Borodkova, S. J. Schwartz et al. (1999). Comprehensive study of the magnetospheric response to a hot flow anomaly. J. Geophys. Res., **104**(A3), 4577–93.

Silin, I., I. R. Mann, R. D. Sydora, D. Summers and R. L. Mace (2011). Warm plasma effects on electromagnetic ion cyclotron wave MeV electron interactions in the magnetosphere. J. Geophys. Res., **116**, A05215, doi: 10.1029/2010JA016398.

Siscoe, G. L., and C. K. Chen (1975). The paleomagnetosphere. J. Geophys. Res., **80**(34), 4675–80.

Southwood, D. J. (1974). Some features of field line resonances in the magnetosphere. Planet. Space Sci., **22**(3), 483–91.

Southwood, D. J. (1976). A general approach to low-frequency instability in the ring current plasma. J. Geophys. Res., **81** (19), 3340–48.

Southwood, D. J., and W. J. Hughes (1983). Theory of hydromagnetic waves in the magnetosphere. Space Sci. Rev., **35**(4), 301–66.

Southwood, D. J., J. W. Dungey and R. J. Etherington (1969). Bounce resonant interaction between pulsations and trapped particles. Planet. Space Sci., **17**(3), 349–61.

Spence, H. E., G. D. Reeves, D. N. Baker, J. B. Blake, M. Bolton, S. Bourdarie, A. A. Chan, S. G. Claudepierre, J. H. Clemmons, J. P. Cravens and S. R. Elkington (2013). Science goals and overview of the radiation belt storm probes (RBSP) energetic particle, composition, and thermal plasma (ECT) suite on NASA's Van Allen probes mission. Space Sci. Rev., **179**(1–4), 311–36.

Stadelmann, A., J. Vogt, K.-H. Glassmeier, M.-B. Kallenrode and G.-H. Voigt (2010). Cosmic ray and solar energetic particle flux in paleomagnetospheres. Earth Planets Space, **62**(3), 333–45.

Subbotin, D. A., Y. Y. Shprits and B. Ni (2011). Long-term radiation belt simulation with the VERB 3-D code: Comparison with CRRES observations. J. Geophys. Res., **116**, A12210, doi: 10.1029/2011JA017019.

Summers, D., B. Ni and N. P. Meredith (2007). Timescales for radiation belt electron acceleration and loss due to resonant wave-particle interactions: 2. Evaluation for VLF chorus, ELF hiss, and electromagnetic ion cyclotron waves. J. Geophys. Res., **112**(A4), doi: 10.1029/2006JA01180.

Surkov, V., and M. Hayakawa (2014b). *Ionospheric Alfven Resonator (IAR) in Ultra and Extremely Low Frequency Electromagnetic Fields*, Springer, Tokyo.

Surkov, V., and M. Hayakawa (2014a). *Ultra and Extremely Low Frequency Electromagnetic Fields*, Springer, Tokyo.

Takahashi, K., and A. Y. Ukhorski (2007). Solar wind control of Pc5 pulsation power at geosynchronous orbit. J. Geophys. Res., **112**, A11205, doi: 10.1029/2007JA012483.

Takahashi, K., C. L. Waters, K.-H. Glassmeier, C. A. Kletzing, W. S. Kurth and C. W. Smith (2015). Multifrequency compressional magnetic field oscillations and their relation to multiharmonic toroidal mode standing Alfvén waves. J. Geophys. Res., **120**, 10384–403, doi: 10.1002/2015JA021780.

Takahashi, K., et al. (2010). Multipoint observation of fast mode waves trapped in the dayside plasmasphere. J. Geophys. Res., **115**, A12247, doi: 10.1029/2010JA015956.

Takahashi, K., M. D. Hartinger, V. Angelopoulos, K. H. Glassmeier and H. J. Singer (2013). Multispacecraft observations of fundamental poloidal waves without ground magnetic signatures. *J. Geophys. Res.*, **118**(7), 4319–34.

Takahashi, K., S. Oimatsu, M. Nosé, K. Min, S. G. Claudepierre et al. (2018). Van Allen Probes observations of second-harmonic poloidal standing Alfvén waves. *J. Geophys. Res.*, **123**, doi: 10.1002/2017JA024869.

Takahashi, K., P. J. Chi, R. E. Denton and R. L. Lysak, eds. (2006). *Magnetospheric ULF Waves: Synthesis and New Directions*, AGU Monograph 169, American Geophysical Union, Washington, DC.

Takahashi, K., R. E. Denton and H. J. Singer (2010). Solar cycle variation of geosynchronous plasma mass density derived from the frequency of standing Alfvén waves. J. Geophys. Res., **115**, A07207, doi: 10.1029/2009JA015243.

Takahashi, K. (1990). Response of energetic particles to magnetospheric ultra-low-frequency waves. Johns Hopkins APL Tech. Digest, **11**, 255–63.

Tamao, T. (1964). The structure of three-dimensional hydromagnetic waves in a uniform cold plasma. J. Geomagn. Geoelectr., **18**, 89–114.

Tamao, T. (1966). Transmission and coupling resonance of hydromagnetic disturbances in the non-uniform Earth's magnetosphere. Sci. Rep. Tohoku Univ. Ser. *5*, 17, 43.

Tarduno, J. A., R. D. Cottrell, W. J. Davis, F. Nimmo and R. K. Bono (2015). A Hadean to Paleoarchean geodynamo recorded by single zircon crystals. Science, **349**(6247), 521–4.

Tarduno, J. A., R. D. Cottrell, M. K. Watkeys, A. Hofmann, P. V. Doubrovine, E. E. Mamajek, D. Liu, D. G. Sibeck, L. P. Neukirch and Y. Usui (2010). Geodynamo, solar wind, and magnetopause 3.4 to 3.45 billion years ago. Science, **327**(5970), 1238–40.

Tauxe, L. (2010). *Essentials of Paleomagnetism*. University of California Press, Berkeley.

Tu, W., S. R. Elkington, X. Li, W. Liu and J. Bonnell (2012). Quantifying radial diffusion coefficients of radiation belt electrons based on global MHD simulation and spacecraft measurements. J. Geophys. Res., **117**, A10210, doi: 10.1029/2012JA017901.

Turner, D. L., Y. Shprits, M. Hartinger and V. Angelopoulos (2012). Explaining sudden losses of outer radiation belt electrons during geomagnetic storms. Nat. Phys., **8**(3), 208.

Ukhorskiy, A. Y., B. J. Anderson, P. C. Brandt and N. A. Tsyganenko (2006). Storm time evolution of the outer radiation belt: Transport and losses. J. Geophys. Res., **111**, A11S03, doi: 10.1029/2006JA011690.

Usanova, M. E., A. Drozdov, K. Orlova, I. R. Mann, Y. Shprits et al. (2014). Effect of EMIC waves on relativistic and ultrarelativistic electron populations: Ground-based and Van Allen Probes observations. Geophys. Res. Lett., **41**, 1375–81.

Usanova, M. E., I. R. Mann, I. J. Rae, Z. C. Kale, V. Angelopoulos, J. W. Bonnell, K.-H. Glassmeier, H. U. Auster and H. J. Singer (2008). Multipoint observations of magnetospheric compression related EMIC Pc1 waves by THEMIS and CARISMA. Geophys. Res. Lett., **35**, L17S25, doi: 10.1029/2008GL034458.

Vallee, M. A., L. Newitt, R. Dumont and P. Keating (2005). Correlation between aeromagntic data rejection and geomagnetic indices. Geophysics, **70**, J33–8, doi: 10.1190/1.2057982.

Walker, A. D. M. (2000). Reflection and transmission at the boundary between two counterstreaming MHD plasmas – active boundaries or negative-energy waves? J. Plasma Phys., **63**(3), 203–19.

Waters, C. L. and S. P. Cox (2009). ULF wave effects on high frequency signal propagation through the ionosphere. Ann. Geophys., **27**, 2779–88, doi: 10.5194/angeo-27-2779-2009.

Waters, C. L., J. W. Gjerloev, M. Dupont and R. J. Barnes (2015). Global maps of ground magnetometer data. J. Geophys. Res., **120**, doi: 10.1002/2015JA021596.

Waters, C. L., K. Kabin, R. Rankin, E. Donovan and J. C. Samson (2007). Effects of the magnetic field model and wave polarisation on the estimation of proton number densities in the magnetosphere using field line resonances. Planet. Space Sci., **55**, 809–19.

Waters, C. L., R. L. Lysak and M. D. Sciffer (2013). On the coupling of fast and shear Alfvén wave modes by the ionospheric Hall conductance. Earth Planets Space, **65**, 385–96, doi: 10.5047/eps.2012.08.002.

Waters, C. L., F. W. Menk and B. J. Fraser (1991). The resonance structure of low latitude Pc3 geomagnetic pulsations. Geophys. Res. Lett., **18**, 2293–6, doi: 10.1029/91GL02550.

Waters, C. L., K. Takahashi, D. H. Lee and B. J. Anderson (2002). Detection of ultralow-frequency cavity modes using spacecraft data. J. Geophys. Res., **107**(A10).

Webb, D. F., and J. H. Allen (2004). Spacecraft and ground anomalies related to the October–November 2003 solar activity. Space Weather, **2**(3).

Wright, A. N., and I. R. Mann (2006). Global MHD eigenmodes of the outer magnetosphere, in *Magnetospheric ULF Waves: Synthesis and New Directions*, ed. K. Takahashi, P. J. Chi, R. E. Denton and R. L. Lysak, pp. 51–72, American Geophysical Union, Washington, DC, doi: 10.1029/169GM06.

Wright, A. N. (1994). Dispersion and wave coupling in inhomogeneous MHD waveguides. J. Geophys. Res., **99**(A1), 159–67.

Wright, D. M., and T. K. Yeoman (1999). High resolution bistatic HF radar observations of ULF waves in artificially generated backscatter. Geophys. Res. Lett., **26**(18), 2825–8.

Wright, D. M., T. K. Yeoman and T. B. Jones (1999). ULF wave occurrence statistics in a high-latitude HF Doppler sounder. Ann. Geophys., **17**, 749–58.

Yeoman, T. K. and D. M. Wright (2001). ULF waves with drift resonance and drift-bounce resonance energy sources as observed in artificially-induced HF radar backscatter. Ann. Geophys., **19**, 159–70, doi: 10.5194/angeo-19-159-2001.

Yeoman, T. K., M. K. James, D. Y. Klimushkin and P. N. Mager (2016). Energetic particle-driven ULF waves in the ionosphere, in *Low-Frequency Waves in Space Plasmas*, ed. A. Keiling, D.-H. Lee and V. Nakariakov, Geophys. Monogr. 216, American Geophysical Union, Washington, DC.

Yeoman, T. K., D. M. Wright, T. R. Robinson, J. A. Davies and M. Rietveld (1997). High spatial and temporal resolution observations of an impulse-driven field line resonance in radar backscatter artificially generated with the Tromsø heater. Ann. Geophys., **15**(6), 634–44.

Yoshikawa, A., and M. Itonaga (2000). The nature of reflection and mode conversion of MHD waves in the inductive ionosphere: Multistep mode conversion between divergent and rotational electric fields. J. Geophys. Res., **105**, 10565–84.

Zhou, X.-Z., Z.-H. Wang, Q.-G. Zong, R. Rankin, M. G. Kivelson, X.-R. Chen, J. B. Blake, J. R. Wygant and C. A. Kletzing (2016). Charged particle behavior in the growth and damping stages of ultralow frequency waves: Theory and Van Allen Probes observations. J. Geophys. Res., **121**, 3254–63, doi: 10.1002/2016JA022447.

Ziegler, L. B., C. G. Constable, C. L. Johnson and L. Tauxe (2011). PADM2 M: Penalized maximum likelihood model of the 0–2 Ma palaeomagnetic axial dipole moment. Geophys. J. Int., **184**(3), 1069–89.

PART IV

Space Weather

14
Physical Processes of Space Weather

The Sun–Earth interaction is a complex system of multi-scale processes. The spatial scales of interest vary from the mega-meter size of solar corona structures to the few hundred kilometres of the terrestrial magnetopause and even less when kinetic effects need to be considered. The temporal variations also span a wide range of scales, from thousands of years for the hydrological ocean cycles driven by the total solar radiation to scales of minutes and below for particle acceleration in magnetic reconnection. In this chapter we introduce the building blocks of the Sun–Earth system and briefly describe its important components.

Solar disturbances such as solar flares and coronal mass ejection (CME) have the largest impact on geomagnetic activity, especially magnetic storms. Magnetic storms are responsible for large depressions in the horizontal (H) component of the Earth's surface magnetic field. The strength of a storm is quantified by the D_{st} index, which is a local time average of the depression measured along the magnetic equator. The depression during a storm is caused by a ring current around the Earth with additional contributions from the magnetopause and tail currents. We review recent developments of empirical prediction algorithms for the D_{st} index using observations made upstream of the Earth, and alternative procedures based on the same concept including neural networks and the NARMAX method. Future improvements in empirical prediction will require more data from extreme events, additional physical insight to identify the role of other processes, and better measurements of the inputs to the system.

14.1 The Sun–Earth Connection

Bogdan Hnat

The Sun's impact on Earth is difficult to overstate. Its gravity determines the trajectory of our planet in the solar system, and its electromagnetic radiation sustains life while its magnetic field and the solar wind generate the heliosphere, which protects us from cosmic radiation. Our close proximity to the Sun also brings with it some hazards. The Sun's dynamic atmosphere produces very energetic particles that can reach the Earth within a few tens of minutes. While we are protected from the direct impact of these particles by Earth's magnetic field, their interaction with the magnetosphere can disrupt power grids as well as terrestrial and satellite communication systems (Pulkkinen, 2007). The Sun–Earth system should be viewed as a complex system far from equilibrium. A defining feature of such systems is the variety of temporal and spatial scales that are important for their description. The following list highlights some of the temporal scales of interest:

- 10^5–10^4 years: quasi-periodic changes of the Earth's orbit
- ~100 years: Gleissberg cycle of variation in sunspot number
- ~11 years: sunspot cycle
- ~27 days: solar rotation
- ~10 hours: transient events on the Sun, such as flares
- ~10 seconds: typical proton gyro-period in the solar wind plasma

14.1.1 Solar Electromagnetic Radiation

The dominant fraction of Sun's energy output is in the form of electromagnetic (EM) radiation. The *total solar luminosity* has been estimated at $L_\odot \approx 3.86 \times 10^{26}$ W. For comparison, the estimated total energy consumption worldwide in 2008 was about 1.5×10^{13} W. All solar energy is produced by fusing protons into deuterium and then into helium in the core of the Sun. The energy is transported by radiation and then convection towards the solar atmosphere, where it is radiated out in different spectral bands at different atmospheric heights. The maximum of the spectrum is at the visible wavelengths, and nearly half of the total integrated power is in the visible range. At these dominant wavelengths (from near-ultraviolet to radio), the emission spectrum is essentially that of a black body at 5710 K, which is the average temperature of the photosphere. In the context of the Sun–Earth connection, the *total solar irradiance* (TSI) and *spectral solar irradiance* (SSI) are of particular interest. TSI is defined as the EM power falling on a unit area at the mean Earth–Sun distance of 1 AU. TSI is the wavelength integral over the SSI. Solar irradiance is the largest energy input for the Earth, but the extent to which small variations in the TSI could impact the Earth's climate is unclear (Haigh, 2007). Different spectral components of solar EM radiation drive atmospheric process on different temporal scales. TSI is dominated by visible light, which is transmitted through the atmosphere to the Earth's surface. TSI variations in this spectral band directly influence the sea surface temperature and hydrological cycle with timescales of 10^2–10^4 years. The impact of short time irradiance variability on Earth's climate is still debated. The observed ~0.1% variation in solar luminosity due to the solar sunspot cycle (~11 years) gives a change in the irradiance of 0.34 Wm^{-2}, which is further reduced by planetary albedo to about 0.24 Wm^{-2}. In comparison, the estimate of the radiative forcing by anthropogenic greenhouse gases is 2.0–2.5 Wm^{-2} (Reid, 2000). The ocean equilibration time is much longer than the sunspot cycle, hence the direct impact of the solar forcing would be substantially reduced by the thermal inertia of the ocean.

While TSI has been continuously measured since 1978, no single instrument has covered a complete solar cycle. The time series spanning more than 30 years has been constructed from seven different instruments: HF on NIMBUS-7, ACRIM, ACRIM I, ACRIM II, VIRGO, ACRIM III and TIM (for details, see Fröhlich, 2012). In 2010 an additional instrument, PICARD, was launched (Conscience et al., 2011). Combining the measurements from these instruments into a single time series requires non-trivial corrections for ageing and exposure degradation as well as calibration uncertainty (Dewitte et al., 2004). Better sensitivity of the newer instruments and improved understanding of the calibration process have led to the steady decline of the measured absolute value of the mean annual TSI at solar minimum from 1371 Wm^{-2} in 1978 to ~1361 Wm^{-2} in 2013. The current value of the TSI from the SORCE mission is TSI = 1360.8 (±0.5) W m^{-2} (Fröhlich, 2012). This lower value, while not universally accepted, is also consistent with the first results from PICARD (Meftah et al., 2014). The TSI value at the current minimum is almost 0.3 Wm^{-2} lower compared to previous ones. This corresponds to 25% of the cycle amplitude, but only 2%–5% difference is observed in chromospheric proxies, such as the Ly-α index.

TSI cycle-to-cycle variation is ~0.1% (Fröhlich, 2012), but changes in the irradiance on scales of days to weeks can exceed 0.3%. These short-term variations have been successfully modelled using photospheric magnetism (Yeo et al., 2014). The magnetic field generated by the dynamo action in the solar interior rises due to its buoyancy and emerges at the photosphere, where it forms *bipolar active regions*. Photospheric field forms flux tubes of various sizes (Solanki et al., 2013). The largest flux tubes form *sunspots*. Smaller flux tubes, called *faculae*, usually surround sunspots. Sunspots are cooler than the surrounding plasma and appear dark, while faculae are brighter (Vögler et al., 2005). As the strength of the solar magnetic field increases, the averaged emission excess from faculae is larger than the emission deficit of the sunspots so that the solar EM output increases with the increasing magnetic field strength (Foukal, 2012). As a consequence, the brightness of the Sun increases from minimum to maximum of the solar cycle (Wenzler et al., 2006). Current models of cyclic TSI and SSI variation are of two types. Proxy models utilise the Photometric Sunspot Index (PSI), radio flux at 10.7 cm (F10.7) and the MgII index (Yeo et al., 2014), of solar surface magnetic activity, using regressions to match observed TSI changes. Physics-based models consider the solar spectrum as a linear superposition of reference spectra originating in different features (spots, faculae), and the brightness of each component is calculated using semi-empirical radiative transfer codes (Shapiro et al., 2010). These models capture the essential variability very accurately and in some cases can reproduce more than 90% of TSI variations for the entire data coverage (Ermolli et al., 2013). Modelling long trends of TSI, which may be of importance for the Earth's climate, is more challenging. Proposed models of the slow variability include the overlap of consecutive solar cycles (Solanki et al., 2002) and a magnetic flux transport model (Wang et al., 2005). Significant progress has been associated with the improved measurements of faculae area in both active regions and active networks (Bertello et al., 2010). Indirect evidence of solar variability on the Earth's climate can be obtained from ice cores and ocean sediments, which contain different levels of cosmogenic isotopes ^{10}Be and ^{14}C, indicative of solar activity. Data from these sources show a good correlation between temperature and solar variations (Bond et al., 2001), with periodicity of about 10^3 years. Similar statistical correlations have been identified between SSI and global temperatures (Usoskin et al., 2005) and between PSI and mean surface temperatures (Wenzler et al., 2006). These statistical features, while robust, cannot establish causality, nor do they reveal any physics fundamental to the Sun–Earth connection.

14.1.2 Solar EUV Radiation

A sharp change in the solar spectrum can be observed at the *extended ultraviolet* (UV) wavelengths between 10 and 400 nm. The power decreases by a factor of ~10^4, and the black body radiation approximation breaks down. The bulk of UV emission comes from the upper photosphere, chromosphere and the transition region, which are far from equilibrium and do not conform to a Planck spectrum. The heat transfer from the hot corona towards the chromosphere is balanced by the radiation of the hydrogen Ly-α line at 121.6 nm, which is a dominant feature of the far UV spectrum.

Shorter wavelengths associated with the extreme UV (EUV) emission from the solar corona, the transition region and the chromosphere, are dominated by strong emission lines. For wavelengths larger than 200 nm, there is a large contribution from absorption lines of heavy elements, Fe I, Fe II, Mg II and Ca II. This contribution is very sensitive to changes in magnetic field topology and is useful as a proxy of the chromospheric magnetic activity. The UV spectral component is a primary energy source for the Earth's atmosphere, driving ionisation, dissociation and excitation of atoms and molecules. These processes set the vertical temperature profile in the upper and middle atmosphere and the dynamical coupling down to the troposphere.

The first continuous record of the EUV variability was obtained by the AE-E/EUVS instrument, from a solar minimum in 1976 to a solar maximum in 1979 (Hinteregger, 1981). The first coverage of a full solar cycle was done by UARS/SUSIM (Brueckner et al., 1993) between 1991 and 2005. Currently, there are three missions providing constant monitoring of the EUV irradiance: SOHO/SEM, TIMED/SEE and the SDO/EVE. The measured variability in the near-UV is much larger than that of TSI, reaching several percent at 200–250 nm and several tens of percent below 200 nm (Floyd et al., 2002). Due to its direct impact on the Earth's atmosphere, the EUV spectral range is of particular interest. While the UV radiation represents less than 8% of the TSI, its variability may impact Earth climate through the complex feedback mechanisms (Gray et al., 2010). A series of SIM and SOHO results stirred some controversy. The SOURCE/SIM instrument measured UV flux decrease during the declining phase of cycle 23 (2004–8), which was twice the decrease measured in TSI (Ermolli et al., 2013; Yeo et al., 2014). Similarly, SOHO reported a significant decrease of EUV irradiance of ~15% between the minima of 1996 and 2008. The lower EUV irradiance has been linked with anomalously low thermospheric neutral density and ionospheric peak density during the minimum of 2007–9 (Didkovsky and Wieman, 2014). This large decrease in the EUV irradiance is not reflected in the common proxies (Chen et al., 2011) and could be due to the instrument degradation.

The lack of long reliable data sets makes empirical models of the UV irradiance critical. Proxy models are constructed similarly to these for TSI. Linear combinations of the 10.7 cm radio and MgII core-to-wing index or nonlinear functions of them are often used (Lean et al., 2011). Semi-empirical models use solar images to identify features on the Sun and each feature type is associated with radiative transfer based on solar heliocentric position. Spectra are calculated for all features, and the sum gives the irradiance estimate. A common feature of these models is a strong correlation of the emission from the different layers of the atmosphere on timescales that exceed the dynamic time of solar events. The EUV range variability is still not fully understood (Suess et al., 2016). Variations in the UV range can be captured more accurately if the solar rotation and the solar cycle variability are modelled separately, as the active network contributes differently in both cases (Woods et al., 2000).

14.1.3 Solar Wind and Interplanetary Magnetic Field (IMF)

Despite its temperature of ~10^6 K, the solar corona is a low β plasma – its dynamics is dominated by its magnetic field. It experiences large energy losses, some mediated by the kinetic energy of the solar wind flow, which is accelerated to a supersonic and super-Alfvénic velocity within ~10 $R_\odot$. The exact mechanisms that govern coronal heating and the solar wind acceleration are still debated (Cranmer et al., 2015). The solar wind flow pulls magnetic field with it, forming the *interplanetary magnetic field* (IMF). The solar wind and the IMF are at the heart of all heliospheric physics. The structure of the solar wind near the Sun has been probed by the Ulysses spacecraft during its heliographic latitude scans of 1993–5 (Balogh et al., 2001) and by the UVCS/SOHO instrument (Kohl et al., 1997). Increased magnetic structuring of the solar corona at solar maximum is clearly visible from the composite images from SOHO/EIT and LASCO/C2 instruments and with the Mauna Loa K coronameter (McComas et al., 2007). The simplest configuration is at the solar minimum, when the solar magnetic field is dominated by its dipolar component, with the magnetic dipole axis tilted by a few degrees to the solar rotation axis. *Coronal holes* are the unipolar regions of the open magnetic flux, that is, with only one foot attached at the Sun while the other stretches into the heliosphere. They are visible as darker areas in EUV and X-ray emissions in the high latitude regions. Near the equator the magnetic flux of opposite magnetic polarities create dense bright *streamer belt*.

Traditionally, solar wind is categorised as fast solar wind with V_{sw} > 500 km/s and a slow solar wind with V_{sw} < 500 km/s based on the bulk speed of the plasma. There is a consensus based on the Ulysses data that the fast solar wind definitively originates from long-lived coronal holes (McComas et al., 2008). It has large temperature anisotropy $T_\perp \gg T_\parallel$, and is accelerated to supersonic speed close to the Sun, possibly at about 2 $R_\odot$ (Antonucci et al., 2000). Two alternative mechanisms for fast wind acceleration have been proposed: wave or turbulence particle interaction (Cranmer et al., 2007) and interchange reconnection (Parker, 1992). The origin of the slow wind is less clear, but there is some evidence that its source is located within the legs of the large streamers at the equatorial region (Strachan et al., 2002). The lack of speed-composition correlation has been recently cited as a compelling evidence for sporadic release of closed field plasma via magnetic reconnection at the boundary between open and closed flux in the Sun's atmosphere (Kepko et al., 2016). At solar minimum, the heliosphere is dominated by fast or high-speed streams (HSS). At solar maximum, when higher order multipole moments of the magnetic field are more pronounced, the slow solar wind extends to high latitudes and dominates the emission. Numerous space missions provided detailed observations of the IMF and plasma in the heliosphere (Schwenn, 2006). Fast wind is nearly incompressible and less variable, while slow wind shows large deviations from its mean values in velocity and the number density. The total energy is very similar for both slow and fast wind.

A radial solar wind outflow with constant speed, independent of distance from the Sun and the latitude, combined with a rigid rotation of magnetic field line footprints at the photosphere and the frozen-in condition, leads to the equatorial magnetic field in a form of a Parker spiral (Borovsky, 2010). For the solar wind velocity of V_{sw} and the angular frequency of the photosphere Ω, a relative angle of the solar wind velocity and the magnetic field is $\tan(\vartheta)=\Omega r/V_{sw}$. At 1 AU this angle is about 45°, but at large heliospheric distances the magnetic field is nearly toroidal. Spacecraft observations show that the angle of the IMF to the radial direction closely follows that predicted by Parker, with a general unwinding of the spiral at higher latitudes (Schwadron, 2002). Observations in the ecliptic at 1 AU also found a simple IMF topology of a few *polarity sectors* in each solar rotation, independent of the complexity of the photospheric and coronal magnetic field (Wilcox and Ness, 1965). These sectors arise from non-dipolar magnetic field and the tilt of the magnetic axis with respect to the rotational one (Schulz, 1973). Different polarities are separated by the *heliospheric current sheet* (HCS) (Smith, 2001). The HCS changes during the solar cycle due to the relative angle between the solar magnetic dipole and the solar rotation axis (Owens and Forsyth, 2013). The HCS is always embedded in the slow wind (Burlaga and Ness, 2012). During the declining part of the solar cycle, the coronal holes extend to low latitudes and the inclination of the solar magnetic axis to the solar rotation axis increases. This leads to fast and slow solar wind entering the heliosphere from neighbouring latitudes. When HSS catches up with slow wind ahead of it, the region of solar wind compression and deflection is called the *stream interaction region* (SIR), and when it co-rotates with the Sun, a *co-rotating interaction region* (CIR) (Richardson, 2004).

There are considerable deviations from the Parker spiral on smaller spatial scales, especially at larger distances from the Sun. These are attributed to solar ejecta, structures of solar origin (Bruno et al., 2001) and turbulent solar wind dynamics (Owens et al., 2011). Variability in the solar wind speed near the solar surface may also produce radial magnetic field (Borovsky, 2010). The IMF fluctuations show all the key features of turbulence (Hnat et al., 2003) and there is evidence that turbulence and reconnection coexist in the solar wind (Gosling, 2007; Phan et al., 2006).

14.1.4 Solar Wind Interactions with a Magnetised Body

The interaction of the solar wind with an object depends on its distance from the Sun, the size of the object, the presence of the atmosphere and the ionosphere and the magnetic properties of the object. The supersonic and super-Alfv énic flow of the solar wind around an obstacle generates a sharp boundary, a *bow shock*, which heats and slows down the solar wind. If the object has its own strong magnetic field, as in the case of the Earth, the solar wind plasma is separated from the planetary plasma by another boundary known as a *magnetopause*. Here, we concentrate on the interaction between the solar wind and the terrestrial magnetic field. The MHD frozen-in approximation prohibits any direct mixing of solar wind and terrestrial plasmas. Thus, the flow of the solar wind around the Earth generates a magnetic cavity, or magnetosphere. The standoff distance of the magnetopause, a boundary between the IMF and the terrestrial field, is about $10R_E$. The magnetosphere has a water drop–like shape. The dipole field is squeezed towards the equatorial plane and extended in the anti-sunward direction to distances of 200–300R_E. A thin current sheet separates the northern and southern lobes of the magnetosphere on the nightside. In the inner magnetosphere, particles can become trapped, populating the Van Allen belts for long periods (Hudson et al., 2004). The ring current located roughly at 4–6R_E radial distance encircles the Earth with highly variable intensity modulated by the level of geomagnetic activity. This complicated structure is maintained by a system of currents, which flow around boundaries and close through the ionosphere (Daglis et al., 1999). The solar wind kinetic energy delivers ~5×10^{-2} Wm^{-2} of power to the magnetosphere. The size of the magnetosphere depends on the solar wind conditions, but for the average parameters, the magnetospheric radius $R_{ms} \approx 15R_E$. This gives total power input of order 10^{15} W, compared with 10^{17} W for the power input from the TSI. While the energy input from the TSI vastly exceeds that from the solar wind, it is the complex interaction of the solar wind with the terrestrial magnetosphere that is a source of most geomagnetic phenomena. Detailed understanding of this process is of fundamental importance for space weather. The overall level of geomagnetic activity varies with the 11 year sunspot cycle and annually due to the changing tilt of the Earth's magnetic dipole with respect to the IMF (Russell and McPherron, 1973). The global energy input from the solar wind into the magnetosphere is measured by the *Akasofu parameter*, $\varepsilon = 10^7 V_{sw} B^2 (7R_E)^2 \sin^4(\Theta/2)$, where V_{sw} is the solar wind speed, B the interplanetary magnetic field magnitude, and $\tan(\Theta) = B_y/B_z$ (in GSM coordinates) (Akasofu, 1981).

Dayside reconnection is by far the most important mechanism for injecting solar wind plasma into the magnetosphere. This is the most efficient when the IMF has a southward B_z component and can be enhanced by solar wind ram pressure changes (Borovsky et al., 2008). The dynamics and the morphology of the terrestrial magnetosphere are dominated by its interaction with the solar wind and are described by the *Dungey cycle* (Dungey, 1961). For the IMF with the southward component, the cycle begins with the solar wind plasma entering the magnetosphere by magnetic reconnection at low latitudes where the IMF is anti-parallel to Earth's dipole field. The reconnection process creates open flux tubes, which are then convected by the solar wind to the magnetotail and eventually reconnect again. The newly closed field lines in the tail convect initially Earthward, then divert along the flanks back to the dayside. In reality, the IMF rarely has a southward orientation, and the location of the reconnection varies with the IMF orientation. Several models have been proposed for the localisation and orientation of the magnetic reconnection line (Moore et al., 2002), and observations suggest that the reconnection line is located where the magnetic shear angle has its maximum (Trattner et al., 2016). Significant progress has been achieved in understanding *dual-lobe reconnection* occurring during the northward orientation of the IMF (Fuselier et al., 2012). Quantifying reconnection rate and understanding the feedback of the injected plasma on this process is fundamental for space weather applications. Historically, the motional electric field component $-V_{sw}B_z$ has been assumed to control the reconnection rate (Gonzalez and Mozer, 1974). Numerical simulations found that the reconnection rate and its scaling with the local parameters are broadly consistent with a 2--D asymmetric reconnection model (Cassak and Shay, 2007). The feedback of the injected plasma on the reconnection rate has been added to numerical models and this appears to slow down the reconnection process (Borovsky and Birn, 2014). In situ observations produced reconnection rate estimates consistent with the Cassak–Shay formula and highlighted the importance of the kinetic physics in the reconnection process (Burch et al., 2016).

The intermittency of the IMF components causes episodic energy injection and dissipation cycles known as *substorms* (Baker et al., 1996). The Earthward directed tail reconnection outflows sheared by the slower ambient flows generate the *substorm current wedge* system (McPherron, 1979), coupling the near-Earth plasma sheet with the ionosphere via downward and upward field-aligned currents. When the enhanced energy input continues for longer than about 3 hours, a magnetic *storm* develops. Storms are accompanied by the formation of an intense current ring, peaking inside the geostationary orbit, at $R = {\sim}4R_E$ (Pulkkinen, 2007). There is currently no consensus on the role of substorms in the ring current development (Ganushkina et al., 2005).

14.1.5 Solar Cycle

The solar activity cycle is traditionally associated with the formation of sunspots and their migration towards the solar equator. This cycle has a well-known periodicity of about 11 years and is referred to simply as the *solar cycle*. Another cycle with a period of 22 years is associated with the solar magnetic field polarity reversal and is called the Babcock–Leighton solar dynamo cycle. The solar cycle is not exactly periodic and varies in the amplitude and the length (Hathaway and Rightmire, 2010). A longer modulation of the 11-year cycle, with a centennial period, is called the Gleissberg cycle. Historical records related to mini ice age periods, Maunder Minimum (1645–1715) and the Dalton Minimum (1790−1830), illustrate a significant variation in the amplitude of the solar cycle (Usoskin et al., 2007). There are also shorter-period oscillations: *quasi-biennial oscillations* (QBO) (1–4 years) (Kolotkov et al., 2015) and *Reiger-type oscillations* with periodicities shorter than 1 year (Chowdhury et al., 2010). The solar activity cycle can be seen in many parameters related directly to the Sun and in the heliospheric observations. Sunspot area provides the most common measure of the solar cycle, related directly to the photospheric magnetic field. The 11 year period is also present in the 10.7 cm radio flux (Tapping and DeTracey, 1990) and in the frequency shift of helioseismic p-modes (Chaplin et al., 2007). These proxies probe different regions of the Sun: the 10.7 cm radio signal originates from the upper chromosphere and lower corona, p-modes probe the interior of the Sun and their frequency shift is sensitive to both the strong and the weak components of the magnetic field (Broomhall and Nakariakov, 2015). In the heliosphere, the complexity of the solar magnetic field at the solar maximum leads to the increased open solar flux realised through the larger number of CIRs and interplanetary coronal mass ejections (ICMEs) (Connick et al., 2010).

The lack of regularity in period and in the amplitude makes solar cycle periodicities difficult to establish with some accuracy. Fourier-based techniques are not adequate, as the prescribed basis of the decomposition is harmonic, while the variability is not. Various alternative approaches are wavelets, the Krylov–Bogolyubov method, proper orthogonal decomposition and the empirical mode decomposition (EMD). These techniques find similar dominant periodicities in the Reiger range, QBO range and 11-year period. Interestingly, the QBO period was also detected in the daily coronal index of Fe XIV (Deng et al., 2015). The Reiger-type oscillations often have periods that are multiples of ~25 day solar rotation (Bai and Sturrock, 1991), which could be related to the evolution of the photospheric field on this temporal scale (Knaack and Stenflo, 2005). While the bulk of research into solar cycle periodicities is aimed at the solar dynamo problem, these periodicities are also found in the solar wind parameters and in the IMF, and it is this heliospheric connection that is of interest for space weather and the Earth–Sun connection (Cho et al., 2014).

Magnetic helicity generation due to the ICMEs also shows an 11 year cycle (Lynch et al., 2016).

14.2 Solar Disturbances

Bogdan Hnat

The solar wind proposed by Parker represents continuous loss of mass from the Sun. During its nonlinear advection away from the Sun, the solar wind fluctuates and may develop some structures, but these are not large enough to disturb the Earth's magnetic field significantly. However, the Sun also produces a number of disturbances, often impulsive in their nature, and these events can have a significant impact on the communication infrastructure, our ability to travel through space and even on life on Earth.

14.2.1 Solar Flares

Solar flares are large and impulsive increases in the solar brightness with the EM radiation emitted across the entire spectrum. The total energy of a flare can vary from a fairly small 10^{19} J to an extremely large 10^{25} J, released on the timescales of a few minutes to a few tens of minutes. The vast majority of the energy is emitted as EM radiation and a small fraction is lost to heating and particle acceleration. One of the most spectacular flares was that observed by R. C. Carrington in 1859, with historical records reporting a doubling of the visible solar emission and global disruption of the telegraph services due to the geomagnetically induced currents. Unprecedented geomagnetic activity has also been observed, with the aurora lights visible at latitudes down to 20° (Tsurutani et al., 2003). The largest solar flare observed by modern spacecraft instrumentation occurred on 4 November 2003. This event has been estimated to be of class X35, compared with class X45 of the Carrington event (Cliver and Dietrich, 2013). The solar flare classification system uses the flare peak radiation of 100 to 800 picometre X-rays near Earth to define classes C1-9, M1-9 and X1-9, corresponding to intensities of $1\text{–}9 \times 10^{-6}$, $1\text{–}9 \times 10^{-5}$ and $1\text{–}9 \times 10^{-4}$ W m^{-2}, respectively. A flare with solar X-ray (SXR) intensity of 45×10^{-4} W m^{-2} is designated X45.

Solar flares originate in the active regions of the Sun and trigger emissions from every layer of the solar atmosphere. It is believed that the plasma motion in and below the photosphere tangles coronal fields and increases magnetic energy until it is released by the reconnection. While the exact chain of events that produces flares is not known, it is very likely that the reconnection powers flare emission. Since this energy and the accelerated particles are largely contained by magnetic loops of the active region, it leads to plasma heating that cools by radiative emission. Many flares, but not all, produce hard X-ray radiation from *non-thermal electrons*, which are accelerated near the flaring site and then impact on the denser layers of the solar atmosphere, producing hard X-rays by the *bremsstrahlung* process (Mann, 2015). Soft X-rays are often observed during the initial phase of the flare. This *thermal emission* consists of the *bremsstrahlung continuum* from the active region heated to about 10^7 K and spectral lines of heavily ionised atoms. The UV band is generated in the chromosphere, with the Ly-α emission of the upper chromosphere dominating the spectrum. At longer wavelengths, near the peak of the visible range, the Balmer H-α emission from the lower parts of the solar atmosphere dominates. The most energetic flares can accelerate protons to energies of a few tens of keV. When these energetic protons penetrate deep into the denser layers of the solar atmosphere, gamma rays and white light are produced. The strongest line in the gamma-ray range is due to neutron capture by protons, which leads to electron–positron annihilation with an emission peak at 511 keV (Lin et al., 2003). *Gyro-synchrotron radiation* is produced by accelerated electrons gyrating in strong chromospheric magnetic field. Radio emission from deci- to kilometric wavelengths is generated by accelerated electrons escaping into space and interacting with the solar wind plasma. These radio bursts are divided into four types (I–IV), depending on their spectral properties.

The impact of solar flares on Earth is broadly similar to any other increase in the solar luminosity. The excess UV radiation from the solar flares is absorbed in the Earth's atmosphere, causing instantaneous heating and expansion. The response to an X30 or X40 flare would produce a ~75% and ~100% density increase, respectively, on the dayside at 400 km (Le et al., 2016). This not only changes the chemical composition, especially the amount of ozone, but also has practical effects on the operation of low orbit satellites. Atmospheric expansion during extreme events produces an enhanced drag on the spacecraft, which can cause unexpected re-entry of satellites, and difficulty in identifying and tracking satellites and other space objects (Nwankwo et al., 2015). The hard X-ray radiation can damage electronic components of the communication satellites, while signals in the radio frequencies can interfere with radar observations. Since the late 1980s it became increasingly clear that many phenomena that were originally associated with flaring activity were in fact due to the coronal mass ejection that often, but not always, accompanies a flare.

14.2.2 Coronal Mass Ejection (CME)

Coronal mass ejections (CMEs) are the most spectacular eruptive events occurring in the solar atmosphere. CME eruptions eject a vast amount of solar material into interplanetary space on a timescale of days and are the source of the largest geomagnetic events. Terrestrial impact of the 1859 event, for example, was due to a CME that followed

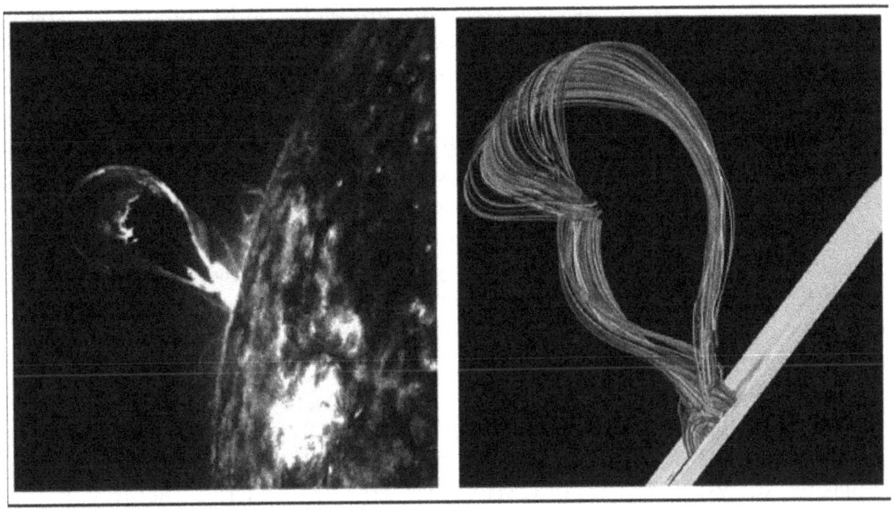

Figure 14.2.1 (left) Typical shape of the erupting flux rope observed by SDO/AIA on 14 October 2012, at 2:32 UT. (right) Late stage of numerical model evolution of the flux rope for the active region AR10930. Image adapted from Amari et al. (2014, Figures 4c and 4d). (A black-and-white version of this figure appears in some formats. For the colour version, please refer to the plate section.)

the flare and arrived at the Earth some 17 hours later. Observations show that flares and CMEs are independent phenomena (Robbrecht et al., 2009), although large flares are usually accompanied by CMEs. This is more likely due to the underlying magnetic reconfiguration process, which is critical to both events.

CMEs have been observed in situ since 1971, but the most recent missions, STEREO and SDO, provided a step-like improvement in our understanding of these events. The longest and the most accurate CME data set has been collected by LASCO on board the SOHO spacecraft. LASCO observed 10^4 CMEs (Kohl et al., 2006) and provided new estimates of the characteristic CME parameters. The average mass expelled during the CME eruption is estimated at 10^{12} kg. Most of the energy extracted from the unstable magnetic field configuration is deposited into kinetic energy of the mass flow, which is on average 10^{23} J. This is an order of magnitude larger than the average flare energy (Emslie et al., 2005). The kinetic energy distribution follows the power law with an index of -1 (Vourlidas et al., 2002). The vast majority of CMEs are accelerated to their final speed of 300–550 km/s before they reach the distance of $2R_{\odot}$ from the Sun (Yashiro et al., 2004). Some CME velocities have been known to exceed 2000 km/s (Gopalswamy, 2010).

The number of observed CMEs follows the solar cycle variation, with about 1 CME per day at the solar minimum to nearly 5, on average, during the solar maximum. It is of great interest to space weather models to detect CMEs, which can directly impact on the Earth magnetosphere. The geo-effectiveness of a CME is directly related to its position on the Sun, with the CMEs near the centre being most likely to perturb the Earth's magnetic field (Gopalswamy, 2010). These CMEs appear as faint halos surrounding the occulting disc of near-Earth coronagraphs. Only 10% of all CMEs are full or partial halo CME. In order to establish that the *halo CME* is Earth directed, and did not erupt on the backside of the solar disc, a point of origin must be identified.

There is no unique form to the observed CME eruption. Often a CME may show three components: a front made of compressed plasma, a flux rope and a cavity surrounded by a bright filament/prominence (Illing and Hundhausen, 1985). There is a growing consensus that despite the lack of clear association of flares with CME, a universal mechanism governs both processes. A key physics is flux rope formation and its stability in the presence of a current sheet (Linton and Moldwin, 2009). There are two competing scenarios: a twisted flux rope moves out of equilibrium or becomes unstable, and the subsequent reconnection then powers the ejection (Amari et al., 2011) or a new flux rope forms as a result of the reconnection of the magnetic lines of an arcade (a group of arches of field lines) during the eruption itself (Antiochos et al., 1999). Observational support for both mechanisms has been claimed (Cheng et al., 2013; Kumar and Innes, 2013). Recent advances in numerical modelling allow three-dimensional simulations of the entire eruption process, and there is some evidence that the first scenario is more likely (Amari et al., 2014). Figure 14.2.1 shows a striking resemblance of typical observations from the SDO/AIA instrument (left panel) and the late stages of flux tube evolution from numerical model (right panel).

CMEs may travel through the surrounding medium at speeds that exceed the local Alfvén and sound speeds. In such circumstances the front of the structure develops into shock and it can accelerate particles to high energies. Particles processed by the shock must have enough energy to return upstream, as the efficient shock acceleration requires multiple particle–shock interactions. The first stage acceleration may be provided by a flare, which can heat electrons to 100 MeV. Then, the fast CME shock provides further acceleration. As much as 10% of the CME kinetic energy might go into particle acceleration (Webb and Howard, 2012).

14.2.3 Solar Energetic Particles (SEP)

There are two different types of SEP events, and the distinction is based on the acceleration mechanism involved. *Impulsive SEP* events are generated directly by solar flares, while *gradual SEP* events are the product of the CME shock acceleration. CMEs that move faster than the surrounding solar wind create a shock wave, which accelerates coronal and solar wind ions and electrons. This classification is not entirely unique, in that the large SEP events are dominated by the CME accelerated particles but may still include a fraction of flare-accelerated species. In fact, it is the fastest particles that are preferentially accelerated by the multiple crossings of the shock front. The different origins of these two types of SEPs are especially clear in the heavy ion abundances. Impulsive SEPs, associated with flaring activity, carry the signature of the heated coronal plasma, showing enhanced ^{3}He/^{4}He ratio and heavy ions, while the composition of gradual SEPs is similar to the ambient solar corona. The impulsive events produce high electron fluxes. Gradual SEPs work more effectively for protons and ions (Cane et al., 1986). A characteristic feature of the impulsive SEPs is their longitudinal localisation on the solar disc. The most relevant sources for space weather are located on the western side of the disc; these are magnetically connected to Earth by the Parker spiral magnetic field lines. The accelerated particles are confined in the flux tube and do not disperse significantly before impacting on the terrestrial atmosphere. In contrast, the sources of gradual SEPs are spread uniformly in solar longitude. The signature of gradual event will be different for the observer that is magnetically well connected to the shock and the observer that is not (Desai and Giacalone, 2016).

Solar flares and CME shocks can accelerate electrons to large (non-thermal) energies. These suprathermal electrons are routinely detected by their emission, or can be directly observed as they escape into interplanetary space along open magnetic field lines (Benz, 2017). For the largest events, electrons with energies of up to 100 MeV have been observed (Moses et al., 1989). An interesting problem of the delayed injection of the near-relativistic electrons from the impulsive events remains open. It refers to the observation that the lower energy electrons arrive at the instrument prior to these highly energetic ones. Various possible explanations have been considered in the literature (Maia and Pick, 2004), but there is no agreement as to the exact mechanism behind this observation. Flares may accelerate protons to several GeV and when these protons interact with atoms they produce a neutron flux. The mean lifetime of the free neutron is about 15 min, and most of the time we do not observe solar neutrons at 1 AU. Due to the large mean free path, nearly relativistic neutrons can reach the Earth and this enhanced neutron flux can be observed on the ground as *ground level events* (GLEs). Reconstruction of such events yields neutron energies of about 500 MeV (Miroshnichenko et al., 2005). It is the relativistic flare and CME accelerated protons, however, that are of the most interest for space weather applications. Travelling near the speed of light, SEPs arrive at Earth 30 min to 1 hour after the event (flare/CME) and travel along the geomagnetic field lines into the upper atmosphere above the North and South Poles. They enhance the ionisation of the lower ionosphere over the polar region. These *polar cap absorption* (PCA) events cause ozone depletions in the polar upper stratosphere (Krivolutsky and Repnev, 2012). The relativistic proton flux is the main concern for the operation of low-orbiting satellites and for manned space missions.

14.2.4 Co-rotating Interacting Regions (CIR)

When the HSS overtakes the slow wind a CIR arise. In the absence of CMEs these large heliospheric structures are the most important source of geomagnetic activity (Denton et al., 2006). At the interface of the fast and the slow solar wind the level of turbulence increases and a shock is formed, which moves inward (reverse shock) and outward (forward shock). The shock does not form instantaneously, and at 1 AU, a non-shock gradual plasma compression is normally observed (Giacalone et al., 2002). When a CIR compression region reaches the Earth's bow shock, a sudden increase of the solar wind speed and the magnetic field (Denton and Borovsky, 2012), lasting up to a few days, is detected. The persistence of low-latitude coronal holes leads to a 27 day recurrence period of the HSS, which is seen in the magnetospheric and ionospheric proxies (Fang and Forbes, 2012).

Different magnetospheric regions respond differently to the HSS and CIR conditions, and there is a positive correlation between the speed of the solar wind and the overall level of the magnetospheric convection. In the *plasmasphere*, a cold plasma torus located above the ionosphere, the HSS/CIR triggers a strong convection of cold and dense plasma structures, called plumes (Moldwin et al., 2016), towards the dayside of the magnetopause. The density enhancement of the plume depends on the strength of the HSS/CIR (Denton and Borovsky, 2012). The injection of the cold dense plasma at the dayside reduces the coupling efficiency of the solar wind to the magnetosphere (Ouellette et al., 2016). Increased convection after the HSS/CIR interaction also leads to dense and hot plasma sheet. Modelling relativistic electron flux from the outer radiation belts is critical for space weather predictions, as these can damage space-borne electronic equipment. It has been found that CIRs lead to the loss of the relativistic electrons from the outer belt, but the exact mechanism is still disputed (Ukhorskiy et al., 2017). In the ionosphere, a response to HSSs/CIRs leads to variation in peak electron number density in different sectors at high latitudes (Grandin et al., 2015) as well as ion heating (Sojka et al., 2009). Besides the direct impact on the Earth's magnetosphere, the gradual plasma compressions of CIRs are also associated with local acceleration of charged particles (Ebert et al., 2012).

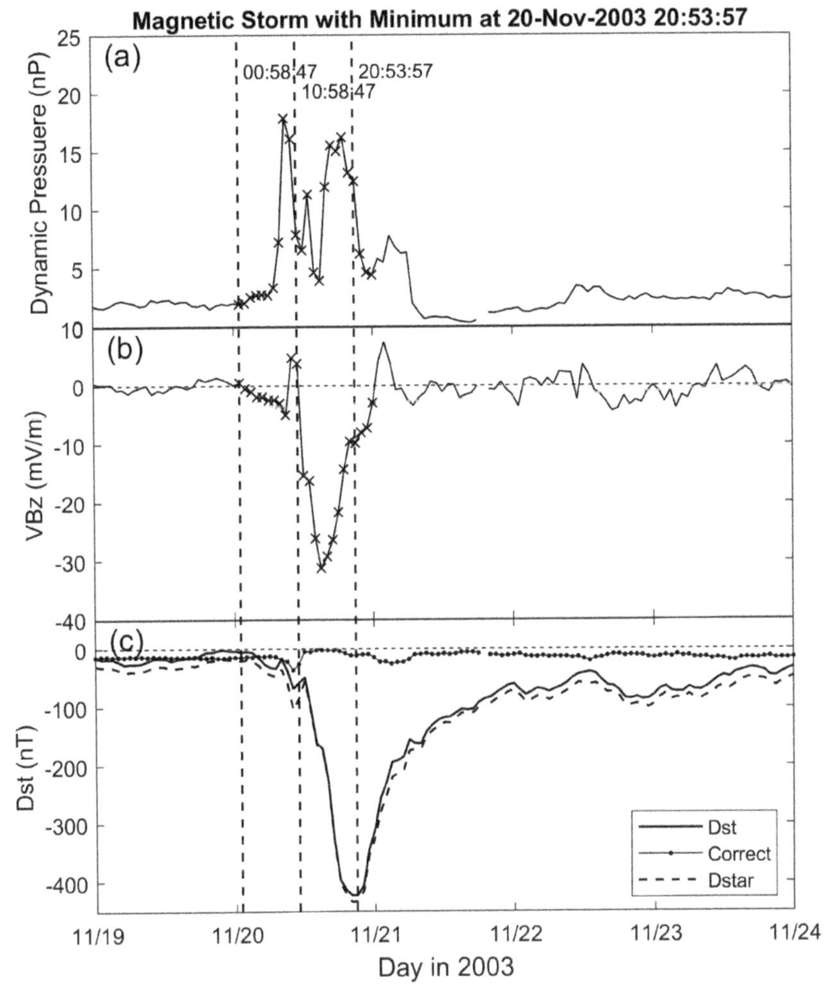

Figure 14.3.1 An example of a magnetic storm as recorded in the hourly D_{st} index and its solar wind drivers in November 2003. (a) The solar wind dynamic pressure. (b) The driver VB_z $(-E_y)$. (c) The D_{st} index (thick black), the dynamic pressure correction (thin dotted) and the corrected D_{st} index, D_{star} (dashed). Three vertical dashed lines are, respectively, start initial phase, start main phase and start recovery phase.

While SEP ions dominate the distribution at solar maximum, the suprathermal heavy ion composition is dominated by CIR-like or solar wind–like material during solar minimum at 1 AU (Dayeh et al., 2009).

14.3 Magnetic Storms

*Robert L. McPherron**

14.3.1 Introduction: Magnetic Storms and the Ring Current

A magnetic storm is an interval of worldwide depression in the surface magnetic field (Chapman and Bartels, 1962). At mid-latitudes, a typical storm exhibits a depression in the horizontal component (H) that ranges from as little as −25 nT to as much as −600 nT. A storm usually has three distinct phases: an initial phase of elevated surface field, a main phase of rapidly decreasing field and a recovery phase of slowly recovering field. Typical storms last for several days. The onset of storms is of two types: gradual commencement and sudden commencement. Many storms exhibit multiple decreases in H with several minima. An example storm from the last solar cycle 23 is plotted in Figure 14.3.1.

Figure 14.3.1a presents the dynamic pressure, which is normally about 2.4 nP, but in an interval of 19 hours on 20 November 2003 rose to a peak of nearly 18 nP. During this interval the product $VB_z = -E_y$ in Figure 14.3.1b reached a value below −30 mV/m. The D_{st} index (defined below) plotted in Figure 14.3.1c decreased to −422 nT. In this storm B_z turned southward at about 01:00 UT, several hours before the increase in dynamic pressure at 07:30, causing D_{st} to decrease without an evident initial phase. Dynamic pressure increased strongly after 07:30, reaching a peak value of about 18 nP. A large drop in B_z began at 11:00, reaching a minimum at 21:00, and returned to positive values at 24:30. D_{st} reached its minimum at 21:00 UT,

* R. L. McPherron's support while preparing this chapter was provided by grants NASA NNX17AB83 and NSF 1602588. Data used in Section 14.3.3 of this report were downloaded from the NASA NSSDC OMNI hourly data file. R. L. M. thanks T. P. O'Brien for helpful advice.

creating a 9 hour long main phase. It then recovered steadily for the next 3 days.

Magnetic storms were first systematically observed in 1806–7 by Alexander von Humboldt through deflections of compass needles. In 1836 Gauss developed a mathematical procedure for representing the magnetic field of the Earth. He subsequently established a number of magnetic observatories that measured the three components of the Earth's vector field allowing him to mathematically describe the field (Stern, 1989). In 1859 an extremely bright solar flare was observed by Carrington, and a day later it was followed by the most severe magnetic storm ever seen (Carrington, 1859). By 1910 sufficient data had been recorded that it was possible to carry out statistical studies of magnetic storms. One of the first used superposed epoch analysis to find the average behaviour of the field as a function of 'storm time' measured from the sudden commencement of the storm (Moos, 1910). A 'new theory of magnetic storms' was introduced in 1930 in the first of a series of papers (Chapman and Ferraro, 1930). It was postulated that the initial phase of elevated field was caused by the arrival from the Sun of a stream of ionised gas that creates a sheet of current in front of the Earth that compresses the field. The fact that this current sheet is always present was not clear until spacecraft observations demonstrated that the solar wind is always present (Snyder et al., 1963). The main phase was attributed to the development of a 'ring current' around the Earth created by charged particles entering from the flanks of the magnetic field. Gold (1959) named the cavity created by this boundary the magnetosphere.

Exploration of the magnetosphere by spacecraft immediately discovered the radiation belts and the fact that the ring current is always present (Davis and Williamson, 1963). It was found that a magnetic storm is created by an enhancement of this current (Stern, 1996). A possible mechanism for the injection of particles into the ring current was put forward by Axford and Hines (1961). They suggested that the solar wind exerts a viscous force on closed field lines near the magnetopause dragging them into a long tail behind the Earth. Eventually these are released and flow towards the Earth carrying charged particles. Particles gain energy by betatron and Fermi acceleration as the field strength increases and field lines shorten. Fluctuations in the strength of convection allow some particles to become trapped. Also at this time Dungey (1961) suggested that an alternative cause of convection is magnetic reconnection. Southward-directed field lines of the interplanetary magnetic field connect with dipole field lines at the subsolar point. The solar wind carries these reconnected field lines over the poles creating a long tail of open field lines. They then reconnect at the centre of the tail and flow towards the Earth as closed field lines. For either cause of convection, particles are carried towards the Earth, and fluctuations in convection trap them in the ring current.

The ring current produces a magnetic perturbation at the centre of the Earth that is parallel to the Earth's magnetic dipole axis. Everywhere on the Earth's surface, the component of this perturbation tangential to the surface is southward, decreasing the horizontal (H) component of the surface field. The growth of the ring current is the cause of the depression seen in the main phase. Eventually the IMF turns northward, allowing convection to decrease and shutting off the energisation and injection of particles. At this point, the interaction of the drifting ions with neutral atoms of the Earth's outer atmosphere causes charge exchange decay of the ring current. This converts them to neutral atoms that are no longer constrained by the magnetic field so that they are lost either to the atmosphere or to the solar wind. This decay of the ring current allows the depressed field at the surface to recover to its quiet-time level. This is the recovery phase of the storm. This hydromagnetic theory of magnetic storms was originally developed by Dessler and Parker (1959) and then extended by Sckopke (1966). The theory demonstrated that the magnetic perturbation of a ring current at the Earth's centre is related to the total energy of the drifting particles by the simple relation $\Delta B_z/B_0 = -2E/3E_m$. In this equation, ΔB_z is the axial field at the Earth's centre, B_0 is the equatorial surface field, E_m is the total energy in the Earth's field external to the Earth and E is the energy of the drifting particles. This relation is now known as the Dessler–Parker–Sckopke (D-P-S) relation. The D_{st} index defined in the next section is a good approximation to ΔB_z.

Spacecraft observations of a magnetic tail were reported by Ness (1965), who also introduced the Geocentric Solar Magnetospheric (GSM) coordinate system that describes observations within the magnetosphere. In this system X is towards the Sun and Z is orthogonal to X in the plane containing X and the Earth's dipole axis. Fairfield and Cahill (1966) soon found that the level of magnetic disturbance in the auroral zone is correlated with the presence of a southward magnetic field in the magnetosheath suggesting reconnection was the more important process. Rostoker and Falthammar (1967) argued that if B_z is the cause of magnetic activity, then it is the Y component of the interplanetary electric field E_y that represents the rate at which the solar wind transports magnetic flux to the dayside. They advocated the use of what is now called solar magnetic (SM) coordinates, with the Sun vector and dipole axis defining the X-Z plane. They showed that auroral zone negative bays and magnetic storms were much more likely when E_y is positive (southward IMF B_z) and quiet conditions were more likely when it is negative. The relation of the *Kp* index of geomagnetic activity to the solar wind was investigated by Schatten and Wilcox (1967), who found that *Kp* increases with both the magnitude of the IMF and the strength of its southward component. Wilcox et al. (1967) established that *Kp* also depends on the solar wind speed but not its density. The most appropriate coordinate system to use in such correlations was investigated statistically by Hirshberg and

Colburn (1969), who concluded that the GSM system organised the relation of *Kp* to solar wind better than did a system based on the solar equator.

14.3.2 *Dst* Index as a Measure of Ring Current Strength

An exhaustive study of 346 magnetic storms during the interval 1905–45 was carried out by Sugiura and Chapman (1960). To perform this study, the authors introduced a new magnetic index that measures the effects of the ring current and quantifies the strength of a storm. This hourly index, called disturbance storm time (D_{st}), is the worldwide average of the equatorial disturbance caused by the ring current. It is not easy to determine because it requires removal of the Earth's main field, its secular variation and the solar quiet day (Sq) variation created by dayside ionospheric currents. The derivation of the original index is described in Sugiura (1964). A modified version of the index using different stations, and an approximation of the equatorial response is described by Sugiura and Kamei (1991). A 1 min resolution version of D_{st} called Sym-H and Asym-H (and also for the perpendicular component, D) has been developed, as explained by Iyemori et al. (1994). Accurate estimation of the ring current effects is an important issue in the determination of models of the Earth's magnetic field and several modified versions of the index have been developed. The history of these developments is described in Kauristie et al. (2017).

In their study of magnetic storms, Chapman and Ferraro (1931) considered the current system induced by the Earth's field in the front of an advancing stream of ionised gas. They showed that this current was a two-celled system with foci on the noon meridian well above and below the magnetic equator. Current flows from dawn to dusk between the foci, producing a positive perturbation in H on the Earth's surface. The return current is westward poleward of the two foci. In the steady solar wind, this is the form of the current that creates the magnetopause. The location of the subsolar magnetopause can be approximated by balancing solar wind dynamic pressure against magnetic pressure. In typical conditions, this is at 10 R_E, where the dipole field is 31 nT. For an infinite sheet current, the field just inside the magnetopause is doubled to 62 nT and constant over the Earth. A realistic model of the magnetopause shows that the actual field is about 41 nT at 10 R_E and decreases to about 25 nT at the Earth (Mead, 1964). The dynamic pressure of the solar wind is constantly changing so that the standoff distance and the strength of the perturbation change as well. Siscoe (1966) showed that this leads to the relation $B_{mp} = \left(B_0/\sqrt{P_0}\right)\sqrt{P_{mp}} = b\sqrt{P_{mp}}$, where B_{mp} is the magnetic field of the magnetopause current at the Earth in disturbed conditions, B_0 is this field in quiet conditions, P_{mp} is the disturbed dynamic pressure and P_0 is the quiet-time pressure.

14.3.3 The Burton Equation for Changes in *Dst*

The perturbation of the magnetopause current is present in the measured D_{st} index, as discussed by Burton et al. (1975). They show that the observed D_{st} index can be corrected for the effects of the disturbed magnetopause and quiet ring current and magnetopause with the relation

$$D^*_{st} = D_{st} - b\sqrt{P_{dyn}} + c. \qquad (14.3.1)$$

In quiet times the ring current decays at a rate proportional to D^*_{st}, as shown by Davis and Parthasarathy (1967). Thus, if dynamic pressure is low and there is no ring current injection, the ring current contribution decays exponentially, $dD^*_{st}/dt = -dD^*_{st}/\tau$. The theory of storms proposes that the rate of injection is controlled by convection which is driven by the solar wind electric field VB_z. (Note that in what follows we have chosen to work with the negative of GSM E_y as it has the same sign as B_z.) Therefore, assume that the ring current injection rate is some function of VB_z, obtaining the 'Burton equation' $dD^*_{st}/dt = F(VB_z) - D^*_{st}/\tau$. Burton et al. (1975) assumed that the injection rate is linearly proportional to the rectified solar wind electric field $F(VB_z) = aVB_s$, where VB_s equals VB_z when IMF B_z is southward and is zero when it is northward. Substituting the relation between corrected and measured D_{st} leads to the following equation:

$$\frac{dD_{st}}{dt} = aE_s - \left(\frac{1}{\tau}\right)D_{st} + b\left(\frac{d\sqrt{P_{\text{dyn}}}}{dt}\right) + \left(\frac{b}{\tau}\right)\sqrt{P_{\text{dyn}}} + \left(\frac{c}{\tau}\right). \qquad (14.3.2)$$

Equation (14.3.2) contains four free parameters $[a, b, c, \tau]$ that must be determined empirically. Given the values for these parameters, it is possible to calculate the change in D_{st} at future times. Adding this change to the current value allows D_{st} to be predicted.

The empirical parameters enter Equation (14.3.2) in a nonlinear manner so a simple linear regression is not possible. Burton et al. (1975) therefore estimated them in a stepwise manner. The constant b was estimated from storm sudden commencements that occurred when the magnetopause and ring current had their quiet values. The quiet-time constant c was determined from Equation (14.3.1), which for a quiet magnetopause and ring current becomes $D_{st} - b\sqrt{P_{\text{dyn}}} + c = 0$. The decay rate was estimated from the recovery phase of storms when VB_z was zero and the ring current was smoothly recovering. Finally, the constant a was estimated from intervals when the dynamic pressure remained constant while VB_s shifted to a large negative value. The values obtained by Burton et al. (1975) are listed in column 2 of Table 14.3.1.

We have used Equation (14.3.2) and the Burton values to predict the hourly change of D_{st} and compared the predictions to the observations used to generate the model parameters. The Burton values predict only 10.3% of the variance in hourly

Table 14.3.1 *Parameters in the Burton Equation*

Parameter	Burton	Lower	Probable	Upper	Units
τ	7.7	10.57	10.81	10.89	hours
a	5.4	3.05	3.07	3.09	(nT/h)/(mV/m)
b	15.87	7.02	7.02	7.26	nT/√nPa
c	20	6.72	7.00	7.33	nT

Note. The first column shows the parameters and the last column their metric units. Column 2 contains the values obtained by Burton et al. (1975). Values obtained by nonlinear inversion of VB_s <0 are presented in column 4 (probable value).

change. Integration of this differential equation discussed below produces a D_{st} time series with a higher prediction efficiency of 49.4%.

In contrast to the stepwise procedure described by Burton et al. (1975), we have used nonlinear regression (described below) to obtain optimum values for the four parameters (probable values in column 4). For comparison, the prediction efficiency using these optimum values is much higher at 37.4%. Multistep integration then yields a prediction efficiency for D_{st} of 68.9%. It can be seen that the Burton values contain more rapid decay than the optimum values but compensate by a stronger injection rate. Note that the pressure constant b obtained by nonlinear regression is about half the Burton value. A similar result was reported by Siscoe et al. (1968). Also the optimum value for c is about one-third the Burton value. If only northward records are used in the regression $VB_z = 0$, then a drops out of the inversion. In this case τ, b and $c = [16.98 \text{ hr}, 9.28 \text{ nT}/\sqrt{\text{nP}_\text{a}}, 6.01 \text{ nT}]$ with a prediction efficiency for changes of 29.5%. The decay rate is much longer for northward B_z, which suggests the ring current decay time might be a function of the driver strength VB_z.

14.3.4 Estimation of Driver-Dependent Burton Parameters with Nonlinear Regression

Because the decay rate obviously changes between north and south B_z, we make the assumption that all parameters in Equation (14.3.2) are a function of the solar wind driver and the form of the injection function is unknown. Then, in terms of instantaneous values and their hourly changes, we have

$$\Delta D_{st} = \left[\left(\Delta\sqrt{P_{dyn}}\right), (-D_{st}), \left(\sqrt{P_{dyn}}\right), 1 \right] * [b, 1/\tau, b/\tau, \mathit{Off}]^T. \quad (14.3.3)$$

The left-hand side is a vector of successive hourly changes of D_{st}. The first bracket on the right-hand side is the design matrix constructed from the four vectors identified by parentheses. The second bracket is a column vector containing the unknowns, which by assumption depend only on VB_z. The parameter $\mathit{Off} = [F(VB_z) - c/\tau]$ contains the dependence of the injection rate on VB_z and the constant c. Equation (14.3.3) has the form $\vec{Y} = (X)\,\vec{\beta}$. We utilise nonlinear regression to solve for the parameters $\vec{\beta}$. To do this, rows of X for which VB_z falls within a narrow bin are selected. To obtain a sufficient number of records to perform an inversion, the width of the bin varies with VB_z from a minimum of 0.01 near zero to 4.0 at ±35 mV/m. Then, in each bin, all parameters are approximately constant, and nonlinear regression provides a solution $\vec{\beta}$. The dependence of the four parameters on the driver is then found by moving the bin across the available range of values of VB_z. We consider the full range of values of VB_z, both positive and negative.

The results of the nonlinear regression using data binned by solar wind VB_z are presented in Figure 14.3.2. Figure 14.3.2a presents the pressure factor b as a function of VB_z. It is immediately apparent that it is not constant and instead varies between 10 nT/$\sqrt{\text{nPa}}$ with positive B_z and near zero at large negative B_z. The heavy line shows a model of the inversion results summarised in Table 14.3.2. McPherron and O'Brien (2001) speculated that this decrease is caused by the tail current. As VB_z becomes stronger, the inner edge of the tail current moves closer to the Earth, becoming more intense as it approaches. Its effect on the Earth is a negative H perturbation, which partially offsets the positive contribution of the magnetopause current.

Figure 14.3.2b presents the ring current decay rate versus VB_z. For values near zero, τ is close to 20 hours and then decays slowly as VB_z becomes more positive to about 14 hours. For negative VB_z the decay time decreases exponentially to about 4 hours. An explanation for the behaviour with negative VB_z was suggested in O'Brien and McPherron (2000). When convection is strong, the ring current is very asymmetric, with ions drifting rapidly through the dusk magnetosphere close to the Earth where charge exchange decay is fast.

Figure 14.3.2c plots the injection function in the expression $\mathit{Off} = [F(VB_z) - c/\tau]$ as a function of VB_z. For positive values the injection rate is zero, allowing us to determine the parameter c from Off and τ. For negative values it decreases linearly, becoming stronger as the strength of the driver increases. Close to zero, the curve bends over in a smooth fashion. Burton et al. (1975) modelled this feature by a slight offset of the linear fit towards negative values (–0.5). A fit which includes the smooth transition and joins the horizontal axis with zero value and zero slope is plotted in the figure. Figure 14.3.2d displays the prediction efficiency obtained in every bin. At the largest values of VB_z, this reaches about 40%. Near zero, where signals are weaker, the efficiency falls to about 20%. Dots in Figure 14.3.2d denote the location of bin centres.

The heavy lines in Figures 14.3.2a–14.3.2c show fits to the dependence of the parameter on VB_z. These fits are purely mathematical functions chosen to summarise the calculated parameters. Initially, each parameter was fit separately with nonlinear regression of a simple function. This model leads to a slight improvement of 39.2% over the optimised Burton model. Subsequently, we developed a single function that

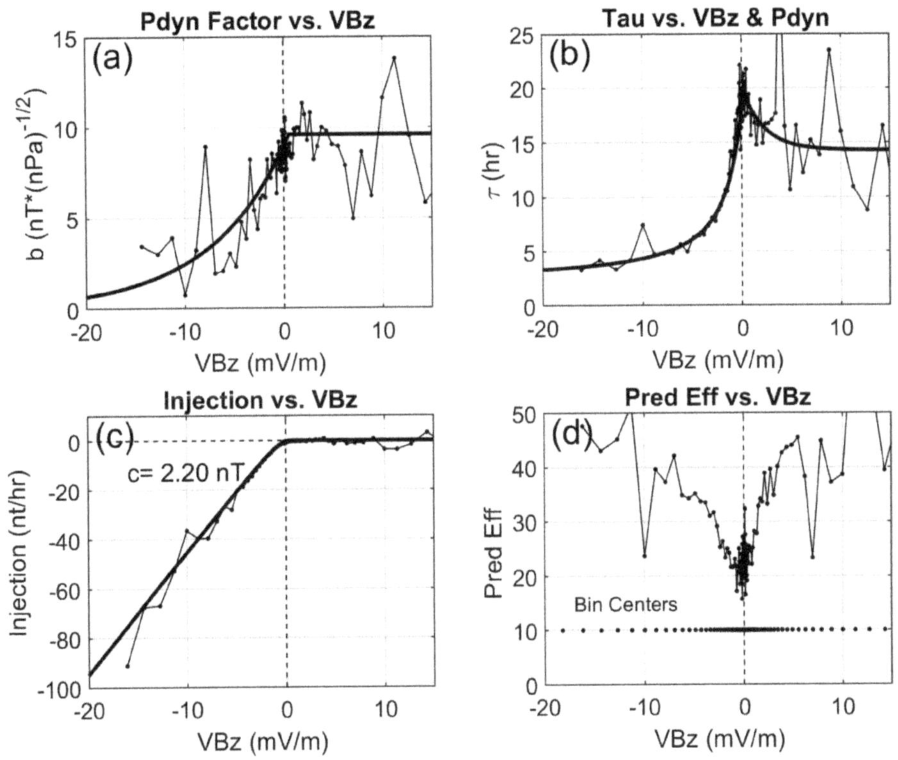

Figure 14.3.2 The results obtained by nonlinear regression of Equation (14.3.3) using narrow bins of solar wind VB_z. (a) Factor in the dynamic pressure equation as a function of VB_z. (b) Dependence of the decay time τ on VB_z. (c) Injection term. The constant c (nT) is the baseline offset in Equation (14.3.1). (d) Prediction efficiency of the model for prediction of the hourly change in D_{st}. Heavy lines are mathematical fits described in Table 14.3.2. Dots on all curves show the location of bin centres.

used the 10 parameters of the model fits to predict the change in D_{st}. All data were accumulated in a single array and input to the nonlinear inversion routine to obtain optimum values for all parameters simultaneously. This procedure increased the prediction efficiency of the fits to 41.7%. It should be emphasised that there are few observations with $abs(VB_z) > 15$ mV/m so the calculated parameters become erratic beyond this value. To obtain the results displayed in Figure 14.3.2, we increased the bin width with increasingly larger values of E_y, as discussed above. In addition, we used a technique described in O'Brien and McPherron (2000) to project the calculated values for each bin to the centre of the bin. There is no guarantee that the mathematical fits are appropriate for these high values of VB_z. The equations incorporated in the final model are summarised in Table 14.3.2. The coefficients in each equation were determined by simultaneous optimisation using all available data.

A different approach to the determination of a model of the changes in D_{st} is described in O'Brien and McPherron (2000). Because the observations ΔD_{st} are so noisy, the authors examined maps of the probability of a given D_{st} as a function of the hourly change in its value and the current value of VB_z. They argue that dynamic pressure changes are likely to average to zero in these maps so that it is not necessary to correct for these effects in the determination of the injection and decay rates. Once these are determined, the pressure multiplier and constant offset can be calculated, provided they are assumed fixed for all values of VB_z. This work was extended by O'Brien and McPherron (2002) by considering the possibility that ring current dynamics also depend on the tilt, ψ, of the dipole towards and away from the Sun, a process generally known as the equinoctial effect. They show that b, c, τ and Q are all modulated by the Svalgaard (1977) function $S(\psi) = 1.15/\left(1 + 3\cos^2(\psi)\right)^{2/3}$.

14.3.5 Prediction of the *Dst* Index with Modified Burton Equation

Every hour, the model parameters are determined by solar wind E_y. The D_{st} index then satisfies the differential equation

$$\frac{dD_{st}}{dt} = -\left[\frac{1}{\tau}\right]D_{st} + \left[b\frac{d\sqrt{P_{\text{dyn}}}}{dt} + \frac{b}{\tau}\sqrt{P_{\text{dyn}}} + Q - \frac{c}{\tau}\right]. \tag{14.3.4}$$

This differential equation has the form $y' = -fy + g$, where f and g depend on the solar wind dynamic pressure and electric field. Given a continuous sequence of hourly measurements, we can calculate f and g for each hour. Their values at an arbitrary time are determined by interpolation and then input to an explicit Runge–Kutta (4, 5) formula. The integration is initialised with the known value of D_{st} at the beginning of the sequence. The solution is obtained at a sequence of time points dictated by the variability of f and g. We fit a spline to these points sampling the solution at the times used in the original data. This provides a predicted sequence of hourly D_{st} values.

Table 14.3.2 *Parameters of the Burton Equation as Functions of VB_z*

Parameter	Equation	Units
c	2.2	nT
b	$b = 8.95\exp(x/7.73)$ $VB_z < 2$ $b = 9.4$ $VB_z >= 2$	$nT/\sqrt{nP}$
τ	$\tau = 2.29 \cdot \exp(9.25/(4.45 - VB_z))$ $VB_z < 0$ $\tau = (20 - 14.29) \cdot \exp(-VB_z/2.22) + 14.29$ $VB_z >= 0$	hours
Q	$Q = 4.94 \cdot (VB_z - 0.47) - 1.27 \cdot \exp((VB_z - 0.47)/1.27) + 1.27$ $VB_z < 0.47$ $Q = 0.0$ $VB_z \geq 0.47$	$\frac{nT}{mV/m}$

Note. All parameters except c behave differently for VB_z above and below zero (B_z northward and southward). The functional forms are simply mathematical curves that approximate the observations. All numerical values were obtained in a non-linear inversion that maximised the prediction efficiency of the calculated hourly changes in D_{st}.

Table 14.3.3 *Prediction Efficiency (PEF) of Various D_{st}. Prediction Models for 297 Segments of Length Greater than Four Days*

Model	Avg PEF	Peak PEF	80 %tile	Mag Storm
Original Burton (Bur)	49.4	52.0	63.9	94.0
Optimum Burton (Opt)	65.6	70.0	73.6	61.1
O'Brien & McP (OBMC)	72.2	73.0	78.7	88.7
McP & OB (MCOB)	68.9	74.0	78.0	75.9
Simultaneous (McP)	71.0	75.0	79.5	91.5

Note. Average is for all segments. Peak is location of maximum of probability distributions; 80 %tile is value of PEF exceeded by 20% of the segments. Mag Storm is the PEF of predictions plotted in Figure 14.3.2.

The prediction fails when solar wind data are missing and must be restarted after data again become available. We have taken this into account by selecting continuous segments of data longer than 4 days. For each segment we start the integration with the known value of D_{st} at the start of the segment. For each segment and using all data, we have determined the prediction efficiency of the models. The results are shown in Table 14.3.3. Averaged over all segments, the original Burton model behaves poorly at 49.4%. In optimised form it does better at 65.6%. The last three models are virtually indistinguishable at about 72%.

Table 14.3.3 characterises the quality of the model predictions by three measures: the average prediction efficiency (PEF) over all data, the peak of the probability distribution for the PEF of each segment and the value of PEF above which 20% of all segments fall. There is virtually no difference between the optimised model with all parameters depending on E_y and the O'Brien and McPherron (2000) model. The main difference is that the optimum model allows the pressure factor to vary with E_y and allows the decay time to decrease as positive E_y increases. These seem to have little effect in the final prediction.

The quality of the predictions from the different models is illustrated in Figure 14.3.3 for the magnetic storm plotted in Figure 14.3.1. The heavy solid line is the observed D_{st} index. The traces produced by the various models are identified in the legend with abbreviations corresponding to entries in Table 14.3.3. The two models with the highest prediction efficiency are the original Burton model (cross-dashed line) at 94% and the simultaneously fitted model (triangle-dotted line) at 91.5%. It should be noted that this ordering of models is very different from their averages over all data. The averages over all data are dominated by weak to moderate storms. When the models are applied to extreme events, they do not work well.

14.3.6 Other Empirical *Dst* Prediction Models

The O'Brien and McPherron (2000) model was extended in O'Brien and McPherron (2002) to include the dependence of the parameters of the Burton equation on the tilt of the dipole towards and away from the Sun. They show that the pressure factor, the baseline correction, the decay constant and the injection rate all depend on the dipole tilt through the Svalgaard function (Svalgaard, 1977) $S(\psi) = 1.15/(1 + 3\cos^2(\psi))^{2/3}$. This function is a mathematical representation of the equinoctial effect (Cliver et al., 2004; McIntosh, 1959) and is distinct from the Russell–McPherron effect (Russell and McPherron, 1973).

A much more complex model than any of the above has been developed by Temerin and Li (2002). This model depends on universal time and season to take account of the orientation of the dipole axis relative to the solar wind. It contains three terms of the form used in the Burton equation – injection and decay –

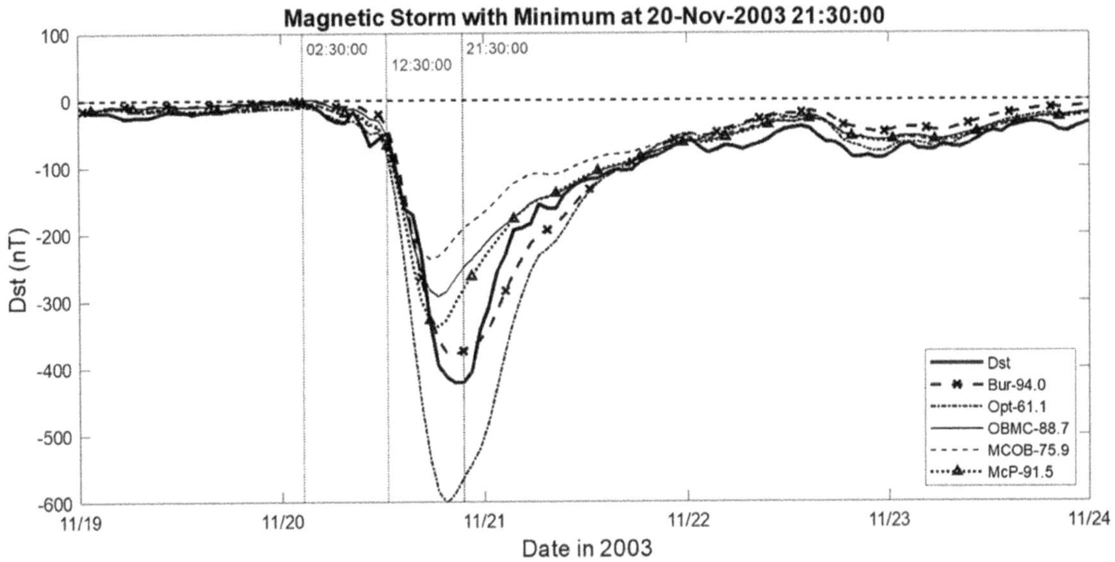

Figure 14.3.3 Multi-step prediction of the magnetic storm presented in Figure 14.3.1 using five different models. The heavy black line is the measured D_{st} index. Other lines are the predictions of D_{st} obtained by the different models. The prediction efficiency of each model is shown in percentage of D_{st} variance.

with each injection and decay term having different functional dependence on the solar wind. In addition, the model includes a pressure correction term and an offset term. The first injection/decay term also includes two thresholds where the behaviour of the function changes when solar wind variables exceed their values. To obtain the change in D_{st}, the five terms are added together with different time delays. The model is initially evaluated with solar wind data at 10 min cadence that is later interpolated to 1 hour resolution. In the authors' presentation, it is impossible to determine which quantities were treated as free parameters and which are fixed by geometry. However, the authors say they have introduced offsets in some of the geometric parameters to improve the prediction efficiency of the model so it is likely all parameters have been adjusted. As best we are able to count, the model has about 50 parameters, but time delays introduced for each term should also be counted. The method by which these parameters were determined is described as iterative least squares. The parameters were manually varied through a range of values to obtain an optimum prediction efficiency. Although it is unclear from the paper, we suspect that this was done in a stepwise manner, similar to what was done in the original Burton et al. (1975) model.

Temerin and Li (2002) report extensive tests of their model over the interval 1995–9 and compare its performance to the Burton model. They find that over all data, their model predicts 81.8% of the D_{st} variance, and that during disturbed times, it reaches 95.8%. By comparison, they state that the original Burton model predicts 59.7% of the variance, a higher value than we find. When the Burton parameters are optimised, we find the model predicts 65.6%. Although their model does extremely well, the authors are unable to provide a physical explanation for its various elements.

An alternative approach to D_{st} prediction is through neural networks (NN). A typical NN will use three layers of nodes: input, hidden and output. The nodes in each layer are connected via weights and sigmoid transfer functions. The networks are trained on real data using a method called back-propagation. The current output of the system and the current weights are adjusted so that the next iteration improves the agreement between output and observations. A typical network may have many hundreds of weights and may take millions of iterations to stabilise at an optimum solution. The technique was first applied to D_{st} prediction by Gleisner et al. (1996), Lundstedt and Wintoft (1994) and Wu and Lundstedt (1996). Using raw solar wind data rather than coupling functions such as VB_s and dynamic pressure, they report a prediction efficiency of 83%. Wu and Lundstedt (1997) compare various coupling functions and different sets of solar wind variables as predictors of D_{st} and conclude that coupling functions do more poorly than original solar wind variables. An operational implementation of four different prediction models including a neural network model is described in Lundstedt et al. (2002). They find that their neural network is the best predictor, but only slightly better than the O'Brien and McPherron (2000) model.

A more recent neural network model for 2 hour ahead prediction of D_{st} is presented in Watanabe et al. (2014). This model on average predicts about 80% of the variance in D_{st}. The authors use the model as an analogue computer to simulate the effects of changing a single variable. In this way they demonstrate that the orientation of IMF B_y has a significant effect on ring current development. The authors also perform a detailed comparison of their results for two storms with predictions from the Burton model and the O'Brien and McPherron model. As we noted above, the

Burton model does poorly, and the O'Brien and McPherron model does better but is limited to storms weaker than −150 nT. They argue that the neural network model is not limited in its range and makes better predictions than the other models.

Another empirical method for modelling the dynamics of D_{st} is the Nonlinear Autoregressive Moving Average Model with Exogenous Inputs (NARMAX). In this method, 'autoregressive' implies that the current output depends on outputs at previous times and 'moving average' means it also depends on previous inputs to the system. An exogenous input is one that is unaffected by the system it drives. In this case the magnetosphere does not change the solar wind input. In the NARMAX method the current output is taken to be a polynomial function of all past inputs, outputs, and errors. For a quadratic function and several past values, there is an enormous number of monomials that could contribute. The NARMAX method includes a procedure for determining how important a given monomial is in explaining the variance of the output. Keeping only the most important of these provides a simplified model for the output.

Several applications of this method have been reported (Boaghe et al., 2001; Boynton et al., 2013; Boynton et al., 2011a, 2011b). In the first of these papers the authors develop a model for the current point k that contains 15 terms. The two most important include the values of the previous input and output. The next most important is $D_{st}(k-2)VB_s(k-1)$, which is a nonlinear contribution from two different time lags. The authors do not provide a quantitative estimate of the model prediction efficiency but do show that the frequency-dependent coherency between predictions and data is 0.8 at low frequencies. A very clear description of the NARMAX procedure is provided in Boynton et al. (2011b), who use it to study solar wind coupling functions. A coupling function is a combination of solar wind variables such as VB_s that attempts to linearise the relation between the solar wind and some magnetospheric variable. They conclude that the best function for D_{st} is $P^{1/2}V^{4/3}B_T\sin^6(\theta/2)$, where P is dynamic pressure, V is solar wind speed, B_T is the GSM magnetic field strength transverse to the X-axis and θ is the clock angle of the IMF around the X-axis. Boynton et al. (2011a) evaluate this function, showing that it predicts 71% of the D_{st} variance.

14.3.7 Discussion

In describing magnetic storms, Chapman noted that during their main phase, there are frequent magnetic fluctuations of ~3 hour duration that he called 'polar substorms' (Chapman, 1961; 1962). This association suggested the possibility that substorms inject ring current particles. In support of this idea, Davis and Parthasarathy (1967) demonstrated that the D_{st} index could be predicted from the sum of 10 previous hourly AE values. Kamide and Fukushima (1971) implemented this idea using an exponentially weighted sum of previous values of the AL index (see section 3.4 in Kamide et al. (1998)). Later, Kamide (1992) encapsulated this idea in the equation Storm $= \sum_i$ Substorm$_i$. This idea appears to be what led Chapman to introduce the concept of a substorm. This approach was extended by Cade et al. (1995) using a slightly different weighting scheme. However, it is well known that both substorms and storms are driven by the solar wind, so McPherron (1997) argued that the AL index was simply a proxy for solar wind driving.

The work of Burton et al. (1975) was based on a simple physical hypothesis that convection injects particles into the ring current and charge exchange eliminates them. Their model had only four free parameters and, as we showed above, predicts hourly D_{st} quite well with an average prediction efficiency of 65.6%. This model was significantly improved by O'Brien and McPherron (2000), who implemented the idea that the charge exchange decay rate depends on the strength of the solar wind electric field. Their model has six free parameters and achieves an average performance of 72.2%. A similar model by McPherron and O'Brien (2001) allows the pressure factor during southward IMF and decay rate during northward IMF to vary with solar wind electric field. This model has a slightly lower average prediction efficiency of 71.0% but with a slightly higher 80 percentile level. Apparently, the additional constants needed to represent these variations do not lead to significant improvements.

We suspect that several other changes in the model are more likely to improve D_{st} prediction efficiency. First is propagation and filtering of the solar wind variables. The hourly averages used in this work are the values determined at the L1 point upstream. Burton et al. (1975) showed that when high cadence data (2.5 min samples) are used, it is necessary to propagate the observations to the subsolar point and low-pass filter them with a 25 min running average. Others have suggested that VB_s is not the best solar wind coupling function to use and that $VB_s\,P^{1/3}$ is better (Murayama, 1982). Alternatively, Wang et al. (2003) demonstrate that the decay rate during northward IMF depends on dynamic pressure and that the injection rate should be multiplied by the factor $(P_{dyn}/3.0)^{0.2}$. Boynton et al. (2011b) used the NARMAX method to conclude that $P^{1/2}V^{4/3}B_T\sin^6(\theta/2)$ is a better input function.

14.3.8 Conclusions

Magnetic storms are evident on the Earth's surface as a rapid decrease (~10 hours) in the horizontal component (H) of the surface magnetic field followed by a slow recovery (>3 days). The strength of storms is quantified by the D_{st} index, which is the local time average of equatorial H. It is normally calculated at 1-hour resolution, but an approximation Sym-H is available at 1 min resolution. The surface perturbations are caused by the magnetopause current, the tail current, field-aligned currents and the ring current. The portion due to the ring current is proportional to the total energy present in particles drifting around the Earth. Empirical models have been developed to predict the D_{st} index. These models are based on the assumption that energy is injected into the ring current by magnetospheric convection driven by the solar wind electric field and lost by charge exchange between the drifting particles and atmospheric neutrals. The first such model had only four constant parameters, two of which described corrections that compensate for the magnetopause and ring current at quiet and disturbed times. The parameters of this model were determined by stepwise approximation that led to a prediction of 49.4% of the variance in D_{st}. When the parameters are determined by non-linear inversion, the original Burton model predicts about 66% of the variance in the index. The model is significantly improved by making the parameters functions of the solar wind electric field, leading to a prediction efficiency of 72% with a model having 10 free parameters. Other empirical models include neural networks predicting up to 85% of the D_{st} variance and the NARMAX model that predicts about 71%. The best model yet developed is that of Temerin and Li (2002), which on average accounts for 88%–91% of its variance. It does this at the expense of more than 50 adjustable parameters and an inability to provide a physical explanation for many parts of the algorithm. Similar difficulties exist for the neural network and NARMAX models. It is unlikely that any model can be constructed that will exceed an average prediction efficiency of 90%, as there are significant errors in the measurements used to drive the models and in the index itself.

References

Akasofu, S. I. (1981). *Space Science Reviews*, **28**(2), 121–90.

Amari, T., Aly, J. J., Luciani, J. F., Mikic, Z. and Linker, J. (2011). *The Astrophysical Journal Letters*, **742** (2), L27–33.

Amari, T., Canou, A. and Aly, J. J. (2014). *Nature*, **514**(7523), 465–9.

Antiochos, S. K., DeVore, C. R. and Klimchuk, J. A. (1999). *The Astrophysical Journal*, **510**(1), 485–93.

Antonucci, E., Dodero, M. A. and Giordano, S. (2000). *Solar Physics*, **197**(1), 115–34.

Axford, W. I. and Hines, C. O. (1961). *Canadian Journal of Physics*, **39**, 1433–64.

Bai, T. and Sturrock, P. A. (1991). *Nature*, **350**(6314), 141–3.

Baker, D. N., Pulkkinen, T. I., Angelopoulos, V., Baumjohann, W. and McPherron, R. L. (1996). *Journal of Geophysical Research: Space Physics*, **101**(A6), 12975–13010.

Balogh, A., Marsden, R. G. and Smith, E. J. (2001). *The Heliosphere Near Solar Minimum: The Ulysses Perspective*. Springer Science & Business Media.

Benz, A. O. (2017). *Living Reviews in Solar Physics*, **14**, 2.

Bertello, L., Ulrich, R. K. and Boyden, J. E. (2010). *Solar Physics*, **264**(1), 31–44.

Boaghe, O. M., Balikhin, M. A., Billings, S. A. and Alleyne, H. (2001). *Journal of Geophysical Research: Space Physics*, **106** (A12), 30047–66. doi:10.1029/2000ja900162.

Bond, G., Kromer, B., Beer, J., Muscheler, R., Evans, M. N., Showers, W., Hoffmann, S., Lotti-Bond, R., Hajdas, I. and Bonani, G. (2001). *Science*, **294**(5549), 2130–36.

Borovsky, J. E. (2010). *Journal of Geophysical Research: Space Physics*, **115**, A09101, 1–33.

Borovsky, J. E. and Birn, J. (2014). *Journal of Geophysical Research: Space Physics*, **119**(2), 751–60.

Borovsky, J. E., Hesse, M., Born, J. and Kuznetsova, M. M. (2008). *Journal of Geophysical Research: Space Physics*, **113**, A08110.

Boynton, R. J., Balikhin, M. A., Billings, S. A. and Amariutei, O. A. (2013). *Ann. Geophys.*, **31**(9), 1579–89. doi:10.5194/angeo-31-1579-2013.

Boynton, R. J., Balikhin, M. A., Billings, S. A., Sharma, S. A. and Amariutei, O. A. (2011a). *Ann. Geophys.*, **29**(5), 965–71. doi:10.5194/angeo-29-965-2011.

Boynton, R. J., Balikhin, M. A., Billings, S. A., Wei, H. L. and Ganushkina, N. (2011b). *Journal of Geophysical Research: Space Physics*, **116**. doi:10.1029/2010ja015505

Broomhall, A. M. and Nakariakov, V. M. (2015). *Solar Physics*, **290**(11), 3095–3111.

Brueckner, G. E., Edlow, K. L., Floyd, L. E., Lean, J. L. and VanHoosier, M. E. (1993). *Journal of Geophysical Research: Atmospheres*, **98**(D6), 10695–711.

Bruno, R., Carbone, V., Veltri, P., Pietropaolo, E. and Bavassano, B. (2001). *Planetary and Space Science*, **49**(12), 1201–10.

Burch, J. L., Moore, T. E., Torbert, R. B. and Giles, B. L. (2016). *Space Science Reviews*, **199**(1–4), 5–21.

Burlaga, L. F. and Ness, N. F. (2012). *The Astrophysical Journal*, **749**(1), 13–18.

Burton, R. K., McPherron, R. L. and Russell, C. T. (1975). *J. Geophys. Res.*, **80**(31), 4204–14. doi:10.1029/JA080i031p04204.

Cade, W. B., III, Sojka, J. J. and Zhu, L. (1995). *J. Geophys. Res.*, **100**(A1), 97–105. doi:10.1029/94JA02347.

Cane, H. V., McGuire, R. E. and Von Rosenvinge, T. T. (1986). *The Astrophysical Journal*, **301**, 448–59.

Carrington, R. C. (1859). *Monthly Notices of the Royal Astronomical Society*, **20**, 13–15. doi:10.1093/mnras/20.1.13.

Cassak, P. A. and Shay, M. A. (2007). *Physics of Plasmas*, **14**(10), 102114. doi:10.1063/1.2795630.

Chapman, S. (1961). *American Scientist*, **49**(3), 249.

Chapman, S. (1962). *Journal of the Physical Society of Japan*, **17** (A-I), 6.

Chapman, S. and Bartels, J. (1962). *Geomagnetism*. Vol. 1, 2nd edn. Clarendon Press.

Chapman, S. and Ferraro, V. C. A. (1930). *Nature*, **126**, 129–30. doi:10.1038/126129a0.

Chapman, S. and Ferraro, V. C. A. (1931). *Terrestrial Magnetism and Atmospheric Electricity*, **36**(3), 171–86. doi:10.1029/TE036i003p00171.

Chaplin, W. J., Elsworth, Y., Miller, B. A., Verner, G. A. and New, R. (2007). *The Astrophysical Journal*, **659**(2), 1749–60.

Chen, Y., Liu, L. and Wan, W. (2011). *Journal of Geophysical Research: Space Physics*, **116**, A04304.

Cheng, X., Ding, M. D., Guo, Y., Zhang, J., Vourlidas, A., Liu, Y. D., Olmedo, O., Sun, J. Q. and Li, C. (2013). *The Astrophysical Journal*, **780**(1), 1–9.

Cho, I. H., Hwang, J. and Park, Y. D. (2014). *Solar Physics*, **289** (3), 707–19.

Chowdhury, P., Khan, M. and Ray, P. C. (2010). *Planetary and Space Science*, **58**(7), 1045–9.

Cliver, E. W. and Dietrich, W. F. (2013). *Journal of Space Weather and Space Climate*, **3**(A31), 1–15.

Cliver, E. W., Svalgaard, L. and Ling, A. G. (2004). *Annals of Geophysics*, **22**(1), 93–100. doi:10.5194/angeo-22-93-2004.

Connick, D. E., Smith, C. W. and Schwadron, N. A. (2010). *The Astrophysical Journal*, **727**(1), 1–6.

Conscience, C., Meftah, M., Chevalier, A., Dewitte, S. and Crommelynck, D. (2011). *Proc. SPIE 8146, UV/Optical/IR Space Telescopes and Instruments: Innovative Technologies and Concepts V*, 814613. doi:10.1117/12.895447.

Cranmer, S. R., Asgari-Targhi, M., Miralles, M. P., Raymond, J. C., Strachan, L., Tian, H. and Woolsey, L. N. (2015). *Philosophical Transactions of the Royal Society, Series A*, 373(2041), 20140148.

Cranmer, S. R., Van Ballegooijen, A. A. and Edgar, R. J. (2007). *The Astrophysical Journal Supplement Series*, **171** (2), 520–51.

Daglis, I. A., Thorne, R. M., Baumjohann, W. and Orsini, S. (1999). *Reviews of Geophysics*, **37**, 407–38.

Davis, L., Jr. and Williamson, J. M. (1963). Low energy trapped protons, paper presented at Space Research III, Proceedings of the Third International Space Science Symposium, North Holland Publishing Company Amsterdam/New York., Washington, DC, 2–8 May.

Davis, T. N. and Parthasarathy, R. (1967). *Journal of Geophysical Research*, **72**(23), 5825–36. doi:10.1029/JZ072i023p05825.

Dayeh, M. A., Desai, M. I., Dwyer, J. R., Rassoul, H. K., Mason, G. M. and Mazur, J. E. (2009). *The Astrophysical Journal*, **693**(2), 1588–1600.

Deng, L. H., Li, B., Xiang, Y. Y. and Dun, G. T. (2015). *Journal of Atmospheric and Solar-Terrestrial Physics*, **122**, 18–25.

Denton, M. H. and Borovsky, J. E. (2012). *Journal of Geophysical Research: Space Physics*, **117**, A00L05.

Denton, M. H., Borovsky, J. E., Skoug, R. M., Thomsen, M. F., Lavraud, B., Henderson, M. G., McPherron, R. L., Zhang, J. C. and Liemohn, M. W. (2006). *Journal of Geophysical Research: Space Physics*, **111**, A07S07.

Desai, M. and Giacalone, J. (2016). *Living Reviews in Solar Physics*, **13**, 3.

Dessler, A. and Parker, E. (1959). *Journal of Geophysical Research*, **64**(12), 2239–52. doi:10.1029/JZ064i012p02239.

Dewitte, S., Crommelynck, D., Mekaoui, S. and Joukoff, A. (2004). *Solar Physics*, **224**(1–2), 209–16.

Didkovsky, L. and Wieman, S. (2014). *Journal of Geophysical Research: Space Physics*, **119**(6), 4175–84.

Dungey, J. W. (1961). *Physical Review Letters*, **6**(2), 47–8.

Ebert, R. W., Dayeh, M. A., Desai, M. I. and Mason, G. M. (2012). *The Astrophysical Journal*, **749**(1), 1–13.

Emslie, A. G., Dennis, B. R., Holman, G. D. and Hudson, H. S. (2005). *Journal of Geophysical Research: Space Physics*, **110**, A11103.

Ermolli, I., Matthes, K., Dudok de Wit, T., Krivova, N. A., Tourpali, K., Weber, M., Unruh, Y. C., Gray, L., Langematz, U., Pilewskie, P. and Rozanov, E. (2013). *Atmospheric Chemistry and Physics*, **13**(8), 3945–77.

Fairfield, D. H. and Cahill, L. J., Jr. (1966). *J. Geophys. Res.*, **71**(1), 155–69.

Fang, T. W. and Forbes, J. M. (2012). *Journal of Geophysical Research: Space Physics*, **117**, A01318.

Floyd, L., Tobiska, W. K. and Cebula, R. P. (2002). *Advances in Space Research*, **29**(10), 1427–40.

Foukal, P. (2012). *Solar Physics*, **279**(2), 365–81.

Fröhlich, C. (2012). *Surveys in Geophysics*, **33**(3–4), 453–73.

Fuselier, S. A., Trattner, K. J., Petrinec, S. M. and Lavraud, B. (2012). *Journal of Geophysical Research: Space Physics*, **117**, A08235.

Ganushkina, N. Y., Pulkkinen, T. I. and Fritz, T. (2005). *Annals of Geophysics*, **23**, 579–91.

Giacalone, J., Jokipii, J. R. and Kóta, J. (2002). *The Astrophysical Journal*, **573**(2), 845–50.

Gleisner, H., Lundstedt, H. and Wintoft, P. (1996). *Annals of Geophysics*, **14**(7), 679–86. doi:10.1007/s00585-996-0679-1.

Gold, T. (1959). *Journal of Geophysical Research*, **64**(9), 1219–24. doi:10.1029/JZ064i009p01219.

Gonzalez, W. D. and Mozer, F. S. (1974). *Journal of Geophysical Research*, **79**(28), 4186–94.

Gopalswamy, N. (2010). Corona mass ejections: A summary of recent results, in *20th National Solar Physics Meeting*, pp. 108–30, Slovak Central Observatory. http://adsabs.harvard.edu/abs/2010nspm.conf.....D.

Gosling, J. T. (2007). *The Astrophysical Journal Letters*, **671**(1), L73–6.

Grandin, M., Aikio, A. T., Kozlovsky, A., Ulich, T. and Raita, T. (2015). *Journal of Geophysical Research: Space Physics*, **120** (12), 10, 669–87.

Gray, L. J., Beer, J., Geller, M., Haigh, J. D., Lockwood, M., Matthes, K., Cubasch, U., Fleitmann, D., Harrison, G., Hood, L. and Luterbacher, J. (2010). *Reviews of Geophysics*, **48**, RG4001, 1–53.

Haigh, J. D. (2007). *Living Reviews in Solar Physics*, **4**(1), 2.

Hathaway, D. H. and Rightmire, L. (2010). *Science*, **327**(5971), 1350–52.

Hinteregger, H. E. (1981). *Advances in Space Research*, **1**(12), 39–52.

Hirshberg, J. and Colburn, D. S. (1969). *Planetary and Space Science*, **17**, 1183–1206.

Hnat, B., Chapman, S. C. and Rowlands, G. (2003). *Physical Review E*, **67**(5), 056404.

Hudson, M. K., Kress, B. T., Mazur, J. E., Perry, K. L. and Slocum, P. L. (2004). *Journal of Atmospheric and Terrestrial Physics*, **66**, 1389–97.

Illing, R. M. E. and Hundhausen, A. J. (1985). *Journal of Geophysical Research*, **90**, 275–82.

Iyemori, T., Araki, T., Kamei, T. and Takeda, M. (1994). Mid-latitude geomagnetic indices ASY and SYM (provisional). Rep. No. 1 (1989–1990), Data Analysis Center for Geomagnetism and Space Magnetism, Faculty of Science, Kyoto University, Kyoto, Japan.

Kamide, Y. (1992). *Journal of Geomagnetism and Geoelectricity*, **44** (2), 109–17. doi:10.5636/jgg.44.109.

Kamide, Y. and Fukushima, N. (1971). *Report of Ionosphere and Space Research in Japan*, **26**, 79.

Kamide, Y., et al.(1998). *Journal of Geophysical Research*, **103** (A8), 17705–28. doi:10.1029/98JA01426.

Kauristie, K., Morschhauser, A., Olsen, N., Finlay, C. C., McPherron, R. L., Gjerloev, J. W. and Opgenoorth, H. J. (2017). *Space Science Review*, **206**(1–4), 61–90. doi:10.1007/s11214-016-0301-0.

Kepko, L., Viall, N. M., Antiochos, S. K., Lepri, S. T., Kasper, J. C. and Weberg, M. (2016). *Geophysical Research Letters*, **43**(9), 4089–97.

Knaack, R. and Stenflo, J. O. (2005). *Astronomy and Astrophysics*, **438**(1), 349–63.

Kohl, J. H., Noci, G., Antonucci, E., Tondello, G., Huber, M. C. E., Gardner, L. D., Nicolosi, P., Fineschi, S., Raymond, J. C., Romoli, M. and Spadaro, D. (1997). *Advances in Space Research*, **20**(1), 3–14.

Kohl, J. L., Noci, G., Cranmer, S. R. and Raymond, J. C. (2006). *The Astronomy and Astrophysics Review*, **13**(1–2), 31–157.

Kolotkov, D. Y., Broomhall, A. M. and Nakariakov, V. M. (2015). *Monthly Notices of the Royal Astronomical Society*, **451**(4), 4360–67.

Krivolutsky, A. A. and Repnev, A. I. (2012). *Geomagnetism and Aeronomy*, **52**(6), 685–716.

Kumar, P. and Innes, D. E. (2013). *Solar Physics*, **288**(1), 255–68.

Le, H., Liu, L., Ren, Z., Chen, Y., Zhang, H. and Wan, W. (2016). *Journal of Geophysical Research: Space Physics*, **121**(1), 832–40.

Lean, J. L., Woods, T. N., Eparvier, F. G., Meier, R. R., Strickland, D. J., Correira, J. T. and Evans, J. S. (2011). *Journal of Geophysical Research: Space Physics*, **116**, A01102.

Lin, R.C., Krucker, S., Hurford, G.J., Smith, D.M., Hudson, H.S., Holman, G.D., Schwartz, R.A., Dennis, B.R., Share, G.H., Murphy, R.J. and Emslie, A.G. (2003). *The Astrophysical Journal Letters*, **595**, L69–L76.

Linton, M.G. and Moldwin, M.B. (2009). *Journal of Geophysical Research: Space Physics*, **114**, A00B09.

Lundstedt, H., Gleisner, H. and Wintoft, P. (2002). *Geophysical Research Letters*, **29**(24), 31–4. doi:10.1029/2002gl016151.

Lundstedt, H. and Wintoft, P. (1994). *Annals of Geophysics*, **12**(1), 19–24. doi:10.1007/s005850050031.

Lynch, B. J., Masson, S., Li, Y., DeVore, C. R., Luhmann, J. G., Antiochos, S. K. and Fisher, G. H. (2016). *Journal of Geophysical Research: Space Physics*, **121**, 10677–97. doi:10.1002/2016JA023432.

Maia, D. J. F. and Pick, M. (2004). *The Astrophysical Journal*, **609** (2), 1082–97.

Mann, G. (2015). *Journal of Plasma Physics*, **81**, 475810601. doi:10.1017/S0022377815001166.

McComas, D. J., Velli, M., Lewis, W. S., Acton, L. W., Balat-Pichelin, M., Bothmer, V., Dirling, R. B., Feldman, W. C., Gloeckler, G., Habbal, S. R. and Hassler, D. M. (2007). *Reviews of Geophysics*, **45**, RG1004. doi:10.1029/2006RG000195.

McComas, D. J., Ebert, R. W., Elliott, H. A., Goldstein, B. E., Gosling, J. T., Schwadron, N. A. and Skoug, R. M. (2008). *Geophysical Research Letters*, **35**, L18103. doi:10.1029/2008GL034896.

McIntosh, D. H. (1959). *Philosophical Transactions of the Royal Society of London. Series A, Mathematical and Physical Sciences*, **251**(1001), 525–52. doi:10.1098/rsta.1959.0010.

McPherron, R. L. (1979). *Reviews of Geophysics*, **17**(4), 657–81.

McPherron, R. L. (1997). Role of substorms in the development of magnetic storms, in *Magnetic Storms*, edited by B. T. Tsurutani, W. D. Gonzalez, Y. Kamide and J. K. Arballo, pp. 131–47, American Geophysical Union. doi:10.1029/GM098p0131.

McPherron, R. L. and O'Brien, T. P. (2001). Predicting geomagnetic activity: The Dst index, in *Space Weather*, edited by P. Song, G. L. Siscoe and H. Singer, pp. 339–45, American Geophysical Union. doi:10.1029/GM125p0339.

Mead, G. D. (1964). *Journal of Geophysical Research*, **69**(7), 1181–95. doi:10.1029/JZ069i007p01181.

Meftah, M., Dewitte, S., Irbah, A., Chevalier, A., Conscience, C., Crommelynck, D., Janssen, E. and Mekaoui, S. (2014). *Solar Physics*, **289**(5), 1885–99.

Miroshnichenko, L. I., Klein, K. L., Trottet, G., Lantos, P., Vashenyuk, E. V., Balabin, Y. V. and Gvozdevsky, B. B. (2005). *Journal of Geophysical Research: Space Physics*, **110**, A09S08. doi:10.1029/2004JA010936.

Moldwin, M. B., Zou, S. and Heine, T. (2016). *Annales Geophysicae*, **34**(12), 1243–53.

Moos, N. A. F. (1910). *Colaba Magnetic Data, 1846 to 1905, 2, The Phenomena and Its Discussion*, Central Government Press.

Moore, T. E., Fok, M. C. and Chandler, M. O. (2002). *Journal of Geophysical Research: Space Physics*, **107**, 1332, doi:10.1029/2002JA009381.

Moses, D., Droege, W., Meyer, P. and Evenson, P. (1989). *The Astrophysical Journal*, **346**, 523–30.

Murayama, T. (1982). *Review of Geophysics*, **20**(3), 623–9. doi:10.1029/RG020i003p00623.

Ness, N. F. (1965). *Journal of Geophysical Research*, **70**(13), 2989. doi:10.1029/JZ070i013p02989.

Nwankwo, V. U., Chakrabarti, S. K. and Weigel, R. S. (2015). *Advances in Space Research*, **56**(1), 47–56.

O'Brien, T. P. and McPherron, R. L. (2000). *Journal of Geophysical Research: Space Physics*, **105**(A4), 7707–19. doi:10.1029/1998ja000437.

O'Brien, T. P. and McPherron, R. L. (2002). *Journal of Geophysical Research*, **107**(A11), 1–10. doi:10.1029/2002JA009435.

Ouellette, J. E., Lyon, J. G., Brambles, O. J., Zhang, B. and Lotko, W. (2016). *Journal of Geophysical Research: Space Physics*, **121**(5), 4111–18.

Owens, M. J. and Forsyth, R. J. (2013). *Living Reviews in Solar Physics*, **10**, 5.

Owens, M. J., Wicks, R. T. and Horbury, T. S. (2011). *Solar Physics*, **269**, 411–20.

Parker, E. N. (1992). *Journal of Geophysical Research: Space Physics*, **97**(A4), 4311–16.

Phan, T. D., Gosling, J. T., Davis, M. S., Skoug, R. M., Øieroset, M., Lin, R. P., Lepping, R. P., McComas, D. J., Smith, C. W., Reme, H. and Balogh, A. (2006). *Nature*, **439** (7073), 175–78.

Pulkkinen, T. (2007). *Living Reviews in Solar Physics*, **4**, 1.

Reid, G. C. (2000). *Space Science Reviews*, **94**(1–2), 1–11.

Richardson, I. G. (2004). *Space Science Reviews*, **111**(3–4), 267–376.

Robbrecht, E., Patsourakos, S. and Vourlidas, A. (2009). *The Astrophysical Journal*, **701**(1), 283–91.

Rostoker, G. and Falthammar, C. G. (1967). *Journal of Geophysical Research*, **72**(23), 5853–63. doi:10.1029/JZ072i023p05853.

Russell, C. and McPherron, R. L. (1973). *Journal of Geophysical Research*, **78**(1), 92–108. doi:10.1029/JA078i001p00092.

Schatten, K. H. and Wilcox, J. M. (1967). *Journal of Geophysical Research*, **72**(21), 5185–91. doi:10.1029/JZ072i021p05185.

Schulz, M. (1973). *Astrophysics and Space Science*, **24**(2), 371–83.

Schwadron, N. A. (2002). *Geophysical Research Letters*, **29**(14), 1663. doi:10.1029/2002GL015028.

Schwenn, R. (2006). *Living Reviews in Solar Physics*, **3**, 2.

Sckopke, N. (1966), *Journal of Geophysical Research*, **71**(13), 3125–30. doi:10.1029/JZ071i013p03125.

Shapiro, A. I., Schmutz, W., Schoell, M., Haberreiter, M. and Rozanov, E. (2010). *Astronomy and Astrophysics*, **517**, A48.

Siscoe, G. L. (1966). *Planetary and Space Science*, **14**(10), 947. doi:10.1016/0032-0633(66)90132-2.

Siscoe, G. L., Formisano, V. and Lazarus, A. J. (1968). *Journal of Geophysical Research*, **73**(15), 4869. doi:10.1029/JA073i015p04869.

Smith, E. J. (2001). *Journal of Geophysical Research: Space Physics*, **106**(A8), 15819–31.

Snyder, C. W., Neugebauer, M. and Rao, U. R. (1963). *Journal of Geophysical Research*, **68**(24), 6361–70.

Sojka, J. J., McPherron, R. L., van Eyken, A. P., Nicolls, M. J., Heinselman, C. J. and Kelly, J. D. (2009). *Geophysical Research Letters*, **36**, L19105.

Solanki, S. K., Krivova, N. A. and Haigh, J. D. (2013). *Annual Review of Astronomy and Astrophysics*, **51**, 311–51.

Solanki, S. K., Schüssler, M. and Fligge, M. (2002). *Astronomy and Astrophysics*, **383**(2), 706–12.

Stern, D. P. (1989). *Review of Geophysics*, **27**(1), 103–14. doi:10.1029/RG027i001p00103.

Stern, D. P. (1996). *Review of Geophysics*, **34**(1), 1–31. doi:10.1029/95RG03508.

Strachan, L., Suleiman, R., Panasyuk, A. V., Biesecker, D. A. and Kohl, J. L. (2002). *The Astrophysical Journal*, 57, 1008–14.

Suess, K., Snow, M., Viereck, R. and Machol, J. (2016). *Journal of Space Weather and Space Climate*, **6**, A10.

Sugiura, M. (1964). *Annals of the International Geophysical Year*, **35**, 9–45.

Sugiura, M. and Chapman, S. (1960). Akademie der Wissenschaften zu Göttingen Mathematisch-Physikalische Klasse, Sonderheft Nr., 4, 51–3.

Sugiura, M. and Kamei, T. (1991). Equatorial Dst index 1957–1986, Rep. IAGA Bulletin No 40, ISGI Publications Office, Saint-Maur-des-Fosses, France.

Svalgaard, L. (1977). Geomagnetic activity: Dependence on solar wind parameters, in *Coronal Holes and High Speed Wind Streams*, edited by J. B. Zirker, Colorado University Press.

Tapping, K. F. and DeTracey, B. (1990). *Solar Physics*, **127**(2), 321–32.

Temerin, M. and Li, X. (2002). *Journal of Geophysical Research: Space Physics*, **107**(A12), 1472. doi:1410.1029/2001JA007532.

Trattner, K. J., Burch, J. L., Ergun, R., Fuselier, S. A., Gomez, R. G., Grimes, E. W., Lewis, W. S., Mauk, B., Petrinec, S. M., Pollock, C. J. and Phan, T. D. (2016). *Geophysical Research Letters*, **43**(10), 4673–82.

Tsurutani, B. T., Gonzalez, W. D., Lakhina, G. S. and Alex, S. (2003). *Journal of Geophysical Research: Space Physics*, **108** (A7), 1268. doi:10.1029/2002JA009504.

Ukhorskiy, A. Y., Sitnov, M. I., Merkin, V. G., Gkioulidou, M. and Mitchell, D. G. (2017). *Journal of Geophysical Research: Space Physics*, **122**(3), 3040–54.

Usoskin, I. G., Schüssler, M., Solanki, S. K. and Mursula, K. (2005). *Journal of Geophysical Research: Space Physics*, **110**, A10102. doi:10.1029/2004JA010946.

Usoskin, I. G., Solanki, S. K. and Kovaltsov, G. A. (2007). *Astronomy and Astrophysics*, **471**(1), 301–9.

Vögler, A., Shelyag, S., Schüssler, M., Cattaneo, F., Emonet, T. and Linde, T. (2005). *Astronomy and Astrophysics*, **429**(1), 335–51.

Vourlidas, A., Buzasi, D., Howard, R. A. and Esfandiari, E. (2002). *Solar Variability: From Core to Outer Frontiers*, **506**, 91–4.

Wang, C. B., Chao, J. K. and Lin, C. H. (2003). *Journal of Geophysical Research*, **108**(A9), SMP5-1–11.

Wang, Y. M., Lean, J. L. and Sheeley, N. R., Jr. (2005). *The Astrophysical Journal*, **625**, 522–38.

Watanabe, S., Sagawa, E., Ohtaka, K. and Shimazu, H. (2014). *Earth, Planets and Space*, **54**(12), e1263. doi:10.1186/bf03352454.

Webb, D. F. and Howard, T. A. (2012). *Living Reviews in Solar Physics*, **9**, 3.

Wenzler, T., Solanki, S. K., Krivova, N. A. and Fröhlich, C. (2006). *Astronomy and Astrophysics*, **460**(2), 583–95.

Wilcox, J. M. and Ness, N. F. (1965). *Journal of Geophysical Research*, **70**(23), 5793–5805.

Wilcox, J. M., Schatten, K. H. and Ness, N. F. (1967). *Journal of Geophysical Research*, **72**(1), 19–26. doi:10.1029/JZ072i001p00019.

Woods, T. N., Tobiska, W. K., Rottman, G. J. and Worden, J. R. (2000). *Journal of Geophysical Research: Space Physics*, 105 (A12), 27195–215.

Wu, J. G. and Lundstedt, H. (1996). *Geophysical Research Letters*, **23**(4), 319–22. doi:10.1029/96GL00259.

Wu, J. G. and Lundstedt, H. (1997). *Journal of Geophysical Research*, **102**(A7), 14255–68. doi:10.1029/97JA00975.

Yashiro, S., Gopalswamy, N., Michalek, G., St. Cyr, O. C., Plunkett, S. P., Rich, N. B. and Howard, R. A. (2004). *Journal of Geophysical Research: Space Physics*, **109**, A07105. doi:10.1029/2003JA010282.

Yeo, K. L., Krivova, N. A., Solanki, S. K. and Glassmeier, K. H. (2014). *Astronomy and Astrophysics*, **570**, A85.

15

Space Weather Effects in the Ionosphere, in the Thermosphere and at Earth's Surface

In the context of space weather effects, magnetosphere–ionosphere coupling is one of the fundamental processes controlling energy transfer and dissipation in geospace. Alfvén waves appear to play a key role in this coupling, specifically in coupling the dynamics of magnetospheric convection to the ionosphere and in generating the region 1 and region 2 global field-aligned current systems. The momentum transport from the magnetosphere to the ionosphere can be described as the result of the generation and propagation of Alfvén waves, for example as arising along newly reconnected magnetic field lines, and in general in terms of their incidence on and reflection from the ionosphere.

The thermosphere experiences dramatic changes in density and composition during magnetic storms. Intense Joule heating and particle precipitation at auroral latitudes cause intense thermal expansion, air upwelling and strong wind circulations. The Joule heating at E-layer altitudes can cause both density enhancements and depletions at higher altitudes and complicate the interpretation of mass density anomalies at high latitudes. The thermospheric response to storms at middle and low latitudes is less complicated, where the averaged density enhancement is linearly proportional to the solar wind input. Magnetic substorms during active periods also cause mass density perturbations.

Magnetic storms and substorms can cause disturbances up to thousands of nT at the Earth's surface. The time derivative of the magnetic field provides a proxy for the associated geoelectric field, which can drive geomagnetically induced currents in Earthed conductors. The geoelectric field is thus a key quantity for space weather effects on technological systems such as power grids, and it can be obtained by modelling the magnetic field using ionospheric currents and model ground conductivity as inputs.

15.1 Magnetosphere–Ionosphere Coupling and Ionospheric Dynamics during Storms and Substorms: Role of Alfvén Waves

*Ian R. Mann, Ivan P. Pakhotin, I. Jonathan Rae, Kyle. R. Murphy, Louis. G. Ozeke, David J. Knudsen, Andy Kale and David K. Milling**

15.1.1 Introduction

Magnetosphere–ionosphere coupling (MIC) is one of the fundamental processes controlling energy transfer and dissipation in geospace. It is well known that southward interplanetary magnetic field (IMF) conditions allow magnetic reconnection to transport energy and momentum from the solar wind into the magnetosphere (Dungey, 1961), creating a convection pattern in the ionosphere (e.g. Lockwood et al., 1990; see also Chapter 14). However, the fact that these field lines are anchored in the (imperfectly) conducting ionosphere ensures that ionospheric impacts are actively fed back into the magnetosphere.

The convection flows in the magnetosphere drive global-scale field-aligned current (FAC) systems, which close in the ionosphere. Finite ionospheric conductivity exerts a drag on the anti-sunward convecting field lines, and this results in the Ohmic dissipation of energy in the ionosphere. Under typical

* The authors thank all optical imagery and ground magnetometer providers. Swarm data were provided by the European Space Agency. The authors thank NRCan for the provision of data from the Canadian magnetic observatory system (CANMOS) Network and CARISMA (www.carisma.ca), operated by the University of Alberta, funded by the CSA. We acknowledge NASA contract NAS5-02099 and V. Angelopoulos for use of data from the THEMIS Mission and NSF support of GIMNAST through grant AGS-1004736. Specifically, we thank S. Mende and E. Donovan for use of the THEMIS ASI data. Deployment and data retrieval of the THEMIS ASIs were partly supported by CSA contract 9 F007-046101 to the University of Calgary. We thank S. Mende and C. T. Russell for the use of the THEMIS ground-based magnetometer data and NSF for support through grant AGS-1004814. IRM is supported by a Discovery Grant from the Canadian Natural Sciences and Engineering Research Council. I. J. R. is funded in part by Science and Technology Facilities (STFC) grant ST/N000749/1 and Natural Environmental Research Council (NERC) grant NE/L007495/1. This activity was undertaken with the financial support of the Canadian Space Agency.

southward IMF conditions, the FACs take the form of region 1 and 2 current systems (Iijima and Potemra, 1976), with the convection patterns and related FAC systems changing in response to changing upstream IMF orientation (e.g. Lockwood et al., 1990; Weimer, 2001). Of course, the Dungey cycle also requires magnetic reconnection in the night-side of the magnetosphere and the closure of open magnetic flux. Milan (2013) summarises how the overall response of the geospace system results from the dynamic balance between the rate of opening and closing of polar cap flux on the dayside and nightside (cf. also Chapter 14). The occurrence rates and intensity of substorms, which close flux on the nightside, are also controlled by the magnetospheric state – with larger open flux in the polar cap at substorm onset leading to larger substorms and more intense FAC (Coxon et al., 2014).

The fact that the system is not steadily driven suggests that temporal dynamics, and hence the related propagation of Alfvén waves (Alfvén, 1942) might play a significant role – not least because in a magnetohydrodynamic (MHD) treatment the only way to change FAC intensity is through field-aligned Alfvén wave propagation (see also e.g. the review by Song and Lysak, 2001, and references therein). In fact, the drag exerted on anti-sunward convecting field lines, as well as the field-aligned currents (FACs) generated, and the rate of energy dissipation in the ionosphere, can all be understood in terms of the generation and propagation of Alfvén waves, and their (partial) reflection at the ionosphere and subsequent inference (e.g. Wright, 1996). Similar Alfvénic wave processes may also impact closed field line flux tube dynamics in the magnetotail (e.g. Strangeway, 2012).

In general, the FACs, which connect magnetospheric disturbances to the ionosphere, are established and modified dynamically through the propagation of Alfvén waves. This can occur as a result of changes in the upstream IMF, or in response to nightside flux closure and/or during periods of bursty bulk flows (BBFs; e.g. Angelopoulos et al., 1992). Alfvén wave propagation between the plasma sheet and the ionosphere also plays an important role in driving FACs linking auroral features with associated tail reconnection and/or tail plasma flows or BBF drivers in the tail (e.g. Nakamura et al., 2001; see also e.g. Allan and Wright, 2000). Similar MIC must presumably also operate more generally in association with nightside convection.

The further energetic significance of Alfvén waves for MIC was emphasised by Wygant et al. (2000) and Keiling et al. (2003), who showed that there is sufficient power in the Earthward Alfvén wave Poynting flux at altitudes of around 6 Earth radii to power the magnetically conjugate aurora below. Overall, it is therefore imperative to develop a better understanding of Alfvén wave exchange in the magnetosphere–ionosphere system, and to assess the impact of low altitude electrodynamics for MIC and related overall energy and momentum transfer. The reader is also referred to Chapter 13 of this volume for a more detailed review of the nature and characteristics of ULF plasma waves, and an examination of their role in the dynamics of some other elements of the coupled geospace system.

15.1.2 Alfvén Waves and Ionospheric Convection

In terms of MHD wave modes (cf. Section 13.2), the fast magnetoacoustic mode is often invoked to transfer energy from the magnetopause across magnetic field lines into the inner magnetosphere (e.g. Mann et al., 1999, and references therein; see also Chapter 13 of this volume). However, and in contrast, the Alfvén mode is constrained to transport energy along magnetic field lines. Alfvén wave MIC develops due to partial reflection at the ionosphere, the magnitude of this reflection depending upon the relationship between the ionospheric Pedersen conductance, Σ_P, and through the Alfvén speed just above the ionosphere (v_{AI}) the Alfvén conductance $\Sigma_A = 1/\mu_0 v_{AI}$. Waves are strongly reflected when Σ_A is either very much greater or very much smaller than Σ_P (e.g. Scholer, 1970). In terms of the dynamics of anti-sunward convection, for example, the momentum transport from the magnetosphere to the ionosphere can be described as the result of the generation and propagation of Alfvén waves along newly reconnected field lines, and their incidence on, and reflection from, the ionosphere and related wave interference (see e.g. Wright, 1996, and references therein, for an elegant and detailed description of this process).

Very recent research, for example by Pakhotin et al. (2018) using data from the three spacecraft ESA Swarm mission in low-Earth orbit (LEO) (Friis-Christensen et al., 2008), has further explored the hypothesis that Alfvén wave propagation may play a very important role in MIC. The study by Pakhotin et al. (2018) examined the magnetic and electric field perturbations associated with FAC at Swarm altitudes during a period of northward IMF. At that time the magnetic perturbations observed on the ground were very small, being primarily associated only with solar quiet (Sq) current systems and with magnitudes ~10 s of nT. Remarkably, at high latitudes during polar passes the in situ Swarm measurements showed magnetic perturbations with >100 nT p-p amplitudes. These disturbances also varied significantly when observed by the side-by-side Swarm A and C spacecraft, even though they were separated by only around 10 s along-track and around 1.4° cross-track.

As has been shown by prior studies (e.g. Gjerloev et al., 2011), the magnetic variations associated with large-scale (greater than ~150 km) FAC current systems in the ionosphere often appear to be well correlated and rather quasi-static between relatively closely spaced spacecraft passes. See also, however, the recent analysis by Forsyth et al. (2017), which challenges the stationarity assumption and therefore the accuracy of FACs estimated using single spacecraft methods in LEO. Magnetic features at small scales (less than ~10 km) are often attributed to Alfvén waves (e.g. Lühr et al., 2015), with recent work by Miles et al. (2018) associating the dynamics of the internal structure of discrete auroral arcs with Alfvén wave reflection and interference, and even a possible role for Alfvén wave trapping inside the ionospheric Alfvén resonator (IAR).

Intermediate-scale (~10–150 km) magnetic structures are often ignored; a number of analyses use filtering at periods of around 20 s to remove such intermediate-scale non-stationary signatures when deriving FAC using dual spacecraft techniques in LEO (e.g. Ritter et al., 2013). However, as described by Pakhotin et al. (2018), these intermediate scales appear to be a potentially significant and important element of MIC associated with Alfvén waves. Indeed, using both electric and magnetic field measurements, Pakhotin et al. (2018) were able to demonstrate a frequency dependence of the wave impedance and phase difference. This was consistent with an Alfvén wave picture, but inconsistent with a static FAC picture.

Here we present a preliminary analysis of the role of Alfvén waves in region 1 and region 2 global FAC structures, which is complementary to the analysis of Swarm satellite data completed by Pakhotin et al. (2018) during a northward IMF interval. See Pakhotin et al. (2018) for additional details relating to the methodology. Figure 15.1.1

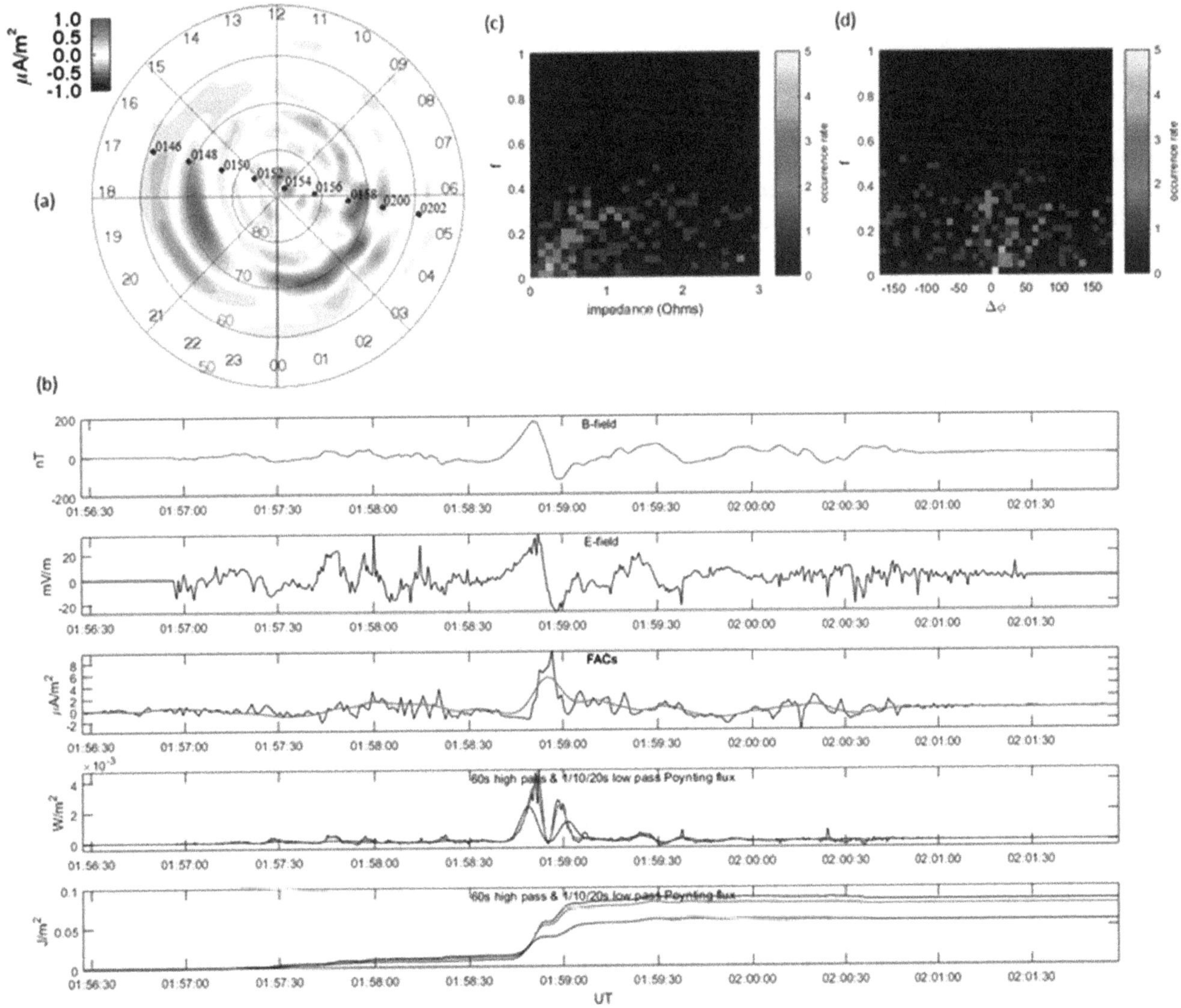

Figure 15.1.1 Field-aligned currents (FACs) and Alfvén waves in the coupled magnetosphere–ionosphere system on 7 January 2016. (a) FAC maps as estimated using the AMPERE experiment (Anderson et al., 2000) superposed with the orbit of the Swarm A satellite from 0146 UT to 0200 UT. (b) Swarm A measurements of detrended magnetic (top panel) and electric (second panel) fields, the Swarm FAC (third panel) derived using the single (blue) and dual (red) spacecraft techniques (cf. Ritter et al., 2013), as well as the instantaneous Poynting flux (fourth panel) for the cases where the electric and magnetic fields used to compute it have been time domain filtered at 1 s (blue), 10 s (red) and 20 s (black), and (bottom panel) the Poynting flux integrated along this section of the Swarm A orbit computed using the same filtering. (c, d) Occurrence distributions of the Alfvén wave impedance magnitude and phase, respectively, as a function of frequency for the polar pass shown in panel (b). See text for details. (A black-and-white version of this figure appears in some formats. For the colour version, please refer to the plate section.)

shows analysis of electric field instrument (EFI; Knudsen et al., 2017) and vector fluxgate magnetometer (VFM; Friis-Christensen et al., 2008) data from Swarm A during a polar pass from 0146 UT to 0202 UT on 7 January 2016. Figure 15.1.1a shows the Swarm A orbit superposed on a global large-scale FAC map derived using data from the attitude control magnetometers on board the Iridium constellation through the Active Magnetosphere and Planetary Electrodynamics Response Experiment (AMPERE) project (Anderson et al., 2000). The structures of the region 1 and region 2 currents are clearly seen at this time, Swarm providing the ability to examine the electric and magnetic fields associated with the MIC electrodynamics. Figure 15.1.1b shows, from top to bottom, the detrended electric and magnetic fields, the FAC (single spacecraft technique in blue, dual spacecraft in red; see Ritter et al., 2013), as well as the instantaneous Poynting flux, and the Poynting flux integrated along the Swarm A pass through this current system. In each of the Poynting flux panels in Figure 15.1.1 (bottom two panels), the blue, red and black traces show the respective Poynting flux when derived using a 1 s, 10 s and 20 s low-pass filter applied to each of the contributing electric and magnetic field time series.

Immediately obvious is the extremely clear and strong correlation between the electric and magnetic field time series (top two panels of Figure 15.1.1b), which show oscillations associated with the FACs derived from the magnetic field data, especially the two-spacecraft FAC product (red trace in the third panel of Figure 15.1.1b). The correlation is especially clear at large scales but extends to smaller amplitude and smaller scales as well. At large scales, the apparent in-phase relationship between the electric and magnetic fields is consistent with the standard picture that large-scale FAC systems close to the ionospheric boundary should obey the boundary condition that $dE/dB = 1/\mu_0\Sigma_P$, with the electric and magnetic fields also being in phase.

However, as discussed by Pakhotin et al. (2018), long-period (low-frequency) Alfvén wave disturbances will have a longer field-aligned wavelength than shorter-period (higher-frequency) waves. At altitudes much less than one field-aligned Alfvén wave wavelength from the reflection point in the conducting ionosphere (such as the E-layer), even if the low-frequency Alfvén waves are characterised by a superposition of interfering and oppositely directed waves as a result of ionospheric reflection, they will still obey the same ionospheric boundary condition as in the static case. However, higher-frequency waves will have shorter field-aligned wavelengths, such that finite field-aligned phase advance between the reflection point and the altitude of the Swarm spacecraft ensures that the ratio of the elctric and magnetic field amplitudes is no longer defined purely by Σ_P. As a result, for Alfvén waves, the ratio dE/dB and the relative phase between dE and dB at the altitudes of Swarm should be frequency dependent. Here dB and dE are the perpendicular magnetic and electric fields of the wave, having polarisations in the eastward (azimuthal) and northward directions, respectively, consistent with the dispersion relation of a parallel propagating Alfvén wave with field-aligned Poynting flux (cf. Pakhotin et al., 2018; see also e.g. Knudsen et al., 1992; Lysak, 1991).

Figures 15.1.1c and 15.1.1d show the wave impedance (μ_0 dE/dB) and the relative phase between dE and dB, respectively, as a function of frequency plotted as an occurrence distribution (Grzesiak, 2000; Pakhotin et al., 2018). As described in detail by Pakhotin et al. (2018), these occurrence distributions of impedance and relative phase as a function of frequency are derived from multiple Fourier spectra using windows of finite length that are stepped through the data along the Swarm pass. This approach provides an assessment of the nature of the MIC electrodynamics, in particular a diagnosis of the potential role of Alfvén waves. Figure 15.1.1c clearly shows a finite impedance at low frequencies, consistent with the expectation that the impedance should approach a value of $1 / \Sigma_P$ as $\omega \rightarrow 0$. In the quasi-static FAC boundary condition, the impedance should be frequency independent and maintain this impedance at higher frequencies as well. However, Figure 15.1.1c shows that the impedance clearly increases with frequency – a clear diagnostic demonstrating the Alfvénic nature of the MIC during this polar pass. Consistent with an Alfvén wave interpretation, Figure 15.1.1d shows that the relative phase between dE and dB spans a range of values between +90° and −90°, rather than being concentrated on zero phase as expected for a quasi-static FAC system. As described by Pakhotin et al. (2018), this is consistent with a model of interfering Alfvén waves with net overall downward Poytning flux as expected for Alfvénic MIC.

In summary, Figure 15.1.1 provides strong support for the hypothesis that Alfvén waves play an important role in the MIC associated with magnetospheric convection and the generation of region 1 and region 2 global FACs. Moreover, the bottom two plots in Figure 15.1.1b highlight the energetic importance of small-scale electromagnetic disturbances which, based on the results from Figure 15.1.1c, we argue are representative of Alfvén waves. In particular, the bottom plot in Figure 15.1.1b indicates that disturbances with timescales in the frame of the Swarm orbit between 1 s and 20 s contribute around 30% of the electromagnetic energy input into the ionosphere as estimated by integrating the instantaneous Poynting flux along the Swarm trajectory. For this pass the vast majority of this intermediate-scale energy input has timescales between 10 s and 20 s, corresponding to spatial scales between ~80 and 160 km if these are interpreted as spatial structures crossed by the Swarm spacecraft. Given the Alfvén wave analysis in Figure 15.1.1, and the non-stationarities in the magnetic fields often observed between the closely spaced Swarm A and C spacecraft (e.g. Pakhotin et al., 2018; Forsyth et al., 2017), we interpret these structures as being spatiotemporal and associated with the fields of ionospherically reflected and interfering Alfvén waves.

Overall, we argue that this case study provides strong evidence for the energetic significance of Alfvén waves for MIC at intermediate scales. Indeed, as discussed by Pakhotin et al. (2018) (see also e.g. Lysak, 1991; Song and Lysak, 2001), even the longer-period, larger-spatial-scale electromagnetic disturbances could be interpreted in the context of Alfvén wave dynamics. Even though such long-period disturbances are typically associated with quasi-static FACs, with equal validity, they could alternatively be interpreted in the context of Alfvén wave dynamics at low frequencies even as $\omega \rightarrow 0$. Certainly based on the results from this case study it would seem reasonable to consider an alternative Alfvén wave MIC interpretation – at least on timescales of the order of the 'there-and-back' Alfvén wave propagation time in the magnetospheric Alfvén resonator, arising from wave reflection between conjugate ionospheric field line footpoints in the Northern and Southern Hemispheres.

15.1.3 Alfvén Wave Magnetosphere–Ionosphere Coupling during Substorms

Waves in the ULF band have long been associated with the dynamical processes active during magnetospheric substorms, including those which connect the magnetotail to the aurora through the propagation of Alfvén waves. For example, waves in the Pi2 band (Jacobs et al., 1964; cf. Table 13.2.1 in Chapter 13) are frequently used to time the onset of substorms, the propagation of Alfvén waves of course being required to establish the field-aligned currents (FACs) that carry a diverted cross-tail current into the ionosphere to form the substorm current wedge (SCW; McPherron et al., 1973; see also the recent review by Kepko et al., 2015). Pi2s can also be driven by the Earthward impact of tail BBFs (e.g. Kepko et al., 2001) and be generated at high latitudes by mode conversion from fast waves trapped in the tail waveguide into Earthward propagating Alfvén waves in the plasmasheet boundary layer (e.g. Wygant et al., 2002; Allan and Wright, 2000; see also the review by Wright and Mann, 2006). Pi2s are also excited in the form of plasmaspheric cavity modes at low and middle latitudes (e.g. Takahashi et al., 2003) and can take the form of short-lived field line resonance oscillations on the flanks (e.g. Kepko et al., 2001). The reader is further directed to the review by Keiling and Takahashi (2011) for more details.

Significant scientific progress has recently been made in relation to distinguishing outside-to-in substorm onset processes, such as would be initiated by tail magnetic reconnection (e.g. Angelopoulos et al., 2008), from inside-to-out processes, such as might arise from ballooning instability at the transition between dipolar and stretched tail-like magnetic fields in the nightside magnetosphere (e.g. Rae et al., 2009b, and references therein). One of the challenges, however, is that the uncertainties often fall within a 2 min range, generating the so-called 2 minute problem (e.g. Ohtani, 2004). Given that the period of Pi2 waves lies by definition from 40 to 150 s (Jacobs et al., 1964; Table 13.2.1 in Chapter 13), deriving more accurate timing using Pi2 waves of such long period is extremely challenging. This has led to recent assessments of the potential use of higher-frequency waves in the Pi1/2 band (typically assumed to be a 24–96 s period band, e.g. Rae et al., 2009a) and Pi1 bands (1–40 s period; see Jacobs et al., 1964, and Table 13.2.1 in Chapter 13 for more accurate timing). For example, recent work has shown how relative arrival times of higher-frequency Pi1 and Pi1/2 waves can be used to identify the epicentre of the substorm onset on the ground (e.g. Milling et al., 2008; Murphy et al., 2009; Rae et al., 2009a).

Figure 15.1.2 shows results taken from the recent paper by Rae et al. (2017), which examines the relationship between substorm onset and westward travelling surge related aurora and the Pi onset epicentres, in particular contours of wave arrival times characterised separately in the Pi2, Pi1/2 and Pi2 bands. The top panel shows the locations of selected magnetometer stations from the Canadian Array for Real-time Investigation of Magnetic Activity (CARISMA; Mann et al., 2008), and the Time History of Events and Macroscale Interactions during Substorms (THEMIS; Angelopoulos et al., 2008) GMAG (Russell et al., 2008) and EPO (Peticolas et al., 2009) magnetometer chains, together with those from the Mid-continent Magnetoseismic Chain (McMAC) and the CANadian Magnetic Observatory System (CANMOS) magnetometer arrays, together with their four-letter station IDs (see Rae et al., 2017, for more details). The three panels underneath show the contours of the arrival times inferred from the network of magnetometer stations by using the time at each station the signal in a given discrete wavelet band rises above the noise (this technique has been named the automated wavelet estimation of substorm onset and magnetic events (AWESOME) algorithm; see Murphy et al., 2009, for more details). The results for three wavelet period bands are shown: Pi2 (48–192 s), Pi1/2 (24–96 s) and Pi1 (12–48 s).

As can be clearly seen, the nature of the first onset timing, its geographic location with respect to the stations in the ground arrays, and the characteristic profiles of the subsequent 2-D expansion of the first arrival time of the signals, are different in each band. However, the location of at least one epicentre in each band is aligned with the onset arc seen in all-sky imager data shown in the top panel at GILL at 06:04:20 UT, and is essentially contemporaneous with the Pi1/2 onset in the middle panel. Rae et al. (2009a, 2009b) have previously shown an association of the Pi1/2 epicentre with the time of the development of auroral beads, and postulated that the onset of Pi2 waves was related to auroral breakup, consistent with the optical analysis in Rae et al. (2017) for this event.

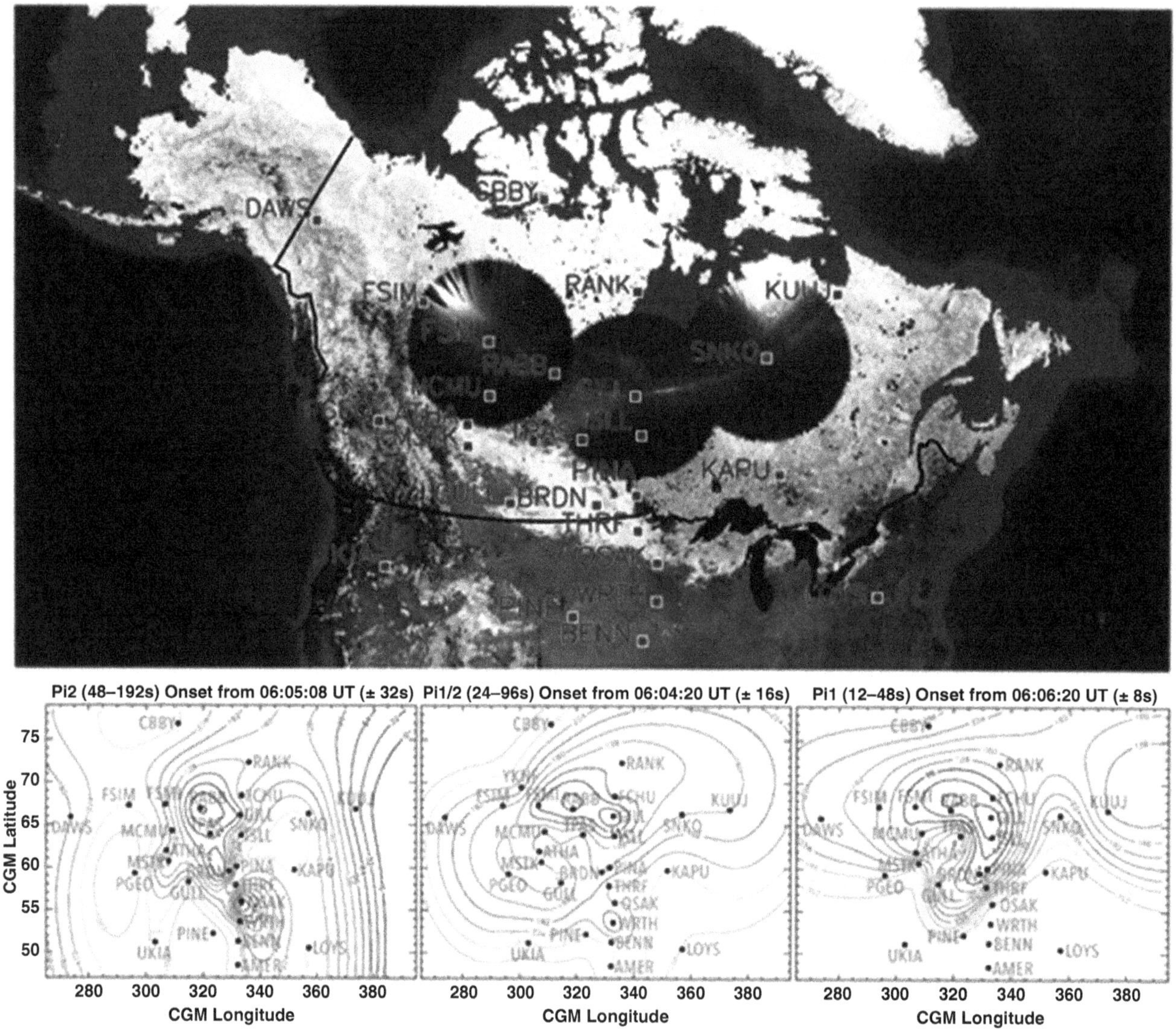

Figure 15.1.2 (top) Selected ground magnetometer stations from the CARISMA, THEMIS, CANMOS and McMAC magnetometer chains. Over-plotted are false colour white light all-sky imager (ASI) data from the GILL, FSMI and SNKQ THEMIS ASIs at 06:04:20 UT. This corresponds to the time of auroral brightening in this small substorm event on 5 March 2008. (bottom) AWESOME timing contours for the Pi2 (48–192 s; left), Pi1/2 (24–96 s; middle) and Pi1 (12–48 s; right) wavelet bands. Adapted from Rae et al. (2017, Figures 1 and 4). (A black-and-white version of this figure appears in some formats. For the colour version, please refer to the plate section.)

Interestingly, in Figure 15.1.2, in the Pi2 band, there are two distinct epicentres arising from the onset of the magnetospheric substorm, one at higher auroral latitudes and another at mid-latitudes. Following the concept of the Tamao (1964) path, which relates to the combined fast-mode and mode-converted Alfvén wave pathways from the magnetotail/magnetosphere to the ionosphere, previous authors such as Chi et al. (2009) have identified two possible latitudinal locations for the minimum travel time: one involving Alfvén wave propagation at high latitudes in the plasmasheet boundary layer and another just outside the plasmapause. As described by Rae et al. (2017), the results shown for the Pi2 band in Figure 15.1.2 appear to be consistent with that interpretation. The Pi1 onset also appears to be at least 2 min after the onset of the waves in the Pi1/2 band and the brightening and/or beading of the onset arc. Most likely, these higher-frequency waves are related to filamentary current structures and/or Alfvén wave aurora during the poleward and westward expansion of the aurora during the process of auroral breakup and the release of the westward travelling surge at the beginning of the expansion phase.

Overall, waves with irregular waveforms in the Pi bands (cf. Table 13.2.1) are closely associated with various substorm onset–related phenomena, and new developments related to their characterisation with wavelet analysis have opened a new window for comparing the spatiotemporal development of ULF wave power in the Pi1, Pi1/2 and Pi2 bands. There appears to be significant promise for using tools such as the AWESOME algorithm for connecting the behaviour in different bands with different physical phenomena active in the magnetosphere and ionosphere, and coupled by MIC, during the onset and expansion phase of substorms.

15.1.4 Conclusion and Future Perspectives

As described in detail in Section 15.1.2, Alfvén waves appear to play a key role in magnetosphere–ionosphere coupling, and specifically in coupling the dynamics of the Dungey convection cycle to the ionosphere and generating the region 1 and region 2 global FAC systems. In the preliminary analysis presented here, Swarm data show not only that Alfvén waves appear to be a ubiquitous element of the coupled MI system but also that they appear to be a significant component of the electromagnetic energy transfer from the magnetosphere to the ionosphere as well. Since Alfvén waves have also been associated with smaller-scale discrete arc dynamics (e.g. Miles et al., 2018) and shown to possess sufficient energy at high altitudes to power the aurora (e.g. Wygant et al. 2000; Keiling et al., 2003), they appear to be a critical aspect of the dynamically coupled geospace system. As argued by Pakhotin et al. (2018), it is also possible that MIC at larger spatial scales corresponding to the structure of global FAC systems described by Iijima and Potemra (1976) might also be described within the same paradigm of Alfvén wave interaction between the magnetosphere and ionosphere (cf. also Song and Lysak, 2001). Further work is required to assess the validity of this Alfvén wave paradigm and to assess the scale dependence of the electromagnetic energy transport and MIC associated with magnetospheric convection.

As described in Section 15.1.3, Alfvén waves also appear to play a key role in substorm dynamics, in relation to auroral brightenings and auroral beads at onset, to substorm-related convection, to establishing the substorm current wedge (SCW) and in general coupling the related magnetospheric dynamics into the ionosphere. As shown here, there is a more complex and broader band ULF wave response at substorm onset than just that in the Pi2 band. Not only can shorter-period Pi1/2 waves be used to more accurately identify the onset epicentre (Milling et al., 2008; Murphy et al., 2009; Rae et al., 2009a, 2017) but these waves might be directly linked to the auroral beads (e.g. Donovan et al., 2006; Rae et al., 2009b), which have now been shown to accompany the majority of substorm onsets (Kalmoni et al., 2017).

Despite significant recent advances in understanding substorm physics, the relatively sparse nature of available in situ satellite observations in the magnetosphere and the strong dependence of the nature of such in situ observations on spacecraft location relative to driving phenomena in the tail, indicate the importance of additional meso- and global-scale monitoring capabilities. For example, as described above, results derived using the AWESOME algorithm applied to ground-based magnetometer data can be used to connect processes in the magnetospheric substorm to auroral features seen in the ionosphere. Such approaches promise further advances in understanding the precise nature of the physical processes active during substorm onset.

Overall, in the context of both the Dungey convection cycle and the substorm sequence, further exploitation of new analysis techniques promises to reveal the role and energetic significance of Alfvén waves. This should apply to the role of Alfvén waves in generating dynamic coupling between the magnetosphere and the ionosphere, both in response to solar wind forcing and in relation to the release of stored magnetic energy in the magnetotail during substorms.

15.2 Thermospheric Density Response to Geomagnetic Storms and Substorms: Heating and Expansion, Composition Changes and Gravity Waves

*Hermann Lühr, Huixin Liu, Ruosi Liu and Patricia Ritter**

15.2.1 Introduction

The mass density of the thermosphere is a highly variable quantity. Its value changes between solar minimum and maximum by typically a factor of 10 around altitudes of 400 km (e.g. Emmert et al., 2010). The knowledge of this quantity is of importance, for example, for understanding thermosphere–ionosphere coupling processes. Even more critical is the air density for orbit disturbances of low-Earth-orbiting (LEO) satellites and for monitoring space debris. In spite of this relevance, detailed knowledge of the thermosphere variability was missing in the past. This was mainly due to sparse measurements of mass density. This situation has improved significantly with the launch of the satellites CHAMP and GRACE in 2000 and 2002, respectively. Sensitive accelerometers enabled these satellites to derive details of the global mass density and wind distributions.

During geomagnetic storms, a lot of energy is dissipated in the upper atmosphere. Many processes contribute to the heating of the air. Precipitating electrons ionise the neutral particles. Strong magnetospheric electric fields drive currents along the field lines that cause Joule heating in the auroral ionosphere. In addition, these mapped E-fields cause rapid plasma convection in the ionosphere, being another source of

* The authors are grateful to Eelco Doornbos for analysing the CHAMP and GRACE mass density data. The CHAMP mission was sponsored by the Space Agency of the German Aerospace Center (DLR) through funds of the Federal Ministry of Economics and Technology, following a decision of the German Federal Parliament (grant code 50EE0944). GRACE is a joint USA (NASA) and German (DLR) mission.

frictional heating due to ion-neutral collision. Practically, the complete ion temperature enhancement during active periods is dumped into the neutral gas of the thermosphere due to close coupling of the species (e.g. Lühr et al., 2012). Furthermore, the solar extreme ultraviolet (EUV) radiation is enhanced during years of solar maximum when many magnetic storms occur. The irradiance in that wavelength range is completely absorbed at altitudes above 100 km. In principle, the thermosphere behaves like a 'black body', absorbing all the energy coming in from space. Monitoring the thermospheric density and probably also the temperature provides a good and independent estimate of the total energy input during magnetically disturbed periods (see Burke et al., 2009).

Atomic oxygen is the most abundant neutral species in the altitude range of ~200–800 km. Above this altitude range, helium becomes more important, and at lower heights, the molecules N_2 and O_2 dominate. Due to the exponential decay of thermospheric density with increasing height the mean free path between collisions becomes larger with increasing altitude. Therefore, the constituents start to separate above 120 km, and their densities follow individual barometric laws. As a consequence, the density scale height increases from about 10 km at 100 km to more than 80 km at 1000 km altitude. However, this average thermostatic picture is not valid during geomagnetic storms, when locally large amounts of energy are dissipated. In those cases, the resultant composition changes complicate the relation between mass density variation and energy input. Furthermore, a transient local Joule heating will cause an upwelling of air. Such a local disturbance cannot be stationary but propagates in horizontal directions, comparable to waves spreading out after a stone falls into water. All these processes have to be considered when describing the response of the thermospheric mass density to geomagnetic storms.

In the sub-sections to follow, we will first outline the general characteristics of the thermospheric density. In particular, we describe the dependence of the mass density on various ambient parameters. In Section 15.2.3, we present different effects that magnetic storms have on the air density. Also some attempts to parameterise the storm effects are described. Section 15.2.4 is devoted to mass density effects during substorms. These transient events on the nightside cause density anomalies that show particular spatial and temporal evolutions. Finally, the important features of mass density response to magnetic activity are summarised.

15.2.2 Dependence of Mass Density on the Various Parameters

The actual thermospheric density depends on a number of controlling parameters. These are ambient conditions like solar EUV intensity or magnetic activity but also time and location, e.g. local time, season and altitude, latitude and longitude. As expected, the dependence on altitude is particularly strong because of the exponential decay of the density following the barometric law. In this article we mainly refer to CHAMP and GRACE measurements within the altitude range of 300–500 km. Here the scale height varies around 60 km (Prölss, 2005). In this chapter we will not discuss the height dependence in detail but mostly present results normalised to a common altitude.

Very important for the thermospheric density is the level of solar EUV flux, commonly approximated by the radio flux at 10.7 cm wavelength and represented by the F10.7 index. This index can vary from about 60 solar flux units (sfu) at solar minimum to more than 250 sfu during solar maximum years (1 sfu = 10^{-22} W m^{-2} Hz^{-1}). Since the thermosphere has a certain inertia, it does not respond fully to the day-to-day variations of F10.7. Density changes correlate much better to a somewhat smoothed version, P10.7 = 0.5 (F10.7 + AF10.7), where AF10.7 is an average over 81 days (three solar rotations) (e.g. Guo et al., 2007). In a dedicated study, Müller et al. (2009) tried to isolate the dependences of the mass density on the various parameters. By analysing CHAMP measurements at 400 km altitude, they found a good linear correlation with the EUV flux. During magnetically quiet days, the density varies as

$$\rho = \rho_0(P10.7 - 59), \tag{15.2.1}$$

where ρ_0 is the mass density at a solar flux level of P10.7 = 60 sfu. This simple relation turned out to be valid both for day and night, and it is only slightly dependent on latitude. Based on Equation (15.2.1), there is a simple way of normalising the observed density values ρ_{amb} to a selected solar flux reference level:

$$\rho_{amb} = \rho_r \frac{P10.7 - 59}{P10.7_r - 59}, \tag{15.2.2}$$

where ρ_r is the density value at a quiet reference day (e.g. before the storm) and $P10.7_r$ the flux level on that day. With the help of Equation (15.2.2), the effect of changing solar EUV during a storm can be separated from the magnetic activity effects.

Mass density values are commonly higher by a factor of 2 at daytime than at night. As expected, maxima and minima of the diurnal variation appear slightly delayed at 13:30 local time (LT) and 01:30 LT, respectively (e.g. Müller et al., 2009). There is also a clear annual variation of the mass density. It has long been known that the air at a given altitude is thinnest around June solstice and highest at equinoxes. Liu et al. (2007) showed the seasonal variation along with hemispheric asymmetry of the mass density at 400 km. Müller et al. (2009) reported a density depletion by a factor of about 1.5 around 1 July. This global annual variation adds to the hemispheric seasonal variation shown in Liu et al. (2007). In the Northern Hemisphere the local summer effect is partly compensated by the global minimum, while in the Southern Hemisphere the winter depletion coincides with the global minimum around June solstice. That means both effects add up causing larger seasonal variation of the mass density in the south.

Emmert and Picone (2010) presented later an independent analysis of average mass density climatology based on orbit analysis of large space debris objects. They confirm in principle the quiet-time characteristics as derived from CHAMP observations. These quiet-time density variations have to be taken into account when interpreting the thermospheric response to geomagnetic storms.

15.2.3 Thermospheric Effects during Geomagnetic Storms

The thermosphere experiences dramatic changes in density and composition during magnetic storms (e.g. Forbes et al., 1996; Prölss, 1997, 2011; Liu and Lühr, 2005). Intense input of Joule heating and particle precipitation at auroral latitudes cause intense thermal expansion, air upwelling and strong wind circulations. For interpreting these effects, it is advisable to discuss the high and middle latitudes separately. The bulk of storm-related energy input occurs at auroral latitudes. Here the intense field-aligned currents (FACs) drive Joule heating in the ionosphere and precipitating charged particles hit the atmosphere. As a consequence, density enhancements will be observed first at high latitudes and subsequently propagate to lower latitudes.

A severe magnetic storm on 20 November 2003 has been studied for example by Liu and Lühr (2005) and Forbes et al. (2005). The related observations exhibit many of the typical mass density evolution features occurring after a sudden activity increase. Figure 15.2.1 shows the density variation during the 3 days around the storm occurrence with respect to the quiet-time level before the storm. In this case CHAMP sampled the thermospheric density close to noon (11 LT) and midnight (23 LT). On the dayside the disturbance amplitudes are larger than on the nightside. Also, larger effects can be observed in the Southern Hemisphere compared to the Northern Hemisphere. Both facts can be associated with the larger diurnal/seasonal-related ambient densities. The figure confirms the delayed propagation of density disturbance to low latitudes from auroral regions. From the slant angle of the contours, a propagation velocity of the order of 600 m/s can be deduced. An equatorward propagation of that kind has been observed by Bruinsma and Forbes (2007) and confirmed by numerical modelling (Fujiwara and Miyoshi, 2006). Both groups relate the equatorward expansion to travelling atmospheric disturbances (TAD). Due to the discrete nature of the TADs and the sampling period of 93 min by the LEO satellites, the contours in Figure 15.2.1 do not always represent properly the actual signal propagation. For example, on the nightside a beating between the two discrete features suggests an apparent propagation from the equator to the poles. This means a dedicated interpretation of satellite data is required to get the correct expansion of the disturbance front. For this event Liu and Lühr (2005) reported local density enhancements of up to a factor of 8 at auroral/subauroral latitudes. After a time of about 24 hours the thermospheric density is restored to pre-event conditions.

Commonly used empirical thermosphere models such as MSIS or JB2008 are not able to represent the density variations during storms. For example, Liu and Lühr (2005) compared density observations during the "Halloween" storm (30 October 2003) with the MSIS model. The predictions by the model had little in common with the data, not in

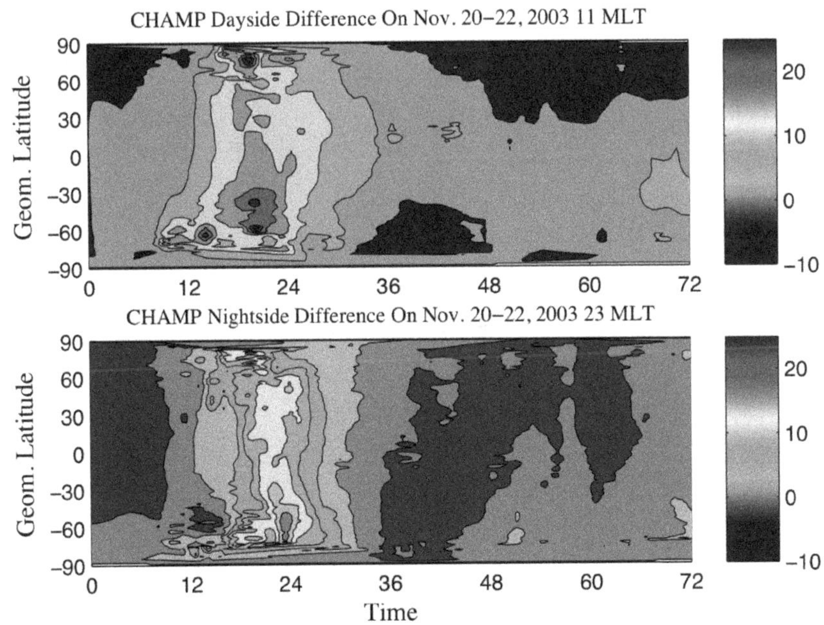

Figure 15.2.1 Storm-time deviation of the thermospheric density during the 20 November 2003 magnetic storm. Values are in units of 10^{-12} kg/m^3 and relative to the pre-storm quiet-time level. Modified after Liu and Lühr (2005, Figure 4). (A black-and-white version of this figure appears in some formats. For the colour version, please refer to the plate section.)

timing, location or amplitude. Similar discrepancies for other storms were reported by Burke et al. (2007). It therefore probably requires a different approach for modelling these very transient effects.

At high latitudes, beyond ±60° magnetic latitude (MLat), the interpretation of mass density anomalies is not so straightforward. Here intense Joule heating at E-layer altitudes can cause both density enhancements and depletions at 400 km altitude (e.g. Lei et al., 2010). This is due to local composition changes. An induced upwelling of heavier molecules like N_2 and O_2 can reduce the average scale height and, with that, reduce the mass density at satellite height. According to Prlöss (1987), the composition disturbance occurs primarily at high latitudes and can expand to 50° MLat during storm-time. In spite of that complication, Kwak et al. (2009) revealed in a statistical analysis the mass density dependence on the interplanetary magnetic field (IMF) orientation. Their study, based on CHAMP observations in the Southern Hemisphere high-latitude region during summer conditions, shows that the thermospheric density poleward of 60° MLat tends to be strongest when IMF B_z is negative and weakest when B_z is positive. For positive IMF B_y, larger density changes appear on the duskside and, for negative B_y, on the dawnside. These observations also confirm that at high latitudes the density reacts on average positively to enhanced solar wind input.

Quite prominent features at high latitudes are the mass density anomalies in the polar cap appearing during magnetic storms. Liu et al. (2010a) have checked systematically all storms during the CHAMP mission exceeding D_{st} amplitudes of 100 nT. Practically all of the 30 events considered occurred during the first half of the mission, during solar maximum years. Encounters of the satellite with polar cap anomalies could be found in all storms. Since the lifetime of such an anomaly is commonly shorter than the orbit period of 93 min (see Liu et al., 2010a), CHAMP may have missed several of them. Figure 15.2.2 shows two examples of mass density peaks in the polar cap. In the top frame the storm of 17 January 2005 is considered. In the top panel the mass density is plotted, and in the bottom panel, the small-scale FAC activity. A polar cap density anomaly appears around 11:00 magnetic local time (MLT). We observe a tripling of the density with respect to background values. Collocated with the anomaly an intense burst of small-scale FACs is recorded. This suggests a direct driving of air upwelling by intense Joule heating.

A different situation is encountered during the storm on 27 July 2004 (Figure 15.2.2, bottom frame). A prominent density peak is observed on the nightside (23 MLT). But this time it is not accompanied by FAC activity. FACs are limited to the auroral zone. Some other process must have been

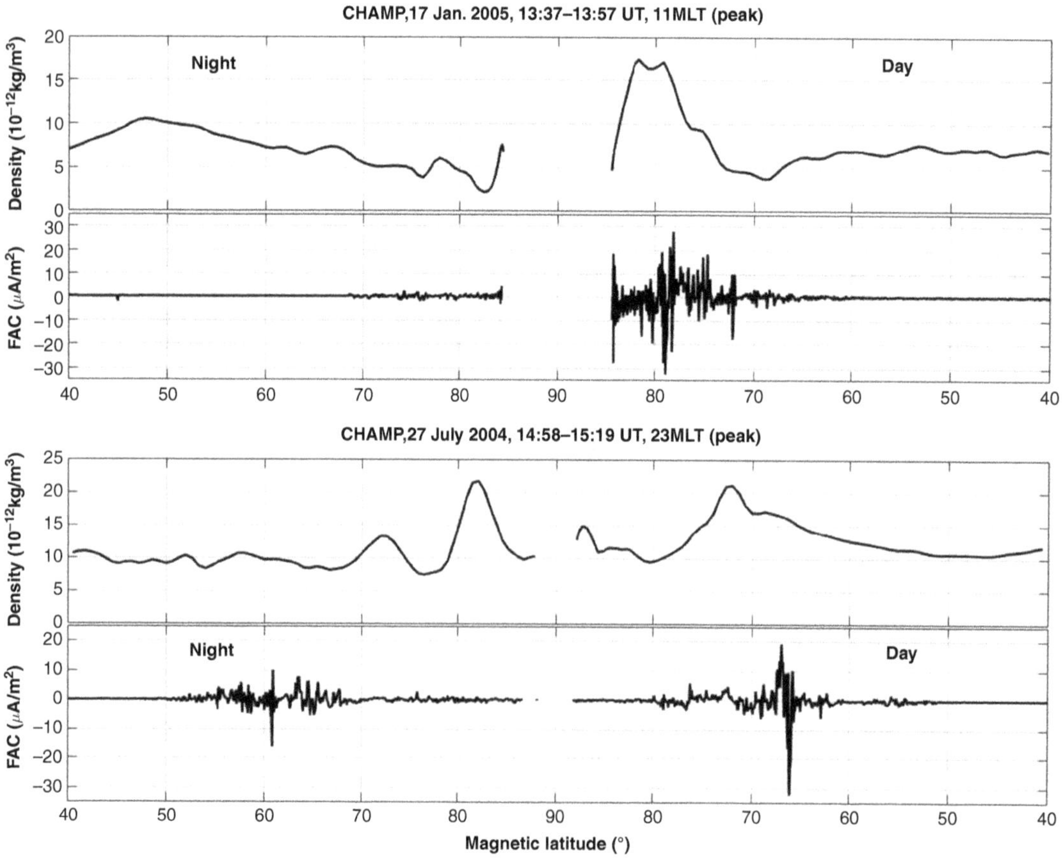

Figure 15.2.2 Observation of a localised mass density peak in the polar cap during storm-time for two events. In the second panel, collocated FAC intensity is shown. About 50% of the peaks are not accompanied by FAC activity. Modified after Liu et al. (2010a, Figures 3 and 4).

responsible for the local doubling of density. When considering the 30 storms investigated, Liu et al. (2010a) found for about half of the polar cap density anomalies collocated small-scale FACs and for the other half an absence of FACs.

In order to get a better understanding of the high-latitude storm effects on the thermospheric density, Lu et al. (2016) performed a dedicated study of the 17 January 2005 event. All available observations were assimilated into the AMIE (Assimilative Mapping of Ionospheric Electrodynamics) framework. The rather realistic high-latitude electrodynamics output of AMIE was coupled with the TIEGCM (Thermosphere–Ionosphere Electrodynamics General Circulation Model) for simulating the processes operating during that storm. Important conclusions regarding the energy input could be drawn, at least for this event. On average, the energy input over the polar cap amounts to about 20% compared to the total input and only occasionally becomes strong, whereas the bulk of energy is dumped into the auroral region. The modelling results of thermospheric density variations are qualitatively in agreement with CHAMP observations. An examination of the simulation results revealed that (1) enhanced temperature lifts up the atmosphere and causes a density increase at satellite altitude; (2) intense Joule heating induces upwelling of heavier molecules, changing locally the scale height in the upper thermosphere; and (3) impulsive Joule heating causes the launch of large-scale atmospheric gravity waves (AGW), which travel poleward or equatorward. These results imply that CHAMP observed density anomalies on successive orbits at different evolutional states, either right at their generation, collocated with intense Joule heating, or later, when they had travelled away from the source.

The thermospheric response to storms at middle and low latitudes is less complicated. Liu et al. (2010b, 2011) investigated mass density variations during storms at magnetic latitudes below 60° MLat. This area plays a dominant role for LEO satellite dynamics because it covers almost 90% of the Earth's surface. The heated thermosphere in the auroral regions is lifted up and then expands over the entire globe. Liu et al. (2010b) found that the enhancement in density at constant altitude is well correlated with the solar wind input occurring some time before. A suitable proxy for the input is the merging electric field (E_m), as defined by Kan and Lee (1979), or even better by the coupling function of Newell et al. (2007). In order to account for the history of the thermosphere, it was necessary to consider the E_m variations during the previous 3 hours. Interestingly, the authors found no saturation of thermospheric density enhancement for strong solar wind driving. In contrast, the saturation of the cross-polar cap potential for large E_m values is well established. Obviously, the upper atmospheric heating is not controlled by that potential.

In particular, for mass density averages over an orbit, Liu et al. (2010) found a simple linear relation with the merging electric field:

$$\rho = aE_m + \rho_{amb}, \tag{15.2.3}$$

where a is a scaling factor, ρ_{amb} is the density reduced to the pre-event quiet-time level as described in Equation (15.2.2), and E_m is the time-integrated merging electric field as defined in Equation (8) of Liu et al. (2010b). For CHAMP the factor a turned out to be close to 0.5 over the whole mission, when ρ is given in 10^{-12} kg m^{-3} and E_m in mV/m. For two storm periods, Figure 15.2.3 shows a comparison of predictions based on Equation (15.2.3) with direct observations. In both cases a good agreement is achieved including even individual features. In the case of the 22 July 2004 storm, the storm lasted almost 1 week. The predictions track rather well the different levels of activity over the whole period. This simple scaling law worked equally well for all levels of solar flux and magnetic activity (including super storms). In essence this observation implies that the mid-latitude density enhancement is an add-on to the prevailing density, and only dependent on the amount of solar wind input. As a consequence, we find much larger relative density variations around solar minimum years than during solar maximum years. In practice this means that magnetic storms driven by co-rotating interaction regions (CIR), which occur mainly during the declining phase of a solar cycle, cause much stronger relative density variations than the prominent coronal mass ejections (CME) during maximum years. This fact has been observed before but was never traced back to a simple relation like Equation (15.2.3). Liu et al. (2011) also checked GRACE density measurements and was able to confirm the simple linear relation with the merging electric field. But for that spacecraft, a scaling factor resulted that was smaller and decreased over the mission time. From the comparison between the two spacecraft the authors concluded that the scaling factor, a, is constant on a given thermospheric pressure level and becomes smaller at higher altitudes. Through coincidence, the orbit of CHAMP decayed at a similar rate as the pressure level during the years 2001 through 2005 from high to low solar activity, resulting in a quasi-constant scale factor. This inference should be tested and eventually considered in atmospheric models.

Interestingly, the solar wind dynamic pressure seems to have only little influence on the overall density enhancement. Only for the storm on 22 January 2005 had dynamic pressure to be considered, in order to achieve a reasonable agreement between prediction and observation (see Liu et al., 2011). In that case the dynamic pressure jumped up from typically 5 nPa to almost 100 nPa. Under such extremely compressed conditions the presented simple model is no longer valid.

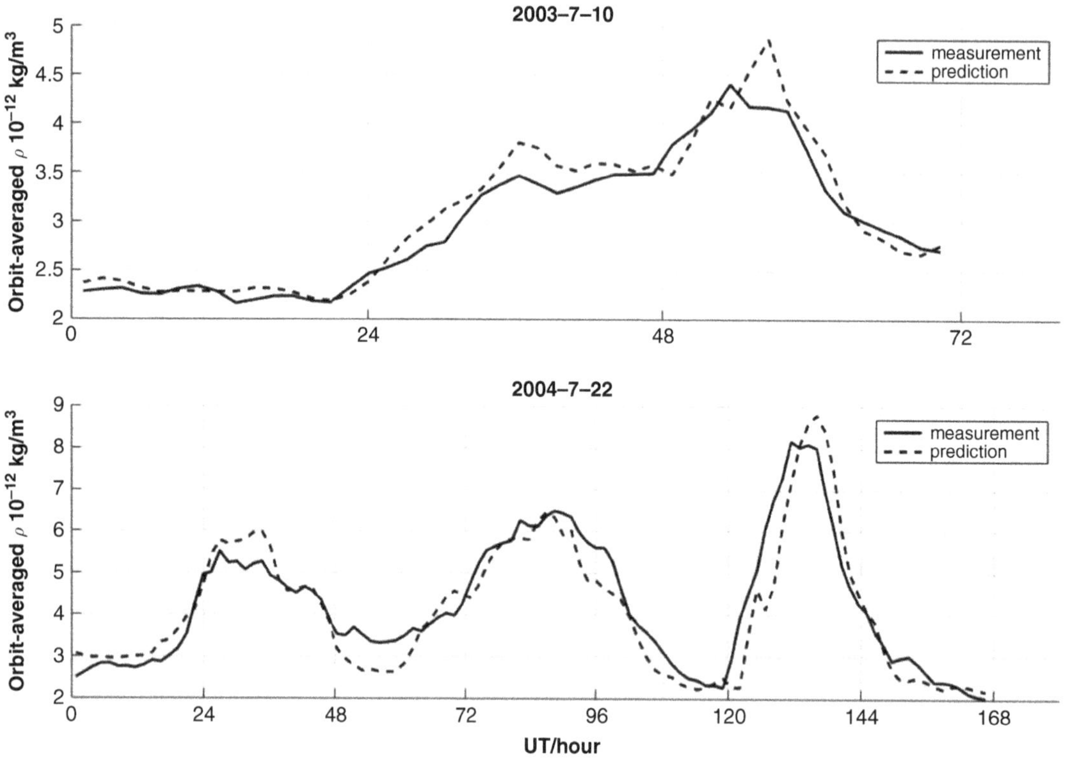

Figure 15.2.3 Comparison between CHAMP orbit-averaged density measurements and predictions of the linear model using a solar wind input integration time of 3 hours and a scaling factor $a = 0.5$ for the two storms. For more details, see Liu et al. (2010b).

There have been other attempts of modelling the global mass density enhancement during storms. For example, Zhou et al. (2009) made use of the quiet-time NRLMSIS-00 mass density values. For characterising the storm-time enhancements a special Joule heating index and the ring current index SYM-H were considered. By properly choosing the parameters, a good match between prediction and observations could be achieved. Unfortunately, the suitable sets of controlling parameters varied from storm to storm.

15.2.4 Density Anomalies during Substorms

Another important process during magnetically active periods is the magnetospheric substorm. During such an event magnetic energy stored in the magnetospheric tail is explosively released and partly deposited in the auroral region on the nightside. The thermospheric response to substorms has been studied by Ritter et al. (2010). Based on a substorm catalogue containing more than 4000 entries from the years 2000 to 2005, the average mass density response has been deduced on the nightside. For this study only CHAMP observations in close proximity, in time and space, to the substorm onsets were used. Due to the large number of events, a meaningful superposed epoch analyses could be performed for retrieving the main features.

One important finding is that the effect of substorms on the thermosphere depends strongly on the level of activity in which they are embedded. Isolated substorms at quiet times ($Kp < 2+$) have no appreciable effect. Only substorms occurring during more active times ($Kp > 4$) are worth being investigated. Obviously, the more energetic plasma in the magnetospheric tail during stronger storms makes the difference, but there is still no consensus about the actual reasons. For this class of more efficient substorms, superposed epoch analyses were performed. The aim was to find out the temporal and spatial evolution of the mass density disturbance. Figure 15.2.4 presents average latitude profiles of density enhancement (above pre-event level), as derived from CHAMP data, taken in the 2 hour local time sector eastward of the onset location. The right panel shows observations at high latitudes, and the left panel visualises enhancements at middle and low latitudes. The three curves represent density profiles at different times after substorm onset; the solid line represents observations right after onset, the dashed curve one orbit later (~1.5 hours) and the dash-dotted line two orbits later. A clear density anomaly appears at high latitudes right after the onset (solid line). At the same time, towards middle latitudes, not much of an effect can be observed. On the next orbit (dashed line) the density enhancement has reached latitudes of 30° MLat. At the equator, however, no effect is seen. This changes completely during the third orbit (dash-dotted curve), when a large density bulge appears

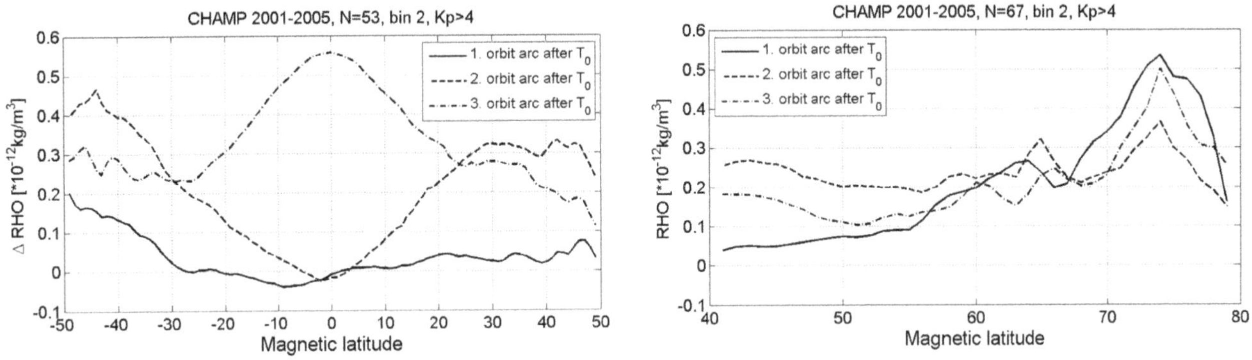

Figure 15.2.4 Spatial and temporal evolution of mass density anomaly with respect to the pre-substorm level. Full line: 0.25–1.25 hours after substorm onset T_0; dashed line: 1.75–2.75 hours after T_0; dashed-dotted line: 3.25–4.25 hours after T_0. Modified after Ritter et al. (2010, Figure 5).

at low latitudes. Density waves from both hemispheres pile up here. The amplitude at the equator is comparable with the initial density anomaly at auroral regions. Ritter et al. (2010) interpreted the time dependent density profiles as signatures of travelling atmospheric disturbances (TAD). From the delayed response at lower latitudes they deduced a velocity of about 650 m/s, which is typical for a TAD. Associated with such TADs are strong equatorward winds (e.g. Fujiwara and Miyoshi, 2006), with velocities up to 300 m/s. Towards lower latitudes, they change direction and turn into more westward zonal winds due to the action of the Coriolis force.

The observations described above show that substorms not only influence the high-latitude mass density but also affect thermosphere dynamics down to the equator at a comparable level. Predicting the disturbances in response of a substorm requires a model that considers both the temporal and spatial evolution of the density anomaly correctly.

15.2.5 Summary

In this section of the chapter we have presented the main responses of the thermospheric density to magnetic storms and substorms. The connection between ionospheric processes and thermospheric dynamics has been known for a long time, but quantitative relations have only been deduced quite recently, mainly based on detailed mass density observations by the CHAMP and GRACE satellites. Further insights may come from the Swarm constellation mission, when a more reliable mass density dataset is available.

At high latitudes, several kinds of mass density anomalies exist. A quite prominent one is the cusp-related density anomaly (e.g. Lühr et al., 2004). It constitutes a local density enhancement, which may reach a factor of 2. This feature is present almost every day. During geomagnetic storms, additional density peaks show up at high latitude. Relating their locations and times of appearance to the source is not straightforward. Impulsive Joule heating at auroral latitudes launches atmospheric gravity waves that cause peaks and troughs, which propagate away from the source. This makes a difference between the transient storm effects and the more stationary cusp-related anomalies.

At middle and low latitudes, a more homogeneous picture emerges from the storm-driven thermospheric modifications. It has been shown that the density enhancement averaged over a CHAMP orbit is linearly proportional to the solar wind input. A suitable parameter for characterising the density enhancement is the merging electric field integrated over the preceding 3 hours. Interestingly, the scaling factor between the two quantities is constant and independent of the solar activity level, as long as measurements are performed on a constant pressure level. This means that magnetic storms are most effective in producing relative density variations in the declining phase of the solar cycle.

Magnetospheric substorms also cause mass density perturbations. But these are only significant when they occur during more active ($Kp > 4$) periods. Right after onset, a density anomaly appears at high latitudes in the midnight sector. The density bulges from both hemispheres propagate as travelling atmospheric disturbances (TAD) equatorward at a speed of around 650 m/s. About 4 hours after onset, the anomalies from the two hemispheres reach the equator and pile up there. The resulting density enhancement is rather comparable with the initial high-latitude anomaly after onset.

The thermospheric response to magnetic activity is not well reflected in existing empirical atmospheric models. During the past decade a lot of progress has been made in characterising the mass density responses to magnetospheric storm and substorms. This additional knowledge should be used for designing improved storm-time models.

15.3 Earth Surface Response to Geomagnetic Storms and Substorms: Geomagnetic and Geoelectric Disturbances

*Ari Viljanen**

There are complex electric current systems in the Earth's magnetosphere and ionosphere, which are primarily maintained by the interaction of the solar wind and the geomagnetic field (Cowley et al., 2000; Ganushkina et al., 2015; Milan et al., 2017). Irregular variations in the solar wind manifest themselves as temporal and spatial changes of the near-space currents and the geomagnetic field. The largest magnetic storms cause disturbances whose amplitudes at the Earth's surface can reach some thousands of nT at the most (Nikitina et al., 2016; Thomson et al., 2011). Typical periods vary from seconds to hours (Campbell, 1973; Jacobs et al., 1964). In this context, the secular variation of the main geomagnetic field originating at the deep interior of the planet has no direct significance. Variations of the ground magnetic field can be modelled empirically (e.g. Juusola et al., 2015; Weigel et al., 2003; Weimer, 2013; Wintoft et al., 2015) or by simulations of the near-space plasma (e.g. Janhunen et al., 2012; Lyon et al., 2004; Ogino et al., 1994; Raeder et al., 2008; Toth et al., 2005), using solar wind data as input in both cases.

Most dynamic and intense magnetic variations occur at high latitudes (Campbell, 1973; Thomson et al., 2011). Figure 15.3.1 illustrates an active event in northern Europe showing the measured geographic north component of the magnetic field. After quiet morning hours, there is a small spike (sudden storm commencement, SSC) occurring simultaneously before 05:00 UT at all magnetometer sites. It is caused by a solar wind shock with an increased interplanetary magnetic field, density and velocity. After the SSC, magnetic variations continue at a moderate level, but in the afternoon the field increases as a signature of an eastward electrojet flowing across the magnetometer chain. At about 17:30 UT, there is a rapid negative excursion at the substorm onset, and the main ionospheric current then flows westward. On top of the large-amplitude positive and negative excursions due to the electrojets, there are also smaller irregular variations such as pulsations. Later in the evening and around the local midnight, the substorm activity continues with the largest amplitudes at high latitudes. This is also the time for the most spectacular auroral displays. For a detailed overview of different high-latitude current structures, see Untiedt and Baumjohann (1993).

Due to the time dependence of the magnetic field, there is always an associated electric field as Faraday's law of induction implies. This geoelectric field drives currents in the conducting ground. It follows that the varying magnetic and electric fields measured on the ground have two sources: primary (external) space currents and secondary (internal) induced currents in the ground, also known as telluric currents. Using ground-based magnetic field measurements, these two parts can in principle be separated (Weaver, 1964). Although this separation is seldom carried out, telluric currents have sometimes a notable contribution to the total variation field (Tanskanen et al., 2001).

Geomagnetically induced currents (GIC) in power grids, pipelines and other technological conductors provide an illustrative example of induction. GIC is basically a direct current driven by the relatively slowly varying geoelectric field that creates a voltage source in a conductor system (Lehtinen and Pirjola, 1985; Pirjola and Lehtinen, 1985). The geoelectric field is thus a key quantity concerning space weather effects on technological conductor systems (Pulkkinen et al., 2017). Figure 15.3.2 shows the measured GIC along the Finnish natural gas pipeline compared to the time derivative of the magnetic field at a nearby observatory. The sign of the time derivative is reversed to provide a proxy for the eastward electric field, as will become evident in Equation (15.3.1). As can be expected, this proxy has quite a similar time dependence to GIC, and such a feature is visible in nearly all reported GIC events everywhere in the world. However, under some special ground conductivity cases, GIC can follow more closely the variation magnetic field instead of its time derivative, as can be argued by theoretical modelling (Pirjola, 2010) and as has been reported (at least) in Japan (Pulkkinen et al., 2010; Watari et al., 2009).

In practice, the geoelectric field is seldom measured by dense arrays for long periods. So it is much more typical to obtain it by modelling from the magnetic field, whose temporal and spatial availability is nowadays quite extensive on continental areas especially in the Northern Hemisphere (Gjerloev, 2012). The main difficulty in the modelling process is the ground conductivity, which can vary greatly in a 2-D and 3-D sense and can thus cause large and local spatial variations in the electric field. An example is any sharp conductivity boundary such as a coastline with a continent on one side and a salty high-conducting ocean on the other side, but corresponding situations exist inland as well. Charges accumulated at the conductivity boundary increase the (horizontal) electric field on the continent side. On the ocean side, the higher surface conductivity decreases the field.

From the theoretical viewpoint, the complete modelling of the electric field requires two inputs: ionospheric currents as a function of time and space and a 3-D model of the ground conductivity. A simple but illustrative model to describe the connection of the electric field to the magnetic field is to consider a plane wave reflecting from the surface of a uniformly conducting ground. Then the time evolution of

* The author thanks the institutes providing data to the IMAGE magnetometer network (http://space.fmi.fi/image/). Gasum Oy is acknowledged for the long-term collaboration in studies of geomagnetically induced currents in the Finnish natural gas pipeline.

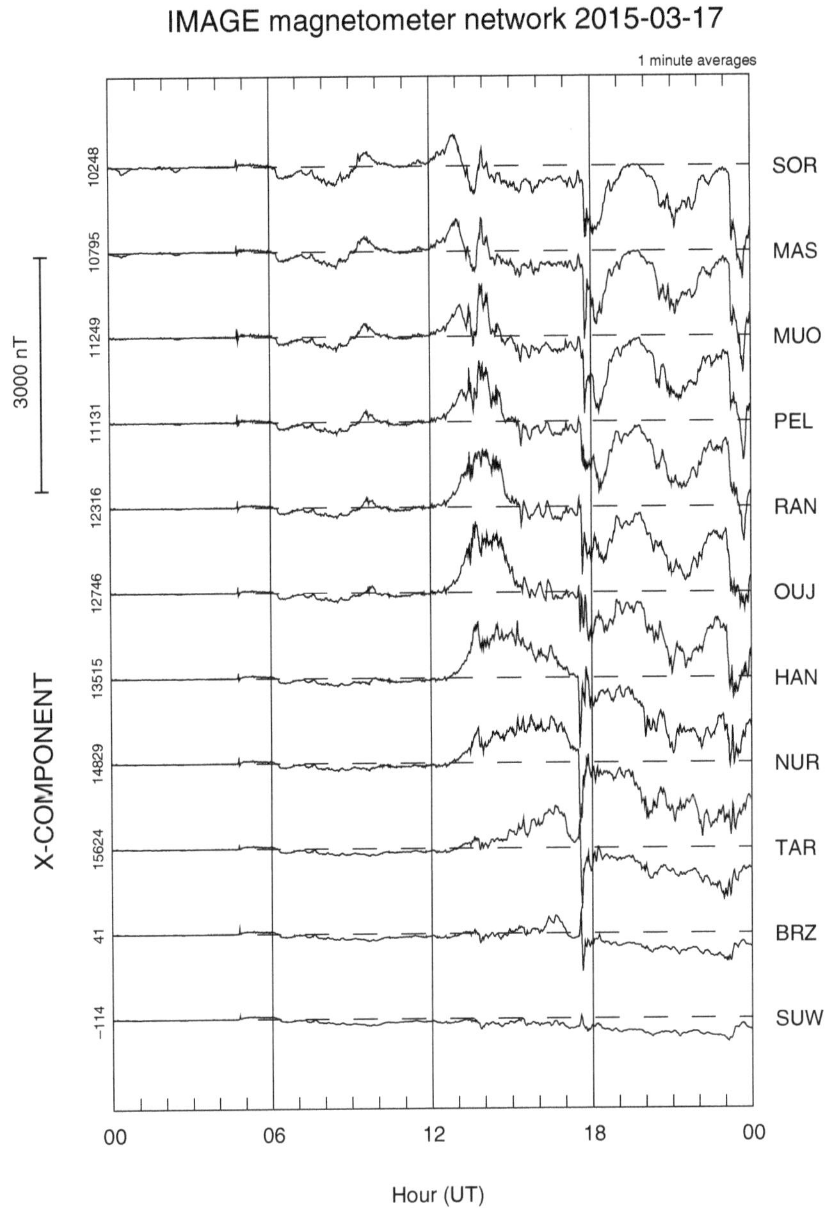

Figure 15.3.1Geographic north component of the magnetic field along a meridional chain from SUW (Suwalki, northeast Poland, 54.0°N) to SOR (Sørøya, north coast of Norway, 70.5°N) on 17 March 2015. In geomagnetic latitudes, the chain extends from about 50° to 67°N. The magnetic local time is approximately UT + 2 hours. Data are from the IMAGE magnetometer network (http://space.fmi.fi/image/).

the horizontal electric field can be expressed as a convolution of the magnetic field as follows:

$$E(t) = -\frac{1}{\sqrt{\pi\mu_0\sigma}}\int_{-\infty}^{t}\frac{g(u)}{\sqrt{t-u}}du, \tag{15.3.1}$$

where E is the eastward electric field, $g = dB/dt$ is the time derivative of the northward magnetic field, μ_0 is the vacuum permeability and σ is the ground conductivity (Cagniard, 1953). The same result holds for the northward electric and eastward magnetic fields when the sign in front of the integral is changed.

Equation (15.3.1) reveals some general features of the temporal behaviour of the electric field. First, there is an explicit dependence of the time derivative of the magnetic field. Second, all previous values of the magnetic field affect the present value of the electric field, but the future values do not, as required by causality. However, the most recent values are the most important ones due to the weighting factor proportional to the inverse square root of time, so the electric field is obviously quite similar to dB/dt. The weighting factor also affects as a filter that makes the electric field curve a little smoother than dB/dt (cf. Figure 15.3.2). Third, this model shows that an increasing ground conductivity decreases the electric field.

It is also easy to show that, for a given frequency, the amplitude of the electric field attenuates exponentially as a function of depth in a uniformly conducting ground. It follows qualitatively that the electric field related to short-period (high-frequency) magnetic variations is mostly

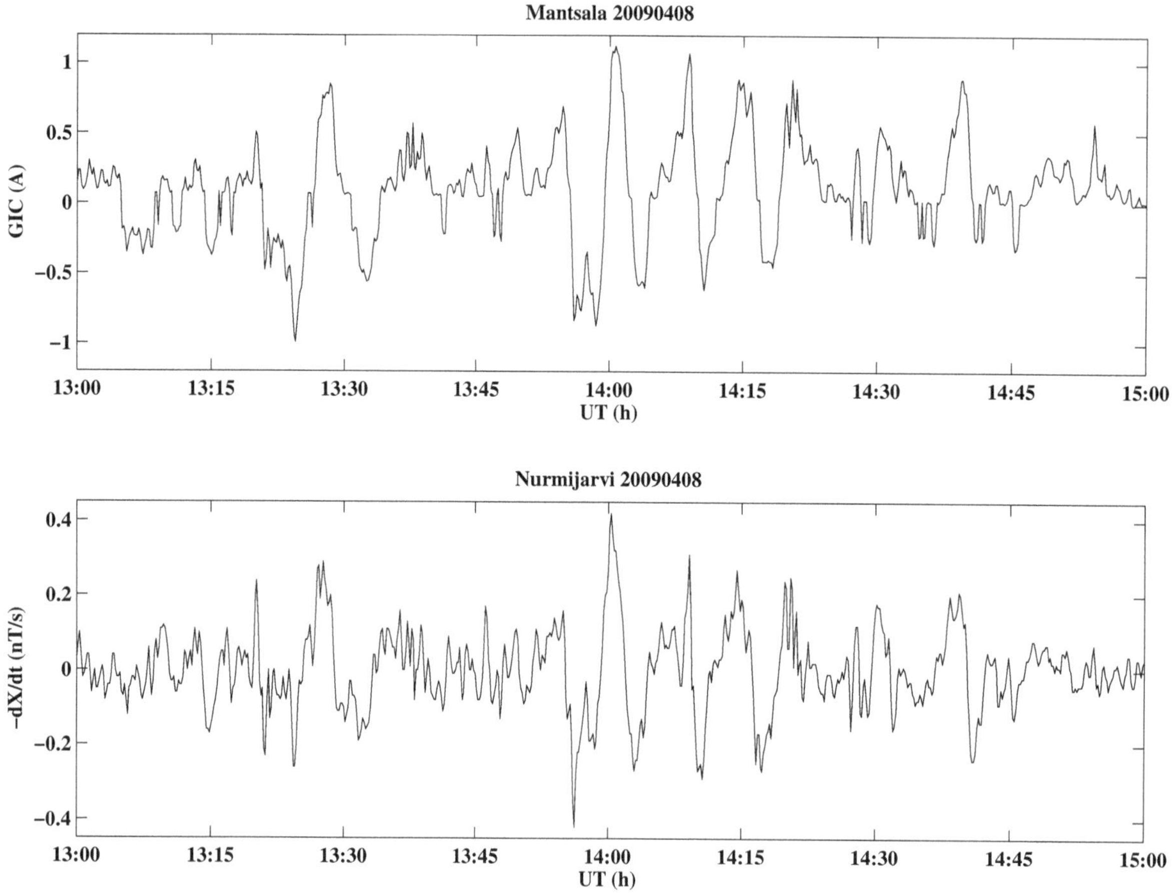

Figure 15.3.2 (top) Geomagnetically induced current flowing along the Finnish natural gas pipeline close to the Mäntsälä compressor station on 8 April 2009. Positive GIC refers to a current flowing to the east. (bottom) Negative of the time derivative of the northward magnetic field at Nurmijärvi Geophysical Observatory at about 40 km from the pipeline. Data are from the Finnish Meteorological Institute, in collaboration with Gasum Oy.

affected by the conductivity of the ground close to the surface. In turn, variations of longer periods are less affected by near-surface geology but are more influenced by deep conductivity structures.

A more realistic ground model consists of uniform layers of different conductivities and thicknesses. Assuming still that the incoming field is a plane wave, a component of the horizontal electric field is related to a transverse magnetic field component in the frequency domain by the surface impedance $Z : E(\omega) \sim Z(\omega)B(\omega)$, where ω is the angular frequency. The surface impedance can be calculated recursively for any model with N layers (e.g. Kaufman and Keller, 1981). Although this formulation nominally assumes a plane wave field, it has been applied in cases with non-uniform fields due to spatially varying ionospheric currents. Furthermore, it is possible to assume a different layered structure for each location on the ground (e.g. Marti et al., 2014; Viljanen et al., 2012; Wei et al., 2013). Such a local plane wave method is computationally very fast, allowing for the use of real-time input (Viljanen et al., 2006). It can be easily combined with input data from space plasma simulations too (Ngwira et al., 2014; Zhang et al., 2012). Although this method only approximates the true 3-D situation, it can produce satisfactory results for the geoelectric field (Figure 15.3.3) and GIC (Viljanen et al., 2012).

A related approach that takes more accurately into account 3-D features caused by the non-uniform ground is based on the impedance tensor from magnetotelluric measurements (Bonner and Schultz, 2017; Kelbert et al., 2017; Weigel, 2017). In such a general case, a horizontal component of the geoelectric field depends on both horizontal components of the magnetic field. When applied for GIC purposes, as outlined by Bonner and Schultz (2017), the first step is to project the horizontal magnetic field variations measured at a set of observatories through a multiple station transfer function to predict magnetic field time series at locations where magnetotelluric impedance tensors are available from previous studies, such as EarthScope in the United States (Bedrosian and Love, 2015; Love et al., 2016). Using

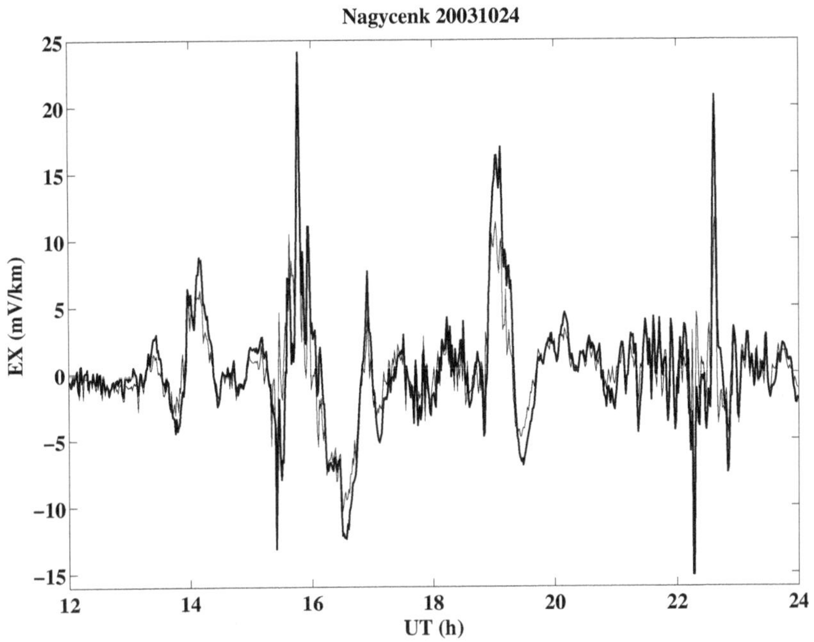

Figure 15.3.3 Measured (thick line) and modelled (thin line) northward electric field at Nagycenk (Hungary, 47.6°N, 16.7°E) on 24 October 2003. A three-layer ground conductivity model is used. Figure modified from Viljanen et al. (2012, Figure 7).

the local impedance tensors, the geoelectric field at each site is then calculated from the predicted magnetic fields. The last step is to interpolate the electric field at any other point, such as the paths of power lines, to calculate GIC. Compared to the full 3-D modelling, one reason for using empirical impedance tensors is the method's capacity for the real-time estimation of the geoelectric field (Kelbert et al., 2017). This method also produces more accurate estimates of the electric field than the local 1-D plane wave method.

An explicit step towards the full 3-D first principles modelling is the thin sheet method. Its basic form assumes a thin surface sheet, which represents the uppermost part of the ground and whose conductance (depth-integrated conductivity) can vary in horizontal directions. Below the sheet, there is a sequence of uniform layers (Vasseur and Weidelt, 1977). It is also possible to set several sheets at different depths (Engels et al., 2002). The external source can have any spatial and temporal dependence. Thanks to continuously improving computational facilities, global-scale modelling has become feasible (Püthe and Kuvshinov, 2013; Püthe et al., 2014), and at local scales, very detailed features can be included (Beggan, 2015). However, the thin sheet method is still too time consuming for real-time applications (Kelbert et al., 2014).

Concerning GIC, the need for the electric field modelling at a very small spatial scale can be relaxed, and larger-scale conductivity model catalogues, such as given by Alekseev et al. (2015), provide a good starting point. When GIC is determined in a given conductor system, such as a power grid or pipeline, the electric field is integrated along the conductor lines to calculate the voltage that drives currents. Typical length scales of power transmission lines and pipelines are of the order of 10–100 km. Integration over such distances filters spatial variations of the order of 1–10 km in the electric field, respectively.

Concerning the magnitude of GIC, the location of ionospheric currents with respect to the region of interest has the primary control (Beggan, 2015). Similarly, the same is true for the geoelectric field. However, local conductivity anomalies can still have a significant role affecting GIC at nearby power grid substations. So a challenge is to find conductivity models that give detailed enough information but are still feasible to be used computationally.

Concluding, the calculation of the geoelectric field and GIC from archived or real-time magnetic field recordings is straightforward. The accuracy will obviously increase with more precise ground conductivity models, which can be implemented with increasing computational power. Concerning operational purposes, the main immediate challenge is in forecasting the geoelectric field and GIC. This returns to the ability to predict ionospheric and magnetospheric currents that primarily produce geomagnetic variations.

There are general ideas of which kind of ionospheric currents can cause large GIC (Viljanen et al., 1999): eastward and westward electrojets and 'breakup events' (substorm onsets) are the most common ones. Pulsations and sudden impulses can also sometimes lead to large GIC. However, there is still a lack of more quantitative classification. As Pulkkinen et al. (2006) have shown for subauroral and auroral latitudes, the spatiotemporal behaviour of the time derivative of the magnetic field above temporal scales of 100 s resembles that of uncorrelated white noise. This obviously sets constraints on the feasible forecasting accuracy of dB/dt. One possibility to circumvent this difficulty is to consider a more general activity measure such as the maximum dB/dt within a given time window (Wintoft et al., 2015). Although it misses detailed information of a time series, it can still provide a proxy for the overall GIC level in a power grid (Viljanen et al., 2015). Secondly, spatial symmetry of the horizontal magnetic field fluctuations increases during substorms, indicating the presence of spatially less ordered ionospheric equivalent currents. In other words, ionospheric currents have a complicated 2-D structure, accompanied by field-aligned currents, which sets further challenges for forecasting. These complications also impact the geoelectric field as indicated by Ngwira et al. (2015) and Pulkkinen et al. (2015).

References

Alekseev, D., Kuvshinov, A. and Palshin, N. (2015). Compilation of 3D global conductivity model of the Earth for space weather applications. *Earth Planets Space*, **67**, 108, doi: 10.1186/s40623-015-0272-5.

Alfvén, H. (1942). Existence of electromagnetic-hydromagnetic waves, *Nature*, **150**, 405.

Allan, W. and Wright, A. N. (2000). Magnetotail waveguide: Fast and Alfvén waves in the plasma sheet boundary layer and lobe. *J. Geophys. Res.*, **105**(A1), 317–28.

Anderson, B. J., Takahashi, K. and Toth, B. A. (2000). Sensing global Birkeland currents with Iridium engineering magnetometer data. *Geophys Res Lett*, **27**(24), 4045–8.

Angelopoulos, V., Baumjohann, W., Kennel, C. F., et al. (1992). Bursty bulk flows in the inner central plasma sheet. *J. Geophys. Res.*, **97**(A4), 4027–39.

Angelopoulos, V., McFadden, J. P., Larson, D., et al. (2008). Tail reconnection triggering substorm onset. *Science*, **321**(5891), 931–5.

Bedrosian, P. A. and Love, J. J. (2015). Mapping geoelectric fields during magnetic storms: Synthetic analysis of empirical United States impedances. Geophys. Res. Lett., **42**, doi: 10.1002/2015GL066636.

Beggan, C. D. (2015). Sensitivity of geomagnetically induced currents to varying auroral electrojet and conductivity models. Earth Planets Space, **67**, 24, doi: 10.1186/s40623-014-0168-9.

Bonner, L. R. and Schultz, A. (2017). Rapid prediction of electric fields associated with geomagnetically induced currents in the presence of three-dimensional ground structure: Projection of remote magnetic observatory data through magnetotelluric impedance tensors. Space Weather, **14**, doi: 10.1002/2016SW001535.

Bruinsma, S. L. and Forbes, J. M. (2007). Global observation of traveling atmospheric disturbances (TADs) in the thermosphere. Geophys. Res. Lett., **34**, L14103, doi: 10.1029/2007GL030243.

Burke, W. J., Huang C. Y., Marcos, F. A. and Wise, J. O. (2007). Interplanetary control of thermospheric densities during large magnetic storms. J. Atmos. Sol. Terr. Phys., **69**, 279–87.

Burke, W. J., Lin, C. S., Hagan, M. P., Huang, C. Y., Weimer, D. R., Wise, J. O., Gentile, L. C. and Marcos, F. A. (2009). Storm time global thermosphere: A driven-dissipative thermodynamic system. J. Geophys. Res., **114**, A06306, doi: 10.1029/2008JA013848.

Cagniard, L. (1953). Basic theory of the magneto-telluric method of geophysical prospecting. Geophysics, **18**, 605–35, doi: 10.1190/1.1437915.

Campbell, W. H. (1973). Spectral composition of geomagnetic field variations in the period range of 5 min to 2 hr as observed at the Earth's surface. Radio Sci., **8**, 929–32.

Chi, P. J., Russell, C. T. and Ohtani, S. (2009). Substorm onset timing via travel time magnetoseismology. Geophys. Res. Lett., **36**(8).

Cowley, S. W. H. (2000). Magnetosphere–ionosphere interactions: A tutorial review, in *Magnetospheric Current Systems*, ed. S.-I. Ohtani, R. Fujii, M. Hesse and R. L. Lysak, American Geophysical Union, Washington, DC, doi: 10.1029/GM118p0091.

Coxon, J. C., Milan, S. E., Clausen, L. B. N., Anderson, B. J. and Korth, H. (2014), A superposed epoch analysis of the regions 1 and 2 Birkeland currents observed by AMPERE during substorms, J. Geophys. Res., **119**(12), 9834–46.

Donovan, E., Mende, S., Jackel, B., et al. (2006). The azimuthal evolution of the substorm expansive phase onset aurora. Proc. ICS, **8**, 55–60.

Dungey, J. W. (1961). Interplanetary magnetic field and the auroral zones. Phys. Rev. Lett., **6**(2), 47.

Emmert, J. T., Lean, J. L. and Picone, J. M. (2010). Record-low thermospheric density during the 2008 solar minimum. Geophys. Res. Lett., **37**, L12102, doi: 10.1029/2010GL043671.

Emmert, J. T. and Picone, J. M. (2010). Climatology of globally averaged thermospheric mass density. J. Geophys. Res., **115**, A09326, doi: 10.1029/2010JA015298.

Engels, M., Korja, T. and the BEAR Working Group (2002). Multisheet modelling of the electrical conductivity structure in the Fennoscandian Shield. Earth Planets Space, **54**, 559–73, doi: 10.1186/BF03353045.

Forbes, J. M., Gonzalez, R., Marcos, F. A., Revelle, D. and Parish, H. (1996). Magnetic storm response of lower thermospheric density, J. Geophys. Res., **101**, 2313–19.

Forbes, J. M., Lu, G., Bruinsma, S., Nerem, R. S. and Zhang, X. (2005). Thermosphere density variations due to the 15–24 April 2002 solar events from CHAMP/STAR accelerometer measurements, J. Geophys. Res., **110**, A12S27, doi: 10.1029/2004JA010856.

Forsyth, C., Rae, I. J., Mann, I. R. and Pakhotin, I. P. (2017). Identifying intervals of temporally invariant field-aligned currents from Swarm: Assessing the validity of single-spacecraft methods. J. Geophys. Res., **122**, 3411–19. doi: 10.1002/2016JA023708.

Fujiwara, H. and Miyoshi, Y. (2006). Characteristics of the large-scale traveling atmospheric disturbances during geomagnetically quiet and disturbed periods simulated by a whole atmosphere general circulation model, Geophys. Res. Lett., **33**, L20108, doi: 10.1029/2006GL027103.

Friis-Christensen, E., Lühr, H., Knudsen, D. and Haagmans, R. (2008). Swarm – an Earth observation mission investigating geospace. Adv. Space Res., **41**, 210–16, doi: 10.1016/j.asr.2006.10.008.

Ganushkina, N. Y., Liemohn, M. W., Dubyagin, S., Daglis, I. A., Dandouras, I., De Zeeuw, D. L., Ebihara, Y., Ilie, R., Katus, R., Kubyshkina, M., Milan, S. E., Ohtani, S., Østgaard, N., Reistad, J. P., Tenfjord, P., Tofoletto, F., Zaharia, S. and Amariutei, O. (2015). Defining and resolving current systems in geospace. Ann. Geophys., **33**, 1369–1402, doi: 10.5194/angeo-33-1369-2015.

Gjerloev, J. W. (2012). The SuperMAG data processing technique. J. Geophys. Res., **117**, A09213, doi: 10.1029/2012JA017683.

Gjerloev, J. W., Ohtani, S., Iijima, T., Anderson, B., Slavin, J. and Le, G. (2011). Characteristics of the terrestrial field-aligned current system. Ann. Geophys., **29**, 1713–29, doi: 10.5194/angeo-29-1713-2011.

Grzesiak, M. (2000). Ionospheric Alfvén resonator as seen by Freja satellite. Geophys. Res. Lett., **27**, 923–6, doi: 10.1029/1999GL010747.

Guo, J., Wan, W., Forbes, J. M., Sutton, E., Nerem, R. S., Woods, T. N., Bruinsma, S. and Liu, L. (2007). Effects of solar variability on thermosphere density from CHAMP accelerometer data. J. Geophys. Res., **112**, A10308, doi: 10.1029/2007JA012409.

Iijima, T. and Potemra, T. A. (1976). The amplitude distribution of field-aligned currents at northern high latitudes observed by Triad. J. Geophys. Res., **81**(13), 2165–74.

Jacobs, J. A., Kato, Y., Matsushita, S. and Troitskaya, V. A. (1964). Classification of geomagnetic micropulsations. J. Geophys. Res., **69**(1), 180–81.

Janhunen, P., Palmroth, M., Laitinen, T., Honkonen, I., Juusola, L., Facsko, G. and Pulkkinen, T. I. (2012). The GUMICS-4 global MHD magnetosphere–ionosphere coupling simulation. J. Atmos. Sol. Terr. Phys., **80**, 48–59, doi: 10.1016/j.jastp.2012.03.006.

Juusola, L., Kauristie, K., van de Kamp, M., Tanskanen, E. I., Mursula, K., Asikainen, T., Andreeova, K., Partamies, N., Vanhamäki, H. and Viljanen, A. (2015). Solar wind control of ionospheric equivalent currents and their time derivatives. J. Geophys. Res., **120**, doi: 10.1002/2015JA021204.

Kalmoni, N. M. E., Rae, I. J., Murphy, K. R., et al. (2017). Statistical azimuthal structuring of the substorm onset arc: Implications for the onset mechanism. Geophys. Res. Lett., **44**(5), 2078–87.

Kan, J. K. and Lee, L. C. (1979). Energy coupling function and solar wind-magnetosphere dynamo. Geophys. Res. Lett., **6**(7), 577–80.

Kaufman, A. A. and Keller, G. V. (1981). *The Magnetotelluric Sounding Method*. Elsevier, New York.

Keiling, A. and Takahashi, K. (2011). Review of Pi2 models. Space Sci. Rev., **161**(1–4), 63–148.

Keiling, A., Wygant, J. R., Cattell, C. A., Mozer, F. S. and Russell, C. T. (2003). The global morphology of wave Poynting flux: Powering the aurora. Science, **299**(5605), 383–6.

Kelbert, A., Balch, C. C., Pulkkinen, A., Egbert, G. D., Love, J. J., Rigler, E. J. and Fujii, I. (2017). Methodology for time-domain estimation of storm-time geoelectric fields using the 3D magnetotelluric response tensors. Space Weather, **15**, doi: 10.1002/2017SW001594.

Kelbert, A., Kuvshinov, A., Velimsky, J., Koyama, T., Ribaudo, J., Sun, J., Martinec, Z. and Weiss, C. J. (2014). Global 3-D electromagnetic forward modelling: A benchmark study. Geophys. J. Int., **197**, 785–814, doi: 10.1093/gji/ggu028.

Kepko, L., Kivelson, M. G. and Yumoto, K. (2001). Flow bursts, braking, and Pi2 pulsations. J. Geophys. Res., **106**(A2), 1903–15, doi: 10.1029/2000JA000158.

Kepko, L., McPherron, R. L., Amm, O., Apatenkov, S., Baumjohann, W., et al. (2015). Substorm current wedge revisited. Space Sci. Rev., **190**(1–4), 1–46.

Knudsen, D. J., Kelley, M. C. and Vickrey, J. F. (1992).Alfvén waves in the auroral ionosphere: A numerical model compared with measurements. J. Geophys. Res., **97**(A1), 77–90, doi: 10.1029/91JA02300.

Knudsen, D. J., Burchill, J. K., Buchert, S. C., et al. (2017). Thermal ion imagers and Langmuir probes in the Swarm electric field instruments. J. Geophys. Res., **122**, 2655–73, doi: 10.1002/2016JA022571.

Kwak, Y.-S., Richmond, A. D., Deng, Y., Forbes, J. M. and Kim, K.-H. (2009). Dependence of the high-latitude thermospheric densities on the interplanetary magnetic field. J. Geophys. Res., **114**, A05304, doi: 10.1029/2008JA013882.

Lei, J., Thayer, J. P., Burns, A. G., Lu, G. and Deng, Y. (2010). Wind and temperature effects on thermosphere mass density response to the November 2002 Geomagnetic Storm. J. Geophys. Res., **115**, A05303, doi: 10.1029/2009JA014754.

Lehtinen, M. and Pirjola, R. (1985). Currents produced in earthed conductor networks by geomagnetically-induced electric fields. Ann. Geophys., **3**, 479–84.

Liu, H. and Lühr, H. (2005). Strong disturbance of the upper thermospheric density due to magnetic storms: CHAMP observations. J. Geophys. Res., **110**, A09S29, doi: 10.1029/2004JA010908.

Liu, H., Lühr, H. and Watanabe, S. (2007). Climatology of the Equatorial Thermospheric Mass Density Anomaly. J. Geophys. Res., **112**, A05305, doi: 10.1029/2006JA012199.

Liu, R., Lühr, H. and Ma, S. Y. (2010a). Storm-time related mass density anomalies in the polar cap as observed by CHAMP. Ann. Geophys., **28**(1), 165–80.

Liu, R., Lühr, H., Doornbos, E. and Ma, S. Y. (2010b). Thermospheric mass density variations during geomagnetic storms and a prediction model based on the merging electric field. Ann. Geophys., **28**, 1633–45, doi: 10.5194/angeo-28-1633-2010.

Liu, R., Ma, S.-Y. and Lühr, H. (2011). Predicting storm-time thermospheric mass density variations at CHAMP and GRACE altitudes. Ann. Geophys., **29**, 443–53, doi: 10.5194/angeo-29-443-2011.

Lockwood, M., Cowley, S. W. H. and Freeman, M. P. (1990). The excitation of plasma convection in the high-latitude ionosphere. J. Geophys. Res., **95**(A6), 7961–72.

Love, J. J., Pulkkinen, A., Bedrosian, P. A., Jonas, S., Kelbert, A., Rigler, E. J., Finn, C. A., Balch, C. C., Rutledge, R., Wagge, R. M., Sabata, A. T., Kozyra, J. U. and Black, C. E. (2016). Geoelectric hazard maps for the continental United States. Geophys. Res. Lett., **43**, 9415–24, doi: 10.1002/2016GL070469.

Lu, G., Richmond, A. D., Lühr, H. and Paxton, L. (2016). High-latitude energy input and its impact on the thermosphere. J. Geophys. Res., **121**, 7108–24, doi: 10.1002/2015JA022294.

Lühr, H., Park, J., Gjerloev, J. W., Rauberg, J., Michaelis, I., Merayo, J. M. G. and Brauer, P. (2015). Field-aligned currents' scale analysis performed with the Swarm constellation. Geophys. Res. Lett., **42**, 1–8, doi: 10.1002/2014GL062453.

Lühr, H., Park, J., Ritter, P. and Liu, H. (2012). In-situ CHAMP observation of ionosphere-thermosphere coupling. Space Sci. Rev., **168**, 237–60, doi: 10.1007/s11214-011-9798-4.

Lühr, H., Rother, M., Köhler, W., Ritter, P. and Grunwaldt, L. (2004). Thermospheric up-welling in the cusp region, evidence from CHAMP observations, Geophys. Res. Lett., **31**, L06805, doi: 10.1029/2003GL019314.

Lyon, J. G., Fedder, J. A. and Mobarry, C. M. (2004). The Lyon-Fedder-Mobarry (LFM) global MHD magnetospheric simulation code. J. Atmos. Sol. Terr. Phys., **66**, 1333–50, doi: 10.1016/j.jastp.2004.03.020.

Lysak, R. L. (1991). Feedback instability of the ionospheric resonant cavity. J. Geophys. Res., **96**, 1553–68, doi: 10.1029/90JA02154.

Mann, I. R., Milling, D. K., Rae, I. J., et al. (2008). The upgraded CARISMA magnetometer array in the THEMIS era. Space Sci. Rev., **141**(1–4), 413–51.

Mann, I. R., Wright, A. N., Mills, K. J. and Nakariakov, V. M. (1999). Excitation of magnetospheric waveguide modes by magnetosheath flows. J. Geophys. Res., **104**(A1), 333–53.

Marti, L., Yiu, C., Rezaei-Zare, A. and Boteler, D. (2014). Simulation of geomagnetically induced currents with piecewise layered-Earth models. IEEE Trans. Power Delivery, **29**, 1886–93, doi: 10.1109/TPWRD.2014.2317851.

McPherron, R. L., Russell, C. T. and Aubry, M. P. (1973). Satellite studies of magnetospheric substorms on August 15, 1968: 9. Phenomenological model for substorms. J. Geophys. Res., **78** (16), 3131–49.

Milan, S. E. (2013). Modeling Birkeland currents in the expanding/contracting polar cap paradigm, J. Geophys. Res., **118**(9), 5532–42.

Milan, S. E., Clausen, L. B. N., Coxon, J. C., Carter, J. A., Walach, M.-T., Laundal, K., Østgaard, N., Tenfjord, P., Reistad, J., Snekvik, K., Korth, H. and Anderson, B. J. (2017). Overview of solar wind-magnetosphere–ionosphere-atmosphere coupling and the generation of magnetospheric currents. Space Sci. Rev., doi: 10.1007/s11214-017-0333-0.

Miles, D. M., Mann, I. R., Pakhotin, I. P., et al. (2018). Alfvénic dynamics and fine structuring of discrete auroral arcs: Swarm and e-POP observations. Geophys. Res. Lett., **45**(2), 545–55.

Milling, D. K., Rae, I. J., Mann, I. R., et al. (2008). Ionospheric localisation and expansion of long-period Pi1 pulsations at substorm onset. Geophys. Res. Lett., **35**(17).

Müller, S., Lühr, H. and Rentz, S. (2009). Solar and magnetospheric forcing of the low latitude thermospheric mass density, as observed by CHAMP. Ann. Geophys., **27**, 2087–99.

Murphy, K. R., Rae, I. J., Mann, I. R., et al. (2009). Wavelet-based ULF wave diagnosis of substorm expansion phase onset. J. Geophys. Res., **114**(A1).

Nakamura, R., Baumjohann, W., Schödel, R., Brittnacher, M., Sergeev, V. A., Kubyshkina, M., Mukai, T. and Liou, K. (2001). Earthward flow bursts, auroral streamers, and small expansions. J. Geophys. Res., **106**(A6), 10791–802, doi: 10.1029/2000JA000306.

Newell, P. T., Sotirelis, T., Liou, K., Meng, C. I. and Rich, F. J. (2007). A nearly universal solar wind-magnetosphere coupling function inferred from magnetospheric state variables. J. Geophys. Res., **112**, A01206, doi: 10.1029/2006JA012015.

Ngwira, C. M., Pulkkinen, A. A., Bernabeu, E., Eichner, J., Viljanen, A. and Crowley, G. (2015). Characteristics of extreme geoelectric fields and their possible causes: Localized peak enhancements. Geophys. Res. Lett., **42**, 6916–21, doi: 10.1002/2015GL065061.

Ngwira, C. M., Pulkkinen, A., Kuznetsova, M. M. and Glocer, A. (2014). Modeling extreme 'Carrington-type' space weather events using three-dimensional global MHD simulations. J. Geophys. Res., **119**, 4456–74, doi: 10.1002/2013JA019661.

Nikitina, L., Trichtchenko, L. and Boteler, D. H. (2016). Assessment of extreme values in geomagnetic and geoelectric field variations for Canada. Space Weather, **14**, doi: 10.1002/2016SW001386.

Ogino, T., Walker, R. J. and Ashour-Abdalla, M. (1994). A global magnetohydrodynamic simulation of the response of the magnetosphere to a northward turning of the interplanetary magnetic field. J. Geophys. Res., **99**, 11027–42, doi: 10.1029/93JA03313.

Ohtani, S. I. (2004). Flow bursts in the plasma sheet and auroral substorm onset: Observational constraints on connection between midtail and near-Earth substorm processes. *Space Sci. Rev.*, **113**(1–2), 77–96.

Pakhotin, I. P., Mann, I. R., Lysak, R. L., et al. (2018). Diagnosing the role of Alfvén waves in magnetosphere-ionosphere coupling: Swarm observations of large amplitude nonstationary magnetic perturbations during an interval of northward IMF. J. Geophys. Res., **123**(1), 326–40.

Peticolas, L. M., Craig, N., Odenwald, S. F., et al. (2009). The Time History of Events and Macroscale Interactions during Substorms (THEMIS) education and outreach (E/PO) program, in *The THEMIS Mission*, pp. 557–83, Springer, New York.

Pirjola, R. (2010). Derivation of characteristics of the relation between geomagnetic and geoelectric variation fields from the surface impedance for a two-layer Earth. Earth Planets Space, **62**, 287–95.

Pirjola, R. and Lehtinen, M. (1985). Currents produced in the Finnish 400 kV power transmission grid and in the Finnish natural gas pipeline by geomagnetically-induced electric fields. Ann. Geophys., **3**, 485–91.

Prölss, G. W. (1987). Storm-induced changes in the thermospheric composition at middle latitudes, Planet. Space Sci., **35**, 807–11.

Prölss, G. W. (1997). Magnetic storm associated perturbations of the upper atmosphere, in *Magnetic Storms*, ed. B. T. Tsurutani, W. D. Gonzalez, Y. Kamide and J. K. Arballo, pp. 227–41, Geophys. Monogr. 98, AGU, Washington, DC.

Prölss, G. W. (2005). *Physics of the Earth's Space Environment*, Springer, Berlin.

Prölss, G. W. (2011). Density perturbations in the upper atmosphere caused by the dissipation of solar wind energy, Surv. Geophys., **32**, 101–95, doi: 10.1007/s10712-010-9104-0.

Pulkkinen, A., Bernabeu, E., Eichner, J., Viljanen, A. and Ngwira, C. (2015). Regional-scale high-latitude extreme geoelectric fields pertaining to geomagnetically induced currents. Earth Planets Space, **67**, 93, doi: 10.1186/s40623-015-0255-6.

Pulkkinen, A., Bernabeu, E., Thomson, A., Viljanen, A., Pirjola, R., Boteler, D., Eichner, J., Cilliers, P. J., Welling, D., Savani, N. P., Weigel, R. S., Love, J. J., Balch, C., Ngwira, C. M., Crowley, G., Schultz, A., Kataoka, R., Anderson, B., Fugate, D., Simpson, J. J. and MacAlester, M. (2017). Geomagnetically induced currents: Science, engineering and applications readiness. Space Weather, **15**, doi: 10.1002/2016SW001501.

Pulkkinen, A., Kataoka, R., Watari, S. and Ichiki, M. (2010). Modeling geomagnetically induced currents in Hokkaido, Japan. Adv. Space Res., **46**, 1087–93, doi: 10.1016/j.asr.2010.05.024.

Pulkkinen, A., Klimas, A., Vassiliadis, D., Uritsky, V. and Tanskanen, E. (2006). Spatiotemporal scaling properties of the ground geomagnetic field variations. J. Geophys. Res., **111**(A3), A03305, doi: 10.1029/2005JA011294.

Püthe, C. and Kuvshinov, A. (2013). Towards quantitative assessment of the hazard from space weather: Global 3-D modellings of the electric field induced by a realistic geomagnetic storm. Earth Planets Space, **65**, 1017–25, doi: 10.5047/eps.2013.03.003.

Püthe, C., Manoj, C. and Kuvshinov, A. (2014). Reproducing electric field observations during magnetic storms by means of rigorous 3-D modelling and distortion matrix co-estimation. Earth Planets Space, **66**, 162, doi: 10.1186/s40623-014-0162-2.

Rae, I. J., Mann, I. R., Murphy, K. R., et al. (2009a). Timing and localization of ionospheric signatures associated with substorm expansion phase onset. J. Geophys. Res., **114**(A1).

Rae, I. J., Mann, I. R., Angelopoulos, V., et al. (2009b). Near-Earth initiation of a terrestrial substorm. J. Geophys. Res., **114**(A7).

Rae, I. J., Murphy, K. R., Watt, C. E., et al. (2017). Using ultra-low frequency waves and their characteristics to diagnose key physics of substorm onset. Geosci. Lett., **4**(1), 23.

Raeder, J., Larson, D., Li, W., Kepko, E. L. and Fuller-Rowell, T. (2008). OpenGGCM simulations for the THEMIS mission. Space Sci. Rev., **141**, 535, doi: 10.1007/s11214-008-9421-5.

Ritter, P., Lühr, H. and Doornbos, E. (2010). Substorm-related thermospheric density and wind disturbances derived from CHAMP observations. Ann. Geophys., **28**, 1207–20, doi: 10.5194/angeo-28-1207-2010.

Ritter, P., Lühr, H. and Rauberg, J. (2013). Determining field-aligned currents with the Swarm constellation mission. Earth Planet Space, **65**(11), 1285–94, doi: 10.5047/eps.2013.09.006.

Russell, C. T., Chi, P. J., Dearborn, D. J., Ge, Y. S., Kuo-Tiong, B., Means, J. D., Pierce, D. R., Rowe, K. M. and Snare, R. C. (2008). THEMIS ground-based magnetometers. Space Sci. Rev., **141**(1–4), 389–412.

Scholer, M. (1970). On the motion of artificial ion clouds in the magnetosphere. Planet. Space Sci., **18**, 977.

Song, Y. and Lysak, R. L. (2001). Towards a new paradigm: From a quasi-steady description to a dynamical description of the magnetosphere. Space Sci. Rev., **95**(1–2), 273–92.

Strangeway, R. J. (2012). The relationship between magnetospheric processes and auroral field-aligned current morphology. Auror. Phenomenol. Magnetos. Process. Earth Planets, **197**, 355–64.

Takahashi, K., Lee, D.-H., Nosé, M., Anderson, R. R. and Hughes, W. J. (2003). CRRES electric field study of the radial mode structure of Pi2 pulsations. J. Geophys. Res., **108**, 1210, doi: 10.1029/2002JA009761, A5.

Tamao, T. (1964). The structure of three-dimensional hydromagnetic waves in a uniform cold plasma. J. Geomagn. Geoelectr., **18**, 89–114.

Tanskanen, E. I., Viljanen, A., Pulkkinen, T. I., Pirjola, R., Häkkinen, L., Pulkkinen, A. and Amm, O. (2001). At substorm onset, 40 % of AL comes from underground. J. Geophys. Res., **106**, 13119–34.

Thomson, A., Dawson, E. and Reay, S. (2011). Quantifying extreme behaviour in geomagnetic activity. Space Weather, **9**, S10001, doi: 10.1029/2011SW000696.

Toth, G., Sokolov, I. V., Gombosi, T. I., Chesney, D. R., Clauer, C. R., DeZeeuw, C. D. L., Hansen, K. C., Kane, K. J., Manchester, W. B., Oehmke, R. C., Powell, K. G., Ridley, A. R., Roussev, I. I., Stout, Q. F., Volberg, O., Wolf, R. A., Sazykin, S., Chan, A., Yu, B. and Kota, J. (2005). Space weather modeling framework: A new tool for the space science community. J. Geophys. Res., **110**. doi: 10.1029/2005JA011126.

Untiedt, J. and Baumjohann, W. (1993). Studies of polar current systems using the IMS Scandinavian magnetometer array. Space Sci. Rev., **63**, 245–390, doi: 10.1007/BF00750770.

Vasseur, G. and Weidelt, P. (1977). Bimodal electromagnetic induction in non-uniform thin sheets with an application to the northern Pyrenean induction anomaly. Geophys. J. R. Astron. Soc., **51**, 669–90, doi: 10.1111/j.1365-246X.1977.tb04213.x.

Viljanen, A., Amm, O. and Pirjola, R. (1999). Modelling geomagnetically induced currents during different ionospheric situations. J. Geophys. Res., **104**, 28059–72, doi: 10.1029/1999JA900337.

Viljanen, A., Pirjola, R., Wik, M., Adam, A., Pracser, E., Sakharov, Ya. and Katkalov, Yu. (2012). Continental scale modelling of geomagnetically induced currents. J. Space Weather Space Clim., **2**, A17, doi: 10.1051/swsc/2012017.

Viljanen, A., Pulkkinen, A., Pirjola, R., Pajunpää, K., Posio, P. and Koistinen, A. (2006). Recordings of geomagnetically induced currents and a nowcasting service of the Finnish natural gas pipeline system. Space Weather, **4**, S10004, doi: 10.1029/2006SW000234.

Viljanen, A., Wintoft, P. and Wik, M. (2015). Regional estimation of geomagnetically induced currents based on the local magnetic or electric field. J. Space Weather Space Clim., **5**, A24, doi: 10.1051/swsc/2015022.

Watari, S., Kunitake, M., Kitamura, K., Hori, T., Kikuchi, T., Shiokawa, K., Nishitani, N., Kataoka, R., Kamide, Y., Aso, T., Watanabe, Y. and Tsuneta, Y. (2009). Measurements of geomagnetically induced current in a power grid in Hokkaido, Japan. Space Weather, **7**, S03002, doi: 10.1029/2008SW000417.

Weaver, J. T. (1964). On the separation of local geomagnetic fields into external and internal parts. Z. Geophys., **30**, 29–36.

Wei, L. H., Homeier, N. and Gannon, J. L. (2013). Surface electric fields for North America during historical geomagnetic storms. Space Weather, **11**, 451–62, doi: 10.1002/swe.20073.

Weigel, R. S. (2017). A comparison of methods for estimating the geoelectric field. Space Weather, **15**, 430–40, doi: 10.1002/2016SW001504.

Weigel, R. S., Klimas, A. J. and Vassiliadis, D. (2003). Solar wind coupling to and predictability of ground magnetic fields and their time derivatives, J. Geophys. Res., **108**(A7), 1298, doi: 10.1029/2002JA009627.

Weimer, D. R. (2001). Maps of ionospheric field-aligned currents as a function of the interplanetary magnetic field derived from Dynamics Explorer 2 data. J. Geophys. Res., **106**(A7), 12889–902.

Weimer, D. R. (2013). An empirical model of ground-level geomagnetic perturbations. Space Weather, **11**, 107–20, doi: 10.1002/swe.20030.

Wintoft, P., Wik, M. and Viljanen, A. (2015). Solar wind driven empirical forecast models of the time derivative of the ground magnetic field. J. Space Weather Space Clim., **5**, A7, doi: 10.1051/swsc/2015008.

Wright, A. N. (1996). Transfer of magnetosheath momentum and energy to the ionosphere along open field lines. J. Geophys. Res., **101**(A6), 13169–78.

Wright, A. N. and Mann, I. R. (2006). Global MHD eigenmodes of the outer magnetosphere, in *Magnetospheric ULF Waves: Synthesis and New Directions*, ed. K. Takahashi, P. J. Chi, R. E. Denton and R. L. Lysak, pp. 51–72, American Geophysical Union, Washington, DC.

Wygant, J. R., Keiling, A., Cattell, C. A., et al. (2002). Evidence for kinetic Alfvén waves and parallel electron energization at 4–6 RE altitudes in the plasma sheet boundary layer. J. Geophys. Res., **105**(A8), doi: 10.1029/2001JA900113.

Wygant, J. R., Keiling, A., Cattell, C. A., et al. (2000). Polar spacecraft based comparisons of intense electric fields and Poynting flux near and within the plasma sheet-tail lobe boundary to UVI images: An energy source for the aurora. J. Geophys. Res., **105**(A8), 18675–92.

Zhang, J. J., Wang, C. and Tang, B. B. (2012). Modeling geomagnetically induced electric field and currents by combining a global MHD model with a local one-dimensional method. Space Weather, **10**, S05005, doi: 10.1029/2012SW000772.

Zhou, Y. L., Ma, S. Y., Lühr, H., Xiong, C. and Reigber, C. (2009). An empirical relation to correct storm-time thermospheric mass density modeled by NRLMSISE-00 with CHAMP satellite air drag data, Adv. Space Res., **43**, 819–28.

16
Technological Impacts of Space Weather

*Michael Hapgood**

Over the past decade, policy makers around the world have recognised that space weather is a natural hazard that poses a significant risk to modern societies. This chapter outlines how this risk has gradually developed over nearly 200 years through impacts on technological systems, and now affects a diverse set of technologies, including systems vital to provision of power, transport and communications. It illustrates the importance of these impacts through detailed examination of some key impact areas, looking at the physics that creates these impacts as well as their societal consequences.

The areas examined include (a) the impact of geomagnetically induced currents on electrically grounded infrastructures such as electric power grids and railway track circuits, (b) the disruption of satellite navigation by space-weather-induced changes in the upper atmosphere, and (c) the impact of space radiation on satellites, on aircraft and on ground systems. A final section reviews our wider knowledge of space weather impacts and discusses how this can be applied, and improved, to meet future challenges. It stresses the need for improved socio-economic assessments of space weather impacts to aid decision-making and, in particular, to assess the value of investment in improved forecasting and other systems for mitigation.

Space weather is not just fascinating science; it is a set of natural phenomena that can have adverse impacts on everyday life in modern technological societies. These adverse impacts arise because the performance of many technologies is influenced by the natural environments in which they operate. Some of these influences arise from fundamental features of the technology. For example, like many other electrical systems, power grids are electrically connected to the body of the Earth to ensure safety; but these 'ground connections' also provide a route by which natural geoelectric fields, such as those induced by space weather, can influence grid operations. But other environmental influences are just inadvertent and simply reflect features in the technology that are sensitive to their operational environment. A key example here is the sensitivity of digital devices to energetic particle radiation. These devices were originally designed to work in the low-radiation environments in which humans live and generally work well in those environments. But we have gradually discovered that they can be disrupted by radiation, most obviously in more intense radiation environments such as those in space and at aircraft cruise altitudes (10–12 km).

In this chapter we will explore the complex web that links space weather environments and their technological impacts. Figure 16.0.1 gives an outline of this web. Here we emphasise the three main natural environments influenced by space weather, namely the electromagnetic environment at Earth's surface, the upper atmosphere (both its neutral and ionised parts) and the radiation environment in the atmosphere and near-Earth space. In the rest of this chapter we will show how these environments drive some of the many different impacts that arise from space weather.

A discussion of all impacts is impracticable for many reasons (not least length), so this chapter focuses on presenting some key impacts in depth, so as to engage a reader interested in the physics behind those impacts. For a wider discussion of space weather impacts, see Hapgood (2017a).

16.1 Geomagnetically Induced Currents

The headline impact of space weather is its ability to disrupt the supply of electricity. This supply is the fundamental infrastructure of modern societies, so anything that can disrupt it will get serious attention from national governments, as we are seeing today with the global interest in space weather.

The risk of space weather disruption to electricity supplies is a consequence of additional electric currents that are injected into power transmission networks, the power lines that carry electricity over hundreds of kilometres using very high voltages (100–1000 kV). These networks have been a central feature in delivery of electrical power since the mid twentieth century, enabling a move away from small-scale local power generation to large power stations built in convenient locations, e.g. close to coal fields. Today transmission networks play an equally important role enabling

* The author thanks the Geological Survey of Sweden and the US Geological Survey for supporting their magnetic observatories and INTERMAGNET for promoting high standards of magnetic observatory practice (www.intermagnet.org).

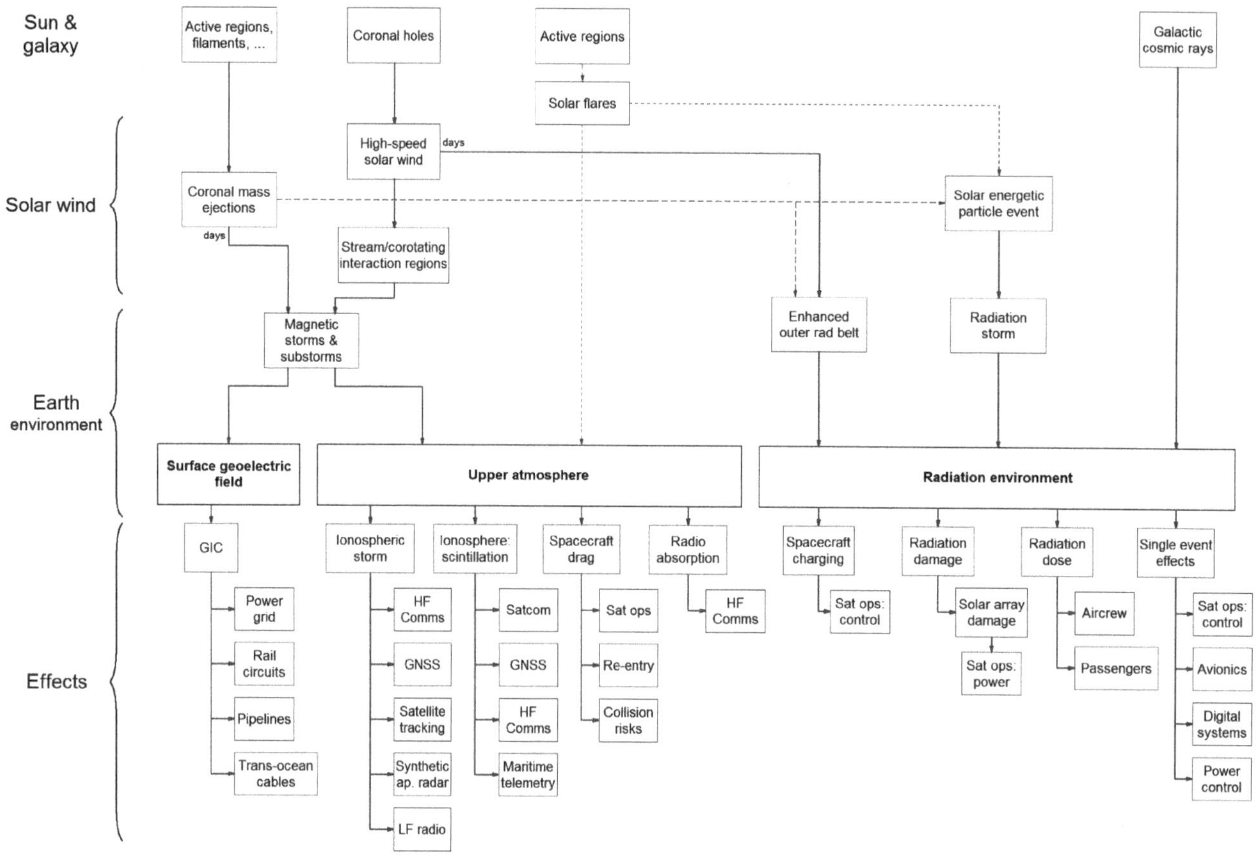

Figure 16.0.1 Overview of how processes on the Sun and in our galaxy influence natural environments on Earth, leading to a wide range of adverse effects on technological systems, the effects of many of which are discussed in this chapter. This web of processes and effects lies at the heart of space weather and is the topic of this chapter. Dashed lines of different weights are sometimes used (a) to avoid confusion when lines cross and (b) to emphasise CMEs as a driver of SEPs. Figure by the author.

delivery of low carbon power generation, as many such sources (nuclear, hydro, tidal, wind and solar) are necessarily located far from users.

Transmission networks are an example of an electrically grounded infrastructure, i.e. a spatially extended system that delivers societally important services, and that is threaded by metal structures that have electrical connections to the surface of the Earth. These ground connections contribute to both electrical safety and electromagnetic compatibility. They ensure that the electrical voltage on exposed metal surfaces matches the voltage of the adjacent Earth surface, so they cannot cause shocks to people. They also reduce the electrical noise generated by the infrastructure so that it can be a good neighbour.

These extended infrastructures need to have multiple ground connections to ensure that their zero electrical voltage tracks that of the Earth's surface. As a result, the geoelectric fields generated by geomagnetic activity (see Section 15.3) can drive electric currents (geomagnetically induced currents or GICs) along the high-conductivity path provided by the infrastructure. These currents have a long history of disrupting grounded infrastructures used in the first wave of telecommunications technologies, with geomagnetic storm impacts on the electric telegraph (Barlow, 1849), telephone (Preece, 1894), communications used for railway traffic control (Maunder, 1900) and transoceanic cables (*New York Times*, 1921a). These particular impacts have now disappeared as optical fibre has replaced twisted copper and coaxial cable as the core component of landline telecommunications, hence removing the ability of space weather to directly interfere with communications signals. But other GIC impacts have arisen as new technologies become important. These include power grids, where the first impacts were reported as early as 1940 (McNish, 1940) but only came to prominence when the great geomagnetic storm of March 1989 led to a major blackout of the Hydro-Québec power grid (Bolduc, 2002; Guillon et al., 2016). Other contemporary systems at risk from GICs include pipelines (Viljanen et al., 2006a) and railway track circuits (Eroshenko et al., 2010; Krausmann et al., 2015).

To understand GIC impacts on power grids, it is important to recognise that GIC variations follow those of the geoelectric fields generated by geomagnetic activity. As a result, GICs have much of their power at millihertz frequencies (see Section 15.3), much lower than the 50 or 60 Hz frequency of the alternating currents that flow in power grids. Thus GICs act as quasi direct currents in those grids, particularly in the transformers that link the transmission networks with generators and with distribution networks. The quasi-DC behaviour of GICs means that they can drive grid transformers into half-cycle saturation, leading to a number of problems:

1 Some of the power passing through the transformer will be dissipated as heat and vibration, leading to increased ageing of the transformer (Gaunt, 2014) and in the worst cases to catastrophic damage (Wrubel, 1992). Fortunately, many transformers have sensors that can detect the gases produced by internal heating, so the operator may respond to alarms from these sensors by switching off the transformer to protect it from further damage. Such a switch-off should ensure that the transformer will be available to provide power once a space weather event is past, but it also reduces grid capacity during the event, increasing the risk of a short-term power outage during the event. But this is much preferred to the risk of long-term outages that can arise if transformers are catastrophically damaged.
2 The non-linear behaviour of a transformer in half-cycle saturation will generate harmonics of the basic grid frequency of 50 or 60 Hz. These harmonics will be transmitted over the network and can disrupt operation of other parts of the system.
3 The amount of reactive power in the grid will be reduced by consumption in transformers, thus requiring the operator to inject additional reactive power into the grid. Reactive power is the power needed to operate the grid, e.g. the power flowing between the capacitances and inductances in the grid. It is distinguished by grid engineers from the power that is delivered to users and is managed carefully because it is vital to the stability of the grid. If not balanced by additional supply, loss of reactive power can lead to a voltage collapse on the grid and hence to power outage.

In summary, space weather impacts on power grids create two primary risks: power outages and damage to grid systems. Grid operators are making major efforts to minimise the former and to prevent the latter, both through a mixture of better engineering and use of space weather forecasts to adjust grid operation. Better engineering includes use of more resilient transformers, designs less prone to damage by GICs. This is a major lesson learned from the 1989 storm; many operators (e.g. National Grid in the UK) require that new transformers are more resilient to GIC. These engineering solutions can be reinforced by operational measures taken in response to space weather forecasts. A key measure would be to cancel scheduled maintenance and otherwise ensure that the maximum number of grid lines are operational, so that the grid has high redundancy against any problems arising during the event. In addition, operations teams would be alerted and key personnel may be deployed ready to fix problems at locations known to be vulnerable. These are vital steps to eliminate the risk of catastrophic damage to transformers and other grid systems, and limit the impact of space weather to that of short localised power outages, similar to those caused by bad tropospheric weather or accidental damage. These are still significant problems, but problems that are entirely manageable once we recognise the need for societal resilience against short outages of electric power, e.g. that individuals, businesses and government organisations should all have a capability to cope with such outages.

We now turn to the issue of GIC impacts on rail systems. These will obviously be disrupted by power outages, especially where electricity is used to power trains, but also more generally by the loss of power to operate vital services at train stations. But there is also a distinct GIC impact on the track circuits that are a key element in control of rail traffic. Track circuits are used to detect the presence of a train in a particular section of rail track. A small electric voltage is applied between the two rails and causes an electric current to flow through a relay that operates the signals and sets them to green. When a train enters that section of track, it provides a short-circuit between the rails that diverts the current, causing the relay to drop out and turn signals to red to warn other train drivers not to enter the section. There are many implementations of this basic principle, not least to cope with the simultaneous presence of other much larger currents when train engines are powered by electricity. Many, possibly all, of these implementations include electrical grounding to Earth at points within the power supply to each track circuit; hence GICs can enter track circuits and interfere with their operation. This risk is supported by circumstantial evidence from major space weather events – in that 'right-side failures' (signals turning red when they should be green) occurred during large geomagnetic storms in 1982, 1989 and 2003 (Wik et al., 2009; Eroshenko et al., 2010). A wider review of the problem (Krausmann et al., 2015) suggested that 'wrong-side failures' (signals turning green when they should be red, and hence creating the danger of a train crash) may also be possible. However, significant research is still needed to better understand how GIC affects the electrical architecture of modern railways (not least modern high-speed systems). These architectures can be complex given that they often have to supply power to trains, as well as detecting the presence of trains and operating signals. Indeed, it is only recently that Liu et al. (2016) demonstrated the first direct measurement of GIC in rail systems. Thus there is considerable scope for work to better understand how GIC affects rail systems.

There are several different processes that generate geomagnetic variations, thus leading to significant geographical variations in the risk from GIC. Most important are the ionospheric currents in the auroral zone (see Section 15.1),

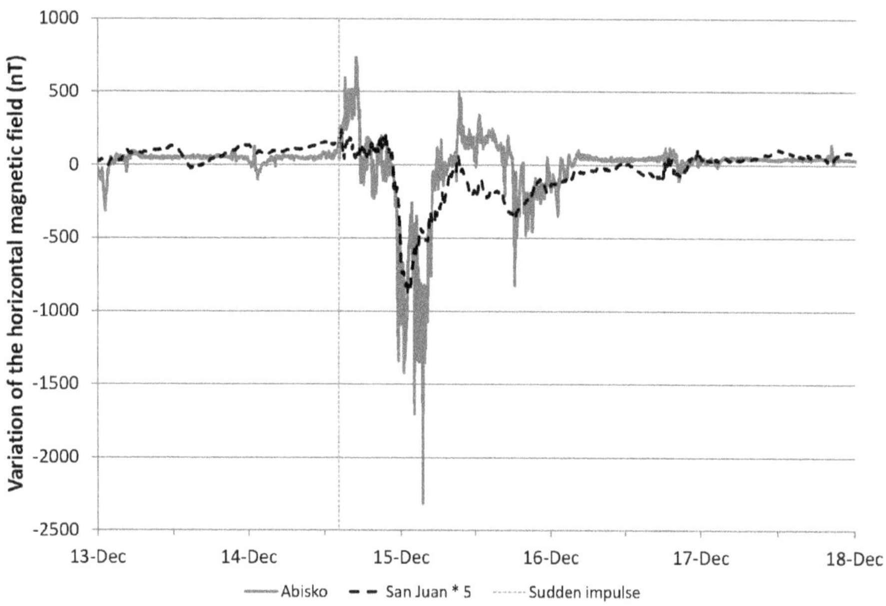

Figure 16.1.1 The auroral electrojet produces much larger variations in the ground magnetic field than does the ring current – as shown in this example from a geomagnetic storm on 14/15 December 2006. This compares the simultaneous variations seen at Abisko in northern Sweden and at San Juan in Puerto Rico, with the latter exaggerated by a factor of five. Abisko is a prime site for monitoring the auroral electrojet and San Juan for the ring current. Abisko magnetometer data courtesy of the Geological Survey of Sweden and San Juan data courtesy of the US Geological Survey. Sudden impulse timing courtesy of the International Service on Rapid Magnetic Variations, hosted by Observatori de l'Ebre. Figure by the author.

which can drive very large magnetic variations (≫1000 nT), as shown in Figure 16.1.1. In everyday space weather conditions, these currents are confined to the polar regions. But as space weather activity increases, the currents move equatorward creating a GIC risk in populous areas, such as Europe, North America, Australia and New Zealand. A key feature of these currents is that the most intense geomagnetic variations are associated with substorms (see Section 15.1; Viljanen et al., 2006b; Pulkkinen et al., 2015) and are often focused into limited spatial scales (perhaps ~500 km) on the ground (Ngwira et al., 2015; Pulkkinen et al., 2015). Thus the extreme risk to mid-latitude power grids could be quite localised, but, when a large space weather event produces a series of substorms, there is likely to be a series of impact footprints gradually moving westwards as the rotation of the Earth brings different areas into the nightside where substorms have their main impact. A good example comes from the great storm of March 1989 (see Figure 16.1.2). A substorm over Canada caused the voltage collapse of the Hydro-Québec power grid, followed some 14–18 hours later by two substorms over Western Europe that led to problems with two transformers in the UK.

At low latitudes, the GIC risk comes mainly from the ring current, a torus of electric current in the equatorial magnetosphere (see Section 14.3). Changes in the ring current produce geomagnetic variations that are weaker than those from auroral currents (see Figure 16.1.1) but that are the dominant variation at lower latitudes. Ring current variations can be significant, as shown by damage to a large number of transformers in South Africa in autumn 2003 (Gaunt and Coetzee, 2007; Thomson and Wild, 2010).

A further GIC risk on the dayside of the Earth arises from sudden impulses, the abrupt increases of the geomagnetic field strength that arise when coronal mass ejections compress the magnetosphere. This can produce a spike of GIC in power grids on the dayside of the Earth (Zhang et al., 2015). There is theoretical evidence that this could be particularly significant in regions around the magnetic equator, as the ionospheric current over these regions (the equatorial electrojet) may magnify the effect of a sudden impulse (Carter et al., 2015). This is an important issue for the future as countries near the magnetic equator build up significant electrical infrastructures.

16.2 Disruption of Satellite Navigation

Satellite navigation technology is one of the wonders of the modern world. Technically known as global navigation satellite systems (GNSS), this technology gives accurate measurements of position anywhere on Earth and also the current time to high accuracy. It underpins a wide range of applications, most obviously for all forms of navigation, from people on foot to cars, ships, aircraft and even satellites in low Earth orbit, but also is central to synchronisation of activities across networked infrastructures, including communications, broadcasting, finance and electric power.

A GNSS receiver works by receiving signals simultaneously from several (at least four, often more) satellites in a constellation around 20,000 km altitude. The US Global Positioning System (GPS) is the best known of these constellations, but others include the Galileo system being deployed by the EU as well as Glonass from Russia and Beidou from China. These satellite signals include a coding that is autocorrelated with an internal reference pattern to determine the travel time of the signal from each satellite. The receiver then converts these travel times into distances

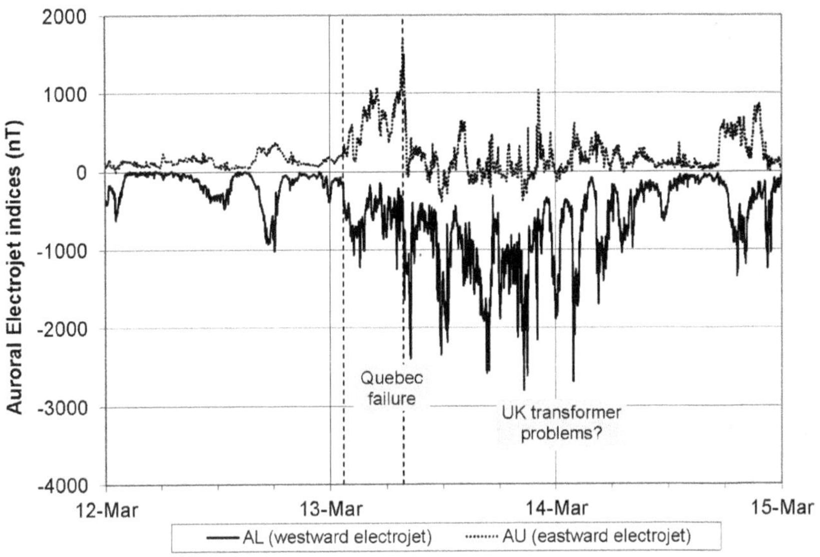

Figure 16.1.2 Auroral electrojet activity during the great geomagnetic storm of March 1989. The auroral electrojet indices are derived from magnetic field measurements at 12 sites around the northern auroral oval (Davis and Sugiura, 1966). The upper trace indicates the maximum positive excursion at any of these sites at each time, whilst the lower indicates the maximum negative excursion. These give an indication of the strength of the eastward and westward electrojets. The series of large dips in the lower trace indicates a series of substorms. The two vertical dashed lines indicate the times of sudden impulses, compressions of the magnetosphere caused by the arrival of coronal mass ejections. The failure of the Hydro-Québec power system (Bolduc, 2002) was caused by the substorm peaking just after the second sudden impulse. Problems on the UK National Grid (noise from, and damage to, transformers, as reported by Smith, 1990, and Erinmez et al., 2002) were associated with both of the large substorms peaking around 2200 on 13 March and 0200 on 14 March. AE index data supplied by the UK Solar System Data Centre. Sudden impulse timings courtesy of the International Service on Rapid Magnetic Variations hosted by Observatori de l'Ebre. Figure by the author.

and, given knowledge of each satellite position (also coded into the signals), calculates the location that best fits this set of distances. Figure 16.2.1 shows an example of location derived using signals from nine satellites.

But to make this work accurately, several corrections must be made. First is to correct for relativistic effects on the clocks on GNSS satellites – GNSS receiver software must correct for the change in clock rate due to satellite motion and lower gravitational field. Second is to correct for the variable rotation rate of the Earth so that locations relative to satellites can be converted into geographic locations. This rate is routinely derived to very high accuracy from observations of pulsars and disseminated via GNSS control centres, so that GNSS positions are accurate to a few metres (or better). Third is to correct for the delay of GNSS signals as they pass through the ionosphere. The plasma in the ionosphere has a group refractive index slightly greater than unity at GNSS signal frequencies (1–2 GHz). The difference is no more than a few parts in 10,000, but this is sufficient to delay the signal by tens of nanoseconds and hence change the position derived by GNSS by many metres.

Whilst all three corrections involve a sophisticated understanding of distinct physical phenomena, the first two can be made accurately. The ionospheric correction is more tricky, since it can vary dramatically with space weather conditions. The delay depends on the total electron content (TEC) along the signal path, i.e. the column density of electrons integrated along that path. In quiet space weather conditions, this follows a regular diurnal pattern increasing during the day as EUV from the Sun creates additional ionisation in the upper atmosphere, and falling during the night as ions and electrons recombine. This pattern also follows a seasonal modulation (due to changing elevation of the Sun) and solar cycle modulation (due to changing EUV brightness of the Sun). The pattern also varies with location on Earth (e.g. due to the effect of the geomagnetic field on plasma transport in the quiet-time ionosphere).

These quiet-time patterns can be modelled fairly well, so simple GNSS systems use those models to estimate the ionospheric corrections (Klobuchar, 1987; Nava et al., 2008). But space weather can cause dramatic changes in the density of the ionosphere, so simple GNSS systems will become inaccurate at such times. In particular, geomagnetic storms can drive both major increases and major decreases in TEC. Major increases occur in the early phase of a storm when solar wind electric fields and thermospheric winds cause uplift of the ionosphere, resulting in reduced recombination rates and hence higher TEC. Major decreases occur in the later phases of a storm, when auroral heating causes injection of molecular species into the thermosphere thereby increasing recombination rates and hence lower TEC, especially at night. It is much more challenging to model these variations, so, in practice, professional GNSS systems use ionospheric corrections derived from near-real-time measurements of TEC. These corrections are now a vital element in accurate navigation for aviation and shipping. Governments and navigation authorities around the world have invested billions of dollars in systems to monitor the

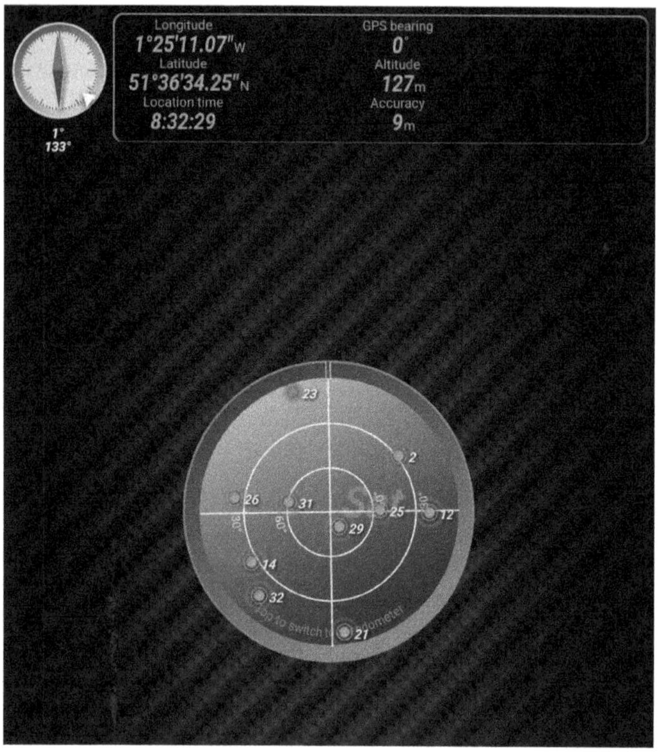

Figure 16.2.1 Screenshot from a GNSS application. This shows a sky view of the satellites visible to the GNSS receiver, zenith at the centre of the circle and horizon at the circumference. The position fix shown at the top uses 9 of the 10 visible satellites; these 9 are marked green. Image taken by the author using an Ulysee Gizmos app running on a Nexus 7 tablet with Android 6.0.1.

ionosphere and distribute corrections. Examples include (1) satellite-based augmentation systems that use satellite links to distribute corrections over wide areas, e.g. WAAS in the US (Loh et al., 1995), EGNOS in Europe (Gauthier et al., 2001) and GAGAN in India (Rao, 2007), and (2) differential GNSS systems that provide local corrections around key facilities, such as ports (Trinity House, 2016). There is strong policy and commercial interest in extending these systems to other countries, e.g. as in current work to deploy satellite-based augmentation systems for Africa (Avanti, 2017).

However, these corrections are at risk during severe geomagnetic storms. These storms can produce rapid spatial as well as temporal variations in TEC, making it very difficult to determine accurate ionospheric corrections, for example as reported in studies of the Halloween storms of 2003 (Jakowski et al., 2008). This possibility is recognised in the design of correction systems through the incorporation of an integrity flag in the data sent to professional GNSS receivers. In a severe event, this flag can be set to advise when GNSS receivers should not be used. In this case GNSS users should switch to alternative systems, e.g. use of inertial navigation systems on aircraft.

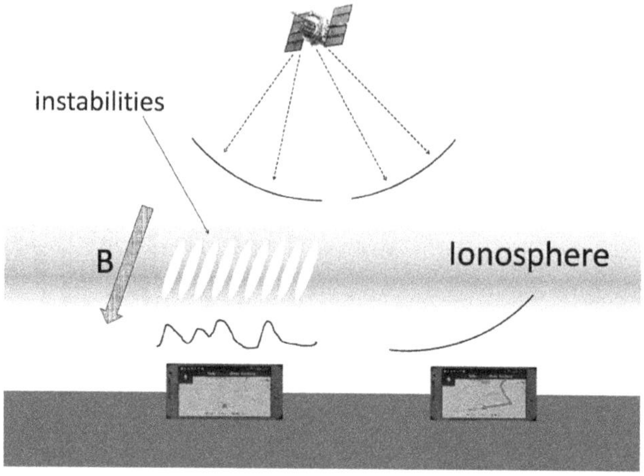

Figure 16.2.2 Schematic showing how instabilities in the ionosphere can disrupt propagation of radio signals. Here the signal from a satellite propagates towards the ionosphere, as shown by the two curved wavefronts above the ionosphere. On the right, the ionosphere is smoothly structured, so the wavefront passes through with little disruption. But on the left, instabilities act as a diffraction screen so that the wavefront is highly modified – leading to rapid fluctuations in amplitude and phase, which are observed as scintillation. Thus the GNSS receiver on the left may be unable to lock on to the satellite signal. Figure by the author.

So far we have considered how space weather affects GNSS by changing TEC. However, space weather has another effect on GNSS, one that is much harder to mitigate, namely the effect of ionospheric instabilities on GNSS signal propagation. As a region of plasma, the ionosphere is prone to instabilities. At high latitudes, these are commonly associated with strong spatial gradients in plasma density, which provide a source of free energy that drives filamentation of the plasma – adjacent regions of high and low density aligned with the magnetic field (see Figure 16.2.2). Similar structures arise at equatorial latitudes around dusk when there is usually an upflow of ionospheric plasma, a flow that naturally breaks down via the Rayleigh–Taylor instability (Ott, 1978). This filamentation of the plasma means that the ionosphere acts as a diffraction screen for the GNSS signals propagating down from the satellites. This diffraction modulates the amplitude and the phase of those signals – an effect that we call scintillation, a radio wave analogue of the twinkling of starlight (itself due to propagation of light through instabilities in the lower atmosphere). Ionospheric scintillation can be a major challenge to reception of GNSS signals; in particular, strong phase scintillation can cause GNSS receivers to lose phase lock on signals. This is a major concern – in a severe space weather event, increased occurrence and strength of ionospheric scintillation, and especially its expansion to mid-latitudes, could cause widespread denial of access to GNSS services (Cannon et al., 2013). In essence, the highly disturbed ionosphere would scramble GNSS signals so they would not be usable.

Given the growing dependence of modern societies on GNSS, this would cause major societal disruption through impacts on road-based transport/logistics (Luntama et al., 2017), possibly also on aviation and shipping, and impacts on systems that use GNSS to synchronise operations of networks (e.g. mobile phone and broadcasting networks in some countries).

Scintillation can be mitigated to some extent by use of high-quality GNSS receivers that are less likely to lose phase lock. Also, there is much research interest in exploiting our scientific understanding of scintillation and the capabilities of modern software radio technologies to develop receivers that can better track signal phase (Kintner et al., 2007; van den IJssel, 2016).

In summary, GNSS is an amazing technology that is bringing great benefits to modern societies, but it is also a technology that is vulnerable to space weather through the need to correct for the ionospheric delay of GNSS signals and through signal disruption by ionospheric scintillation. It is vital that users of GNSS systems have some awareness of space weather (Hapgood, 2017b), in particular that it can on occasion deny access to GNSS services. Thus users should have alternative ways of working in the absence of GNSS services; if they do not, they may have to halt their work and wait out a severe space weather event.

16.3 Radiation Effects on Satellites

Satellites are inevitably exposed to energetic particle radiation – electrons, protons and heavier ions at a wide range of energies from keV to GeV and above. The fluxes of these particles vary considerably with both time and location and lead to a range of effects.

16.3.1 Single-Event Effects

A proton or ion with an energy above 10 MeV can penetrate a significant distance into solid materials and produce ionisation inside those materials by knocking electrons out of their orbits around the nuclei in the material. If that material is an electronic device, such as a computer memory or a digital processor, then the electrons released by that single ion can have a significant effect on the state of the device, disrupting its digital state or even damaging discrete elements within the device. These single-event effects (SEEs) have become a significant issue over the past few decades; the gradual reduction in size of features inside digital devices (now below 100 nm) has increased the sensitivity of these devices to SEEs whilst delivering ever more computer power (Moore, 1965). This increased sensitivity has manifested as a growing range of effects (Cannon et al., 2013; Pisacane, 2016) as well as more frequent occurrence of those effects. Some key examples of SEEs follow:

- *Single-event upsets.* These were the first class of SEEs to be recognised and arise when the electrons released by the passing ion flip the value of a computer bit (i.e. 0 changes to 1, or 1 to 0). Thus the ion corrupts the data or code held in a device, but does no damage. Fortunately, these corruptions can readily be detected and fixed by modern error correction codes. Single-event upsets will only be a significant problem during intense solar energetic particle (SEP) events when the rate of upsets outstrips the rate at which the correction code can fix them. Thus intense events pose a significant challenge for satellite operators; they will often need to take manual action to fix problems caused by single-event upsets, e.g. by rebooting satellite systems.
- *Single-event transients.* These are another class of SEE that does not cause damage: the electrons released by the passing ion create a strong electric field that is misinterpreted as a signal and thereby cause the device to behave in an anomalous fashion. This may require action by satellite operators to review and fix the anomaly.
- *Single-event gate rupture and single-event burnout.* These are classes of SEE that cause actual damage to a digital device. In the case of gate rupture, the passing ion creates a conducting path through a feature on the device, whereas burnout leads to a conducting path through the whole device. Thus gate rupture eliminates the capability provided by one feature on the device, but burnout leads to the catastrophic failure of the device. Both classes of SEE are likely to require significant action by satellite operators, first to understand the scope of the damage and then to explore ways of working around that damage.
- *Single-event latch-up.* This is an intermediate class of SEE in which devices become locked into a particular state. It may require action by satellite operators to restore the device to a normal state, e.g. by switching power off and on or through operation of additional functionality, external to the device, that can enable unlocking. But there may be cases in which it proves impossible to unlock the device and hence operators will need to explore ways of working around loss of the device.

Thus SEEs are an important issue for the design and operation of satellites. Good satellite design will consider how to minimise the impact of SEEs, e.g. by shielding susceptible devices to reduce the flux of energetic ions and by provision of parallel redundancy so that operators can work around problems with any one device. Good practice will recognise that anomaly management is a major and critical part of the work of satellite operators.

16.3.2 Radiation Damage

Protons and ions passing through solid materials can also cause damage to the structure of the material. This is a non-ionising effect that arises from Coulomb interactions between the ions and the nuclei of atoms in material. The electric force exerted by passing ions can displace some of those nuclei from their normal positions. These displacements can degrade performance of materials in critical satellite sub-systems, e.g. reducing the power available from solar arrays or causing a gradual increase in the currents that flow through digital devices. These changes will accumulate over the years in line with the fluence of damaging ions, i.e. the time integral of the flux of those ions.

Radiation damage is therefore a major constraint on satellite lifetime as it can lead to a situation where there is insufficient power to run the satellite or where key subsystems have failed. It is an important consideration in satellite design, e.g. solar arrays are designed with allowance for decay across the planned lifetime of the satellite, so they deliver excess power immediately after launch and just enough power at end of life. Designers must consider how different types of radiations contribute to the decay of power from the arrays: a fairly steady decay due to cosmic rays and occasional abrupt changes due to SEP events. Thus satellite designers need statistical models of how SEP fluence is likely to accumulate over many years. Then they can size solar arrays such that there is only a small probability (maybe 5%) that this fluence will result in insufficient power after the design lifetime of the satellite.

16.3.3 Surface Charging

Satellites are also affected by lower-energy (~keV) particles that form the bulk of hot plasmas in the magnetospheres of Earth and other planets such as Jupiter: the surfaces of satellites in these regions are continually being bombarded by electrons and ions with keV energies.

However, the flux of electrons bombarding the satellite will be much greater than the ion flux so that electrons will be the dominant factor in depositing charge on the surfaces of satellites. This is a consequence of the electron mass being much lower than the mass of the proton or other ions. Electrons and ions in plasmas usually have similar mean energies, so that the mean velocity of electrons will be greater than that of the ions by a factor equal to the square root of the ion-electron mass ratio (~43 for protons to electrons). Given that the density of ions and electrons must be the same to ensure the overall electrical neutrality of the plasma, the flux of electrons will be greater than the flux of ions by a similar factor.

Thus, when a hot plasma fills the space around the satellite, the surface of that satellite will acquire a negative charge due to accumulation of electrons. The potential of the surface, relative to that plasma, will tend to a value in volts close to the mean electron energy in electron-volts, a value which is sufficient to suppress further accumulation of electrons. Thus surface charging in Earth's outer magnetosphere (e.g. in geostationary orbit) can lead to satellite surfaces acquiring potentials of thousands of volts. This can become a problem if different parts of the satellite surfaces acquire different potentials, so that there is a risk of electrical discharges between those different parts. At best, such discharges can create false signals that may lead to anomalous behaviour, and at worst, they can cause catastrophic damage. Satellite manufacturers seek to mitigate this problem by ensuring that all exposed satellite surfaces are electrically interconnected, e.g. by physically making such connections and by applying a thin conductive coating to dielectric materials. However, this mitigation is not always perfect, e.g. the performance of some dielectric surfaces may be degraded by application of conductive coating, so there needs to be a trade-off between that performance and the need to reduce charging risks. Thus surface charging remains a significant challenge that needs to be taken into account by satellite operators. Good design can greatly reduce the risk, but operations teams need to be aware of the risk and have the tools to recognise when it is occurring and be ready to resolve any anomalies that arise as a result of surface charging.

One natural factor that can mitigate surface charging is sunlight, more specifically solar ultraviolet. This will stimulate photoelectron emission from the satellite surface, causing that surface to acquire a positive charge. Photoemission will dominate over electron bombardment if, as is often the case, the flux of UV photons exceeds the flux of plasma electrons. In this case, the potential of the satellite surfaces will tend to a positive value of a few volts, approximately equal to the work function, the threshold for photoemission, of the surface materials of the satellite. This potential is sufficient to suppress further photoemission. Thus photoemission greatly limits the risk of surface charging. That risk is confined to regions where satellites are in darkness or where the electron fluxes exceed the UV photon flux from the Sun.

16.3.4 Internal (Deep Dielectric) Charging

Satellites are also vulnerable to a second and very different charging mechanism. Electrons at MeV energies can penetrate deep inside satellite systems and deposit charge inside dielectric materials such as circuit boards and insulation on wires. This charge can accumulate if high fluxes of MeV electrons continue for several days (as often occurs in geostationary orbit) and create strong electric fields inside the dielectric. Eventually, this will lead to a breakdown of the dielectric and to electric discharges within the dielectric (a vivid benchtop demonstration of such breakdown can be viewed on YouTube at https://youtu.be/eCz7BL74D4Y). Such discharges can create false signals that cause satellite anomalies and, in some cases, can cause damage to satellite subsystems. This potential for damage has sometimes led to MeV electrons being nicknamed 'killer electrons', a piece of hyperbole that gets attention in the wider media.

Deep dielectric charging is a difficult problem for satellite design and operations. Unlike surface charging, it cannot easily be mitigated by providing conductive paths to spread the accumulated electric charge. Shielding around sensitive systems can help to reduce the electron fluxes, but will always require a trade-off against the mass budget of the satellite. It is therefore important that satellite operators are aware of space weather conditions that lead to a significant risk of deep dielectric charging, so they can be ready to deal with anomalies that arise during these conditions. The occurrence of MeV electron fluxes in near-Earth space is episodic, typically showing a sharp rise (hours, sometimes minutes) and a slow decay over many weeks. Thus awareness of adverse conditions can most obviously be provided by near-real-time monitoring of electron fluxes, e.g. as done by the GOES satellites in geostationary orbit. But forecasting is also possible, based on our knowledge of solar wind conditions that are likely to lead to high MeV electron fluxes, in particular high solar wind speeds. High speeds may arise during and following the passage of a coronal mass ejection, but they can also arise when the Earth is enveloped by a high-speed solar wind stream, originating from a coronal hole on the Sun (see Section 14.2). The latter case commonly occurs during the declining phase of the solar cycle, as coronal holes are then more likely to extend to low-latitude regions of the Sun's surface. There are now several examples of forecasting systems built upon a detailed understanding of the science outlined above (Horne et al., 2013; Boynton et al., 2015).

Satellite operators can also benefit from understanding of how MeV electron fluxes (and hence charging risk) vary with different orbit locations within Earth's magnetosphere. These fluxes are ordered by the McIlwain L parameter (McIlwain, 1961). This can roughly be considered as the geocentric distance, measured in Earth radii, at which an energetic particle crosses the magnetic equator when guided by the geomagnetic field (the full definition is physically more sophisticated, see the reference above). MeV electron fluxes typically peak at an L-value of 4 and fall off towards higher and lower values. Thus navigation satellites such as GPS and Galileo are particularly exposed to these electrons as their orbits include substantial portions around L-value of 4. Satellites in geostationary orbits, around L-value of 6–7, have lower, but still significant, exposure. A key feature is that L-value varies around geostationary orbit (see Figure 16.3.1), reaching a primary maximum over the western Atlantic, so satellites in that region will be least exposed. Exposure increases (because L-value decreases) to both west and east, so that satellites over the eastern Pacific and over Europe and Africa will be more exposed. Statistical studies of available MeV electron data by Meredith et al. (2015) have confirmed that there is a significant westward increase, with electron fluxes around 135°W some 2.5 times higher than at 75°W. This highlights the importance of satellite location as a factor in determining the risks that killer electrons pose for satellite operations.

Deep dielectric charging is one of the most important space weather impacts on satellites as it is hard to mitigate by engineering methods. We are gradually developing good means to forecast when it is likely to occur, so that operators can be ready to respond to problems.

16.4 Atmospheric Radiation Effects

Radiation from space makes a significant contribution to the radiation environment in Earth's atmosphere. Even in quiet space weather conditions, it provides 8% of the

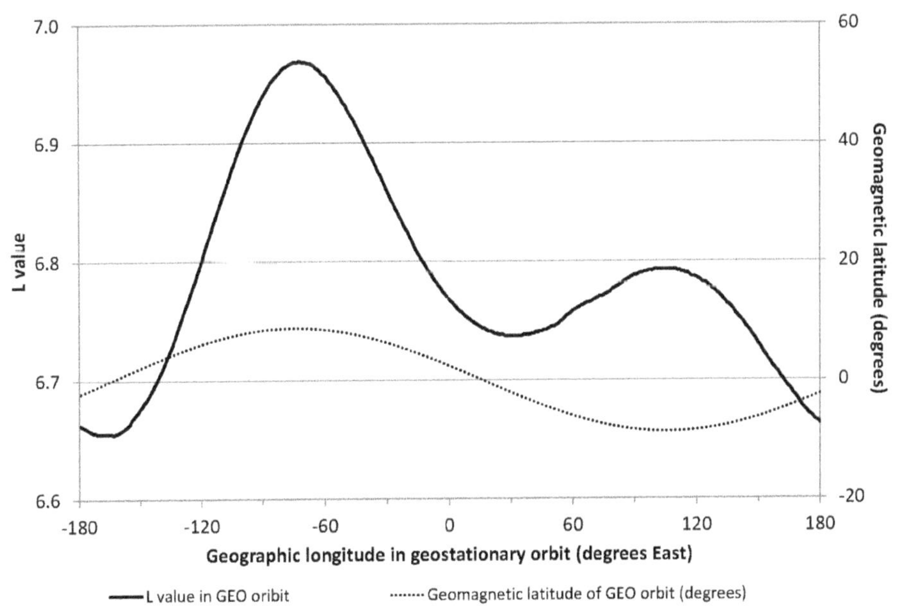

Figure 16.3.1 The variation of L-value around geostationary orbit. The modulation is mainly driven by the variation of geomagnetic latitude around the orbit, with L-value peaking when geomagnetic latitude reaches its greatest values north and south. The differences in the two peaks are due to the offset of the geomagnetic dipole from the centre of the Earth (BIRA-IASB, 2016). The shift of the dipole towards East Asia reduces the maximum near that region and increases the maximum over the western Atlantic. Figure by the author. L-values calculated using the MAGLIB Fortran library (Kosik, 1994).

radiation observed at the surface and is the dominant source above 3 km altitude. Modern aircraft cruising at 12 km altitude fly in a significant radiation environment, so aviation safety must take account of the risk that this poses to humans and to aircraft systems. During a severe solar energetic particle event, the contribution of radiation from space can be greatly enhanced. The size of this enhancement depends markedly on the energy spectrum of the SEP event – with atmospheric radiation enhanced only if the event has a 'hard' spectrum with significant fluxes above 400 MeV. At least 70 such events have been observed by ground-based radiation detectors since observations started in the 1940s.

These 'ground-level enhancements' (GLEs) can pose a significant risk to aviation and perhaps also to ground-based systems. The strongest observed GLE (back in 1956) produced a 50-fold increase in the ground-level radiation environment over a few hours (Gold and Palmer, 1956; Marsden et al., 1956) and is estimated to have produced a 300-fold increase at aircraft altitudes (Dyer et al., 2007). A similar event today could deliver a significant radiation dose (2 millisieverts (mSv)) to any aircrew and passengers in flight during the event (Dyer et al., 2007). Other studies suggest that even larger events are possible and that we should plan for a reasonable worst-case exposure of 20 mSv during an extreme space weather event (Cannon et al., 2013; Hapgood et al., 2016; Dyer et al., 2018).

However, it is important to appreciate that these doses do not pose any immediate risk to human health. They fall well below the 1000 mSv level at which short-term radiation sickness is likely. But they are significant in terms of the long-term cancer risks that arise from accumulated exposure to radiation. Thus the response to a severe radiation event is likely to comprise an assessment of the radiation dose received by people in flight during the event and provision of advice on how they should manage their future radiation exposure, especially if their work makes them more exposed to radiation. Examples of such work include, very obviously, aircrew, but also workers in nuclear industry, underground facilities (via exposure to radon gas) and medical diagnostics.

At the present time it is difficult for the aviation industry to take action to mitigate radiation exposure due to a GLE, because we lack good forecasts of GLE onset. Once an event has been detected, it would be possible to protect people by keeping their aircraft on the ground until the event has subsided. But it is much more difficult to take action in respect of aircraft already in flight. In principle, these could gain protection by descending to a lower altitude (or even landing) so that they are better shielded from radiation by the atmosphere. But a change of flight plan would bring many other problems, including decreased aircraft range and collision risks. The risks associated with a change of flight plan, especially if aircraft communications are also at risk from other space weather effects (e.g. disruption of satellite signals by ionospheric scintillation as discussed in Section 16.2), generally outweigh the risks from increased radiation doses.

But radiation effects on aircraft are by no means limited to human radiation dose. Another, probably more serious issue is disruption of electronic systems via single-event effects, as we discussed for satellites. As aircraft control has become more dependent on electronics, there is considerable concern that atmospheric radiation can induce SEEs that disrupt aircraft control (Cannon et al., 2013). The risk is illustrated by a 2008 incident in which anomalous operation of a flight control system caused serious injuries to 12 passengers and aircrew. Whilst the cause of the anomaly was never fully resolved, the accident report (ATSB, 2011) emphasised that aircraft control systems must be able to deal with a wide range of anomalies, not least those caused by SEEs. There is a clear need to

1 design aircraft control systems so that they are resilient to SEEs, in particular to understand how SEE impacts will evolve with the continuing reduction in feature sizes in the digital devices at the heart of those systems
2 verify those design by tests of actual control systems in simulated radiation environments (Hambling, 2014; STFC, 2017)

SEEs can also occur in digital systems on the ground. This is clear from theoretical considerations: the radiation that causes SEEs on aircraft is primarily neutrons generated by collisions between space radiation and atmospheric species. Those neutrons also reach the surface of the Earth, and some will inevitably enter and disrupt digital systems on the ground. However, the scale of SEE impacts on ground systems is poorly understood. They are likely responsible for some fraction of cases where computers malfunction and require rebooting to clear the problem, but the fraction is unknown. Good evidence on specific cases is hard to come by. One well-known example occurred in the Belgian town of Schaerbeek during a 2003 local election, when a voting machine recorded an extra 4096 votes. This was attributed to an SEE changing the thirteenth bit in a counter (PourEVA, 2003). Another example, from the 1990s, was a series of premature failures of solid state power control systems on trains. These were attributed to single-event burnout and fixed by redesign of the power control system (Normand, 1997). There is also concern that the growing use of electronics to control cars creates a risk that SEEs could cause cars to malfunction (Wood and Caustin, 2006). This is a significant concern today with the advent of driverless cars, where reliable operation of software is obviously critical to safety – as has sadly been demonstrated when test vehicles have crashed (Stilgoe, 2017).

Thus it is clear that the atmospheric radiation environment poses risks to both human health and technological systems. The risk to human health is probably modest and can be linked with wider management of long-term cancer risks for people whose homes or work places expose them to higher than normal levels of radiation. The risk to our technological systems is probably greater since SEEs have the potential to disrupt the operation of systems that are critical to human safety. Our biggest gap in understanding these risks is our knowledge of what is the reasonable worst-case radiation environment that we should plan for. The 1956 event already mentioned (Gold and Palmer, 1956; Marsden et al., 1956) provides a clear baseline on which to build, but there is evidence that more severe events have happened in the past, e.g. radiocarbon records show that there was a severe global atmospheric radiation event around 774/775 AD (Miyake et al., 2012; Usoskin et al., 2013). Thus there is wide scientific interest in establishing what is the worst-case atmospheric radiation event, and doing so over a range of timescales from the usual 1-in-100 year case, as in most risk management schemes, to a 1-in-10,000 year case, as required for the operation of critical technologies such as nuclear power (HSE, 1992; Dyer et al., 2018).

16.5 Impact on Technological Society

The natural phenomena underlying space weather (e.g. the solar wind, geomagnetic activity and solar energetic particle events) have existed since the Earth and the Sun were born 5 billion years ago. Furthermore, prior to the nineteenth century, human civilisation had developed over many millennia without any adverse impacts from the space weather phenomena discussed above. It is only in the nineteenth century that we began to see adverse impacts and then in the late twentieth century that these impacts have expanded to become a major societal risk.

This highlights the critical role that technology plays in creating the risk from space weather. Unlike most other natural hazards, such as flooding, bad weather and disease, space weather has no direct impact on humans in their everyday lives on Earth. Its impact is instead mediated by our increasing dependence on technology. The early impacts of space weather (e.g. on the electric telegraph [Barlow, 1849], on telephones [Preece, 1894] and on HF radio communications [Anderson, 1928]) affected systems that were then used only occasionally in everyday life. Space weather could sometimes impact strategically important communications and surveillance systems, such as those used to support military activities (Knipp et al., 2016), thereby gaining governmental attention. But its wider societal impact was limited until the latter part of the twentieth century.

This situation has changed dramatically since the 1960s as wider society has become critically dependent on a number of key infrastructures that are vulnerable to the adverse impacts of space weather:

1 *use of power grids for the long distance transmission of electricity*: This enabled a gradual phase-out of use of coal as a direct local power source in homes, offices and factories (as this author experienced in his childhood). Modern societies now rely on the continuous supply of electricity, and any significant interruption of that supply will quickly become a major problem for the people affected.
2 *use of civil aviation to quickly transport people and goods over long distances*: Aviation is the main means of conveying passengers over long distances in most parts of the world. Aviation also plays a key role in the supply chains for some goods, e.g. the shipping of fresh flowers from Kenya to Europe.
3 *ability to communicate instantly almost anywhere on Earth using low-cost telephone and internet services, whether for work or personal purposes*: The use of paper-based mail services to exchange information, with exchanges taking days or even weeks, now seems quaint. Yet these services were a mainstay of communications just a few decades ago.

Our vulnerability appears to be growing as technological innovation enhances the way that many infrastructures are run, and sometimes also opens up new vulnerabilities. For example, road transport and logistics in many countries are now heavily dependent on use of GNSS to allow drivers to reach their destinations quickly, without the need for prior personal knowledge of the route. Thus severe space weather could disrupt logistics chains leading to wide societal impact. Other infrastructures where innovation may create space weather vulnerabilities include future rail systems (GIC; GNSS impacts on control systems), wider energy generation and distribution (SEE impacts on control systems) and driverless cars (GNSS and SEE impacts on sensors and control systems).

It is important to recognise that technological innovation has greatly improved the standard of living for billions of people around the world but that this innovation can sometimes create new vulnerabilities to all kinds of natural hazards. A prime example from the wider world was the disruption of civil aviation in Northern Europe as a result of ash clouds from the 2010 eruption of the Eyjafjallajökull volcano in Iceland. This event was a wake-up call; it showed policy makers that there were natural hazards that could cause major societal disruption, and that were well known to science but not to policy makers. It triggered policy interest in a wider assessment of natural hazards that could cause major societal disruption at the 1-in-100 year-level. This wider assessment has been one of the reasons why space weather has gained the attention of policy makers in recent years. There is much evidence that severe space weather at the 1-in-100 year-level could cause major societal disruption. This follows in particular from studies of the 1859 Carrington event (Carrington, 1859; Hodgson, 1859;

Clauer and Siscoe, 2006) but is reinforced by studies of more recent events. These include the reports of strong geoelectric fields (Hapgood, 2019) and GIC-induced fires and faults in communications systems (*New York Times*, 1921a, 1921b; Karsberg et al., 1959) during the geomagnetic storm of May 1921. They also include the July 2012 observation of a Carrington-class CME (Russell et al., 2013). Fortunately this missed our planet, but simulations indicate that a hit by this event would have caused severe societal disruption (Baker et al., 2013). The weight of this evidence has demonstrated to policy makers that space weather is a risk that they need to take seriously (e.g. BIS, 2015; OSTP, 2015). There are still many uncertainties about the scale and probability of extreme events, so further research is clearly needed (and scientists are encouraged to pursue such research). But that research should proceed in synergy with wider policy work to mitigate space weather risks.

It is important that policies to mitigate space weather be developed as part of an integrated framework for managing all natural hazards. Then it is possible to identify where generic mitigation measures can help to deal with space weather impacts, and also where measures specific to space weather are needed. A prime example of generic measures is the procedures and capabilities needed to restore electric power, irrespective of whether the power loss was caused by space weather, normal tropospheric weather or some other cause. In contrast, an example of specific measures is the support of space weather forecasting and of the observations needed to derive forecasts.

It is now becoming clear that this policy development needs to be underpinned by a good understanding of the socioeconomic impacts of space weather and of how these are likely to change in response to measures to mitigate space weather. Over the past 10–15 years, there have been a number of reports suggesting the potential for enormous economic impacts, e.g. the much-cited summary of talks given at a 2008 US National Research Council workshop includes claims that extensive power outages following damage to power grids during a severe event could result in costs of trillions of dollars (Space Studies Board, 2009). In contrast, studies by the US power industry suggest that such an event would cause short outages and little damage, leading to costs of only a few billion dollars (NERC, 2012).

These total value-at-risk figures are primarily of use to the insurance industry, but of limited use to policy makers. Insurers need to understand the value-at-risk from space weather, so that they can cost that risk and thereby decide whether to offer insurance for the risk. Policy makers need to understand how the value-at-risk is reduced following investment in mitigation. In particular, does the reduction in that value justify the investment? (As a rule of thumb, the investment is justified if the reduction is at least three times the investment.) A number of recent and current studies are now looking at this problem. A study for ESA (Luntama et al., 2017) provided evidence that investment in space weather instruments at the L1 and L5 Lagrange points (for monitoring solar activity, solar wind disturbances and solar energetic particles) could provide substantial cost savings for a number of vulnerable infrastructures including power grids, aviation, transport logistics and resource exploitation. These savings could more than justify investment in the missions. Similar results were obtained in a study funded by the UK Space Agency (Eastwood et al., 2018) and in a recent study by Oughton et al. (2018).

However, it is important to note that these socioeconomic studies are very much the first steps in the field. It is important to encourage further collaboration between space weather scientists and economists in order to deepen our understanding of the economic value of specific investments, e.g. in the monitoring and forecasting of space weather conditions. Experience suggests that this emphasis on economic value will be important for the development of services to mitigate space weather. Policy makers see space weather very much as an issue that needs to be justified in economic terms and not as a strategic issue of political value (unlike space survey and tracking (SST) activities, where there is strong political interest in maintaining situational awareness of space objects). Thus, at present, it is often harder to justify investment in space weather than in SST, but conversely, it is much easier to justify and promote international collaboration on space weather. These distinctions may be important in future as both space weather and SST activities develop in parallel.

In summary, space weather is increasingly recognised as a significant source of risk for technological societies, and a risk whose good management needs engagement from policy makers. Policy development needs to be underpinned not just by good scientific understanding of the risk but also by a good socioeconomic assessment of the costs that could arise from space weather and of how these can be reduced by taking measures to mitigate space weather risks.

References

Anderson, C. N. (1928). Correlation of long wave transatlantic radio transmission with other factors affected by solar activity. Proc. Inst. Radio Eng. 16, 297–347. doi: 10.1109/JRPROC.1928.221400

ATSB (Australian Transport Safety Bureau) (2011). In-flight upset, 154 km west of Learmonth, Western Australia, 7 October 2008, VH-QPA, Airbus A330-303. Report AO-2008–070.

Avanti (2017). SBAS-AFRICA. http://sbas-africa.avantiplc.com/.

Baker, D. N., et al. (2013). A major solar eruptive event in July 2012: Defining extreme space weather scenarios. Space Weather 11, 585–91. doi: 10.1002/swe.20097.

Barlow, W. H. (1849). On the spontaneous electrical currents observed in wires of the electric telegraph. Philos. Trans. R. Soc. London, **139**, 61–72. doi: 10.1098/rstl.1849.0006

BIRA-IASB (2016). The eccentric dipole model. www.spenvis.oma.be/help/background/magfield/cd.html#ED.

BIS (2015). Space weather preparedness strategy. www.gov.uk/government/publications/space-weather-preparedness-strategy.

Bolduc, L. (2002). GIC observations and studies in the Hydro-Québec power system. J. Atmos. Sol. Terr. Phys. 64, 1793–1802. doi: 10.1016/S1364-6826(02)00128-1.

Boynton, R. J., et al. (2015). Online NARMAX model for electron fluxes at GEO. Ann. Geophys. 33, 405–11. doi: 10.5194/angeo-33-405-2015.

Cannon, P., et al. (2013). *Extreme Space Weather: Impacts on Engineered Systems and Infrastructure*. UK Royal Academy of Engineering, London.

Carrington, R. C. (1859). Description of a singular appearance seen in the Sun on September 1. Mon. Not. Roy. Astron. Soc. 20, 13–15. doi: 10.1093/mnras/20.1.13

Carter, B. A., et al. (2015). Interplanetary shocks and the resulting geomagnetically induced currents at the equator. Geophys. Res. Lett. 42, 6554–9. doi: 10.1002/2015GL065060.

Clauer C. R. and G. Siscoe. (2006). The great historical geomagnetic storm of 1859: A modern look. Adv. Space Res. 38, 115–388.

Davis, T. N. and M. Sugiura (1966). Auroral electrojet activity index AE and its universal time variations. J. Geophys. Res. 71(3), 785–801, doi: 10.1029/JZ071i003p00785.

Dyer, C. S., et al. (2007). Solar particle events in the QinetiQ atmospheric radiation model. IEEE Trans. Nucl. Sci. 54, 1071–5. doi: 10/1109/TNS.2007.893537.

Dyer, C., et al. (2018). Extreme atmospheric radiation environments & single event effects. IEEE Trans. Nucl. Sci. doi: 10.1109/TNS.2017.2761258.

Eastwood, J. P., et al. (2018). Quantifying the economic value of space weather forecasting for power grids: An exploratory study. Space Weather 16, 2052–67. doi: 10.1029/2018SW002003.

Erinmez, I. A., et al. (2002). Management of the geomagnetically induced current risks on the national grid company's electric power transmission system. J. Atmos. Sol. Terr. Phys. 64, 743–56. doi: 10.1016/S1364-6826(02)00036-6.

Eroshenko, E. A., et al. (2010). Effects of strong geomagnetic storms on Northern railways in Russia. Adv. Space Res. 46, 1102–10. doi: 10.1016/j.asr.2010.05.017.

Gaunt, C. T. (2014). Reducing uncertainty – responses for electricity utilities to severe solar storms. J. Space Weather Space Clim. 4, A01. doi: 10.1051/swsc/2013058

Gaunt, C. T. and Coetzee, G. (2007). Transformer failures in regions incorrectly considered to have low GIC-risk, presented at the IEEE Powertech Conference, Lausanne, Switzerland, July. doi: 10.1109/PCT.2007.4538419.

Gauthier, L., et al. (2001). EGNOS: The first step in Europe's contribution to the global navigation satellite system. ESA Bull. 105, 35–42. http://esamultimedia.esa.int/multimedia/publications/ESA-Bulletin-105/.

Gold, T. and Palmer, D. R. (1956). The solar outburst, 23 February 1956 – Observations by the Royal Greenwich Observatory. J. Atmos. Terr. Phys. 8, 287–90.

Guillon, S., Toner, P., Gibson, L. and Boteler, D. (2016). A colorful blackout: The havoc caused by auroral electrojet generated magnetic field variations in 1989. IEEE Power Energy Mag. 14(6), 59–71. doi: 10.1109/MPE.2016.259176.

Hambling, D. (2014). Burnout. New Scientist 223, 42–5. doi: 10.1016/S0262-4079(14)61863-7.

Hapgood, M. (2017a). *Space Weather*. IOP Publishing, Bristol. doi: 10.1088/978-0-7503-1372-8.

Hapgood, M. (2017b). Satellite navigation – Amazing technology but insidious risk: Why everyone needs to understand space weather. Space Weather 15, 545–8. doi: 10.1002/2017SW001638.

Hapgood, M. (2019). The Great Storm of May 1921: an Exemplar of a Dangerous Space Weather Event. Space Weather 17. https://doi.org/10.1029/2019SW002195.

Hapgood, M., et al. (2016). Summary of space weather worst-case environments. Revised edition. RAL Technical Report RAL-TR-2016-06. http://purl.org/net/epubs/work/25015281.

Hodgson, R. (1859). On a curious appearance seen in the Sun. Mon. Not. Roy. Astron. Soc. 20, 15–16. doi: 10.1093/mnras/20.1.15.

Horne, R. B., et al. (2013). Space weather impacts on satellites and forecasting the Earth's electron radiation belts with SPACECAST. Space Weather 11, 169–86. doi: 10.1002/swe.20023.

HSE (Health and Safety Executive) (1992). The tolerability of risk from nuclear power stations. www.onr.org.uk/documents/tolerability.pdf.

Jakowski, N., et al. (2008). Large-scale ionospheric gradients over Europe observed in October 2003, J. Atmos. Sol. Terr. Phys. 70, 1894–1903. doi: 10.1016/j.jastp.2008.03.020.

Karsberg, A., et al. (1959). *The Influences of Earth Magnetic Currents on Telecommunication Lines*. English edn. Televerket, Stockholm.

Kintner, P. M., B. M. Ledvina and E. R. de Paula (2007). GPS and ionospheric scintillations. Space Weather 5, S09003. doi: 10.1029/2006SW000260.

Klobuchar, J. A. (1987). Ionospheric time-delay algorithm for single-frequency GPS users. IEEE Trans. Aerosp. Electron. Syst. AES-23, 325–31. doi: 10.1109/TAES.1987.310829

Knipp, D. J., et al. (2016). The May 1967 great storm and radio disruption event: Extreme space weather and extraordinary responses. Space Weather 14, 614–33. doi: 10.1002/2016SW001423.

Kosik, J. C. (1994). *Maglib User's Guide*. Version 2. CNES, Toulouse.

Krausmann, E., et al. (2015). Space weather and rail: Findings and outlook. EU Joint Research Centre report 98155. doi: 10.2788/211456

Liu, L., et al. (2016). Analysis of the monitoring data of geomagnetic storm interference in the electrification system of a high-speed railway. Space Weather 14, 754–63. doi: 10.1002/2016SW001411.

Loh, R., et al. (1995). The U.S. Wide-Area Augmentation System (WAAS). Navigation 42, 435–65. doi: 10.1002/j.2161-4296.1995.tb01900.x

Luntama J., et al. 2017. Report on the ESA Space-Weather Socio-Economic Study. www.ukssdc.ac.uk/meetings/L5InTandemWithL1/talks/session01/04_SSA%20SWE%20CBA%20Study.pptx.

Marsden, P.L., et al. (1956). Variation of cosmic-ray nucleon intensity during the disturbance of 23 February 1956. J. Atmos. Terr. Phys. 8, 278–81.

Maunder, E. W. (1900). *The Royal Observatory Greenwich: A Glance at Its History and Work*. Reprinted by Cambridge University Press, Cambridge, 2013.

McIlwain, C. E. (1961). Coordinates for mapping the distribution of magnetically trapped particles. J. Geophys. Res. 66, 3681–91. doi: 10.1029/JZ066i011p03681.

McNish, A. G. (1940). The magnetic storm of March 24, 1940. Terr. Magn. Atmos. Electr. 45, 359–64. doi: 10.1029/TE045i003p00359.

Meredith, N. P., R. B. Horne, J. D. Isles and J. V. Rodriguez (2015), Extreme relativistic electron fluxes at geosynchronous orbit: Analysis of GOES E >2 MeV electrons. Space Weather 13. doi: 10.1002/2014SW001143.

Miyake, F., et al. (2012). A signature of cosmic-ray increase in AD 774–775 from tree rings in Japan. Nature 486, 240–42. doi: 10.1038/nature11123.

Moore, G. E. (1965). Cramming more components onto integrated circuits. Electronics 38, 114–17.

Nava, B., P. Coisson and S. M. Radicella (2008). A new version of the NeQuick ionosphere electron density model. J. Atmos. Sol. Terr. Phys. 70(15), 1856–62. doi: 10.1016/j.jastp.2008.01.015.

NERC (North American Electrical Reliability Corporation) (2012). GMD Task Force Report: Effects of GMD on the bulk power system. www.nerc.com/pa/Stand/Geomagnetic%20Disturbance%20Resources%20DL/2012_GMD_Report_112012.pdf.

New York Times (1921a). Sunspot credited with rail tie-up, 16 May, p. 2. http://query.nytimes.com/mem/archive-free/pdf?res=9E05E7D61E3FEE3ABC4E52DFB366838A639EDE.

New York Times (1921b). Cables damaged by sunspot aurora. 17 May. http://query.nytimes.com/mem/archive-free/pdf?res=9407E2D61E3FEE3ABC4F52DFB366838A639EDE.

Ngwira, C. M., et al. (2015). Characteristics of extreme geoelectric fields and their possible causes: Localized peak enhancements. Geophys. Res. Lett. 42, 6916–21. doi: 10.1002/2015GL065061.

Normand, E. (1997). Neutron-induced single event burnout in high voltage electronics. IEEE Trans. Nucl. Sci. 44, 2358–66. doi: 10.1109/23.659062.

OSTP (2015). National Space Weather Strategy. https://obamawhitehouse.archives.gov/sites/default/files/microsites/ostp/final_nationalspaceweatherstrategy_20151028.pdf.

Ott, E. (1978). Theory of Rayleigh–Taylor bubbles in the equatorial ionosphere. J. Geophys. Res. 83, 2066–70. doi: 10.1029/JA083iA05p02066.

Oughton, E., et al. (2018). A risk assessment framework for the socio-economic impacts of electricity transmission infrastructure failure due to space weather: An application to the UK. Risk Analy.. doi: 10.1111/risa.13229.

Pisacane, V. L. (2016). *The Space Environment and Its Effects on Space Systems*. 2nd edn. AIAA, Reston, VA. doi: 10.2514/4.103537.

PourEVA (2003). Le Ministre DEWAEL reconnait la faillibilité du vote électronique grâce à un rayon cosmique complice! www.poureva.be/article.php3?id_article=36.

Preece, W. H. (1894). Earth currents. Nature 49, 554. doi: 10.1038/049554b0.

Pulkkinen, A., et al. (2015). Regional-scale high-latitude extreme geoelectric fields pertaining to geomagnetically induced currents. Earth Planets Space 67, 93. doi: 10.1186/s40623-015-0255-6.

Rao, K. N. (2007). GAGAN – The Indian satellite based augmentation system. Ind. J. Radio Space Phys. 36, 293–302. http://nopr.niscair.res.in/handle/123456789/4707.

Russell, C. T., et al. (2013). The very unusual interplanetary coronal mass ejection of 2012 July 23: A blast wave mediated by solar energetic particles. Astrophys. J. 770, 38. doi: 10.1088/0004-637X/770/1/38.

Smith, P. M. (1990). Effects of geomagnetic disturbances on the national grid system. Presented at the Universities Power Engineering Conference (UPEC).

Space Studies Board and National Research Council (2009). *Severe Space Weather Events – Understanding Societal and Economic Impacts: A Workshop Report*. National Academies Press, Washington, DC.

STFC (2017). ChipIR: Instrument for rapid testing of effects of high energy neutrons. www.isis.stfc.ac.uk/instruments/chipir/chipir8471.html.

Stilgoe, J. (2017). Tesla crash report blames human error – this is a missed opportunity. The Guardian, 21 January. www.theguardian.com/science/political-science/2017/jan/21/tesla-crash-report-blames-human-error-this-is-a-missed-opportunity.

Thomson, A. and J. Wild (2010). When the lights go out ... Astron. Geophys. 51, 5.23–4. doi: 10.1111/j.1468-4004.2010.51523.x.

Trinity House (2016). Satellite navigation ground based augmentations. www.trinityhouse.co.uk/dgps.

Usoskin, I. G., et al. (2013). The AD775 cosmic event revisited: The Sun is to blame. Astronomy and Astrophysics 552, L3.

van den IJssel, J., Forte, B. & Montenbruck, O. (2016). Impact of Swarm GPS receiver updates on POD performance. Earth Planet and Space 68, 85. doi: 10.1186/s40623-016-0459-4.

Viljanen, A., et al. (2006a). Recordings of geomagnetically induced currents and a nowcasting service of the Finnish natural gas pipeline system. Space Weather 4, S10004. doi: 10.1029/2006SW000234.

Viljanen, A., et al. (2006b). Relation between substorm characteristics and rapid temporal variations of the ground magnetic field. Ann Geophys 24, 725–33. doi: 10.5194/angeo-24-725-2006.

Wik, M., et al. (2009). Space weather events in July 1982 and October 2003 and the effects of geomagnetically induced currents on Swedish technical systems. Ann. Geophys. 27, 1775–87. doi: 10.5194/angeo-27-1775-2009.

Wood, J. and E. Caustin (2006). Timely testing avoids cosmic ray damage to critical auto electronics. www.eetimes.com/document.asp?doc_id=1272752.

Wrubel, J. N. (1992). Monitoring program protects transformers from geomagnetic effects. IEEE Comput. Appl. Power 5, 10–14. doi: 10.1109/67.111465.

Zhang, J. J., et al. (2015). GIC due to storm sudden commencement in low-latitude high-voltage power network in China: Observation and simulation. Space Weather 13, 643–55. doi: 10.1002/2015SW001263.

PART V

Magnetic Fields beyond the Earth and beyond Today

17
Magnetic Field Evolution in Terrestrial Bodies from Planetesimals to Exoplanets

Doris Breuer

17.1 Introduction

Magnetic fields generated in the interior of terrestrial bodies are common in our solar system – either they are present today or have been active in the early evolution of a body. Of the terrestrial planets and satellites Earth, Mercury and Ganymede are known to have largely dipolar internally generated magnetic fields, while for instance Venus and Mars apparently lack such fields. For Earth, Moon, Mars and Mercury, there is also evidence for an earlier self-generated field in the form of remanently magnetised crust that formed up to billions of years ago. Magnetised rock can provide information on the magnetic field present at the time when the material cooled below its Curie temperature. Similarly, magnetised meteorites suggest that planetesimals, i.e. small planetary bodies that are the building blocks of the terrestrial planets, may have had their own self-generated fields (e.g. Weiss et al., 2008). For Venus, we have no information on the existence of a previous magnetic field, both because of a lack of data and missions and because the high surface temperature is near or above the Curie temperature for most minerals thereby preventing magnetisation of the crust.

Generation of a magnetic field requires an electrically conducting and convecting layer within a planet. In the terrestrial planets and the satellites this region is believed to be the fluid iron-rich core at the centre – although for icy satellites a weak magnetic field can be also generated in the salty ocean layers. Convection can result from thermal buoyancy in the case where there is a of a sufficiently large temperature difference between the core and the rocky mantle surrounding the core. Alternatively, solidification of the iron-rich material provides a buoyancy flux that may drive the dynamo. The buoyancy in this case derives from a difference in composition between the solid and the fluid material. In a eutectic system, which is typical for the assumed core compositions, light alloying elements such as sulphur and oxygen tend to be expelled from the solidifying core and concentrate in the fluid core. For Earth, freezing starts at the core centre and continues with an outward solidification by forming a solid inner core. In recent years, it has been recognised that for planetary bodies smaller than Earth, the solidification can start at the core–mantle boundary and initiate so-called iron snow – a process that might generate the dynamos in Ganymede (Hauck et al., 2006) and Mercury (Chen et al., 2008) but could also be relevant for the Moon (Laneuville et al., 2014), Mars (Stewart et al., 2007) and planetesimals (Scheinberg et al., 2016). Interestingly, crystallisation starting at the core–mantle boundary has also been suggested for massive exoplanets larger than 2.5 Earth masses (Gaidos et al., 2010). The crystallisation scenario may further differ from the Earth case if the content of the light alloying element in the core is greater than that at the eutectic.

Regardless of whether thermal or chemical buoyancy drive core convection, the existence of a dynamo and the magnetic field strength are strongly dependent on the heat transfer rate through the mantle of the planet (e.g. Stevenson et al., 1983). The thermal buoyancy flux is directly related to the heat flow extracted by the mantle from the core. In a chemically driven dynamo, the solidification rate and thus the rate of buoyancy release depends on the core cooling rate and thus on the heat flow from the core. The heat flow from the core is directly dependent on the mantle heat transfer rate. The latter can be calculated as a function of time from planet thermal history models (e.g. Stevenson et al., 1983; Breuer et al., 2010). Some common features of the magnetic field evolution can be assumed for all bodies. A thermal dynamo is more likely for the early evolution of a planetary body but is – if it existed at all – in general short-lived. Long-standing dynamo action and a present-day field is mostly related to the core solidification process. Nevertheless, each body has its individual magnetic field history, which is mainly related to its core composition, planet and core size and the prevailing heat transport mechanism in the mantle.

In the present chapter, we discuss the subject of the magnetism of the rocky bodies. We first describe the mechanisms for dynamo generation by thermal or thermochemical convection. This is followed by a discussion of the magnetic field evolution of Mercury, Moon, Mars, Ganymede and planetesimals. We do not take Venus into account in this discussion because there are no data available for this planet on its early magnetic field in the form of magnetised rock. Instead, we will address the possibility of dynamo generation in massive rocky exoplanets, as space missions like Kepler and CoRoT have detected a growing number of rocky exoplanets up to 10 Earth masses.

17.2 Mechanism for Dynamo Generation in Iron-Rich Cores

The generation of a self-sustained magnetic field requires convection within the electrically conductive, metallic core. Convection can be obtained by large heat flows at the core–mantle boundary or by chemical buoyancy due to crystallisation of the iron-rich material.

17.2.1 Thermal Dynamo

Thermal convection in the core, like thermal convection in the mantle, is driven by a sufficiently large super-adiabatic temperature difference between the core and the mantle. It occurs if the core heat flux exceeds that conducted along the core adiabat. The latter heat flux, therefore, serves as a criterion for the existence of thermally driven convection in the core:

$$q_c > q_{ad} = k_c dT/dr|_{ad} = k_c \alpha_c g_c T/C_p, \tag{17.2.1}$$

where k_c is the thermal conductivity of the core, $dT/dr|_{ad}$ is the adiabatic temperature gradient in the core, α_c is the thermal expansivity of the core, C_p is the core heat capacity at constant pressure and g_c and T are the gravitational acceleration and temperature at the core–mantle boundary (CMB). Values for the critical heat flux that have been used in the literature vary between 5 and 20 mW m^{-2} for Mars (Nimmo and Stevenson, 2000) and Mercury (Stevenson et al., 1983; Schubert et al., 1988), and between 1 and 10 mW m^{-2} for the Moon (Zhang et al., 2013; Laneuville et al., 2014; Evans et al., 2014) and Ganymede (Hauck et al., 2006; Bland et al., 2009; Kimura et al., 2009; Rückriemen et al., 2015). The large range of values for the critical heat flux stems mostly from uncertainties in the thermal conductivity of the iron-rich alloy at core pressures (Stacey and Anderson, 2001). Recent ab initio calculations of transport properties suggest that the thermal conductivity in the Earth's core is a factor of 2 higher than previously thought, and thus upward revisions to the thermal conductivity in other terrestrial planets are also being discussed. Revised values of the thermal conductivity in the Earth's core range from 90 to 150 W m^{-1} K^{-1} (de Koker et al., 2012; Pozzo et al., 2012). Similar values also result from measurements of the electrical conductivity and from using the Wiedemann–Franz law to calculate the thermal conductivity (Gomi et al., 2013). However, direct measurements of solid iron using a dynamically laser heated diamond-anvil cell place the thermal conductivity of Earth's core near the low end of previous estimates, at 18–44 W m^{-1} K^{-1} (Konôpková et al., 2016).

Irrespective of the conflicting estimates of the critical heat flow, the core cooling is strongly controlled by the thermal evolution of the mantle above. If the mantle removes heat from the core at a rate that exceeds the critical heat flow, then the core will convect. If the mantle removes heat at a rate less than the critical heat flow, the core is thermally stably stratified and dynamo action by thermal convection is not possible.

17.2.2 Chemical Dynamo

Compositional convection in the core is driven by crystallisation of iron-rich fluid with non-eutectic composition (Braginsky, 1964). The existence of light elements in the core has been assumed for all terrestrial planets (Sohl and Schubert, 2015). Potential candidates that are mixed with the iron are S, Si, O, C and H, and their concentration depends on pressure, temperature and oxidation state during core formation. In any case, in addition to reducing the average core density, these elements will significantly reduce the melting temperature of the core alloy (e.g. Usselman, 1975; Fei et al., 1997). Crystallisation will occur if the temperature in the fluid core lies between that of the solidus and the liquidus of the core alloy.

Although the effects of a variety of light elements are conceivable in terrestrial cores, the physics of crystallisation has been studied mostly for Fe–FeS systems, both experimentally and theoretically. In the following, we describe for this specific system crystallisation scenarios that might be of relevance for terrestrial bodies.

17.2.2.1 Inner-Core Growth by Outward Solidification

The conventional model of inner-core growth on the Fe-rich side of the eutectic is applicable when the positive slope of the core liquidus is steeper than that of the core temperature, $dT_m/dP > dT_c/dP$ (e.g. Stevenson et al., 1983), as is the case for Earth (Figure 17.2.1). As a consequence, during cooling, the liquidus is reached first at the centre. At this stage, iron will crystallise and an inner solid core will form – this solid core will freeze from the inside out upon further cooling. Due to crystallisation of iron, the outer core will be increasingly depleted in iron and enriched in sulphur until the eutectic composition is reached. At that point, the core will continue to freeze in an eutectic composition.

During crystallisation, the concentration of sulphur will not be homogeneous; instead, a boundary layer will form just above the inner core, where the sulphur concentration is above average. The fluid in this layer will be lighter than in the remainder of the fluid core, and compositional convection will be induced to homogenise the outer-core composition. Latent heat release at the inner-core/outer-core boundary will also contribute to the strength of convection. Inner-core growth will induce outer-core convection even if the heat flow through the CMB is lower than the critical heat flow for a thermal dynamo, as discussed above (e.g. Stevenson et al., 1983). Thus, the temperature gradient may be even shallower than adiabatic, and compositional convection may carry heat downward, against the temperature gradient. The buoyancy release associated with this compositional change exceeds the work done against the stabilising thermal stratification. However, if core cooling continues to decrease and not enough power is available to obtain sufficient magnetic advection against magnetic diffusion, the dynamo stops, although the inner core may still grow slowly over time. Nevertheless, it is generally expected that the compositional dynamo by outward solidification will be sustained for a long time on the order of billions of years.

17.2.2.2 *Iron Snow Regime*

Compositional convection can be very different from the scenario described above in smaller planetary bodies at core pressures lower than those in Earth's core. Experimental studies (e.g. Fei et al., 1997, 2000; Li et al., 2001; Chudinovskikh and Boehler, 2007; Stewart et al., 2007; Chen et al., 2008; Morard et al., 2008; Buono and Walker, 2011) have revealed two important aspects of the Fe–FeS phase diagram (1) at pressures below 14 GPa, the eutectic melting temperature decreases with increasing pressure and (2) the sulphur content at the eutectic decreases with increasing pressure up to a pressure of 60 GPa. Note that the exact decline of the eutectic concentration with pressure is somewhat uncertain.

A melting-temperature profile with a negative slope (or a shallower melting-temperature profile than that of the core temperature, $dT_m/dP < dT_c/dP$) suggests that upon cooling Fe will precipitate at the CMB rather than in the centre (Figure 17.2.1). Since the solid iron particles are heavier than the surrounding Fe–FeS fluid, it may fall as iron snow. A snow zone forms and is limited from below at a depth where the core temperature is higher than the melting temperature and the iron snow particles remelt. Due to the solidification of iron, the concentration of sulphur increases in the snow zone, and the liquidus of the fluid decreases there (e.g. Buono and Walker, 2011). Under the assumption of thermodynamic equilibrium, iron continues to solidify, and the local sulphur concentration in the liquid increases until the liquidus reaches the core temperature. Thus, in the snow zone, the liquidus becomes collinear with the core temperature and reverses its slope, so that the liquidus temperature increases with increasing depth. The increase of the liquidus with depth in the snow zone is inevitably accompanied by a decrease in the sulphur concentration with depth, which

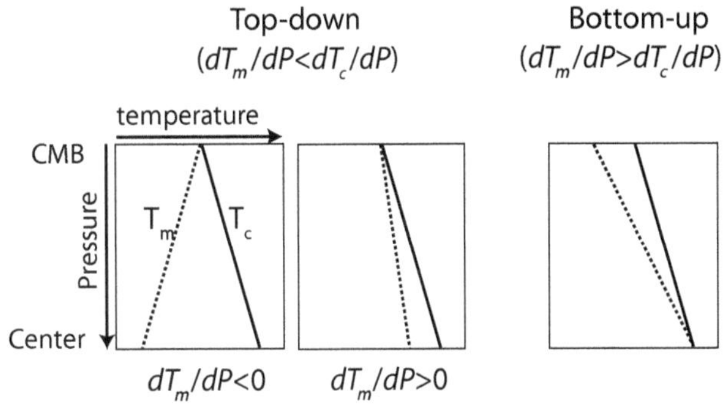

Figure 17.2.1 Schematic views of core temperature, T_c, (solid line) and liquidus, T_l (dashed line), for top-down (two left panels) and bottom-up freezing (right panel). Pressure increases from the CMB towards the centre of the core.

implies the presence of a stable chemical gradient across the snow zone (Hauck et al., 2006; Williams, 2009; Rückriemen et al., 2015).

The remelting of iron particles below the snow zone results in a local enrichment of iron in the Fe–FeS melt. This creates a gravitationally unstable situation, where a heavier layer is formed on top of a lighter one. The Fe-rich layer initiates compositional convection, which leads to a well-mixed state of the deeper, entirely liquid core below the snow zone. With ongoing cooling, the snow zone grows at the expense of the deeper, well-mixed fluid core. Finally, an inner core begins to grow when the snow zone encompasses the entire core – the sinking iron particles can no longer remelt. It is suggested that the dynamo may be located in the iron-rich layer below the snow zone as the stable chemical gradient in the snow zone may prevent large-scale convection that is necessary for dynamo action (Christensen and Wicht, 2007). Vigorous convection below the snow zone caused by the release of buoyancy upon remelting of iron may drive the dynamo (Vilim et al., 2010; Christensen, 2015; Rückriemen et al., 2015; Olson et al., 2017). This dynamo would only be active during the period between the formation of the snow zone and the time when it reached the centre of the core and is shorter than the convectional dynamo described above. Note, however, that the iron snow scenario described above assumes thermodynamic equilibrium. It might also be possible that the iron particles sink so rapidly that they fail to remelt (completely) before reaching the centre. Alternatively, iron particles may stick to the CMB and form dendritic structures before sinking into the centre (Scheinberg et al., 2016). Whether or not a dynamo may then still be possible is an open question.

This so-called iron snow regime has already been suggested to apply to the present cores of Ganymede (Hauck et al., 2006; Christensen, 2015; Rückriemen et al., 2015) and Mercury (Chen et al., 2008; Vilim et al., 2010; Wicht and Heyner, 2014). It has also been suggested that Mars' core may enter the iron snow regime in the future (Stewart et al.,

2007) and that the lunar core or cores of planetesimals may have gone through an iron snow episode (Zhang et al., 2013; Laneuville et al., 2014; Scheinberg et al., 2016). Furthermore, Gaidos et al. (2010) propose this crystallisation mechanism for massive exoplanets larger than 2.5 Earth masses.

17.2.2.3 Precipitation of FeS, Fe_3S_2 or Fe_3S

If the core sulphur content is greater than the eutectic concentration, solid FeS (or Fe_3S_2 for pressures between 14 GPa and 20 GPa, and Fe_3S for even higher pressures) will precipitate within the core (Fei et al., 2000; Hauck et al., 2006; Campbell et al., 2007; Stewart et al., 2007; Morard et al., 2008). Different crystallisation scenarios can arise due to the difference in density of the solid FeS (or Fe_3S_2 for pressures between 14 GPa and 20 GPa, and Fe_3S for even higher pressures) relative to the liquid. Solid FeS will be less dense than the residual liquid because of the high sulphur content of FeS (~36.5 wt% S). In contrast, Fe_3S is denser than the residual liquid (Stewart et al., 2007). As a consequence, either a solid Fe_3S inner core or a solid FeS layer can form below the core–mantle boundary. Depending on the relative slope between the melting and core temperature, solid Fe_3S or solid FeS start crystallising either at the core–mantle boundary or at the core centre. A detailed description of the different precipitation scenarios on the sulphur-rich side of the eutectic can be found in Breuer et al. (2015). It is suggested that apart from the scenario with Fe_3 S crystallising at the core centre, all other scenarios may in principle produce a dynamo. Dynamos caused by FeS, Fe_3S_2 or Fe_3S precipitation have been suggested as a possibility for Ganymede (Hauck et al., 2006; Zhan and Schubert, 2012; Rückriemen et al., 2018) and Mars (Stewart et al., 2007).

17.3 Magnetic Field Evolution of Rocky Bodies

Whether a terrestrial body features core convection and whether it can have a thermal or a chemical dynamo depends mainly on the cooling rate of the core, the core temperatures and the core melting temperatures. In addition to the thermal condition and the composition of the core, the efficiency of heat transport out of the interior is of importance. The Earth for instance removes efficiently heat by plate tectonics, while the other terrestrial planets have a stagnant lid on top of a convecting mantle – heat transport between surface and interior is less efficient. As a consequence, a planet in the so-called stagnant lid regime cools slower and its temperature evolution differs from a planet with plate tectonics. To better illustrate this difference on the magnetic field evolution, consider two generic planets with the same core composition and same size and structure. The planet in the stagnant lid regime would start crystallising the core later than the planet in the plate tectonics regime. This would have consequences on the onset of chemical convection and an associated dynamo. The magnetic field evolution of any planetary body is therefore strongly coupled to the temperature evolution of the mantle, and a coupled study is required (Breuer and Moore, 2015).

In general, thermal evolution models suggest that the existence of a present-day magnetic field in a terrestrial planet requires a crystallising core. This finding is independent of the heat transport mechanism of the mantle. A purely thermally driven dynamo is unlikely for any present-day terrestrial body, since the models typically predict a slowly cooling core with a core heat flux that is smaller than the heat flux along the core adiabat (Breuer et al., 2010). In fact, a thermal dynamo is in general difficult to generate and only likely if the core is superheated with respect to the mantle due to the core formation process. The thermal dynamo typically shuts off very early during the evolution, since the heat flow from the core decreases rapidly during the first few hundred million years. Note that with plate tectonics, this time can be prolonged by a few hundred million years but cannot expand to the present time. A chemical dynamo can then start if the core cools below the core liquidus.

Figure 17.3.1 shows two possible scenarios of core evolution and associated magnetic field generation for a planet with either bottom-up crystallisation (Earth case) or top-down crystallisation (Fe snow) at different states. Both scenarios start with a strongly convecting mantle, which removes efficiently heat out of the core. The fully liquid core is thermally convecting and a dynamo can be generated (1). In the second stage, the mantle cooling declines and core convection stops. The dynamo ceases although the core is still entirely fluid (2). For the bottom-up crystallisation, an inner solid core starts to form when the core temperatures fell below the melting temperature. This initiates compositional convection in the outer liquid core, which can power a dynamo (3). Over time, as cooling efficiency decreases, compositional convection in the outer liquid core weakens and is not strong enough to overcome resistive dissipation. Alternatively, the eutectic composition is reached. For both cases, there is no dynamo generation (4). In the case of top-down crystallisation, iron starts to crystallise at the core–mantle boundary. Remelting of sinking iron below the iron snow zone initiates compositional convection and dynamo action in the inner liquid core (5). With further cooling, dynamo action stops as soon as the iron snow region comprises the entire core and an inner solid core forms (6).

To estimate the magnetic field strength of a thermal or chemical dynamo, scaling laws are used that have been derived for hydromagnetic dynamos which relate magnetic induction to the buoyancy flux, a measure of the convective power available for the dynamo, and the dimensions of the dynamo region (e.g. Christensen and Tilgner, 2004; Christensen and Aubert, 2006; Olson and Christensen,

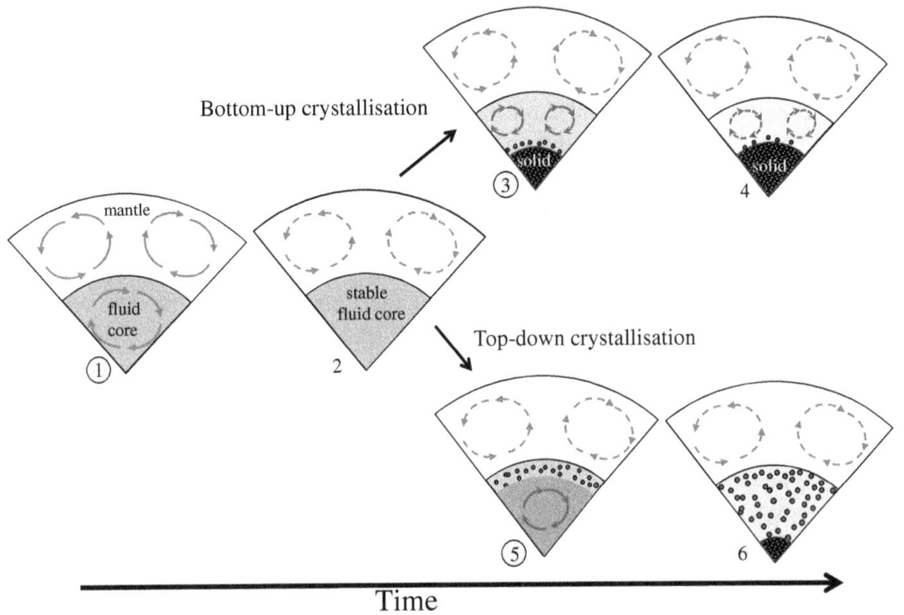

Figure 17.3.1 Possible scenarios of core crystallisation and magnetic field evolution. Time is increasing from left to right. The upper row is representative for bottom-up (Earth case) freezing and the lower row for top-down freezing (iron snow). White areas show solid silicate mantle. Grey areas indicate liquid Fe–FeS, and the darker the grey, the higher the iron content. Dark grey area with pattern represents a solid inner iron core, and small black circles show sinking iron crystals. Solid lines with arrows indicate strong convection and cooling. Dashed lines with arrows indicate weak convection and cooling in the mantle and dashed lines with arrow indicate weak convection that is not sufficient to overcome ohmic dissipation (no dynamo generation) in the core. The numbers 1–6 denote the different states of a cooling planet, which may change in time and transit from one to another. The circle around a number indicates a state with dynamo action. Possible paths of the evolution are 1–4 or 1,2,5,6. Depending on the core composition, there is no step 2, and the scenarios go from 1,3,4 or 1,5,6.

2006; Aubert et al., 2009). These recent scaling laws suggest that the magnetic field strength is related to the convective power available to balance ohmic dissipation. It is not determined by a force balance between the Coriolis and Lorentz forces, as in the older scaling laws (e.g. Busse, 1976; Curtis and Ness, 1986). The convective power available to drive a dynamo is related to the rate of entropy production, which is given by (e.g. Nimmo, 2015)

$$E_O = E_s + E_r + E_L + E_G - E_k, \tag{17.3.1}$$

where E_O is the rate of entropy production by ohmic dissipation, E_S the entropy production rate due to secular cooling, E_r is the production rate due to radioactive heating, E_k is the entropy production rate due to conduction along the adiabat, E_G the production rate due to the release of chemical buoyancy and E_L is the entropy production rate due to latent heat release. E_G and E_L only occur for core solidification. A detailed discussion of the rate of entropy production can be found in Nimmo (2015) and of the various scaling laws in Christensen (2010).

In addition to constraints on the power available to generate a dynamo, the magnetic Reynolds number must be greater than a numerically defined critical value of about 50–100 (Christensen and Aubert, 2006). The magnetic Reynolds number characterises the importance of magnetic advection versus magnetic diffusion and is defined as $\mathrm{Re}_m = vL/\eta$, where v is the characteristic velocity, L the characteristic length (the thickness of the convective shell, $D = L$) and η is the magnetic diffusivity. Planetary rotation rate is thought to have little influence on dynamo strength, although it may control the relative strength of the dipole and multipolar components (Christensen et al., 2009). The reason why Venus does not have a magnetic field at present is therefore not due to its slow rotation, as is often assumed.

In the following, we describe the thermal and magnetic field evolution for Mercury, Mars, Moon and Ganymede as well as for planetesimals and rocky exoplanets in more detail. For each body, we first introduce the available data of the observed magnetic field and discuss their magnetic field evolution. Figure 17.3.2 shows the times of the magnetic field generation of Mercury, Venus, Earth, Moon, Mars and Ganymede, as suggested from observation data, i.e. remanent magnetisation of the old crust and today's self-generated magnetic field. These data are used to constrain models of the magnetic field evolution of the bodies.

17.3.1 Mercury

The internally generated magnetic field of Mercury had been detected during close encounters of Mariner 10 in 1974 and 1975 (Ness et al., 1974, 1975) and was confirmed by the MESSENGER mission, which orbited Mercury between 2011 and 2015 (Anderson et al., 2011). Mercury's magnetic field possesses a small dipole moment, which is a factor of about 10^4 smaller than that of the Earth. Its equivalent dipole is offset towards the north by about 0.2 planetary radii and shows only a small deviation from axial symmetry (Anderson et al., 2012; Johnson et al., 2012). Moreover, there is evidence

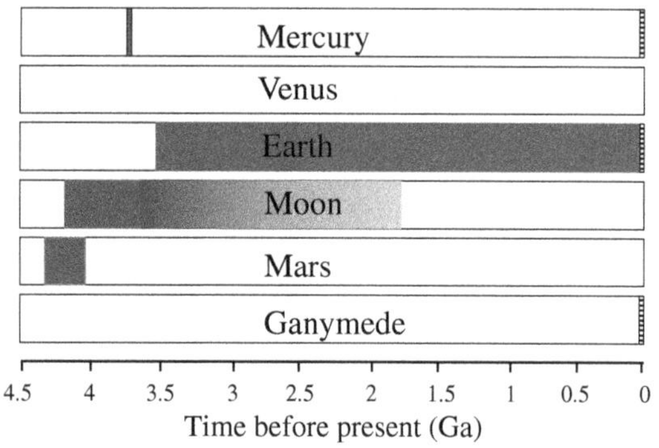

Figure 17.3.2 Times of magnetic field generation of Mercury, Venus, Earth, Moon, Mars and Ganymede, as suggested from observation data, i.e. crustal remanent magnetisation and today's self-generated magnetic field. The dark grey surface shows remanent magnetised crust. In case of the Moon, dark grey shows a strong field and light grey a weaker field, indicating the strong decrease of the magnetic field strength with time. The dashed area shows today's self-generated magnetic field. It is important to note that the lack of observed remanent magnetised crust is no clear sign that also an internally generated magnetic field was lacking at that time. For Ganymede, for example, the dynamo action probably started earlier, but no data are available for the remanent magnetisation. See text for further information about the proposed magnetic field evolution of these bodies.

for remanent magnetisation of crustal rock with an age of 3.7–3.9 Ga ago suggesting that a dynamo has existed early in the evolution of the planet (Johnson et al., 2015).

Mercury is peculiar among the terrestrial planets due to its high mean density. It suggests that the core must be comparatively large, about 0.82 planetary radii, *Rp* (*Rp* = 2400 km). The core radius range has been estimated to be between 1840 km and 2040 km (Smith et al., 2012; Hauck et al., 2013; Rivoldini and Van Hoolst, 2013), using recent MESSENGER gravity data (Smith et al., 2012) and the amplitude of libration measurements with Earth-based radar (Margot et al., 2007, 2012). The data also support the presence of a fluid outer core (Peale, 1976; Margot et al., 2012). As for the other terrestrial planets, sulphur has been the most likely candidate for the light element in the iron core (e.g. Ringwood, 1977; McCammon et al., 1983). However, the model of an Fe–FeS core has recently been questioned, since the MESSENGER X-ray spectrometer (XRS) data (Nittler et al., 2011; Weider et al., 2012) indicate strongly reducing conditions, which has been used to suggest a core containing Si as well as S (e.g. Malavergne et al., 2010; Chabot et al., 2014). The presence of silicon may have major implications for the core's structure and evolution (Smith et al., 2012; Hauck et al., 2013), and for the magnetic field evolution (Knibbe and van Westrenen, 2018).

For the present-day magnetic field, it has mostly been argued that a hydromagnetic dynamo driven by chemical convection in an iron-rich core provides the best explanation. Thermal evolution models (Schubert et al., 1988; Hauck et al., 2004; Grott et al., 2011; Tosi et al., 2013) suggested that a typical evolution of the Hermean magnetic field starts with a thermally driven dynamo, followed by a chemically driven dynamo after core crystallisation. Thermally driven dynamo action may cease before the inner core begins to freeze, in which case there would be a gap in the magnetic field history. Whether such a gap occurs and its length in time would depend on the value of the heat flow along the core adiabat and the core's volatile content. The larger the heat flux along the core adiabat becomes, the shorter the lifetime of a thermally driven dynamo. Estimates range between a few tens to few hundred million years (Grott et al., 2011; Tosi et al., 2013). On the other hand, the higher the volatile content in the core, the later the onset of inner-core growth as the core needs to cool below the melting temperature that decreases with the volatile content (e.g. Schubert et al., 1988; Hauck et al., 2004; Grott et al., 2011; Tosi et al., 2013). For example, the core starts to freeze after 0.75, 0.8 and 2.5 Gyr for an initial concentration of 0.2, 1 and 5 wt% S, respectively, assuming the same mantle parameters (Schubert et al., 1988). Models that consider the combined thermal, magmatic and tectonic evolution of Mercury (Hauck et al., 2004; Grott et al., 2011; Tosi et al., 2013) suggest a bulk core sulphur content between 6.5 wt% and 10 wt% and a late formation of the inner core at 1 Ga. An important constraint for these models is the observed surface contraction of ~7 km during the last ~4 Ga derived from the distribution, lengths and heights of lobate scarps (Byrne et al., 2014). This small radial contraction suggests a rather small solid inner core, as the solidification of large parts of the core would imply a too large volume change that is not consistent with observation. Thus, dynamo action by compositional convection in Mercury to produce the present field may be recent and the observed early magnetic field episode by purely thermal convection was likely brief in particular if the core consists of Fe–FeS.

It is important to note that the classical hydromagnetic dynamo, as assumed in the thermal evolution models above, typically suggest a much stronger field than observed (Stevenson et al., 1983; Schubert et al., 1988; Wicht et al., 2007). To explain the weak field, several models have been suggested. Among those are thin-shell dynamo models which require a large inner core (relative inner-core size larger than about 0.8) (Stanley et al., 2005). As described above, these models are not consistent with geodetic and tectonic observations (Grott et al., 2011; Dumberry and Rivoldini, 2015). Alternatively, a deep dynamo with a small inner core (of less than 1000 km in radius) has been proposed by Heimpel et al. (2005), Christensen (2006) and Christensen and Wicht (2008). In the latter two models, the dynamo operates below a stable conducting layer in the outermost core. The field is then strongly attenuated by the electromagnetic skin effect of the stagnant (i.e. no vertical movement) but electrically conducting layer. However, the stagnant layer may be destabilised by double diffusive effects, which can only be avoided if the sulphur concentration in Mercury's core is only a fraction of a per cent (Manglik et al., 2010). A core with such a low sulphur concentration is, as a result of the early core crystallisation, again unsuitable for explaining the geodetic and tectonic observations (Hauck et al., 2004; Grott et al., 2011; Tosi et al., 2013).

Most dynamo models assume an Earth-like dynamo with inner-core freezing starting at the centre. However, as discussed before, at moderate pressures the gradient of the liquidus temperature of Fe–FeS is smaller than the gradient of the core temperature in the outer layers of Mercury's core. The present core would then most likely be precipitating solid iron in the form of snow at a single depth or in two distinct zones (Chen et al., 2008). Such a scenario suggests a stably stratified layer in the outermost core, as required to explain the geometry and magnetic field strength through the electromagnetic skin effect (Christensen, 2006; Wicht and Christensen, 2008; Vilim et al., 2010). As a drawback, dynamo models that include an iron snow layer are strongly time-dependent and generate Mercury-like magnetic fields only for a fraction of the total simulation time (Wicht and Heyner, 2014). Large-scale variations in the CMB heat flux may help to stabilise strong Mercury-like quadrupole terms (Cao et al., 2014; Tian et al., 2015), but these do not coincide with the current understanding of Mercury's lower mantle, which shows rather small heat flux variations at the CMB (e.g. Tosi et al., 2013).

Alternatively to a Fe–FeS core, Si has been suggested as a core phase in addition to S (Malavergne et al., 2010) as a consequence of the reducing conditions under which Mercury likely formed (Nittler et al., 2011). Partitioning of Si into the core has important implications for the structure and evolution of the planet. The Fe–S–Si ternary system shows an immiscibility of S- and Si-rich liquids up to a modest pressure of ~14 GPa (Sanloup and Fei, 2004; Morard and Katsura, 2010). This would lead to a segregation of S-rich liquids near the top of the core with more Si-rich materials deeper. Hauck et al. (2013) even suggest that the outer FeS layer is solid and part of the solid outer shell (crust, mantle and upper core). To fit the observed moment-of-inertia data, this would imply a mantle even thinner than 400 km (Hauck et al., 2013). Furthermore, core–mantle temperatures need to be lower than that for the liquidus of pure FeS but these temperatures are not consistent with the most recent thermal evolution models (Tosi et al., 2013). The implications of a Fe–S–Si system for the dynamo are, however, unknown. Thermal evolution models with a pure Fe–Si core suggest that thermal buoyancy in a deep liquid core layer may generate Mercury's magnetic field (Knibbe and van Westrenen, 2018) – the upper liquid core being stable due to a sub-adiabatic core–mantle heat flow. In contrast to a Fe–S core, Si fractionates non-preferentially between solid and liquid metal implying that core solidification does not produce compositional buoyancy, but it releases more latent heat, and that the inner core solidification may have started early (Knibbe and van Westrenen, 2018).

The present magnetic field data and constraints on core composition and interior structure do not allow us to conclusively prove or reject the various dynamo models and thermal history calculations. More details about the magnetic field are expected from the upcoming BepiColombo mission to be launched in 2018. BepiColombo consists of two individual orbiters: the Mercury Planetary Orbiter (MPO) to map the planet and the Mercury Magnetospheric Orbiter (MMO) to investigate its magnetosphere. Both spacecraft may help to better disentangle the magnetic field and learn more about the thermal and magnetic field evolution of the planet.

17.3.2 Mars

The Mars Global Surveyor (MGS) mission (e.g. Acuña et al., 1999; Connerney et al., 1999) confirmed that there is no active dynamo on Mars but revealed that part of its crust is remanently magnetised, indicating that the planet must have had an active magnetic field. The discovered strongly magnetised regions in the crust are closely associated with the ancient, cratered terrain of the highlands in the Southern Hemisphere. Magnetic anomalies are weak or absent within younger impact basins. On the basis of the surface distribution of the magnetic anomalies, it has been proposed that this magnetisation event predated the formation of the Hellas and Argyre basins (Connerney et al., 1999; Lillis et al., 2013), roughly at the end of the Early Noachian (Head et al., 2001). The latter impact basins do not show any signs of magnetisation, which implies that the dynamo action in Mars stopped before their formation. Crater counting of the magnetised crustal regions further suggests an end of dynamo action at ~4.1 Ga (e.g. Connerney et al., 2004; Vervelidou et al., 2017) (Figure 17.3.2). The existence of an early magnetic field has been supported by the ancient Martian meteorite ALH84001, which shows a magnetisation that was acquired at 4 Ga or even earlier (Weiss et al., 2002). However, the lack of an appreciable magnetisation in two ancient areas on the highlands, the South Province and the Tempe Terra, further suggests that no strong dynamo existed in the first 100 Ma of Mars' history (Arkani-Hamed and Boutin, 2012).

The large-scale spatial pattern of magnetisation on Mars is of particular interest and essentially coincides with the so-called crustal dichotomy. This dichotomy describes the topographic division between the young and smooth plains in the Northern Hemisphere and the old cratered terrain in the Southern Hemisphere. Remarkable in the pattern of magnetisation are also linear structures running in east–

west direction, which extend over 2000 kilometres and suggest an analogy to plate tectonics (Connerney et al., 1999, 2004). Other models, however, favour the thermal remanence acquired during the period when these old rocks were cooled down in conjunction with dike intrusions (Nimmo, 2000). The derived crustal magnetisations, which amount up to 10–30 Am^{-1}, are an order of magnitude stronger than those of the Earth's continents (Arkani-Hamed and Dyment, 1996) and comparable to the remanent magnetisation of fresh extrusive basalt near the oceanic ridge axes of Earth (Bleil and Petersen, 1983). These magnetic anomalies thus indicate a strong ancient intrinsic magnetic moment of Mars, which is 0.1–10 times as large as that of today's Earth's (Ness et al., 1999). The highest values require both very strongly magnetic rocks and magnetisation through a large section of the crust, possibly to a depth of 30 km for magnetisation intensity ±20 A/m (Connerney et al., 1999).

To explain the early magnetic field for Mars and the absence of a present dynamo field, most models suggest that the Martian core may be lacking an inner core and that an early thermally driven dynamo had ceased to operate after a few hundred million years of evolution (e.g. Stevenson et al., 1983; Spohn et al., 1998, Hauck and Phillips, 2002; Williams and Nimmo, 2004; Breuer and Spohn, 2003, 2006). These models further assume that Mars was in a stagnant lid regime since its early evolution and that the core was initially superheated with respect to the mantle. The lack of an inner core is further supported by geochemical evidence for the composition of the Martian core and by interior structure models (e.g. Rivoldini et al., 2011). A molten core for the entire evolution is possible if the core contains at least ~5–10 wt% of sulphur (Williams and Nimmo, 2004; Stewart et al., 2007). A core sulphur concentration of 14.2 wt% has been suggested (Dreibus and Wänke, 1985; McSween, 1985) based on chemical analysis of SNC meteorites. Furthermore, the most recent estimates of the moment of inertia and tidal Love number suggest a core with a size of 1794 ± 65 km and a sulphur concentration of 16 ± 2 wt% (Rivoldini et al., 2011). This would suggest a composition close to the eutectic (even on the sulphur-rich side of the eutectic) for the pressure range of the Martian core. A composition close to the eutectic, on the other hand, has low melting temperatures of less than ~ 1500 K (e.g. Chudinovskikh and Böhler, 2007). Thermal evolution models show that the core typically does not cool below these temperatures (Plesa et al., 2016). Thus, a completely liquid Martian core is likely to be present today. However, recent experiments showed that it is impossible to simultaneously account for the Martian mantle depletion of moderately siderophile elements if the Martian core sulphur content exceeds 10.5 wt% (Rai and van Westrenen, 2013). In addition, chalcophile elements in Martian meteorites suggest sulphur content in the core of less than 5–10 wt% (Wang and Becker, 2017). To explain the geophysical data (Rivoldini et al., 2011), these latter findings may imply that the Martian core also contains some other light elements in addition to sulphur.

Alternative scenarios that explain the magnetic field history suggest changes in the tectonic style and in the heat transport mechanism of the mantle, from plate tectonics to stagnant lid (Nimmo and Stevenson, 2000; Stevenson, 2001; Breuer and Spohn, 2003), or consider the effects of impacts. Some of the latter models point out that a late impact may have heated the mantle to such an extent that convection and dynamo action in the core were stopped (e.g. Roberts et al., 2009; Arkani-Hamed and Olson, 2010), while Reese and Solomatov (2010) argued that a late large impact may have triggered the onset of the dynamo. Which of these conflicting scenarios is more likely depends mainly on the dynamics in the core during fusing an iron diapir with a pre-existing core, a process that is barely understood (Monteux et al., 2013). Although these models suggest a different thermal history of Mars, they have in common that the Mars dynamo is only generated by thermal convection and that there is no inner solid core. It is further expected that with ongoing cooling, a dynamo can be initiated in the future due to core crystallisation in the form of iron snow or FeS precipitation (Stewart et al., 2007). Davies and Pommier (2018) even suggest a scenario for which iron snow may have started already and models that match the available magnetic and geodetic constraints have an initial sulphur concentration of 10% and snow zones that occupy approximately the top 100 km of the present-day Martian core.

To explain the observed dichotomy in the remnant magnetisation with the Southern Hemisphere being more strongly magnetised than the Northern Hemisphere (e.g. Langlais et al., 2004; Morschhauser et al., 2014), two general types of models have been suggested. A homogeneous magnetisation may have been destroyed in the Northern Hemisphere by, for example, hydrothermal alteration (Solomon et al., 2005) or impacts (e.g. Rochette et al., 2001; Mohit and Arkani-Hamed, 2004). The alternative explanation suggests that the dynamo itself produced a hemispherical field owing to lateral heat flow variations at the CMB (Stanley et al., 2008; Amit et al., 2011). This hemispherical dynamo would produce strong magnetic fields in the Southern Hemisphere if the heat flow from the core below the Southern Hemisphere were stronger than that below the Northern Hemisphere. However, these hemispherical dynamos may reverse on timescales of about 10 kyr (Dietrich and Wicht, 2013). These reversals are too quick to allow for the unidirectional magnetisation of a thick crustal layer and to explain the strong magnetisation in the Southern Hemisphere.

17.3.3 Moon

Today the Moon does not have an internally generated global magnetic field but shows remanent magnetisation in the crust and returned Apollo samples, which proves that a lunar magnetic field was present billions of years ago (e.g. Weiss and Tikoo, 2014). It has now been established that a core dynamo likely existed in the Moon from at least 4.25 billion to 3.56 billion years ago, with an intensity of ~100 μT similar to that at the surface of Earth today. The field then declined by at least an order of magnitude by 3.3 billion years ago and possibly persisted until 2.25–1 Ga b.p. (Tikoo et al., 2014, 2017) or even until 200 Ma b.p. (Wieczorek et al., 2006) at a low level of a few μT. Alternative explanations to a dynamo-generated global field for the lunar magnetic record suggest that at least part of the remanent magnetisation was acquired from plasma clouds that had been generated by impacts antipodal to the magnetic anomaly (e.g. Hide, 1972; Hood et al., 2001).

The observations and measurements of the remnant magnetisation require that any magnetic field generation model needs to explain the characteristic features of the palaeomagnetic data. First, remanently magnetised rock is younger than 4.25 Ga. Two alternative explanations have been suggested. (1) The remanent magnetisation of older crustal rock could have been destroyed by impact gardening and/or volcanic activity in the first 250 Ma. Alternatively, (2) the internal field may have started only about 250 Ma after Moon formation. Second, the inferred palaeo-intensity between 3.56 Ga and 4.25 Ga b.p. was about 2 orders of magnitude higher than suggested by estimates from magnetic field scaling laws for typical thermal or chemical dynamos (e.g. Evans et al., 2014; Laneuville et al., 2014; Weiss and Tikoo, 2014; Evans et al., 2017). Finally, there is a rapid decline of the field intensity with time at ~3.6 Ga b.p., but no lunar dynamo model exists to date that is capable of explaining such a decline while continuing to produce a weak field thereafter.

A lunar dynamo requires an iron-rich core, and its existence has long been controversially discussed also because the leading hypothesis for lunar origin, the giant impact model, predicts a moon composed primarily of material from Earth's mantle and the impactor's mantle and, therefore, little or no metallic core (Cameron, 1997). If the giant impact occurred before completion of Earth accretion, then this could possibly explain the presence of a small core (Cameron and Canup, 1998). The lunar polar moment of inertia (Konopliv et al., 1998) and electromagnetic sounding data from the Lunar Prospector (Hood et al., 1999; Shimizu et al., 2013) are consistent with a core radius of 250 km to 450 km. Furthermore, reanalyses of Apollo seismic data (Garcia et al., 2011; Weber et al., 2011) have confirmed the existence of a core with a radius between 340 km and 420 km. Weber et al. (2011) even suggests the existence of a solid inner core with a radius of 240 km radius.

Most previous thermal and magnetic field evolution models have considered a thermally driven dynamo during the early evolution of the Moon (Konrad and Spohn, 1997; Spohn et al., 2001; Laneuville et al., 2014). These models show that a thermally driven dynamo would have been active since core formation and may have lasted a maximum of 500 Myr, until 4 Ga b.p., if the core was superheated by a few hundred degrees with respect to the mantle. A recent study by Evans et al. (2014) showed a long-standing thermal dynamo of up to 2 Gyr if the lower mantle rheology is weakened by water, but they also assume a low thermal conductivity of the iron-rich core of 25 W m^{-1} K^{-1} and an adiabatic heat flux of 2.4 mW m^{-2} – models predicting a shorter thermal dynamo duration assume an adiabatic heat flux closer to a value of 10 mW m^{-2}. All these thermal evolution models neglect any chemical stratification of the mantle. The latter, however, is suggested as a consequence of the solidification of the lunar magma ocean (e.g. Snyder et al., 1992).

Models of a late onset of dynamo action invoked the thermal and chemical consequences of a magma ocean (Stegman et al., 2003; Zhang et al., 2013). After crystallisation of the magma ocean, a dense ilmenite and pyroxene cumulate layer, located just below the enriched KREEP layer, sinks to the deep lunar mantle and thermally insulates the core. (The KREEP layer is enriched in potassium (K), rare earth elements (REE) and phosphorous (P).) The core is thus prevented from cooling convectively and also from developing a dynamo. Eventually, because of the heating by radiogenic elements, the dense lower mantle becomes buoyant and rises towards the surface. With the removal of the thermal blanket, the core begins to convect vigorously, cooling and producing a short-lived thermal dynamo. The period it takes the thermal blanket to heat up and rise back towards the surface is essential in these models and is broadly consistent with the time of the eruption of the mare basalts and the magnetisation of lunar rocks between 4.25 Ga and 3.55 Ga b.p. However, newer models of Zhang et al. (2017) show that the ilmenite-rich cumulates remain stable at the core–mantle boundary and do not rise towards the surface as a result of their weaker rheology. The associated efficient cooling of this layer by convection prevents its heating and thus the rise of the dense material.

Other models explain the magnetic signatures by mechanically stirring the core through either precession (Williams et al., 2001; Dwyer et al., 2011) or changes in the lunar rotation rate from large impacts (Le Bars et al., 2011). In particular, the precession dynamo may be capable of producing the strong surface magnetisation, since the power caused by differential motion can be much larger compared with the power driving thermal or chemical convection (Lin et al., 2016). However, the persistence of such a dynamo beyond ~2 Ga would likely require that the Moon's orbit receded from the Earth at a rate far slower than expected (Tikoo et al., 2014).

If the presence of a solid inner core is correct – as suggested by seismic data (Weber et al., 2011) – a compositionally driven dynamo may be an alternative. Previous studies (e.g. Konrad and Spohn, 1997) have discarded this possibility because the absence of a present-day magnetic field is difficult to reconcile with a growing inner core as a chemical dynamo with inner-core growth is difficult to stop in a continuously cooling planet. It has recently been shown that even with a growing inner core, power is indeed not sufficient to drive a present-day dynamo (Scheinberg et al., 2015). The growth rate of the core and, therefore, the compositional buoyancy becomes too small to overcome the stable temperature gradient for subadiabatic core cooling. These models assume Earth-like core crystallisation but for a sulphur content above about 8 wt%, iron snow is a likely mechanism for core solidification (Laneuville et al., 2014) and the formation and growth of the inner core would mark the demise of dynamo action (Christensen, 2015; Rückriemen et al., 2015). In any case, a chemical dynamo also suggests a field strength that is two orders of magnitude too low to explain the observed early strong magnetisation and one order too low to explain the weaker, younger magnetisation (Laneuville et al., 2014; Scheinberg et al., 2015; see Figure 17.3.3). This finding also implies that the dynamo may have been powered by at least two distinct mechanisms operating during early and late lunar history.

17.3.4 Ganymede

In 1996, the Galileo mission measured a magnetic field around the Jovian moon Ganymede (Kivelson et al., 1996), which was interpreted as a field of internal origin embedded in the Jovian magnetic field. The field would have an equatorial surface field strength of approximately 719 nT (Kivelson et al., 2002). The magnetic field has an anomalously low quadrupole moment relative to its dipole moment, placing a strong constraint on the dynamo mechanism (Kivelson et al., 2002). The discovery of a self-sustained magnetic field at Ganymede came as a surprise, since it was commonly believed that the icy moon – with an average density of 1936 kg m^{-3} – was composed of a rocky core surrounded by a large mantle of water or water ice with an ice surface.

As for the other terrestrial planets and the Moon, the origin of Ganymede's magnetic field is most likely found in an iron-rich core, because alternative mechanisms such as remanent magnetisation of a silicate layer containing magnetic minerals below the ice mantle or a dynamo in an electrically conducting, salty subsurface ocean are implausible (Schubert et al., 1996). For remanent magnetisation, both the concentration of magnetic minerals and the external field (i.e. the present Jovian magnetic field at Ganymede's orbit, ~100 nT) are too low to reasonably explain the observed field. On the other hand, the flow velocity in its ocean must be implausibly large, on the order of 1 m s^{-1}, to explain the observed magnetic field through a dynamo in a subsurface ocean. The interpretation of the measured field, however, requires the consideration of a superimposed induced magnetic dipole in a subsurface

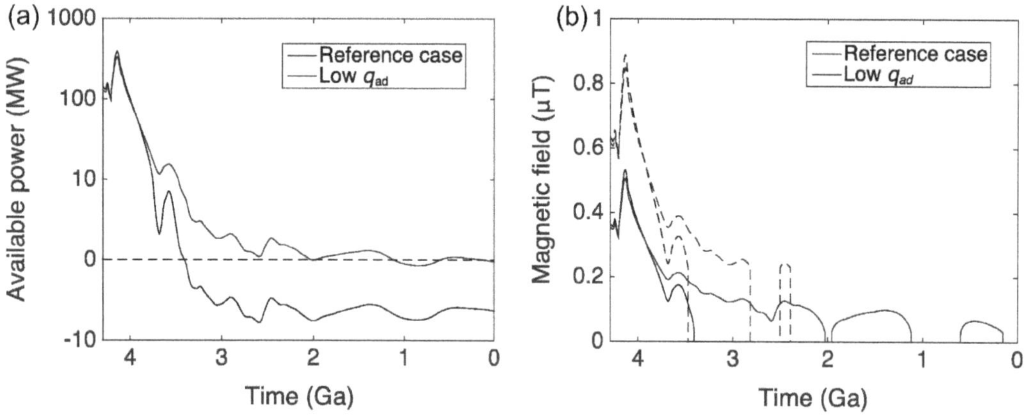

Figure 17.3.3 (a) Power available to a lunar dynamo and (b) its magnetic field strength for two different values of the heat flow along the core adiabat. The reference model is calculated using $\alpha_c = 9{\cdot}10^{-5}$ K^{-1} and $k_c = 50$ Wm^{-1} K^{-1} and the low q_{ad} case is calculated using $\alpha_c = 7{\cdot}10^{-5}$ K^{-1} and $k_c = 25$ Wm^{-1} K^{-1}. In both models, an inner core starts to solidify early, i.e. after 250 Ma. The results demonstrate that the calculated field strength of such a compositional dynamo is much lower than the inferred palaeo-intensity of ~70 μT between 3.56 Ga and 3.85 Ga b.p. Knowledge of the critical heat flow along the adiabat is thus crucial as the lifetime of the dynamo can extend nearly to present day. After Scheinberg et al. (2015).

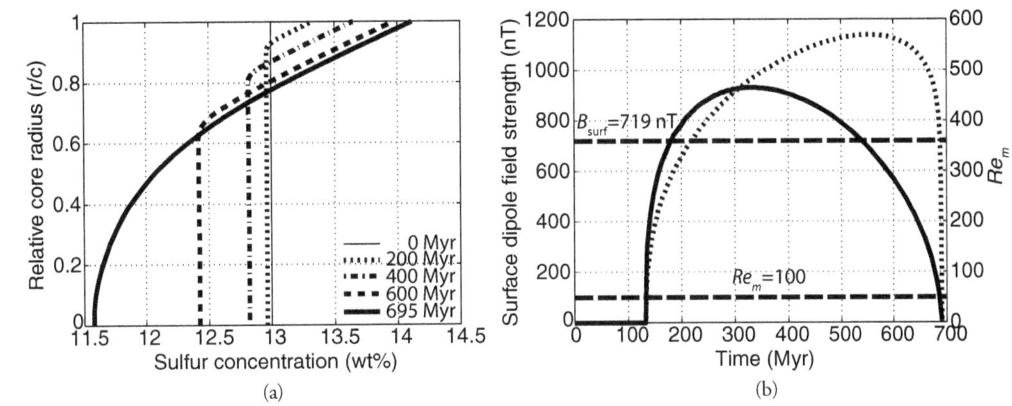

Figure 17.3.4 (a) The sulphur concentration as a function of relative core radius (*r*/*c*) at distinct points in time. The range of the snow zone can be identified where the sulphur gradient varies with depth. Note that iron starts to solidify at the CMB for 0 Myr and that the snow zone encompasses the entire core for 695 Myr. (b) Predicted surface dipole magnetic field strength (solid line) and the magnetic Reynolds number (dotted line) for the nominal model with an initial sulphur content in the core of 13 wt.%, a core heat flux of 4 mW/m^2 and a thermal conductivity in the core of 40 W/mK. Horizontal dashed lines indicate the observed magnetic field B_{surf} of 719 nT and the critical magnetic Reynolds number Re_m of 100. The dipole magnetic field strength at the surface is calculated with b_{dip} = 7 (the relative strength of the internal field to the dipole field at the core–mantle boundary). The duration of the dynamo is less than 700 Myr. After Rückriemen et al. (2015).

ocean driven by the time varying component of the externally imposed magnetic field of Jupiter's magnetosphere (Kivelson et al., 2002).

The existence of an iron-rich core is further supported by the small mean moment-of-inertia factor (MoI = 0.3105) derived from Galileo gravity and rotation data (e.g. Anderson et al., 1996; Sohl et al., 2002; Schubert et al., 2004). Fixing the silicate and ice-shell densities to plausible values and allowing for core compositions ranging from pure iron to pure FeS, Sohl et al. (2002) found core radii between one-fourth and one-third Ganymede radii for a three-layer model. As a light and siderophile element, sulphur is the main candidate (e.g. Schubert et al., 2004). Assuming a composition identical to a type I carbonaceous chondrite (CI), experiments by Scott et al. (2002) indicate that the alloying of iron and sulphur may form FeS-dominated cores. However, because of the unknown oxidation state of Ganymede's interior during core differentiation, it is difficult to say whether the core composition is likely to be of Fe- or FeS-rich eutectic composition.

Compositional-driven convection is the preferred mechanism for driving Ganymede's dynamo (Hauck et al., 2006; Bland et al., 2009), although Kimura et al. (2009) have suggested a model including a present-day thermal dynamo by using a heat flux along the adiabat as small as 1 mW m^{-2} to meet the necessary condition for a thermal dynamo. The melting behaviour of Fe–FeS alloys at Ganymede's core pressure, between 6 GPa and 10 GPa (Sohl et al., 2002), prompted Hauck et al. (2006) to consider the iron snow model for Ganymede. Depending on the iron liquidus and the Fe–FeS eutectic temperature used, the iron snow regime works at bulk sulphur concentrations above 0–6 wt% up to about 22 wt.% the sulphur concentration at the eutectic composition (Hauck et al., 2006; Rückriemen et al., 2015). Because of the pressure gradient in the core, it is possible that the iron snow regime is established in the outer core, while an inner core begins to freeze independently at the centre for certain ranges of sulphur concentration (Rückriemen et al., 2015; see Figure 17.3.4).

The implications of the Fe snow regime for the core dynamics and magnetic field evolution are controversial. Hauck et al. (2006) proposed that the sinking iron particles may cause motions within the fluid of the stable snow zone and, thereby, provide the driving force for the magnetic field of Ganymede. By applying the magnetic field strength scaling law of Christensen and Aubert (2006), they found surface magnetic field strengths between 600 nT and 1600 nT. However, it not clear whether large-scale fluid motions that are necessary for a dynamo (Christensen and Wicht, 2007) can be expected for particle settling in a stably stratified fluid. Bland et al. (2009) argued that the conditions for magnetic field generation cannot be met in the iron snow regime, because the latent heat released by the solidification of iron would readily be removed by the mantle above and thus be unavailable to power the dynamo. However, the assumption of latent heat being released close to the CMB is only correct initially, when the snow zone starts to grow. For continuous cooling, the snow zone extends into deeper core regions, thereby triggering the release of latent heat in the deeper core as well. This deep and entirely fluid core below the zone snow could be the locale of the dynamo (Christensen, 2015; Olson et al., 2017). Compositional convection would be driven by the sinking and remelting of iron crystals in the lower core region underlying an electrically conducting snow zone with a strongly stabilising density gradient. For plausible values of the buoyancy flux from Rückriemen et al. (2015), the model can reproduce Ganymede's observed dipole. Such a dynamo would be restricted in its lifetime by the time the snow zone takes to grow through the core. Rückriemen et al. (2015) found values of 320–800 Myr, depending mainly on the initial core sulphur content and the heat flow from the core (Figure 17.3.4). The dynamo would then start late in the evolution and could easily last to the present day. Moreover, the model would predict the absence of a solid inner core.

Magnetic field generation for a core composition on the sulphur-rich side of the eutectic has been studied less often. FeS floating, where FeS crystals form first in the core centre and rise, is considered a viable mechanism for dynamo action in Ganymede (Hauck et al., 2006; Bland et al., 2009; Zhan and Schubert, 2012). However, at Ganymede's core pressures, FeS crystallisation at the CMB is more likely than in the centre, and a solid outer-core layer would grow inward, similar to the Earth's inner-core growth, but in reverse (Rückriemen et al., 2018). Such a dynamo could be long lasting (at least longer than the dynamo in the Fe snow regime), because compositional convection in a cooling Ganymede would only stop when the eutectic composition is reached in the liquid inner core. To solve the question about Ganymede's magnetic field generation, better constraints on the thermal properties, the melting relations, as well as the compositions of the cores are required through experimental work on phase diagrams and transport properties, and numerical modelling. Furthermore, data from the ESA JUICE (Jupiter Icy moons Explorer) mission to Jupiter and its moons will contribute to this issue. To do this, some of the mission's instruments will measure Ganymede's magnetic field, both its intrinsic component and that induced by Jupiter.

17.3.5 Planetesimals

The discovery of palaeomagnetism recorded in a variety of meteorites suggests that strong magnetic fields were even common in early solar system bodies. Remanent magnetisation has been found in both chondritic meteorites, which never underwent melting, and achondritic meteorites, which underwent at least partial melting. Evidence for past core dynamos has been identified in plutonic angrites (Wang et al., 2017), eucrites (Fu et al., 2012), and main group pallasites (Bryson et al., 2015; Tarduno et al., 2012), but also in the carbonaceous chondrite Allende, both in bulk samples and in some chondrules (Carporzen et al., 2011).

It is suggested that these palaeofields originated from an early internal dynamo in a rapidly formed iron-rich core. Bodies that accreted to more than ~10 km radius before ~2.5 Ma after the formation of calcium–aluminium-rich inclusions (CAIs) likely contained sufficient 26Al, a short-lived radiogenic element, to significantly heat and melt the interior (e.g. Hevey and Sanders, 2006; Neumann et al., 2012). These early-accreting bodies would have melted from the interior outward, resulting in a separation of metal and silicates to form the core. Models show that a typical structure consists of a differentiated interior with an iron core and a silicate mantle that is covered by an undifferentiated upper layer with its lower part being sintered and compacted and its upper part still porous (e.g. Hevey and Sanders, 2006; Sahijpal et al., 2007; Neumann et al., 2012). This outer region could be the origin of the magnetised chondritic meteorites (Weiss et al., 2010; Elkins-Tanton et al., 2011). A core dynamo is not the only proposed mechanism to explain the observed magnetisation in meteorites. Alternatively, solar magnetic field during the active T-Tauri phase has been suggested but the meteorites and their acquired magnetic fields are likely younger (e.g. Weiss et al., 2008).

Studies of planetesimal dynamos have been limited (Nimmo, 2009; Weiss et al., 2008, 2010; Elkins-Tanton et al., 2011; Sterenborg and Crowley, 2013; Bryson et al., 2015, Scheinerg et al., 2016). Nimmo (2009) examined the energetics of planetesimal cores and evaluated the minimum cooling rate required to produce a dynamo with core cooling rates of ~1–100 K/Ma in the absence of compositional convection, and <0.1 K/Ma if compositional convection occurs. The strong cooling rates in absence of compositional convection can be obtained if a magma ocean formed above the iron-rich core. It may have sufficiently cooled the core to initiate thermal convection and thus powered a dynamo to generate the magnetic field (Weiss et al., 2008). Elkins-Tanton et al. (2011) found out that the core heat flow in bodies with a radius of only 100 km is sufficient to maintain a dynamo for more than 10 Ma and for more than 50 Ma in bodies with a radius of 300 km. However, an analysis by Sterenborg and Crowley (2013) showed that convection in a purely fluid core in bodies with a radius of less than 500 km quickly fell below the critical magnetic Reynolds number required for dynamo action and would only be capable to produce a weak dynamo for less than 10 Myr. A further problem with dynamo generation in a fluid core caused by a rapidly cooling magma ocean stems from differentiation models of planetesimals. These show that considering magma heat transport and redistribution of radiogenic heat sources with the melt towards the surface inhibits the formation of a global magma ocean above the core (Moskovitz and Gaidos, 2011; Neumann et al., 2014).

Alternatively, Scheinberg et al. (2016) considered the magnetic field generation of a planetesimal by core solidification and only a thin silicate shell. Such a structure could be analogous to some M-type asteroids including Psyche, i.e. the target of a future NASA discovery mission to be launched in 2022 (Elkins-Tanton et al., 2014). They examined two scenarios: inward solidification (both concentric and dendritic) and cumulate inner-core growth. The cumulate inner-core growth considers iron snow or destabilised dendrites that crystallise at the core–mantle boundary and sink to form an inner solid core. They found that this cumulate inner core solidification would be energetically capable of sustaining a dynamo during solidification.

However, it may further require that iron snow or dendrites do not remelt during sinking to avoid the stable density stratification in the snow region (Rückriemen et al., 2015). Less power is available and therefore less likely is the magnetic field generation in an inward concentric or dendritic solidification scenario. In all scenarios, dynamo action during solidification is possible if the body is sufficiently large, and the assumed critical magnetic Reynolds number is sufficiently small. In an inwardly solidifying body, such a field could have been captured by the overlying solid iron shell. Solidification scenarios can therefore be distinguished by magnetic field measurements of the future Psyche Mission.

17.3.6 Rocky Exoplanets

With the vast array of extrasolar rocky planets now being discovered, an obvious question is whether these planets also possess an internally generated magnetic field and what dynamo behaviour can be expected. This question became particularly interesting because it is assumed that maintaining a habitable surface (i.e. having stable surface water) over long periods of time may require a magnetic field to shield the atmosphere from mass loss, retain large amounts of water and protect the surface from charged particles (Dehant et al., 2007; Lammer et al., 2008). Detectability of magnetic fields of rocky exoplanets are not yet possible but in principle possible when electrons from the stellar wind interact with the dynamo-generated magnetic field from the planet, emitting cyclotron radiation. However, predicted frequencies for Earth-like dynamos are below the 10 MHz cut-off of Earth's ionosphere and only anomalous boosts and modulations to the magnetic field intensity and cyclotron emission may allow for their detection in the future (e.g. Driscoll and Olson, 2011). Detection might also be possible with radio telescopes in space, which would not suffer from the low-frequency ionospheric cut-off.

So far, most models have investigated the magnetic field evolution of rocky exoplanets for planets that are equal or more massive than the Earth and have an Earth-like composition. Furthermore, it is assumed that these planets either have an active surface deformation such as plate tectonics or stagnant lid convection (e.g. Gaidos et al., 2010; Driscoll and Olson, 2011; Tachinami et al., 2011; van Summeren et al., 2013). Other studies also examine the effect of tidal heating on the dynamo generation, as a lot of these planets are detected close to their star where tidal effects can be very large (Driscoll and Barnes, 2015) or the difference in magnetic protection of tidally locked and unlocked planets (Zuluaga et al., 2013). The lack of observation, against which we can compare the predictions of these models, leave much room for uncertainties. This is especially true regarding mantle rheology, core composition and thermodynamic properties. Despite these severe limitations, a global picture of the thermal and magnetic history of the rocky exoplanets is emerging.

The thermal evolution model by van Summeren et al. (2013) predict a long-lived (~8 Gyr) field for Earth and similar dynamo evolution for Earth-mass exoplanets with plate tectonics. This dynamo is generated first by purely thermal and later by compositional convection in the core, as it is assumed for the Earth. It is further suggested that stagnant lid planets that have a relatively lower CMB heat flow produce more long-lived (~10.5 Gyr) but weaker dynamos. These weaker dynamos also start with purely thermal convection and may cease for several billions of years before they are subsequently reactivated due to inner core growth.

With increasing mass of the planets, the discussion about their thermal and magnetic field evolution becomes more controversial. Firstly, the pressure dependence of the mantle rheology for super-Earths is a topic of controversy. If viscosity increases with pressure (Stamenkovic et al., 2011; Tackley et al., 2013), then this will suppress convective vigour in the deep mantles of massive planets, which in turn will reduce heat flow across the CMB. Tachinami et al. (2011) predict that, above a critical planetary mass that also depends on the initial temperature jump across the CMB, enhanced viscosity terminates dynamo operation. If viscosity is almost constant or decreases with pressure by 2 to 3 orders of magnitude due to changes in the diffusion mechanism (from vacancy to interstitial) and a series of chemical and electronic phase transitions over 0.1–1 TPa as suggested by Karato (2011), then the deep mantles of super-Earths will be less viscous. In that case, the thermal and magnetic field evolution is more similar to Earth possibly showing a long-standing dynamo activity. Planets with higher surface temperatures can also support a dynamo more easily, as the resulting higher mantle temperatures lead to a lower viscosity and thus to more efficient convection and core cooling (Gaidos et al., 2010). Secondly, the higher pressure for more massive rocky exoplanets may also influence the melting and adiabatic curves in their cores. Gaidos et al. (2010) found an inversion of gradients of the melting and adiabatic curves in the core for rocky planets larger than $2.5M_{Earth}$. They suggest the presence of iron snow that crystallises at the core–mantle boundary – a crystallisation process that is already suggested for planetary bodies smaller than Earth (see Section 17.2). As described above, the dynamo action for this solidification process is more limited in time. Either massive rocky planets cannot sustain magnetic field when the solidification of the core begins (Bland et al., 2009; Gaidos et al., 2010) or their magnetic field ceases as soon as a solid inner core begins to form (Rückriemen et al., 2015). The presence of a long-standing dynamo may thus be less likely for more massive rocky planets.

17.4 Summary and Conclusion

Present-day self-generated magnetic fields in terrestrial bodies have been detected on Mercury and Jupiter's moon Ganymede, in addition to the Earth. Indications as to whether planets have already generated a magnetic field in their early evolution can be found in the remanent magnetisation of their silicate crusts. Data from planetary missions show remanent magnetised crust on Mercury, the Moon, Mars and also on some planetesimals. However, it is important to note that the lack of data does not necessarily imply the non-existence of past dynamo action. First, there must be sufficient magnetisable material that cools below its Curie temperature at the time of the dynamo action. This can be obtained with volcanism which typically decreases significantly over time for stagnant lid planets such as Mars and Mercury, making late dynamo action more difficult to detect. Second, remanent material can be later destroyed for instance by impacts or weathering. These effects are particularly relevant for the oldest crust. Third, the magnetisation must be strong enough to be detected at orbiter height. In the case of Ganymede with its thick ice layer of about 900 km on top the silicate interior, this can be very difficult. In addition, unfavourable eccentric orbits can also severely restrict the detection of remanent magnetised crust, as in the case of the MESSENGER mission to Mercury.

It is generally accepted that magnetic fields of the terrestrial bodies are triggered by dynamo action via thermal and/or compositional convection in the iron-rich cores although alternative mechanisms of dynamo generation such as mechanical stirring of the core due to mantle precession and impacts are discussed. Models have shown that thermal dynamos are mainly active in the early evolution of a planet when the core is strongly cooling. These dynamos are typically short-lived. Chemical dynamos are generated when the iron-rich core begins to crystallise. Normally they are active much longer than thermal dynamos and are most likely responsible for today's magnetic fields of Earth, Mercury and Ganymede. Despite some similarities in the magnetic field generation and its effects on the planet, such as the crust magnetisation, each terrestrial body has some special characteristics. Mercury with its large iron core shows a weak present-day magnetic field. Its equivalent dipole is offset towards the north by about 0.2 planetary radii and shows only a small deviation from axial symmetry. The detection of an old remanently magnetised crust suggests an early dynamo but it is not known whether a dynamo was continuously produced or whether it restarted later – may be even only recently – in the evolution. Mercury may have formed under very reducing conditions, which implies that silicon is also incorporated in the iron-rich core as a light element. However, how core crystallisation works and how a magnetic field is generated under these conditions is not known. The lunar magnetic field is long-lived with a lifetime of more than 2 billion years and possibly even 3 billion years. The early magnetisation indicates a magnetic field of great strength that later strongly declined. Probably there was a change in the mechanism of magnetic field generation, since a mechanism alone cannot explain the observations made so far. For instance, a precession driven dynamo has been proposed as an alternative mechanism for the early strong field, which was then replaced by a compositional dynamo. However, more work is needed in this direction to better understand the various mechanisms to generate a magnetic field. Mars shows a dichotomy in the crustal magnetisation with a strong magnetisation of the Southern Hemisphere, while the Northern Hemisphere has only a weak or no magnetisation. Whether this dichotomy has its origin in the core, in which a semi-spherical dynamo was produced, or whether it is rather connected with other processes, e.g. the remanent magnetisation in the northern crust was later destroyed, is debated. Ganymede shows a relatively strong present-day magnetic field, but it is uncertain when the core dynamo started. It is expected that a chemical dynamo is responsible for the magnetic field, and in the case of a Fe–FeS core with a composition on the iron-rich side of the eutectic, the self-generated magnetic field is probably a very young feature. In contrast to the Earth, another crystallisation mechanism is in work, in which iron crystallises in Ganymede's core near the core–mantle boundary as iron snow and sinks into the centre.

In conclusion, the study of the magnetic field evolution of terrestrial bodies has advanced significantly in the last decades, driven by improvements in modelling not only of the dynamo generation in the core but also of the thermochemical evolution of the planets, increasing computing power, and new observations. In particular, our perception of the iron-rich cores and their crystallisation has changed substantially in the past few years. Improved Fe–FeS phase diagrams and melting relations have motivated novel models of core dynamics and dynamos, in particular for planetary bodies with masses significantly smaller than Earth's. All in all, a rich variety of crystallisation scenarios and dynamos have become apparent and it has become clear that the Earth-like dynamo is probably not that typical. These new findings also leave many new questions open, some of them have already been mentioned in this chapter and which wait to be answered. To this end, coming mission such as BepiColombo to Mercury, Juice to Jupiter's icy Moons and Psyche to a M-type asteroid will help to better understand the magnetic field history of the planets. Not to forget the new research field of exoplanets, which has just emerged and will help to answer fundamental questions as to whether and under what conditions other rocky planets can generate magnetic fields, but also how important is a magnetic field for a habitable world.

References

Acuña, M. H., Connerney, J. E. P., Ness, N. F., Lin, R. P., Mitchell, D., Carlson, C. W., McFadden, J., Anderson, K. A., Rème, H., Mazelle, C., Vignes, D., Wasilewski, P. and Cloutier, P. (1999). Global distribution of crustal magnetism discovered by the Mars Global Surveyor MAG/ER experiment, *Science*, 284, 790–93.

Amit, H., Christensen, U. R. and Langlais, B. (2011). The influence of degree-1 mantle heterogeneity on the past dynamo of Mars, *Phys. Earth Planet. Inter.*, 189, 63–79.

Anderson, J. D., Jurgens, R. F., Lau, E. L., Slade, M. A., III and Schubert, G. (1996). Shape and orientation of Mercury from radar ranging data, *Icarus*, 124(2), 690–97.

Anderson, B. J., Johnson, C. L., Korth, H., Purucker, M. E., Winslow, R. M., Slavin J.A. and Zurbuchen, T.H. (2011). The global magnetic field of Mercury from MESSENGER orbital observations, *Science*, 333(6051), 1859–62.

Anderson, B. J., Johnson, C. L., Korth, H., Winslow, R. M., Borovsky, J. E., Purucker, M. E., Slavin, J. A., Solomon, S. C., Zuber, M. T. and McNutt, R. L., Jr. (2012). Low-degree structure in Mercury's planetary magnetic field, *J. Geophys. Res.*, 117, doi: 10.1029/2012JE004159.

Arkani-Hamed, J. and Boutin, D. (2012). Is the primordial crust of Mars magnetized?, *Icarus*, 221(1), 192–207.

Arkani-Hamed, J. and Dyment, J. (1996). Magnetic potential and magnetization contrasts of Earth's lithosphere, *J. Geophys. Res.*, 101(B5), 11401–26.

Arkani-Hamed, J. and Olson, P. (2010). Giant impact stratification of the Martian core, *Geophys. Res. Lett.*, 370(L02), 201, doi: 10.1029/2009GL041417.

Aubert, J., Labrosse, S. and Poitou, C. (2009). Modelling the palaeo-evolution of the geo-dynamo, *Geophys. J. Int.*, 179, 1414–28.

Bland, M. T., Showman, A. P. and Tobie, G. (2009). The orbital–thermal evolution and global expansion of Ganymede, *Icarus*, 200(1), 207–21.

Bleil, U. and Petersen, N. (1983). Variations in magnetization intensity and low-temperature titanomagnetite oxidation of ocean floor basalts, *Nature*, 301, 384–8.

Braginsky, S. I. (1964). Magnetohydrodynamics of the Earth core, *Geomag. Aeron.*, 4, 698–712.

Breuer, D., Labrosse, S. and Spohn, T. (2010). Thermal evolution and magnetic field generation in terrestrial planets and satellites, *Space Sci. Rev.*, 152(1), 449–500.

Breuer, D. and Moore, B. (2015). Dynamics and thermal history of the terrestrial planets, the Moon, and Io, in *Treatise on Geophysics*, 2nd edn., vol. 10, ed. T. Spohn, pp. 255–305, Elsevier, Oxford.

Breuer, D. and Spohn, T. (2003). Early plate tectonics versus single-plate tectonics: Evidence from the magnetic field history and crust evolution, *J. Geophys. Res.*, 108(E7), 5072, doi: 10.1029/20002JE001999.

Breuer, D. and Spohn, T. (2006). Viscosity of the Martian mantle and its initial temperature: Constraints from crust formation history and the evolution of the magnetic field, *Planet. Space Sci.*, 54, 153–69.

Bryson, J. F., Nichols, C. I., Herrero-Albillos, J., Kronast, F., Kasama, T., Alimadadi, H. and Harrison, R. J. (2015). Long-lived magnetism from solidification-driven convection on the pallasite parent body, *Nature*, 517(7535), 472–5.

Buono, A. S. and Walker, D. (2011). The Fe-rich liquidus in the Fe–FeS system from 1 bar to 10GPa, *Geochim. Cosmochim. Acta*, 75(8), 2072–87.

Busse, F. H. (1976). Generation of planetary magnetism by convection, *Phys. Earth Planet. Inter.*, 12(4), 350–58.

Byrne, P. K., Klimczak, C., Şengör, A. C., Solomon, S. C., Watters, T. R. and Hauck, S. A. (2014). Mercury's global contraction much greater than earlier estimates, *Nat. Geosci.*, 7(4), 301–7.

Cameron, A. G. W. (1997). The origin of the Moon and the single impact hypothesis V, *Icarus*, 126(1), 126–37.

Cameron, A. G. W. and Canup, R. M. (1998). The giant impact and the formation of the Moon, *Origin Earth Moon*, 957, 3.

Campbell, A. J., Seagle, C. T., Heinz, D. L., Shen, G. and Prakapenka, V. B. (2007). Partial melting in the iron–sulfur system at high pressure: A synchrotron X-ray diffraction study, *Phys. Earth Planet. Inter.*, 162(1), 119–28.

Cao, H., Aurnou, J. M., Wicht, J., Dietrich, W., Soderlund, K. M. and Russell, C. T. (2014). A dynamo explanation for Mercury's anomalous magnetic field, *Geophys. Res. Lett.*, 41 (12), 4127–34.

Carporzen, L., Weiss, B. P., Elkins-Tanton, L. T., Shuster, D. L., Ebel, D. and Gattacceca, J. (2011). Magnetic evidence for a partially differentiated carbonaceous chondrite parent body, *Proc. Natl. Acad. Sci.*, 108(16), 6386–9.

Chabot, N. L., Wollack, E. A., Klima, R. L. and Minitti, M. E. (2014). Experimental constraints on Mercury's core composition, *Earth Planet. Sci. Lett.*, 390, 199–208.

Chen, B., Li, J. and Hauck, S. A. (2008). Non-ideal liquidus curve in the Fe-S system and Mercury's snowing core, *Geophys. Res. Lett.*, 35, L07201, doi: 10.1029/2008GL033311.

Christensen, U. R. and Aubert, J. (2006). Scaling properties of convection-driven dynamos in rotating spherical shells and application to planetary magnetic fields, *Geophys. J. Int.*, 166 (1), 97–114.

Christensen, U. R., Holzwarth, V. and Reiners, A. (2009). Energy flux determines magnetic field strength of planets and stars, *Nature*, 457(7226), 167–9.

Christensen, U. R. and Wicht, J. (2007). Numerical dynamo simulations, in *Core Dynamics, Treatise on Geophysics*, vol. 8, ed. G. Schubert, pp. 254–82, Elsevier, Oxford.

Christensen, U. R. and Wicht, J. (2008). Models of magnetic field generation in partly stable planetary cores: Applications to Mercury and Saturn, *Icarus*, 196(1), 16–34.

Christensen, U. R. and Tilgner, A. (2004). Power requirement of the geodynamo from ohmic losses in numerical and laboratory dynamos, *Nature*, 429, 169–71, doi: 10.1038/nature02508.

Christensen, U. R. (2010). Dynamo scaling laws and applications to the planets, *Space Sci. Rev.*, 152, 565–90.

Christensen, U. R. (2015). Iron snow dynamo models for Ganymede, *Icarus*, 247, 248–59.

Chudinovskikh, L. and Boehler, R. (2007). Eutectic melting in the system Fe–S to 44 GPa, *Earth Planet. Sci. Lett.*, 257(1), 97–103.

Connerney, J. E. P., Acuña, M. H., Wasilewski, P. J., Ness, N. F., Reme, H., Mazelle, C. and Cloutier, P. A. (1999). Magnetic lineations in the ancient crust of Mars, *Science*, 284, 794–8.

Connerney, J. E. P., Acuña, M. H., Ness, N. F., Spohn, T. and Schubert, G. (2004). Mars crustal magnetism, in *Mars' Magnetism and Its Interaction with the Solar Wind*, pp. 1–32, Springer, Netherlands.

Curtis, S. A. and Ness, N. F. (1986). Magnetostrophic balance in planetary dynamos: Predictions for Neptune's magnetosphere, *J. Geophys. Res.*, 91(A10), 11003–8.

Davies, C. J. and Pommier, A. (2018). Iron snow in the Martian core?, *Earth Planet. Sci. Lett.*, 481, 189–200.

Dehant, V., Lammer, H., Kulikov, Y. N., Grießmeier, J. M., Breuer, D., Verhoeven, O., Karatekin, Ö., Van Hoolst, T., Korablev, O. and Lognonné, P. (2007). Planetary magnetic dynamo effect on atmospheric protection of early Earth and Mars, *Space Sci. Rev.*, 129(1–3), 279–300.

de Koker, N., Steinle-Neumann, G. and Vlček, V. (2012). Electrical resistivity and thermal conductivity of liquid Fe alloys at high P and T, and heat flux in Earth's core, *Proc. Natl. Acad. Sci.*, 109(11), 4070–73.

Dietrich, W. and Wicht, J. (2013). A hemispherical dynamo model: Implications for the Martian crustal magnetization, *PEPI*, 217, 10–21.

Dreibus, G. and Wänke, H. (1985). Mars: A volatile rich planet, *Meteoritics*, 20, 367–82.

Driscoll, P. and Olson, P. (2011). Optimal dynamos in the cores of terrestrial exoplanets: Magnetic field generation and detectability, *Icarus*, 213(1), 12–23.

Driscoll, P. E. and Barnes, R. (2015). Tidal heating of Earth-like exoplanets around M stars: thermal, magnetic, and orbital evolutions, *Astrobiology*, 15(9), 739–60.

Dumberry, M. and Rivoldini, A. (2015). Mercury's inner core size and core-crystallization regime, *Icarus*, 248, 254–68.

Dwyer, C. A., Stevenson, D. J. and Nimmo, F. (2011). A long-lived lunar dynamo driven by continuous mechanical stirring, *Nature*, 479, 212–14, doi: 10.1038/nature10564.

Elkins-Tanton, L. T., Weiss, B. P. and Zuber, M. T. (2011). Chondrites as samples of differentiated planetesimals, *Earth Planet. Sci. Lett.*, 305(1), 1–10.

Elkins-Tanton, L. T., Asphaug, E., Bell, J., Bercovici, D., Bills, B. G., Binzel, R. P., Bottke, W. F., Jun, I., Marchi, S., Oh, D. and Polanskey, C. A. (2014). Journey to a metal world: Concept for a discovery mission to Psyche, *Lunar Planet. Sci. Conf.*, 45, 1253.

Evans, A. J., Zuber, M. T., Weiss, B. P. and Tikoo, S. M. (2014). A wet, heterogeneous lunar interior: Lower mantle and core dynamo evolution, *J. Geophys. Res.*, 119(5), 1061–77.

Evans, A. J., Tikoo, S. M. and Andrews-Hanna, J. C. (2017). The case against an early lunar dynamo powered by core convection, *Geophys. Res. Lett.*, 44, doi: 10.1002/2017GL075441.

Fei, Y, Bertka, C. M. and Finger, L. W. (1997). High-pressure iron-sulfur compound, Fe_3S_2, and melting relations in the Fe-FeS system, *Science*, 275(5306), 1621–3.

Fei, Y., Li, J., Bertka, C. M. and Prewitt, C. T. (2000). Structure type and bulk modulus of Fe3S, a new iron-sulfur compound, *Am. Min.*, 85(11–12), 1830–33.

Fu, R. R., Weiss, B. P., Shuster, D. L., Gattacceca, J., Grove, T. L., Suavet,C., Lima, E. A., Li, L. and Kuan, A. T. (2012). An ancient core dynamo in asteroid Vesta, *Science*, 338, 238–41.

Gaidos, E., Conrad, C. P., Manga, M. and Hernlund, J. (2010). Thermodynamic limits on magnetodynamos in rocky exoplanets, *Astrophys. J.*, 718, 596–609.

Garcia, R., Gagnepain-Beyneix, J., Chevrot, S. and Lognonne, P. (2011). Very preliminary reference Moon model, *Phys. Earth. Planet. Inter.*, 188, 96–113, doi: 10.1016/j.pepi.2011.06.015.

Gomi, H., Ohta, K., Hirose, K., Labrosse, S., Caracas, R., Verstraete, M. J. and Hernlund, J. W. (2013). The high conductivity of iron and thermal evolution of the Earth's core, *Phys. Earth. Planet. Inter.*, 224, 88–103.

Grott, M., Breuer, D. and Laneuville, M. (2011). Thermo-chemical evolution and global contraction of Mercury, *Earth Planet. Sci. Lett.*, 307, 135–46, doi: 10.1016/j.epsl.2011.04.040.

Hauck, S. A. and Phillips, R. J. (2002). Thermal and crustal evolution of Mars, *J. Geophys. Res.*, 107(E7), 6–1.

Hauck, S. A., Aurnou, J. M. and Dombard, A. J. (2006). Sulfur's impact on core evolution and magnetic field generation on Ganymede, *J. Geophys. Res.*, 111(E9).

Hauck, S. A., Dombard, A. J., Phillips, R. J. and Solomon, S. C. (2004). Internal and tectonic evolution of Mercury, *Earth Planet. Sci. Lett.*, 222(3–4), 713–28.

Hauck, S. A., Margot, J.-L., Solomon, S. C., Lemoine, F. G., Mazarico, E., Peale, S. J., Perry, M. E., Phillips, R. J., Smith, D. E. and Zuber, M. T. (2013). The curious case of Mercury's internal structure, *J. Geophys. Res.*, 118, 1303–22, doi: 10.1002/jgre.20052.

Head, J. W., Greeley, R., Golombek, M. P., Hartmann, W. K., Hauber, E., Jaumann, R. and Carr, M. H. (2001). Geological processes and evolution, *Space Sci. Rev.*, 96(1/4): 263–92.

Heimpel, M. H., Aurnou, J. M., Al-Shamali, F. M. and Gomez Perez, N. (2005). A numerical study of dynamo action as function of spherical shell geometry, *Earth Planet. Sci. Lett.*, 236, 542–57.

Hevey, P. J. and Sanders, I. S. (2006). A model for planetesimal meltdown by 26Al and its implications for meteorite parent bodies, *Meteoritics Planet. Sci.*, 41(1), 95–106.

Hide, R. (1972). Comments on the Moon's magnetism, *Moon*, 4, 39.

Hood, L. L., Mitchell, D. L., Lin, R. P., Acuna, M. H. and Binder, A. B. (1999). Initial measurements of the lunar induced magnetic dipole moment using Lunar Prospector magnetometer data. *Geophys. Res. Lett.*, 26, 2327–30.

Hood, L. L., Zakharian, A., Halekas, J., Mitchell, D. L., Lin, R. P., Acuña, M. H. and Binder, A. B. (2001). Initial mapping and interpretation of lunar crustal magnetic anomalies using Lunar Prospector magnetometer data, *J. Geophys. Res.*, 106, 27825–39.

Johnson, C. L., Purucker, M. E., Korth, H., Anderson, B. J., Winslow, R. M., Al Asad, M. M. and Solomon, S. C. (2012). MESSENGER observations of Mercury's magnetic field structure, *J. Geophys. Res.*, 117(E12).

Johnson, C. L., Phillips, R. J., Purucker, M. E., Anderson, B. J., Byrne, P. K., Denevi, B. W. and Solomon, S. C. (2015). Low-altitude magnetic field measurements by MESSENGER reveal Mercury's ancient crustal field, *Science*, 348, 892–5.

Karato, S. I. (2011). Rheological structure of the mantle of a super-Earth: Some insights from mineral physics, *Icarus*, 212 (1), 14–23.

Kimura, J., Nakagawa, T. and Kurita, K. (2009). Size and compositional constraints of Ganymede's metallic core for driving an active dynamo, *Icarus*, 202(1), 216–24.

Kivelson, M. G., Khurana, K. K., Russell, C. T., Walker, R. J., Warnecke, J., Coroniti, F. V., Polanskey, C., Southwood, D. J. and Schubert, G. (1996). Discovery of Ganymede's magnetic field by the Galileo spacecraft, *Nature*, 384, 537–41.

Kivelson, M. G., Khurana, K. K. and Volwerk, M. (2002). The permanent and inductive magnetic moments of Ganymede, *Icarus*, 157, 507–22.

Knibbe, J. and van Westrenen, W. (2018). The thermal evolution of Mercury's Fe-Si core, *Earth Planet. Sci. Lett.*, 482, 147–59. doi: 10.1016/j.epsl.2017.11.006

Konôpková, Z., McWilliams, R. S., Gómez-Pérez, N. and Goncharov, A. F. (2016). Direct measurement of thermal conductivity in solid iron at planetary core conditions, *Nature*, 534(7605), 99–101.

Konopliv, A. S., Binder, A. B., Hood, L. L., Kucinskas, A. B., Sjogren, W. L. and Williams J. G. (1998). Improved gravity field of the Moon from Lunar Prospector, *Science*, 281(5382), 1476–80.

Konrad, W. and Spohn, T. (1997). Thermal history of the Moon: Implications for an early core dynamo and post-accretional magmatism, *Adv. Space Res.*, 19(10), 1511–21.

Lammer, H., Kasting, J. F., Chassefière, E., Johnson, R. E., Kulikov, Y.N. and Tian, F. (2008). Atmospheric escape and evolution of terrestrial planets and satellites, *Space Sci. Rev.*, 139(1–4), 399–436.

Laneuville, M., Wieczorek, M. A., Breuer, D., Aubert, J., Morard, G. and Rückriemen T. (2014). A long-lived lunar dynamo powered by core crystallization, *Earth Planet. Sci. Lett.*, 401, 251–260.

Langlais, B., Purucker, M. E. and Mandea, M. (2004). Crustal magnetic field of Mars, *J. Geophys. Res.*, 109(E2).

Le Bars, M., Wieczorek, M. A., Karatekin, O., Cebron, D. and Laneuville, M. (2011). An impact-driven dynamo for the early Moon, *Nature*, 479, 215–218, doi: 10.1038/nature10565.

Li, J., Fei, Y., Mao, H. K., Hirose, K. and Shieh, S. R. (2001). Sulfur in the Earth's inner core, *Earth Planet. Sci. Lett.*, 193, 509–14.

Lillis, R. J., Robbins, S., Manga, M., Halekas, J. S. and Frey, H. V. (2013). Time history of the Martian dynamo from crater magnetic field analysis, *J. Geophys. Res.*, 118(7), 1488–1511, doi: 10.1002/jgre.20105.

Lin, Y., Marti, P., Noir, J. and Jackson, A. (2016). Precession-driven dynamos in a full sphere and the role of large scale cyclonic vortices, *Phys. Fluids*, 28, 066601, doi: 10.1063/1.4954295.

Malavergne, V., Toplis, M. J., Berthet, S. and Jones, J. (2010). Highly reducing conditions during core formation on Mercury: Implications for internal structure and the origin of a magnetic field, *Icarus*, 206(1), 199–209.

Manglik, A., Wicht, J. and Christensen, U. R. (2010). A dynamo model with double diffusive convection for Mercury's core, *Earth Planet. Sci. Lett.*, 289(3), 619–28.

Margot, J. L., Peale, S. J., Jurgens, R. F., Slade, M. A. and Holin, I. V. (2007). Large longitude libration of Mercury reveals a molten core, *Science*, 316(5825), 710–14, doi: 10.1126/science.1140514.

Margot, J. L., Peale, S. J., Solomon, S. C., Hauck, S. A., Ghigo, F. D., Jurgens, R. F. and Campbell, D. B. (2012). Mercury's moment of inertia from spin and gravity data, *J. Geophys. Res.*, 117(E12).

McCammon, C. A., Ringwood, A. E. and Jackson, I. (1983). Thermodynamic of the system Fe-FeO-MgO at high pressure and temperature and a model for the formation of the Earth's core, *Geophys. J. R. Astron. Soc.*, 72, 577–95.

McSween, H. Y., Jr (1985). SNC meteorites: Clues to Martian petrologic evolution?, *Rev. Geophys.*, 23, 391–416.

Mohit, P. S. and Arkani-Hamed, J. (2004). Impact demagnetization of the Martian crust, *Icarus*, 168(2), 305–17.

Monteux, J., Jellinek, A. M. and Johnson, C. L. (2013). Dynamics of core merging after a mega-impact with applications to Mars' early dynamo, *Icarus*, 226(1), 20–32.

Morard, G., Andrault, D., Guignot, N., Sanloup, C., Mezouar, M., Petitgirard, S. and Fiquet, G. (2008). In situ determination of Fe-Fe3S phase diagram and liquid structural properties up to 65 GPa, *Earth Planet. Sci. Lett.*, 272, 620–26.

Morard, G. and Katsura, T. (2010). Pressure–temperature cartography of Fe–S–Si immiscible system, *Geochim. Cosmochim. Acta*, 74(12), 3659–67.

Morschhauser, A., Lesur, V. and Grott, M. (2014). A spherical harmonic model of the lithospheric magnetic field of Mars, *J. Geophys. Res.*, 119(6), 1162–88.

Moskovitz, N. and Gaidos, E. (2011). Differentiation of planetesimals and the thermal consequences of melt migration, *Meteorit. Planet. Sci.*, 46(6), 903–18.

Ness, N. F., Acuna, M. H., Connerney, J., Wasilewski, P. and Bauer, S. J. (1999). MGS magnetic fields and electron reflectometer investigation: Discovery of paleomagnetic fields due to crustal remanence, *Adv. Space Res.*, 23(11), 1879–86.

Ness, N. F., Behannon, K. W., Lepping, R. P., Schatten, K. H. and Whang, Y. C. (1974). Magnetic field observations near Mercury: Preliminary results from Mariner 10, *Science*, 185, 151–60.

Ness, N. F., Behannon, K. W., Lepping, R. P. and Whang, Y. C. (1975). Magnetic field of Mercury, *J. Geophys. Res.*, 80, 2708–16.

Neumann, W., Breuer, D. and Spohn, T. (2012). Differentiation and core formation in accreting planetesimals, *Astron. Astrophys.*, 543, A141.

Neumann, W., Breuer, D. and Spohn, T. (2014). Differentiation of Vesta: Implications for a shallow magma ocean, *Earth Planet. Sci. Lett.*, 395, 267–80.

Nimmo, F. (2000). Dike intrusion as a possible cause of linear Martian magnetic anomalies, *Geology*, 28, 391–4.

Nimmo, F. (2009). Energetics of asteroid dynamos and the role of compositional convection, *Geophys. Res. Lett.*, 36(10).

Nimmo, F. (2015). Thermal and compositional evolution of the core, in *Treatise on Geophysics*, 2nd edn., ed. D. Stevenson and G. Schubert, pp. 201–19, Elsevier, Amsterdam.

Nimmo, F. and Stevenson, D. (2000). Influence of early plate tectonics on the thermal evolution and magnetic field of Mars, *J. Geophys. Res.*, 105, 11969–79.

Nittler, L. R., Starr, R. D., Weider, S. Z., McCoy, T. J., Boynton, W. V., Ebel, D. S., Ernst, C. M., Evans, L. G., Goldsten, J. O., Hamara, D. K., Lawrence, D. J., McNutt, R. L., Schlemm, C. E., Solomon, S. C. and Sprague, A. L. (2011). The major-element composition of Mercury's surface from MESSENGER X-ray spectrometry, *Science*, 333, 1847–1950, doi: 10.1126/science.1211567.

Olson, P., Landeau, M. and Hirsh, B. H. (2017). Laboratory experiments on rain-driven convection: Implications for planetary dynamos, *Earth Planet. Sci. Lett.*, 457, 403–11.

Olson, P. and Christensen, U. R. (2006). Dipole moment scaling for convection-driven planetary dynamos, *Earth Planet. Sci. Lett.*, 250(3), 561–71.

Peale, S. J. (1976). Inferences from the dynamical history of Mercury's rotation, *Icarus*, 28(4), 459–67.

Plesa, A.-C., Grott, M., Tosi, N., Breuer, D., Spohn, T. and Wieczoreck, M. (2016). How large are present-day heat flux variations across the surface of Mars?, *J. Geophys. Res.*, 121 (12), 2386–2403, doi: 10.1002/2016JE005126.

Pozzo, M., Davies, C., Gubbins, D. and Alfe, D. (2012). Thermal and electrical conductivity of iron at Earth's core conditions, *Nature*, 485, 355–359.

Rai, N. and Westrenen, W. (2013). Core-mantle differentiation in Mars, *J. Geophys. Res.*, 118(6), 1195–1203.

Reese, C. C. and Solomatov, V. S. (2010). Early Martian dynamo generation due to giant impacts, *Icarus*, 207(1), 82–97.

Ringwood, A. E. (1977). Composition of the core and implications for the origin of the earth, *Geochem. J.*, 11, 111–35.

Rivoldini, A. and Van Hoolst, T. (2013). The interior structure of Mercury constrained by the low-degree gravity field and the rotation of Mercury, *Earth Planet. Sci. Lett.*, 377, 62–72.

Rivoldini, A., Van Hoolst, T., Verhoeven, O., Mocquet, A. and Dehant, V. (2011). Geodesy constraints on the interior structure and composition of Mars, *Icarus*, 213(2), 451–72.

Roberts, J. H., Lillis, R. J. and Manga, M. (2009). Giant impacts on early Mars and the cessation of the Martian dynamo, *J. Geophys. Res.*, 114(E4).

Rochette, P., Lorand, J.-P., Fillion, G. and Sautter, V. (2001). Pyrrhotite and the remanent magnetization of SNC meteorites: A changing perspective on Martian magnetism, *Earth Planet. Sci. Lett.*, 190(1–2), 1–12.

Rückriemen, T., Breuer, D. and Spohn, T. (2015). The Fe snow regime in Ganymede's core: A deep-seated dynamo below a stable snow zone, *J. Geophys. Res.*, 120(6), 1095–1118.

Rückriemen, T., Breuer, D. and Spohn, T. (2018). Top-down freezing in a Fe–FeS core and Ganymede's present-day magnetic field, *Icarus*, 307, 172–96.

Sahijpal, S., Soni, P. and Gupta, G. (2007). Numerical simulations of the differentiation of accreting planetesimals with 26Al and 60Fe as the heat sources, *Meteoritics Planet. Sci.*, 42(9), 1529–48.

Sanloup, C. and Fei, Y. (2004). Closure of the Fe-S-Si liquid miscibility gap at high pressure, *Phys. Earth Planet. Inter.*, 147, 57–65, doi: 10.1016/j.pepi.2004.06.008.

Scheinberg, A., Soderlund, K. M. and Schubert, G. (2015). Magnetic field generation in the lunar core: The role of inner core growth, *Icarus*, 254, 62–71.

Scheinberg, A., Elkins-Tanton, L. T., Schubert, G. and Bercovici, D. (2016). Core solidification and dynamo evolution in a mantle-stripped planetesimal, *J. Geophys. Res.*, 121 (1), 2–20.

Schubert, G., Ross, M. N., Stevenson, D. J. and Spohn, T. (1988). Mercury's thermal history and the generation of its magnetic field, in *Mercury*, ed. F. Viulas, C. R. Chapman and M. S. Matthews, pp. 514–61, University Press of Arizona, Tucson.

Schubert, G., Zhang, K., Kivelson, M. G. and Anderson, J. D. (1996). The magnetic field and internal structure of Ganymede, *Nature*, 384(6609), 544–5.

Schubert, G., Anderson, J. D., Spohn, T. and McKinnon, W. B. (2004). Interior composition, structure and dynamics of the Galilean satellites, in *Jupiter: The Planet, Satellites and Magnetosphere*, pp. 281–306. Cambridge University Press, Cambridge.

Scott, H. P., Williams, Q. and Ryerson, F. J. (2002). Experimental constraints on the chemical evolution of large icy satellites, *Earth Planet. Sci. Lett.*, 203, 399–412.

Shimizu, H., Matsushima, M., Takahashi, F., Shibuya, H. and Tsunakawa, H. (2013). Constraint on the lunar core size from electromagnetic sounding based on magnetic field observations by an orbiting satellite. *Icarus*, 222(1), 32–43.

Snyder, G. A., Taylor, L. A. and Neal, C. R. (1992). A chemical model for generating the sources of mare basalts: Combined equilibrium and fractional crystallization of the lunar magmasphere, *Geochim. Cosmochim. Acta*, 56(10), 3809–23.

Smith, D., Zuber, M., Phillips, R., Hauck, S. A., Lemoine, F., Mazarico, E., Neumann, G., Peale, S., Margot, J.-L., Johnson, C. L., Torrence, M. H., Perry, M. E., Rowlands, D. D., Goossens, S., Head, J. W. and Taylor, A. H. (2012). Gravity field and internal structure of Mercury from MESSENGER, *Science*, 336(6078), 214–17, doi: 10.1126/science.1218809.

Sohl, F., Spohn, T., Breuer, D. and Nagel, K. (2002). Implications from Galileo observations of the interior structure and chemistry of the Galilean satellites, *Icarus*, 157, 104–19.

Sohl, F. and Schubert, G. (2015). Interior structure, composition, and mineralogy of the terrestrial planets, in *Physics of Terrestrial Planets and Moons, Treatise on Geophysics*, 2nd edn., vol. 10, pp. 23–64, Elsevier, New York.

Solomon, C. S., Aharonson, O., Aurnou, J. M., Banerdt, W. B., Carr, M. H., Dombard, A. J., Frey, H. V., Golombek, M. P., Hauck, S. A., Head, J. W., Jakosky, B. M., Johnson, C. L., McGovern, P. J., Neumann, G. A., Phillips, R. J., Smith, D. E. and Zuber, M. T. (2005). New perspectives on ancient Mars, *Science*, 307, 1214–20.

Spohn, T., Acuña, M. A., Breuer, D., Golombek, M., Greeley, R., Halliday, A., Hauber, E., Jaumann, R. and Sohl, F. (2001). Geophysical constraints on the evolution of Mars, *Space Sci. Rev.*, 96, 231–62.

Spohn, T., Sohl, F. and Breuer, D. (1998). Mars, *Astron. Astrophys. Rev.*, 8, 181–235.

Stacey, F. D. and Anderson, O. L. (2001). Electrical and thermal conductivities of Fe–Ni–Si alloy under core conditions, *Phys. Earth Planet. Inter.*, 124(3), 153–62.

Stamenković, V., Breuer, D. and Spohn, T. (2011). Thermal and transport properties of mantle rock at high pressure: Applications to super-Earths, *Icarus*, 216(2), 572–96.

Stanley, S., Bloxham, J., Hutchinson, W. E. and Zuber, M. T. (2005). Thin shell dynamo models consistent with Mercury's weak observed magnetic field, *Earth Planet. Sci. Lett.*, 234, 27–38.

Stanley, S., Elkins-Tanton, L., Zuber, M. T. and Parmentier, E. M. (2008). Mars' paleomagnetic field as the result of a single-hemisphere dynamo, *Science*, 321(5897), 1822–5.

Stegman, D. R., Jellinek, A. M., Zatman, S. A., Baumgardner, J. R. and Richards, M. A. (2003). An early lunar core dynamo driven by thermochemical mantle convection, *Nature*, 421(6919), 143–6.

Sterenborg, M. G. and Crowley, J. W. (2013). Thermal evolution of early solar system planetesimals and the possibility of sustained dynamos, *Phys. Earth Planet. Inter.*, 214, 53–73.

Stevenson, D. J. (2001). Mars core and magnetism, *Nature*, 412, 214–19.

Stevenson, D. J., Spohn, T. and Schubert, G. (1983). Magnetism and thermal evolution of the terrestrial planets, *Icarus*, 54, 466–89.

Stewart, A. J., Schmidt, M. W., van Westrenen, W. and Liebske, C. (2007). Mars: A new core-crystallization regime, *Science*, 316 (5829), 1323–5.

Tachinami, C., Senshu, H. and Ida, S. (2011). Thermal evolution and lifetime of intrinsic magnetic fields of super-Earths in habitable zones, *Astrophys. J.*, 726(2), 70.

Tackley, P. J., Ammann, M., Brodholt, J. P., Dobson, D.P. and Valencia, D. (2013). Mantle dynamics in super-Earths: Post-perovskite rheology and self-regulation of viscosity, *Icarus*, 225(1), 50–61.

Tarduno, J. A., Cottrell, R. D., Nimmo, F., Hopkins, J., Voronov, J., Erickson, A., Blackman, E., Scott, E. R. and McKinley, D. R. (2012). Evidence for a dynamo in the main group pallasite parent body, *Science*, 338, 939–42.

Tian, Z., Zuber, M. T. and Stanley, S. (2015). Magnetic field modeling for Mercury using dynamo models with a stable layer and laterally variable heat flux, *Icarus*, 260, 263–8.

Tikoo, S. M., Weiss, B. P., Cassata, W., Shuster, D. L., Gattacceca, J., Lima, E. A., Suavet, C., Nimmo, F. and Fuller, M. (2014). Decline of the lunar core dynamo, *Earth Planet. Sci. Lett.*, 404, 89–97.

Tikoo, S. M., Weiss, B. P., Shuster, D. L., Suavet, C., Wang, H. and Grove, T. L. (2017). A two-billion-year history for the lunar dynamo, *Sci. Adv.*, 3, e1700207, doi: 10.1126/sciadv.1700207s.

Tosi, N., Grott, M., Plesa, A.-C. and Breuer, D. (2013). Thermo-chemical evolution of Mercury's interior, *J. Geophys. Res.*, doi: 10.1002/jgre.20168.

Usselman, T. M. (1975). Experimental approach to the state of the core; Part I, The liquidus relations of the Fe-rich portion of the Fe-Ni-S system from 30 to 100 kb, *Am. J. Sci.*, 275(3), 278–90.

Van Summeren, J., Gaidos, E. and Conrad, C. P. (2013). Magnetodynamo lifetimes for rocky, Earth-mass exoplanets with contrasting mantle convection regimes, *J. Geophys. Res.*, 118(5), 938–51.

Vervelidou, F., Lesur, V., Grott, M., Morschhauser, A. and Lillis, R. J. (2017). Constraining the date of the Martian dynamo shutdown by means of crater magnetization signatures, *J. Geophys. Res.*, doi: 10.1002/2017JE005410.

Vilim, R., Stanley, S. and Hauck, S. A. (2010). Iron snow zones as a mechanism for generating Mercury's weak observed magnetic field, *J. Geophys. Res.*, 115(E11).

Wang, H., Weiss, B. P., Bai, X. N., Downey, B. G., Wang, J., Wang, J., Suavet, C., Fu, R. R. and Zucolotto, M. E. (2017). Lifetime of the solar nebula constrained by meteorite paleomagnetism, *Science*, 355, 623–7.

Wang, Z. and Becker, H. (2017). Chalcophile elements in Martian meteorites indicate low sulfur content in the Martian interior and a volatile element-depleted late veneer, *Earth Planet. Sci. Lett.*, 463, 56–68.

Weber, R. C., Lin, P. Y., Garnero, E. J., Williams, Q. and Lognonne, P. (2011). Seismic detection of the lunar core, *Science*, 331, 309–12, doi: 10.1126/science.1199375.

Weider, S. Z., Nittler, L. R., Starr, R. D., McCoy, T. J., Stockstill-Cahill, K. R., Byrne, P. K. and Solomon, S. C. (2012). Chemical heterogeneity on Mercury's surface revealed by the MESSENGER X-Ray Spectrometer, *J. Geophys. Res.*, 117(E12).

Weiss, B. P., Berdahl, J. S., Elkins-Tanton, L., Stanley, S., Lima, E. A. and Carporzen, L. (2008). Magnetism on the angrite parent body and the early differentiation of planetesimals, *Science*, 322(5902), 713–16.

Weiss, B. P., Gattacceca, J., Stanley, S., Rochette, P. and Christensen, U. R. (2010). Paleomagnetic records of meteorites and early planetesimal differentiation, *Space Sci. Rev.*, 152(1–4), 341–90.

Weiss, B. P., Shuster, D. L. and Stewart, S. T. (2002). Temperatures on Mars from 40Ar/39Ar thermochronology of ALH84001, *Earth Planet. Sci. Lett.*, 201(3–4), 465–72.

Weiss, B. P. and Tikoo, S. M. (2014). The lunar dynamo, *Science*, 346, 1198–1209, doi: 10.1126/science.1246753.

Wicht, J., Mandea, M., Takahashi, F., Christensen, U. R., Matsushima, M. and Langlais, B. (2007). The origin of Mercury's internal magnetic field, *Space Sci. Rev.*, 132, 261–90, doi: 10.1007/s11214-007-9280-5.

Wicht, J. and Heyner, D. (2014). Mercury's magnetic field in the MESSENGER era, *Planet. Geod. Remote Sens.*, 223.

Wieczorek, M. A., Jolliff, B. L., Khan, A., Pritchard, M. E., Weiss, B. P., Williams, J. G., Hood, L. L., Righter, K., Neal, C. R., Shearer, C. K., McCallum, I. S., Tompkins, S., Hawke, B. R., Peterson, C., Gillis, J. J. and Bussey, B. (2006). The constitution and structure of the lunar interior, *Rev. Mineral. Geochem.*, 60, 221–364.

Williams, Q. (2009). Bottom-up versus top-down solidification of the cores of small solar system bodies: Constraints on paradoxical cores, *Earth Planet. Sci. Lett.*, 284(3), 564–9.

Williams, J. G. and Nimmo, F. (2004). Thermal evolution of the Martian core: Implications for an early dynamo, *Geology*, 32 (2), 97–100.

Williams, J. G., Boggs, D. H., Yoder, C. F., Ratcliff, J. T. and Dickey, J. O. (2001). Lunar rotational dissipation in solid body and molten core, *J. Geophys. Res.*, 106(E11), 27933–68.

Zhan, X. and Schubert, G. (2012). Powering Ganymede's dynamo, *J. Geophys. Res.*, 117(E8).

Zhang, N., Parmentier, E. M. and Liang, Y. (2013). Effects of lunar cumulate mantle overturn and megaregolith on the expansion and contraction history of the Moon, *Geophys. Res. Lett.*, 40(19), 5019–23.

Zhang, N., Dygert, N., Liang, Y. and Parmentier, E. M. (2017). The effect of ilmenite viscosity on the dynamics and evolution of an overturned lunar cumulate mantle, *Geophys. Res. Lett.*, 44, 6543–52, doi: 10.1002/2017GL073702.

Zuluaga, J. I., Bustamante, S., Cuartas, P. A. and Hoyos, J. H. (2013). The influence of thermal evolution in the magnetic protection of terrestrial planets, *Astrophys. J.*, 770, doi: 10.1088/0004-637X/770/1/23.

18

Solar Variability

Causes, Current Understanding, Prospects for the Future

André Balogh

18.1 Total Solar Irradiance (TSI) as a Function of the Solar Activity Cycle

The total energetic output of the Sun is calculated to be 3.83 × 10^{33} J/s. This corresponds to a total solar irradiance (TSI) at the orbit of the Earth of 1360.8 ± 0.5 W/m^2 (Kopp, 2016), called the *solar constant*. Until accurate measurements of the solar constant could be made in space, no reliable variations of TSI could be detected. In the past 40 years, the accuracy and, very importantly, the cross-calibration of instruments on different space missions measuring the solar constant have continually improved to yield generally agreed results (Fröhlich, 2006, 2013; Zacharias, 2014; Dudok de Wit et al., 2017). The variations of TSI could then be studied quantitatively, on different timescales, in particular in relation to the 11 year solar activity cycle (Solanki et al., 2013; Yeo et al., 2017a). Figure 18.1.1 shows the composite TSI over the past nearly 40 years for which observations were made in space, with instruments on a range of spacecraft that could, after painstaking cross-calibration, provide a consistent time series. There is a modulation of the TSI with the 11 year solar activity cycle, but at a very small amplitude compared to the average irradiance. Yeo et al. (2014) reviewed the current state of the variations observed and modelled in solar irradiance as a function of the solar activity cycle. The total variability is less than one part per thousand; this needs to be compared to many other solar parameters that vary over a much broader range, as can be seen below in the more frequently used parameters to describe solar variability, such as the sunspot number. The variations in TSI have been recently shown to be very largely due to the variability of solar surface magnetism (Yeo et al., 2017b).

Even if the fraction of the energy in the more variable components is a very small proportion of the TSI, the effects of the solar UV radiation on the Earth's space environment are significant and vary with the solar activity cycle (Haigh, 2007). The variations in TSI are largely due to the variable contribution of the short-wavelength UV radiation below wavelengths of ~400 nm (Krivova et al., 2006). The measurement of the solar spectral irradiance (SSI) shows that at the shorter wavelengths, for instance around Lyman-alpha, 121.6 nm, the variations in the irradiance can be as much as 50%–100%, as opposed to the nearly constant value of the TSI.

The two results, the dependence of the solar cycle variation of TSI on the short-wavelength UV output of the Sun and on the surface magnetic field, are effectively two aspects of the same causative agent. The short-wavelength output of the Sun is dominated by photospheric magnetic structures that present both higher intensities of the field and more diverse surface features that contribute to the increase of the intensity of short-wavelength radiation.

18.2 Predictability and Forecasting: Tests of Understanding Solar Phenomena

Variability is the most characteristic property of all solar phenomena. For solar–terrestrial physics (the science underpinning space weather applications), causal relations that link the different phenomena on the Sun to their impact on the near-Earth environment need to be understood for effective predictions. In the following, solar variability is described, primarily in the context of the 11 year activity cycle. There is a plethora of solar parameters and observations, some on very long timescales, some on short timescales. There are equally many correlations between the parameters, generally dominated by correlated variations (and generally in phase with, or nearly so, the dominant solar activity index – the sunspot numbers). Sunspots (Solanki, 2003) remain the most prominent indicators of both solar magnetism and solar activity.

However, correlations do not imply direct physical causality. The main thrust of solar research now and in the future is to investigate the relationships among the solar observables, to try to determine the physical processes that can be used to construct causal chains of phenomena. At present, no such chains are identified, despite considerable efforts over the past decades, underpinned by increasingly detailed observations.

Improving the modelling of solar phenomena, using both physical insights and observational constraints, is the objective of most solar research. The main test is to what extent

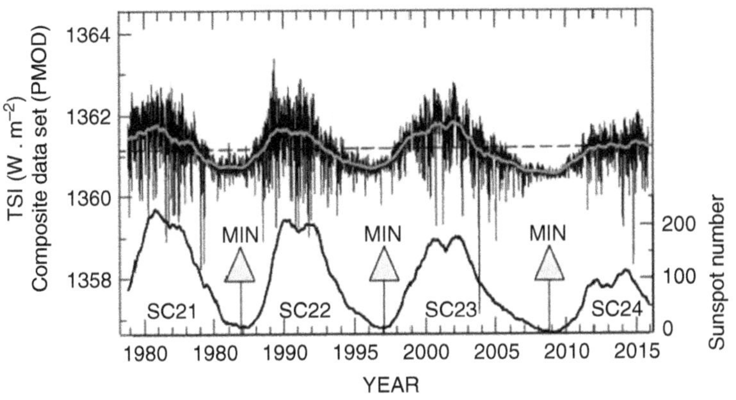

Figure 18.1.1 Composite total solar irradiance, from cross-calibration of six instruments by the Physikalisch-Meteorologisches Observatorium Davos (PMOD), Switzerland (Fröhlich, 2013). TSI is modulated on short timescales primarily by bright areas on the Sun (plages) and sunspot groups, but a large fraction of variation on solar cycle timescales is due to variations in the UV part of the spectrum. (Figure based on Kopp, 2016.)

models can predict future observations, given current sets of parameters. Looked at from this point of view, using a range of empirical relationships, there are some satisfying successes. A good test case, described in detail below, is the prediction of the solar cycle, in terms of the maximum number of sunspots in the next cycle. This was attempted for the current cycle 24. The predictions ranged over values with a factor 2.5 between the smallest and the largest. Although several predictions proved to be very good, this cannot be considered a success until a better physical relationship is elaborated between the phenomena involved in the prediction process. In the past 60 years or so, the vastly increased observational database, both from the ground and from space platforms, has provided observations that should enable a firmer foundation to be built for predictive solar science. While two decades ago, there was general optimism that the forecasting ability of solar models was improving, the Sun provided an unexpected surprise in two unusual solar cycles, 23 and 24, which placed question marks on many of the correlations previously envisaged. In particular, by bringing up the shadow of a deep solar minimum activity phase, as experienced in the seventeenth century, the previously trusted correlations have been subjected to much more demanding scrutiny.

This is the situation currently. In the following, some of the more prominent correlations between solar phenomena are described, with a view to identifying possible ways forward. This is a fast moving target. Currently, predictions are made concerning the forthcoming solar cycle 25, expected to start late in 2019 or early in 2020. However, this research is conducted with less confidence than a cycle ago, given the disappointing results of forecasting cycle 24. Nevertheless, it is expected that lessons have been learned and that research will concentrate on the more promising threads that have emerged from the previous attempts.

18.3 Sunspots and the Solar Activity Cycle

The (approximately) 11 year solar activity cycle and its general development have been well observed in many solar phenomena. The cycles are based on sunspot numbers, which have reasonably well-defined rising and declining phases separated by peak numbers at solar maximum activity. Solar cycles are conventionally defined as the intervals between successive minima in sunspots. For an introductory overview, see Balogh et al. (2014). For a detailed review that presents a comprehensive list of what the observations have shown about the solar cycle, including the challenges created by the observations of solar cycle 23 and the significantly less active solar cycle 24, see Hathaway (2015). The underlying periodicity, however, is based on the Sun's magnetic field, which has a 22 year cycle, as the polarity of the main, dipole component of the Sun reverses close to the maximum in each 11 year sunspot cycle (Babcock, 1961). The 22 year magnetic cycle is of fundamental importance in the dynamics of the solar interior – however, the 11 year periodicity of most of the observed activity indicators remains the target of solar activity research.

Samuel Schwabe, Alexander von Humboldt and Rudolf Wolf identified the near-11-year cyclic nature of sunspots in the middle decades of the nineteenth century (for a brief summary of the historical context, see e.g. Balogh and Thompson, 2009, and references therein). This led to the identification of the modern sequence of sunspot cycles from the mid eighteenth century; the first numbered cycle in the contemporary series started in 1755. Cycles in the twenty-first century of interest so far are cycle 23 (ended in December 2008), cycle 24 (from December 2008 to late 2019 or early 2020) and cycle 25 (expected to start by 2020).

Between the discovery of sunspots around 1610 and the mid eighteenth century, the Sun entered a nearly spotless state, lasting about 70 years (from 1645 to 1715, as defined by Eddy, 1976), now called the Maunder minimum. This state of the Sun has been investigated extensively, for both a better understanding of the Sun and the effects of its low activity levels on the Earth (Lockwood et al., 2017). The possibility of a cooling of the climate in response to low solar activity remains a subject of both research and speculation. This has been prompted by the unusually long and deep sunspot minimum between cycles 23 and 24 and the general decrease in the Sun's magnetic field. Understanding how solar activity can decrease to Maunder minimum levels is one of the questions that remain open.

Difficulties in observing and counting sunspots and sunspot groups in a consistent manner have led to recent efforts to achieve a consistency that could be extrapolated to historical data (Clette et al., 2014). In July 2015, a new, internationally agreed sunspot number list (now version V2.0) was introduced. The data are held in and maintained by the Solar Data Analysis Center (www.sidc.be/silso/). The previously used lists of sunspot numbers, now referred to as the "American" sunspot numbers, are still available and held by NOAA, National Centers for Environmental Information (www.ngdc.noaa.gov/stp/solar/solar-indices.html). Solar literature even now often uses the American lists; publications prior to 2015, including all the papers and studies about sunspot number prediction, use these same lists. In this work, both are used, as it is not possible to recalibrate previous studies to the two standards. However, in all cases when sunspot numbers are used, including figures based on earlier publications, explicit reference is made to the source of the data.

Even though the sunspot cycle has now been reconstructed over millennia (Usoskin, 2017), its underlying physical causes are little understood. One thing is clear: the variability of the Sun, including the quasi-periodic cycles in sunspot numbers, is a manifestation of internal processes best observed in solar magnetic phenomena. Sunspots are only the most visible, longest observed magnetic manifestations of internal dynamics. The origin of the solar magnetic field is the solar dynamo, operating in the solar interior. Its drivers are turbulent convective motion of the plasma in the outer third of the Sun and latitude- and depth-dependent differential rotation.

The operation of the dynamo remains, in several of its details, insufficiently understood, despite the progress made in probing the solar interior using helioseismology and in the spatial and temporal details of surface magnetic observations. The model originally proposed by Babcock (1961) and Leighton (1969) is still a valid basic template for modern developments of the solar dynamo. Many refinements have been added to this model, of which the addition of the meridional circulation observed through the solar cycle has had the most lasting effect (Wang and Sheeley, 1991), leading to the formulation of the flux transfer dynamo that has been in the mainstream of dynamo investigations. Modelling and computational advances have also brought increased understanding of Leighton's dynamo concept through the exploration of the relevant parameter space (Cameron and Schüssler, 2017). Recent, detailed reviews of the current state and future prospects of the solar dynamo models can be found in Brun et al. (2004), Dikpati and Gilman (2009), Weiss and Thompson (2009), Jiang et al. (2013), Charbonneau (2014) and Brun et al. (2015). It is generally agreed that the solar activity cycle and its variability, following the operation of the dynamo that drives it, are chaotic: the solar activity cycle is driven deterministically rather than stochastically (Weiss and Tobias, 2000). This limits currently the prospects of its predictability, which will improve only with better dynamo models.

Investigating the interior of the Sun has been made possible by helioseismology, the study of acoustic oscillations in the photosphere, which can be used for determining internal physical and dynamical parameters (Deubner and Gough, 1984). Since the discovery of solar oscillations about 50 years ago, the advances made in the physics of the interior of the Sun have enabled the building of a very detailed picture of its dynamic behaviour (Hanasoge et al., 2015). In particular, helioseismology has brought a new perspective on solar magnetism by enabling the formulation of processes, increasingly constrained by observational data that contribute to the generation and cyclical evolution of the magnetic field (Brun et al., 2015).

The continued accumulation of helioseismic observations, their increasing resolution, their quality and their analysis over the past three solar cycles have confirmed the complexity of the solar dynamo (including its multi-polar nature) that current models still cannot represent fully (Fletcher et al., 2010; Simoniello et al., 2016). While progress is being made to overcome the computational complexities of modelling the solar dynamo (e.g. Charbonneau, 2013), this is a domain in which future high-resource computing capabilities will be required (Stejko et al., 2017).

The asymmetry in the activity between the hemispheres, the possible existence of a second dynamo and the persistence of the modulation of solar activity with a period between 1 and 3 years (the Quasi-Biennial Oscillation, QBO), all discussed below, strongly imply the need for further studies, observations and modelling of the causative agents of solar activity.

The Sun is not the only star with activity cycles. A range of other stars have been identified with starspots, activity cycles and even flares (Strassmeier, 2009). The underlying magnetic field processes have also been identified in some cases. Despite the orders of magnitude fewer observational data available compared to the Sun, stellar activity cycles, driven by dynamos in different parameter regimes, may contribute to understanding the solar dynamo. For the current state of different aspects of the research, concerning the solar–stellar activity cycle connection, see Brun et al. (2015), Pipin and Kosovichev (2016), Korhonen (2016) and Choudhuri (2017).

18.4 Activity Indicators and Forecasting the Solar Cycles

Most solar phenomena vary with the solar cycle, even if the sunspot numbers remain the baseline against which other indicators are considered. However, the variability associated with the different measures of solar activity is not unique, nor is 'activity' the only component of variability. Different measures are not always in phase with sunspot numbers and with each other. Causally justified correlations remain generally tentative.

The composite presentation of the key observations in cycles 21 (in part), 22, 23 and 24 (up to mid 2017) is shown in Figure 18.4.1. Figure 18.4.1a is the butterfly diagram, showing the progression of sunspots in heliolatitude as the activity cycles progress; originally identified by Maunder (1904), the butterfly diagram has become a key presentation of solar activity cycles. The colour scheme represents a way of representing the evolution in the area occupied by the sunspots. Sunspots of a new cycle first appear in mid-latitudes; as the cycle progresses, sunspots appear at lower latitudes. The last spots of a cycle are close to the equator. Solar cycles overlap in activity measures; the first spots of new cycles appear before the near-equatorial spots of the current cycle disappear. As seen in Figure 18.4.1, the overlap was smaller between cycles 23 and 24 than between cycles 22 and 23, owing to the much extended minimum that concluded cycle 23.

Figure 18.4.1b shows the sunspot numbers (using the new version V2.0, source: WDC-SILSO, Royal Observatory of Belgium, Brussels; see Clette et al., 2014) that remain the most used parameter to characterise solar activity. The figure also shows the solar polar magnetic field, in anti-phase with the sunspot number (data courtesy of the Wilcox Solar Observatory, http://wso.stanford.edu/). The polar fields have a determinant role in the solar magnetic cycle (Petrie et al., 2014; see Section 18.5).

In Figure 18.4.1c, the magnetic butterfly diagram is shown (based on Petrie, 2017, Figure 7) to illustrate the progression of the Sun's surface magnetic field through the activity cycles. The diagram shows the drift of magnetic fields of opposite polarity towards the polar regions from solar maximum to solar minimum, then their gradual decay and reformation and drift to the opposite poles as the Sun's magnetic field reverses. Superimposed, the axial dipole component of the Sun is shown to be of maximum magnitude around minimum activity, when the polar fields are at their strongest.

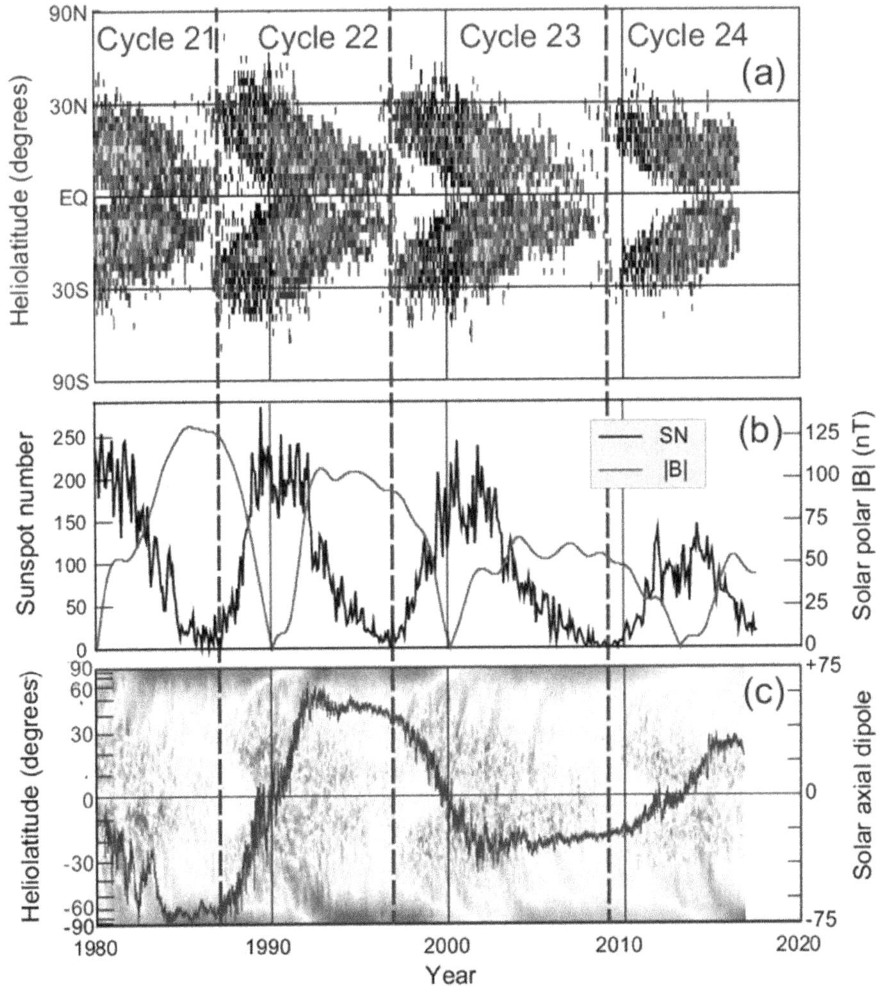

Figure 18.4.1 Signatures of the solar activity cycle. (a) The butterfly diagram of sunspots as a function of latitude. (b) The sunspot number (Version 2, WDC-SILSO, Royal Observatory of Belgium, Brussels) and the magnitude of the solar polar magnetic fields (Wilcox Solar Observatory). (c) The magnetic butterfly diagram (based on Petrie, 2017, Figure 7) and the axial dipole of the Sun (Wilcox Solar Observatory). This figure is discussed in detail in the text.

The parameter most often used in lieu of sunspot numbers is the solar radio noise flux at 2800 MHz or at a wavelength of 10.7 cm. It is very close in terms of phase and goodness of correlation to sunspot numbers. Now generally referred to as the $F_{10.7}$ measure of solar activity, it has been measured consistently since 1947 (Tapping, 2013). The physical mechanisms underlying the measure of the solar radio noise are bremsstrahlung in the corona and gyromagnetic resonance emission in strong magnetic fields above sunspots (Schonfeld et al., 2015). The bremsstrahlung component of the radio noise observations has been used for modelling terrestrial upper atmospheric effects as it has been considered a good proxy for the extreme ultraviolet (EUV) solar radiation that affects the upper atmosphere. Recent results from the deep solar minimum between cycles 23 and 24 have shown that the correlation between the $F_{10.7}$ index and EUV has changed, so that upper atmospheric modelling needs to take into account this varying relationship (Chen et al., 2011; Girazian and Withers, 2015).

The gyroresonant component (about 90% of the total intensity for the $F_{10.7}$ index; see Dudok de Wit et al., 2015) is related to the sunspot magnetic fields. This had provided the historically close correlation to sunspot numbers. However, from about 2000, an excess in the radio flux was noted (Tapping, 2013), compared to the sunspot numbers, using the earlier correlations. The cause of this deviation from the earlier correlation has not as yet been convincingly identified. The probable cause is a diminishing of the strength of umbral magnetic fields in sunspots (Livingston et al., 2012), reducing the contrast with the surrounding photosphere so that smaller sunspots may not be counted, but the strength of the magnetic field is still sufficient to generate the gyroresonant interactions leading to the radio emission. This interpretation of the comparative change between the sunspot numbers and the $F_{10.7}$ index (Tapping and Morgan, 2017) is contested by the findings of de Toma et al. (2013) and Rezaei et al. (2015), who find a solar cycle variation but no long-term trend in sunspot umbral magnetic field intensity. A possible resolution of the relative decrease of sunspot numbers with respect of the $F_{10.7}$ index may be due to two distinct populations of sunspots (Nagovitsyn and Pevtsov, 2016): a population of smaller, transient sunspots with weaker umbral magnetic field and a population of large sunspots, more mature and with higher-strength umbral fields. The relative distribution of these populations varies with the phase of the solar cycle, and there could be an underlying longer-term trend that would explain the changing relationship between sunspot numbers and the $F_{10.7}$ index (Nagovitsyn et al., 2017).

A superposed plot of sunspot numbers for cycles 1–24 is shown in Figure 18.4.2. It is clear that the shapes of the cycles show great variability and a highly variable peak (as measured by the maximum in sunspot numbers), but the near-11-year length of the cycles is one of the fundamental parameters of all solar models. The cycle duration is

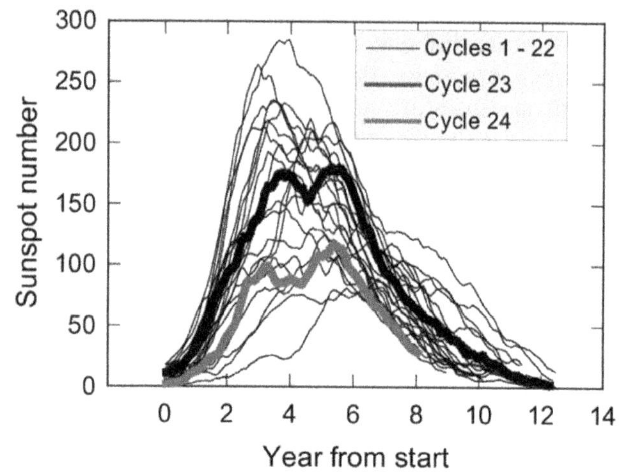

Figure 18.4.2 Superimposed plot of 13-month smoothed sunspot numbers in solar cycles 1–14. Cycles 23 and 24 are highlighted to show the significant reduction in the height of the successive maxima that has been used to investigate the underlying causal chain in the Sun's magnetic variability. (Version 2 sunspot numbers are from WDC-SILSO, Royal Observatory of Belgium, Brussels.)

sensitive to how the activity minima are determined, because of the overlap between cycles (see below). Hathaway et al. (2002) found a near-normal distribution for the first 22 cycles, with an average duration of 131.7 months (11 years = 132 months), with individual cycles ranging from 9 to 13.6 years.

Forecasting solar cycles has in general proved to be very difficult. The main goal of forecasts is to predict the number of sunspots at maximum; secondary objectives are to predict the time of the next maximum and the duration of the cycle. These are seemingly modest goals, but even these have proved to be challenging. The state of the art up to and including cycle 24 (1996–2008) was reviewed in detail by Petrovay (2010). Available forecasts for cycle 24, started in December 2008, were reviewed by Pesnell (2012, 2016); these can be used to assess how well the phenomenological or modelling techniques can forecast the main parameters of activity in the next solar cycle. A large number of methods have been developed and assessed for identifying reliable precursors and other potential predictors, as reviewed by Hathaway et al. (1999), Hathaway (2009), Petrovay (2010) and Hathaway (2015).

Recent solar cycles, especially since the availability of space observations and the parallel increase in the capabilities of ground observatories, covering effectively cycles 20 (beginning in October 1964) to 24 (expected to end in 2019–20), have provided vast amounts of data and have led to increasingly refined theoretical models for all aspects of solar variability. The unexpectedly long and deep minimum in sunspot numbers in 2008–9, concluding cycle 23, and the subsequent weak cycle 24 have exposed the limits of our understanding of solar activity. In particular, attempts at forecasting sunspot numbers at solar maximum in cycle 24 proved that causal relationships among solar phenomena and activity indicators were poorly understood. However, cycle 24 now serves as the best documented test case where the basis of the different forecasts before the cycle has been confronted with the actual realisation of the activity cycle (Sun et al., 2015).

Solar cycle forecasts use historical correlations between the sunspot numbers and phenomena and parameters that may be causally related to the sunspot cycle. Many such correlations have been discovered. These empirically derived correlations between different phenomena and sunspot numbers have been subjected to physical scrutiny to try and explain their functional relationships. This remains possibly the most important, if complex and multifaceted, research topic relating solar variability to its terrestrial effects.

Remarkably, some geomagnetic activity indices have proved to be among the more reliable predictors of solar activity cycles (see Pesnell, 2014, and references therein). However, explaining the forecasting ability of geomagnetic indices prior to the activity cycle involves trying to identify the causal chain in reverse. A very large number of geomagnetic indices have been correlated with solar activity with a view to test their potential use for predicting the key parameters of future solar activity cycles (see Hathaway, 2015).

The differences between cycles 23 and 24 have already been noted. The weak cycle 24 was anticipated based a significant weakening of the polar fields (Schatten, 2005) and the magnitude of the axial dipole moment already during cycle 23 (Wang et al., 2009). Two predictions are shown in Figure 18.4.3. The first was made by Jiang et al. (2007), using a dynamo model adjusted with a high diffusivity in the convection zone, using the poloidal field at minimum. The second, quoted by Pesnell (2016), was the prediction of Schatten (2005). Both predictions proved to be close to the observed sunspot number, as did that published by Svalgaard et al. (2004) – also based on the strength of polar fields – indicating the reliability of the use of the polar field of the Sun as a predictor. Following the maximum activity in cycle 24 that occurred in April 2014, Pesnell (2016) assessed the predictions made prior to the cycle and compared these to how the cycle actually evolved. These successful predictions contrast with many others: Pesnell (2016) quotes 105 different predictions for cycle 24, ranging from ~40 to 185 at sunspot maximum against the observed ~80 (using the American sunspot numbers).

The physical basis of the relationships between predictors and sunspot numbers has been examined, among others, by Wang and Sheeley (2009) and Jiang (2013). The minimum level of geomagnetic activity, as measured by the *aa* index near solar minimum, was found by Wang and Sheeley (2009) to be a better precursor than the maximum geomagnetic activity level near minimum solar activity, as used by, among others, Hathaway and Wilson (2006).

Maximum geomagnetic activity near sunspot minimum is related to the high-speed stream structure of the solar wind,

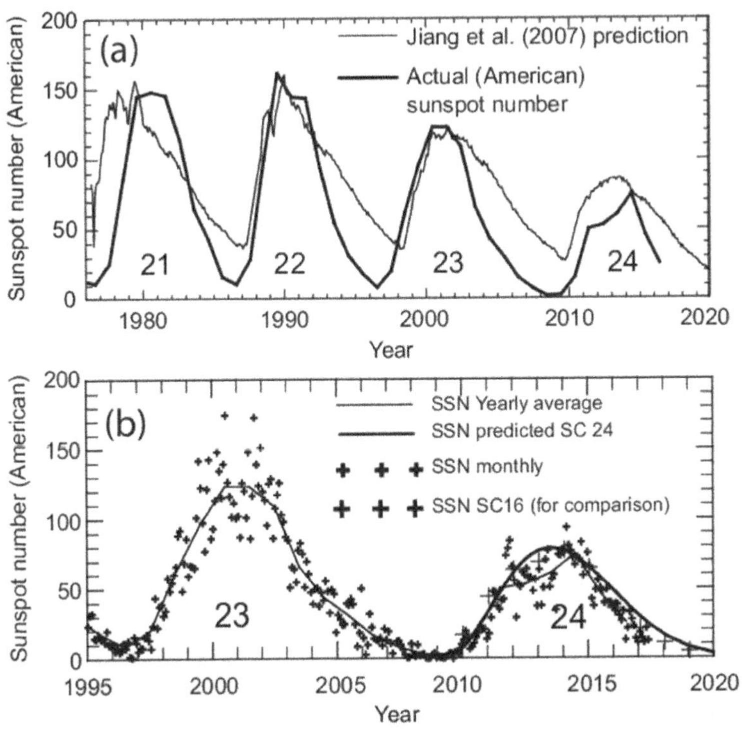

Figure 18.4.3 Two forecasts for solar cycle 24, published before the start of the cycle in December 2008. (a) The prediction of Jiang et al. (2007) based on the assumption of a dynamo model incorporating high diffusivity in the convection zone. This version of the dynamo model predicted successfully a weak cycle 24. (b) Figure based on Pesnell (2016, Figure 1), with the actual sunspot numbers extended into 2017. The predicted curve for cycle 24 is based on Schatten (2005), who correctly predicted a weak cycle 24, using the Sun's polar field as predictor.

itself related to the equatorial dipole component of the solar magnetic field, which, once decaying at solar minimum, has no influence on the next cycle. The strength of the solar polar field, directly related to the axial dipole component of the Sun's magnetic field, is arguably the best predictor of activity in the next cycle, and so is its geomagnetic equivalent, the minimum value of the activity index *aa*.

Wang and Sheeley (2009) argued that the total open flux, a direct measure of the solar polar field, is the equivalent predictor; that, in turn, can be directly evaluated from the radial component of the interplanetary magnetic field (IMF). Following the determination of the heliolatitude independence of the magnetic flux density by the Ulysses solar polar probe (Smith and Balogh, 1995), it was shown by Erdõs and Balogh (2014) that the magnetic flux density only depends on time, so that the radial component of the IMF, measured at 1 AU, can be used to calculate the total open flux of the Sun. The conclusion of Wang and Sheeley (2009) that the IMF radial component at solar minimum can be used for predicting the strength of the subsequent solar cycle is illustrated in Figure 18.4.4. Jiang (2013) reviewed the links between the geomagnetic precursor *aa*, the Sun's polar fields, the radial component of the IMF (as a measure of the open magnetic flux of the Sun) and their implications for dynamo models. The solar and geomagnetic precursors are closely correlated – this is a phenomenological rather than a deterministic correlation, as the mechanisms which link the polar field or open flux to the geomagnetic phenomena underlying the *aa* index have not been causally elaborated.

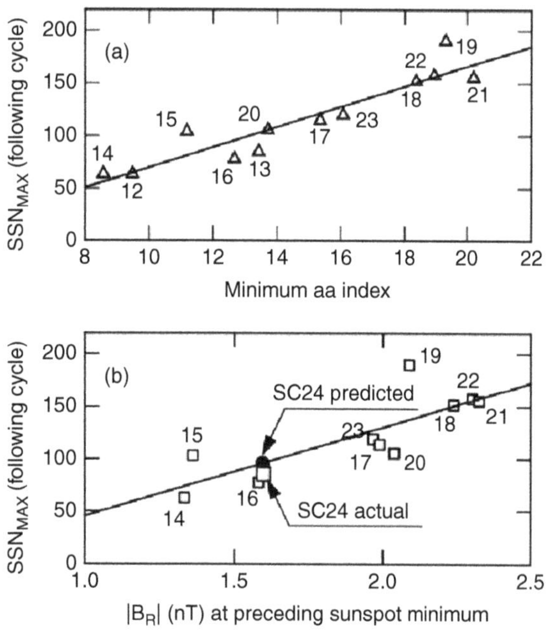

Figure 18.4.4 (a) Correlation between the geomagnetic disturbance index *aa* at solar cycle minimum and sunspot number maximum in the following activity cycle. (b) Correlation between the absolute value of the radial component of the interplanetary magnetic field at solar cycle minimum and sunspot number maximum in the following activity cycle. This figure is a combined version of Figures 2 and 5 of Wang et al. (2009). In panel (b), the predicted and actual values of maximum sunspot numbers in cycle 24 are highlighted.

18.5 Hemispheric Asymmetries of Solar Activity

Differences in the manifestations of solar activity have been observed and noted between the two hemispheres. This is most visible in sunspot numbers, described by Babcock (1959; and see e.g. Temmer et al., 2006, and references within, as illustrated in Figure 18.5.1a). The excess of sunspots in one hemisphere over the other changes within each activity cycle, but there are periodicities on timescales longer than the 11 year activity cycle. It appears that following four cycles in which there is an excess in the North followed by an excess in the South in each cycle, the next four cycles show the excess first in the South, then in the North (Li et al., 2002; Muraközy and Ludmány, 2010). However, the peak numbers in the two hemispheres do not differ by more than 20%, nor are the phase lags between the hemispheres longer than 2 years (<20% of the cycle length). This, taken together with the consistency of the recurrence pattern of the asymmetry every four cycles, reinforces the view that coupling of the underlying dynamo processes between the two hemispheres is not stochastic (Norton et al., 2014).

The solar dynamo generates higher-order magnetic moments, beyond the dipole (Muñoz-Jaramillo et al., 2013). Departure from a simple axial dipole leads to the asymmetries of the photospheric field, with respect to the rotational equator. The global solar magnetic field is most dipole-like at solar minimum activity; away from minimum, higher-order terms increase their contribution to the global field (Petrie, 2013; Wang, 2014). The emergence of sunspots, as the activity cycle is initiated, is related to the increase in complexity of the dynamics of the magnetic flux bundles below the photosphere; this is the basic mechanism that leads to the increase in the higher-order magnetic moments.

There are other important manifestations of the operation of the solar dynamo that provide constraints to dynamo models. Sunspots are known to emerge in pairs, with the two spots showing opposite magnetic polarities and the pairs generally oriented east–west. As observed already by Hale et al. (1919), the polarities in bipolar sunspots are always ordered so that the preceding (westward) spot has one polarity in the Southern Hemisphere and the opposite polarity in the Northern Hemisphere. The following (eastward) spots also have opposite polarities in the two hemispheres. However, the hemispheric ordering of polarities in bipolar groups reverses between solar cycles (Hale's law). In addition, leading polarity sunspots in sunspot pairs are closer to the equator than the following polarity spots (Joy's law). The tilt of sunspot pairs is smaller when they emerge closer

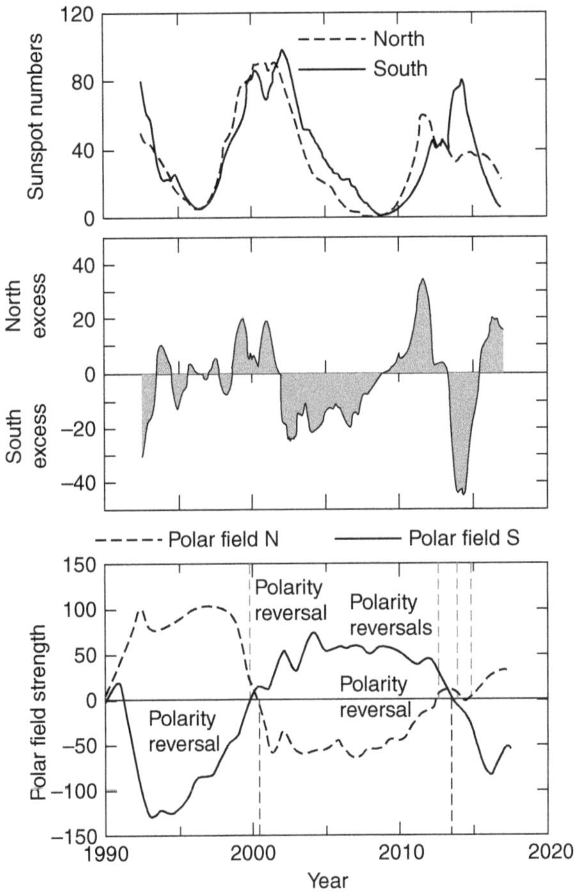

Figure 18.5.1 (a) Hemispheric asymmetry of sunspot numbers. (Version 2 sunspot numbers are from WDC-SILSO, Royal Observatory of Belgium, Brussels.) The asymmetry between the hemispheres is well documented for each solar cycle, with the relative strengths of the Northern and Southern Hemispheres modulated on timescales of several solar cycles. (b) Heliospheric asymmetry of the strength of the polar fields of the Sun, with phase differences up to a year or longer in the reversal of the magnetic polarity of the two polar regions. (Data from the Wilcox Solar Observatory, Stanford, CA, courtesy of J. Todd Hoeksema.)

to the equator. As the emergence of sunspots moves closer to the equator as the activity cycle progresses (see the butterfly diagram in Figure 18.4.1), the decreasing tilt has been found to be related to the strength of the activity cycle (Dasi-Espuig et al., 2010), providing yet another indicator for forecast models.

Dynamo models that include meridional transport of magnetic flux had originally considered flux moving away from the solar equator towards the poles, generally symmetrically around the equator (Hathaway and Upton, 2014). However, observationally, large bipolar magnetic regions – sunspot groups – that emerge close to the solar equator can straddle it, and their effect on the flux transported in the two hemispheres can be significant and can contribute to a large fraction of cross-equator magnetic flux (Cameron et al., 2014). As such events are discrete and not frequent, their effect on the formation of polar fields and the open magnetic flux at solar minimum leads to the imbalance between the two hemispheres, as shown in Figure 18.5.1b. This also leads to the polarity reversals occurring at different times in the two hemispheres. For the polarity reversal (as measured by the change of sign of the polar field, independently in the two hemispheres) near the maximum of cycle 24, Mordvinov et al. (2016) found multiple 'reversals' from south to north in the northern polar region and a single one from north to south in the southern polar region, as illustrated in Figure 18.5.1b.

The emergence of sunspots and the subsequent flux transport are subject to stochastic processes that lead to hemispheric asymmetries (Chen et al., 2017). While the general processes of flux emergence, sunspot formation, decay of activity complexes and the subsequent flux transport towards the solar poles are a well-understood phenomenological chain in the evolution of solar activity, the stochasticity of the individual events limits the statistical predictability of the quantitative measures of solar activity (Cameron et al., 2013).

In future, the increased ability to measure and quantify the evolution of surface magnetic fields, and better modelling of the complete process from flux emergence to flux transport, will lead to improved capability to move to quantitative forecasting (Jiang et al., 2014). The importance of this chain in the functioning of the solar dynamo needs to be better understood and integrated into detailed dynamo modelling (Cameron and Schüssler, 2015).

18.6 Magnetic Flux Emergence, Solar Flares and Coronal Mass Ejections

Solar flares and coronal mass ejections (CMEs), the two classes of the largest solar eruptive phenomena, are primary causes of space weather events. They are frequently, if not always, linked to each other and to other manifestations of fast – indeed, explosive – changes in the solar atmosphere. Yet the relationships are not sufficiently understood to allow confident forecasting of the occurrence and development of events, their impact on each other and the propagation of the resulting energetic phenomena into interplanetary space. The acceleration of energetic charged particles, both ions and electrons, accompanies both flares and CMEs; it is indeed an integral part of their phenomenology. The emission of X-rays, a consequence of the acceleration of electrons, is an often used direct measure of the energy of the solar eruptions.

Both solar flare and CME occurrence rates vary with the solar activity cycle and are generally correlated with sunspot numbers (see Figure 18.6.1). For cycles 23 and 24, see the rate of flares and CMEs in Figure 18.6.2. However, determining occurrence rates is prone to bias, mostly due to the difficulty in determining the

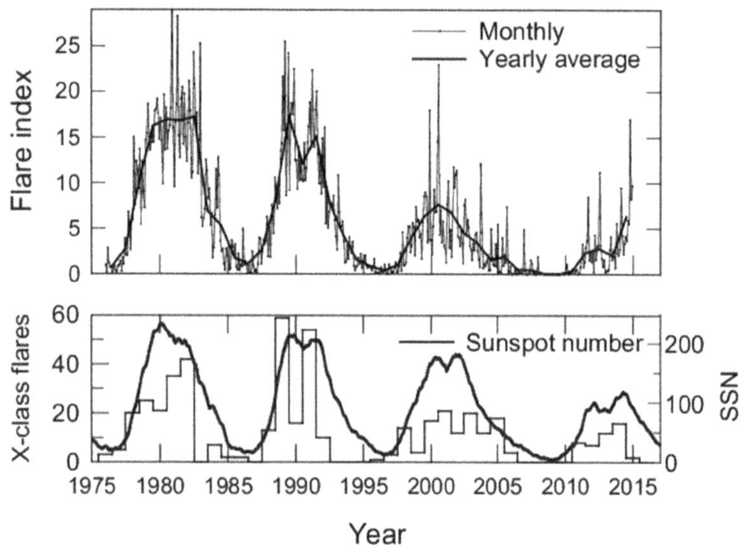

Figure 18.6.1 (a) Monthly and yearly Flare Index calculated by Bogazici University Kandilli Observatory, Istanbul, Turkey, courtesy of T. Ataç, A. Özgüç and O. Dogan, www.koeri.boun.edu.tr/astronomy/findex.html. (b) Yearly occurrence rate of the largest X-ray flares, showing the association with high solar activity, but without a consistent correlation: large flares can occur within years of the maximum activity.

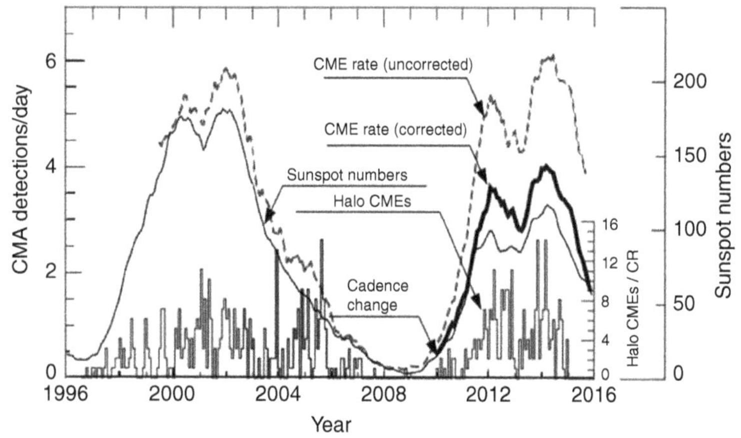

Figure 18.6.2 Daily rates of coronal mass ejections. Given the very broad range of CMEs in total mass, initial velocity and angular width, automatic detection schemes are widely used on the optical observations from spacecraft (primarily the LASCO instrument on the SOHO spacecraft; Brueckner et al., 1995). This figure, based on the analysis of Hess and Colaninno (2017), shows that when instrumental effects (cadence rate) are taken into account, the CME rate tracks the sunspot number as for previous solar cycles. At least a portion of the higher, 'uncorrected' rate may be due to the weakened solar polar fields and their effects on the CME rate (Petrie, 2015) or propagation characteristics (Gopalswamy et al., 2015).

appropriate lower threshold of events. For flares, the observational database is long, but the detection and identification of events have evolved over time, so a consistent long-term database, stretching back to the past as with sunspots, is not available.

While studies of the solar cycle dependence of flare and CME rates have established broad correlations and phenomenological scenarios, reaching the goal of causal functional relations remains elusive. Both flares and CMEs are the product of instabilities in multi-scale dynamic magnetic structures. However, the complexity of the structures, their observational variety and contextual dependences need to be taken into account when physical relationships are assessed.

Until cycle 24, the rate of CMEs tracked the sunspot number; this meant that the ratio of the average rate of CMEs and averaged sunspot numbers remained approximately constant over cycles 21–23 – not a particularly long baseline, due to the unavailability of continuous observations of CMEs. However, the apparent rate of CMEs compared to the sunspot numbers in cycle 24 showed an excess number of CMEs compared to the previous cycles. What is shown in Figure 18.6.2 is the observed rate of CMEs for both cycles 23 and 24 (Wang and Colaninno, 2014), together with their assessment of the correction to the observed rate in cycle 24, once a change in the image cadence from the SOHO instrument in 2010 is taken into account.

However, alternative explanations for the increase in the CME to sunspot number ratio have been put forward that have a more physical basis, with important implications for the effect of the polar magnetic fields on the initiation and launch of CMEs. It has been argued by Gopalswamy et al. (2015) that the decrease in the strength of the solar polar magnetic fields in cycle 24 (Petrie, 2012), and the subsequent decrease in interplanetary pressure led to a greater expansion of CMEs and thus to a higher detection rate. Petrie

(2015) related the changed properties of coronal fields to the generation and coronal propagation of CMEs. This observation provides a potentially useful tool to understand better the role of the coronal and interplanetary context in the formation and evolution of CMEs.

The driving motivation to assess how the eruptions of flares and initiation of CMEs can be predicted is their central role in defining and affecting space weather, the condition of the near-Earth space environment. In recent years, considerable effort has been made on a broad front to investigate the solar observations for potential signatures that could be used in predictive algorithms. Even given the wealth of observations and their increasing resolution, the search for causal signatures remains a very difficult process. Improvements are expected also to come from modelling and theory, not just from further increases in the temporal and spatial resolution of observations. A greater effort to combine observations of different phenomena in models is required to generate a comprehensive picture of eruptions.

Progress on modelling the initiation of eruptions from complex magnetic structures has suggested the existence of some basic underlying mechanisms which may in fact be common to both flares and CMEs (Wyper et al., 2017). There is diversity in the description of how flares and CMEs are initiated, how they develop and what mechanisms relate them (see Schmieder et al., 2015; Green et al., 2018).

Magnetic complexity manifests itself in the morphology of sunspot groups, of the active regions (ARs) associated with them and of the chromospheric and coronal magnetic structures. The source of the complexity is magnetic flux emergence from the solar convection zone to the photosphere and the overlying layers of the solar atmosphere (Schmieder et al., 2014). ARs feature sunspots as the most intense magnetic features but comprise a larger area of the photosphere that is also characterised by still intense magnetic features, resulting in a broad range of energetic phenomena from X-rays to energetic particles (van Driel-Gesztelyi and Green, 2015). The formation of sunspots follows the emergence of intense magnetic flux tubes from the solar convection zone through the photosphere (see e.g. Cheung et al., 2017). It is the variability in the emerging magnetic flux that leads to the formation of different size sunspots and sunspot groups; the variety of sunspot groups and their associated ARs are a direct consequence of the complexity of the flux emergence process. The complexity of the magnetic field configuration in the ARs and around sunspots indicates a build-up of stored magnetic energy.

Models of solar flares always invoke sudden breakdowns in complex, unstable magnetic structures in the solar atmosphere, resulting in the explosive release of energy, heating the plasma and accelerating energetic particles. When attempting to find common ground in magnetic field or magnetic structure changes in the photosphere associated with flaring activity, limits of observational resolution make it difficult to identify unambiguous signatures. Schrijver et al. (2005) showed that the probability of flaring is enhanced by the formation of non-potential magnetic field structures, following the emergence of new flux. This is one aspect of structures building up magnetic energy that can be released in the flaring process. Another perspective on the role complex magnetic field structures leading to solar flares is the existence and characteristics of magnetic polarity inversion lines (PILs) in active regions. It has been known since the observations of Severnyi (1958) that solar flares preferentially erupt on the two sides of the line in the photosphere that separate opposite polarity magnetic fields. Current theories and observations are still built largely on those early notions, even though the details and resolution of observations have improved orders of magnitude since the early discoveries.

The large amount of increasingly refined observations, together with increased computational capabilities, has made the initiation of solar flares better understood in detail and case-to-case complexity (e.g. Melrose, 2017). Schrijver (2016) reviewed the evidence for non-potential fields of different scales in the region where the solar flare is initiated. He concluded that the small-scale non-potentiality (field gradients and currents) in proximity of the PIL is the significant predictive signature of flaring potential. Modelling on small scales has led to the detailed matching of observed flaring regions with models incorporating the phenomenological building blocks of solar flares (Aulanier et al., 2010).

The ability to forecast flares based on the characteristics of PILs and the non-potentiality of active region was assessed by Sadykov and Kosovichev (2017). They found that forecasts based on the PIL are more effective than those using the whole active region; in general, the forecasting ability based on the magnetic characteristics around the PIL is improved if vector magnetograms are used, rather than simply line-of-sight measurements.

Magnetic helicity is a quantity characteristic of magnetic fields with complex topologies, such as force-free fields, kinks, twisted loops and sheared layers of flux (Berger, 1999; Blackman, 2015). Filament channels (Martin, 1998) are highly sheared magnetic structures that form above the PIL and contain high values of magnetic helicity. They also store large amounts of relatively unstable magnetic energy, ready to initiate eruptive events (Wang and Muglach, 2007). Given the high-level non-potentiality of the active region structures around the PIL, several instabilities can trigger magnetic reconnection that quickly leads to an explosive release of stored magnetic energy. This leads to structural changes in the field configuration and the electric current systems associated with the field structure in the close vicinity of the PIL (Sharykin et al., 2017). This process can be tracked through the conservation of magnetic helicity (Berger, 2005) as magnetic energy is released in the eruption, resulting in structural changes to the magnetic field configuration. In this model of the eruptive process, the magnetic energy and magnetic helicity emerge through the photosphere (Demoulin and Berger, 2003) from structures that will eventually lead to unstable and unsustainable levels of magnetic energy which are released by magnetic reconnection, triggered by instabilities. To understand the basic steps, simple models have tracked how the energy release leads to new structures, while conserving (exactly in the magnetohydrodynamics, but approximately because of the finite conductivity of the chromospheric and coronal plasma) the values of magnetic helicity (Priest et al., 2016).

The importance of forecasting solar flares and CMEs has led to numerous alternative studies of potentially usable observational scenarios as predictors. Predictive algorithms have used many different observables and potential relationships; however, owing to the correlations among relevant parameters that are usually difficult to quantify, even in a statistical sense, the model predictions rarely lead to a generally applicable forecasting technique. The current status of flare forecasting methods and the comparison of their success in forecasting flares have been reviewed by Barnes et al. (2016) and Green et al. (2018). Although the modelling and forecasting efforts have been considerable, it has been generally concluded that successful prediction models and algorithms cannot improve significantly on 'climatological models' based on past statistics (Barnes et al., 2016; Murray et al., 2017).

It seems likely that little progress will be made in predicting solar eruptive events, be they flares or CMEs, until a much better causal understanding can be reached of the physical processes that control the storage and sudden release of magnetic energy in complex coronal structures. In particular, it is unlikely that increasingly sophisticated machine-learning algorithms, based on large observational databases, can bring other than incremental improvements to the forecast programmes.

The clear role magnetic fields play in eruptive events and the complex structures that evolve in active regions following the emergence of flux from below the photosphere that can both store magnetic energy and dissipate it suddenly are the key to understanding the initiation of flares and CMEs. The concept of magnetic helicity appears to be the best suited to assess the likelihood of eruptions in any concrete magnetic region, as already introduced above. Deterministic criteria based on magnetic helicity may lead to better predictions of the conditions that lead to eruptions in solar active regions, as discussed by Pariat et al. (2017). Their simulations have highlighted the importance of relative helicity rather than its magnitude: the helicity in the current-carrying magnetic structures when compared to the total helicity (that includes the potential component of the magnetic field) has been found to be best correlated to eruptivity. Simulations such as these will provide the framework for a phenomenological assessment of solar active regions, isolating those features that are the most reliable predictors of eruptions.

18.7 The Origin and Dynamic Evolution of the Interplanetary Medium

Dynamic processes on the Sun and in the solar atmosphere propagate into interplanetary space to the Earth's neighbourhood to form space weather. The most prominent space weather events are the near-Earth phenomena that have their origins in solar energetic events: solar flares and CMEs. The medium of propagation is the solar wind: the supersonic plasma that expands from the solar corona to fill the heliosphere, as first proposed by Parker (1958).

Parker's solar wind model was isotropic, with a spherically symmetric plasma flow – however, this model was only to illustrate how the supersonic flow would arise from the million degree solar corona. Parker himself recognised that the plasma flow was non-uniform with consequences for the dynamic evolution of interacting plasma streams of different speeds (Parker, 1959). Following the five decades of interplanetary exploration since Parker's proposals, the complex dynamic structures of the solar wind and the embedded interplanetary magnetic field (IMF) have been extensively described and documented (cf. Gosling, 1996; McComas et al., 2003; Balogh and Erdõs, 2013; Owens and Forsyth, 2013).

In the inner heliosphere, the solar wind is clearly composed of two components: slow and fast. Parker's original (uniform, isotropic) idea is a theoretical model that proved remarkably prescient and fruitful. It did not, however, consider the physical processes underlying the formation of the solar wind. While the variability of the solar wind has been known since the first observations in the 1960s (Neugebauer and Snyder, 1962), the vast range of variability in all its observed parameters and its causes have continued to be discovered observationally and explored through theory, modelling and simulations.

18.7.1 Slow and Fast Solar Wind Streams

The most widely used measured parameter is the speed of the solar wind. It can range between about 250 km/s to in excess of 1000 km/s and up to 2000 km/s or more in transient flows. Based simply on speed – not the best indicator, as discussed below – above about 600 km/s, the solar wind is considered to be fast, and below about 500 km/s, the solar wind is considered to be slow. Observations over the past decades have confirmed the bimodality of the solar wind and its general properties, including the parameter ranges encountered as a function of solar cycle and heliolatitude (e.g. Richardson et al., 1996; Ebert et al., 2009; Zerbo and Richardson, 2015).

With the discovery of coronal holes in the 1970s and their association with high-speed solar wind streams (see e.g. Cranmer, 2009) a clear solar origin was identified for the fast solar wind. With the Ulysses spacecraft observations, the delineation between fast and slow wind streams became well understood (Geiss et al., 1995; McComas et al., 2003):

coronal holes, in particular the solar polar coronal holes, provided a nearly uniform, fast solar wind (at speeds about 700–800 km/s), and other regions of the Sun and the corona were the source of a much more variable, significantly slower solar wind component. The questions concerning the origin and acceleration of the fast solar wind have been reviewed by McComas et al. (2007), among others. It is clear that future observations, as these authors argue, will be needed to constrain theories and models of the fast solar wind despite being relatively simple in its properties.

An important empirical finding concerning the solar wind was by Wang and Sheeley (1990) and related the speed of fast solar wind to the expansion of the magnetic flux tubes rooted in the base of the corona in the coronal hole from which the wind originates. They derived a relationship between ratio of the cross-sectional areas of magnetic flux tubes in coronal holes at the surface where magnetic field vectors are normal to it (the so-called source surface) at about 2.5 solar radii and at the base of the corona and the speed of the solar wind. The expansion factor can be calculated from models of coronal field, generally based on the potential field source surface (PFSS) approach used to compute the magnetic field configuration in the corona. For a review, see Wiegelmann et al. (2017), who provide a critical discussion of the different coronal magnetic field models that have been proposed and are generally used. They conclude that, despite difficulties caused by the observational shortcomings, PFSS models can be successfully used for a range of purposes, including the estimation of flux tube expansion factors, hence the prediction of the fast solar wind speed from coronal holes.

All available observational tests over different phases of the solar cycle and over (the so-far explored) three-dimensional heliosphere have shown that flux tube expansion is the main controlling cause of the speed of the fast solar wind. This evidence is particularly striking in the case of the solar polar observations of the Ulysses spacecraft. Over three polar orbits around the Sun from 1990 to 2006, Ulysses uniquely explored the solar wind as a function of heliolatitude and solar activity levels. Using the flux tube expansion model for modelling the solar wind speed, Wang and Sheeley (2006) were able to reproduce in great detail the actual observations of speed profiles in the heliosphere. They were able to verify that the speed of the solar wind is inversely correlated with the rate of flux tube expansion in the corona, extrapolated from the measured photospheric field, using the current-free PFSS model. In particular, they identified in the coronal observations the sources of the long-lived, high-speed wind streams observed by Ulysses.

One particular characteristic of the slow solar wind is its very high variability in all its parameters: speed, density and temperature (see e.g. Abbo et al., 2016). The origin of the slow solar wind remains disputed (see the recent review of Cranmer et al., 2017). The most likely reason for the difficulty in identifying both the region(s) of origin of the slow wind and the underlying processes that generate it is that plasma of the slow wind originates in the closed field line regions of the corona and that its acceleration and escape are complex, multi-scale processes, depending on the coronal structures. A comprehensive study (Kilpua et al., 2016) of the kinetic and compositional properties of the slow solar wind through the solar minimum between cycles 23 and 24 concluded that, based on interplanetary measurements of the slow wind, no definitive conclusion can be drawn on the relative contributions of the five or six different possible source regions and source mechanisms that had been proposed. It seems very likely that the variability associated with the slow wind is due to the origin of the plasma in closed coronal structures, subject to many complex processes, with neighbouring magnetic field lines and loops populated with plasma subjected to different processes.

The distinction between slow and fast wind has been challenged by Stakhiv et al. (2015), who, based on the unique observations at mid-heliolatitudes by Ulysses, noted observations of solar wind streams with a charge state distribution (corresponding to the temperature of the coronal region of origin) corresponding to the slow wind, although its elemental composition would identify it as originating in a coronal hole. A possible explanation (instead of evoking a new type of solar wind) could be that at the boundaries of coronal holes, plasma is caught up in boundary reconnection processes that contribute to heating the plasma, even before release from the corona, to high coronal temperatures. This explanation is at least indirectly supported by observations of the signatures of reconnections in the slow solar wind (Kepko et al., 2017; Zhao et al., 2017). Although modelling of reconnecting magnetic structures in the corona and their probable contribution to the generation of the slow solar wind remain one, possibly major source process, further observations and their interpretations will be necessary to confirm the contribution of this process to slow wind streams observed in interplanetary space (Edmondson, 2012).

The dual nature of the solar wind flows is best observed in the composition of the solar wind plasma. Geiss et al. (1995) showed that the two kinds of winds originated in different regions of the solar atmosphere and had markedly different ionic composition. An important aspect of the different natures of the fast and slow solar wind streams is the markedly different temperatures in the coronal regions in which they originate. The coronal temperatures of the regions of origin are determined by the ratios of differently charged ions that are characteristic of the ionic charge distribution in the plasma in which the ions collisionally interact and hence can be taken to be the origin of the collisionless plasma flow of the solar wind. A still uniquely capable space instrument, SWICS, which was flown on the Ulysses and ACE space missions, measured with high accuracy the charge states of ions of different solar wind species. From the charge state distribution, the temperature at which the distribution was 'frozen into' the solar wind flow could be determined (see Gloeckler et al., 1992). In particular, taking a measure of the temperature characteristic of the charge distribution, using the ratio of the densities of two charge states of oxygen, $O7+/O6+$, allows the determination of the coronal freezing-in temperature. This has proved to be a very powerful observation that has allowed, on the two space missions Ulysses and ACE, the determination of the coronal region of origin of the observed solar wind plasma. Although solar wind plasma compositions have been included on other space missions, the accurate ionic charge state composition measurements that are needed for the freezing-in temperature determination have not been available other than those two missions.

A relatively low (1–1.2 MK) coronal freezing-in temperature indicates fast solar wind from coronal holes; a higher (>1.4 MK) freezing-in temperature corresponds to slow solar wind from the hotter, generally loop-filled corona. The observed solar wind velocities and the correlation between speed and coronal freezing-in temperatures are shown for the two space missions on which these observations were made in Figure 18.7.1. The upper panels show the in-ecliptic observations by ACE; the lower panel shows the observations around the solar polar orbit of Ulysses.

18.7.2 Solar Wind Stream Interaction Regions

An important conclusion from the observations and correlations in Figure 18.7.1 is that the bimodality of the solar wind is best observed at mid- to high heliolatitudes, where the two kinds of solar winds have less chance to interact and form, through dynamic evolution, solar wind streams, which, while distinguishable on the basis of compositional measurements, show less contrast in their kinetic (velocity, density, kinetic temperature) parameters.

The decisive contribution of the uniquely distinguishable coronal freezing-in temperature (as derived from the in situ charge state composition of oxygen or carbon ions) to the analysis of solar wind observations is illustrated in Figure 18.7.2. These observations show a stream interaction region (SIR), formed when a fast solar wind stream follows a slow stream along the radial direction. The fast wind stream, with a speed of ~650 km/s, compresses the slow solar wind ahead, which has a speed <400 km/s. This results in an increase in the magnetic field strength at the leading edge of the interaction region and a gradual increase in the solar wind speed to that of the fast stream. The plot also shows the freezing-in temperature of the plasma, indicating the two different types of solar wind streams, with the slow wind originating in the hot corona (at 1.5–1.8 MK), and the fast solar wind from a much cooler coronal hole (0.8–1.0 MK). The interface between the two kinds of solar wind is more abrupt than the speed differential, and showing (in this case) three abrupt drops in the coronal temperature, presumably corresponding to the structure of the boundaries of the coronal hole (Wimmer-Schweingruber and Hassler, 2016). In general terms, the interfaces between fast and slow solar wind streams are discernible by their plasma composition characteristics: elemental compositions and ionic charge state distributions (Wimmer-Schweingruber et al., 1999). In turn, this distinction allows constraining the coronal acceleration and heating processes and structures in the regions of origin of different solar wind streams.

Stream interaction regions (SIRs) are ubiquitous structures to be found in the heliosphere when both fast and slow streams are emitted by the Sun at the heliolatitude of the observer. For an up-to-date and very comprehensive review, see Richardson (2018). When the coronal structure is stable over several solar rotations, the interaction pattern is repeated in successive solar rotation, resulting in a corotating pattern of the interaction regions (CIRs). SIRs exist at all heliolatitudes and at all phases of the solar activity cycle. Even at solar maximum, there are small, transient coronal holes, emitting fast solar wind that interacts with the (mostly) slow solar wind from the complex loop structures dominating the corona at maximum activity (Balogh and Smith, 2001).

An important feature of SIRs is the formation of leading and trailing shock waves at the leading and trailing boundaries of the compression region, due to the increased plasma pressure. However, shock waves are generally formed beyond 1 AU, although numerous CIRs have been observed at Earth. Shock waves cause geomagnetic effects (Sudden Commencements, or SCs) by the sudden compression of the magnetosphere. They are also the site of energetic particle acceleration.

Forecasting the evolution of SIRs is an important goal in space weather studies. Magnetohydrodynamic modelling of the stream formation and dynamic evolution of the interaction between slow and fast stream has steadily progressed, particularly through increasing computing capabilities and their exploitation using simulation and modelling software of increasing resolution. Operational models follow the method outlined in Odstrčil (2003), although considerably more sophisticated models have been applied to specific individual, mostly transient events such as CMEs. These have been discussed in the review by Richardson (2018). The need for long-term forecasting of solar wind conditions at Earth has prompted looking for new, statistical modelling techniques (Owens and Riley, 2017) which promise useful results, even if the probabilistic approach that has been found necessary is not deeply rooted in the causal chain of physical processes from the Sun to the Earth.

18.7.3 Propagation of CMEs from the Sun to the Earth

Once launched, CMEs propagate through the corona and are embedded in the solar wind flow. Sheeley et al. (1999) found that there were two kinds of CMEs: gradual and

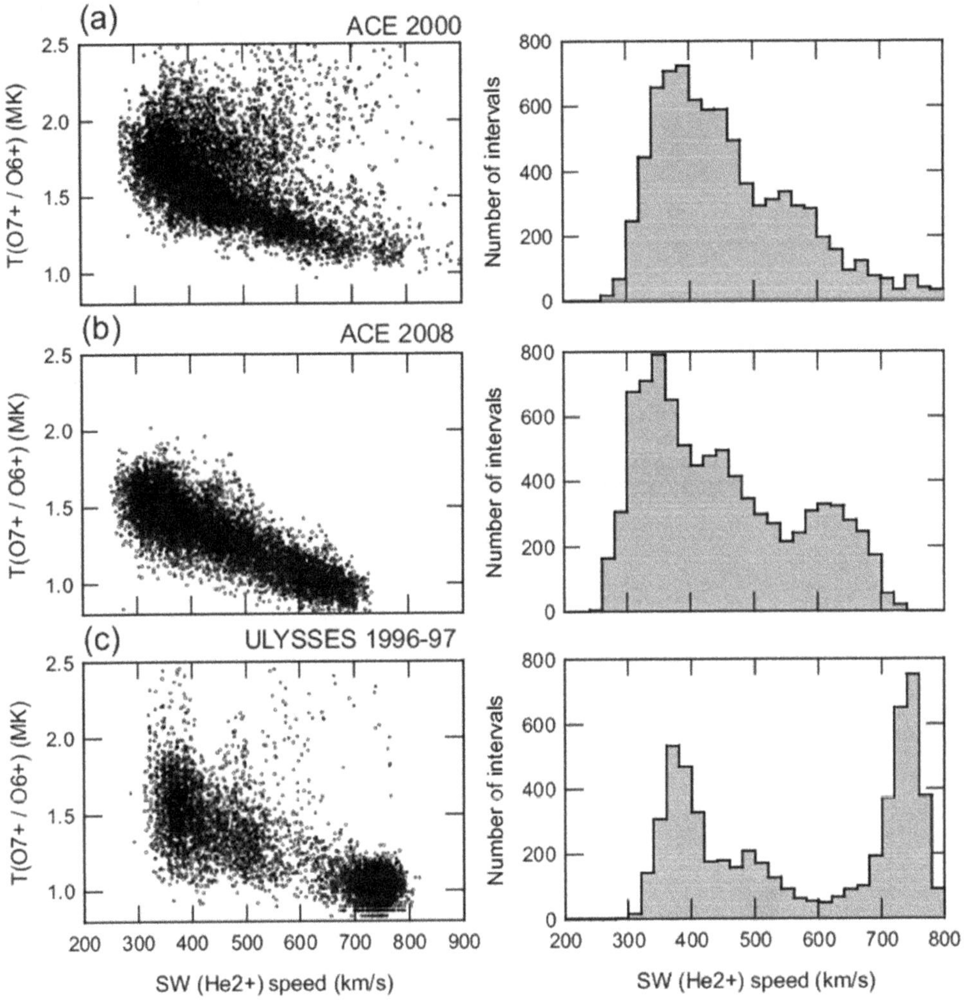

Figure 18.7.1 The histograms of solar wind speeds, correlated with the temperature of the corona calculated from the freezing-in temperature of oxygen, determined from the ratio of ionisation states of oxygen ions O^{7+}/O^{6+}, measured by the SWICS instrument on the ACE and Ulysses spacecraft (Gloeckler et al., 1992; plots based on Erdős and Balogh, 2012). (a, b) ACE observations of the solar wind during solar maximum in 2000 and solar minimum 2008, respectively. (c) Ulysses observations around its nearly polar solar orbit between 1990 and 2008. In the left-hand panels, the anti-correlation between coronal temperature and solar wind speed is clearly displayed. In the right-hand panels, the difference between the in-ecliptic and the three-dimensional heliographic measurements shows the strong peak of the relatively unevolved high-speed wind from the polar coronal holes (McComas et al., 2003). (Figure from Erdős and Balogh, 2012.)

impulsive. Gradual CMEs tend to be slower, reaching speeds of 400–600 km/s, and their implied shape is that of a balloon, with a central core that is of lower speed than the leading edge. Impulsive CMEs are much faster, with speeds in excess of 750 km/s, although they generally decelerate as they move away from the Sun. Impulsive CMEs are also more generally associated with flares. Their shape is less regular and is suggestive of being made up of several components. So, in general, CMEs can propagate in the solar wind at speeds comparable to the solar wind but can also be significantly faster. Yurchyshyn et al. (2005), based on a study of 4315 CMEs observed between 1998 and 2001 (at the peak of solar cycle 23), found that it is not possible to describe the statistical distribution of CME speeds by any simple distribution. There is a long tail to the distribution, suggestive of a log-normal distribution. Yurchyshyn et al. (2005) speculated that the underlying complexity and multi-scale nature of the eruption process leads to this distribution. In this view, eruptions leading to the launching of CMEs are caused by magnetic reconnection at multiple sites on a large range of spatial scales.

Studies of the propagation of CMEs from their formation in the corona to their arrival at Earth have yielded insights into the complex range of interacting phenomena that are responsible for shaping the observations. Such studies include the eruptive process itself, such as by Török et al. (2018) and Temmer et al. (2017a), as well as a modelling of the interplanetary propagation and evolution of the CME. The study by Temmer et al. (2017a) also showed that magnetic reconnection processes had been initiated at the eruption site before the flare became visible, reinforcing the view that CMEs and flares are related but that the relationship is dependent on the topology of the magnetic structures, their stability and the way the magnetic instabilities initiate the reconnection processes.

A main contributor to the effectiveness of CMEs to cause disturbances in the Earth's space environment is the large-scale, often very fast shock wave that is formed at its leading surface as it propagates away from the Sun. CME-associated propagating shocks are also the primary sources of accelerated energetic particles. This has been known for several decades, but owing to the complexity of

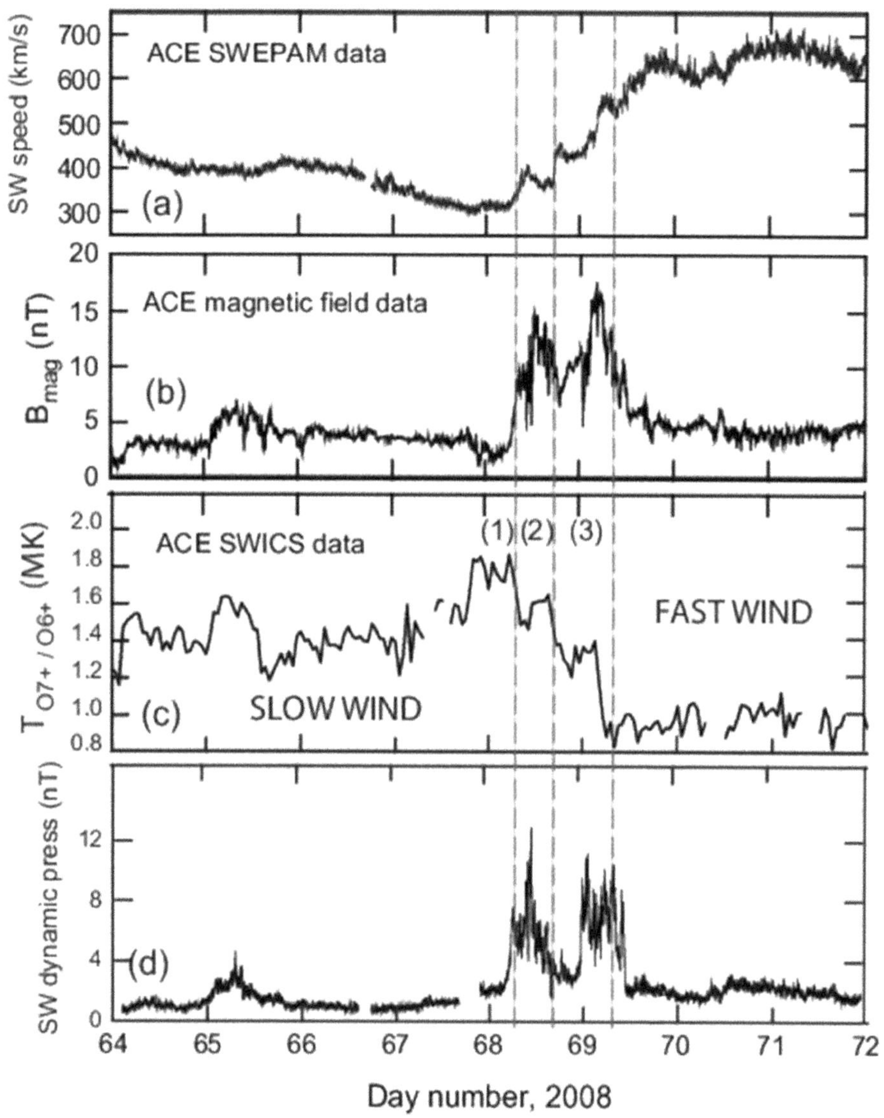

Figure 18.7.2 A solar wind stream interaction in the ecliptic, as observed by the ACE spacecraft in March 2008. (a) The solar wind speed (red) and (b) the magnitude of the magnetic field (blue). On days 68 and 69, the solar wind speed gradually increases from ~400 km/s to near 700 km/s, indicating the arrival of a high-speed stream at the spacecraft. At the leading edge of the high-speed stream, the magnetic field shows a region of compression of the plasma. (d) The dynamic pressure of the solar wind (black), also indicating a pressure pulse that shows the compression of the slow wind ahead of the fast wind. (c) The freezing-in temperature of the solar wind plasma (green) is calculated from the ion charge state ratios $O7^{+}/O6^{+}$, which shows that the fast wind originated in a coronal hole with temperature ~1 MK, while the slow wind originated from coronal regions with a more variable temperature ~1.3–1.8 MK. There are three abrupt decreases in the coronal freezing-in temperature, marked (1), (2) and (3), that are likely to correspond to the boundaries of the coronal holes from which the fast stream originates.

understanding the shock acceleration process itself and the propagation of the energetic particles that pose a difficult three-dimensional problem on the scale of the inner heliosphere, studies generally concentrate on analysing individual events. This helps in understanding the associated phenomena but generally cannot be used for forecasts. Up-to-date studies help put the problem in context; such is the study of three events by Xie et al. (2017).

About 40% of CMEs contain a recognisable flux rope carried into interplanetary space by the CME. Their structure is rarely simple and requires considerable care in its identifications. This is difficult for single-point measurements but becomes more tractable if the observations are available from different vantage points. The achievement of this goal was one of the primary aims of the two-spacecraft STEREO mission. The review of the observations of magnetic clouds by Wood et al. (2017) is therefore the most authoritative for use in reconstructing this central feature of many CMEs.

The irregular structure of CMEs observed at Earth is mainly the result of their interaction with the ambient solar wind, although it is also, to a degree, dependent on their original formation in the corona (Manchester et al., 2017). The concept of CME propagation into a 'pre-conditioned' environment (Temmer et al., 2017b) allows taking into account the 'history' of the medium into which the CME is propagating. The methodology to assess quantitatively the change in the solar wind medium by the passage of CMEs was to compare actual solar wind conditions to those modelled simultaneously, based solely on the solar boundary conditions. It was found that the passage of CMEs through the solar wind affected its characteristic parameters several days longer than the size of the CME itself.

In more general terms, CMEs may not be considered to be coherent structures, owing to the complex history and their propagation through the corona and the interplanetary medium (Owens et al., 2017). The obvious way CMEs are disrupted during propagation away from the Sun is through interaction with other CMEs. Such interacting CMEs have been observed at high solar activity, when their rate is high. Lugaz et al. (2017) found a range of effects that could be attributed to the interaction of CMEs, both in the corona and in interplanetary space. Such effects include a higher than expected accelerated energetic particle population, increased wave activity and higher levels of turbulence. In addition, when the leading shock wave propagates through a preceding CME, significant distortion of the magnetic structure can be observed.

Even if some of the effects of solar wind irregularities on CME propagation may be only of minor consequence, the larger-scale gradients due to the presence of interacting fast- and slow-speed solar wind streams cannot be neglected (Owens et al., 2017). Observations of CMEs in interplanetary space are usually restricted to single vantage points, naturally so for in situ observations of the CME plasma and even for imaging remotely. The interpretation of the measurements will remain quite challenging: CMEs are complex, evolving, three-dimensional structures, and their reconstruction on the available measurements and observations needs to be handled with considerable care.

18.8 The Solar Origin of the Heliospheric Magnetic Field

In his first presentation of his solar wind model, Parker (1958) also addressed the question of the extension of the solar magnetic field from the corona into the heliosphere. Given the parameters of the solar wind plasma, in particular its near-infinite electrical conductivity, the magnetic field from the corona is carried into the heliosphere by the solar wind. The field is 'frozen into' the plasma. The general equations of the heliospheric magnetic field (HMF) follow, in Parker's simple, spherically symmetric model, the geometric entrainment of the field lines by the spherically (and uniformly) flowing solar wind. Although just as for the solar wind, the real magnetic field departs from Parker's description due to the variable solar magnetic field, nevertheless, the overall Parker geometry of spiral field lines of constant pitch (equal to the colatitude of the origin of the field line in the corona) remains the basic reference frame for describing the HMF. Time-dependent departures, such as caused by dynamic processes in interplanetary space – SIRs, for example – can still be described against the basic Parker geometry. In the solar equatorial plane, the field lines follow an Archimedean (or Parker) spiral.

Parker's (1958) field line equations assume the existence and knowledge of the magnetic field vector on a 'source surface', close to the Sun, from which the equations are extrapolated into the heliosphere. The source surface magnetic field can be a complex function of the location in general. In addition, the source surface field is obviously time dependent; this has significant consequences for even the simple extension of the field into the heliosphere. The concept of the source surface, not elaborated in detail by Parker, has evolved considerably since it was first introduced. It remains a key concept that underlies the linking of the HMF with its origins in the corona.

Solar magnetic field lines that extend into the heliosphere are rooted in open regions in the corona; these appear dark in the high-temperature spectrum and in EUV and X-rays and are commonly called coronal holes. As has been observed in X-ray and EUV images of the Sun, coronal holes generally cover the polar regions around solar minimum activity but become smaller, even evanescent, around solar maximum. Coronal holes can, however, extend towards the equator, thus modifying the overall structure of the corona. Coronal holes are the source of fast solar wind streams (see e.g. Cranmer, 2002, 2009) and are threaded by the open magnetic field lines that extend into the heliosphere.

Several coronal models have been proposed that calculate the topology of magnetic fields near the Sun, generally out to a source surface at ~2.5 solar radii (R_S). The most generally used model, the PFSS model, was originally proposed by Schatten et al. (1969) and Altschuler and Newkirk (1969) and elaborated for current use by Hoeksema et al. (1982), Wang and Sheeley (1992, 1994), Petrie et al. (2011) and Wiegelmann et al. (2017). A particularly good illustrative example of the application of the PFSS model is shown in Figure 18.8.1, courtesy of G. Petrie (NSO), following Petrie et al. (2011). This model has been, despite its rather restrictive assumptions, quite successful when compared to extrapolations with observed magnetic fields in the heliosphere. In the PFSS model, the coronal magnetic field is calculated by assuming that the corona is current-free from the photosphere to a spherical source surface concentric with the Sun, so that the magnetic field is potential and can be determined by solving Laplace's equation. The inner boundary condition is the photospheric magnetic field measured from the ground and the outer boundary condition is that the magnetic field is radial on the source surface. A much used data record, available from 1976, consists of the magnetic field contours on the source surface, calculated for each Carrington rotation, using the photospheric measurements by the Wilcox Solar Observatory (Hoeksema et al., 1982; see http://wso.stanford.edu/synsourcel.html). The assumptions of the model have been criticised on a number of grounds, such as the likely non-sphericity of the source surface (Schulz, 1997) and the existence of currents in the corona. Nevertheless, it has retained its practical importance through its relative simplicity and generally reasonable agreement with observables in the heliosphere, in particular with the magnetic sector structure.

Global MHD models of the coronal magnetic field have been developed that use measured photospheric fields as inner boundary conditions and calculate the magnetic field and solar wind configuration to beyond the corona out to 30 solar radii (e.g. Linker et al., 1999). An earlier comparison between the MHD models and the PFSS model has been made by Riley et al. (2006); see also the up-to-date review of coronal field models by Wiegelmann et al. (2017). The conclusion at present is that different models give insights into different aspects of the configuration of coronal magnetic fields; for global modelling and comparison with heliospheric observations, the PFSS model remains the most useful. However, the use of other models in parallel can highlight shortcomings of the PFSS model when the corona evolves significantly on the timescale of the solar rotation or when generally local non-potential effects weaken the use of the current-free solution. The PFSS model performs best around solar minimum, when the global solar field is dominated by the axial dipole term and the corona (including the coronal holes in the polar regions) is stable over many solar rotations. Figure 18.8.1 shows a three-dimensional snapshot of coronal and heliospheric magnetic fields inside the source surface, using PFSS modelling (courtesy of Gordon Petrie). A generally prominent feature of PFSS models at the low solar activity phase of the solar cycle is the 'magnetic equator' of the Sun, the contour line on the source surface separating the large-scale outward and inward pointing magnetic polarities. The extension of this contour line, also called the *neutral line*, into interplanetary space is the heliospheric current sheet (HCS) separating the two dominant polarities of the solar magnetic field in interplanetary space (e.g. Smith, 2001). The PFSS model does least well around solar maximum, when the corona is evolving rapidly and the Sun's large-scale magnetic field least resembles an axial dipole (Balogh and Smith, 2001; Wang et al., 2002). Around solar maximum, generally only small and evanescent coronal holes are observed (Zhang et al., 2003; Miralles et al., 2004); together with the complex large-scale magnetic topology, these conditions lead to a complex topology of the HCS that may in fact lead to disconnected and/or multiple current sheets (Schrijver and DeRosa, 2003). However, there is no evidence for such complex surfaces in the heliosphere.

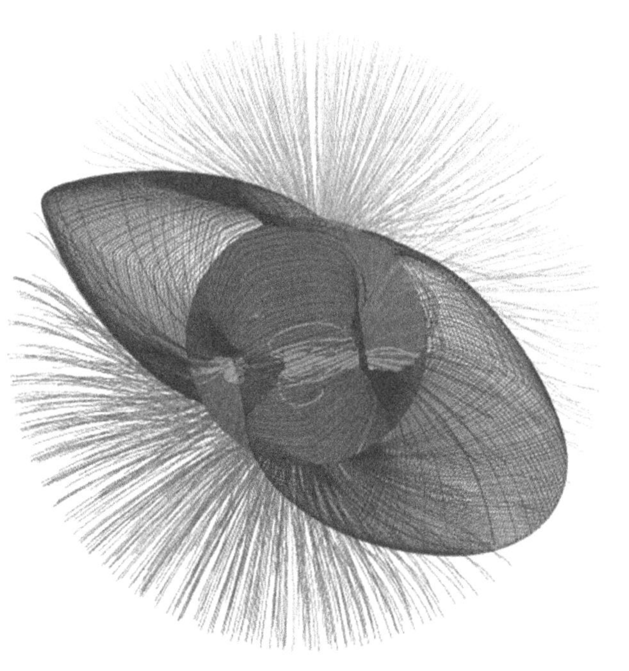

Figure 18.8.1 Potential field source surface model of the corona. Open magnetic fields that extend into the heliosphere are shown in green (outward or positive fields) and red (inward or negative fields). Closed magnetic field lines are shown in blue. The outer limit of the model is the spherical source surface at 2.5 solar radii. The open magnetic field lines are orthogonal to the source surface. (From http://gong.nso.edu/science/gongpfss/gongpfss.html, courtesy Gordon Petrie, National Solar Observatory.) (A black-and-white version of this figure appears in some formats. For the colour version, please refer to the plate section.)

During the unusually long declining phase of cycle 23, due to major changes in the Sun's polar fields, the applicability of the PFSS model was tested under conditions different from previous periods of low solar activity (Luhmann et al., 2009). The polar magnetic fields of the Sun were much weaker than in previous minima, and therefore the axial dipole term was much weakened with respect to the multipole terms (Wang et al., 2009). Following this minimum, the recovery of the solar magnetic fields in cycle 24 has added to the complexity of the relationship between solar photospheric fields and their extension into the corona and the heliosphere (Sheeley and Wang, 2015). An overview of the large-scale magnetic field of the Sun, presented from the point of view of a solar physicist, is to be found in Wang (2004), who links the global solar field and its characteristics with the magnetic field in the heliosphere. This linkage is further explored in the following section.

18.9 The Open Magnetic Flux of the Sun in the Heliosphere

The total open magnetic flux of the Sun is directly related to the strength of the solar magnetic field in the open coronal regions (Wang et al., 2000). It can be calculated from integrating the open flux on the source surface (see the previous section). On the other hand, in situ measurements of the heliospheric magnetic field provide a measure of the in situ magnetic flux density: it is taken to be the radial component of the magnetic field vector. To deduce the total magnetic flux, assumptions must be made about the heliographic dependence of the flux density. Integrating the flux density (as measured by the radial component) over a spherical (or simply just closed) surface around the Sun must yield zero, as the magnetic field is divergence-free. This is taken to be one of the boundary conditions in the PFSS models of coronal fields.

Early spacecraft data confirmed the heliocentric dependence of the radial component as r^{-2}, in agreement with Parker's model (Behannon, 1978. The Ulysses out-of-ecliptic observations (e.g. Balogh and Erdõs, 2013) have confirmed this result in the three-dimensional heliosphere (see Figure 18.9.1a). The amplitude of the fluctuations (deviations from the modelled r^{-2} fit) of daily averages remains relatively constant, as is also shown in Figure 18.9.1a. The data used in Figure 18.9.1 correspond, as shown in the upper panel of Figure 18.9.1b, to solar wind from the large northern polar coronal hole in 1993–5, while Ulysses was moving to high heliolatitudes during the approach to solar minimum in fast solar wind, as shown by the low coronal temperatures. The value of the radial component, normalised to 1 AU, is seen to be independent of heliocentric distance in the lower panel of Figure 18.9.1b. During this interval, the radial component was unipolar, corresponding to the northern solar polar polarity, even though there were large fluctuations around the mean; the fluctuations were significantly more prominent at the data resolution of 1 hour than when averaged over a day. The same point is made in Figure 18.9.2c, where the histograms of the normalised radial component of the field are shown. The importance of the knowledge of the total open magnetic flux is its relation to solar irradiance and its variability for space weather studies and also to other solar activity indicators, such as cosmic ray modulation (Lean et al., 2002).

Observations by the solar polar Ulysses spacecraft discovered (Smith and Balogh, 1995) that the flux density, the radial component of the field scaled to 1 AU, is independent of heliolatitude. This was most apparent in the unipolar high-speed streams from the large polar coronal holes at the time of low solar activity. In Figure 18.9.2a, the four portions of the out-of-ecliptic orbit of Ulysses that corresponded to the long-lasting high-speed solar wind streams from the polar coronal holes are shown. The flux density, measured along these orbit portions, is plotted as a function of heliocentric distance in Figure 18.9.2b. The histograms of the flux density for the four orbit portions are shown in Figure 18.9.2c. It is clear that, taking one portion at a time, the flux density is constant as a function of both heliolatitude and heliocentric distance. The flux densities in fact change between orbit portions 1 and 2, on one hand, and orbit portions 3 and 4, on the other hand, as the latter two portions correspond to the solar activity minimum closing solar cycle 23, when the polar magnetic flux from the Sun was about one-third less than in the previous activity minimum (Wang et al., 2009). In a more general way, the independence of the solar magnetic flux density on heliolatitude and heliographic distance was found to be the case even in the dynamically more disturbed solar wind flows near the ecliptic (Erdõs and Balogh, 2012, 2014). The flux density, hence the total flux, depends only on time, in

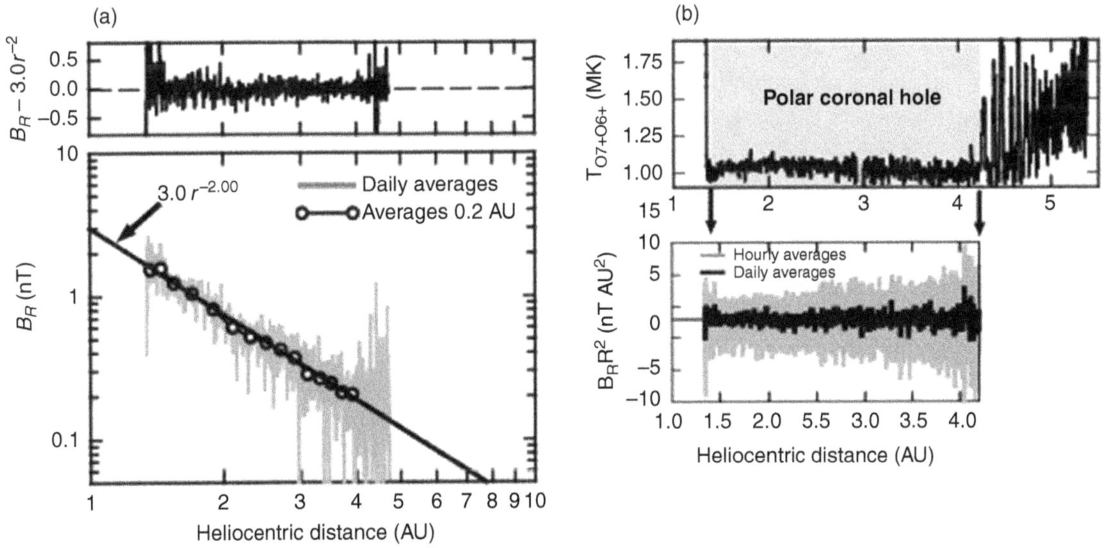

Figure 18.9.1 Heliocentric radial dependence of the magnetic field in the high-speed, coronal hole solar wind in 1993–5 as determined on the Ulysses solar polar space probe. (a) In the upper panel, deviation of the measured radial component of the magnetic field, referenced to 1 AU, from the inverse square dependence on heliocentric distance. In the lower panel, the inverse square dependence of the radial magnetic field component. (b) In the upper panel, the freezing-in temperature of the solar wind measured in the large-scale polar coronal hole as Ulysses travelled from the ecliptic plane to high heliolatitudes over a heliocentric distance from 4.2 AU to 1.4 AU. In the lower panel, the heliospheric magnetic flux density remained constant when referred to 1 AU, over the range of heliocentric distance observations.

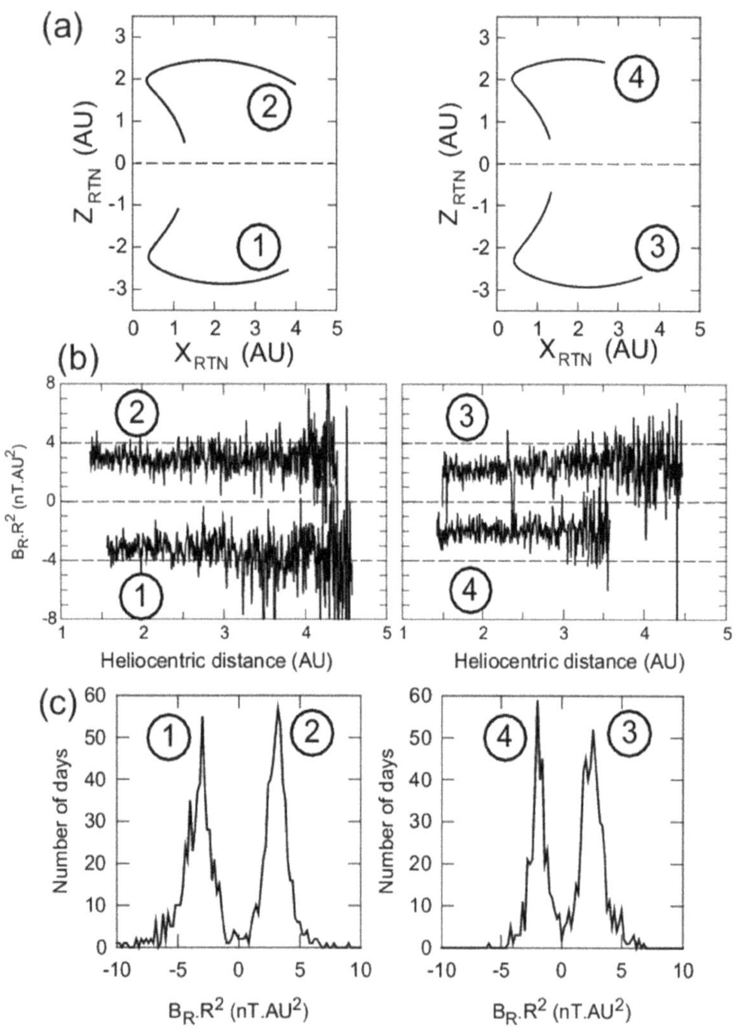

Figure 18.9.2 (a) The four out-of-ecliptic orbit portions of Ulysses in unipolar high-speed solar wind streams. (b) The normalised radial magnetic field component, representing the magnetic flux density, during the four orbit portions shown in (a). (c) Histograms of the normalised magnetic field radial component during the orbit intervals in (a).

particular on the evolution of the strength of the polar fields of the Sun with the phase of the solar cycle.

Although the open solar flux calculated from coronal modelling has been quite reasonably matched to that inferred from interplanetary measurements (Wang, 2009), this has been questioned by Linker et al. (2017), who find that trying to match the open flux to the coronal holes leads to the conclusion that the interplanetary flux implies that some of the open solar flux is not rooted in the observed coronal holes. At present it is unclear how this conundrum will be resolved.

18.10 Current Status and Future Prospects

Understanding of the Sun and the solar phenomena that affect the Earth's environment has of course progressed enormously in the past decades, thanks in large part to the increasingly sophisticated space missions, providing increasingly high-resolution data dedicated to the subject. More recently, ground-based observational facilities have provided equally high-quality complementary data. In addition, the development of new tools, such as helioseismology, has opened new windows on the complexities of solar processes both inside the Sun and in its atmosphere. In the above sections, key topics in solar physics have been highlighted, based on the current state of knowledge.

Nevertheless, despite the significant progress in observations, modelling and theoretical investigations, the obviously complex causal chains of phenomena have generally resisted a sufficiently quantitative formulation to provide deterministic predictions, given observational data. Examples evoked in this chapter include the most fundamental solar process, the dynamo that generates the solar magnetic field, and, through its variability, controls effectively all relevant solar phenomena. The recent review of the dynamo by Charbonneau (2014) has already been quoted. New developments, such as by Lemerle and Charbonneau (2017), effectively present new and more detailed constraints on the likely working of the solar dynamo, notably still in the general framework of the so-called Babcock–Leighton phenomenological model, already mentioned in Section 18.3, that dates back to the 1960s. It is expected that the increasingly detailed simulations of alternative dynamo models will make good use of the evolution of helioseismologic observations as well as of the observations on photospheric circulation to constrain the solar dynamo. The question remains if these developments will lead to more confident predictions of the cyclic evolution of the dynamo that are needed for forecasting solar activity cycles. The range of predictions for solar cycle 24 is a sobering reminder that much progress remains to be made.

It is equally sobering that it is not known what sets the length of the solar activity cycle to be about 11 years. As pointed out by Charbonneau (2014), it is possible to adjust parameters in most dynamo models to achieve a solar-cycle-like periodicity, essentially unconstrained by observations. Realistic computer simulations of the solar dynamo action are not yet within reach of even the most powerful computing systems. In addition, the smaller the scales that are included in the simulation, their (likely non-linear) interactions need to be investigated and, if possible, better understood.

The key role of the evolution of polar magnetic fields needs better observations: from ground-based observatories, there are clear limitations. The already mentioned Ulysses solar polar mission (1990–2009) did not carry remote solar imaging instruments, and even if its contribution to solar wind studies will remain unique for the foreseeable future, the next space mission which will provide high-heliolatitude solar observation will be the European Space Agency's Solar Orbiter mission, currently due for launch in 2020, which could reach 7 years after launch to a heliolatitude of 25° and even 34° at the end of an extended mission phase.

Solar eruptive phenomena, flares and CMEs, are equally difficult to forecast – even though forecasts are of a very different nature, on very different timescales, from forecasting the solar activity cycles. Many of the necessary observational constraints that will be needed for flares and CMEs have been identified and many specific cases of eruptive activity analysed and presented. These relate to the existence and characteristics of magnetically active regions, with complex magnetic topologies as constrained by magnetic helicity, including topologies of magnetic polarity inversion lines.

One such complex event was studied by Park et al. (2013), in many ways typical of the complex event chain leading to flares and CMEs. The observations showed the injection of oppositely signed helicity in the sunspots and in the flaring regions, respectively, with large amounts of magnetic energy added in the flare regions. Furthermore, a number of small bipolar magnetic fragments emerged, with their connecting arcades antiparallel with existing overlying magnetic loops connecting the main sunspots of the active region. These conditions led to a dynamic evolution of the geometry of the magnetic fields, gathering the ingredients to the magnetic eruptions, manifesting themselves as both flares and associated CMEs.

Given the increasing database of solar observations, it is expected that such detailed case studies will continue to be carried out in the future. The quoted event described by Park et al. (2013) was typical in its ingredients, but other events provide a range of different conditions, while retaining the same phenomenological features. Deterministic descriptions of solar eruptive events will become possible, thanks to a distillation of such event descriptions, with their measured characteristics, which will allow more quantitative associations among the solar measurable, such as magnetic field characteristics, topology and helicity.

For the larger picture, from the solar eruption to its interplanetary effects, combining observations from a range of sources are of course essential (Kilpua et al., 2017). Even then, tools for the study of individual events to include the initiation of the eruption (flare and CME) and its propagation through interplanetary space are still largely in development. Their potential importance for space weather applications enables considerable resources to be devoted to improving the propagation models for operational purposes (Zheng et al., 2013). The most comprehensive study (at the time of writing) of the actual solar context of a very large eruption and its propagation to Earth, using the best available magnetohydrodynamics simulation code, is that of one of the largest and best observed eruptions of the past decades, the so-called Bastille Day (14 July 2000) event by Török et al. (2018). Such studies are essential to test our understanding of the phenomenological causal chains that are operationally important for effective space weather forecasting.

Major progress is expected from the observations of the Parker Solar Probe, a spacecraft instrumented for the study of all aspects of the solar wind and its solar sources, at an unprecedented small distance of the Sun (Fox et al., 2016). By filling the observational gap between remote sensing of the solar corona and the near-Sun interplanetary space and more distant in situ observations, the mission is expected to provide vital constraints for modelling the relationship between solar and interplanetary phenomena. The probe was launched on 12 August 2018. It will approach the Sun gradually, on succeeding orbits, finally to less than 10 solar radii in 2024 (to be compared to the mean distance of the Earth to the Sun of 213 solar radii).

Clearly, many events are only partially observed, and the recorded chain of phenomena may not be fitted precisely in a causative model. But the richness of the available observations, increased by new observing facilities, will lead, even if in an incremental way, to a more confident deterministic description of solar eruptions. Combined with the expected increase in computational and simulation resources, such events (and solar phenomena, more generally) are likely to be much better understood in the decades to come.

References

Abbo, L., Ofman, L., Antiochos, S. K., Hansteen, V. H., Harra, L., Ko, Y.-K., Lapenta, G., Li, B., Riley, P., Strachan, L., von Steiger, R. & Wang, Y.-M. Slow solar wind: Observations and modeling, Space Sci. Rev., 201, 55–108 (2016). doi: 10.1007/s11214-016-0264-1

Altschuler, M. D. & Newkirk, G., Jr. Magnetic fields and the structure of the solar corona I: Methods of calculating coronal fields, Sol. Phys., 9, 131–49 (1969).

Aulanier, G., Török, T., Démoulin, P. & DeLuca, E. E. Formation of torus-unstable flux ropes and electric currents in erupting sigmoids, Astrophys. J., 708, 314–33 (2010). doi: 10.1088/0004-637X/708/1/314

Babcock, H. D. The Sun's polar magnetic field, Astrophys. J., 130, 364–5 (1959). doi: 10.1086/146726

Babcock, H. W. The topology of the Sun's magnetic field and the 22-year cycle, Astrophys. J., 133, 572–87 (1961).

Balogh, A. & Erdõs, G. The heliospheric magnetic field, Space Sci. Rev., 176, 177–215 (2013). doi: 10.1007/s11214-011-9835-3

Balogh, A. & Smith, E. J. The heliospheric magnetic field at solar maximum: Ulysses observations, Space Sci. Rev., 97, 147–60 (2001). doi: 10.1023/A:101185490

Balogh, A. & Thompson, M. J. Introduction to solar magnetism: The early years, *Space Sci. Rev.*, 144, 1–14 (2009). doi: 10.1007/s11214-009-9493-x

Balogh, A., Hudson, H. S., Petrovay, K. & von Steiger, R. Introduction to the solar activity cycle: Overview of causes and consequences, Space Sci. Rev., 186, 1–15 (2014). doi: 10.1007/s11214-014–0125

Barnes, G., Leka, K. D., Schrijver, C. J., Colak, T., Qahwaji, R., Ashamari, O. W., Yuan, Y., Zhang, J., McAteer, R. T. J., Bloomfield, D. S., Higgins, P. A., Gallagher, P. T., Falconer, D. A., Georgoulis, M. K., Wheatland, M. S., Balch, C., Dunn, T. & Wagner, E. L. A comparison of flare forecasting methods. I. Results from the 'All-Clear' workshop, Astrophys. J., 829(2), Article 89 (2016). doi: 10.3847/0004-637X/829/2/89

Behannon, K. W. Heliocentric distance dependence of the interplanetary magnetic field, Rev. Geophys. Space Phys., 16, 125–45 (1978). doi: 10.1029/RG016i001p00125

Berger, M. A. Introduction to magnetic helicity, Plasma Phys. Control. Fusion, 41, B167–75 (1999). doi: 10.1088/0741-3335/41/12B/312

Berger, M. Magnetic helicity conservation, Highlights Astron., 13, 85–8 (2005). doi: 10.1017/S1539299600015148

Blackman, E. G. Magnetic helicity and large scale magnetic fields: A primer, Space Sci. Rev., 188, 59–91 (2015). doi: 10.1007/s11214-014-0038-6

Brueckner, G. E., Howard, R. A., Koomen, M. J., et al. The Large Angle Spectroscopic Coronagraph (LASCO), Sol. Phys., 162, 357–402 (1995). doi: 10.1007/BF00733434

Brun, A. S., Miesch, M. S. & Toomre, J., Global-scale turbulent convection and magnetic dynamo action in the solar envelope, Astrophys. J., 614, 1073–98 (2004). doi: 10.1086/423835

Brun, A. S., Browning, M. K., Dikpati, M., Hotta, H. & Strugarek, A. Recent advances on solar global magnetism and variability, Space Sci. Rev., 196, 101–36 (2015). doi: 10.1007/s11214-013-0028-0

Cameron, R. H. & Schüssler, M. The crucial role of surface magnetic fields for the solar dynamo, Science, 347, 1333–5 (2015). doi: 10.1126/science.1261470

Cameron, R. H. & Schüssler, M. An update of Leighton's solar dynamo model, Astron. Astrophys., 599, Article A52 (2017). doi: 10.1051/0004-6361/201629746

Cameron, R. H., Dasi-Espuig, M., Jiang, J., Işik, E., Schmidt, D. & Schüssler, M. Limits to solar cycle predictability: Cross-equatorial flux plumes, Astron. Astrophys., 557, Article A141 (2013). doi: 10.1051/0004-6361/201321981

Cameron, R. H., Jiang, J., Schüssler, M. & Gizon, L. Physical causes of solar cycle amplitude variability, J. Geophys. Res., 119, 680–88 (2014). doi: 10.1002/2013JA019498

Charbonneau, P. Where is the solar dynamo? J. Phys. Conf. Ser., 440, Article 012014 (2013). doi: 10.1088/1742-6596/440/1/012014

Charbonneau, P. Solar dynamo theory, Ann. Rev. Astron. Astrophys., 52, 251–90 (2014). doi: 10.1146/annurev-astro-081913-040012

Chen, F., Rempel, M. & Fan, Y., Emergence of magnetic flux generated in a solar convective dynamo. I: Formation of sunspots and active regions, and origin of their asymmetries, Astrophys. J., 846, Article 149 (2017). doi: 10.3847/1538-4357/aa85a0

Chen, Y., Liu, L. & Wan, W., Does the $F_{10.7}$ index correctly describe solar EUV flux during the deep solar minimum of 2007–2009? J. Geophys. Res., 116, Article A04304 (2011). doi: 10.1029/2010JA016301

Cheung, M. C. M., van Driel-Gesztelyi, L., Martínez Pillet, V. & Thompson, M. J., The life cycle of active region magnetic fields, Space Sci. Rev., 210, 317–49 (2017). doi: 10.1007/s11214-016-0259-y

Clette, F., Svalgaard, L., Vaquero, J. M. & Cliver, E. W. Revisiting the sunspot number: A 400-year perspective on the solar cycle, Space Sci. Rev., 186(1–4), 35–103 (2014). doi: 10.1007/s11214-014-0074-2

Choudhuri, A. R. Starspots, stellar cycles and stellar flares: Lessons from solar dynamo models, Sci. Chin. Phys. Mech. Astron., 60, Article 1:019601 (2017). doi: 10.1007/s11433-016-0413-7

Cranmer, S. R. Coronal holes and the high speed solar wind, Space Sci. Rev., **101**, 229–94 (2002). doi: 10.1023/A:1020840004535

Cranmer, S. R., Coronal holes, Living Rev. Sol. Phys., 6, Article 3 (2009). doi: 10.12942/lrsp-2009-3

Cranmer, S. R., Gibson, S. E. & Riley, P. Origins of the ambient solar wind: Implications for space weather, Space Sci. Rev., 212, 1345–84 (2017). doi: 10.1007/s11214-017-0416-y

Dasi-Espuig, M., Solanki, S. K., Krivova, N. A., Cameron, R. & Peñuela, T. Sunspot group tilt angles and the strength of the solar cycle, Astron. Astrophys., 518, Article A7 (2010). doi: 10.1051/0004-6361/201014301

Demoulin, P. & Berger, M. A. Magnetic energy and helicity fluxes at the photospheric level, Sol. Phys., 215, 203–15 (2003). doi: 10.1023/A:1025679813955

de Toma, G., Chapman, G. A., Cookson, A. M. & Preminger, D. Temporal stability of sunspot umbral intensities: 1986–2012, Astrophys. J. Lett., 771, Article L22 (2013). doi: 10.1088/2041-8205/771/2/L22

Deubner, F.-L. & Gough, D. Helioseismology: Oscillations as a diagnostic of the solar interior, Annu. Rev. Astron. Astrophys., 22, 593–619 (1984). doi: 10.1146/annurev.aa.22.090184.0031

Dikpati, M. & Gilman, P. A. Flux-transport solar dynamos, Space Sci. Rev., 144, 67–75 (2009). doi: 10.1007/s11214-008-9484-3

Dudok de Wit, T., Bruinsma, S. & Shibasaki, K. Synoptic radio observations as proxies for upper atmosphere modelling, J. Space Weather Space Clim., 4, Article A06 (2015). doi: 10.1051/swsc/2014003

Dudok de Wit, T., Kopp, G., Fröhlich, C. & Schöll, M. Methodology to create a new Total Solar Irradiance record: Making a composite out of multiple data records, Geophys. Res. Lett., 44, 1196–1203 (2017). doi: 10.1002/2016GL071866

Ebert, R. W., McComas, D. J., Elliott, H. A., Forsyth, R. J. & Gosling, J. T. Bulk properties of the slow and fast solar wind and interplanetary coronal mass ejections measured by Ulysses: Three polar orbits of observations, J. Geophys. Res., 114, Article A01109 (2009). doi: 10.1029/2008JA013631

Eddy, J. A. The Maunder minimum, Science, 192(4245), 1189–1202 (1976).

Edmondson, J. K. On the role of interchange reconnection in the generation of the slow solar wind, Space Sci. Rev., 172, 209–25 (2012). doi: 10.1007/s11214-011-9767-y

Erdõs, G. & Balogh, A. Magnetic flux density measured in fast and slow solar wind streams, Astrophys J., 753, Article 130 (2012). doi: 10.1088/0004-637X/753/2/130.

Erdõs, G. & Balogh, A. Magnetic flux density in the heliosphere through several solar cycles, Astrophys. J., 781(1), Article 50 (2014). doi: 10.1088/0004-637X/781/1/50

Fletcher, S. T., Broomhall, A.-M., Salabert, D., Basu, S., Chaplin, W. J., Elsworth, Y., Garcia, R. A. & New, R. A seismic signature of a second dynamo? Astophys. J. Lett., 718, L19–22 (2010). doi: 10.1088/2041-8205/718/1/L19

Fox, N. J., Velli, M. C., Bale, S. D., Decker, R., Driesman, A., Howard, R. A., Kasper, J. C., Kinnison, J., Kusterer, M., Lario, D., Lockwood, M. K., McComas, D. J., Raouafi, N. E. & Szabo, A. The Solar Probe Plus mission: Humanity's first visit to our star, Space Sci. Rev., 204, 7–48 (2016). doi: 10.1007/s11214-015-0211-6

Fröhlich, C. Solar irradiance variability since 1978: Revision of the PMOD composite during Solar Cycle 21, Space Sci. Rev., 125, 53–65 (2006). doi: 10.1007/s11214-006-9046-5.

Fröhlich, C. Total solar irradiance: What have we learned from the last three cycles and the recent minimum? Space Sci. Rev., 176, 237–52 (2013). doi: 10.1007/s11214-011-9780-1.

Geiss, J., Gloeckler, G. & von Steiger, R. Origin of the solar wind from composition data, Space Sci. Rev., 72, 49–60 (1995). doi: 10.1007/BF00768753

Girazian, Z. & Withers, P. An empirical model of the extreme ultraviolet solar spectrum as a function of the $F_{10.7}$ index, J. Geophys. Res., 120, 6779–94 (2015). doi: 10.1002/2015JA021436.

Gloeckler, G., Geiss, J., Balsiger, H., Bedini, P., Cain, J. C., Fischer, J., Fisk, L. A., Galvin, A. B., Gliem, F., Hamilton, D. C., Hollweg, J. V., Ipavich, F. M., Joos, R., Livi, S., Lundgren, R. A., Mall, U., McKenzie, J. F., Ogilvie, K. W., Ottens, F., Rieck, W., Tums, E. O., von Steiger, R., Weiss, W. & Wilken, B. The solar wind ion composition spectrometer, Astron. Astrophys. Suppl. Ser., 92, 267 (1992).

Gopalswamy, N., Xie, H., Akiyama, S., Mäkelä, P., Yashiro, S. & Michalek, G. The peculiar behavior of halo coronal mass ejections in Solar Cycle 24, Astrophys. J. Lett., 804(1), Article L23 (2015). doi: 10.1088/2041-8205/804/1/L23

Gosling, J. T. Corotating and transient solar wind flows in three dimensions, Annu. Rev. Astron. Astrophys., 34, 35–73 (1996). doi: 10.1146/annurev.astro.34.1.35

Green, L. M., Török, T., Vršnak, B., Manchester, B., IV & Veronig, A. The origin, early evolution and predictability of solar eruptions, Space Sci. Rev., 214, Article 46 (2018). doi: 10.1007/s11214-017-0462-5

Haigh, J. D. The Sun and the Earth's climate, Living Rev. Sol. Phys., 4, Article 2 (2007). doi: 10.12942/lrsp-2007-2

Hale, G. E., Ellerman, F., Nicholson, S. B. & Joy, A. H. The magnetic polarity of sunspots, Astrophys. J., 49, 153–78 (1919).

Hanasoge, S., Miesch, M. S., Roth, M., Schou, J., Schüssler, M. & Thompson, M. J. Solar dynamics, rotation, convection and overshoot, Space Sci. Rev., 196, 79–99 (2015). doi: 10.1007/s11214-015–0144–0

Hathaway, D. H. Solar cycle forecasting, Space Sci. Rev., 144, 401–12 (2009).

Hathaway, D. H. The solar cycle, Living Rev. Sol. Phys., 12, Article 4 (2015). doi: 10.1007/lrsp-2015-4

Hathaway, D. H. & Upton, L. The solar meridional circulation and sunspot cycle variability, J. Geophys. Res., 119, 3316–24 (2014). doi: 10.1002/2013JA019432

Hathaway, D. H. & Wilson, R. M. Geomagnetic activity indicates large amplitude for sunspot cycle 24, Geophys. Res. Lett., 33, L18101 (2006). doi: 10.1029/2006GL027053

Hathaway, D. H., Wilson, R. M. & Reichmann, E. J. A synthesis of solar cycle predictions, J. Geophys. Res., 104, 22375–88 (1999). doi: 10.1029/1999JA900313

Hathaway, D. H., Wilson, R. M. & Reichmann, E. J. Group sunspot numbers: Sunspot cycle characteristics, Sol. Phys., 211, 357–70 (2002). doi: 10.1023/A:1022425402664

Hess, P. & Colaninno, R. C. Comparing automatic CME detections in multiple LASCO and SECCHI catalogs, Astrophys. J., 836, Article 134 (2017). doi: 10.3847/1538-4357/aa5b85

Hoeksema, J. T., Wilcox, J. M. & Scherrer, P. H. Structure of the heliospheric current sheet in the early portion of Sunspot Cycle 21, J. Geophys. Res., 87, 10331–8 (1982). doi: 10.1029/JA087iA12p10331

Jiang, J. Solar-cycle precursors and predictions, in *Solar and Astrophysical Dynamos and Magnetic Activity*, ed. A. G. Kosovichev, E. M. de Gouveia Dal Pino & Y. Yan, pp. 49–60, Proceedings IAU Symposium 294 (2013). doi: 10.1017/S1743921313002196

Jiang, J., Chatterjee, P. & Choudhuri, A. R. Solar activity forecast with a dynamo model, Mon. Not. R. Astron. Soc., 381, 1527–42 (2007). doi: 10.1111/j.1365-2966.2007.12267.x

Jiang, J., Cameron, R. H., Schmitt, D. & Işık, E. Modelling solar cycles 15 to 21 using a flux transport dynamo, Astron. Astrophys., 553, Article A128 (2013). doi: 10.1051/0004-6361/201321145

Jiang, J., Hathaway, D. H., Cameron, R. H., Solanki, S. K., Gizon, L. & Upton, L. Magnetic flux transport at the solar surface, Space Sci. Rev., 186, 491–523 (2014). doi: 10.1007/s11214-014-0083-1

Kepko, L., Viall, N. M., Antiochos, S. K., Lepri, S. T., Kasper, J. C. & Weberg, M. Implications of L1 observations for slow solar

wind formation by solar reconnection, Geophys. Res. Lett., 43, 4089–97 (2017). doi: 10.1002/2016GL068607
Kilpua, E. K. J., Madjarska, M. S., Karna, N., Wiegelmann, T., Farrugia, C., Yu, W. & Andreeova, K. Sources of the slow solar wind during the Solar Cycle 23/24 minimum, Sol. Phys., 291, 2441–56 (2016). doi: 10.1007/s11207-016-0979-x
Kilpua, E. K. J., Balogh, A., von Steiger, R. & Liu, Y. D. Geoeffective properties of solar transients and stream interaction regions, Space Sci. Rev., 212, 1271–1314 (2017). doi: 10.1007/s11214-017-0411-3
Kopp, G. Magnitudes and timescales of total solar irradiance variability, J. Space Weather Space Clim., 6, Article A30 (2016). doi: 10.1051/swsc/2016025
Korhonen, H. Properties of stellar activity cycles, Proc. IAU, 11 (A29A), 354–9 (2016). doi: 10.1017/S1743921316003276
Krivova, N. A., Solanki, S. K. & Floyd, L. Reconstruction of solar UV irradiance in cycle 23, Astron. Astrophys., 452, 631–9 (2006). doi: 10.1051/0004-6361:20064809
Lean, J. L., Wang, Y.-M. & Sheeley, N. R., Jr. The effect of increasing solar activity on the Sun's total and open magnetic flux during multiple cycles: Implications for solar forcing of climate, Geophys. Res. Lett., 29(24), Article 2224 (2002). doi: 10.1029/2002GL015880
Leighton, R. B. A magneto-kinematic model of the solar cycle, Astrophys. J., 156, 1–26 (1969). doi: 10.1086/149943
Lemerle, A. & Charbonneau, P. A coupled 2 × 2D Babcock-Leighton solar dynamo model. II. Reference dynamo solutions, Astrophys. J., 834(2), Article 133 (2017). doi: 10.3847/1538-4357/834/2/133
Li, K. J., Wang, J. X., Xiong, S. Y., Liang, H. F., Yun, H. S. & Gu, X. M. Regularity of the north-south asymmetry of solar activity, Astron. Astrophys., 383, 648–52 (2002). doi: 10.1051/0004-6361:20011799
Linker, J. A., Mikic, Z., Biesecker, D. A., Forsyth, R. J., Gibson, S. E., Lazarus, A. J., Lecinski, V., Riley, P., Szabo, A. & Thompson, B. J. Magnetohydrodynamic modeling of the solar corona during Whole Sun Month, J. Geophys. Res., 104, 9809–30 (1999). doi: 10.1029/1998JA900159
Linker, J. A., Caplan, R. M., Downs, C., Riley, P., Mikic, Z., Lionello, R. Henney, C. J., Arge, C. N., Liu, Y., Derosa, M. L., Yeates, A. & Owens, M. J. The open flux problem, Astrophys. J., 848(1), Article 70 (2017). doi: 10.3847/1538-4357/aa8a70
Livingston, W., Penn, M. J. & Svalgaard, L. Decreasing sunspot magnetic fields explain unique 10.7 cm radio flux, Astrophys. J. Lett., 757, Article L8 (2012). doi: 10.1088/2041-8205/757/1/L8
Lockwood, M., Owens, M., Hawkins, E., Jones, G. S. & Usoskin, I. Frost fairs, sunspots and the Little Ice Age, Astron. Geophys., 83(2), 17–23 (2017). doi: 10.1093/astrogeo/atx057
Lugaz, N., Temmer, M., Wang, Y. & Farrugia, C. J. The interaction of successive coronal mass ejections: A review, Sol. Phys., 292(4), Article 64 (2017). doi: 10.1007/s11207-017-1091-6
Luhmann, J. G., Lee, C. O., Li, Y., Arge, C. N., Galvin, A. B., Simunac, K., Russell, C. T., Howard, R. A. & Petrie, G. Solar wind sources in the late declining phase of Cycle 23: Effects of the weak solar polar field on high speed streams, Sol. Phys., 256, 285–305 (2009). doi: 10.1007/s11207-009-9354-5
Manchester, W., IV, Kilpua, E. K. J., Liu, Y. D., Lugaz, N., Riley, P., Török, T. & Vršnak, B. The physical processes of CME/ICME evolution, Space Sci. Rev., 212, 1159–1219 (2017). doi: 10.1007/s11214-017-0394-0
Martin, S. F. Conditions for the formation and maintenance of filaments, Sol. Phys., 182, 107–37 (1998). doi: 10.1023/A:100502681
Maunder, E. W. Note on the distribution of sunspots in heliographic latitude, 1874 to 1902, Mon. Nat. R. Astron. Soc., 64, 747–61 (1904).
McComas, D. J., Elliott, H. A., Schwadron, N. A., Gosling, G. T., Skoug, R. M. & Goldstein, B. E. The three-dimensional solar wind around solar maximum, Geophys. Res. Lett., 30(10), 1517–20 (2003). doi: 10.1029/2003GL017136
McComas, D. J., Velli, M., Lewis, W. S., Acton, L. W., Balat-Pichelin, M., Bothmer, V., Dirling, R. B., Feldman, W. C., Gloeckler, G., Habbal, S. R., Hassler, D. M., Mann, I., Matthaeus, W. H., McNutt, R. L., Mewaldt, R. A., Murphy, N., Ofman, L., Sittler, E. C., Smith, C. W. & Zurbuchen, T. H. Understanding coronal heating and solar wind acceleration: Case for in situ near-Sun measurements, Rev. Geophys., 45(1), Article RG1004 (2007). doi: 10.1029/2006RG000195
Melrose, D. B. Current-driven flare and CME models, J. Geophys. Res., 122, 7963–78 (2017). doi: 10.1002/2017JA024035
Miralles, M. P., Cranmer, S. R. & Kohl, J. L. Low-latitude coronal holes during solar maximum, Adv. Space Res., 33, 696–700 (2004). doi: 10.1016/S0273-1177(03)00239-4
Mordvinov, A. V., Pevtsov, A. A., Bertello, L. & Petrie, G. J. D. The reversal of the Sun's magnetic field in Cycle 24, Sol. Terr. Phys., 2(1), 3–18 (2016). doi: 10.12737/19856
Muñoz-Jaramillo, A., Balmaceda, L. A. & DeLuca, E. E. Using the dipolar and quadrupolar moments to improve solar-cycle predictions based on the polar magnetic fields, Phys. Rev. Lett., 111, Article 041106 (2013). doi: 10.1103/PhysRevLett.111.041106
Muraközy, J. & Ludmány, A. North-south differences of solar cycles, Centr. Eur. Astrophys. Bull., 34, 99–107 (2010).
Murray, S. A., Bingham, S., Sharpe, M. & Jackson, D. R. Flare forecasting at the Met Office Space Weather Operations Centre, Space Weather, 15, 577–88 (2017). doi: 10.1002/2016SW001579
Nagovitsyn, Y. A. & Pevtsov, A. A. On the presence of two populations of sunspots, Astrophys. J., 833, Article 94 (2016). doi: 10.3847/1538-4357/833/1/94
Nagovitsyn, Y. A., Pevtsov, A. A. & Osipova, A. A. Long-term variations in sunspot magnetic field–area relation, Astron. Nachr., 338, 26–34 (2017). doi: 10.1002/asna.201613035
Neugebauer, M. & Snyder, C. W. Solar plasma experiment, Science, 138, 1095–7 (1962).
Norton, A. A., Charbonneau, P. & Passos, D. Hemispheric coupling: Comparing dynamo simulations and observations, Space Sci. Rev., 186, 251–83 (2014). doi: 10.1007/s11214-014-0100-4
Odstrčil, D. Modeling 3-D solar wind structure, Adv. Space Res., 32, 497–506 (2003). doi: 10.1016/S0273-1177(03)00332-6
Owens, M. J. & Forsyth, R. J. The heliospheric magnetic field, Living Rev. Sol. Phys., 10(5), 1–52 (2013). doi: 10.12942/lrsp-2013-5
Owens, M. J., Lockwood, M. & Barnard, L. A. Coronal mass ejections are not coherent magnetohydrodynamic structures,

Sci. Rep., 7, Article 4152 (2017). doi: 10.1038/s41598-017-04546-3

Owens, M. J. & Riley, P. Probabilistic solar wind forecasting using large ensembles of near-Sun conditions with a simple one-dimensional 'upwind' scheme, Space Weather, 15, 1461–74 (2017). doi: 10.1002/2017SW001679

Pariat, E., Leake, J. E., Valori, G., Linton, M. G., Zuccarello, F. P. & Dalmasse, K. Relative magnetic helicity as a diagnostic of solar eruptivity, Astron. Astrophys., 601, Article A125 (2017). doi: 10.1051/0004-6361/201630043

Park, S.-H., Kusano, K., Cho, K.-S., Chae, J., Bong, S.-C., Kumar, P., Park, S.-Y. Kim, Y.-H. & Park, Y.-D. Study of magnetic helicity injection in the active region NOAA 9236 producing multiple flare-associated coronal mass ejection events, Astrophys. J., 778(1), Article 13 (2013). doi: 10.1088/0004-637X/778/1/13

Parker, E. N. Dynamics of the interplanetary gas and magnetic fields, Astrophys. J., 128, 664–76 (1958).

Parker, E. N. Extension of the solar corona into interplanetary space, J. Geophys. Res., 64, 1675–81 (1959).

Pesnell, W. D. Solar cycle predictions, Sol. Phys., 281, 507–32 (2012). doi: 10.1007/s11207-012-9997-5

Pesnell, W. D. Predicting Solar Cycle 24 using a geomagnetic precursor pair, Sol. Phys., 289, 2317–31 (2014). doi: 10.1007/s11207-013-0470-x

Pesnell, W. D. Predictions of solar cycle 24: How are we doing? Space Weather, 14, 10–21 (2016). doi: 10.1002/2015SW001304

Petrie, G. J. D. Evolution of active and polar photospheric magnetic fields during the rise of cycle 24 compared to previous cycles, Sol. Phys., 281, 577–98 (2012). doi: 10.1007/s11207-012-0117-3

Petrie, G. J. D. Solar magnetic activity cycles, coronal potential field models and eruption rates, Astrophys. J., 768, Article 12 (2013). doi: 10.1088/0004-637X/768/2/162

Petrie, G. J. D. On the enhanced coronal mass ejection detection rate since the Solar Cycle 23 polar field reversal, Astrophys. J., 812(1), Article 74 (2015). doi: 10.1088/0004-637X/812/1/74

Petrie, G. High-resolution vector magnetograms of the Sun's poles from Hinode: Flux distributions and global coronal modelling, Sol. Phys., 292, Article 13 (2017). doi: 10.1007/s11207-016-1034-7

Petrie, G. J. D., Canou, A. & Amari, T. Nonlinear force-free and potential-field models of active-region and global coronal fields during the Whole Heliosphere Interval, Sol. Phys., 274, 163–94 (2011). doi: 10.1007/s11207-010-9687-0

Petrie, G. J. D., Petrovay, K. & Schatten, K. Solar polar fields and the 22-year activity cycle: Observations and models, Space Sci. Rev., 186, 325–57 (2014). doi: 10.1007/s11214-014-0064-4

Petrovay, K. Solar cycle prediction, Living Rev. Sol. Phys., 7, Article 6 (2010). doi: 10.12942/lrsp-2010-6#

Pipin, V. V. & Kosovichev, A. G. Dependence of stellar magnetic activity cycles on rotational period in a nonlinear solar-type dynamo, Astophys. J., 823(2), Article 133 (2016). doi: 10.3847/0004-637X/823/2/133

Priest, E. R., Longcope, D. W. & Janvier, M. Evolution of magnetic helicity during eruptive flares and coronal mass ejections, Sol. Phys., 291, 2017–36 (2016). doi: 10.1007/s11207-016-0962-6

Rezaei, R., Beck, C., Lagg, A., Borrero, J. M., Schmidt, W. & Collados, M. Variation in sunspot properties between 1999 and 2014, Astron. Astrophys., 578, Article A43 (2015). doi: 10.1051/0004-6361/201425557

Richardson, I. G. Solar wind stream interaction regions throughout the heliosphere, Liv. Rev. Sol. Phys., 15(1), Article 1 (2018). doi: 10.1007/s41116-017-0011-z

Richardson, J. D., Belcher, J. W., Lazarus, A. J. & Paularena, K. I. Statistical prpperties of the solar wind, AIP Conf. Proc., 382, 483–6 (1996). doi: 10.1063/1.51433

Riley, P., Linker, J. A., Mikic, Z., Lionello, R., Ledvina, S. A. & Luhmann, J. G. A comparison between global solar magnetohydrodynamic and potential field source surface model results, Astrophys. J., 653, 1510–16 (2006). doi: 10.1086/508565

Sadykov, V. M. & Kosovichev, A. G. Relationships between characteristics of the line-of-sight magnetic field and solar flare forecasts, Astrophys. J., 849, Article 148 (2017). doi: 10.3847/1538-4357/aa9119

Schatten, K. Fair space weather for solar cycle 24, Geophys. Res. Lett., 32, Article L21106 (2005). doi: 10.1029/2005GL024363

Schatten, K. H., Wilcox, J. M. & Ness, N. F. A model of the interplanetary and coronal magnetic fields, Sol. Phys., 6, 442–55 (1969).

Schmieder, B., Archontis, V. & Pariat, E. Magnetic flux emergence along the solar cycle, Space Sci. Rev., 186, 227–50 (2014). doi: 10.1007/s11214-014-0088-9

Schmieder, B., Aulanier, G. & Vršnak, B. Flare-CME models: An observational perspective (invited review), Sol. Phys., 290, 3457–86 (2015). doi: 10.1007/s11207-015-0712-1

Schonfeld, S. J., White, S. M., Henney, C. J., Arge, C. N. & McAteer, R. T. J. Coronal sources of the F10.7 radio flux, Astrophys. J., 808, Article 29 (2015). doi: 10.1088/0004-637X/808/1/29

Schrijver, C. J. The nonpotentiality of coronae of solar active regions, the dynamics of the surface magnetic field, and the potential for large flares, Astrophys. J., 820, Article 103 (2016). doi: 10.3847/0004-637X/820/2/103

Schrijver, C. J. & DeRosa, M. L. Photospheric and heliospheric magnetic fields, Sol. Phys., 212, 165–200 (2003). doi: 10.1023/A:1022908504100

Schrijver, C. J., De Rosa, M. L., Title, A. M. & Metcalf, T. R. The nonpotentiality of active-region coronae and the dynamics of the photospheric magnetic field, Astrophys. J., 628, 501–13 (2005). doi: 10.1086/430733

Schulz, M. Non-spherical source-surface model of the heliosphere: A scalar formulation, Ann. Geophys., 15, 1379–87 (1997). doi: 10.1007/s00585-997-1379-1

Severnyi, A. B. Nonstationary processes in solar flares as a manifestation of the pinch effect, Sov. Astron., 2, 310–25 (1958).

Sharykin, I. N., Sadykov, V. M., Kosovichev, A. G., Vargas-Dominguez, S. & Zimovets, I. V. Flare energy release in the lower solar atmosphere near the magnetic field Polarity Inversion Line, Astrophys. J., 840, Article 84 (2017). doi: 10.3847/1538-4357/aa6dfd

Sheeley, N. R., Walters, J. H., Wang, Y.-M. & Howard, R. A. Continuous tracking of coronal outflows: Two kinds of coronal mass ejections, J. Geophys. Res., 104(A11), 24739–68 (1999). doi: 10.1029/1999JA900308

Sheeley, N. R., Jr & Wang, Y.-M. The recent rejuvenation of the Sun's large-scale magnetic field: A clue for understanding past and future sunspot cycles, Astrophys. J., 809(2), Article 113 (2015). doi: 10.1088/0004-637X/809/2/113

Simoniello, R., Tripathy, S. C., Jain, K. & Hill, F. A new challenge to solar dynamo models from helioseismic observations: The latitudinal dependence of the progression of the solar cycle, Astrophys. J., 828, Article 41 (2016). doi: 10.3847/0004-637X/828/1/41

Smith, E. J. The heliospheric current sheet, J. Geophys. Res., 106, 15819–31 (2001). doi: 10.1029/2000JA000120

Smith, E. J. & Balogh, A. Ulysses observations of the radial magnetic field, Geophys. Res. Lett., 22, 3317–20 (1995). doi: 10.1029/95GL02826

Solanki, S. K. Sunspots: An overview, Astron. Astrophys. Rev., 11(2–3), 153–286 (2003). doi: 10.1007/s00159-003-0018-4

Solanki, S. K., Krivova, N. A. & Haigh, J. D. Solar irradiance variability and climate, Annu. Rev. Astron. Astrophys., 51, 311–51 (2013). doi: 10.1146/annurev-astro-082812-141007

Stakhiv, M., Landi, E., Lepri, S. T., Oran, R. & Zurbuchen, T. H. On the origin of mid-latitude fast wind: Challenging the two-state solar wind paradigm, Astrophys. J., 801, Article 100 (2015). doi: 10.1088/0004-637X/801/2/100

Stejko, A. M., Guerrero, G. G. & Kosovichev, A. G. 3D global modelling of the solar dynamo, eprint arXiv:1701.08450 (2017).

Strassmeier, K. G. Starspots, Astron. Astrophys. Rev., 17, 251–308 (2009). doi: 10.1007/s00159-009-0020-6

Sun, X., Hoeksema, J. T., Liu, Y. & Zhao, J. On polar magnetic field reversal and surface flux transport during Solar Cycle 24, Astrophys. J., 798, Article 114 (2015). doi: 10.1088/0004-637X/798/2/114

Svalgaard, L., Cliver, E. W. & Kamide, Y. Sunspot cycle 24: Smallest cycle in 100 years? Geophys. Res. Lett., 32, Article L01104 (2004). doi: 10.1029/2004GL021664

Tapping, K. F. The 10.7 cm solar radio flux ($F_{10.7}$), Space Weather, 11, 394–406 (2013). doi: 10.1002/swe.20064

Tapping, K. F. & Morgan, C. Changing relationships between sunspot number, total sunspot area and F10.7 in Cycles 23 and 24, Sol. Phys., 292, Article 73 (2017). doi: 10.1007/s11207-017-1111-6

Temmer, M., Rybák, J., Bendík, P., Veronig, A., Vogler, F., Otruba, W., Pötzi, W. & Hanslmeier, A. Hemispheric sunspot numbers R_n and R_s from 1945–2004: Catalogue and N-S asymmetry analysis for solar cycles 18–23, Astron. Astrophys., 447, 735–43 (2006). doi: 10.1051/0004-6361:200 54060

Temmer, M., Thalmann, J. K., Dissauer, K., Veronig, A. M., Tschernitz, J., Hinterreiter, J. & Rodriguez, L. On flare-CME characteristics from Sun to Earth combining remote-sensing image data with in situ measurements supported by modelling, Sol. Phys., 292, Article 93 (2017a). doi: 10.1007/s11207-017-1112-5

Temmer, M., Reiss, M. A., Nikolic, L., Hofmeister, S. J. & Veronig, A. M. Preconditioning of interplanetary space due to transient CME disturbances, Astrophys. J., 835, Article 141 (2017b). doi: 10.3847/1538-4357/835/2/141

Török, T., Downs, C., Linker, J. A., Lionello, R., Titov, V. S., Mikić, Z., Riley, P., Caplan, R. M. & Wijaya, J. Sun-to-earth MHD simulation of the 2000 July 14 'Bastille Day' eruption, Astrophys. J., 856, Article 75 (2018). doi: 10.3847/1538-4357/aab36d

Usoskin, I. G. A history of solar activity over millennia, Living Rev. Sol. Phys., 14, Article 3 (2017). doi: 10.1007/s41116-017-0006-9

van Driel-Gesztelyi, L. & Green, L. M. Evolution of active regions, Living Rev. Sol. Phys., 12, Article 1 (2015). doi: 10.1007/lrsp-2015-1

Wang, Y. M. The Sun's large-scale magnetic field and its long-term evolution, Sol. Phys., 224, 21–35 (2004). doi: 10.1007/s11207-005-4982-x

Wang, Y.-M. Coronal holes and open magnetic flux, Space Sci. Rev., 144, 383–99 (2009). doi: 10.1007/s11214-008-9434-0

Wang, Y.-M. Solar cycle variation of the Sun's low-order magnetic multipoles: Heliospheric consequences, Space Sci. Rev., 186, 387–407 (2014). doi: 10.1007/s11214-014-0051-9

Wang, Y.-M. & Colaninno, R. Is Solar Cycle 24 producing more coronal mass ejections than Cycle 23? Astrophys. J. Lett., 784, Article L27 (2014). doi: 10.1088/2041-8205/784/2/L27

Wang, Y.-M. & Muglach, K. On the formation of filament channels, Astrophys. J., 666, 1284–95 (2007). doi: 10.1086/520623

Wang, Y.-M. & Sheeley, N. R. Jr. Solar wind speed and coronal flux-tube expansion, Astrophys. J., 355, 726–32 (1990). doi: 10.1086/168805

Wang, Y.-M. & Sheeley, N. R., Jr. Magnetic flux transport and the sun's dipole moment – New twists to the Babcock-Leighton model, Astrophys. J., 375, 761–70 (1991). doi: 10.1086/170240

Wang, Y.-M. & Sheeley, N. R., Jr. On potential field models of the solar corona, Astrophys. J., 392, 310–19 (1992). doi: 10.1086/171430

Wang, Y.-M. & Sheeley, N. R., Jr., Global evolution of interplanetary sector structure, coronal holes, and solar wind streams during 1976–1993: Stackplot displays based on solar magnetic observations, J. Geophys. Res., 99(A4), 6597–608 (1994). doi: 10.1029/93JA02105

Wang, Y.-M., Sheeley, N. R., Jr & Lean, J. Understanding the evolution of the Sun's open magnetic flux, Geophys. Res. Lett., 27, 505–8 (2000). doi: 10.1029/1999GL010744

Wang, Y.-M., Sheeley, N. R., Jr & Andrews, M. D. Polarity reversal of the solar magnetic field during cycle 23, J. Geophys. Res., 107(A12), Article 1465 (2002). doi: 10.1029/2002JA009463

Wang, Y. M. & Sheeley, N. R., Jr. Sources of the solar wind at Ulysses during 1990–2006, Astrophys. J., 653, 708–18 (2006). doi: 10.1086/508929

Wang, Y.-M. & Sheeley, N. R., Jr. Understanding the geomagnetic precursor of the solar cycle, Astrophys. J., 694, L11–15 (2009). doi: 10.1088/0004-637X/694/1/L11

Wang, Y.-M., Robrecht, E. & Sheeley, N. R., Jr. On the weakening of the polar magnetic fields during solar cycle 23, Astrophys. J., 707, 1372–86 (2009). doi: 10.1088/0004-637X/707/2/1372

Weiss, N. O. & Tobias, S. M. Physical causes of solar activity, Space Sci. Rev., 94, 99–112 (2000). doi: 10.1023/A:1026790416627

Weiss, N. O. & Thompson, M. J. The solar dynamo, Space Sci. Rev., 144, 53–66 (2009).

Wiegelmann, T., Petrie, G. J. D. & Riley, P. Coronal magnetic field models, Space Sci. Rev., 210, 249–74 (2017). doi: 10.1007/s11214-015-0178-3

Wimmer-Schweingruber, R. F. & Hassler, D. M. Tracing heliospheric structures to their solar origin, AIP Conf. Proc., 1720, Article 100002 (2016). doi: 10.1063/1.4943857

Wimmer-Schweingruber, R. F., von Steiger, R. & Paerli, R. Solar wind stream interfaces in corotating interaction regions: New SWICS/Ulysses results, J. Geophys. Res., 104, 9933–46 (1999). doi: 10.1029/1999JA900038

Wood, B. E., Wu, C. C., Lepping, R. P., Nieves-Chinchilla, T., Howard, R. A., Linton, M. G. & Socker, D. G. A STEREO survey of magnetic cloud coronal mass ejections observed at Earth in 2008–2012, Astrophys. J. Suppl. Ser., 229(2), Article 29 (2017). doi: 10.3847/1538-4365/229/2/29

Wyper, P. E., Antiochos, S. & DeVore, C. R. A universal model for solar eruptions, Nature, 544, 452–9 (2017). doi: 10.1038/nature22050

Xie, H., Mäkelä, P., St. Cyr, O. C. & Gopalswamy, N. Comparison of the coronal mass ejection shock acceleration of three widespread SEP events during solar cycle 24, J. Geophys. Res., 122, 7021–41 (2017). doi: 10.1002/2017JA024218

Yeo, K. L., Krivova, N. A. & Solanki, S. K. Solar cycle variation in solar irradiance, Space Sci. Rev., 186, 137–67 (2014). doi: 10.1007/s11214-014-0061-7

Yeo, K. L., Krivova, N. A. & Solanki, S. K. EMPIRE: A robust empirical reconstruction of solar irradiance variability, J. Geophys. Res., 122, 3888–3914 (2017a). doi: 10.1002/2016JA023733

Yeo, K. L., Solanki, S. K., Norris, C. M., Beeck, B., Unruh, Y. C. & Krivova, N. A. Solar irradiance variability is caused by the magnetic activity on the solar surface, Phys. Rev. Lett., 119(9), Article 091102 (2017b). doi: 10.1103/PhysRev Lett.119.091102

Yurchyshyn, V., Yashiro, S., Abramenko, V., Wang, H. & Gopalswamy, N. Statistical distributions of speeds of coronal mass ejections, Astrophys. J., 619, 599–603 (2005). doi: 10.1086/426129

Zacharias, P. An independent review of existing total solar irradiance records, Surv. Geophys., 35, 897–912 (2014). doi: 10.1007/s10712-014-9294-y

Zerbo, J.-L. & Richardson, J. D. The solar wind during current and past solar minima and maxima, J. Geophys. Res., 120, 10250–56 (2015). doi: 10.1002/2015JA021407.

Zhang, J., Woch, J., Solanki, S. K., von Steiger, R. & Forsyth, R. Interplanetary and solar surface properties of coronal holes observed during solar maximum, J. Geophys. Res., 108, Article 1144 (2003). doi: 10.1029/2002JA009538

Zhao, L., Landi, E., Lepri, S. T., Kocher, M., Zurbuchen, T. H., Fisk, L. A. & Raines, J. M. An anomalous composition in slow solar wind as a signature of magnetic reconnection in its source region, Astrophys. J. Suppl. Ser., 228(1), Article 4 (2017). doi: 10.3847/1538-4365/228/1/4

Zheng, Y., Macneice, P., Odstrcil, D., Mays, M. L., Rastaetter, L., Pulkkinen, A., Taktakishvili, A., Hesse, M., Kuznetsova, M., Lee, H. & Chulaki, A. Forecasting propagation and evolution of CMEs in an operational setting: What has been learned, Space Weather, 11, 557–74 (2013). doi: 10.1002/swe.20096

19

Long- and Short-Term Geomagnetic Prediction

Weijia Kuang, Andrew Tangborn, Terence Sabaka and Robert Tyler

Prediction of geomagnetic variability depends on the accuracy of geomagnetic field modeling, dynamical modeling of source regions that contribute to geomagnetic signals and advanced assimilation algorithms that combine effectively the results of geomagnetic field and dynamic models to make accurate estimates of the dynamic states of the sources and, therefore, accurate forecast of geomagnetic variations. In this chapter we provide an overview of recent research efforts in these three research areas, focusing primarily on geomagnetic variations from the dynamic outer core and from solar and lunar tidal effects, but also including a review of relevant research results and developments. Prediction of weak but periodic tidal phenomena, and of strong but chaotic secular variation (SV), showcases two very important new developments which will lead to new opportunities in geomagnetic research and application.

19.1 Introduction

Geomagnetic data measured on the Earth's surface and in the low Earth orbits (LEO) include contributions from various sources within the Earth system and of external origins. Several examples can be found in other chapters of this book. These contributions depend on chemical and dynamical properties of the source regions. Therefore, observed geomagnetic variations provide important information for understanding the Earth system, and thus a means to monitor magnetically changes of the Earth system in space and in time.

Generally speaking, prediction of geomagnetic variations from a particular source region, e.g. the Earth's outer core, requires assimilation of observations into relevant dynamical models. Thus it involves parallel efforts in dynamical models of the source region, geomagnetic field models to extract contributions of the source region from surface observations and efficient assimilation algorithms (in both information acquisition and computational need). But they also complement each other: for example, results from the field models are used to constrain the dynamical models so that they can make better estimation of the 'true' state (more accurately, the partial truth specified by the observations), and the improved dynamical models can be also used to assist the field models for more accurate source separations. This very simple description of geomagnetic prediction should therefore be fully explored for future geomagnetic research and application.

In this chapter, we focus on prediction of geomagnetic variation from two very different systems: geomagnetic secular variation (SV) from the convective outer core, and magnetic anomalies from tidal effects. They represent two extreme cases of geomagnetic variability: the former is dominantly strong (over 95% of energy) in the geomagnetic measurements, but is chaotic (thus difficult to predict) in nature, while the latter is extremely weak but highly regular (thus predictable) in its temporal variation patterns. In particular, prediction of SV from the core and of geomagnetic fluctuations from ocean tidal flow are new developments in geomagnetic studies and applications and therefore deserve special attention from the IAGA community.

Observed SV plays a crucial role in understanding the core dynamics: small changes observed at the surface are the manifestation of variations in the outer core dynamic state, such as the fluid velocity field (e.g. Kuang and Tangborn, 2015). Therefore, its complex spatial/temporal variation patterns are not predictable over extended periods of time (e.g. Kuang et al., 2009; Hulot et al., 2010). By relative contrast, geomagnetic fluctuations due to tidal phenomena are approximately 10^{-5} order in magnitude of that of the core field. They are very predictable since they are generated at phase-locked astronomical periods. However, it is expected or observed that solar heating effects (also included as 'tidal' geomagnetic variations) are much less stationarity. Their contrasting variation patterns require very different approaches in analyzing and interpreting the observed signals.

Prediction of SV could greatly help our understandings of fundamental dynamical properties for interpretation of the SV observed at the surface, the geodynamo and the dynamic state in the outer core and the properties of the deep interior. The principle of geodynamo is very simple: the core field is generated and maintained by the outer core fluid motion, most likely the convection driven by gravitational energy released from secular cooling and differentiation of the Earth over geological timescales. But the dynamic details are still very unresolved. Numerical geodynamo simulation can provide a partial depiction of the core dynamics (Christensen et al., 2010) but is hampered by lack of resolution (due to limited computing resource) to reach Earth-like parameter regimes. On the other hand, geomagnetic observations provide direct evidences of complex spatial and temporal variation of the intrinsic field. But the observational evidence is also partial, limited by short measurement records and long distances to the outer core (the source region of SV). As suggested by recent research results (Aubert and Fournier, 2011; Aubert, 2014; Kuang and Tangborn, 2015), geomagnetic data assimilation can integrate partial information from both studies to obtain better estimates of the core state, thus improving our understandings of the core dynamics and its implications for interactions and responses between the outer core and other components of the Earth.

Predictability of global geomagnetic signals due to ocean tides is supported by the agreement shown in independent descriptions derived from theoretical forward models and field models of observations (Tyler et al., 2003; Sabaka et al., 2015, 2016). These studies have recently opened an opportunity for using the tidal magnetic signals to probe ocean electrical conductivity, variability in ocean tidal processes and mantle conductivity (Schnepf et al., 2015; Grayver et al., 2016, 2017; Tyler et al., 2017).

This chapter is organized as follows: an overview of geomagnetic field modeling is given in the next section, followed by sections discussing the tidal magnetic signals, geodynamo and geomagnetic data assimilation using an ensemble Kalman filter, SV prediction on decadal and longer timescales and, finally, a concluding discussion.

19.2 Source Separation in Geomagnetic Field Modelling

This chapter concerns the forecasting of long- and short-timescale geomagnetic fields. Provided with initial conditions, the partial differential equations (PDEs) and boundary conditions (BCs) governing a particular constituent of the geomagnetic field would produce a time series of magnetic fields that would likely diverge from what could be or is observed. This is due to several reasons, including the inadequacy of the PDEs and BCs to properly describe reality and the sensitivity of these to initial conditions. If realistic forecasting is to be achieved, then the PDEs and BCs must interact with actual observations, which is the basis of data assimilation described in Section 19.4. However, near the Earth, one finds a rich variation of current systems, and any measurement of the magnetic field will be composed of a superposition of these sources. It should be clear that if one is interested in predicting the Earth's core field, then any geomagnetic data assimilation system (GDAS) should be presented with measurements of the core field, not, for example, with a measurement of the night-side ionospheric-induced field. It is crucial, then, that some type of source separation mechanism be developed before any meaningful forecasting can take place.

This section will present a particular philosophy and methodology of field modeling that provides separation of field sources in an optimal way. The general classification of these models is of the 'inverse' variety, which is to say that the geophysical parameters of a numerical model are inferred from measurements of the magnetic field. Furthermore, these models are mostly 'empirical' in that they describe the variations and patterns of the observed magnetic field with generic mathematical functions while abiding by some physical restrictions. This is in contrast to the usual 'forward' models, which are composed of the PDEs and BCs alluded to above and run in free form given some initial conditions, independent of measurements. The fusing of inverse and forward models provides the basis for GDAS.

There are two concepts of field modeling that should be understood: correlation and bias. The first occurs when the mathematical functions corresponding to each parameter or coefficient can describe some portion of the measurements or data space that other functions do. If the part of the data space described by the function, or its span, can also be described by a combination of the spans of all other functions, then the model is ill posed and does not possess a unique solution. If the other extreme occurs where no other function can describe any part of the data space described by any other function, then the functions are orthogonal over the data space with a zero correlation, and each parameter may be estimated independently. Usually there is a non-empty intersection between each pair of spans, but there is also part of each span that is unique, thus rendering a unique solution. This non-empty intersection of spans leads to correlation between the associated parameters. Bias, on the other hand, is a discrepancy between the estimated and true values of a parameter that cannot be eliminated by resampling and averaging over the data space and is usually due to errors of omission or commission. A successful estimation approach will attempt to mitigate bias while taking correlation into account.

Two major methodologies have emerged in dealing with correlation in geomagnetic field modeling. For discussion purposes, consider the following simple linear model consisting of two sets of parameter vectors, $\mathbf{x}$ and $\mathbf{y}$, associated with basis functions that comprise the columns of matrices $\mathbf{A}$ and $\mathbf{B}$, respectively, describing the data vector of measurements $\mathbf{d}$ in the presence of a vector of additive error $\boldsymbol{\nu}$ whose mean is zero and covariance is the identity matrix, such that

$$\mathbf{d} = \mathbf{A}\mathbf{x} + \mathbf{B}\mathbf{y} + \nu. \tag{19.2.1}$$

The first method, 'sequential estimation', attempts to estimate the parameters in an arbitrary order, say $\mathbf{x}$ followed by $\mathbf{y}$, where $\mathbf{x}$ is estimated directly from $\mathbf{d}$ followed by $\mathbf{y}$ estimated from the residuals $\mathbf{d} - \mathbf{A}\tilde{\mathbf{x}}$, where the tilde indicates the estimate of $\mathbf{x}$. It can be shown that if the cycle is repeated with subsequent $\mathbf{x}$ being estimated from the latest residuals $\mathbf{d} - \mathbf{B}\tilde{\mathbf{y}}$ and $\mathbf{y}$ estimated from the latest residuals $\mathbf{d} - \mathbf{A}\tilde{\mathbf{x}}$, then this constitutes a dual-block Gauss–Seidel method that converges as long as the entire system is non-singular. However, there are pitfalls with this method in that (1) in practice, only a single iteration is often performed; (2) convergence can be very slow; and (3) the method can mask co-linearities between basis functions that lie in separate blocks.

The second method is 'co-estimation', which attempts to estimate all parameters in a single inversion. This method can be understood by considering a data space D and the spans of the columns of matrices $\mathbf{A}$ and $\mathbf{B}$ within this space. The regions of intersection between the spans of $\mathbf{A}$ and $\mathbf{B}$, denoted '$A \cap B$', represent areas of dispute where both sets can describe the data and thus lead to non-uniqueness. In co-estimation, these regions are eliminated from consideration, and the $\mathbf{x}$ and $\mathbf{y}$ parameters are rather determined from the unique regions of '$A \setminus B$' and '$B \setminus A$', respectively.

The error between a true model state $\mathbf{x}$ and its estimate $\tilde{\mathbf{x}}$ as measured by the mean-squared error (MSE) may be decomposed into two terms, as shown in the following equation, one involving bias $\mathbf{b}$ and the other involving variance $\mathbf{C}$:

$$\begin{aligned} \mathrm{MSE}(\tilde{\mathbf{x}}) &= E[(\tilde{\mathbf{x}} - \mathbf{x})^{\mathrm{T}}(\tilde{\mathbf{x}} - \mathbf{x})] \\ &= \mathbf{b}^{\mathrm{T}}\mathbf{b} + Tr[\mathbf{C}], \end{aligned} \tag{19.2.2}$$

where $E[\cdot]$ is the expectation operator, $Tr[\cdot]$ is the trace operator, $\mathbf{b} = \mathbf{x} - \bar{\mathbf{x}}$ and $\bar{\mathbf{x}}$ and $\mathbf{C}$ are the expected value and covariance of $\tilde{\mathbf{x}}$, respectively. These two terms represent orthogonal contributions over the space of all estimates. In the case of least squares estimation, the bias term is *assumed* zero, and thus the model error is often reported only as the variance term. However, when modeling real magnetic fields, this is surely never the case, and in fact the bias term may actually dominate.

As with correlation, basically two approaches are used to minimize the effects of bias. The first is data selection, in which subsets are chosen that do not contain an undesirable field constituent. For example, external fields exhibit a range of intensities, and their presence can be detrimental to the resolution of fields from the core and lithosphere (see Finlay et al., 2017; Olsen and Stolle, 2017). Therefore, most models will use data from so-called quiet times, when external fields are relatively weak. In another example, since the primary ionospheric field is mostly absent during the night sector, most models interested in the core and lithosphere will choose data during this time. However, because of the complexities of the magnetic environment, it is difficult to make data selections that eliminate all unwanted contaminates. For instance, when night sector data is chosen to filter out primary ionospheric signals, it turns out that the associated secondary or induced field is still present and this indeed biases estimates of the core field. This particular contaminate affects the odd, low-degree zonal terms of a spherical harmonic expansion of the core field.

The second approach entails co-estimating the biases, or nuisance parameters, along with the nominal parameters. The assumption is that there exists a subset of measurements whose signal-to-noise ratio is large enough to accurately resolve a particular subset of parameters. All other data subsets are not used to resolve these parameters. If this holds true, then a functional representation of the actual contaminate is not needed, since only its projection onto the nominal basis functions comes into play leaving only nuisance versions of the nominal parameters of interest to be estimated. Using the example of night sector contamination by induced ionospheric fields, if day sector data are introduced, then a representation of this induced field may be estimated from the data. However, since the day sector data will be contaminated with respect to core and lithospheric fields, they are used to estimate nuisance versions of these. The result is that core and lithospheric field derived from night sector data are now not contaminated by ionospheric induction.

Some level of the co-estimation approach is followed by all modeling efforts, including the very successful POMME (Maus et al., 2006; Maus, 2017), GRIMM (Lesur et al., 2008, 2015a, 2015b) and CHAOS (Olsen et al., 2014; Finlay et al., 2016) suites. However, the suite of Comprehensive Models (CMs) (Sabaka et al., 2002, 2004, 2015, 2016) in general attempts to co-estimate more sources. Aside from the usual core and lithospheric fields, these models have been able to simultaneously resolve the solar quiet (Sq) fields as well as the oceanic M_2 tidal field. As for biases, the CMs use limited data selection and employ the selective infinite variance weighting (SIVW) scheme of Sabaka and Olsen (2006) and Sabaka et al. (2013), which essentially treat the biases in the manner described above. It is envisioned that an estimation scheme like that used in the CMs will be combined with geodynamo simulation algorithms as a front-end filter to provide a GDAS algorithm for improved long-term geomagnetic forecasting.

19.3 Predictability of Tidal Phenomena in the Geomagnetic Record

When considering rapid temporal fluctuations around the slowly changing main field, one may anticipate high predictability resulting from phenomena driven by the Sun and Moon. The spin/orbits and gravitational potentials of the Earth, Moon, Sun and even the radiation from the Sun are all highly predictable. Modulations in the associated geomagnetic fluctuations may be expected then to primarily reflect temporal variations in the Earth's response to these 'tidal' driving forces.

To describe the predictability of the tidal effects in the geomagnetic field, one can consider the long record of observations from the Honolulu (station HON) geomagnetic observatory. Specifically, the hourly sampled records are differenced in time to produce the eastward (E), northward (N) and radial (R) vector-component time series.

In Figure 19.3.1 the power spectral density is shown (the spectrum is produced using the Welch averaging method on 24 × 1024-hour Hamming windowed segments). While one sees a red spectrum for periods longer than about a week, shorter-period power density increases and shows narrow spectral peaks. The larger of these peaks appear to be multiples of the Earth's rotation frequency with respect to the Sun, as well as frequencies derived from adding or subtracting pairs of such frequencies. This suggests that the variability may be predominantly due to solar radiation and gravity tidal forces under amplitude and phase modulation by the parameters involved in the Earth's response.

Empirical orthogonal functions (EOFs) compactly describe the variability in these time series in terms of statistical modes. The EOF analyses can be applied to the data matrix created as follows. Start by regarding $[E, N, R]$ as a three-column matrix, each column representing a time series of hourly samples. It is then replaced with a 72-column matrix in which each row presents one day of data (the $1 \cdots 24$ hourly samples of the E, N, R, sequentially). In this case, the data matrix presents 72 time series of daily samples. While simply averaging this data matrix along the columns would give a 1 × 72 row vector describing the time-averaged behavior of E, N, R over the course of the solar day, such averaging could remove, or even cancel, components of regular signals with strong modulations due to seasons or other factors. The EOF analyses remove this limitation.

In the EOF analysis, the original 72 time series are linearly combined to produce 72 new time series. Each linear combination is referred to as an EOF mode, and the associated time series modulating the mode is referred to as the expansion coefficient series. The EOF modes and expansion coefficients are both orthogonal sets and have the useful property that the amount of total variance explained decreases monotonically from mode 1 to the higher modes. They are statistically most efficient at compactly describing the variance presented in the data matrix described above. The construction of the data matrix implicitly imposes a request for regular structures organized with respect to the solar day, as is the proximate goal.

In Figure 19.3.2 the results of the EOF analysis are shown. Remarkably, the first mode explains a fraction 0.48 of the variance, and the second mode explains 0.14 such that more than half of the variance is captured in only two of the 72 EOF modes. Also important, there is a large drop and gradual decline in the variance explained by the higher modes, suggesting that other fluctuations in the data that are not particularly organized with respect to the solar day are simply being represented through this remaining part of the orthonormal expansion.

The EOF spectrum, on its own, suggests much of the solar daily variability is captured in two modes. This is

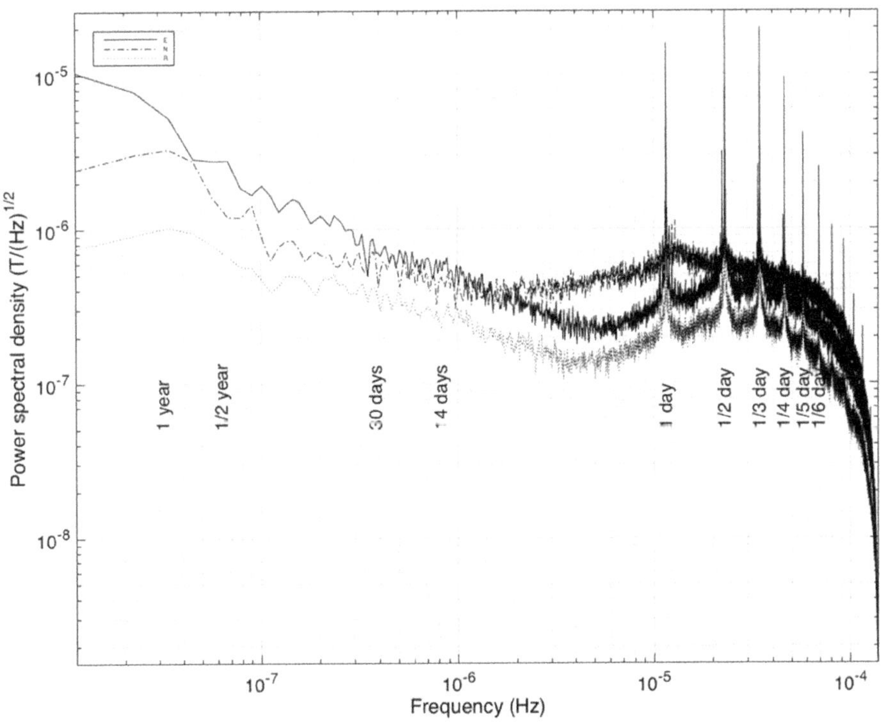

Figure 19.3.1 Eastward (E), northward (N) and radial (R) vector components of the time rate of change of geomagnetic observations at the Honolulu observatory.

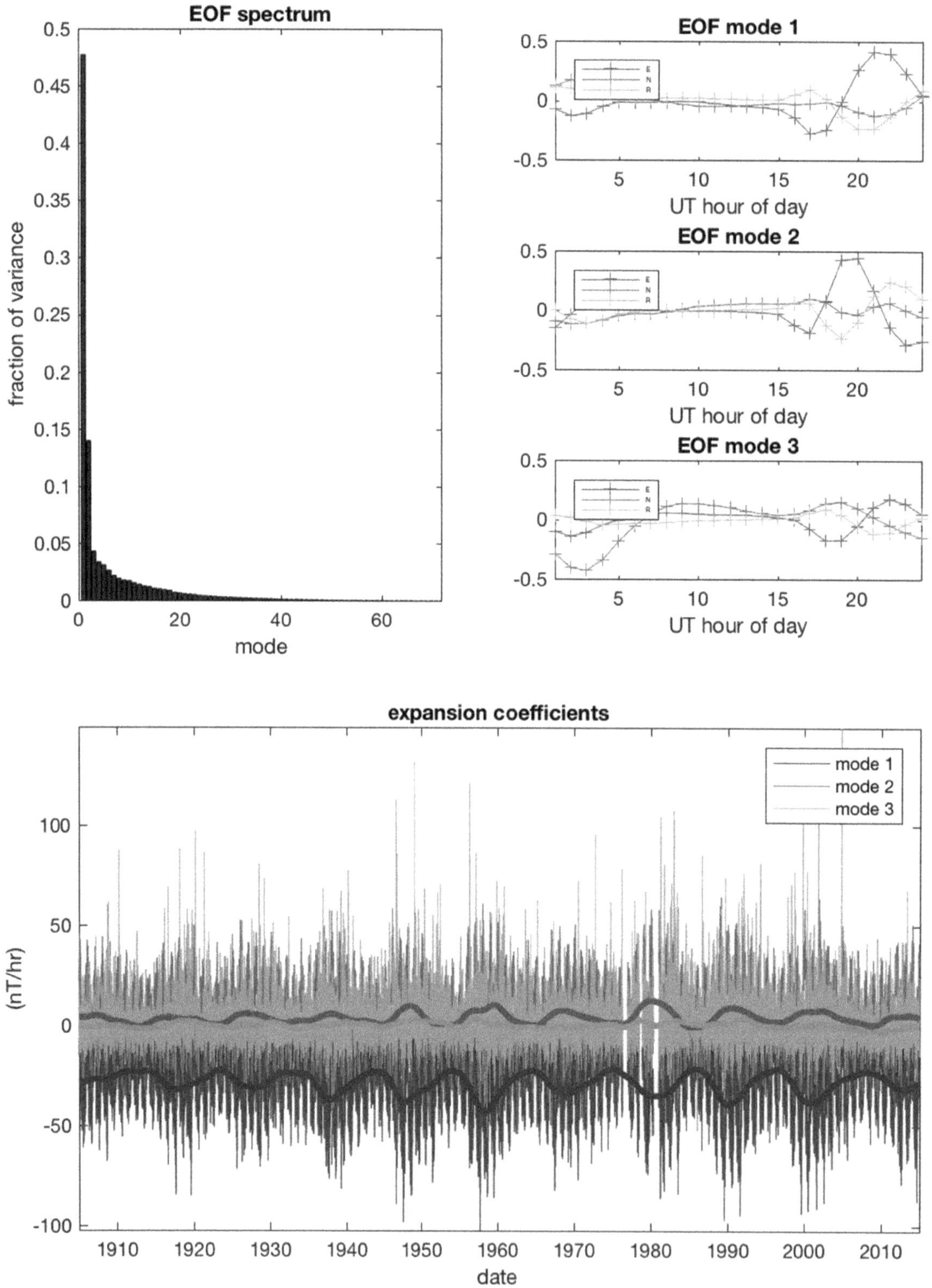

Figure 19.3.2 EOF analyses of the day vs hour data matrix of geomagnetic data from Honolulu. The EOF spectrum (upper left panel) shows that more than half of the variance is explained by the first two EOF modes; the EOFs (daily structures of *E, N, R*) for the first three modes are shown in the upper right panel. The associated expansion coefficients (lower panel) of the first two modes primarily show a seasonal cycle modulated by the solar cycle (as indicated by the thicker curves obtained by low-pass filtering the series with a three-year moving window). Because seasons as well as the solar cycle can be predicted, around half of the variance *E, N, R* is expected also to be predictable using only the first two EOF modes. (A black-and-white version of this figure appears in some formats. For the colour version, please refer to the plate section.)

further supported by the EOFs shown in the upper right panels of Figure 19.3.2. The first two modes show well-studied solar-quiet-type curves with larger amplitudes during the local day time and larger amplitudes in the horizontal components consistent with large-scale sheet currents overhead. The first two modes also appear in quadrature, suggesting one common process with phase migration.

The associated expansion coefficients (which modulate the EOF structures shown in the upper right panels) are also highly intuitive. Magnification shows a strong seasonal cycle in modes 1 and 2, which also appear in quadrature, suggesting again a common process with phase modulation. The low-pass curves show more clearly that there is also amplitude modulation with the solar cycle in modes 1 and 2. EOF mode 3 appears to also include influences due to the seasonal and solar cycles, but the correspondence is less clear, and other influences are also present.

While solar influences appear to dominate the spectral peaks, lunar periodicities are also clear on closer examination of the spectra. The M_2 semi-diurnal tide with a period of 12.42 hours is an example that can be seen to the left of the 1/2-day peak in Figure 19.3.1. Love and Rigler (2014) have conducted a careful and thorough spectral study of the HON time-differenced records using maximum-entropy and Lomb periodogram methods. They show and discuss solar, lunar and luni-solar periodicities corresponding to astronomically established periods ranging from a couple hours to a couple decades. While both the Moon and the Sun are clearly involved in these tidal fluctuations in the magnetic field, from a single observatory, it is not immediately clear if the tides involved are primarily in the ocean or upper atmosphere.

In the case of the M_2 constituent, comparisons of observations with forward-modeled simulations have established that the magnetic field is primarily generated through motional induction by the ocean tidal flow. This can be seen in Figure 19.3.3, which shows the first frame of a movie (available at https://neptune.gsfc.nasa.gov/gngphys/index.php?section=455, and see Tyler et al., 2003; Sabaka et al., 2015, 2016, for further information). One may note that while the M_2 map derived from observations (middle panel) has spherical harmonic base functions that independently attempt to separate internal and external sources, both the ionospheric tides as well as the oceanic tides are below the satellite observatories, and the spatial coverage of land observatories is poor over the ocean regions. The simplest independent indication of the oceanic origin is in the correlation of amplitudes with ocean basins, and certainly the rough correspondence with the M_2 surface displacement is also indicative. The primary disagreement seen between the theoretical and observational M_2 fields is in the difference in spatial resolution. Whereas the theoretical simulations are conducted using the one sixth-degree lon./lat. resolution of the ocean tidal model data, the observations extend only to a much coarser resolution (up to the spherical harmonic degree 18). With continued collection of satellite data, and improved modeling, it is expected that the agreements between the models will either increase or reveal new physical effects.

Because only the gravitational (and not radiational) tidal forces by the Moon on the Earth are expected to be important, and because the gravitational tidal forces are known quite precisely, the lunar tidal forces can be regarded as highly predictable. Of course, the tides that respond to these forces also depend on internal parameters and processes that may vary in time, with effects on the predictability of the response. On large scales, at least, the ocean fluid's tidal response can also be regarded as highly predictable (Egbert and Ray, 2017). Because the ocean electrical conductivity is also known from observations (Tyler et al., 2017), and is seen to have only very small temporal variations when considering offshore depth integrals, it is expected that ocean tidal magnetic fields may have a similarly high degree of predictability. In this case, the forward simulations such as shown in the middle panel of Figure 19.3.3 may be expected to reliably predict the tidal magnetic fields, including the high-spatial variations not yet confirmed by lower-resolution observational results, such as shown in the bottom panel.

The magnetic fields motionally generated by lunar tidal forces on the ocean may present the most predictable, large-scale, naturally occurring geomagnetic fluctuations. In this case, it is potentially an important source for sounding and/or monitoring the electrical conductivity of the oceans as well as the upper mantle. But the variability or even the stationarity of the ocean tidal magnetic fields has not yet been adequately examined, despite many observations demonstrating appreciable fluctuations in at least the tides of coastal observatories. In fact, the small observed variations in the M_2 surface height amplitude at coastal observatories, notably Honolulu (Colosi and Munk, 2006), have been closely studied, as they may indicate changes in ocean parameters, processes and stratification. Tyler (2013) presented a corresponding description of the history of M_2 variations seen in the magnetic observations at HON and showed that the fractional amplitude of the variability in the magnetic field is much higher than that seen in the sea surface displacement. This must be further studied to determine the stability of the tidal magnetic fields used as a sounding source.

19.4 Mathematics of Geomagnetic Data Assimilation

Currently, there are many geodynamo models based on different physics and numerical schemes (Christensen et al., 2001; Jones et al., 2011; Matsui et al., 2016) and many different data assimilation algorithms (see Kalnay, 2011). The mathematical description in this section is based on the formulations used in our geomagnetic data assimilation system (Jiang and Kuang,

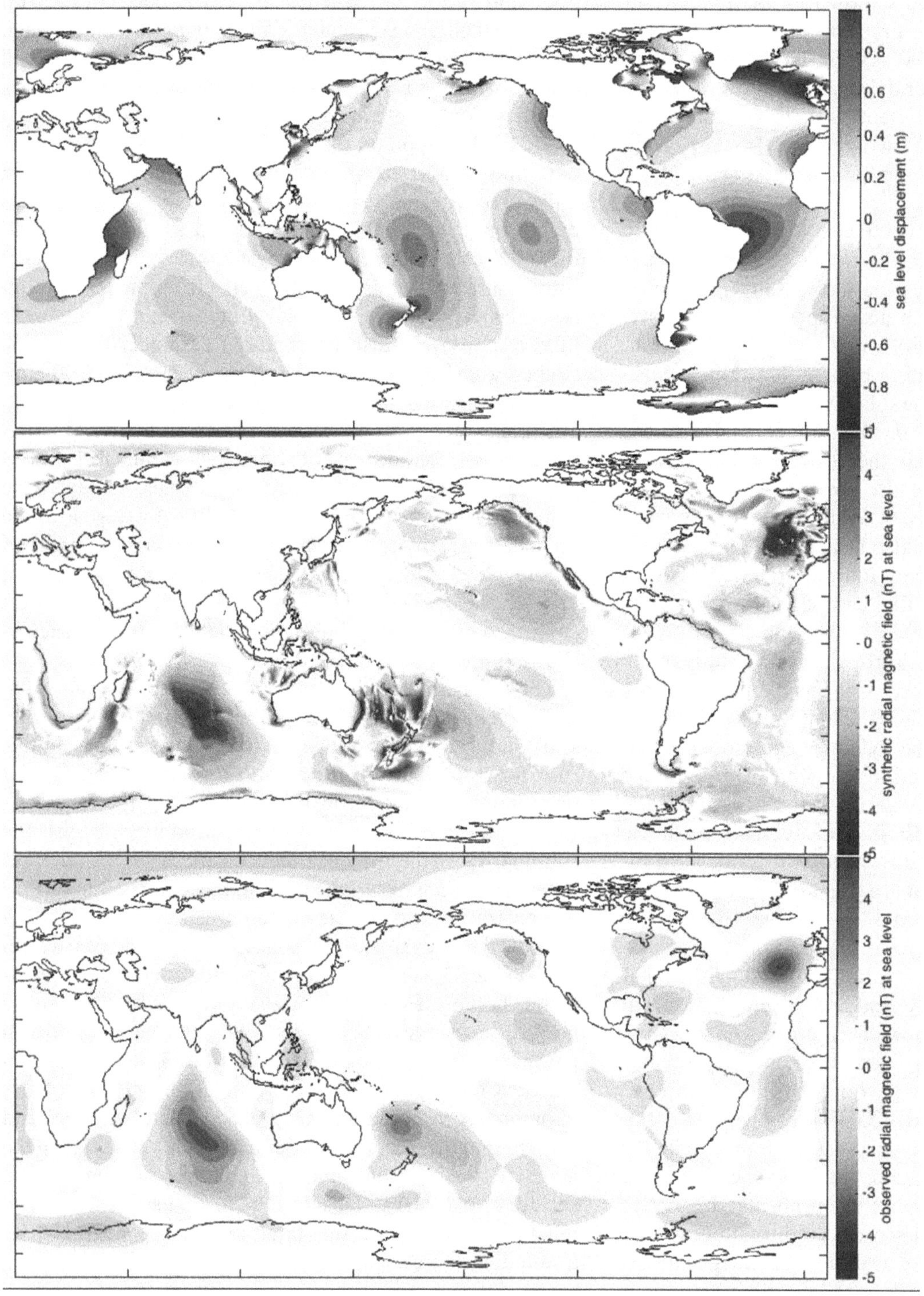

Figure 19.3.3 Movie frame of M_2 tidal surface displacement (top frame) and the radial component of magnetic field generated by the tidal flow as calculated in a numerical model (middle frame) and as seen in observations (bottom frame). The agreement between the bottom two frames strongly supports the assumption that the observed M_2 magnetic fluctuations are primarily due to ocean tides. (A black-and-white version of this figure appears in some formats. For the colour version, please refer to the plate section.)

2008; Kuang et al., 2008; Sun and Kuang, 2015). But they can be easily adapted to other geodynamo models. Note that all variables, unless otherwise specified, are non-dimensional in the rest of this chapter.

In our geodynamo model, the Earth's fluid outer core is approximated as a Boussinesq, electrically conducting fluid shell between the inner core boundary (ICB) with the mean radius r_{icb}, and the core–mantle boundary (CMB) with the mean radius r_{cmb}. The core state is described by the velocity field $\mathbf{v}$, the magnetic field $\mathbf{B}$ and the relative density anomaly $\delta\rho$. Since $\mathbf{v}$ and $\mathbf{B}$ are solenoidal, they can each be described by two independent scalar fields, e.g.

$$\mathbf{B} = \mathbf{B}_T + \mathbf{B}_P \equiv \nabla \times (T_b \hat{\mathbf{r}}) + \nabla \times \nabla \times (P_b \hat{\mathbf{r}}) \quad (19.4.1)$$

where T_b and P_b are called the toroidal and poloidal scalars, respectively. These five scalar fields are approximated by spherical harmonic expansions, with the spectral coefficients defined on discrete radial grid points, e.g.

$$\begin{bmatrix} T_b \\ P_b \end{bmatrix} = \sum_{0 \le m \le l}^{L} \begin{bmatrix} j_l^m(r_i) \\ b_l^m(r_i) \end{bmatrix} Y_l^m(\theta, \phi) + C.C., \quad (19.4.2)$$

where Y_l^m are the spherical harmonic functions of degree l and order m, r_i are the radial grids and $C.C.$ are the complex conjugate parts. The core state can be symbolically defined by a state vector $\mathbf{x}_D$ and is governed by the following initial-value problem:

$$\begin{aligned} \frac{\partial \mathbf{x}_D}{\partial t} &= \mathrm{M}(\mathbf{x}_D, \Lambda) \\ \mathbf{x}_D(t_0) &= \mathbf{x}_D^0. \end{aligned} \quad (19.4.3)$$

where Λ denotes the dynamo parameters arising from model scaling rules and describing the outer core properties, such as the small magnetic Rossby number R_o (for the fluid inertia) and the small Ekman number E (for the viscous effect). For the details of (19.4.3) and the dynamo parameters, we refer the reader to e.g. Kuang and Bloxham (1999) and Kuang and Chao (2003).

In geomagnetic data assimilation, the model output $\mathbf{x}_D$ is called the forecast (and denoted by $\mathbf{x}^f$ in the rest of the section). In sequential data assimilation, the forecast $\mathbf{x}^f$ is modified at the time t_a when observations $\mathbf{y}$ are made:

$$\mathbf{x}^a = \mathbf{x}^f + \mathbf{K} \cdot (\mathbf{y} - \mathbf{H} \cdot \mathbf{x}^f), \quad (19.4.4)$$

where $\mathbf{K}$ is called the gain matrix and $\mathbf{H}$ is the observation operator which projects the model space to the observation space (the two are in general different). The modified state $\mathbf{x}^a$, called the analysis, is then used as the initial state for making the forecast at $t > t_a$ via (19.4.3). The gain matrix $\mathbf{K}$ is chosen to minimize the (weighted) differences between $\mathbf{x}^a$ and $\mathbf{x}^f$, and between $\mathbf{x}^a$ and $\mathbf{y}$, and is of the form

$$\mathbf{K} = \mathbf{P}^f \mathbf{H}^T (\mathbf{H}\mathbf{P}^f \mathbf{H}^T + \mathbf{R})^{-1}, \quad (19.4.5)$$

where $\mathbf{P}^f$ and $\mathbf{R}$ are the error covariances of the forecasts and of the observations, respectively. In the ensemble Kalman filter (EnKF) approach (e.g. Kalnay, 2011), $\mathbf{P}^f$ is determined with an ensemble $\{\mathbf{x}_i^f | i \le N\}$ of forecasts:

$$\mathbf{P}^f = \left\langle (\mathbf{x}^f - \bar{\mathbf{x}}^f)(\mathbf{x}^f - \bar{\mathbf{x}}^f)^T \right\rangle, \; \bar{\mathbf{x}}^f = \frac{1}{N} \sum_{i=1}^{N} \mathbf{x}_i^f, \quad (19.4.6)$$

where $\langle \cdot \rangle$ implies the ensemble average. Through repeated processes, it is expected to pull the geodynamo model (19.4.3) closer to the truth, i.e. to reduce the difference between the forecast $\mathbf{x}^f$ and the true core state $\mathbf{x}^t$ (not really known) in time.

Often approximations are made to $\mathbf{P}^f$ to reduce computational cost or to deal with unknown state properties, e.g. modeled (or parameterized) $\mathbf{P}^f$ in the optimal interpolation (OI) approach (e.g. Kalnay, 2011). Both approaches are used in our geomagnetic data assimilation system (e.g. Kuang et al., 2008; Sun and Kuang, 2015; Tangborn and Kuang, 2015).

In geodynamo simulation, the truncations are typically of order 10^6 (or 5×10^6 parameters to define the core state). But surface geomagnetic measurements can obtain the poloidal coefficients b_l^m up to the degree $L_{\text{obs}} \le 13$ (e.g. Langel, 1987), or slightly better $L_{\text{obs}} \le 20$ for SV (e.g. Langel, 1987; Finlay et al., 2016), mainly due to the spatial attenuation between the CMB and the surface, and due to the contamination of the crustal field, leading to approximately an order 10^5 differences between the coefficients from the dynamo models and from the observations. Therefore, the observation operator $\mathbf{H}$ is very simple and very sparse. If denoting by H_{ij} the elements of $\mathbf{H}$, then,

$$H_{ij} = \delta_{ip}\delta_{jq}, \quad (19.4.7)$$

where the index pairs (p, q) are defined by $\{b_l^m(r_d) | 0 \le m \le l \le 20\}$.

19.5 Geomagnetic Secular Variation Prediction

19.5.1 An Overview

There has been a long history of SV prediction in geomagnetic research and application. Perhaps the best examples are the 5-year predictive models included in the International Geomagnetic Reference Field (IGRF). These models are defined by the SV Gauss coefficients that are time-invariant in the forecast period. Prior to IGRF-11, they were produced via some optimal mathematical extrapolation of the Gauss coefficients obtained from prior geomagnetic measurements (e.g. Maus et al., 2005). Starting from IGRF-11, the predictive models also include contributions from two new approaches: the forecast based on the core flow inverted from geomagnetic observations (Maus et al., 2008; Beggan and Whaler, 2009; Gillet et al., 2013, 2015; Whaler and Beggan, 2015; Barrois et al., 2017) and the forecast made by assimilation of past geomagnetic observations into geodynamo models (Kuang et al., 2010; Fournier et al., 2015).

In the former approach, the forecast is made based on the non-diffusive magnetic induction at the CMB, with the fluid velocity field also derived from the observed SV. Therefore this approach can be characterized as 'data driven' and 'kinematic' (or partially dynamic) in nature. In the latter approach, core dynamics models are determined independently from any observations. These models and surface geomagnetic observations provide two independent estimates of the SV. And the assimilation algorithm employed in this approach works to optimally utilize the two independent assessments for making improved SV forecast. Therefore, the latter is called the geomagnetic data assimilation (GDAS) and is the focus of this section. Since the two approaches share many common methodologies and concerns, such as the stochastic EnKF analysis, the discussion in this section is also applicable to the first approach.

19.5.2 Decadal SV Prediction

Initial work in GDAS began with observing system simulation experiments (OSSEs), in which synthetic observations are created by a model and are assimilated into the same model with some parameter differences. The first two were performed with very similar one-dimensional MHD models that mimic the nonlinear geodynamo process (Sun et al., 2007; Fournier et al., 2007), with sparse observations taken from the full system. These two experiments, one based on an ensemble Kalman filter algorithm (Sun et al., 2007) and the other on the variational approach (Fournier et al., 2007), demonstrated that assimilation of the sparsely distributed data can bring the forecasts closer to the true state of the system. They also highlighted the differences between sequential and variational data assimilation algorithms. The former has the advantage of relatively simple implementation, since they do not require the construction of a tangent linear model (TLM) or its adjoint. The latter has the potential to improve estimates of earlier states of the core, when there were relatively sparse observations available for assimilation. Most of recent development has involved sequential algorithms. But there has been recent progress in variational methods as well (Canet et al., 2009; Li et al., 2011, 2014).

Liu et al. (2007) carried out OSSEs with two different fully nonlinear and three-dimensional geodynamo models separated by two different Rayleigh numbers. In their experiments, the synthetic data were generated at the CMB (i.e. the boundary) from one dynamo solution, while the other model was used for assimilation. Their results further confirmed that similar conclusions can be drawn for fully non-linear geodynamo models. These experiments partially addressed a fundamental concern on whether geomagnetic observations, only available to a portion of the poloidal magnetic field at the CMB, could be used to improve model estimation (i.e. the full geodynamo solution) of the true core state, which can then be used to make accurate geomagnetic forecasts.

Since then, numerous studies have furthered our knowledge of GDAS, and our capabilities to probe core dynamical processes and to make accurate SV forecasts. For example, Kuang et al. (2009), Fournier et al. (2010) and Hulot et al. (2010) worked on understanding the geomagnetic forecast time limit. Aubert and Fournier (2011), Aubert (2014), Fournier et al. (2011), Kuang and Tangborn (2015) and Tangborn and Kuang (2015) used GDAS to probe dynamical processes in deep outer core that are consistent with geomagnetic observations and to understand statistical properties of geodynamo solutions. In particular, Kuang et al. (2010) and Fournier et al. (2015) used GDAS to provide 5-year predictive geomagnetic candidate models for IGRF. More recently, efforts have been made on prediction of longer-term geomagnetic variability, including reversals (Morzfeld et al., 2017; Tangborn and Kuang, 2018). It should be pointed out that most of the GDAS systems currently in use are based on EnKF algorithms (Fournier et al., 2011; see also Section 19.4 for the mathematical details). This approach retains covariances between different state variables, thus enabling corrections to unobserved variables.

Despite the progress made thus far, we are still facing many challenges in GDAS. One is the excessive computational expense arising from large ensemble sizes used in the EnKF, such as those computations with as many as 500 model simulations by Fournier et al. (2013), which is prohibitive for experiments that might encompass thousands of years of Earth core dynamical processes. An alternative is to develop modeled covariances. Tangborn and Kuang (2015) validated these using observation minus forecast (O-F) statistics, though the changes to the unobserved are made indirectly through the model and therefore take longer. A compromise approach that could utilize the best of the two is perhaps a hybrid of the ensemble-generated and modeled covariances (Hammill and Snyder, 2000), which would allow a much smaller ensemble size while retaining the cross covariances. It should be also pointed out that approaches utilizing simplified core dynamics models (e.g. Jault, 2008; Canet et al., 2009) can also help bring down the computing expense. But, as discussed in the following, simplified models could potentially produce larger model biases.

Another critical issue is the sparsity of the geomagnetic data (see Section 19.4). This will remain for a very long time. Utilization of archeomagnetic and paleomagnetic data (e.g. Korte and Constable, 2011; Licht et al., 2013; Nilsson et al., 2014) will help and will be discussed in the following subsection. Efforts should be also on assimilation of derived products from geomagnetic measurements and non-geomagnetic data relevant to core dynamics. For example, observed SV could potentially increase the available data by an order of magnitude (Kuang and Tangborn, 2015) and length-of-day (LOD) and polar motion variation on decadal timescales that are from core–mantle angular momentum exchanges (Jault et al., 1988; Holme and Whaler, 2001; Gillet et al., 2010; Kuang et al., 2017) could provide additional constraints on the (unobserved) toroidal field $\mathbf{B}_T$ at the CMB. Utilization of these new data will bring complex and even non-linear observation operator $\mathbf{H}$ to GDAS.

The EnKF algorithm depends on the error statistics of geomagnetic observations (e.g. geomagnetic field models) and of core dynamics (e.g. geodynamo models). Geodynamo model error statistics are approximated by the forecast covariance matrix $\mathbf{P}^f$ in (19.4.5) (Fournier et al., 2013, 2015; Sun and Kuang, 2015). Similar covariances are also provided in recent geomagnetic field models (e.g. Gillet et al., 2013; Licht et al., 2013; Nilsson et al., 2014). However, these are far from satisfactory, because $\mathbf{P}^f$ could only provide knowledge of the distribution but not the bias of the ensemble. Such biases can exist in both the field modeling (see Section 19.2) and in geodynamo models (Kuang et al., 2010), due to different physics and mathematics employed in the models. For example, geodynamo model biases are from different physics/hypotheses (e.g. boundary conditions discussed by Kuang and Bloxham, 1997, and Sakuraba and Roberts, 2009) and different parameter regimes (e.g. Christensen et al., 2010; Aubert et al., 2017) which affect the scaling rules (e.g. Kuang et al., 2008; Aubert, 2014) used in GDAS. Comprehensive investigations may require large ensembles of the models with different physics and mathematics in different parameter domains.

Rescaling numerical dynamo models for GDAS will remain for many years since, due to computational constraint, the numerical parameters in dynamo simulation are several orders of magnitude away from those appropriate for the Earth's core. Mathematically, this implies that the scaled magnetic field $\mathbf{B}/B$ will be used for assimilation. The scaling factor B can be obtained in different means. One approach, first adopted by Aubert (2014), is to define B with the asymptotic scaling rules derived from geodynamo solutions (Christensen et al., 2010). Kuang et al. (2008) took a much simpler approach by making B the time-averaged magnitude of $\mathbf{B}$. This approach performs surprisingly well, as it is not sensitive to model parameters (similar to the asymptotic scaling rules). To better explain this, consider the following quantity:

$$\delta B \equiv (|\mathbf{B}| - B)/B, \qquad (19.5.1)$$

where $|\mathbf{B}|$ is the *rms* magnetic field in the outer core (and B is therefore the time average of $|\mathbf{B}|$). As shown in Table 19.5.1, the mean B varies strongly with the Rossby number R_o: it decreases by more than a factor of 4 from $R_o = 2.5 \times 10^{-6}$ to $R_o = 3.125 \times 10^{-7}$. However, the standard deviation $\sigma(\delta B)$ remains nearly unchanged in the all Rossby numbers. Similar properties are also found for the poloidal field $\mathbf{B}_P$ at the CMB (but with slightly smaller R_o).

But the time scalings are more complicated. From the power spectra of δB and of δB_P for the poloidal field at the CMB (in Figure 19.5.1), one can find that the power spectra for $R_o \geq 1.25 \times 10^{-6}$ (gray solid and dotted curves) are significantly different from those for $R_o \leq 6.25 \times 10^{-7}$ (black solid and dotted curves). In particular, the latter two are very similar, suggesting asymptotic convergences in the frequency domain in the poloidal field at the top of the D''-layer, bringing a hope to find a parameter-independent, dynamical consistent timescale for all dynamo solutions used in GDAS with $R_o \leq 6.25 \times 10^{-7}$. Similar properties are also shown in the total magnetic field anomaly δB of $\mathbf{B}$ in the outer core (but are not shown here). It should be pointed out that the frequency domain in 19.4 is controlled by the time steps and the output settings of our simulation. Further numerical analysis is needed to investigate variations with even higher frequencies. However, these figures are sufficient for our current discussion. We should point out also that similar analysis on timescales was done by others, e.g. Bouligand et al. (2016) analyzed statistical properties of dynamo fields with different parameter values and compared to those from observations.

19.5.3 Millennial and Longer SV Prediction

Many of the important physical processes in the Earth's core take place on timescales of thousands of years or longer. This means that in order to capture core dynamics in any meaningful way in a geomagnetic data assimilation system, it is essential to assimilate data from epochs dating long before direct measurements of the geomagnetic field. An extreme example of this involves using a scaler stochastic model over millions of years that assimilates paleomagnetic data over this time frame. Morzfeld et al. (2017) showed that this can lead to prediction of dipole reversals as far as 4000 years in the future. But this type of model does not give any insight into the physical processes within the Earth's core that trigger the reversals. This requires a fully three-dimensional geodynamo model, which is at present computationally too expensive to be feasible on these very long timescales. But intermediate assimilation experiments on the order of thousands of years are now possible, and recent work has been carried out on these. For example, Sanchez et al. (2016) used a geodynamo model to produce a geomagnetic field model over an 800 year period with archeomagnetic data as observations in a data assimilation system. The resulting field model is constrained by the physics of the geodynamo

Table 19.5.1 *The Mean* **B** *and the Standard Deviation* $\sigma(\delta B)$ *for Different Rossby Numbers* R_o

R_o	2.5×10^{-5}	1.25×10^{-6}	6.25×10^{-7}	3.125×10^{-7}
B	1.2356	0.4635	0.3225	0.2869
$\sigma(\delta B)$	0.0697	0.0877	0.0602	0.0767

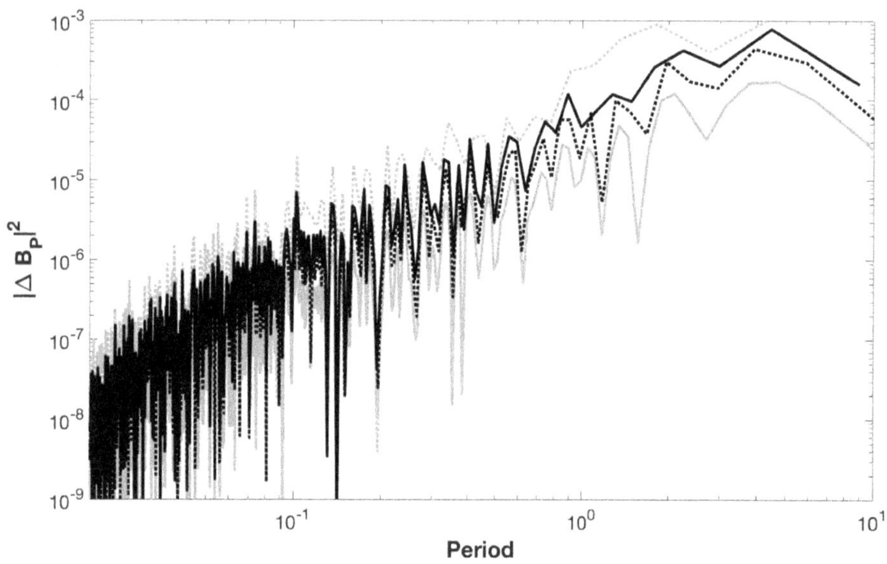

Figure 19.5.1 The power spectra of the time series δB_P of the poloidal field $\mathbf{B}_P$ at the top of the D''-layer defined in (10) for $R_o = 2.15 \times 10^{-6}$ (dotted gray), 1.25×10^{-6} (solid gray), 6.25×10^{-7} (dotted black) and 3.125×10^{-7} (black). The horizontal axis is the non-dimensional period (the unit period is equivalent to the magnetic free-decay time $\tau_d \approx 20,000$ years for the Earth).

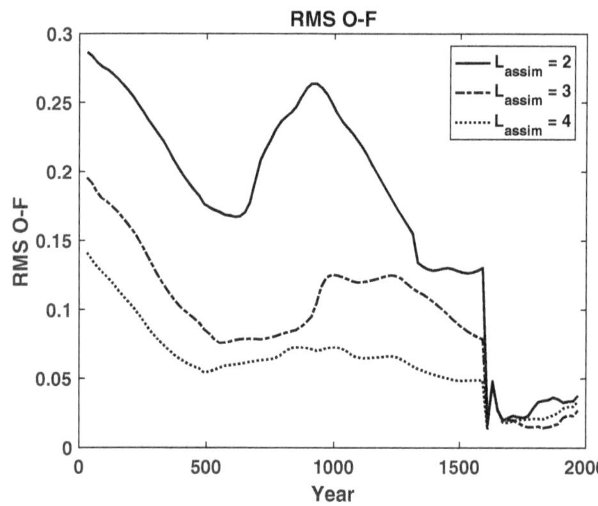

Figure 19.5.2 $RMS(\boldsymbol{O}\text{-}\boldsymbol{F})$ of 20 year forecasts for degrees up to $L = 3$, for assimilation runs starting in 10 CE and finishing in 1990 CE. In all cases the observation error is set to $\sigma^o = 1.0|b_L^m|$ (so $\alpha^o = 1.0$), and coefficients are assimilated up to degree $L_{assim} = 2$ (black), $L = 3$ (dashed) and $L = 4$ (dotted).

model rather than the regularization and smoothing used in traditional field models.

Another approach is to assimilate traditional field model outputs into a geodynamo model over longer periods of time. For this we need to include the indirect paleo- and archeomagnetic measurements. These are made from lava flows and marine sediments, as well as fired clay bricks or other human artifacts, respectively (Donadini et al., 2009). This type of data contains one or more of magnetic inclination, declination and intensity values and is relatively more frequent during the last 3000 years. All of these data sources are parametrized in terms of spherical harmonic expansions spatially, and B-splines temporally. Field models are generally divided by the type of data that they incorporate, e.g. the CALS3 k.4 model (Korte and Constable, 2011), which is based on both and paleo- and archeomagnetic data available for the last 3000 years, but is also constrained by gufm1 (Jackson et al., 2000) for the last 400 years of the model.

The information available from this early era in the CALS3 k.4 model makes it valuable for the study of longer-timescale changes. The more direct measurements are incorporated into the gufm1 model (1590 to 1990), which makes use of the sailing ship compass records and magnetic observatory data starting in the mid nineteenth century until present. The CM4 model (Sabaka et al., 2004) (1960 to 2002) uses satellite measurements from a variety of missions, allowing it to include smaller spatial-scale structures of the field. The accuracy of these models naturally improves during later times, which can be attributed to both a better spatial distribution and data precision, as well as the progress in separating the different sources that contribute to the observed field. Tangborn and Kuang (2018) have used data assimilation to connect these three field models via a geodynamo model. The purpose of this work was to show how data assimilation can be used to obtain independent estimates of errors in the CALS3 k.4 model from a long assimilation run that ends with the CM4 model in 1990.

It presented a series of nearly 2000-year geomagnetic data assimilation experiments that use the CALS3 k.4 field model as observations from 10 CE to 1590 CE, followed by gufm1 (1590–1950) and CM4 (1970–90). In each epoch, the field models were assimilated into the geodynamo model every 20 years. The goal was to use the higher accuracy of gufm1 and

CM4 to gain additional insight (beyond the estimates provided with the field model) into the precision of CALS3 k.4, particularly as the maximum degree L assimilated is varied. These independent estimates of the uncertainty in the field models used a simple relationship for the Gauss coefficients. The error standard deviation, $(\sigma^o)_l^m$, in each was modeled as being proportional to its magnitude, $(\sigma^o)_l^m = \alpha^o||b_l^m$, where α^o is the observation error coefficient up to $L =$. For higher degrees, larger errors were assigned that essentially removed the observations from the assimilation. Smaller errors are assigned after 1590, when sailing ship data become available, along with observatory data in the nineteenth century and satellite data after 1960.

Tangborn and Kuang (2018) carried out a series of assimilation experiments that run from 10 to 1990 CE, with data from the field models assimilated every 20 years. The goal of these experiments is to determine whether the CALS3 k.4 field model can have a positive impact on the forecasting of the later fields, particularly CM4. About 20 numerical experiments were done, in which the maximum degree (L_{assim}) assimilated and/or the observation error coefficient (α^o) for CALS3 k.4 were varied. In all cases the assimilation is done every 20 years, and geodynamo model parameters used are $R_{th} = 3622$ (or 30 times the critical Rayleigh number for the onset of thermal convection) and an Ekman number $E = 1.25 \times 10^{-6}$.

Here we summarize some of the experimental results from Tangborn and Kuang (2018) in the form of the RMS of the observed minus forecast (O-F) statistics several of the simulations over 2000 years (Figure 19.5.2), and the RMS O-F at the end of the simulations in 1990 for all of the simulations (Figure 19.6.1). In all experiments, observations up to degree $L = 8$ are assimilated after 1590 (when gufm1 begins) with much smaller observation error estimates. Figure 19.5.2 shows trajectories of the RMS (O-F) over 2000 years for three different maximum degree (L) of Gauss coefficients assimilated ($L = 2, 3, 4$). And it shows that, for the observation model used, the most accurate case is with a maximum of $L = 3$ assimilated. This implies that there is information in the CALS3 k.4 field model prior to 1590 that is useful in making geomagnetic forecasts in 1990. The RMS O-F values from 1990 for the complete set of assimilation experiments are shown in Figure 19.6.1. Here both the maximum degree assimilated for CALS3 k.4 and the observation error coefficient, α^o, are varied. This indicates that the maximum assimilated degree of $L = 3$ produces the best forecast in 1990, from the experiments carried out, even though the gufm1 and CM4 observation errors used are the same in all cases.

19.6 Concluding Remarks

In this chapter we have reviewed recent research efforts for prediction of geomagnetic variations on different timescales. We used the two very contrasting examples, the weak periodic geomagnetic anomalies from tides and the strong chaotic SV from the outer core, to explain the physics and the mathematics used for the prediction, the progress made in SV forecasts and the challenges for future development.

The observables and their error-covariance matrices in (19.4.4) and (19.4.5) come from field modeling efforts in which the poloidal field coefficients are separated from other sources. This is a key to provide the assimilation algorithm with the proper data and realistic uncertainties. There are several ways in which this field separation can interact with GDAS. One is to parameterize the poloidal field coefficients in terms of empirical temporal basis functions and subsequently provide the assimilation with the resulting coefficient time series and error covariance. Another approach is to use magnetic measurements as the observables and include non-core sources in the state variables. Either way helps to mitigate bias in the core poloidal field estimates.

While tidal magnetic fluctuations are very weak in amplitude relative to that of SV, they dominate the variability seen in the diurnal to semi-diurnal frequencies. Agreement between theoretical forward models and field models based on observations present at least the primary semi-diurnal lunar tidal fluctuations as highly predictable in at least the coarse features. It is similarly expected that forward modeling of weaker ocean tidal magnetic fields is predictive, though this has yet to be confirmed. The 'tidal' magnetic fields associated with solar heating of the upper atmosphere are expected and observed to be less stationary. Despite this, analyses of pre-whitened ground magnetic observatory data show that much if not most of the variance is organized with respect to the solar day and is modulated by the solar cycle. It seems then that much of the geomagnetic variability at quasi-diurnal frequencies and the associated harmonics can be deterministically related to astronomical spin/orbit parameters and solar variability. Claims in this regard must also consider regional variations, as competing unresolved signals may pose limitations at high latitudes and equatorial regions.

As described in Section 19.5, the relative geomagnetic anomalies defined in (19.5.1) from geodynamo simulation have displayed asymptotically convergence in both the magnitude (see the mean and the standard deviation in Table 19.5.1) and the timescales (see the power spectra in Figure 19.5.1). These can be used to define the scaling relations between the non-dimensional magnetic field from geodynamo simulation and the core field derived from geomagnetic measurements. In particular, asymptotic convergence of the spectra of the magnetic anomalies (in Figure 19.5.1) can be used to estimate the time resolutions of geodynamo models, and thus the limit of forecast

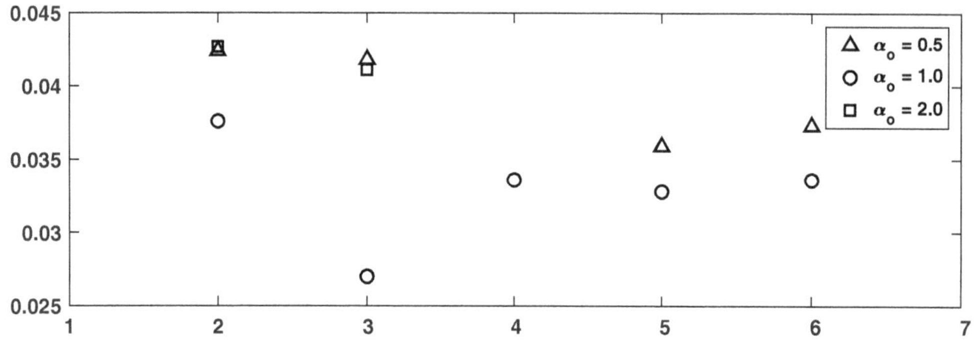

Figure 19.5.3 $RMS(\boldsymbol{O}\text{-}F)$ of 20 year forecasts in 1990, for all of the assimilation experiments starting in 10 CE and finishing in 1990 CE. The x-axis shows the maximum degree L assimilated, and the observation error coefficient used up to 1590, σ^o, is indicated by the symbols in the legend (from 0.5 to 2).

shortest geomagnetic variations from the outer core, such as occurrence of geomagnetic jerks (on sub-annual timescales).

Further improvement in SV prediction relies on improving knowledge of error statistics of geomagnetic field models and geodynamo models, in particular the biases that cannot be described by covariance matrices. But the biases in both modeling are affected by the knowledge of the core dynamics, which is only partially built into geodynamo modeling, and partially (and empirically) implemented in geomagnetic field modeling. The two fragments may not be consistent. And assimilation of the two models could certainly lead to erroneous estimation of the core state. An attractive approach to reduce such inconsistencies (and therefore biases) is to integrate both modeling efforts, such that the field models are used to make initial forecasts; the forecasts are then fed back to improve the field models, which will be used again for improved forecasts. The iterations will continue until an optimal solution is reached. This approach is conceptually similar to the parameter 'co-estimation' used in geomagnetic field modeling (see Section 19.2), but on model 'co-development'. Therefore we name this approach the 'dynamic field modeling'. However, a large ensemble of field models and geodynamo models may be needed to make this work, thus requiring a community effort.

The long-term (2000 years) assimilation experiments point to the conclusion that over more than a millennium, observational information from field models is retained in the numerical geodynamo model so as to improve modern-day geomagnetic forecasts. And because this information is not lost from the model over these long time periods, there remains some hope that longer-term forecasts can also be made. The work of Morzfeld et al. (2017) using a simple scalar model adds to the evidence that this may be possible. However, this may require a numerical model that lies in between these two approaches in complexity and computational cost.

References

Aubert, J. 2014. Earth's core internal dynamics 18402010 imaged by inverse geodynamo modelling. *Geophys. J. Int.*, **197**, 1321–34.

Aubert, J. and Fournier, A. 2011. Inferring internal properties of Earth's core dynamics and their evolution from surface observations and a numerical geodynamo model. *Nonlinear Process. Geophys.*, **18**, 657–74.

Aubert, J., Gastine, T. and Fournier, A. 2017. Spherical convective dynamos in the rapidly rotating asymptotic regime. *J. Fluid. Mech.*, **813**, 558–93.

Barrois, O., Gillet, N. and Aubert, J. 2017. Contributions to the geomagnetic secular variation from a reanalysis of core surface dynamics. *Geophys. J. Int.*, **211**, 50–68.

Beggan, C. D. and Whaler, K. A. 2009. Forecasting change of the magnetic field using core surface flows and ensemble Kalman filtering. *Geophys. Res. Lett.*, doi: 10.1029/2009GL039927.

Bouligand, C., Gillet, N., Jault, D., Schaeffer, N. Fournier, A. and Aubert, J. 2016. Frequency spectrum of the geomagnetic field harmonic coefficients from dynamo simulations. *Geophys. J. Int.*, **207**, 1142–57.

Canet, E., Fournier, A. and Jault, D. 2009. Forward and adjoint quasi-geostrophic models of the geomagnetic secular variation. *J. Geophys. Res.*, doi: 10.1029/2008JB006189.

Christensen, U. R., Aubert, J., Cardin, P., Dormy, E., Gibbons, S., Glatzmaier, G. A., Grote, E., Honkura, Y., Jones, C., Kono, M., Matsushima, M., Sakuraba, A., Takahashi, F., Tilgner, A., Wicht, J. and Zhang, K. 2001. A numerical dynamo benchmark. *Phys. Earth Planet. Inter.*, **128**, 25–34.

Christensen, U. R., Aubert, J. and Hulot, G. 2010. Conditions for Earth-like geodynamo models. *Earth Planet. Sci. Lett.*, **296**, 487–96.

Colosi, J. A. and Munk, W. 2006. Tales of the Venerable Honolulu Tide Gauge. *J. Phys. Oceanogr.*, **36**, 967–96.

Donadini, F., Korte, M. and Constable, C. G. 2009. Geomagnetic field for 0–3 ka: 1. New data sets for global modeling. *Geochem. Geophys. Geosys.*, **10**, Q06007, doi: 10.1029/2008GC002295.

Egbert, G. P. and Ray, R. D. 2017. Tidal prediction. *J. Mar. Res.*, **189–237**, 1–49.

Finlay, C. C., Lesur, V., Thébault, E., Vervelidou, F., Morschhauser, A. and Shore, R. 2017. Challenges handling magnetospheric and ionospheric signals in internal geomagnetic field modeling. *Space Sci. Rev.*, **206**, 157–89, doi: 10.10007/s11214-016-0285-9.

Finlay, C. C., Olsen, N., Kotsiaros, S., Gillet, N. and Toffner-Clausen, L. 2016. Recent geomagnetic secular variation from Swarm and ground observatories as estimated in the CHAOS-6 geomagnetic field model. *Earth Planets Space*, **68**, doi: 10.1186/s40623-016-0486-1.

Fournier, A., Eymin, C. and Alboussiere, T. 2007. A case for variational geomagnetic data assimilation: insights from a one-dimensional, nonlinear, and sparsely observed MHD system. *Nonlin. Process. Geophys.*, **14**, 163–80.

Fournier, A., Hulot, G., Jault, D., Kuang, W., Tangborn, A., Gillet, N., Canet, E., Aubert, J. and Lhuillier, F. 2010. An introduction to data assimilation and predictability in geomagnetism. *Space Sci. Rev.*, **155**, 247–91.

Fournier, A., Aubert, J. and Thébault, E. 2011. Inference on core surface flow from observations and 3-D dynamo modelling. *Geophys. J. Int.*, **186**, 118–36.

Fournier, A., Nerger, L. and Aubert, J. 2013. An ensemble Kalman filter for the time-dependent analysis of the geomagnetic field. *Geochem. Geophys. Geosyst.*, doi: 10.1002/ggge.20252.

Fournier, A., Aubert, J. and Thébaut, E. 2015. A candidate secular variation model for IGRF-12 based on Swarm data and inverse geodynamo modeling. *Earth Planets Space*, **67**, doi: 10.1186/s40623-015-0245-8.

Gillet, N., Jault, D., Canet, E. and Fournier, A. 2010. Fast torsional waves and strong magnetic field within the Earth's core. *Nature*, doi: 10.1038/nature09010.

Gillet, N., Jault, D., Finlay, C. C. and Olsen, N. 2013. Stochastic modeling of the Earth's magnetic field: Inversion for covariances over the observatory era. *Geochem. Geophys. Geosyst.*, doi: 10.1002/ggge.20041.

Gillet, N., Barrois, O. and Finlay, C. C. 2015. Stochastic forecasting of the geomagnetic field from the COV-Obs.x1 geomagnetic field model, and candidate models for IGRF-12. *Earth Planets Space*, doi: 10.1186/s40623-015-0225-z.

Grayver, A. V., Schnepf, N. R., Kuvshinov, A. V., Sabaka, T. J., Manoj, C. and Olsen, N. 2016 Satellite tidal magnetic signals constrain oceanic lithosphere-asthenosphere boundary. *Sci. Adv.*, doi: 10.1126/sciadv.1600798.

Grayver, A. V., Munch, F. D., Kuvshinov, A. V., Khan, A., Sabaka, T. J. and Tøffner-Clausen, L. 2017. Joint inversion of satellite-detected tidal and magnetospheric signals constrains electrical conductivity and water content of the upper mantle and transition zone. *Geophys. Res. Lett.*, doi: 10.1002/2017GL073446.

Hammill, T. M., Snyder, C. 2000. A Hybrid Ensemble Kalman Filterâ€"3D Variational Analysis Scheme. *Mon. Weather Rev.*, **128**, 2905–15.

Holme, R. and Whaler, K. A. 2001. Steady core flow in an azimuthally drifting reference frame. *Geophys. J. Int.*, **145**, 560–69.

Hulot, G., Lhuillier, F. and Aubert, J. 2010. Earth's dynamo limit of predictability. *Geophys. Res. Lett.*, **37**, doi: 10.1029/2009GL041869.

Jackson, A., Jonkers, A. R. T. and Walker, M. R. 2000. Four centuries of geomagnetic secular variation from historical records. *Philos. Trans. R. Soc. London*, **A358**, 957–90.

Jault, D., Gire, C. and LeMouël, J.-L. 1988. Westward drift, core motions and exchanges of angular momentum between core and mantle. *Nature*, **333**, 353–6.

Jault, D. 2008. Axial invariance of rapidly varying diffusionless motions in the Earth?s core interior. *Phys. Earth Planet. Inter.*, **166**, 67–76.

Jiang, W. and Kuang, W. 2008. An MPI-based MoSST core dynamics model. *Phys. Earth Planet. Inter.*, **170**, 46–51.

Jones, C. A., Boronski, P., Brun, A. S. Glatzmaier, G. A., Gastine, T., Miesch, M. S. and Wicht, J. 2011. Anelastic convection-driven dynamo benchmarks. *Icarus*, **216**, 120–35.

Kalnay, E. 2011. *Atmospheric Modeling, Data Assimilation and Predictability*. Cambridge University Press, Cambridge.

Korte, M. and Constable, C. G. 2011. Improving geomagnetic field reconstructions for 0–3 ka. *Phys. Earth Planet. Inter.*, **188**, 247–59.

Kuang, W. and Bloxham, J. 1997. An Earth like numerical dynamo model. *Nature*, **389**, 371–4.

Kuang, W. and Bloxham, J. 1999. Numerical modeling of magnetohydrodynamic convection in a rapidly rotating spherical shell: Weak and strong dynamo action. *J. Comput. Phys.*, **153**, 51–81.

Kuang, W. and Chao, B. F. 2003. Geodynamo modeling and core-mantle interactions, in *Earth's Core: Dynamics, Structure, Rotation*, Geodynamics Series **31**, ed. Dehant, V., Kreager, K. C., Karato, S. and Zatman, S. American Geophysical Union, Washington, DC.

Kuang, W. and Tangborn, A. 2015. Dynamic responses of the Earth's outer core to assimilation of observed geomagnetic secular variation. *Prog. Earth Planet. Sci.*, doi: 10.1186/s40645-015-0071-4.

Kuang, W., Tangborn, A., Jiang, W., Liu, D., Sun, Z., Bloxham, J. and Wei, Z. 2008. MoSSTDAS: The first generation geomagnetic data assimilation framework. *Commun. Comput. Phys.*, **3**, 85–108.

Kuang, W., Wei, Z., Holme, R. and Tangborn, A. 2010. Constraining a numerical geodynamo model with 100 years of surface observations. *Geophys. J. Int.*, doi: 10.1111/j.1365-246X.2009.04376.x.

Kuang, W., Tangborn, A., Wei, Z. and Sabaka, T. 2009. Prediction of geomagnetic field with data assimilation: a candidate secular variation model for IGRF-11. *Earth Planets Space*, **62**, 775–85.

Kuang, W., Chao, B. F. and Chen, J. 2017. Decadal polar motion of the Earth excited by the convective outer core from geodynamo simulations. *J. Geophys. Res.*, in press.

Langel, R. A. 1987. The main geomagnetic field, in *Geomagnetism*, vol. 1, ed. Jacobs, J. A., Academic Press, London.

Lesur, V., Wardinski, I., Rother, M. and Mandea, M. 2008. GRIMM: The GFZ Reference Internal Magnetic Model based on vector satellite and observatory data. *Geophys. J. Int.*, **173**, 382–94.

Lesur, V., Rother, M., Wardinski, I., Schachtschneider, R., Hamoudi, M. and Chambodut, A. 2015a. Parent magnetic field models for the IGRF-12: GFZ-candidates. *Earth Planets Space*, doi: 10.1186/s40623-015-0239-6.

Lesur, V., Whaler, K. and Wardinski, I. 2015b. Are geomagnetic data consistent with stably stratified flow at the core–mantle boundary? *Geophys. J. Int.*, DOI: 10.1093/gji/ggv031.

Li, K., Jackson A. and Livermore, P. W. 2011. Variational data assimilation for the initial-value dynamo problem. *Phys. Rev. E*, **84**, doi: 10.1103/PhysRevE.84.056321.

Li, K., Jackson A. and Livermore, P. W. 2014. Variational data assimilation for a forced, inertia-free magnetohydrodynamic dynamo model. *Geophys. J. Int.*, **199**, 1662–76.

Licht, A., Hulot, G., Gallet, Y. and Thébault, E. 2013. Ensembles of low degree archeomagnetic field models for the past three millennia. *Phys. Earth Planet. Inter.*, **224**, 38–67.

Liu, D., Tangborn, A. and Kuang, W. 2007. Observing system simulation experiments in geomagnetic data assimilation. *J. Geophys. Res.*, **112**, doi: 10.1029/2006JB004691.

Love, J. J. and Rigler, E. J. 2014. The magnetic tides of Honolulu. *Geophys. J. Int.*, **197**, 1335–53.

Matsui, H., Heien, E., Aubert, J., Aurnou, J. M., Avery, M., Brown, B., Buffett, B. A., Busse, F., Christensen, U. R., Davies, C. J., Featherstone, N., Gastine, T., Glatzmaier, G. A., Gubbins, D., Guermond, J.-L., Hayashi, Y., Hollerbach, R., Hwang, L. J., Jackson, A., Jones, C. A., Jiang, W., Kellogg, L. H., Kuang, W., Landeau, M., Marti, P., Olson, P., Ribeiro, A., Sasaki, Y., Schaeffer, N., Simitev, R. D., Sheyko, A., Silva, L., Stanley, S., Takahashi, F., Takehiro, S., Wicht, J. and Willis, A. P. 2016. Performance benchmarks for a next generation numerical dynamo model. *Geochem. Geophys. Geosyst.*, doi: 10.1002/2015GC006159

Maus, S. 2017. POMME-9 magnetic model of the Earth. http://geomag.colorado.edu/pomme-9-magnetic-model-of-the-earth.html.

Maus, S., Macmillan, S., Lowes, F. and Bondar, T. 2005. Evaluation of candidate geomagnetic field models for the 10th generation of IGRF. *Earth Planets Space*, **57**, 1173–81.

Maus, S., Rother, M., Stolle, C., Mai, W., Choi, S., Lühr, H., Cooke, D. and Roth, C. 2006. Third generation of the Potsdam Magnetic Model of the Earth (POMME). *Geochem. Geophy. Geosyst.*, doi: 10.1029/2006GC001269.

Maus, S., Silva, L. and Hulot, G. 2008. Can core-surface flow models be used to improve the forecast of the Earth's main magnetic field? *J. Geophys. Res.*, doi: 10.1029/2007JB005199.

Morzfeld, M., Fournier, A. and Hulot, G. 2017. Coarse predictions of dipole reversals by low-dimensional modeling and data assimilation. *Phys. Earth Planet. Inter.*, **262**, 8–27.

Nilsson, A., Holme, R., Korte, M., Suttie, N. and Hill, M. 2014. Reconstructing Holocene geomagnetic field variation: new methods, models and implications. *Geophys. J. Int.*, doi: 10.1093/gji/ggu120.

Olsen, N. and Stolle, C. 2017. Magnetic signatures of ionospheric and magnetospheric current systems during geomagnetic quiet conditions – an overview. *Space Sci. Rev.*, **206**, 5–25, doi: 10.1007/s11214-016-0279-7.

Olsen, N., Lühr, H., Finlay, C. C., Sabaka, T. J., Michaelis, I., Rauberg, J. and Tøffner-Clausen, L. 2014. The CHAOS-4 geomagnetic field model. *Geophys. J. Int.*, **197**, 815–27.

Sabaka, T. J. and Olsen, N. 2006. Enhancing comprehensive inversions using the *Swarm* constellation. *Earth Planets Space*, **58**, 371–95.

Sabaka, T. J., Olsen, N. and Langel, R. A. 2002. A comprehensive model of the quiet-time, near-Earth magnetic field: phase 3. *Geophys. J. Int.*, **151**, 32–68.

Sabaka, T. J., Olsen, N. and Purucker, M. E. 2004. Extending comprehensive models of the Earth's magnetic field with Ørsted and CHAMP data. *Geophys. J. Int.*, **159**, 521–47.

Sabaka, T. J., Tøffner-Clausen, L. and Olsen, N. 2013. Use of the comprehensive inversion method for Swarm satellite data analysis. *Earth Planets Space*, **65**, 1201–22.

Sabaka, T. J., Olsen, N., Tyler, R. H. and Kushinov, A. 2015. CM5, a pre-Swarm comprehensive geomagnetic field model derived from over 12 yr of CHAMP, Ørsted, SAC-C and observatory data. *Geophys. J. Int.*, **200**, 1596–1626.

Sabaka, T. J., Tyler, R. H. and Olsen, N. 2016. Extracting ocean-generated tidal magnetic signals from Swarm data through satellite gradiometry. *Geophys. Res. Lett.*, doi: 10.1002/2016GL068180.

Sakuraba, A. and Roberts, P. H. 2009. Generation of a strong magnetic field using uniform heat flux at the surface of the core. *Nat. Geosci.*, doi: 10.1038/NGEO643.

Sanchez, S., Fournier, A., Aubert, J. and Gallrt, Y. 2016. Modelling the archaeomagnetic field under spatial constraints from dynamo simulations: A resolution analysis. *Geophy. J. Int.*, **207**, 983–1002.

Schnepf, N. R., Kuvshinov, A. and Sabaka, T. J. 2015. Can we probe the conductivity of the lithosphere and upper mantle using satellite tidal magnetic signals? *Geophys. Res. Lett.*, doi: 10.1002/2015GL063540.

Sun, Z., Tangborn, A. and Kuang, W. 2007. Data assimilation in a sparsely observed one-dimensional modeled MHD system. *Nonlinear Process. Geophys.*, **14**, 181–92.

Sun, Z. and Kuang, W. 2015. An ensemble algorithm based component for geomagnetic data assimilation. *Terr. Atmos. Ocean. Sci.*, **26**, 53–61.

Tangborn, A. and Kuang, W. 2015. Geodynamo model and error parameter estimation using geomagnetic data assimilation. *Geophys. J. Int.*, **200**, 664–75.

Tangborn, A. and Kuang, W. 2018. Impact of archeomagnetic field model data on modern era geomagnetic forecasts. *Phys. Earth Planet. Inter.*, **276**, 2–9.

Tyler, R. H. 2013. Magnetic remote sensing of ocean flow variability, presented at IAGA 12th Scientific Assembly, Living on a Magnetic Planet, Merida, 26–31 August.

Tyler, R. H., Maus, S. and Lühr, H. 2003. Satellite observations of magnetic fields due to ocean tidal flow, *Science*, **299**, 239–41.

Tyler, R. H., Boyer, T. P., Minami, T., Zweng, M. M. and Reagan, J. R. 2017. Electrical conductivity of the global ocean. *Earth Planets Space*, doi: 10.1186/s40623-017-0739-7

Whaler, K. A. and Beggan, C. D. 2015. Derivation and use of core surface flows for forecasting secular variation. *J. Geophys. Res.*, doi: 10.1002/2014JB011697.

INDEX

For EU product safety concerns, contact us at Calle de José Abascal, 56–1°,
28003 Madrid, Spain or eugpsr@cambridge.org.

www.ingramcontent.com/pod-product-compliance
Lightning Source LLC
LaVergne TN
LVHW082114200926
848484LV00008B/1949

* 9 7 8 1 1 0 8 4 1 8 4 8 5 *